Keio University Symposia
for Life Science and Medicine 1

**Springer**

*Tokyo*
*Berlin*
*Heidelberg*
*New York*
*Barcelona*
*Budapest*
*Hong Kong*
*London*
*Milan*
*Paris*
*Santa Clara*
*Singapore*

Y. Ishimura, H. Shimada,
M. Suematsu (Eds.)

# Oxygen Homeostasis and Its Dynamics

With 264 Figures, Including 14 in Color

 Springer

Yuzuru Ishimura, M.D., Ph.D.
Professor and Chairman, Department of Biochemistry, School of Medicine,
Keio University, 35 Shinanomachi, Shinjuku-ku, Tokyo 160, Japan

Hideo Shimada, Ph.D.
Associate Professor, Department of Biochemistry, School of Medicine,
Keio University, 35 Shinanomachi, Shinjuku-ku, Tokyo 160, Japan

Makoto Suematsu, M.D., Ph.D.
Associate Professor, Department of Biochemistry, School of Medicine,
Keio University, 35 Shinanomachi, Shinjuku-ku, Tokyo 160, Japan

ISBN 978-4-431-68478-7

Library of Congress Cataloging-in-Publication Data

Oxygen homeostasis and its dynamics / Y. Ishimura, H. Shimada, M.
    Suematsu, eds.
        p.  cm.—(Keio University symposia for life science and
    medicine series ; 1)
        Papers presented in the 1st Keio International Symposium for Life
    Sciences and Medicine, held in Tokyo, Dec. 8–13, 1996.
        Includes bibliographical references and index.
        Papers presented in the 1st Keio International Symposium for Life
    Sciences and Medicine, held in Tokyo, Dec. 8–13, 1996.
        ISBN 978-4-431-68478-7        ISBN 978-4-431-68476-3 (eBook)
        DOI 10.1007/978-4-431-68476-3
        1. Oxygen—Metabolism—Congresses. 2. Respiration—Congresses.
    3. Cytochrome oxidase—Congresses.   4. Cytochrome P-450—Congresses.
    I. Ishimura, Yuzuru, 1935–   .  II. Shimada, H. (Hideo), 1948–   .
    III. Suematsu, M. (Makoto), 1957–   .   IV. Keio International
    Symposium for Life Sciences and Medicine (1st : 1996 : Tokyo, Japan)
    V. Series.
        [DNLM: 1. Oxygen—metabolism—congresses. 2. Homeostasis—
    congresses. 3. Oxygenases—congresses.   QV 312 O975   1998]
    QP177.O946   1998
    572'.534—DC21
    DNLM/DLC
    for Library of Congress                                                97-41378

Printed on acid-free paper

© Springer-Verlag Tokyo 1998
Softcover reprint of the hardcover 1st edition 1998

Typesetting: Best-set Typesetter Ltd., Hong Kong

SPIN: 10731263          9 8 7 6 5 4 3 2

# Foreword

This first volume in a projected series contains the proceedings of the first of the Keio University International Symposia for Life Sciences and Medicine under the sponsorship of the Keio University Medical Science Fund. As stated in the address by the President of Keio University at the opening of the 1996 symposium, the fund was established by the generous donation of Dr. Mitsunada Sakaguchi. The Keio University International Symposia for Life Sciences and Medicine constitute one of the core activities of the fund. The objective is to contribute to the international community by developing human resources, promoting scientific knowledge, and encouraging mutual exchange. Every year, the Executive Committee of the International Symposia for Life Sciences and Medicine selects the most interesting topics for the symposium from applications received in response to a call for papers to the Keio medical community. The publication of these proceedings is intended to publicize and distribute information arising from the lively discussions of the most exciting and current issues during the symposium.

We are grateful to Dr. Mitsunada Sakaguchi, who made the symposium possible, the members of the program committee, and the office staff whose support guaranteed the success of the symposium. Finally, we thank Springer-Verlag, Tokyo, for their assistance in publishing this work.

Akimichi Kaneko, M.D., Ph.D.
Chairman
Executive Committee
on the Keio University International Symposia
for Life Sciences and Medicine

# Preface

This volume contains the collected papers presented at the Keio University International Symposium for Life Sciences and Medicine: 1996 Conference on Oxygen Homeostasis and Its Dynamics held at the New North Building (Kita-shinkan), Mita campus of Keio University, and the Miyako Hotel Tokyo, Minato-ku, Tokyo, December 8–13, 1996.

The symposium, which was the first of the Keio University International Symposia for Life Sciences and Medicine, was organized to open the latest frontiers of ideas and experience in the field of oxygen metabolism and to develop an overall picture through discussion. It is also the aim of the Keio University Medical Science Fund, the sponsor of the meeting, to contribute to global societies by promoting life sciences and medicine through discussions and mutual exchange of opinions. With these aims in mind, the organizers invited a group of 42 specialists and more than 70 young scientists who were interested in the stated objectives of the symposium. All the participants lived together during the symposium: we ate in the same room in the morning and evening, participated in the lecture sessions at Kita-shinkan, and enjoyed the poster sessions at night. Indeed, the discussions often continued until midnight following the poster sessions.

A total of 79 papers have been organized in this volume according to the reactions through which molecular oxygen and its derivatives exhibit biological actions. These have been subdivided into (1) Cytochrome oxidases, (2) Cytochrome P-450 monooxygenases, (3) Various types of oxidases and oxygenases, (4) Oxygen sensing and regulation of blood flow, and (5) Pathophysiology and physiology of gaseous monoxides. In this manner, we intend to demonstrate the concept of oxygen homeostasis and its dynamics, and its importance in biology and medicine.

The editors wish to thank all participants for their contributions. We offer our thanks and gratitude to the members of the Organizing Committee, to the Scientific Program Committee, and especially to Hiroshi Ohin, Junko Miyai, and Hajime Ebihara, the Conference Secretariats. Without their excellent work and diligence, the conference could not have been held. Finally, it is a pleasure to express our indebtedness to the staff of Springer-Verlag, Tokyo, for their support for this book and their forbearance and unfailing courtesy.

The Editors:
Yuzuru Ishimura
Hideo Shimada
Makoto Suematsu
August 1997

# Contents

Foreword .................................................................. V

Preface ................................................................... VII

List of Contributors ...................................................... XV

Opening Remarks ......................................................... XIX

## Part 1  Cytochrome Oxidases

Structure and Possible Mechanism of Action of Cytochrome *c* Oxidase
from the Soil Bacterium *Paracoccus denitrificans*
  H. Michel, S. Iwata, and C. Ostermeier ............................    3

Crystal Structure and Reaction Mechanism of Bovine Heart Cytochrome *c*
Oxidase
  S. Yoshikawa, K. Shinzawa-Itoh, and T. Tsukihara ..................   13

Cooperation of Two Quinone-Binding Sites in the Oxidation of Substrates
by Cytochrome *bo*
  M. Sato-Watanabe, T. Mogi, H. Miyoshi, and Y. Anraku .............   24

Rapid Formation of a Semiquinone Species on Oxidation of Quinol by the
Cytochrome *bo₃* Oxidase from *Escherichia coli*
  J.P. Osborne, S.M. Musser, B.E. Schultz, D.E. Edmondson, S.I. Chan,
  and R.B. Gennis ..................................................   33

Coupling of Ion and Charge Movements: From Peroxidases to
Protonmotive Oxidases
  P.R. Rich, B. Meunier, and S. Jünemann ...........................   40

Ligand Dynamics in the Binuclear Site in Cytochrome Oxidase
  G.T. Babcock, G. Deinum, J. Hosler, Y. Kim, M. Pressler,
  D.A. Proshlyakov, H. Schelvis, C. Varotsis, and S. Ferguson-Miller ........   47

Time-Resolved Resonance Raman Study of Dioxygen Reduction by
Cytochrome c Oxidase
    T. Kitagawa, T. Ogura, S. Hirota, D.A. Proshlyakov, J. Matysik,
    E.H. Appelman, K. Shinzawa-Itoh, and S. Yoshikawa ................    57

Human Cytochrome c Oxidase Analyzed with Cytoplasts
    Y. Kagawa ........................................................    72

Redox Behavior of Copper A in Cytochrome Oxidase in the Brain In Vivo:
Its Clinical Significance
    Y. Hoshi, H. Eda, O. Hazeki, Y. Nomura, Y. Kakihana, S. Kuroda, and
    M. Tamura .......................................................    84

Crystallization of Bovine Heart Mitochondrial Cytochrome c Oxidase for
X-Ray Diffraction at Atomic Resolution (2.8 Å)
    K. Shinzawa-Itoh, R. Yaono, R. Nakashima, H. Aoyama, E. Yamashita,
    T. Tomizaki, T. Tsukihara, and S. Yoshikawa ......................    98

Mechanism of Dioxygen Reduction by Cytochrome c Oxidase as Studied by
Time-Resolved Resonance Raman Spectroscopy
    T. Ogura, D.A. Proshlyakov, J. Matysik, E.H. Appelman,
    K. Shinzawa-Itoh, S. Yoshikawa, and T. Kitagawa ..................    102

Coupling of Proton Transfer to Oxygen Chemistry in Cytochrome Oxidase:
The Roles of Residues I67 and E243
    B. Meunier and P.R. Rich .......................................    106

Respiration of Helicobacter pylori, cb-Type Cytochrome c Oxidase, and
Inhibition of NADH Oxidation by $O_2$
    N. Sone, S. Tsukita, K. Nagata, and T. Tamura ...................    112

The Regulation by ATP of the Catalytic Activity and Molecular State of
Thiobacillus novellus Cytochrome c Oxidase
    K. Shoji, K. Hori, M. Tanigawa, and T. Yamanaka ..................    120

## Part 2   Cytochrome P-450 Monooxygenases

Structure–Function Studies of P-450BM-3
    S.E. Graham-Lorence and J.A. Peterson .........................    127

Oxygen Binding to $P-450_{cam}$ Induces Conformational Changes of
Putidaredoxin in the Ferrous $P-450_{cam}$–Reduced Putidaredoxin Complex
    H. Shimada, M. Unno, Y. Kimata, R. Makino, F. Masuya, T. Obata,
    H. Hori, and Y. Ishimura ........................................    139

Crystal Structure of Nitric Oxide Reductase Cytochrome $P-450_{nor}$ from
Fusarium oxysporum
    S.-Y. Park, H. Shimizu, S. Adachi, Y. Shiro, T. Iizuka, and H. Shoun ....    147

Analysis of NAD(P)H Binding Site of Cytochrome $P-450_{nor}$
    N. Takaya, T. Kudo, and H. Shoun ..............................    156

Substitutions of Artificial Amino Acids $O$-Methyl-Thr, $O$-Methyl-Asp,
$S$-Methyl-Cys, and 3-Amino-Ala for Thr-252 of Cytochrome P-450$_{cam}$:
Probing the Importance of the Hydroxyl Group of Thr-252 for
Oxygen Activation
    Y. Kimata, H. Shimada, T. Hirose, and Y. Ishimura .................. 161

From a Monooxygenase to an Oxidase: The P-450 BM3 Mutant T268A
    G. Truan and J.A. Peterson ...................................... 166

Thiolate Adducts of the Myoglobin Cavity Mutant H93G as Models for
Cytochrome P-450
    M.P. Roach, S. Franzen, P.S.H. Pang, W.H. Woodruff, S.G. Boxer, and
    J.H. Dawson ..................................................... 172

Pronounced Effects of Axial Thiolate Ligand on Oxygen Activation by
Iron Porphyrin
    T. Higuchi, Y. Urano, M. Hirobe, and T. Nagano .................... 181

Superoxide via Peroxynitrite Blocks Prostacyclin Synthesis
    V. Ullrich and M. Zou ........................................... 189

The N=N Bond Cleavage of Angeli's Salt Is Markedly Enhanced by
Cytochrome P-450 1A2: Effects of Distal Amino Acid Mutations on the
Formation of Nitric Oxide Complexes
    Y. Shibata, H. Sato, I. Sagami, and T. Shimizu .................... 199

Langmuir–Blodgett Films of Cytochrome P-450$_{scc}$: Molecular Organization
and Thermostability
    O.L. Guryev, A.V. Krivosheev, and S.A. Usanov .................... 204

Kinetic Analysis of Successive Reactions Catalyzed by Cytochromes
P-450$_{17\alpha,lyase}$ and P-450$_{11\beta}$
    T. Yamazaki, H. Tagashira-Ikushiro, T. Ohno, T. Imai, and
    S. Kominami ..................................................... 214

Changing Substrate Specificity and Product Pattern in Adrenal
Cytochrome P-450-Dependent Steroid Hydroxylases
    B. Böttner, P. Cao, and R. Bernhardt ............................. 221

Inhibition Studies of Steroid Conversions Mediated by Human CYP11B1
and CYP11B2 Expressed in Cell Cultures
    K. Denner and R. Bernhardt ...................................... 231

Effects of ACTH and Angiotensin II on the Novel Cell Layer Without
Corticosteroid-Synthesizing Activity in Rat Adrenal Cortex
    H. Miyamoto, F. Mitani, K. Mukai, and Y. Ishimura ................ 237

Cell-Specific and Hormonally Regulated Gene Expression Directed by the
CYP11B1 Gene Promoter in Rat Adrenal Cortex
    K. Mukai, F. Mitani, and Y. Ishimura ............................. 244

Gene Organization and Genetic Defects of Bilirubin
UDP-Glucuronosyltransferase
    Y. Emi, S. Ikushiro, and T. Iyanagi .............................. 248

Imaging of Calcium Oscillations and Activity of Cytochrome P-450$_{scc}$ in
Adrenocortical Cells
   T. Kimoto, H. Mukai, R. Homma, T. Bettou, D. Nishimura, Y. Ohta, and
   S. Kawato .................................................................. 252

## Part 3    Various Types of Oxidases and Oxygenases

Fundamentally Divergent Strategies for Oxygen Activation by $Fe^{2+}$ and $Fe^{3+}$
Catecholic Dioxygenases
   J.D. Lipscomb, A.M. Orville, R.W. Frazee, M.A. Miller, and
   D.H. Ohlendorf ........................................................... 263

Three-Dimensional Structure of an Extradiol-Type Catechol Ring Cleavage
Dioxygenase BphC derived from *Pseudomonas* sp. strain KKS102: Structural
Features Pertinent to Substrate Specificity and Reaction Mechanisms
   T. Senda, K. Sugimoto, T. Nishizaki, M. Okano, T. Yamada, E. Masai,
   M. Fukuda, and Y. Mitsui ................................................. 276

Probing the Reaction Mechanism of Protocatechuate 3,4-Dioxygenase
with X-Ray Crystallography
   A.M. Orville, J.D. Lipscomb, and D.H. Ohlendorf ..................... 282

Nitric Oxide Synthases: Structure, Function, and Control
   D. Harris, S.M.E. Smith, C. Brown, and J.C. Salerno ................. 289

Electron Paramagnetic Resonance Studies on Substrate Binding to the
NO Complex of Neuronal Nitric Oxide Synthase
   C.T. Migita, J.C. Salerno, B.S.S. Masters, and M. Ikeda-Saito ........... 298

Heme Oxygenase: A Central Enzyme of Oxygen-Dependent Heme
Catabolism and Carbon Monoxide Synthesis
   M. Ikeda-Saito, H. Fujii, K.M. Matera, S. Takahashi, C.T. Migita,
   D.L. Rousseau, and T. Yoshida ........................................... 304

Heme Degradation Mechanism by Heme Oxygenase: Conversion of
α-*meso*-Hydroxyheme to Verdoheme $IX_\alpha$
   H. Fujii, K.M. Matera, S. Takahashi, C.T. Migita, H. Zhou, T. Yoshida,
   and M. Ikeda-Saito ....................................................... 315

A Spectroscopic Study on the Intermediates of Heme Degradation by
Heme Oxygenase
   Y. Omata and M. Noguchi ............................................... 322

Expression of Heme Oxygenase and Inducible Nitric Oxide Synthase mRNA
in a Human Glioblastoma Cell Line
   E. Hara, K. Takahashi, H. Fujita, and S. Shibahara .................... 328

The Mechanism of Conversion of Xanthine Dehydrogenase to
Xanthine Oxidase
   T. Nishino, K. Okamoto, S. Nakanishi, H. Hori, and T. Nishino ........ 333

Mechanism-Based Molecular Design of Peroxygenases
  Y. Watanabe, S. Ozaki, and T. Matsui ............................  340

Catalytic Roles of the Distal Site Hydrogen Bond Network of Peroxidases
  S. Nagano, M. Tanaka, K. Ishimori, I. Morishima, Y. Watanabe,
  M. Mukai, T. Ogura, and T. Kitagawa ...........................  354

Catalytic Intermediates of Polyethylene-Glycolated Horseradish Peroxidase
in Benzene
  S. Ozaki, Y. Inada, and Y. Watanabe ............................  359

Formation of a Hydroperoxy Complex of Heme During the Reaction of
Ferric Myoglobin with $H_2O_2$
  T. Egawa, H. Shimada, and Y. Ishimura .........................  363

Photooxidation in Ternary System Human Serum Albumin–Chlorin
$e_6$–Tryptophan
  E.V. Petrotchenko, G.A. Kochubeev, S.A. Usanov, and P.A. Kiselev .....  367

## Part 4    Oxygen Sensing and Regulation of Blood Flow

Tissue Oxygen Pressure and Oxygen Sensing by the Carotid Body
  D.F. Wilson, S.A. Vinogradov, A. Mokashi, A. Pastuszko, S. Lahiri,
  and M.W. Dewhirst .............................................  377

Molecular Mechanisms of Hypoxia-Induced Angiogenesis
  E. Ikeda, A. Damert, and W. Risau ..............................  388

Mechanisms of Pulmonary Vasodilatation and Ductus Arteriosus
Constriction by Normoxia
  E.K. Weir, H.L. Reeve, S. Tolarova, D.N. Cornfield, D.P. Nelson, and
  S.L. Archer ....................................................  400

Biological Impediment to Oxygen Sensing in Injured Pulmonary
Microcirculation Exposed to a High-Oxygen Environment
  K. Yamaguchi, K. Suzuki, K. Nishio, T. Aoki, Y. Suzuki, A. Miyata,
  N. Sato, K. Naoki, and H. Kudo ................................  410

Transcriptional Responses Mediated by Hypoxia-Inducible Factor 1
  G.L. Semenza, F. Agani, N. Iyer, B.-H. Jiang, E. Laughner, S. Leung,
  R. Roe, C. Wiener, and A. Yu .................................  421

Differences in Particle Size and Oxygen-Binding Affinity Between
Cross-Linked Hemoglobin and Hemoglobin Vesicle
  S. Takeoka, Y. Mano, and E. Tsuchida .........................  428

Activation of Gene Expression of Collagenase and ICAM-1 by
UVA Radiation and by Exposure to Singlet Oxygen
  K. Briviba, M. Wlaschek, K. Scharffetter-Kochanek, S. Grether-Beck,
  J. Krutmann, and H. Sies ......................................  434

Redox Regulation of the Nuclear Factor Kappa B (NF-κB) Signaling
Pathway and Disease Control
  T. Okamoto, S. Sakurada, J.-P. Yang, and N. Takahashi ..............  438

Thioredoxin and Its Involvement in the Redox Regulation of Transcription
Factors, NF-κB and AP-1
  T. Ohno, K. Hirota, H. Nakamura, H. Masutani, T. Sasada, and
  J. Yodoi .................................................................. 450

Inhibition of Cytokines and ICAM-1 Induction in Rheumatoid Fibroblasts
by anti-NF-κB Reagents
  S. Sakurada, T. Kato, K. Mashiba, J.-P. Yang, and T. Okamoto ......... 457

Detection of a Nuclear 60-kDa Protein Coimmunoprecipitated with
Human Thioredoxin
  A. Nishiyama, K. Furuke, K. Hirota, H. Masutani, and J. Yodoi ........ 464

Ito Cells: A Putative Cellular Component Responsible for Carbon Monoxide-
Mediated Microvascular Relaxation in the Rat Liver
  Y. Wakabayashi, S. Kashiwagi, N. Goda, Y. Ishimura, and M. Suematsu .. 469

K⁺ Channels and the Normoxic Constriction of the Rabbit Ductus Arteriosus
  H.L. Reeve, M. Tristani-Firouzi, S. Tolarova, S.L. Archer, and
  E.K. Weir ................................................................ 473

Modulation of Adhesion Molecule Expression in Pulmonary Vascular
Endothelium by Oxygen
  Y. Suzuki, T. Aoki, K. Nishio, O. Takeuchi, K. Toda, K. Suzuki,
  A. Miyata, N. Sato, K. Naoki, H. Kudo, and K. Yamaguchi ............. 479

Cross Talk Between Nitric Oxide and Cyclooxygenase Pathways in
Glomerular Mesangial Cells
  T. Tetsuka and A.R. Morrison ....................................... 484

## Part 5    Pathophysiology and Physiology of Gaseous Monoxides

Pathophysiological Reactivities of Nitric Oxide
  A.J. Gow, R. Foust III, M. McClelland, S. Malcolm, and
  H. Ischiropoulos ..................................................... 493

Defenses Against Peroxynitrite
  H. Sies, H. Masumoto, V. Sharov, and K. Briviba .................... 505

Bioactive 6-Nitronorepinephrine Formation Requiring Nitric Oxide
Synthase in Mammalian Brain
  T. Nakaki, F. Shintani, S. Kanba, E. Suzuki, and M. Asai ........... 510

Role of Nitric Oxide in the Regulation of Cerebral Blood Flow in
Conscious Rats
  S. Takahashi and L. Sokoloff ....................................... 518

Physiological Implication of Induction of Heme Oxygenase-1
Gene Expression
  S. Shibahara ....................................................... 537

Carbon Monoxide: Toxic Waste or Endogenous Modulator of
Hepatobiliary Function?
    M. Suematsu, T. Sano, S. Takeoka, Y. Wakabayashi, T. Yonetani,
    E. Tsuchida, and Y. Ishimura ...................................    544

Microspectrophotometry of Nitric Oxide-Dependent Changes in
Hemoglobin in Single Red Blood Cells Incubated with
Stimulated Macrophages
    T. Shiraishi, K. Tsujita, and K. Kakinuma .......................    550

An Antioxidant Role of Nitric Oxide in Modulation of Oxidative Stress in
Human Placental Trophoblastic Cells
    N. Goda, M. Natori, M. Suematsu, K. Kiyokawa, Y. Ishimura,
    Y. Yoshimura, and S. Nozawa ....................................    557

Liberation of Nitric Oxide from S-Nitrosothiols
    M. Kashiba-Iwatsuki, K. Kitoh, M. Nishikawa, E.F. Sato, and
    M. Inoue .......................................................    562

Change of Nitric Oxide Concentration in Exhaled Gas After Lung Resection
    H. Horinouchi, M. Kohno, M. Gika, A. Tajima, K. Kuwabara,
    A. Yoshizu, M. Naruke, Y. Izumi, M. Kawamura, K. Kikuchi, and
    K. Kobayashi ...................................................    569

Induction of NADPH Cytochrome P-450 Reductase in Kupffer Cells After
Chronic Ethanol Consumption Associated with Increase of Superoxide
Anion Formation
    H. Yokoyama, Y. Akiba, M. Fukuda, Y. Okamura, T. Mizukami,
    M. Matsumoto, H. Suzuki, and H. Ishii ..........................    576

In Vivo Measurement of Superoxide in the Cerebral Cortex Utilizing
Cypridina luciferin analog (MCLA) Chemiluminescence
    D. Uematsu, Y. Fukuuchi, N. Araki, S. Watanabe, Y. Itoh, and
    K. Yamaguchi ...................................................    580

Effects of Zinc Protoporphyrin IX, a Heme Oxygenase Inhibitor, on
Mitochondrial Membrane Potential in Rat Cultured Hepatocytes
    Y. Shinoda, M. Suematsu, Y. Wakabayashi, and Y. Ishimura ........    585

Structure and Function of NO Reductase with Oxygen-Reducing Activity
    T. Fujiwara, T. Akiyama, and Y. Fukumori .......................    591

Oxidative Modification of Apolipoprotein E3 and Its Biological Significance
    S. Hara, T. Tanaka, M. Yamada, Y. Nagasaka, and K. Nakamura .....    597

Sequential Multistep Mechanisms for Leukocyte Adhesion: Applicable to
Lung Microcirculation?
    T. Aoki, Y. Suzuki, K. Suzuki, A. Miyata, K. Nishio, N. Sato,
    K. Naoki, H. Kudo, H. Tsumura, and K. Yamaguchi ................    603

Role of Nitric Oxide in Autoregulation of Cerebral Blood Flow in the Rat
    K. Tanaka, Y. Fukuuchi, T. Shirai, S. Nogawa, H. Nozaki, E. Nagata,
    T. Kondo, S. Koyama, and T. Dembo .............................    609

Key Word Index ....................................................    617

# List of Contributors

Adachi, S.    147
Agani, F.    421
Akiba, Y.    576
Akiyama, T.    591
Anraku, Y.    24
Aoki, T.    410, 479, 603
Aoyama, H.    98
Appelman, E.H.    57, 102
Araki, N.    580
Archer, S.L.    400, 473
Asai, M.    510

Babcock, G.T.    47
Bernhardt, R.    221, 231
Bettou, T.    252
Böttner, B.    221
Boxer, S.G.    172
Briviba, K.    434, 505
Brown, C.    289

Cao, P.    221
Chan, S.I.    33
Cornfield, D.N.    400

Damert, A.    388
Dawson, J.H.    172
Deinum, G.    47
Dembo, T.    609
Denner, K.    231
Dewhirst, M.W.    377

Eda, H.    84

Edmondson, D.E.    33
Egawa, T.    363
Emi, Y.    248

Ferguson-Miller, S.    47
Foust III, R.    493
Franzen, S.    172
Frazee, R.W.    263
Fujii, H.    304, 315
Fujita, H.    328
Fujiwara, T.    591
Fukuda, Masahiko    576
Fukuda, Masao    276
Fukumori, Y.    591
Fukuuchi, Y.    580, 609
Furuke, K.    464

Gennis, R.B.    33
Gika, M.    569
Goda, N.    469, 557
Gow, A.J.    493
Graham-Lorence, S.E.    127
Grether-Beck, S.    434
Guryev, O.L.    204

Hara, E.    328
Hara, S.    597
Harris, D.    289
Hazeki, O.    84
Higuchi, T.    181
Hirobe, M.    181
Hirose, T.    161

Hirota, K.      450, 464
Hirota, S.      57
Homma, R.      252
Hori, H.      139, 333
Hori, K.      120
Horinouchi, H.      569
Hoshi, Y.      84
Hosler, J.      47

Iizuka, T.      147
Ikeda, E.      388
Ikeda-Saito, M.      298, 304, 315
Ikushiro, S.      248
Imai, T.      214
Inada, Y.      359
Inoue, M.      562
Ischiropoulos, H.      493
Ishii, H.      576
Ishimori, K.      354
Ishimura, Y.      139, 161, 237, 244,
    363, 469, 544, 557, 585
Itoh, Y.      580
Iwata, S.      3
Iyanagi, T.      248
Iyer, N.      421
Izumi, Y.      569

Jiang, B.-H.      421
Jünemann, S.      40

Kagawa, Y.      72
Kakihana, Y.      84
Kakinuma, K.      550
Kanba, S.      510
Kashiba-Iwatsuki, M.      562
Kashiwagi, S.      469
Kato, T.      457
Kawamura, M.      569
Kawato, S.      252
Kikuchi, K.      569
Kim, Y.      47
Kimata, Y.      139, 161
Kimoto, T.      252
Kiselev, P.A.      367
Kitagawa, T.      57, 102, 354

Kitoh, K.      562
Kiyokawa, K.      557
Kobayashi, K.      569
Kochubeev, G.A.      367
Kohno, M.      569
Kominami, S.      214
Kondo, T.      609
Koyama, S.      609
Krivosheev, A.V.      204
Krutmann, J.      434
Kudo, H.      410, 479, 603
Kudo, T.      156
Kuroda, S.      84
Kuwabara, K.      569

Lahiri, S.      377
Laughner, E.      421
Leung, S.      421
Lipscomb, J.D.      263, 282

Makino, R.      139
Malcolm, S.      493
Mano, Y.      428
Masai, E.      276
Mashiba, K.      457
Masters, B.S.S.      298
Masumoto, H.      505
Masutani. H.      450, 464
Masuya, F.      139
Matera, K.M.      304, 315
Matsui, T.      340
Matsumoto, M.      576
Matysik, J.      57, 102
McClelland, M.      493
Meunier, B.      40, 106
Michel, H.      3
Migita, C.T.      298, 304, 315
Miller, M.A.      263
Mitani, F.      237, 244
Mitsui, Y.      276
Miyamoto, H.      237
Miyata, A.      410, 479, 603
Miyoshi, H.      24
Mizukami, T.      576
Mogi, T.      24

Mokashi, A.    377
Morishima, I.    354
Morrison, A.R.    484
Mukai, H.    252
Mukai, K.    237, 244
Mukai, M.    354
Musser, S.M.    33

Nagano, S.    354
Nagano, T.    181
Nagasaka, Y.    597
Nagata, E.    609
Nagata, K.    112
Nakaki, T.    510
Nakamura, H.    450
Nakamura, K.    597
Nakanishi, S.    333
Nakashima, R.    98
Naoki, K.    410, 479, 603
Naruke, M.    569
Natori, M.    557
Nelson, D.P.    400
Nishikawa, M.    562
Nishimura, D.    252
Nishino, Takeshi    333
Nishino, Tomoko    333
Nishio, K.    410, 479, 603
Nishiyama, A.    464
Nishizaki, T.    276
Nogawa, S.    609
Noguchi, M.    322
Nomura, Y.    84
Nozaki, H.    609
Nozawa, S.    557

Obata, T.    139
Ogura, T.    57, 102, 354
Ohlendorf, D.H.    263, 282
Ohno, Takashi    214
Ohno, Tetsuya    450
Ohta, Y.    252
Okamoto, K.    333
Okamoto, T.    438, 457
Okamura, Y.    576
Okano, M.    276

Omata, Y.    322
Orville, A.M.    263, 282
Osborne, J.P.    33
Ostermeier, C.    3
Ozaki, S.    340, 359

Pang, P.S.H.    172
Park, S.-Y.    147
Pastuszko, A.    377
Peterson, J.A.    127, 166
Petrotchenko, E.V.    367
Pressler, M.    47
Proshlyakov, D.A.    47, 57, 102

Reeve, H.L.    400, 473
Rich, P.R.    40, 106
Risau, W.    388
Roach, M.P    172
Roe, R.    421
Rousseau, D.L.    304

Sagami, I.    199
Sakurada, S.    438, 457
Salerno, J.C.    289, 298
Sano, T.    544
Sasada, T.    450
Sato, E.F.    562
Sato, H.    199
Sato, N.    410, 479, 603
Sato-Watanabe, M.    24
Scharffetter-Kochanek, K.    434
Schelvis, H.    47
Schultz, B.E.    33
Semenza, G.L.    421
Senda, T.    276
Sharov, V.    505
Shibahara, S.    328, 537
Shibata, Y.    199
Shimada, H.    139, 161, 363
Shimizu, H.    147
Shimizu, T.    199
Shinoda, Y.    585
Shintani, F.    510
Shinzawa-Itoh, K.    13, 57, 98, 102
Shirai, T.    609

Shiraishi, T.    550
Shiro, Y.    147
Shoji, K.    120
Shoun, H.    147, 156
Sies, H.    434, 505
Smith, S.M.E.    289
Sokoloff, L.    518
Sone, N.    112
Suematsu, M.    469, 544, 557, 585
Sugimoto, K.    276
Suzuki, E.    510
Suzuki, H.    576
Suzuki, K.    410, 479, 603
Suzuki, Y.    410, 479, 603

Tagashira-Ikushiro, H.    214
Tajima, A.    569
Takahashi, K.    328
Takahashi, N.    438
Takahashi, Satoshi    304, 315
Takahashi, Shin-ichi    518
Takaya, N.    156
Takeoka, S.    428, 544
Takeuchi, O.    479
Tamura, M.    84
Tamura, T.    112
Tanaka, K.    609
Tanaka, M.    354
Tanaka, T.    597
Tanigawa, M.    120
Tetsuka, T.    484
Toda, K.    479
Tolarova, S.    400, 473
Tomizaki, T.    98
Tristani-Firouzi, M.    473
Truan, G.    166
Tsuchida, E.    428, 544
Tsujita, K.    550
Tsukihara, T.    13, 98
Tsukita, S.    112

Tsumura, H.    603

Uematsu, D.    580
Ullrich, V.    189
Unno, M.    139
Urano, Y.    181
Usanov, S.A.    204, 367

Varotsis, C.    47
Vinogradov, S.A.    377

Wakabayashi, Y.    469, 544, 585
Watanabe, S.    580
Watanabe, Y.    340, 354, 359
Weir, E.K.    400, 473
Wiener, C.    421
Wilson, D.F.    377
Wlaschek, M.    434
Woodruff, W.H.    172

Yamada, M.    597
Yamada, T.    276
Yamaguchi, Kazuhiro    410, 479, 603
Yamaguchi, Keiji    580
Yamanaka, T.    120
Yamashita, E.    98
Yamazaki, T.    214
Yang, J.-P.    438, 457
Yaono, R.    98
Yodoi, J.    450, 464
Yokoyama, H.    576
Yonetani, T.    544
Yoshida, T.    304, 315
Yoshikawa, S.    13, 57, 98, 102
Yoshimura, Y.    557
Yoshizu, A.    569
Yu, A.    421

Zhou, H.    315
Zou, M.    189

# Opening Remarks*

PROFESSOR YASUHIKO TORII
PRESIDENT, KEIO UNIVERSITY
CHAIRMAN, KEIO UNIVERSITY MEDICAL SCIENCE FUND

Ladies and Gentlemen, Distinguished Guests:

I have great pleasure in extending to you a cordial welcome on behalf of Keio University and the Keio University Medical Science Fund. I am particularly grateful to those scientists who traveled such far distances from every part of the world in order to participate in this symposium, the first Keio University International Symposium for Life Sciences and Medicine. The special topic chosen for this occasion is "Oxygen Homeostasis and Its Dynamics," which, I believe, is essential to understand the principle of all forms of life.

There are several reasons for us at Keio University to host such an International Symposium for Life Sciences and Medicine, an occasion for international scientific exchange. To explain the reasons, I would like to give you a short history of Keio University and of the Keio University Medical Science Fund, though I will be brief.

Keio Gijuku, now Keio University, was founded in 1858 by Yukichi Fukuzawa, a pioneer of modern civilization in Japan. I assume some of you are already familiar with his personal appearance, because his portrait is on the 10 thousand-yen note of Japanese currency. In the more than 138 years since its establishment, we are proud that Keio, as Japan's oldest among 587 universities, has played a major role in developing human resources including academic, business, and political leaders. The present prime minister, Ryutaro Hashimoto, is one of our alumni. At Keio University we now have eight faculties and nine graduate schools, and among the faculties, the school of medicine is one of the most highly regarded medical schools in Japan.

We have also been carrying out, for many years, a wide range of international exchanges with people from various countries. In this connection, I would like to tell you that Yukichi Fukuzawa, the founder, was a member of the very first mission of the Tokugawa Shogunate government to the United States in 1860. Before that year, Japan had closed its door to the world for almost 300 years until Admiral Perry (Matthew Calbraith Perry) knocked on our door in 1853.

---

* This opening address was given by Professor Yasuhiko Torii, President of Keio University, at the opening session of the Keio University International Symposium for Life Sciences and Medicine: 1996 Conference on Oxygen Homeostasis and Its Dynamics on the morning of Monday, December 9, 1996, in the conference room on the 4th floor of the New North Building of the Mita campus, Keio University (Y.I.)

During his visits to the United States and Europe as a member of the Japanese Official Mission, Mr. Fukuzawa realized that education was most important to the future of Japan, and therefore, after coming back to Japan, he established Keio Gijuku—now we call it Keio University—in Tokyo. Thus Keio has its origin in international exchanges and has long aimed for international exchanges of culture and science with many countries. Please understand that international exchanges such as this occasion have been one of the most important academic and social missions of Keio University from its birth.

In the fall of 1994, Dr. Mitsunada Sakaguchi, a 1940 alumnus of the medical school, donated 5 billion yen to the university expressing his wish that it be used to encourage research in life sciences and medicine at Keio University and to promote world-wide advances in biomedical sciences. Being a political economist especially interested in the nation's health-care policy, I totally sympathized and agreed with his wishes, and thus launched the Keio University Medical Science Fund on April 1995 in order to fully reflect Dr. Sakaguchi's unwavering commitment to the cause of medical progress. The International Symposium for Life Sciences and Medicine has thus been organized as one of the several projects of the Keio University Medical Science Fund whose objective is, let me stress again, to contribute to the international community by developing human resources, promoting scientific knowledge, and encouraging mutual exchange.

Time flies. The year 1996 is passing by, and as we witness the dawn of the 21st century, we realize that our society faces many problems from this century which will be carried over into the next. In the field of life sciences and medicine alone, we are still unable to completely cure cancer or AIDS. In addition, many new and unknown problems await us in the new century. We will have to overcome numerous obstacles, including diseases and problems that arise with over-sophisticated civilization and the aging of our population.

I believe that exploring new horizons in life sciences is one of the most vital tasks that we face at the dawn of the 21st century. It is equally important to ensure that the knowledge obtained through these horizons is used in ways which bring genuine happiness to humankind. Conceived in the belief and philosophy I have described, Keio University has organized this first Keio University International Symposium for Life Sciences and Medicine. It is therefore more than a pleasure, and indeed an honor, for me to meet you distinguished medical researchers from world-renowned institutions, and to share and exchange views and opinions in the field of medicine and the life sciences. I also am grateful for the efforts of the organizing committee, chaired by Professor Yuzuru Ishimura, who devoted themselves to making the symposium a high-quality and enjoyable one.

Finally I do hope that this symposium will be both fruitful and productive for all of you. Let me close this address now by wishing you the best of health and further success in your research. Thank you very much for your attention.

# Part 1
# Cytochrome Oxidases

# Structure and Possible Mechanism of Action of Cytochrome c Oxidase from the Soil Bacterium Paracoccus denitrificans

Hartmut Michel[1], So Iwata[2], and Christian Ostermeier[3]

*Summary.* The four protein subunits containing cytochrome c oxidase from the soil bacterium *Paracoccus denitrificans* were crystallized with the help of antibody $F_v$ fragments. The structure, determined at 2.8-Å resolution by X-ray crystallography, is reported. This structure forms the basis for understanding the mechanism of this redox-coupled transmembrane proton pump, which is the key component of the respiratory chain of most aerobic organisms.

*Key words.* Membrane protein crystallization—Cytochrome c oxidase—X-ray crystallography—Structure determination—*Paracoccus denitrificans*

## Introduction

Cellular respiration is one of the most fundamental processes of life. Most of the energy available to animals is generated by respiration and its coupling to the synthesis of adenosine 5′-triphosphate (ATP). In the so-called respiratory chain, four large membrane protein complexes act together to oxidize substrates and finally to reduce oxygen. In the respiratory chains of mitochondria and in many bacteria, either NADH or succinate, both formed preferentially in the citric acid cycle, are oxidized by complex I or complex II, respectively, and ubiquinol is generated. Ubiquinol is oxidized by complex III, also known as the cytochrome $bc_1$ complex, and the electrons are transferred to cytochrome c. Cytochrome c is oxidized by complex IV, the cytochrome c oxidase.

The electrons of cytochrome c are used to reduce molecular oxygen, and water is formed. Complexes I, III, and IV are able to transport (or "pump") protons across the membrane, in addition to those protons that are released from ubiquinol on the

[1] Max-Planck-Institut für Biophysik, Heinrich-Hoffmann-Str. 7, D-60528 Frankfurt am Main, Germany
[2] Department of Biochemistry, Uppsala University, Biomedical Centre, P.O. Box 576, S-75123 Uppsala, Sweden
[3] Department of Molecular Biophysics and Biochemistry, Yale University, 266 Whitney Avenue, New Haven, CT 06520-8114, USA

periplasmic side of the bacterial membrane (or in the intermembrane space of mito-chondria) by complex III or consumed on the cytoplasmic (or matrix) side in complex IV on water formation. The electrochemical potential difference of protons is used to drive ATP synthesis by the $H^+$-translocating ATPases. In some sense the respiratory chain catalyzes the detonating gas reaction, but it must ensure that energy is stored in the electrochemical proton gradient and that no dangerous side products are formed, especially in the reaction catalyzed by cytochrome $c$ oxidase. Generation and release of superoxides, peroxides, or singlet oxygen would be dangerous.

The cytochrome $c$ oxidases are members of a large superfamily of heme-copper-containing terminal oxidases (see [1,2] for review; for general reviews see [3,4]), which also includes the cytochrome $bo$ ubiquinol oxidase from *Escherichia coli*. The se-quences of subunits I and II especially are well conserved. Subunit II of cytochrome $c$ oxidases contains the binuclear $Cu_A$ center, which receives the electrons from cyto-chrome $c$ and transfers them to heme $a$ and finally to the binuclear heme $a_3$–$Cu_B$ center. The $Cu_A$ center is absent in the ubiquinol oxidases, but could be restored by genetic engineering [5]. The heme groups whose chemical identity can be different (hemes A, B, or O have been found) and $Cu_B$ are bound to subunit I. The total number of subunits varies from 2 or 3 in some bacteria to 13 in mammalian mitochondria. During the past 10 years site-directed mutagenesis studies combined with spectrosco-py have provided much structural information [2], most of which was correct.

In this chapter, we describe the structure of cytochrome $c$ oxidase from the soil bacterium *Paracoccus denitrificans* as determined by X-ray crystallography [6]. This cytochrome $c$ oxidase has the advantage of being well suited for site-directed mu-tagenesis. Quite surprisingly, at the same time the structure of the metal sites of cytochrome $c$ oxidase from bovine heart mitochondria [7] and, 1 year later, its com-plete protein structure, were also published [8].

# Crystallization

Membrane protein structure determination is limited by the lack of well-ordered crystals. Up to now atomic models based on X-ray or electron crystallographic struc-ture determinations have been available for members of only seven membrane pro-tein families, namely bacterial photosynthetic reaction centers, bacteriorhodopsin, bacterial porins, prostaglandin $H_2$ synthase-1, plant light harvesting complex II, bac-terial light harvesting complexes, and now cytochrome $c$ oxidases. It is remarkable that it took about 20 years to obtain suited crystals of the bovine cytochrome $c$ oxidase using a conventional crystallization strategy. As had been the case for photosynthetic reaction centers, the choice of the detergent was critical.

Crystallization trials with the bacterial enzyme proceeded for about 6 years, but succeeded within 2 years when a novel strategy, namely cocrystallization with an $F_v$ fragment of a conformation-specific monoclonal antibody, was used [9]. The anti-body fragment binds to a discontinuous epitope at the periplasmic side of subunit II, affecting neither cytochrome $c$ oxidation nor proton pumping (Kannt and Michel, unpublished data). The $F_v$ fragment is the only part of the cytochrome $c$ oxidase-$F_v$ fragment complex involved in protein–protein contacts in the a,b-plane of the crystal lattice. It acts by enhancing the polar surfaces of cytochrome $c$ oxidase, but it does not cover any hydrophobic surfaces. The antibody fragment could also be used to isolate the cytochrome *c* oxidase in a rapid and mild way [10].

## The Structure of the *Paracoccus* Cytochrome *c* Oxidase

A view parallel to the membrane of the entire cytochrome *c* oxidase is shown in Fig. 1. The part integrated into the membrane has a trapezoid-like appearance from the direction shown. The width at the cytoplasmic surface is about 90 Å and at the periplasmic surface approximately 75 Å; the height of the trapezoid, which is formed by 22 transmembrane helices, is 75 Å. The globular domain of subunit II is attached to the trapezoid from the periplasmic side. The central part of the complex is made up of subunit I, which binds both hemes and $Cu_B$. Subunit I is associated with subunit II on one side and subunit III on the other. The amino and carboxy termini of subunit II protrude into the periplasmic space and form a globular domain that contains $Cu_A$; the antibody $F_v$ fragment binds to this globular domain. In a view perpendicular to the membrane (Fig. 2), cytochrome *c* oxidase has an oval shape with its largest dimensions being 90 and 60 Å.

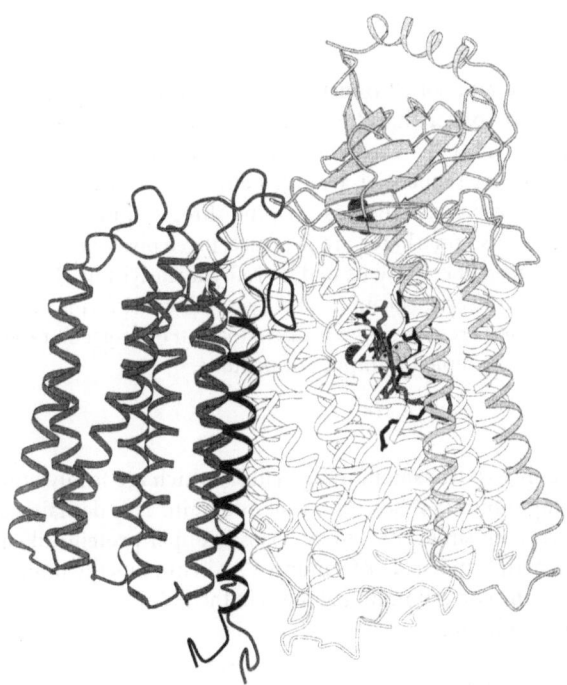

FIG. 1. Ribbon representation of the cytochrome *c* oxidase from *Paracoccus denitrificans* in a view parallel to the membrane. Subunit I, *white*; subunit II, *light gray*; subunit III, *dark gray*; subunit IV, *black*. Heme *a* (*black*), heme $a_3$ (*gray*) and $Cu_B$ bound to subunit I, as well as $Cu_A$ bound to subunit II, are barely visible

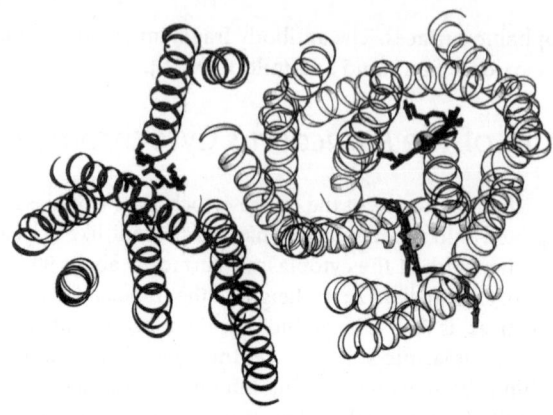

FIG. 2. Major secondary structure elements of cytochrome $c$ oxidase in a view from the periplasmic space, color coded as in Fig. 1. The lipid (*black*) bound into the V-shaped cleft of subunit III is clearly visible.

## Subunit I

Subunit I consists mainly of 12 transmembrane helices. In general, the loops on the periplasmic side are longer than those on the cytoplasmic side. The helices show a fascinating and unexpected arrangement: the 12 closely packed helices, which are in a simple anticlockwise sequential arrangement when viewed from the periplasmic side (see Fig. 2), can be described as forming three symmetry-related semicircular arcs. As a result of this remarkable architecture, three pores are formed. Two of these are blocked by heme $a$, including its hydroxyethylfarnesyl side chain, or heme $a_3$, and the third by mostly conserved aromatic residues. Subunit I forms the core of the entire oxidase complex and is responsible for the oxidation of molecular oxygen to water as well as for the redox-coupled pumping of protons.

## Subunit II

Subunit II has only two transmembrane helices, which are firmly bound to subunit I [11]. It is the only subunit possessing a polar domain. This domain consists of a ten-stranded β-barrel with similarities to the class I copper proteins like plastocyanin. It is bound to the periplasmic side of subunit I and contains the binuclear $Cu_A$ center. The $Cu_A$ center of the corresponding domain in the quinol oxidases is thought to have been lost during evolution.

## Subunits III and IV

Subunit III possesses seven transmembrane helices, arranged in an irregular manner. They form two bundles, one consisting of the first two helices and the other of helices III–VII. Both bundles are in a V-shaped arrangement. The association of subunit III with subunit I is weak [11], and subunits III and IV are removed from the core complex by many detergents. In some members of the oxidase family the first two helices seem to be fused to subunit I, and subunit III possesses only five helices [12]. Subunits II and I can be fused at the DNA level. The resulting construct of the *Escherichia coli* ubiquinol oxidase was found to be active [13]. The idea that subunit III is involved directly in proton pumping has been ruled out; it may play a more indirect role during assembly or as a linker to other proteins.

Furthermore, the structure of the bacterial oxidase indicated a possible channel for diffusion of oxygen leading from the cleft between the two helix bundles of subunit III directly to the binuclear center [14]. In the cleft, subunit III contains at least one firmly bound lipid molecule [6]. The role of subunit III therefore might be to prevent blockage of the channel entrance by other proteins. Because of the high solubility of oxygen in the hydrophobic interior of the membrane generated by the alkyl chains of the lipids, it is likely that the lipids bound to subunit III might aid in the rapid diffusion of oxygen to the heme $a_3$-$Cu_B$ center.

The crystal structure of the *Paracoccus* cytochrome *c* oxidase provided the final proof for the existence of a fourth subunit, which consists of only one transmembrane helix with a small N-terminal extension at the cytoplasmic side. Its function is unknown; deletion of its gene does not cause any phenotype [15]. A number of additional lipid and detergent molecules could be localized around subunit IV.

## The $Cu_A$ Center

As an important difference from the type I copper proteins, $Cu_A$ has been suggested to be a mixed-valence [Cu(1.5)–Cu(1.5)] complex [16–19]. This suggestion agrees with the crystal structure. The ligands of the two copper atoms are the $N_\delta$-atoms of two histidine residues, one methionine sulfur, one backbone carbonyl oxygen from a glutamate residue (which could not be predicted to be a ligand of $Cu_A$), and two cysteine thiolates: the latter bridge the two copper atoms (Fig. 3). Distances between the two copper atoms of 2.6 Å in the bacterial [6], 2.7 Å in the beef heart [7], and 2.5 Å in the reengineered *bo* oxidase [5] have been published.

In the interface region between subunits I and II, a non-redox-active metal can be assigned. A Mg site was modeled in the bovine oxidase structure [7,8], and a Mg/Mn site is found in a similar location in the bacterial oxidase. The function of this metal center is still unclear.

## Heme *a* and Heme $a_3$-$Cu_B$ Center

Heme *a* is a low-spin heme with two axial histidine ligands, whereas heme $a_3$ is a high-spin heme with one histidine ligand. The shortest distance between the two hemes is only 4.7 Å. Both are located 15 Å from the periplasmic surface in the hydrophobic core of subunit I. The heme planes are perpendicular to the membrane, and the interplanar angle between the two heme groups is 108° in the bacterial cytochrome *c* oxidase [6]. The electronic coupling between heme *a* and the binuclear heme $a_3$-$Cu_B$ center is very strong, and fast electron transfer between these two redox centers was recently used to explain the high operational oxygen affinity of the oxidase by "kinetic trapping" of bound oxygen [20].

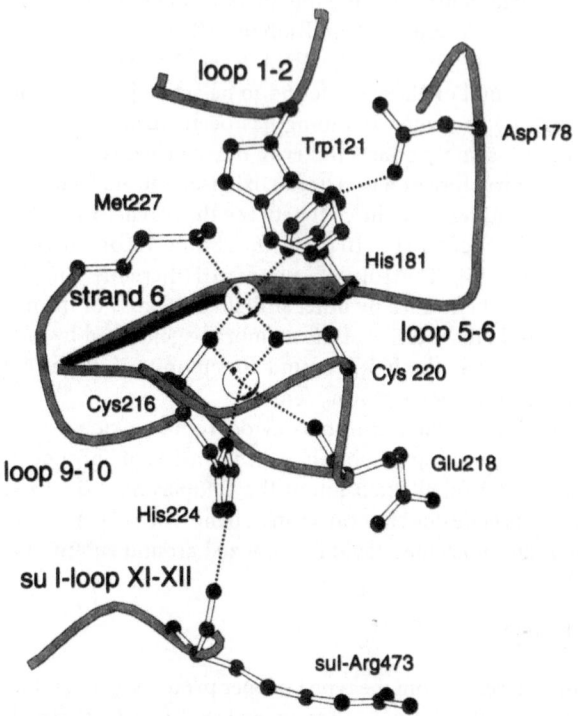

FIG. 3. The $Cu_A$ center. The two $Cu_A$ atoms are shown as *white spheres* bridged by two thiolate sulfurs from cysteine residues. All four atoms lie in one plane. The further $Cu_A$ ligands, all from subunit II, are shown. Tryptophan-121 may form an entry pathway for electrons. Histidine-224 forms a hydrogen bond to the backbone carbonyl oxygen atom of arginine-473 from subunit I and may form the exit pathway for electrons. (From [37], with permission)

The heme $a_3$–$Cu_B$ center is the catalytic core for $O_2$ reduction. The high-spin iron of heme $a_3$ is coordinated by histidine-411 as an axial ligand. The free coordination site of the iron points toward the free coordination site of $Cu_B$, which is ligated by the three histidines: 276, 325, and 326. In the bacterial oxidase, one of the three histidines, His-325, seems to be disordered when the oxidized enzyme has been crystallized in the presence of azide [6]. There is spectroscopic evidence that a bridging ligand may exist between the two metals [21], but there was no evidence for a ligand in the bovine cytochrome $c$ oxidase structure [7]. In contrast, electron density between iron and copper was observed in the bacterial enzyme. The identity of this electron density is unclear. It might be caused by the presence of one or two water molecules, or one water plus one hydroxyl ion, which would be bound to $Cu_B$. The distance between iron and copper is about 5.2 Å in bacterial cytochrome $c$ oxidase. However, the central heme $a_3$ iron atom was found to be 0.7 Å out of plane. The question as to whether there exists a hydroxyl group as a ligand to $Cu_B$ had to remain unanswered at the present resolution of the structure determination.

## Possible Mechanisms

Protein structures are not determined to provide beautiful textbook figures but to understand the function and mechanism of action of the protein under investigation. In the case of cytochrome *c* oxidase, three important mechanistic questions have to be answered: (i) the precise pathway for electron transfer, (ii) the proton transfer pathways, and (iii) the mechanism of coupling proton transfer and proton pumping to oxygen reduction. $Cu_A$ is the first acceptor for electrons delivered from cytochrome *c*. The electrons are then transferred to heme *a*, and heme *a* passes them further to the binuclear center [22]. It is very likely that the endergonic proton pumping reaction is directly coupled to the exergonic redox reactions at the binuclear heme $a_3$–$Cu_B$ center [23].

Many models and hypotheses have been put forward to explain this coupling [23–32]. Inspection of the structure of the bacterial cytochrome *c* oxidase shows two possible proton transfer pathways, which are in agreement with the results of site-directed mutagenesis experiments. Pathway (i) leads along the conserved polar face of transmembrane helix VIII [33]. A prominent residue at this face is lysine-354; this pathway is therefore called the K-pathway. In addition, it includes the hydroxyl group of the hydroxyethylfarnesyl side chain of heme $a_3$ and tyrosine-280. Replacement of

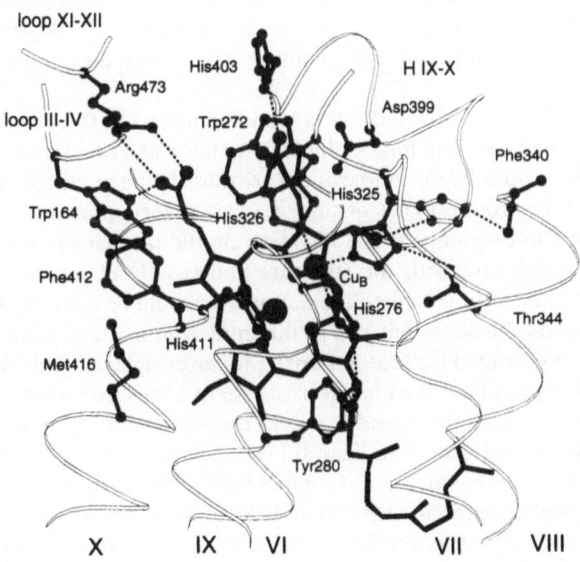

FIG. 4. The binuclear center in subunit I. Heme $a_3$ and some important neighboring residues are shown as *black ball-and-stick models*. The transmembrane helices of subunit I involved are indicated. The heme $a_3$ iron atom and the $Cu_B$ atom are shown as *black spheres*. Histidine-276 and histidine-326 are clear ligands of $Cu_B$. There is no electron density for the side chain of histidine-326, which can be modeled in two different conformations: in one, it is a $Cu_B$ ligand; in the other (*shown in white*) it is not. Switching of the side chain between these two positions might be an essential part of the proton pump mechanism. (From [37], with permission)

the residues of the polar face of transmembrane helix VIII leads to a loss of the enzymatic activity.

Pathway (ii) starts at a conserved aspartate (124) (it is therefore called the D-pathway) at the entrance of the pore formed by helices II–VI (see Fig. 2) and leads into a partly polar cavity containing several solvent molecules. The conserved glutamate residue 278 is found at the end of the cavity. Beyond this glutamate the proton pathway becomes rather speculative. It may lead to histidine-325, which is not visible in the electron density map. Replacing the conserved Asp-124 by Asn, or two Asn residues nearby (113,131) by hydrophobic residues, abolishes proton pumping, but oxygen reduction and water formation still occur albeit at a reduced rate [34–36]. These findings suggest at a first glance that the D-pathway is the one used for protons to be pumped and the K-pathway is that for protons consumed in water formation. However, such an assignment depends on the precise mechanism of coupling proton pumping and redox reactions.

Two papers [28,30] seem to be of special importance for formulating a mechanism of redox-linked proton pumping. Rich [30] postulated the electroneutrality of redox changes around the heme-copper center. It is an attractive hypothesis considering the low polarity of the surrounding hydrophobic membrane environment; this means that on a single reduction of one of the metals a proton has to be taken up for charge compensation and two protons for reduction of both crystals. When in the catalytic cycle protons are taken up from the cytoplasmic side later and consumed in water formation, those protons taken up first are expelled to the periplasmic side by electrostatic interactions and thus pumped. This paper does not present chemical details.

Morgan et al. [28] postulated that a histidine ligand of $Cu_B$ "cycles" between the imidazolate, imidazole, and imidazolium states twice on reduction of one molecule of dioxygen. The protons of the imidazolium are those being pumped. Also in this model the uptake of the protons to be consumed on water formation leads to the expulsion of the protons to be pumped. The histidine shuttle mechanism presented by Iwata et al. [6] is compatible with the structure and strictly obeys the electroneutrality principle. The key residue is histidine-325, which would be a $Cu_B$ ligand in the imidazolate and imidazole states but not in the imidazolium state. In the latter it might assume a position suited for proton transfer to the periplasmic side, which is realized upon arrival of protons at the binuclear site needed for water formation. However, if a hydroxy group were a $Cu_B$ ligand (a special case of the finding of one oxygen with an exchangeable proton(s) as a $Cu_B$ ligand) [21], the first proton taken up on reduction of the binuclear center would unavoidably lead to the formation of water, and the histidine cycle/shuttle mechanism would have to be reformulated.

The idea of having at least one hydroxyl ion between the heme $a_3$ iron and $Cu_B$ in the oxidized form of the enzyme is attractive, because otherwise a strong electrostatic repulsion between the iron atom with one positive charge and $Cu_B$ with two positive charges would exist. Knowledge of the protonation states of the histidine ligands to $Cu_B$, especially of histidine-325, and the pK values for protonation and deprotonation is required to support or to exclude a histidine cycle mechanism. If the histidine ligands stay neutral during the catalytic cycle, alternative mechanisms have to be sought. Of particular interest is the answer to the question of what are the proton acceptors for those protons that are taken up on reduction of cytochrome $c$ oxidase [27].

What is also clearly needed is the determination of the precise structure of the intermediates [23] of the redox reactions of the heme-copper oxidases. There is good reason to believe that the structures of some of these intermediates can be determined by X-ray crystallography using the available cytochrome *c* oxidase crystals.

*Acknowledgments.* We thank Dr. C.R.D. Lancaster and A. Harrenga for preparing some of the figures.

# References

1. Garcia-Horsman JA, Barquera B, Rumbley J, et al. (1994) The superfamily of heme-copper respiratory oxidases. J Bacteriol 176:5587–5600
2. Calhoun MW, Thomas JW, Gennis RB (1994) The cytochrome oxidase superfamily of redox-driven proton pumps. Trends Biochem Sci 19:325–330
3. Saraste M (1990) Structural features of cytochrome oxidase. Q Rev Biophys 23:331–366
4. Malatesta F, Antonini G, Sarti P, et al. (1995) Structure and function of a molecular machine: cytochrome *c* oxidase. Biophys Chem 54:1–33
5. Wilmanns M, Lappalainen P, Kelly M, et al. (1995) Crystal structure of the membrane-exposed domain from a respiratory quinol oxidase complex with an engineered dinuclear copper center. Proc Natl Acad Sci USA 92:11955–11959
6. Iwata S, Ostermeier C, Ludwig B, et al. (1995) Structure at 2.8 Å resolution of cytochrome *c* oxidase from *Paracoccus denitrificans*. Nature 376:660–669
7. Tsukihara T, Aoyama H, Yamashita E, et al. (1995) Structure of metal sites of oxidized bovine heart cytochrome *c* oxidase at 2.8 Å. Science 269:1069–1074
8. Tsukihara T, Aoyama H, Yamashita E, et al. (1996) The whole structure of the 13-subunit oxidised cytochrome *c* oxidase at 2.8 Å. Science 272:1136–1144
9. Ostermeier C, Iwata S, Ludwig B, et al. (1995) $F_v$ fragment-mediated crystallization of the membrane protein bacterial cytochrome *c* oxidase. Nat Struct Biol 2:842–846
10. Kleymann G, Ostermeier C, Ludwig B, et al. (1995) Engineered $F_v$ fragments as a tool for the one-step purification of integral multisubunit membrane protein complexes. Biotechnology 13:155–160
11. Haltia T, Semo N, Arrondo JL, et al. (1994) Thermodynamic and structural stability of cytochrome *c* oxidase from *Paracoccus denitrificans*. Biochemistry 33:9731–9740
12. Castresana J, Lübben M, Saraste M, et al. (1994) Evolution of cytochrome *c* oxidase, an enzyme older than atmospheric oxygen. EMBO J 13:2516–2525
13. Ma J, Lemieux L, Gennis RB (1993) Genetic fusion of subunits I, II, and III of the cytochrome *bo* ubiquinol oxidase from *Escherichia coli* results in a fully assembled and active enzyme. Biochemistry 32:7692–7697
14. Riistama S, Puustinen A, Garcia-Horsman A, et al. (1996) Channelling of dioxygen into the respiratory enzyme. Biochim Biophys Acta 1275:1–4
15. Witt H, Ludwig B (1997) Isolation, analysis, and deletion of the gene coding for subunit IV of cytochrome *c* oxidase in *Paracoccus denitrificans*. J Biol Chem 272:5514–5517
16. Larsson S, Källebring B, Wittung P, et al. (1995) The $Cu_A$ center of cytochrome-*c* oxidase: electronic structure and spectra of models compared to the properties of $Cu_A$ domains. Proc Natl Acad Sci USA 92:7167–7171
17. Farrar JA, Lappalainen P, Zumft WG, et al. (1995) Spectroscopic and mutagenesis studies on the $Cu_A$ centre from the cytochrome-*c* oxidase complex of *Paracoccus denitrificans*. Eur J Biochem 232:294–303
18. Blackburn NJ, Barr ME, Woodruff WH, et al. (1994) Metal-metal bonding in biology: EXAFS evidence for a 2.5 Å copper–copper bond in the $Cu_A$ center of cytochrome oxidase. Biochemistry 33:10401–10407

19. Henkel G, Müller A, Weissgräber S, et al. (1995) The active sites of the native cytochrome-c oxidase from bovine heart mitochondria: EXAFS-spectroscopic characterization of a novel homobinuclear copper center ($Cu_A$) and of the heterobinuclear $Fe_{a3}$-$Cu_B$ Center. Angew Chem Int Ed Engl 34:1488–1492

20. Verkhovsky MI, Morgan JE, Puustinen A, et al. (1996) Kinetic trapping of oxygen in cell respiration. Nature 380:268–270

21. Fann YC, Ahmed I, Blackburn NJ, et al. (1995) Structure of $Cu_B$ in the binuclear heme-copper center of the cytochrome $aa_3$-type quinol oxidase from *Bacillus subtilis*: an ENDOR and EXAFS study. Biochemistry 34:10245–10255

22. Hill BC (1994) Modeling the sequence of electron transfer reactions in the single turnover of reduced, mammalian cytochrome c oxidase. J Biol Chem 269:2419–2425

23. Babcock GT, Wikström M (1992) Oxygen activation and the conservation of energy in cell respiration. Nature 356:301–309

24. Einarsdóttir Ó (1995) Fast reactions of cytochrome oxidase. Biochim Biophys Acta 1229:129–147

25. Einarsdóttir Ó, Geogiadis KE, Sucheta A (1995) Intramolecular electron transfer and conformational changes in cytochrome c oxidase. Biochemistry 34:496–508

26. Hallén S, Brzezinski P, Malmström BG (1994) Internal electron transfer in cytochrome c oxidase is coupled to the protonation of a group close to the bimetallic site. Biochemistry 33:1467–1472

27. Mitchell R, Rich PR (1994) Proton uptake by cytochrome c oxidase on reduction and on ligand binding. Biochim Biophys Acta 1186:19–26

28. Morgan JE, Verkhovsky MI, Wikström M (1994) The histidine cycle: a new model for proton translocation in the respiratory heme-copper oxidases. J Bioenerg Biomembr 26:599–608

29. Musser SM, Chan SI (1995) Understanding the cytochrome c oxidase proton pump: thermodynamics of redox linkage. Biophys J 68:2543–2555

30. Rich PR (1995) Towards an understanding of the chemistry of oxygen reduction and proton translocation in the iron-copper respiratory oxidases. Aust J Plant Physiol 22:479–486

31. Verkhovsky MI, Morgan JE, Wikström M (1995) Control of electron delivery to the oxygen reduction site of cytochrome c oxidase: a role for protons. Biochemistry 34:7483–7491

32. Wikström M (1989) Identification of the electron transfers in cytochrome oxidase that are coupled to proton-pumping. Nature 338:776–778

33. Thomas JW, Lemieux LJ, Alben JO, et al. (1993) Site-directed mutagenesis of highly conserved residues in helix VIII of subunit I of the *bo* ubiquinol oxidase from *Escherichia coli*. An amphipathic transmembrane helix that may be important in conveying protons to the binuclear center. Biochemistry 32:11173–11180

34. Thomas JW, Puustinen A, Alben JO, et al. (1993) Substitution of asparagine-135 in subunit I of the cytochrome *bo* ubiquinol oxidase of *Escherichia coli* eliminates proton-pumping activity. Biochemistry 32:10923–10928

35. Garcia-Horsman JA, Puustinen A, Gennis RB, et al. (1995) Proton transfer in cytochrome $bo_3$ ubiquinol oxidase of *Escherichia coli*: second-site mutations in subunit I that restore proton pumping in the mutant Asp 135 → Asn. Biochemistry 34:4428–4433

36. Fetter JR, Qian J, Shapleigh J, et al. (1995) Possible proton relay pathways in cytochrome c oxidase. Proc Natl Acad Sci USA 92:1604–1608

37. Ostermeier C, Iwata S, Michel H (1996) Cytochrome c oxidase. Curr Opin Struct Biol 6:460–466

# Crystal Structure and Reaction Mechanism of Bovine Heart Cytochrome *c* Oxidase

Shinya Yoshikawa[1], Kyoko Shinzawa-Itoh[1], and Tomitake Tsukihara[2]

*Summary.* Crystal structure of bovine heart cytochrome *c* oxidase at fully oxidized state at 2.8 Å resolution shows that this protein consists of 13 different subunits, each in one copy, and 8 phospholipids in addition to 7 metal ions, 2 irons, 3 coppers, 1 magnesium, and 1 zinc. Three redox active sites, $Cu_A$, heme *a*, and the $O_2$ reduction site containing heme $a_3$ and $Cu_B$, are connected by three possible pathways for facile electron transfers. The pathways between $Cu_A$ and heme *a* and between heme *a* and heme $a_3$ are consistent with the rapid electron transfers determined kinetically. However, the role of the direct pathway between $Cu_A$ and heme $a_3$ is unknown. The coordination geometry of $Cu_B$ together with the proximity between the two hemes suggest that heme *a*, not $Cu_B$, donates electrons to initiate the reduction of $O_2$ in the two electron process. $Tyr^{244}$ is identified as the proton donor for producing water from the intermediates during $O_2$ reduction. Possible proton-pumping sites are mapped well separated from the $O_2$ reduction site. No possible proton-pumping site involving the $O_2$ reduction site has been identified, suggesting an indirect coupling between $O_2$ reduction and proton pumping.

*Key words.* Cytochrome *c* oxidase—Proton pump—$O_2$ reduction—Membrane protein—X-ray crystal structure

## Introduction

Cytochrome *c* oxidase is the terminal oxidase that reduces molecular oxygen ($O_2$) to water coupled with proton pumping across the mitochondrial inner membrane. Since the discovery of this enzyme, many structural and functional studies have been done to understand the reaction mechanism of this intriguing enzyme [1]. The amino acid sequences of this enzyme from more than 80 species have been determined following the pioneering work of Buse et al. for the sequence determination of the bovine heart enzyme [2,3]. The ingenious determination of the subunit composition by Kadenbach

---

[1] Department of Life Science, Faculty of Science, Himeji Institute of Technology, 1479-1 Kanaji, Kamigoricho, Akoh-gun, Hyogo 678-12, Japan
[2] Institute for Protein Research, Osaka University, 3-2 Yamada-oka, Suita, Osaka 565, Japan

13

et al. also should not be ignored [4]. (We followed the terminology of Kadenbach for the subunits of cytochrome *c* oxidase.)

On the other hand, the function of this enzyme was initiated by the landmark investigation of the internal electron transfer with the flow flash method by Gibson and Greenwood [5]. A powerful technique was later introduced for identifying the intermediate species during the course of $O_2$ reduction by the enzyme, such as $Fe^{2+}$–$O_2$, $Fe^{4+}=O$, and $Fe^{3+}$–OH [6,7]. Furthermore, the reaction between the enzyme and cytochrome *c* has been studied steadily for more than 30 years with the initial steady-state analysis [8–10]. A cytochrome *c*-binding site with an extremely high affinity to cytochrome *c* has been proposed in addition to the site with a physiologically relevant affinity [9]. However, the complete analysis including four cytochromes *c* and an $O_2$ molecule as the substrates has not been completed because of the difficulty in the accurate analysis of the initial steady-state measurement and also of the extremely high apparent affinity of the enzyme for $O_2$ [11].

In spite of these structural and functional investigations, the reaction mechanism of this enzyme is still essentially unknown. For example, the roles of metal sites, especially $Cu_B$, in the dioxygen reduction by the enzyme and the mechanism of proton pumping are the two greatest problems. The most important information for solving these problems is the three-dimensional structure of this enzyme at atomic resolution. However, purification and crystallization of such a large multicomponent membrane protein as this enzyme are extremely difficult. Thus, as recently as 1995, crystal structures of the enzyme isolated from beef heart and a soil bacteria, *Paraccocus denitrificans* were reported at atomic resolution, suggesting a new era of this enzyme research [12,13]. This review presents a summary of the crystal structure of beef heart cytochrome *c* oxidase and the possible contribution of the crystal strucuture for understanding the reaction mechanism of this enzyme.

## Structures of the Protein and Phospholipids

Cytochrome *c* oxidase isolated from beef heart and stabilized with a nonionic detergent, decyl maltoside, provides tetragonal crystals that diffract X-rays up to 2.8 Å resolution at 8°C [12]. The crystal structure in the fully oxidized state indicates that this enzyme is in a dimer state. Each monomer has 13 different subunits and 6 metal centers (hemes *a* and *a₃*, $Cu_A$, $Cu_B$, Mg, and Zn) with a total molecular weight of 210 kDa (Fig. 1) [12,14]. The middle part of each monomer consists of 28 α-helices assignable to the transmembrane region in the mitochondrial membrane. A view perpendicular to the membrane surface shows the assembly of the two monomers, with a fairly large "intermonomer space."

Each monomer does not look like the tooth model given by the electron diffraction analysis [15]. The biggest three subunits, which are encoded by mitochondrial genes, aggregate tightly to form a core part of the monomer. The other 10 nuclear-coded subunits surround the core. The biggest subunit, subunit I [16], has the two hemes and $Cu_B$ and Mg sites in the peptide with 12 transmembrane helices. The third biggest subunit with 2 transmembrane helices, subunit II, contains the $Cu_A$ site, which is a dinuclear copper center. Subunit III, the second largest one, has 7 helices. Three nuclear-coded subunits, one on the cytosolic side and the other two on the matrix

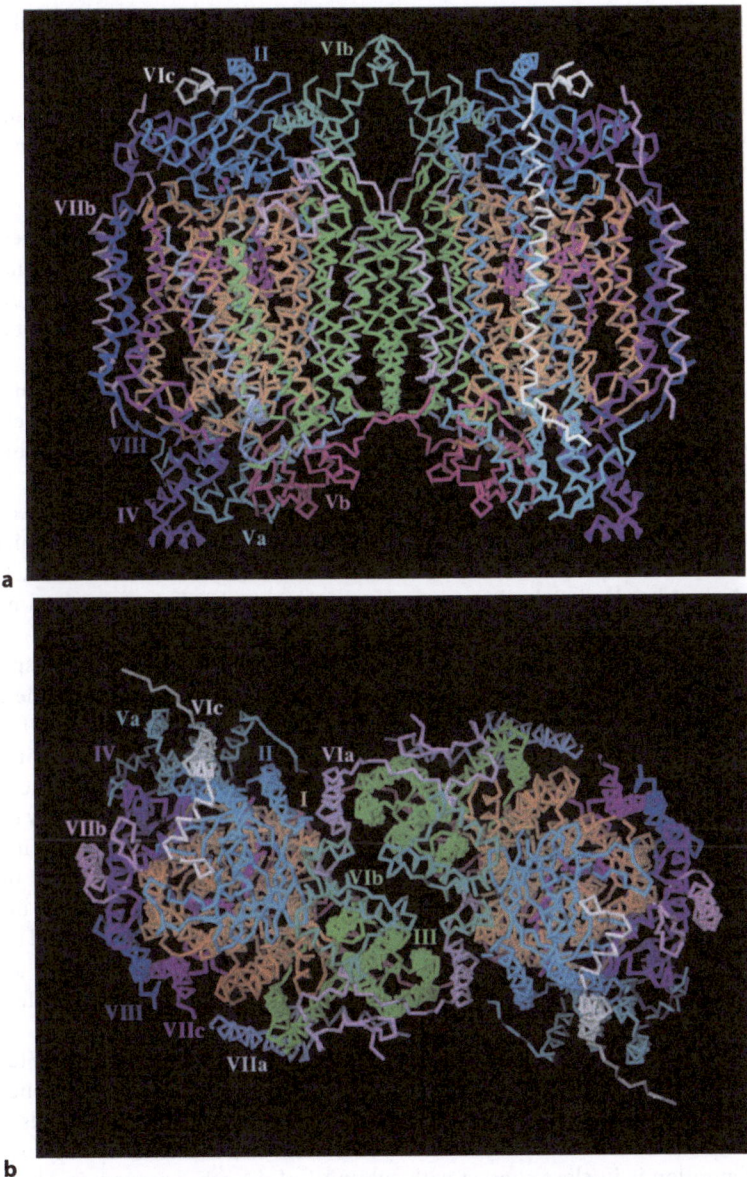

FIG. 1a,b. Crystal structure of dimer of beef heart cytochrome *c* oxidase is shown in the Cα backbone trace. Each monomer consists of 13 different subunits, each shown in a different color with the subunit name in the color of the subunit. **a** Side view against the transmembrane helices. **b** Top view from the cytosolic side

side, have no transmembrane helix. Each of the other seven nuclear-coded subunits contains a single transmembrane helix. Subunit Vb, one of the extramembrane subunits, holds a zinc atom [12,14].

In addition to these peptides and metal sites, five phosphotidyl ethanolamines, three phosphatidyl glycerols, and two cholates are observed in the crystal structure. However, no cardiolipin is detectable, which has long been thought to be indispensable for the enzymic activity. Amino acid residues forming hydrogen bonds or salt bridges with the head groups of the phospholipids are identified in the electron density distribution. Some terminal ends of the fatty acyl groups near the surface of the monomer show the electron density distributions too flat to evaluate the number of carbon atoms [14]. Identification of the unsaturated sites in the fatty acyl tails is impossible at this resolution of the electron density distribution.

The two cholate molecules in the crystal structure are likely to be contaminated during isolation from the mitochondrial membrane using cholate as a detergent. In our isolation procedure, cholate is used as the solubilizing agent, and the solibilized preparation is fractionated with ammonium sulfate in the presence of cholate. Then, the cholate is replaced with a nonionic detergent, decylmaltoside, with repeated ammonium sulfate fractionations in the presence of decyl maltoside. The critical micellar concentration of decyl maltoside is about one order lower than that of cholate. Thus, the cholate molecules involved in formation of the protein–detergent mixed micell must be completely replaced with decyl maltoside.

The two cholate molecules detected in the crystal structure indicate very specific and strong binding to the site. The cholate molecule is closely similar to the ADP molecule in size and shape. The atomic model of ADP fits well to either of the two cholate-binding sites, which suggests that ADP regulates enzymic function by binding to the cholate-binding sites. Actually, Kadenbach et al. demonstrated that the ATP/ADP ratio influences the efficiency of proton pumping [17]. Furthermore, they determined immunochemically one of the ADP-binding sites at the amino terminal moiety of a nuclear coded subunit (VIa), which is one of the cholate-binding sites in the crystal structure. The amino terminal moiety of subunit VIa is in contact with helices VII and VIII of subunit I of the other monomer. The $O_2$ reduction site is placed on the other side of the array of the helices VII and VIII. Thus, ADP binding to the amino terminal moiety could influence the conformation of the $O_2$ reduction site to affect the reactivity.

The dimer state is stabilized mainly by the two nuclear-coded subunits, VIa and VIb. The extramembrane domain of subunit VIa is in contact with the cytosolic side of subunit III, and it serves as a lid of the large crevice of subunit III that holds three phospholipids. The amino-terminal moiety of subunit VIa in essentially fully extended configuration is in close contact with subunit I of the other monomer. Thus, the two subunits VIa bridge the two monomers in two points (Fig. 1). An extramembrane subunit, VIb, is placed near the quasi twofold symmetric axis on the cytosolic surface to form a close contact between the two corresponding segments of the subunits containing 14 amino acids. The intermonomer space could receive two molecules of cardiolipin.

A preliminary analysis of phospholipid contents in this crystalline preparation showed one molecule of cardiolipin per monomer of the enzyme. Thus, two cardiolipin molecules are likely to be placed somewhat loosely to make the electron density distribution in the intermonomer space too flat for detection of the phospholipid molecules. The two cardiolipins in the space would contribute the stability of the dimer. Even bacterial cytochrome *c* oxidase without the nuclear-coded subunits could be in the dimer state stabilized by the phospholipid.

The other possible role of the nuclear-coded subunits, each with a transmembrane helix, is to stabilize the core subunits by giving the stable helix–helix contacts. The helix–helix contact is most stable at the angles of 0°, 20°, and 50° [18]. All the helices of this enzyme are in contact with each other at one of these angles. One of the most mysterious subunits may be subunit Vb, which holds the zinc site. Conformation of the peptide is quite peculiar and is stable only in the enzyme complex, suggesting a potential reactivity.

## Metal Site Structures

$Cu_A$ is a dinuclear copper center with six ligands. The center is similar to the iron–sulfur center of the 2Fe/2S type in which the iron atoms are replaced with copper atoms and the inorganic sulfurs with the SH of cysteine residues. The similarity in the electron spin resonance spectrum between the $Cu_A$ site in cytochrome *c* oxidase and the copper site of nitrous oxide reductase strongly suggests that $Cu_A$ in the fully oxidized state has two cupric copper ions with one electron equivalent delocalized between the two copper atoms, that is, $[Cu^{1.5+} \ldots Cu^{1.5+}]$ [19]. Heme *a* is in a six-coordinated low-spin state with two histidines. The fifth ligands of hemes *a* and $a_3$ are separated by only one amino acid residue. The shortest distance between the two heme porphyrins is only 4 Å. These structures promote rapid electron transfer between the two hemes.

The second copper site, $Cu_B$, which is a mononuclear center, is placed on the opposite site of the heme plane where histidine ligates. Three histidines coordinate to the $Cu_B$, forming an equilateral triangle with the $Cu_B$ at the center. The triangle plane is in parallel with the heme $a_3$ plane. The $Cu_B$ is placed 1 Å off the heme normal at the heme $a_3$. The distance between the two metals is 4.7 Å. The magnesium is bound to Glu[198] of subunit II. The peptide carbonyl of Glu[198] coordinates to one of the copper atoms of $Cu_A$. The amino acids of subunit I and a water are the other three ligands of the magnesium, placed about midway between $Cu_A$ and heme $a_3$ on the interface between subunits I and II. Zinc is coordinated by four cystein sulfurs to form essentially a regular tetrahedron about 40 Å from heme $a_3$ [12,14]. Thus, this metal is unlikely to participate in the basic function of this enzyme, which occurs at the redox active metal sites.

## Electron Transfer Reactions Within the Monomer

As shown of Fig. 2, a very effective electron-transfer path between $Cu_A$ and heme *a* is obvious, including His[204] of subunit II, one of the ligands of $Cu_A$, a peptide bond between Arg[438] and Arg[439] of subunit I, and a propionate group of heme *a*, which are connected with hydrogen bonds. The double bond character of the peptide promotes the through-bond electron transfer. A network including Glu[198], the magnesium, His[368], and a propionate group of heme $a_3$, which are connected by

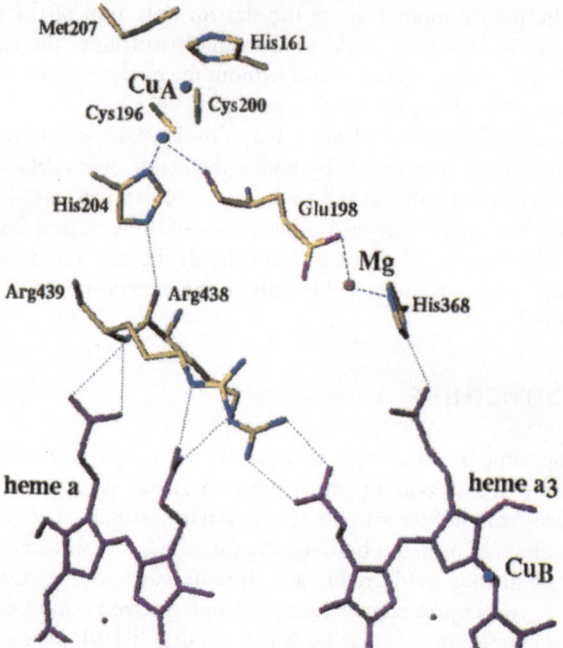

FIG. 2. Hydrogen bond network between the redox active metal sites in beef heart cytochrome *c* oxidase. *Dotted lines,* possible hydrogen bonds; *broken lines,* coordination bonds; *blue, red,* and *purple balls,* copper, magnesium, and iron atoms, respectively; *purple structures* with or without a *blue ball* are heme $a_3$ or *a*, respectively. Three amino acid residues—His[368], Arg[438], and Arg[439]—belong to subunit I; the others belong to subunit II

coordination and hydrogen bonds, can be an effective electron-transfer path between $Cu_A$ and heme $a_3$. The internal electron transfer of this enzyme has been extensively studied since the work of Gibson and Greenwood with the flow flash technique [15].

The electron transfer path has been established as cytochrome $c$–$Cu_A$–heme $a$–heme $a_3$ [20]; that is, no direct elctron transfer between $Cu_A$ and heme $a_3$ has been detected kinetically, indicating that the direct electron transfer between $Cu_A$ and heme $a_3$, if any, is about two orders of magnitude lower than the rate between $Cu_A$ and heme *a*. The distance between $Cu_A$ and heme $a_3$, 3 Å longer than that between $Cu_A$ and heme *a*, may provide such a difference in the electron-transfer rate. Then, what is the role of the network including the Mg site? Furthermore, if the purpose of these redox metal site systems of this enzyme is to convey electrons from cytochrome *c* to the $O_2$ reduction site, why is such a large prosthetic group as heme *a* required? Perhaps the electron-transfer path including Mg may be the most economical for synthesis in the cell. Thus, heme *a* could have an important unknown role, such as that related to proton pumping.

The internal electron transfers within cytochrome *c* oxidase are usually much slower than the elementary process of electron transfer. Thus, the kinetic behaviors of these redox metal sites are not directly related to the electron-transfer paths themselves but to the affinities of these melt sites to electrons. That is, heme $a$ could be reduced with the electrons transferred via Mg and heme $a_3$, but heme a is reduced earlier than heme $a_3$. In other words, both networks given in Fig. 2 could be active under turnover conditions of this enzyme. The redox equilibrium behavior of this enzyme so far examined, although not so extensively as in the case of the kinetic investigation, shows a closely related repulsive interaction between the metal sites, so that at any overall oxidation state all the redox site are in an identical fractional reduction (i.e., the concentration ratio of the reduced form to the oxidized form) [21,22]. The foregoing two networks and the dinuclear center of $Cu_A$ may contribute to the promotion of these metal site interactions.

## Machanism of $O_2$ Reduction by Cytochrome $c$ Oxidase

It has long been well accepted that $Cu_B$, placed very near the $O_2$ reduction site, initiates the two-electron reduction of the bound $O_2$ to the peroxide level [23]. Then, the $O_2$ reduction rate, limited by the electron transfer from $Cu_B$ to one of the two oxygen atoms placed within 3 Å, must be extremely rapid, perhaps on the order of a picosecond. However, recent resonance Raman investigations show that the lifetime of the oxygenated form is unexpectedly long, about 0.1 ms at 4°C. Thus, a special conformation or structure that lowers the electron transfer from $Cu_B$ must be placed on the $O_2$ reduction site. Only the crystal structure at the atomic resolution could solve the control mechanism.

A preliminary analysis of the electron density between the two metals in the $O_2$ reduction site indicates a bridging ligand between $Fe_{a_3}^{3+}$ and $Cu_B^{2+}$. The electron density distribution between the two metals is larger than a single oxygen or nitrogen atom. Also, the distance of 4.7 Å between the two metals is too long for a μ-oxo bridge (Fe-O-Cu). On the other hand, the triangle planar coordination of $Cu^{2+}$ is extremely unstable [24]. Thus, $Cu_B^{2+}$ is likely to have the fourth ligand. The crystal structure of beef heart cytochrome $c$ oxidase at the atomic resolution so far reported is the one in the fully oxidized state. However, crystal structures of many metalloproteins so far reported indicate that the conformational changes induced by the change in the redox state of the metal site are too small to detect in most cases even at atomic resolution. Thus, the conformation of the $O_2$ reduction site in the fully reduced form to which $O_2$ binds is essentially identical to the one in the fully oxidized state. However, no ligand is likely to be on either the two metals in the fully reduced state. That is, heme $a_3$ is in a ferrous high-spin state (five coordinated), and $Cu_B$ is in a triangle planar coordination with the cuprous ion at the center of the triangle. The configuration of $Cu_B^{1+}$ is quite stable [24]; in other words, the $Cu_B^{1+}$ site is not a good electron donor.

As shown in Fig. 3, the $O_2$ reduction site has $Tyr^{244}$, which could form a hydrogen bond with the bound dioxygen although a small conformational change is required. This tyrosine, which is connected to the matrix side with a hydrogen bond network, is the sole possible proton donor to the bound dioxygen to form water. Furthermore, heme a could be a very effective electron donor to the $O_2$ at heme $a_3$, as was suggested. Thus, $O_2$ reduction is initiated possibly by the formation of the hydrogen bond between the bound $O_2$ and $Tyr^{244}$.

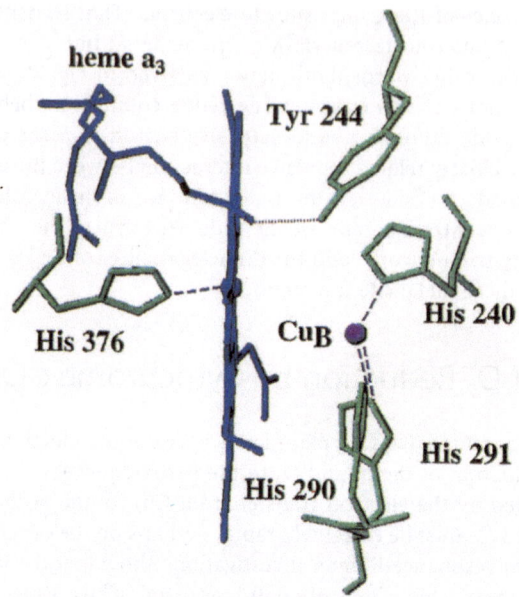

FIG. 3. The $O_2$ reduction site viewed along the heme $a_3$ plane. *Dotted lines*, hydrogen bond; *broken lines*, coordination bonds; *blue structure*, heme $a_3$; *green structures*, amino acids

Formation of a hydrogen bond to the bound $O_2$ may stimulate the electron entry to the $O_2$ by increasing the positive charge on the $O_2$. The hydrogen bond formation cannot be very rapid because it is accompanied by a conformational change. The resulting hydroperoxo intermediate is likely to be very unstable, as in the case of peroxidase, to form the intermediate species at an unusually high oxidation state, corresponding to compounds I and II [25]. The role of $Cu_B^{1+}$, with high redox potential, may be to reduce these species.

## Mechanism of Proton Pumping

The crystal structures of cytochrome $c$ oxidase from beef heart and the soil bacteria *Paraccoccus denitrificans* reveal a striking similarity in the redox active metal site system in the largest two subunits except for the $O_2$ reduction site. As stated earlier, beef heart enzyme at the fully oxidized state free from any respiratory inhibitor has three histidines at the $Cu_B$ site, placed 4.7 Å apart from $Fe_{a_3}^{3+}$ [12]. On the other hand, one of the three histidines at the $Cu_B$ site is missing in the crystal structure of the bacterial enzyme in the fully oxidized azide bound state. The $Cu_B^{2+}$–$Fe_{a_3}^{3+}$ distance is 5.2 Å in the presence of azide. Iwata et al. proposed the missing histidine as the proton-pumping site [13], as has been suggested by Wikström et al. [26].

One of the critical requirements of this cycle, however, is the complete insulation of the protons on the imidazole of the pumping site from the dioxygen reduction site. Most of the $O_2$ reduction intermediates formed on $Fe_{a_3}$ have extremely strong affinity to protons. The trapping of the protons to be pumped by the $O_2$ reduction intermediate to form water results in the short circuiting of the proton pump. The electron

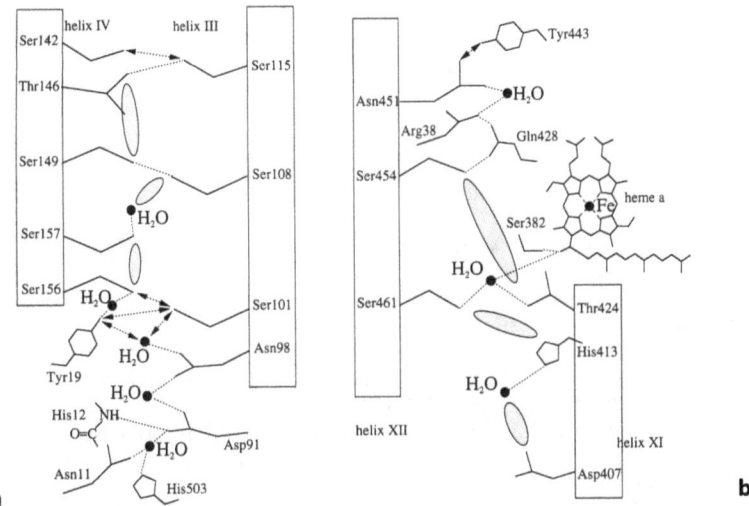

FIG. 4a,b. Possible channels for proton pumping, shown schematizally, between helices III and IV (a) and between helices XI and XII (b). *Dark ovals, dotted lines,* and *dotted lines with arrows* denote cavities, hydrogen bonds, and possible hydrogen bond configurations, respectively. The cavities are spaces that contain no electron density but are large enough to retain randomly oriented or mobile water molecules. A pair of amino acid side chains in a possible hydrogen bond configuration are placed too far apart to form a hydrogen bond. However, a small conformational change in the pair without movement of peptide bonds could provide a new hydrogen bond. The matrix side of each network is placed on the bottom and the cytosolic side on the top. Each figure shows only the structure connecting from the matrix port to the cytosolic port; no dead-end branch is shown except for a branch leading to heme *a* in the channel b. Both channels are in subunit I, and all the amino acid residues belong to subunit I

density of the imidozole is missing in the crystal structure of bacteria, which indicates that the imidazole is not fixed tightly and so is freely mobile. It is not at all easy for the histidine at such a state to keep the protons on the imidazole completely away from the $O_2$ reduction site.

On the other hand, the $O_2$ reduction site in the crystal structure of beef heart enzyme in the fully oxidized state does not have any possible proton-transfer path for pumping protons [14]. The hydrogen bond network from the matrix surface to $Tyr^{244}$ in the dioxygen reduction site cannot be the proton-pumping site because the $Try^{244}$ is placed so close to the bound $O_2$ and the reduction intermediate. The crystal structure of the enzyme in the fully oxidized state has two possible networks for proton pumping (Fig. 4) [14].

Thus, the crystal structure of beef heart cytochrome *c* oxidase suggests proton pumping occurs far from the $O_2$ reduction site [14]. To identify the proton-pumping site, comparison of the crystal structures at different oxidation and ligand-binding states is indispensable. The crystallographic analysis of beef heart enzyme in the fully reduced and the fully reduced CO-bound states is under way in our group.

# References

1. Ferguson-Miller S, Babcock GT (1996) Heme/copper terminal oxidases. Chem Rev 96:2889–2890
2. Meinecke L, Buse G (1986) Studies on cytochrome-c oxidase. XIII. Biol Chem Hoppe-Seyler 367:67–73
3. Hensel S, Buse G (1990) Studies on cytochrome-c oxidase, XIV. Biol Chem Hoppe-Seyler 371:411–422
4. Kadenbach B, Ungibaver U, Jaraush J, et al. (1983) The complexity of respiratory complexes. Trends Biochem Sci 8:398–400
5. Gibson QH, Greenwood C (1963) Reactions of cytochrome oxidase with oxygen and carbon monoxide. Biochem J 86:541–554
6. Kitagawa T, Ogura T (1997) Oxygen activation mechanism at the binuclear site of heme-copper oxidase superfamily as revealed by time-resolved resonance Raman spectroscopy. Prog Inorg Chem 45:431–479
7. Ogura T, Takahashi S, Shinzawa-Itoh K, et al. (1991) Time-resolved resonance Raman investigation of cytochrome oxidase catalysis: observation of a new oxygen-isotope sensitive Raman band. Bull Chem Soc Jpn 64:2901–2907
8. Minnaert K (1961) The kinetics of cytochrome c oxidase. I. The system: cytochrome c-cytochrome oxidase-oxygen. Biochim Biophys Acta 50:23–34
9. Ferguson-Miller S, Brautigan DL, Margoliaoh E (1976) Correlation of the kinetics of electron transfer activity of various eukaryotic cytochromes c with binding to mitochondrial cytochrome c oxidase. J Biol Chem 251:1104–1115
10. Ortega-Lopez J, Robinson NC (1995) Cytochrome c oxidase: biphasic kinetic result from incomplete reduction of cytochrome a by cytochrome c bound to the high-affinity site. Biochemistry 34:10000–10008
11. Petersen LC, Nichols P, Degan H (1976) The effect of oxygen concentration on the steady-state kinetics of the solubilized cytochrome c oxidase. Biochim Biophys Acta 452:59–65
12. Tsukihara T, Aoyama H, Yamashita E, et al. (1995) Structures of metal sites of oxidized bovine heart cytochrome c oxidase at 2.8 Å. Science 269:1069–1074
13. Iwata S, Ostermeier C, Ludwig B, et al. (1995) Structure at 2.8 Å resolution of cytochrome c oxidase from Paracoccus denitrificans. Nature 376:660–669
14. Tsukihara T, Aoyama H, Yamashita E, et al. (1996) The whole structure of the 13-subunit oxidized cytochrome c oxidase at 2.8 Å. Science 272:1136–1144.
15. Fuller SD, Capaldi RA, Henderson R (1979) Structure of cytochrome c oxidase in deoxycholate-derived two-dimensional crystals. J Mol Biol 134:305–327
16. Kadenbach B, Merle P (1981) On the function of multiple subunits of cytochrome c oxidase from higher eukaryotes. FEBS Lett 135:1–11
17. Frank V, Kadenbach B (1996) Regulation of the $H^+/e^-$ stoichiometry of cytochrome c oxidase from bovine heart by intramitochondrial ATP/ADP ratios. FEBS Lett 382:121–124
18. Chotia C, Levitt M, Richardson D (1997) Structure of proteins: packing of α-helices and pleated sheets. Proc Natl Acad Sci USA 74:4130–4134
19. Kroneck PMH, Antholine WA, Riester J, et al. (1988) The cupric site in nitrous oxide reductase contains a mixed-valence [Cu(II), Cu(I)] binuclear center: a multifrequency electron paramagnetic resonance investigation. FEBS Lett 242:70–74
20. Hill BC (1994) Modeling the sequence of electron transfer reaction in the single turn-over of reduced, mammalian cytochrome c oxidase. J Biol Chem 269:2419–2425
21. Wikström M, Krab K, Saraste M (1981) Cytochrome oxidase—a synthesis. Academic, London, pp 88–116
22. Nichols P, Wrigglesworth JM (1988) Routes of cytochrome $a_3$ reduction. Ann NY Acad Aci 550:59–67

23. Caughey WS, Wallace WJ, Volpe JA, et al. (1976) Cytochrome *c* oxidase. In: Boyer PD (ed) The enzymes, 3d edn. Academic, New York, pp 299–344
24. Cotton FA, Wilkinson G (1980) Advanced inorganic chemistry, 4th edn. Wiley, New York, pp 798–821
25. Kitagawa T, Mizutani Y (1994) Resonance Raman spectra of highly oxidized metalloporphyrins and heme proteins. Coord Chem Rev 135/136:685–735
26. Wikström M, Bogacher A, Finel M, et al. (1994) Mechanism of proton translocation by the respiratory oxidases. The histidine cycle. Biochim Biophys Acta 1187:106–111

# Cooperation of Two Quinone-Binding Sites in the Oxidation of Substrates by Cytochrome *bo*

Mariko Sato-Watanabe[1], Tatsushi Mogi[1], Hideto Miyoshi[2], and Yasuhiro Anraku[1]

*Summary.* Cytochrome *bo* is a terminal quinol oxidase of the aerobic respiratory chain of *Escherichia coli* and catalyzes not only the scalar protolytic reactions but also redox-coupled proton pumping. Structure–function studies of the quinol oxidation site ($Q_L$) using systematically selected quinone analogues, 1,4-benzoquinones, substituted phenols, and ubiquinone-2 derivatives revealed the structural features of the quinol oxidation site. We found further that bacterial quinol oxidases share common features of the quinol oxidation site irrespective of their structural similarities. In addition, we identified the presence of a tightly bound ubiquinone-8 ($Q_H$) and examined the possible roles of the $Q_H$ site in the two-electron oxidation of substrates at the $Q_L$ site and in mediating sequential one-electron transfer from $Q_L$ to the low-spin heme *b*. Based on these observations, we discuss molecular mechanism of the substrate oxidation by cytochrome *bo*.

*Key words.* Cytochrome *bo*—Intramolecular electron transfer—Semiquinone radical—Quinol oxidation site—Quinone analogues

## Introduction

In the aerobic respiratory chain of *Escherichia coli*, a four-subunit cytochrome *bo* is a predominant terminal oxidase under high oxygen tension while a two-subunit cytochrome *bd* is expressed predominantly under oxygen-limiting growth conditions [1]. They are not structurally related; however, both catalyze the two-electron oxidation of ubiquinol-8 ($Q_8H_2$) at the periplasmic side and the four-electron reduction of molecular oxygen at the cytoplasmic side. Thus, they establish a proton electrochemical gradient across the cytoplasmic membrane via the scalar protolytic reactions. In addition, cytochrome *bo* can vectorially translocate protons via redox-coupled proton pumping [2–4].

---

[1] Department of Biological Sciences, Graduate School of Science, University of Tokyo, 7-3-1 Hongo, Bunkyo-ku, Tokyo 113, Japan
[2] Department of Agricultural Chemistry, Kyoto University, Kitashirakawa oiwake-cho, Sakyo-ku, Kyoto 606, Japan

Recent molecular biological and physicochemical studies indicate that cytochrome *c* oxidases and bacterial quinol oxidases, except cytochrome *bd*, are classified into the heme-copper respiratory oxidase superfamily [2–4]. Subunits I, II, and III are conserved in bacterial and mitochondrial enzymes. Subunit I binds a low-spin heme and a high-spin heme-$Cu_B$ binuclear center and serves as a reaction center for dioxygen reduction and proton pumping. X-ray crystallographic studies have recently determined atomic structures of bacterial and mammalian $aa_3$-type cytochrome *c* oxidases [5,6] and a water-soluble C-terminal domain of subunit II from cytochrome *bo* [7]. Axial ligands for the redox metal centers were found to be identical to those determined by molecular biological studies [3,4]. Structural biology now provides new insights into the molecular mechanism of redox-coupled proton pumping.

Subunit II of $aa_3$-type cytochrome *c* oxidases contains the $Cu_A$ binuclear copper center in the C-terminal hydrophilic domain and oxidizes ferrous cytochrome *c*, a water-soluble one-electron carrier. Then electrons are transferred to the heme $a_3$–$Cu_B$ binuclear center through heme *a*, present in transmembrane regions of subunit I. Similarly, the counterpart of cytochrome *bo* seems to catalyze the oxidation of quinols [8], lipid-soluble two-electron, two-proton redox components (Fig. 1), although it lacks $Cu_A$ [1–4,7]. Subsequently, it transfers electrons to the heme *o*-$Cu_B$ center through heme *b*. Cytochrome *bo* is closely related to bacterial cytochrome *c* oxidases [9]; however, the location and structural features of the substrate oxidation site remain obscure.

In this chapter, we discuss the structural features of a low-affinity quinol oxidation site ($Q_L$) of cytochrome *bo* and the functional role of a high-affinity quinone binding site ($Q_H$) that connects electron flow from the two-electron redox component (quinols) to a one-electron transfer system (heme irons). We suggest that cooperation of two quinone/quinol-binding sites is essential for the oxidation of substrates by the heme-copper quinol oxidases.

## Structural Features of the $Q_L$ Site

Enzymatic studies have demonstrated that the $Q_L$ site has an apparent $K_m$ value for ubiquinol-1 ($Q_1H_2$) of 10–50 μM, and the oxidase activity was inhibited by the quinone analogues, piericidin A [10], HHQNO (2-heptyl-4-hydroxyquinoline N-oxide) [10,11], UHDBT (5-*n*-undecyl-6-hydroxy-4,7-dioxobenzothiazole) [11], aurachin C, and tridecylstigmatellin [12] (Fig. 1). Antimycin A, rotenone, myxothiazol, mucidin, funiculosin [12], 2,4-dinitro-6-alkylphenols, DCMU (3-(3,4-dichlorophenyl)-1,1-dimethylurea), and simazin (2-chloro-4,6-bis(ethylamine)-*s*-triazine) [13] had little effect on cytochrome *bo*. These results suggest that the $Q_L$ site differs considerably from the quinone redox sites of NADH dehydrogenase, the cytochrome $bc_1$ complex (i.e., $Q_i$ and $Q_o$), and the photosythetic reaction center (RC) (i.e., $Q_B$).

To further characterize structural features of the $Q_L$ site, we carried out structure-inhibitory potency analyses with systematically selected compounds, namely 33 substituted phenols and seven 1,4-benzoquinones (BQ) [13] (see Fig. 1). In substituted phenols, chlorine was the most effective substituent at both *ortho*-positions irrespective of substituent pattern at the *para*-position. Replacing one or both chlorine

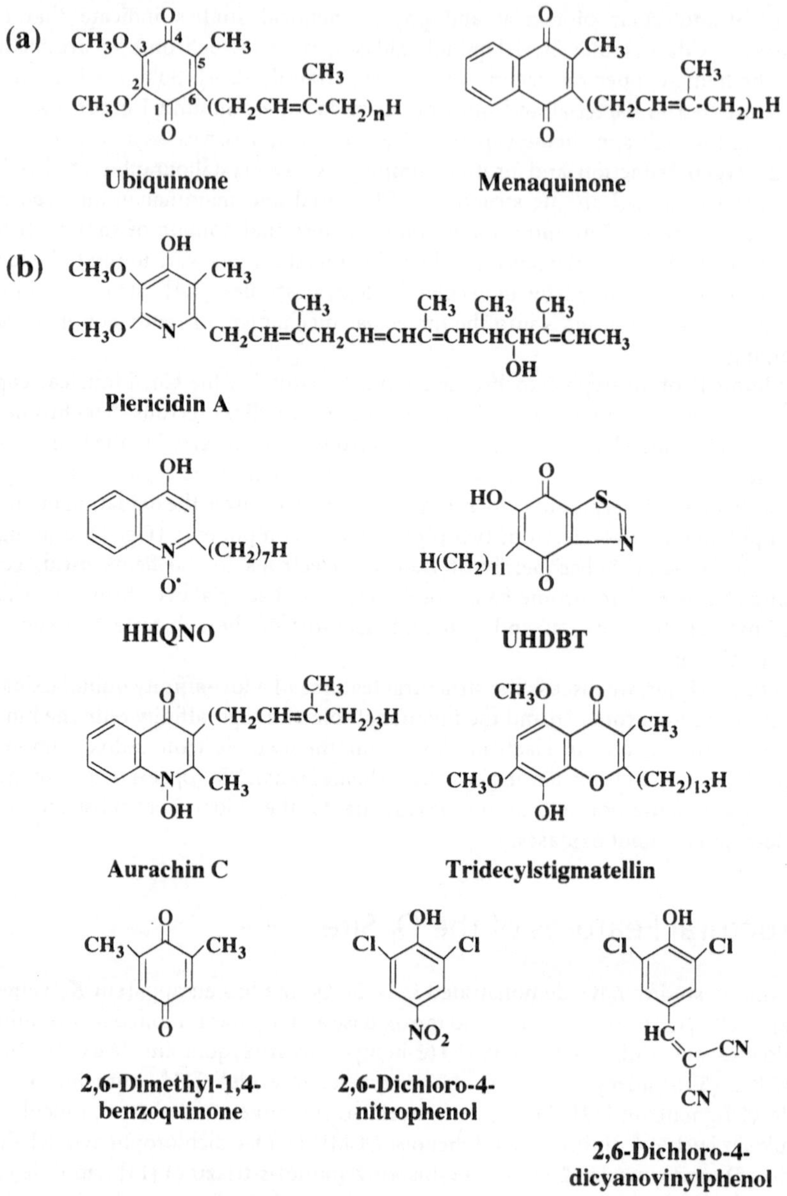

**(a)**

Ubiquinone

Menaquinone

**(b)**

Piericidin A

HHQNO

UHDBT

Aurachin C

Tridecylstigmatellin

2,6-Dimethyl-1,4-
benzoquinone

2,6-Dichloro-4-
nitrophenol

2,6-Dichloro-4-
dicyanovinylphenol

FIG. 1a,b. Structures of quinones and inhibitors reported for the quinol oxidation site of the *E. coli* cytochrome *bo*

atom(s) by another halogen or a methyl group results in a decrease of the inhibitory potency by a considerable extent. Physicochemical characteristics of chlorine and bromine substituents are similar; for instance, the electronic effects in terms of the Hammett-type substituent constant are almost identical [14], and the difference in van der Waals radius is just 0.12 Å [15]. Therefore, a steric structure or bulkiness of the *ortho*-substituents seems to be strictly recognized by the $Q_L$ site.

In addition, we noticed that the electron-withdrawing propensity ($\sigma^-_{para}$) of the *para*-substituents determined an inhibitory potency of the substituted phenols [13]. The inhibitory activity of 2,6-dichlorophenols increases as the $\sigma^-_{para}$ index of the *para*-substituents increases. 2,6-Dichloro-4-dicyanovinylphenol (see Fig. 1), which fulfills these requirements, was the most potent inhibitor among the substituted phenols tested. It suggests that the stability of an anionic form of inhibitor in a lower dielectric binding pocket is an important factor determining the inhibitory activity. Considering the fact that the dicyanovinyl group is bulky and conformationally rigid, the *para*-substituent of the molecule may be out of a binding pocket for the benzoquinone ring.

Among the BQ derivatives examined, 2,6-dimethyl-BQ (see Fig. 1) was found to be the most potent competitive inhibitor, and 2-methyl-BQ and 2,6-dichloro-BQ were comparable to HHQNO and piericidin A [13]. Replacing a methyl group of 2-methyl-BQ and 2,6-dimethyl-BQ by an ethyl group causes a marked decrease in the inhibitory activity. In addition, comparing the inhibition between two isomers, 2,6-substituted BQs have significantly higher inhibitory activity than 2,5-substituted BQs. These results, along with the poor activity of tetramethyl-BQ, suggest that the $Q_L$ site is capable of binding quinone-related compounds in an asymmetric manner, but also recognizes a portion of the ligand molecule corresponding to the 2,6-substituting groups of BQs, as well as the whole molecule in a strict sense.

Subsequently, we synthesized a series of the ubiquinol-2 ($Q_2H_2$) derivatives to identify structural requirements for binding of ubiquinones to the $Q_L$ site [16]. The $V_{max}$ values of 2-methoxy-3-ethoxy analogues were greater than those of 2-ethoxy-3-methoxy analogues irrespective of the side-chain structure. This indicates not only that a methoxy group in the 2-position is recognized more strictly than the 3-position by the $Q_L$ site but also that the side-chain structure does not affect binding of the quinone ring moiety. Similarly, a methyl group was found to be crucial for the 5-position. The apparent $K_m$ values of the $Q_2H_2$ derivatives were much lower than those of corresponding 6-*n*-decyl and 6-(3′,7′-dimethyloctyl) derivatives [16]. The isoprenoid structure is less hydrophobic than the saturated *n*-alkyl group with the same carbon number. Therefore, the π-electron system of the isoprenoid side chain appears to increase binding affinity to the $Q_L$ site (Fig. 2). In contrast, Yu and his colleagues reported that ubiquinone analogues with *n*-alkyl side chain longer than that of $Q_1H_2$ exhibit oxidase activity comparable to $Q_1H_2$ [8].

These observations indicate that the $Q_L$ site of cytochrome *bo* more strictly recognizes the 1-OH, 2-methoxy, 4-OH, and 5-methyl groups on the quinone ring than the 3- and 6-substituents (Fig. 2). Such an asymmetric nature in molecular recognition of the $Q_L$ site may account for the sequential electron tansfer to low-spin heme *b*, which accepts electrons one at a time. Finally, we would like to point out that bacterial quinol oxidases share common structural properties for the quinol oxidation site irrespective of their structural similarities. We carried out structure-

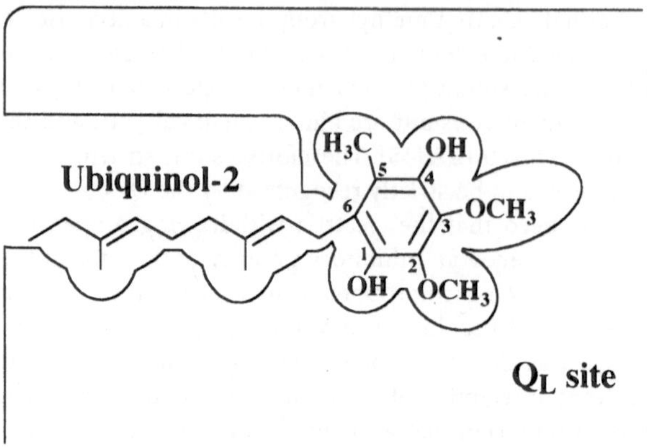

FIG. 2. Binding model for ubiquinol-2 to the quinol oxidation ($Q_L$) site. 2-Methoxy and 5-methyl groups of ubiquinols are strictly recognized by the $Q_L$ site, and the isoprenoid side chain increases the binding affinity [16]

inhibitory potency studies using the BQ derivatives and the substituted phenols, and found that the variations in the residual $Q_1H_2$ oxidase activities were almost linearly correlated between cytochromes *bo* and *bd* from *E. coli* and between cytochrome *bo* from *E. coli* and cytochrome *o* (*bo*-type heme-copper quinol oxidase) from *Acetobacter aceti* [13]. Furthermore, we found that the effects of substituents at all positions of the ubiquinol ring on the electron-donating activity are similar between cytochromes *bo* and *bd* [16]. Recognition of the quinone ring by the two enzymes is not affected by the 6-substituent structure, indicating that the quinol ring and the side-chain moieties independently contribute to the binding of substrates to the quinol oxidation site. Recently, Meunier et al. reported that tridecylstigmatellin and UHDBT more specifically inhibit oxidase activity of cytochrome *bo* whereas aurachin D and antimycin A are relatively specific for cytochrome *bd* [12]. However, structural features that determine the specificity of quinone analogues remain to be identified.

## Role of the $Q_H$ Site in Intramolecular Electron Transfer

In general, many redox proteins that use quinones as mobile electron carriers have two quinone/quinol-binding sites, even though their quinone redox mechanisms are different. Photosynthetic RC and cytochrome $bc_1$ complex are the representatives that have been so far studied in detail. In the RC, $Q_A$ acts as a tightly bound one-electron carrier between pheophytin and $Q_B$ and does not undergo protonation changes [17]. $Q_B$ is a mobile two-electron, two-proton redox component and can leave the $Q_B$ site when the second electron produces $Q_B^{2-}$ and two protons immediately bind to form the final product, $Q_BH_2$. In the cytochrome $bc_1$ complex, the $Q_o$ and $Q_i$ sites are both in dynamic equilibrium with the quinone pool in the membrane and the coordinated redox reactions are tightly coupled to vectorial proton translocation via the protonmotive Q cycle originally postulated by Mitchell [18]. The $Q_o$ site may bind two ubiquinone molecules in a stacked configuration to facilitate bifurcated electron transfer [19].

Unlike cytochrome *c* oxidases, cytochrome *bo* contains only three redox metal centers. Therefore, completion of the reaction of the fully reduced enzyme with dioxygen requires an extra electron donor, either a bound quinone molecule [20] or tryptophan or tyrosine residues near the binuclear center [21]. We examined the presence of a tightly bound $Q_8$ molecule(s) in cytochromes *bo* and *bd* purified in a mild nonionic detergent, sucrose monolaurate, and discovered a high-affinity quinone-binding site ($Q_H$) in cytochrome *bo* distinct from the $Q_L$ site [22]. Later, the presence of the bound quinone in quinol oxidases was confirmed in the *E. coli* cytochrome *bo* [23,24] and cytochrome $ba_3$ from *Paracoccus denitrificans* [25]. We found that $Q_H$ is hardly exchanged with exogenous substrates and the $Q_L$ site competitive inhibitors and cannot be removed by gel filtration chromatography or precipitation of the enzyme [22].

Subsequently, we examined the role of $Q_H$ using a bound $Q_8$-free oxidase isolated from a ubiquinone biosynthesis mutant [22]. The absence of $Q_H$ lowers the affinities for both substrates and the inhibitors at the $Q_L$ site. Reconstitution of the $Q_H$ site with substituted phenols resulted in a marked decrease of the $V_{max}$ value of the $Q_LH_2$ oxidase activity. The dissociation constant for $Q_l$ was determined to be about $2\,\mu M$ much smaller than the $K_m$ value for the $Q_lH_2$ oxidation ($10$–$50\,\mu M$). UV-visible and resonance Raman spectroscopies demonstrated that the environment of low-spin heme *b* was perturbed in the air-oxidized $Q_H$-free oxidase. Accordingly, the $Q_H$ site is required for catalytic activity and seems to be close to the $Q_L$ site and low-spin heme *b*. Thus, we postulated that $Q_H$ mediates electron transfer from $Q_L$ to low-spin heme *b* [22] (Fig. 3).

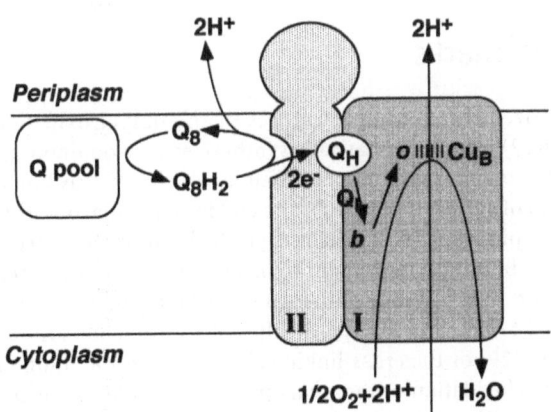

FIG. 3. Model for cooperation of two quinone/quinol-binding sites in electron transfer from the quinol oxidation site to the metal centers. A bound ubiquinone-8 at the high-affinity quinone-binding site ($Q_H$) can mediate the two-electron transfer from the $Q_L$ site, like $Q_B$ of the photosynthetic reaction center, and then the sequential one-electron transfer to low-spin heme *b* similar to $Cu_A$ of cytochrome *c* oxidase

Redox titration studies using electron paramagnetic resonance (EPR) revealed that cytochrome $bo$ can stabilize a ubisemiquinone radical and its $g = 2.0$ signal titrates as a bell shaped curve [26,27]. The $E_m'$ value of a $QH_2/Q$ couple was lower than that of low-spin heme $b$ and decreased sharply with increased medium pH with a slope of about $-60\,mV/pH$ unit, indicative of a $2H^+/2e^-$ stoichiometry. We demonstrated that this signal was originated from a bound $Q_8$ molecule at the $Q_H$ site and that a $E_m'$ value of the heme $b$ was lowered $23$–$30\,mV$ by $Q_H$ [26]. The absence of $Q_H$ would cause conformational change around the $Q_H$ site and consequently affect the electronic state of the metal center itself and/or heme–heme interaction at the air-oxidized state.

These observations are consistent with our proposal [22,26] that $Q_H$ mediates electron transfer from the $Q_L$ site to heme $b$ (see Fig. 3). Cooperation of two quinone/quinol-binding sites ($Q_H$ and $Q_L$) seems crucial for the oxidation of substrates by quinol oxidases. Flow-flash studies on the reaction of the fully reduced wild-type and $Q_H$-free enzymes with dioxygen indicated that $Q_H$ is involved in intramolecular electron transfer [Orii et al, unpublished results; 24]. Recently, on oxidation of $Q_2H_2$, the rapid formation of the ubisemiquinone, which is accompanied by reduction of the heme $b$, was demonstrated by stopped-flow UV/vis and rapid freeze/quench EPR spectroscopies [28].

Both $Q_L$ and $Q_H$ can undergo protonation changes coupled to two-electron oxidoreduction; therefore, they are similar to $Q_B$ of the photosynthetic RC and $Q_o$ and $Q_i$ of the cytochrome $bc_1$ complex [26]. However, $Q_H$ can stabilize the ubisemiquinone and does not leave the binding site when protonated to $Q_HH_2$ [22]. Thus, we conclude that a molecular mechanism for the oxidation of substrates by bacterial quinol oxidases is different from those of the photosynthetic RC and cytochrome $bc_1$ complex. $Q_H$ could not only act as a transient electron reservoir for $Q_L$, a mobile two-electron, two-proton redox component, but also a gate for the low-spin heme through sequential one-electron transfer similar to $Cu_A$ of cytochrome $c$ oxidase.

## Concluding Remarks

We have characterized the structural features of the quinol oxidation site ($Q_L$) of cytochrome $bo$ (Fig. 2) using systematically synthesized benzoquinones, ubiquinones, and substituted phenols, and revealed the possible roles of a tightly bound $Q_8$ at the $Q_H$ site in the oxidation of substrates at the $Q_L$ site and in sequential one-electron transfer from $Q_L$ to the low-spin heme (Fig. 3). However, the locations and structures of these two quinol/quinone-binding sites remain to be determined. Besides crystallographic efforts, identification of the quinone analogue-resistant mutations using the $Q_L$ site inhibitors such as 2,6-dimethyl-BQ, 2,6-dichloro-4-nitrophenol, and 2,6-dichloro-4-dicyanovinylphenol [29] or the cross-linking site(s) of azido-ubiquinone derivatives [8,30] will facilitate elucidation of the structure and function of the $Q_L$ site. In addition, determination of the structural features and location of the $Q_H$ site is also required for understanding of the cooperation of two quinone/quinol-binding sites in bacterial quinol oxidases.

*Acknowledgment.* This work was supported in part by grants-in-aid for scientific research on priority areas (08249106 and 08268216), for scientific research (A)

(07558221 and 07309006) and (B) (08458202), and for exploratory research (08878097) from the Ministry of Education, Science, Sports and Culture, Japan. This is paper XXIV in the series "Structure-function studies on the *E. coli* cytochrome *bo* complex."

# References

1. Anraku Y, Gennis RB (1987) The aerobic respiratory chain of *Escherichia coli*. Trends Biochem Sci 12:262–266
2. Saraste M (1990) Structural features of cytochrome oxidase. Q Rev Biophys 23:331–336
3. Mogi T, Nakamura H, Anraku Y (1994) Molecular structure of a heme-copper redox center of the *Escherichia coli* ubiquinol oxidase: evidence and model. J Biochem (Tokyo) 116:471–477
4. García-Horsman JA, Barquera B, Rumbley J, et al. (1994) The superfamily of heme-copper respiratory oxidases. J Bacteriol 176:5587–5600
5. Iwata S, Ostermeier C, Ludwig B, et al. (1995) Structure at 2.8 Å resolution of cytochrome *c* oxidase from *Paracoccus denitrificans*. Nature 376:660–669
6. Tsukihara T, Aoyama H, Yamashita E, et al. (1996) The whole structure of the 13-subunit oxidized cytochrome *c* oxidase at 2.8 Å. Science 272:1136–1144
7. Wilmanns M, Lappalainen P, Kelly M, et al. (1995) Crystal structure of the membrane-exposed domain from a respiratory quinol oxidase complex with an engineered dinuclear copper center. Proc Natl Acad Sci USA 92:11955–11959
8. Welter R, Gu LQ, Yu L, et al. (1994) Identification of the ubiquinol-binding site in the cytochrome *bo*₃-ubiquinol oxidase of *Escherichia coli*. J Biol Chem 269:28834–28838
9. Castresana J, Lübben M, Saraste M, et al. (1994) Evolution of cytochrome oxidase, an enzyme older than atomospheric oxygen. EMBO J 13:2516–2525
10. Kita K, Konishi K, Anraku Y (1984) Terminal oxidases of *Escherichia coli* aerobic respiratory chain. I. Purification and properties of cytochrome $b_{562}$-*o* complex from cells in the early exponential phase of aerobic growth. J Biol Chem 259:3368–3374
11. Matsushita K, Patel L, Kaback HR (1984) Cytochrome *o* type oxidase from *Escherichia coli*. Characterization of the enzyme and mechanism of electrochemical proton gradient generation. Biochemistry 23:4703–4714
12. Meunier B, Madgwick SA, Reil E, et al. (1995) New inhibitors of the quinol oxidation sites of bacterial cytochromes *bo* and *bd*. Biochemistry 34:1076–1083
13. Sato-Watanabe M, Mogi T, Miyoshi H, et al. (1994) Structure-function studies on the ubiquinol oxidation site of the cytochrome *bo* complex from *Escherichia coli* using *p*-benzoquinones and substituted phenols. J Biol Chem 269:28899–28907
14. Hansch C, Leo A (1979) In: Substituent constants for correlation analysis in chemistry and biology. Wiley, New York
15. Bondi A (1964) van der Waals volumes and radii. J Phys Chem 68:441–451
16. Sakamoto K, Miyoshi H, Takegami K, et al. (1996) Probing substrate binding site of the *Escherichia coli* quinol oxidases using synthetic ubiquinol analogues. J Biol Chem 271:29897–29902
17. Okamura MY, Feher G (1992) Proton transfer in reaction center from photosynthetic bacteria. Annu Rev Biochem 61:861–896
18. Mitchell P (1976) Possible molecular mechanisms of the protonmotive function of cytochrome system. J Theor Biol 62:327–367
19. Brandt U (1996) Energy conservation by bifurcated electron-transfer in the cytochrome-$bc_1$ complex. Biochim Biophys Acta 1275:41–46
20. Orii Y, Mogi T, Sato-Watanabe M, et al. (1995) Facilitated intramolecular electron transfer in the *Escherichia coli* *bo*-type ubiquinol oxidase requires chloride. Biochemistry 34:1127–1132

21. Wang J, Rumbley J, Ching YC, et al. (1995) Reaction of cytochrome $bo_3$ with oxygen: extra redox center(s) are present in the protein. Biochemistry 34:15504–15511

22. Sato-Watanabe M, Mogi T, Ogura T, et al. (1994) Identification of a novel quinone-binding site in the cytochrome $bo$ complex from Escherichia coli. J Biol Chem 269:28908–28912

23. Svensson-Ek M, Thomas JW, Gennis RB, et al. (1996) Kinetics of electron and proton transfer during the reaction of wild type and helix VI mutants of cytochrome $bo_3$ with oxygen. Biochemistry 35:13673–13680

24. Puustinen A, Verkhovsky MI, Morgan JE, et al. (1996) Reaction of the Escherichia coli quinol oxidase cytochrome $bo_3$ with dioxygen: the role of a bound ubiquinone molecule. Proc Natl Acad Sci USA 93:1545–1548

25. Zickermann I, Anemüller S, Richter OMH, et al. (1996) Biochemical and spectroscopic properties of the four-subunit quinol oxidase (cytochrome $ba_3$) from Paracoccus denitrificans. Biochim Biophys Acta 1277:93–102

26. Sato-Watanabe M, Itoh S, Mogi T, et al. (1995) Stabilization of a semiquinone radical at the high-affinity quinone-binding site ($Q_H$) of the Escherichia coli bo-type ubiquinol oxidase. FEBS Lett 374:265–269

27. Ingledew WJ, Ohnishi T, Salerno JC (1995) Studies on a stabilization of ubisemiquinone by Escherichia coli quinol oxidase, cytochrome $bo$. Eur J Biochem 227:903–908

28. Osborn JP, Musser SM, Scultz BE, et al. (1996) Rapid formation of a ubisemiquinone species upon oxidation of quinol by the cytochrome $bo_3$ from Escherichia coli. Keio J Med 45:S52

29. Sato-Watanabe M, Mogi T, Miyoshi H, et al. (1995) Characterization of the quinol oxidation site of the E. coli cytochrome $bo$ complex. Biophys J 68:A318

30. Yang FD, Yu L, Yu CA, et al. (1986) Use of an azido-ubiquinone derivative to identify subunit I as the ubiquinol-binding site of the cytochrome $d$ terminal oxidase complex of Escherichia coli. J Biol Chem 261:14987–14990

# Rapid Formation of a Semiquinone Species on Oxidation of Quinol by the Cytochrome *bo₃* Oxidase from *Escherichia coli*

Jeffrey P. Osborne[1], Sigfried M. Musser[2], Brian E. Schultz[2],
Dale E. Edmondson[3], Sunney I. Chan[2], and Robert B. Gennis[1]

*Summary.* Many bacterial oxidases utilize dihydroquinols, such as ubiquinol or menaquinol, rather than cytochrome *c* as a substrate. The best-characterized ubiquinol oxidase is cytochrome *bo₃* from *Escherichia coli*. In this work, the initial oxidation of ubiquinol by this ubiquinol oxidase is examined. Stopped-flow UV-visible spectroscopy and rapid freeze-quench electron paramagnetic resonance (EPR) spectroscopies were used to examine the oxidation of ubiquinol-2 ($UQ_2H_2$) by cytochrome *bo₃* under multiple turnover conditions. The results show the rapid appearance of the semiquinone radical, coincident with the reduction of the low-spin heme *b* component of the enzyme. The rate of formation of the semiquinone radical is consistent with the proposition that this is a kinetically relevant intermediate in the reaction sequence. As $UQ_2H_2$ is depleted, the radical decays and the enzyme forms a "peroxy," or P, complex with dioxygen. No detectable protein radical is associated with the P complex.

*Key words.* Semiquinone—Quinol oxidase—Heme—*E. coli*—Peroxy state

## Introduction

The cytochrome *bo₃* ubiquinol oxidase, from *Escherichia coli*, is an integral membrane protein. It is a member of the superfamily of heme-copper terminal oxidases, as is cytochrome *c* oxidase [1], for which two crystal structures have recently been solved [2–4]. Cytochrome *bo₃* catalyzes the two-electron oxidation of ubiquinol-8 in the bacterial membrane and reduces dioxygen to water. For each dioxygen, four protons are pumped electrogenically across the membrane bilayer, contributing to the generation of a protonmotive force. In addition, four protons are released into the bacterial periplasm and another four protons (per $O_2$) are taken up from the bacterial cytoplasm to form water ($2H_2O$).

[1] School of Chemical Sciences, University of Illinois, 600 S. Goodwin, Urbana, IL 61801, USA
[2] Noyes Laboratory, 127-72, Department of Chemistry, California Institute of Technology, Pasadena, CA 91125, USA
[3] Department of Biochemistry, Emory University School of Medicine, Atlanta, GA 30322, USA

Cytochrome $bo_3$ contains one low-spin heme $b$ and a heme-copper bimetallic center, consisting of a high-spin heme $o$ and copper ($Cu_B$), where oxygen is activated and reduced. Instead of receiving its electrons from the one-electron carrier cytochrome $c$, as in the cytochrome $c$ oxidases, cytochrome $bo_3$ is reduced by lipid-soluble, two-electron dihydroquinols such as ubiquinol-8 or menaquinol. Additionally, the $Cu_A$ redox component located within subunit II of the cytochrome $c$ oxidases, which functions to transfer electrons from cytochrome $c$ to the low-spin heme, is missing in cytochrome $bo_3$ [5] and the other quinol oxidases within the superfamily. A photoreactive ubiquinol analogue has been covalently cross-linked to the cytochrome $bo_3$, indicating that at least one quinol-binding site is located in subunit II [6]. Previous work has demonstrated that the enzyme probably contains two binding sites for ubiquinol [7]. One, $Q_H$, has a high affinity for ubiquinone, and it has been postulated that this bound quinone species acts as a cofactor during turnover of the enzyme. The other, $Q_L$, has a lower affinity for ubiquinol and is presumably the site where ubiquinol is oxidized to ubiquinone during turnover of the enzyme and which is in rapid exchange with the quinol pool in the membrane. A schematic of these sites in cytochrome $bo_3$ is shown in Fig. 1.

One can consider two types of binding sites where the two-electron oxidation of quinol occurs: pair-splitting, where there are two separate electron acceptors, and sequential, where there is a single electron acceptor and the oxidation occurs in two one-electron steps. A pair-splitting site is illustrated by the quinol oxidation site ($Q_o$) of the $bc_1$ complex. In this case, the quinol donates electrons to two different acceptor groups (low-potential cytochrome $b$ and Rieske Fe-S center) with broadly disparate redox potentials [8]. The initial one-electron oxidation of quinol (by the Fe-S center) generates a very unstable semiquinone with a sufficiently low redox potential to reduce the second electron acceptor (cytochrome $b_L$) at virtually the same time. In contrast, a "simple" oxidation site donates electrons sequentially to the same electron acceptor, requiring at least transient stabilization of the semiquinone

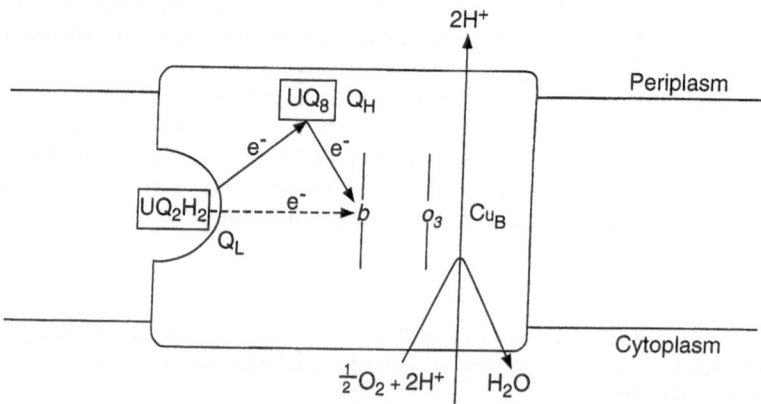

FIG. 1. Cytochrome $bo_3$ contains two ubiquinone-binding sites, designated as $Q_L$ and $Q_H$. Possible electron-transfer pathways are shown between $Q_L$, $Q_H$, and heme $b$. $UQ$, is ubiquinone

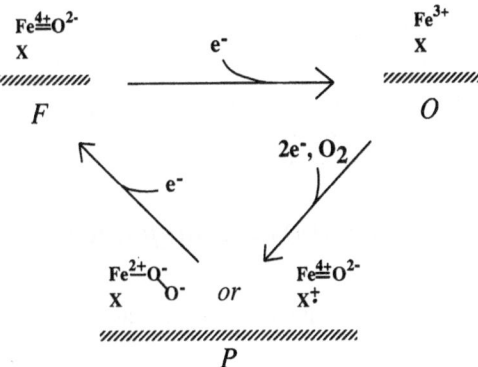

FIG. 2. A simplified version of the catalytic cycle showing the reduction of dioxygen by the heme-copper oxidases. Intermediates are designated $O$ (oxidized), $P$ (peroxy), and $F$ (oxoferryl). $X$ is iron, porphyrin, copper, or protein

while the electron acceptor is reoxidized following its initial reduction by the first electron [9].

The chemistry of dioxygen reduction by cytochrome $bo_3$ (Fig. 2) is the same as that in cytochrome $c$ oxidase [10,11], implying that electrons first arrive at the low-spin heme $b$, and then are transferred to the heme $o/Cu_B$ bimetallic center where oxygen reduction occurs. The sequence of electron transfer from $Q_L$ to the low-spin heme $b$ is not known. A thermodynamically stabilized ubisemiquinone radical has been identified in cytochrome $bo_3$ poised at appropriate redox potentials [9,12]. In principle, a stabilized semiquinone, competent to transfer an electron to the low-spin heme $b$, means that the ubiquinol can donate its electrons sequentially from the $Q_L$ site to heme $b$. The question addressed in this work is whether this ubisemiquinone species is kinetically significant. This is part of a larger study to determine how cytochrome $bo_3$ initially oxidizes ubiquinol by deciphering the order of redox events. It is important to decipher this to understand how the oxygen chemistry is coupled to proton pumping.

It is shown that the previously reported, thermodynamically stabilized ubisemiquinone radical signal is observed under multiple turnover conditions, indicating that this ubisemiquinone is also a kinetic intermediate in the enzyme reaction.

## Materials and Methods

### Purification

Cytochrome $bo_3$ was purified from an overproducing E. coli strain grown aerobically in a 200-l fermentor. The enzyme was solubilized in dodecyl maltoside, the detergent used for the entire preparation. A genetically engineered, six-histidine tag on the protein enabled affinity chromatography with a nickel column to effect a one-step purification (manuscript in preparation).

## UV-Visible Stopped Flow

UV-visible (UV-vis) kinetic experiments were performed using an Applied Photophysics (Leatherhead, UK) SX-18MV stopped-flow spectrophotometer equipped with a photodiode array for recording whole spectra. Results were analyzed using Pro-K global analysis software from Applied Photophysics.

## Rapid Freeze-Quench Electron Paramagnetic Resonance (EPR) Spectroscopy

An illustration of the rapid freeze-quench setup is shown in Fig. 3 and has been described previously [13].

# Results

## Stopped-Flow Reaction of $UQ_2H_2$ with Cytochrome bo$_3$

A ubiquinol-8 analogue, ubiquinol-2 ($UQ_2H_2$), was reacted with cytochrome bo$_3$ using stopped flow, and the reaction was followed in the UV-vis region with a diode array. The low-spin heme $b$ component of the enzyme was reduced rapidly, and on reoxidation a species was formed that absorbed at 580 nm. This species was stable over the course of several seconds.

The spectrum obtained for the reduced low-spin heme $b$ after global kinetic fitting matches the spectrum of the heme $b$ component of the oxidase obtained in the presence of dithionite. Quantitation indicates that the majority of heme $b$ is reduced

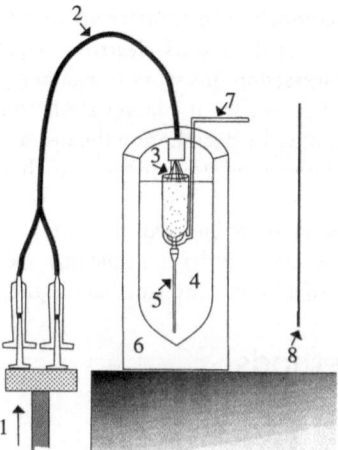

FIG. 3. The rapid freeze-quench setup for making electron paramagnetic resonance (EPR) samples. Numbered items are as follows: (1) constant velocity drive ram for syringes; (2) aging loop controlling the reaction time; (3) sample aerosol; (4) Isopentane at −140°C for freezing the sample; (5) EPR tube; (6) liquid nitrogen in a double dewar; (7) handle for holding the EPR tube; (8) packing plunger

during the reaction with $UQ_2H_2$. Decay of the reduced heme $b$ species occurred before $UQ_2H_2$ was consumed completely, indicating that the enzyme turned over multiple times and that the amount of the reduced heme $b$ intermediate is maximal when the ratio of $UQ_2H_2$ to cytochrome $bo_3$ is high.

Dithiothreitol (DTT), a slow reductant of $UQ_2$ ($28\,M^{-1}s^{-1}$), always was present in the reaction mixture to maintain $UQ_2H_2$ in the reduced state before the reaction was initiated. Reduced heme $b$ formed at a rate much faster than the rereduction of $UQ_2H_2$ by DTT. However, the appearance of the 580-nm species occurred at longer times, such that the reaction with DTT could be contributing. Essentially, during slow turn-over of the oxidase, which is limited by DTT rereducing $UQ_2H_2$, the 580-nm species predominates. A species with a similar absorbance has been reported as the "peroxy" form of oxygenated cytochrome $bo_3$ [10].

## Rapid Freeze-Quench Analysis

The one-electron reduction of heme $b$ in the initially observed species suggested that the second electron from $UQ_2H_2$ might be present within an enzyme-bound ubisemiquinone. To test this hypothesis, rapid freeze-quench EPR spectroscopy was performed on samples at different times after mixing cytochrome $bo_3$ with $UQ_2H_2$. A ubisemiquinone signal was observed near $g = 2$. The line shape and splittings of the signal are the same as that reported for the thermodynamically stabilized ubisemiquinone radical [9,12]. The time course of the appearance of the ubisemiquinone EPR signal is the same as that of reduced heme $b$ determined by UV-vis stopped-flow spectroscopy under the same reaction conditions. Quantitation of the amount of the ubisemiquinone radical indicated that it was present within a significant fraction of the enzyme population.

## Discussion

Sequence homologies make it appear certain that the structure of the cytochrome $bo_3$ quinol oxidase must be very similar to those of the cytochrome $c$ oxidases from bovine heart [3,4] and from *Paracoccus denitrificans* [2]. One exception, of course, is the quinol binding sites, which are unique to the quinol oxidases and are not present in the related cytochrome $c$ oxidases. Much work has demonstrated that the oxygen chemistry and proton pumping mechanisms in the cytochrome $c$ and quinol oxidases are probably the same [10,11]. A question that remains is how the electrons are transferred from substrate quinol to the hemes. In this chapter, it has been established that electrons are transferred from quinol in a sequence that involves the rapid formation of a ubisemiquinone radical which is thermodynamically stabilized within the enzyme.

A high-affinity quinone-binding site, $Q_H$, has recently been described in cytochrome $bo_3$. Structural and functional roles were both postulated for this tightly bound quinone [14], and it was proposed that this thermodynamically stabilized semiquinone was bound at the $Q_H$ site [12]. The unique splittings of the EPR spectrum of the thermodynamically stabilized, enzyme-bound ubisemiquinone that have been observed using equilibrium conditions [9,12] are also observed in the current work in the EPR spectrum of the ubisemiquinone produced during the rapid freeze-quench

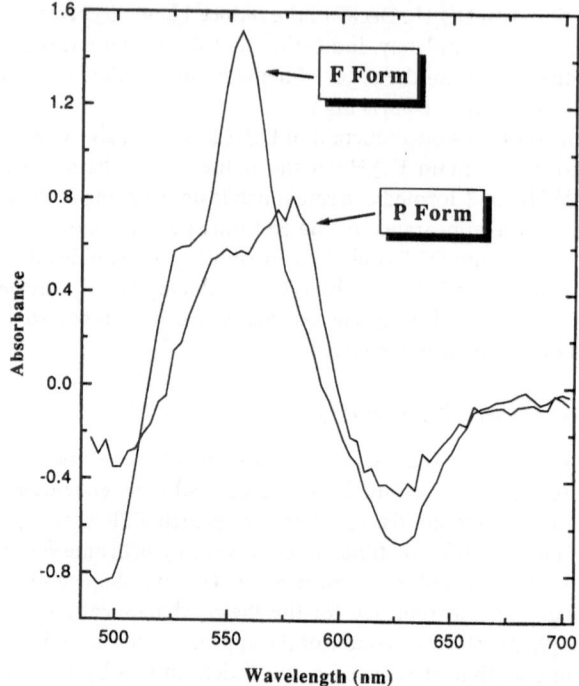

FIG. 4. UV-vis absorption spectra of the *P* and *F* intermediate forms of cytochrome *bo₃* produced by reacting oxidized cytochrome *bo₃* with $H_2O_2$. The P form has a peak at 580 nm, as previously reported [10]; the spectrum of the F form has also been previously reported [18]

experiments (not shown), indicating that the species observed under the different experimental conditions are identical. Hence, it is likely that the kinetically trapped ubisemiquinone is bound at the $Q_H$ site. Although electron transfer from $Q_L$ directly to the low-spin heme *b* cannot be ruled out, the data suggest that electron transfer from the quinol bound at $Q_L$ proceeds through $Q_H$. $Q_H$ may function analogously to $Cu_A$ in the cytochrome *c* oxidases.

The 580-nm species that is observed on decay of the reduced heme *b* and ubisemiquinone appears similar to the P complex which has been produced by reacting oxidized enzyme with $H_2O_2$, binding dioxygen to the two-electron reduced enzyme [10] or during turnover in steady state with ascorbate and mediators [15]. UV-vis absorption spectra of the two intermediates formed during the reaction of oxidized cytochrome *bo₃* with $H_2O_2$ are shown in Fig. 4. Structures for the P intermediate have been postulated that include a protein radical [16,17]. However, in the EPR freeze-quench experiments, at time points when the 580-nm species is formed maximally, there is no observed protein-associated free radical other than some remaining ubisemiquinone.

# References

1. García-Horsman JA, Barquera B, Rumbley J, et al. (1994) The superfamily of heme-copper respiratory oxidases. J Bacteriol 176:5587–5600
2. Iwata S, Ostermeier C, Ludwig B, et al. (1995) Structure at 2.8 Å resolution of cytochrome c oxidase from *Paracoccus denitrificans*. Nature 376:660–669
3. Tsukihara T, Aoyama H, Yamashita E, et al. (1995) Structures of metal sites of oxidized bovine heart cytochrome c oxidase at 2.8 Å. Science 269:1069–1074
4. Tsukihara T, Aoyama H, Yamashita E, et al. (1996) The whole structure of the 13-subunit oxidized cytochrome c oxidase at 2.8 Å. Science 272:1136–1144
5. Minghetti KC, Goswitz VC, Gabriel NE, et al. (1992) Modified, large-scale purification of the cytochrome o complex of *Escherichia coli* yields a two heme/one copper terminal oxidase with high specific activity. Biochemistry 31:6917–6924
6. Welter R, Gu L-Q, Yu L, et al. (1994) Identification of the ubiquinol-binding site in the cytochrome $bo_3$-ubiquinol oxidase of *Escherichia coli*. J Biol Chem 269:28834–28838
7. Sato-Watanabe M, Mogi T, Ogura T, et al. (1994) Identification of a novel quinone binding site in the cytochrome bo complex from *Escherichia coli*. J Biol Chem 269:28908–28912
8. Brandt U, Trumpower B (1994) The protonmotive Q cycle in mitochondria and bacteria. Crit Rev Biochem Mol Biol 29:165–197
9. Ingledew WJ, Ohnishi T, Salerno JC (1995) Studies on a stabilisation of ubisemiquinone by *Escherichia coli* quinol oxidase, cytochrome bo. Eur J Biochem 227:903–908
10. Morgan JE, Verkhovsky MI, Puustinen A, et al. (1995) Identification of a "peroxy" intermediate in cytochrome $bo_3$ of *Escherichia coli*. Biochemistry 34:15633–15637
11. Puustinen A, Verkhovsky MI, Morgan JE, et al. (1996) Reaction of the *Escherichia coli* quinol oxidase cytochrome $bo_3$ with dioxygen: the role of a bound ubiquinone molecule. Proc Natl Acad Sci USA 93:1545–1548
12. Sato-Watanabe M, Itoh S, Mogi T, et al. (1995) Stabilization of a semiquinone radical at the high-affinity quinone-binding site ($Q_H$) of the *Escherichia coli* bo-type ubiquinol oxidase. FEBS Lett 374:265–269
13. Ravi N, Bollinger JM, Huynh BH, et al. (1994) Mechanism of assembly of the tyrosyl radical-diiron (III) cofactor of *E. coli* ribonucleotide reductase. 1. Mössbauer characterization of the diferric radical precursor. J Am Chem Soc 116:8007–8014
14. Sato-Watanabe M, Mogi T, Miyoshi H, et al. (1994) Structure-function studies on the ubiquinol oxidation site of the cytochrome bo complex from *Escherichia coli* using p-benzoquinones and substituted phenols. J Biol Chem 269:28899–28907
15. Moody AJ, Rumbley JN, Ingledew WJ, et al. (1993) The reaction of hydrogen peroxide with cytochrome bo from *E. coli*. Biochem Soc Trans 21:255
16. Ogura T, Hirota S, Proshlyakov DA, et al. (1996) Time-resolved resonance Raman evidence for tight coupling between electron transfer and proton pumping of cytochrome c oxidase upon the change from the $Fe^V$ oxidation level to the $Fe^{IV}$ oxidation level. J Am Chem Soc 118:5443–5449
17. Weng L, Baker GM (1991) Reaction of hydrogen peroxide with the rapid form of resting cytochrome oxidase. Biochemistry 30:5727–5733
18. Cheesman MR, Watmough NJ, Gennis RB, et al. (1994) Magnetic-circular-dichroism studies of *Escherichia coli* cytochrome bo identification of high-spin ferric, low-spin ferric and ferryl [Fe(IV)] forms of heme o. Eur J Biochem 219:595–602

# Coupling of Ion and Charge Movements: From Peroxidases to Protonmotive Oxidases

Peter R. Rich, Brigitte Meunier, and Susanne Jünemann

*Summary.* The energy cost of introduction of charges into regions of low dielectric strength in proteins can be reduced by associated binding of a proton to appropriately placed residues. Such protonations can be important in peroxidases and other soluble proteins, and are likely to be central to the protonmotive mechanism of oxidases. Three residues in subunit I of cytochrome oxidase that are likely to interfere with such protonations have been examined by mutagenesis, and the data are discussed in the light of the known crystal structures.

*Key words.* Cytochrome oxidase—Peroxidase—Proton translocation—Protonation

## Introduction

To understand the protonmotive chemistry of terminal oxidases, we are examining the link between charge changes of their reaction cycle intermediates and the associated movement and binding of protons within the protein structure. Protonations can provide a means for charge compensation of the otherwise net negative charge changes as the enzyme proceeds through some of its reaction cycle steps, thus decreasing the high energy cost of introduction of charges into a region of low dielectric strength [1].

In many proteins, the energetic cost is decreased by solvent and ion rearrangements, by movement of local charges and hydrogen bonds in the protein, by charge delocalization, or by covalent bond changes. In these cases, no cation uptake is observed. However, if these factors are inadequate, and if there are appropriate sites in the protein structure, specific counterions may instead bind in response to the introduction of the charges; that is, the affinities of the sites for the counterions are increased in response to the charges. In the special case of protonation, the result is redox-linked p$K$ shifts. The strength of the effect is given by the p$K$ differences in the oxidized and reduced conditions [2].

Such charge-associated protonations can be important factors that are additional to the microscopic electron transfer rate constants in controlling equilibrium

Department of Biology, University College, Gower Street, London WC1E 6BT, UK

constants and observed rates of reaction. For example, we have shown the importance of the cobinding of a proton in controlling the binding and rate constants of anionic ligands in the distal heme pocket of horseradish peroxidase under some conditions [3,4]. Control of reaction rates through limitations of proton movement rates might be particularly important if the proton has to move through the protein structure to arrive at its binding site, as is likely to be the case for compensation of charge changes in the deeply buried binuclear center of the protonmotive oxidases [5,6].

## Charge-Linked Protonations in Protonmotive Oxidases

Currently, we are considering two general types of mechanism, based on ideas in [7–9], by which proton translocation is coupled to the catalytic reaction cycle of the oxidases. Two protons are known to be bound when the peroxy state is formed from the oxidized enzyme, and we assume that these protons are destined to be translocated, in response to uptake of substrate protons for water generation, rather than acting as the "substrate" protons themselves. One variant assumes that the translocated protons are bound to the conserved glutamate and other moieties in its surroundings (for example, the heme propionates), and are taken up into this region through a channel [9,10] that is physically separated from the binuclear center and its "substrate" proton channel. We propose to call this model the glutamate trap. A second model, introduced by Wikström and termed the histidine cycle, assumes that the two protons are bound to one of the histidine ligands of $Cu_B$ [8]. Hence, they are quite close to the oxygen intermediates and could reach their binding sites either through the same channel used by substrate protons or, perhaps, through a separate route.

We have tested empirically the degree to which the binuclear center balances charge changes by protonation changes . To date, we have found no exception to the general rule that all stable redox or ligand-binding changes in the binuclear center are counterbalanced by protonation changes [11]. The pH dependency of midpoint potentials of the metals during classical redox potentiometry is also consistent with this view (reviewed in [12]). A model for proton translocation has been presented that takes into account this need for net charge balance of the stable intermediates [1,7].

The structures of the four subunit cytochrome oxidase from *Paracoccus denitrificans* [9] and the 13 subunit enzyme from beef heart mitochondria [10,13] have been solved by X-ray crystallography to 2.8 Å resolution, and coordinates for the latter are available. Possible protonation sites and routes in subunit I of the protein can be identified [14,15]. The most obvious candidates for stable sites of charge-linked protonation in subunit I include the histidine ligands to the metal centers, the heme propionates, and conserved protonatable residues lysine-319, glutamate-242, and tyrosine-245 (unless otherwise stated, all numbering refers to the bovine sequence). We have begun to analyze mutant enzymes with a view to assessing their importance in this respect. Figure 1 illustrates the positions of three residues that might influence protonation properties, and our initial analyses of the primary effects of these mutations are described here and also in the chapter by B. Meunier and P.R. Rich, this volume.

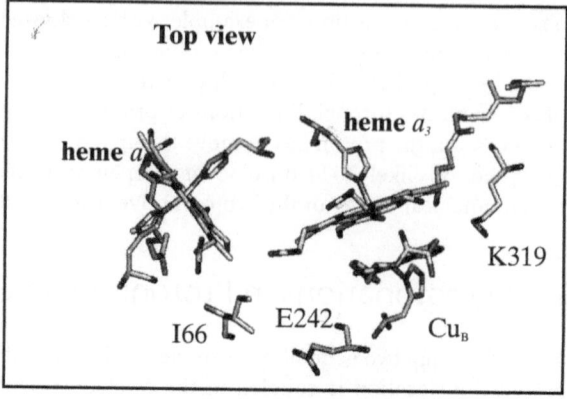

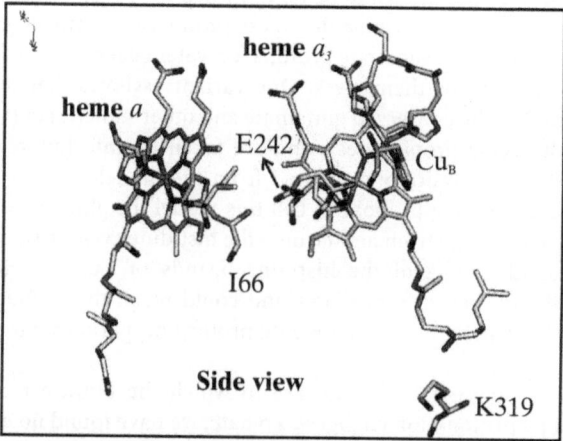

FIG. 1. Positions of residues affecting protonation processes. The structures have been drawn from the coordinates of the bovine enzyme published by Tsukihara et al. [10], and numbering of residues is also according to the bovine subunit I sequence. *Top view*, from the cytochrome *c* side of the membrane; *side view*, a perspective from the membrane

## Mutations in the Region of the Conserved Glutamate, E-242

We have studied a mutant form of cytochrome *c* oxidase that we have isolated from *Saccharomyces cerevisiae* [16] which has a single mutation of I67N in subunit I (equivalent to I66 in the bovine structure in Fig. 1), a position that from the crystallographic data is close to heme *a* and to a conserved glutamate on helix VI (see Fig. 1). The enzyme has a very low catalytic turnover number compared to the wild-type form. This mutation lowers the midpoint potential of heme *a* by 60 mV, but is without significant effect on the midpoint potentials or ligand reactivity of the binuclear center metals. Steady-state spectra indicate that at least some of the oxygen intermediates are readily formed. Analysis of the pH dependencies of midpoint potentials indicates a weakening of the redox-linked protonations associated with heme *a* reduction [17], and this may be the primary cause of its lowered midpoint potential.

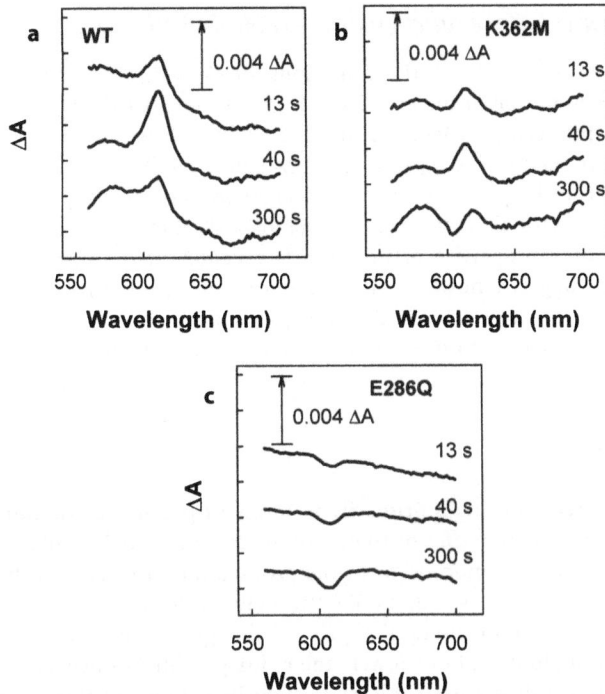

FIG. 2a–c. Reaction of hydrogen peroxide with wild-type (*WT*) and mutant forms of cytochrome *c* oxidase from *Rhodobacter sphaeroides*. Purified enzymes were dissolved to about 0.5 μM in 50 mM potassium phosphate at pH 8.0 and containing 0.1% lauryl maltoside. Samples were incubated with 100 μM potassium ferricyanide until no further oxidation could be discerned; 2 mM hydrogen peroxide was then added and spectra were recorded at the times indicated after this addition. Formation of the peroxy state at 607 nm, followed by formation of the ferryl state at 585 nm, can clearly be observed in both wild-type (a) and K319 (b) forms, whereas only a small heme degradation was observed in the E286Q form (c). (Experiments were performed in collaboration with the laboratory of Prof. R.B. Gennis)

It has already been shown that mutation of the glutamate residue itself leads to a complete loss of catalytic activity and to a reaction of the reduced enzyme with oxygen that proceeds only partially, perhaps to the peroxy form, and without uptake of protons from the medium [18]. Our preliminary analysis of the mutant E284Q isolated from *Rhodobacter sphaeroides* has indicated that the enzyme as prepared may be already primarily in the peroxy state and is unable to proceed past this step even in the presence of a respiratory substrate. This conjecture is based on the absence of a charge transfer band around 655 nm, a significant visible band absorbance in the 600- to 610-nm region, lack of ligand reactivity of the as prepared form (data not shown), and a lack of reaction with hydrogen peroxide to form the characteristic 607-nm peak [15] indicative of transformation into the peroxy form (Fig. 2).

## Mutation of a Conserved Lysine on Helix VIII

In contrast to the effects of the mutations in the conserved glutamate region just described, mutation of the conserved lysine, K-319, is without significant effect on the redox properties of heme $a$. Instead, reduction of the binuclear center to a form that can react with oxygen is no longer possible. Reduction of a major part of the heme $a_3$ in the presence of reduced cytochrome $c$ or dithionite is extremely slow ([19]; also see the chapter by B. Meunier and P.R. Rich, this volume). Analysis of the ligand-binding properties of the binuclear center in the presence of reduced cytochrome $c$ suggests, however, that it does not quantitatively remain in the fully oxidized state when reduced substrate is present. On the basis of our preliminary data, we suggest that up to one charge may be able to enter the binuclear center readily.

# Discussion

We have interpreted our data primarily in terms of possible disruption of charge-compensating protonation sites or routes. In the case of the I67N mutant in yeast, we propose that this substitution perturbs the protonation properties of the conserved glutamate, which in turn is redox-linked to heme $a$. Specifically, the introduction of the asparagine group alters the p$K$s of the glutamate residue, perhaps by hydrogen bonding to it, and this lowers the redox potential of heme $a$. Inhibition of catalytic turnover is then caused at least partially by a difficulty of reduction of heme $a$. At least part of the catalytic cycle of the binuclear center appears to be possible. We have not yet ascertained whether removal of the glutamate residue in the E286Q mutant of *R. sphaeroides* also impairs redox-linked protonation of heme $a$ and thereby lowers its midpoint potential. However, the available data indicate that the binuclear center may be able to reach the peroxy state but no further in the catalytic cycle. Thus, disruption of charge-linked protonation in this region may influence the energetics both of reduction of heme $a$ and of conversion between some of the oxygen intermediates.

Lack of catalytic turnover in the K362M mutant of *R. sphaeroides* may arise for a quite different reason, although again associated with electrostatic factors. In this case, the midpoint potential and reduction of heme $a$ by substrate appear to be normal. Instead, difficulty is encountered in formation of one or more of the catalytic intermediates of the binuclear center. Preliminary experiments indicate that up to one electron transfer into the binuclear center may be facile but transfer of two electrons is not. Because both metal centers of the binuclear center must be reduced (the "R" state) for reaction with oxygen to form the peroxy state, it is the inability to form this R state that prevents catalytic activity. The defect is presumably electrostatic in origin, with loss of the lysine residue preventing appropriate charge-compensating protonation reaction(s). This may arise because the lysine itself is a static redox-linked protonation site (so that the equilibrium constant is affected) or because the lysine provides kinetic access to such a site (so that the rate of formation is affected). Interestingly, direct formation of the two-electron-reduced peroxy state by addition of hydrogen peroxide [20] can occur in the mutant (see Fig. 2). In this reaction, formation of the doubly reduced R state of the binuclear center is not an intermediate, and the hydrogen peroxide effectively carries protons with it, so that formation of the P state now becomes possible.

*Acknowledgments.* We are grateful to EPSRC (GR/J28148), BBSRC (CO2632), HFSP (RG-464/95 M) and The European Community (BIO2-CT-94-8197) for financial support. Parts of this work are being undertaken collaboratively with the laboratories of Prof. R. Gennis (Urbana, IL, USA) and Prof. U. Brandt and Dr. C. Ortwein (Frankfurt, Germany).

# References

1. Rich PR (1996) Electron transfer complexes coupled to ion translocation. In: Bendall DS (ed) Protein electron transfer. BIOS, Oxford, pp 217–248
2. Moore GR, Harris DE, Leitch FA, Pettigrew GW (1984) Characterisations of ionisations that influence the redox potential of mitochondrial cytochrome $c$ and photosynthetic bacterial cytochrome $c_2$. Biochim Biophys Acta 764:331–342
3. Meunier B, Rodriguez-Lopez JN, Smith AT, et al. (1995) Laser photolysis behaviour of ferrous horseradish peroxidase with carbon monoxide and cyanide: effects of mutations in the distal heme pocket. Biochemistry 34:14687–14692
4. Meunier B, Rich PR (1997) Photolysis of the cyanide adduct of the ferrous horseradish peroxidase. Biochim Biophys Acta 1318:235–245
5. Hallén S, Brzezinski P, Malmström BG (1994) Internal electron transfer in cytochrome $c$ oxidase is coupled to the protonation of a group close to the bimetallic site. Biochemistry 33:1467–1472
6. Verkhovsky MI, Morgan JE, Wikström M (1995) Control of electron delivery to the oxygen reduction site of cytochrome $c$ oxidase: a role for protons. Biochemistry 34:7483–7491
7. Rich PR (1995) Towards an understanding of the chemistry of oxygen reduction and proton translocation in the iron-copper respiratory oxidases. Aust J Plant Physiol 22:479–486
8. Morgan JE, Verkhovsky MI, Wikström M (1994) The histidine cycle: a new model for proton translocation in the respiratory heme-copper oxidases. J Bioenerg Biomembr 26:599–608
9. Iwata S, Ostermeier C, Ludwig B, Michel H (1995) Structure at 2.8 Å resolution of cytochrome $c$ oxidase from *Paracoccus denitrificans*. Nature 376:660–669
10. Tsukihara T, Aoyama H, Yamashita E, et al. (1996) The whole structure of the 13-subunit oxidized cytochrome $c$ oxidase at 2.8 Å. Science 272:1136–1144
11. Rich PR, Meunier B, Mitchell RM, Moody AJ (1996) Coupling of charge and proton movement in cytochtrome $c$ oxidase. Biochim Biophys Acta 1275:91–95
12. Rich PR, Moody AJ (1996) Cytochrome $c$ oxidase. In: Gräber P, Milazzo G (eds) Bioelectrochemistry: principles and practice. Birkhäuser, Basel, pp 419–457
13. Tsukihara T, Aoyama H, Yamashita E, et al. (1995) Structures of metal sites of oxidized bovine heart cytochrome $c$ oxidase at 2.8 Å. Science 269:1069–1074
14. Hosler JP, Ferguson-Miller S, Calhoun MW, et al. (1993) Insight into the active-site structure and function of cytochrome oxidase by analysis of site-directed mutants of bacterial cytochrome $aa_3$ and cytochrome $bo$. J Bioenerg Biomembr 25:121–136
15. Babcock GT, Wikström M (1992) Oxygen activation and the conservation of energy in cell respiration. Nature 356:301–309
16. Meunier B, Colson A-M (1994) Random deficiency mutations and reversions in the cytochrome $c$ oxidase subunits I, II and III of *Saccharomyces cerevisiae*. Biochim Biophys Acta 1187:112–115

17. Moody AJ, Rich PR (1990) The effect of pH on redox titrations of heme *a* in cyanide-liganded cytochrome-*c* oxidase: experimental and modelling studies. Biochim Biophys Acta 1015:205–215
18. Svensson-EM, Thomas JW, Gennis RB, et al. (1996) Kinetics of electron and proton transfer during the reaction of wild type and helix VI mutants of cytochrome *bo*$_3$ with oxygen. Biochemistry 35:13673–13680
19. Svensson M, Hallén S, Thomas JW, et al. (1995) Oxygen reaction and proton uptake in helix VIII mutants of cytochrome *bo*$_3$. Biochemistry 34:5252–5258
20. Wrigglesworth JM (1984) Formation and reduction of a "peroxy" intermediate of cytochrome *c* oxidase by hydrogen peroxide. Biochem J 217:715–719

# Ligand Dynamics in the Binuclear Site in Cytochrome Oxidase

GERALD T. BABCOCK[1], GEURT DEINUM[2], JON HOSLER[3], YOUNKYOO KIM[4], MICHELLE PRESSLER[1], DENIS A. PROSHLYAKOV[1], HANS SCHELVIS[1], CONSTANTINOS VAROTSIS[5], and SHELAGH FERGUSON-MILLER[6]

*Summary.* The dioxygen-reduction mechanism in cytochrome oxidase relies on proton control of the electron-transfer events that drive the process. Recent work on proton delivery and efflux channels in the protein that are relevant to substrate reduction and proton pumping is considered, and the current status of this area is summarized. Carbon monoxide photodissociation and the ligand dynamics that occur subsequent to photolysis have been valuable tools in probing possible coupling schemes for linking exergonic electron-transfer chemistry to endergonic proton translocation. Our picosecond-time-resolved Raman results show that the heme $a_3$–proximal histidine bond remains intact following CO photodissociation but that the local environment around the heme $a_3$ center in the photoproduct is in a nonequilibrium state. This photoproduct relaxes to its equilibrium configuration on the same time scale as ligand release occurs from $Cu_B$, which suggests a coupling between the two events and a potential signaling pathway between the site of $O_2$ binding and reduction and the putative element, $Cu_B$, that links the redox chemistry to the proton pump.

*Key words.* Oxygen activation—Proton pump—Ligand shuttle—Ligand photolysis—Respiration

[1] Department of Chemistry, 320 Chemistry Building, Michigan State University, East Lansing, MI 48824, USA
[2] G. R. Harrison Spectroscopy Laboratory, Massachusetts Institute of Technology, 77 Massachusetts Avenue, Cambridge, MA 02139, USA
[3] Department of Biochemistry, University of Mississippi Medical Center, 2500 North State Street, Jackson, MS 39216, USA
[4] Department of Chemistry, Hankuk University of Foreign Studies, Yongin, Kyungki DO 449-791, Korea
[5] Department of Chemistry, University of Crete, 71 409 Iraklion, Crete, Greece
[6] Department of Biochemistry, Michigan State University, East Lansing, MI 48824, USA

47

## Introduction

Cytochrome oxidase catalyzes the final step in the respiratory process whereby electrons exit the electron-transport chain during the reduction of dioxygen to water (for a review, see [1]). This process is critical to the ATP-forming energy transduction reactions that occur in respiration, as it allows a continuous electron current through the redox cofactors and a sustained buildup of the chemiosmotic gradient that is the immediate precursor to ADP phosphorylation. The oxidation of the functional electron donor to cytochrome oxidase, cytochrome $c$, during the $O_2$ reduction is thermodynamically driven by more than 0.5 eV per electron under "no load" conditions across the respiratory membrane. Rather than dissipate this free energy as heat, a mechanism has evolved that allows oxidases to use the free energy made available during the redox chemistry to pump protons against their chemiosmotic gradient, thereby contributing directly to the $\Delta\mu_{H^+}$ [2].

A considerable amount of effort has been devoted to understanding the coupling of dioxygen reduction to proton translocation. Figure 1 summarizes a number of insights that have resulted from this work [1,3]. Beginning with the oxidized form of the enzyme, indicated as the hydroxy species in Fig. 1, two electrons are delivered from cytochrome $c$ through the $Cu_A$ and heme $a$ cofactors of the enzyme to the binuclear center that comprises heme $a_3$ and $Cu_B$. These electron transfers produce the ferrous/cuprous reduced form of the active site, which then binds $O_2$ to yield the initial oxy-intermediate. Subsequent intra- and intermolecular electron and proton transfers generate "peroxy" and "ferryl" intermediates that are ultimately converted back to the hydroxy form to initiate additional cycles of the dioxygen-reducing/proton-pumping process.

A critical feature of the mechanism in Fig. 1 is that only the late intermediates, those that are formally at the peroxy and ferryl levels of $O_2$ reduction, are coupled to proton

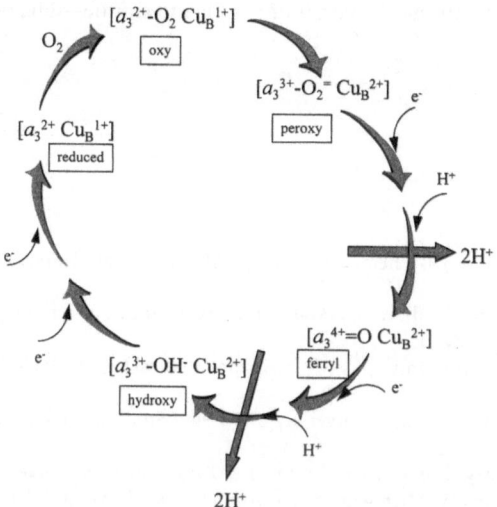

FIG. 1. Simplified scheme for the reduction of $O_2$ by cytochrome oxidase and the coupling of energy released in this process to proton translocation

translocation. That is, even though the overall stoichiometry of the reaction is one proton pumped per electron delivered to $O_2$, only the last two one-electron steps in this process—the reduction of peroxy to ferryl and the reduction of ferryl to hydroxy—are coupled to the proton-pumping machinery in the enzyme. As we have argued in detail elsewhere [1,3], efficient coupling of the pump with the driving redox chemistry is achieved in oxidase by allowing the proton-transfer reactions that immediately precede the pumping steps to become overall rate determining in the process. Operatively, this mechanism requires that the individual steps in the $O_2$-reduction cycle become slower as the reaction proceeds and that the later steps in the cycle, those that involve reduction of the peroxy and ferryl species, become pH dependent. These expectations have been confirmed in a variety of time-resolved optical and resonance Raman experiments [1,3,4–6].

The proton-controlled reduction of $O_2$ by oxidase has a number of important consequences, particularly that the process slows down as the reaction proceeds and that intermediate species build up to substantial concentrations during the catalytic cycle. As we have discussed elsewhere [1], such a mechanism is unusual in the enzymatic activation and reduction of dioxygen. It affords a unique opportunity to understand, in detail, the molecular steps involved in the binding of $O_2$ at a metalloenzyme site and the subsequent electron- and proton-transfer reactions that lead to cleavage of the O=O bond and release of product water or hydroxide. Considerable insight into this chemistry has developed, as reviewed elsewhere [1,5].

## Proton Pumping: General Considerations

A second important consequence of the proton-controlled chemistry in oxidase is that it provides an opportunity to study the pumping mechanism by a variety of spectroscopic, biochemical, molecular biological, and X-ray crystallographic techniques. The crystal structures of the enzyme, obtained independently by Yoshikawa and coworkers [7,8] and Michel and coworkers [9], have been especially useful in this effort, particularly in conjunction with molecular biological methods that can now be guided by the structure [2]. A considerable amount of progress has been made recently on various aspects of the pump mechanism, and we review that progress here before proceeding to some recent experiments that we have carried out to understand ligand dynamics in the binuclear center, which are likely to be of relevance to the molecular basis of the proton translocation process.

For a proton pump to operate effectively, two components are necessary: (a) an element that couples the redox chemistry at the metal centers in the enzyme to the proton translocation process and (b) proton pathways through the protein that are switchable, in terms of proton conductivity, as dictated by the status of the coupling element. At present, good progress has been made on the proton pathway part of this problem. Because protons are both a substrate in reducing $O_2$ to water and the translocated species, we expect that there may be separate pathways for transporting substrate and pumped protons. This somewhat naive expectation is reinforced by current models that have been proposed for this operation of the pump [9–11] and by directed mutagenesis work that has shown that the pumping process can be frustrated by point mutations without abolishing $O_2$ reduction (reviewed in [1]).

## Proton Pump - Coupling Mechanism

1. Proton conduction pathways
2. e⁻/H⁺ coupling element and mechanism

FIG. 2. Proton channels in cytochrome oxidase

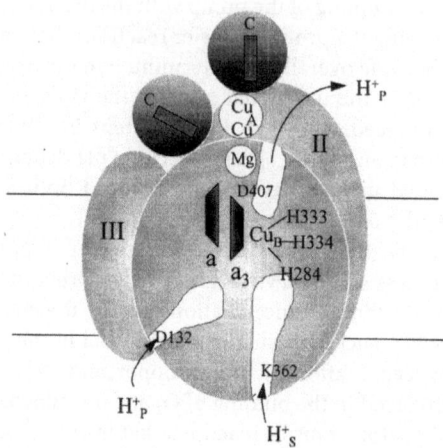

These results are summarized pictorially in Fig. 2, which shows that separate pathways converge on the binuclear center, where the $Cu_B$ site is currently the leading candidate for the coupling element. For proton efflux from this site, of course, only a single pathway is necessary, as shown in Fig. 2, to complete the route for pumped protons through the enzyme. For at least two of the channels in Fig. 2, that is, those that lead to the binuclear center, there is now a reasonably good database to suggest specific residues that participate. Thus, one of the two is likely to originate near D132 in subunit I; the second appears to involve at least K362, T359, and T352, also in subunit I, as critical elements [1]. At present, the pathway involving D132 has been implicated in steps that involve the pumping process, whereas the K362 channel has been implicated in some steps that deliver substrate protons to the binuclear center.

The proton efflux pathway shown in Fig. 2 is somewhat more problematic. Although recent work has shown the occurrence of a $Mg^{2+}/Mn^{2+}$-binding site [12] that has a suggestive position in the vicinity of the binuclear center [7], the role of this metal center in promoting proton translocation remains unclear. Moreover, D407, which looks to be ideal for proton efflux [9], can be mutated to a nonprotonatable residue without interrupting proton translocation [13]. Efforts to identify the route of pumped protons to the C-side of the membrane will undoubtedly continue.

For the second component of the pump, the coupling element, current models [9–11,14,15] favor a direct coupling scheme that involves a ligand shuttle mechanism at the $Cu_B$ site in the binuclear center. Chan and coworkers [16] have laid out the principles of ligand shuttle operation, which are summarized in the generic shuttle mechanism shown in Fig. 3. In the input state of the pump, the metal (M) is ligated by

Generic Ligand Shuttle Mechanism for
Redox-Linked Proton Pump

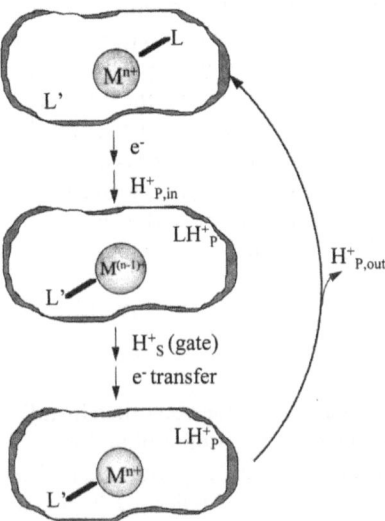

FIG. 3.  A generic model for ligand-shuttle-driven proton pumping (see text for details)

ligand L; reduction of M is accompanied by protonation of L through the proton input channel of the pump and its replacement in the M coordination sphere by L'. Subsequent delivery of substrate protons to active oxygen species at the dioxygen-reduction site forces M oxidation and deprotonation of $LH^+$ through the efflux channel. The requirement for charge neutrality at the M site is critical to coupled proton/electron uptake at the M site [10]. Moreover, such a site can be fairly easily modified to incorporate the $2H^+/e^-$ stoichiometry that occurs during the reduction of the late intermediates in the cytochrome oxidase/dioxygen reaction cycle (see Fig. 1) [9,11].

## Proton-Pumping Models and Ligand Dynamics in the Binuclear Center

At present, there are specific proposals for a distal $Cu_B$-based histidine shuttle in cytochrome oxidase [9,11], for a distal ligand exchange process that has been postulated to function in the proton-pumping cycle [15], and for a proximal ligand exchange process that involves a tyrosine/histidine shuttle as the proximal ligand to heme $a_3$ [14]. Experimental methods are available to test the latter two of these three proposals, and recent work we have carried out to this end is described next.

Carbon monoxide photodissociation from heme $a_3$ and the events that follow photolysis figure prominently in this work, and we review this briefly. Findsen et al. [17] originally studied CO photolysis from heme $a_3$ and concluded that a simple relaxation process occurred in the binuclear center on photodissociation:

$$\text{His} - \text{Fe}_{a_3}^{2+} - \text{CO} \xrightarrow{hv} \left[\text{His} - \text{Fe}^{2+}\right]* + \text{CO} \xrightarrow{1-10\mu s} \text{His} - \text{Fe}^{2+}$$

where *His* is the proximal H419 and the state denoted by an asterisk represents a nonequilibrium state of the heme $a_3$ state characterized by an upshifted Fe-His stretching vibration. This species relaxes to the equilibrium reduced species on the microsecond time scale following photodissociation. Woodruff and coworkers [18] subsequently characterized the photolysis in more detail and showed that the photo-detached CO migrates to $Cu_B$ on the picosecond time scale following photolysis. They also suggested that the movement of CO to $Cu_B$ triggered a ligand exchange such that a distal pocket residue ligated the heme $a_3$ iron and displaced the proximal histidine to produce a new deoxy species, as follows:

$$\text{His} - \text{Fe}^{2+} - \text{CO} \xrightarrow{hv} \text{His} \quad \text{Fe} - \text{X} \quad \text{CO} \xrightarrow{1-5\mu s} \text{His} - \text{Fe}$$
$$Cu_B^{1+} \qquad\qquad Cu_B \qquad\qquad Cu_B$$

The species X, which they assigned as a residue other than histidine, was subsequently invoked in a ligand shuttle mechanism for proton pumping.

Following the Woodruff work, Rousseau and coworkers [14] suggested that the CO photolysis data could best be interpreted as reflecting ligand shuttle activity in the proximal heme $a_3$ pocket:

$$\text{His} - \text{Fe} - \text{CO} \xrightarrow{hv} \text{Tyr} - \text{Fe} \xrightarrow{\mu s} \text{His} - \text{Fe}$$

where Tyr in the intermediate species was suggested to be Y422 of subunit I. With this proposal, they then went on to postulate that the His–Tyr shuttle formed the basis for the pumping activity in the enzyme.

We have recently explored these proposals in some detail. For the Rousseau model, we mutagenized Y422 to phenylalamine and showed that this substitution did not perturb structure in the binuclear center; moreover, proton pumping in the Y422 F mutant was not disrupted. Accordingly, we can eliminate this proposed mechanism for the proton pump [19].

For the distal ligand shuttle advanced by Woodruff et al., a key prediction is that the Fe-His vibration should be absent immediately following CO photodissociation and that this coordination should only reform on the μs time scale. We have tested this proposal recently by carrying out picosecond-time-resolved resonance Raman during the CO photodissociation and subsequent relaxation processes [20]. Figure 4 shows typical data obtained during these experiments with the beef heart enzyme (left panel) and with the *Rhodobacter sphaeroides* protein (right panel). The lower spectra in each panel show static Raman spectra of the resting reduced enzyme and of the carbon-monoxy-ligated form of the enzyme. In the deoxy species, the $214\,\text{cm}^{-1}$ mode, which arises from the $Fe_{a_3}^{2+}$-His vibration is prominent; ligation by CO forms the six-coordinate carbon-monoxy species, and intensity from the Fe-His mode is absent in the spectrum. In the time-resolved spectra in the upper part of both panels, we photodissociated the CO, with the goal of determining the temporal behavior of the Fe-His vibration, that is, to identify the time scale upon which the Fe-His vibration reappears. The data show that the Fe-His vibration is observed at $220\,\text{cm}^{-1}$ in all spectra in which the probe pulse follows the pump or is overlapped with it. These results demonstrate histidine ligation to the $a_3^{2+}$ Fe at the earliest times following CO

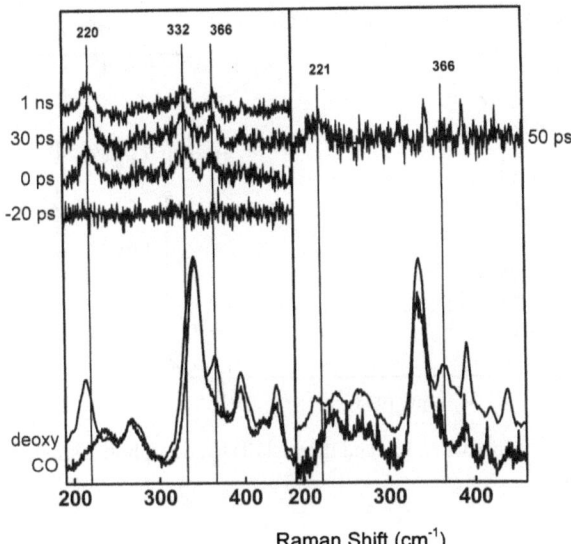

FIG. 4. Static and picosecond (ps) time-resolved resonance Raman spectra for reduced, carbon-monoxy, and transient photoproduct species. Data for this beef heart enzyme are shown in the *left panel*; for *Rhodobacter sphaeroides*, in *right panel*

photodissociation. The ~6 cm$^{-1}$ upshift in $\nu$(Fe-His) in the photoproduct species is consistent with that reported earlier by Findsen et al. [17]; that is, the initial state created following photolysis is not the equilibrium deoxy species but rather one in which the Fe-His mode is perturbed. The earlier work by Findsen et al. showed that the 221 cm$^{-1}$ species relaxes to the equilibrium 215 cm$^{-1}$ form with a time constant on the microsecond time scale, which dovetails well with Woodruff's result that CO dissociates from Cu$_B$ in the same time range [18].

An important issue in interpreting the data in Fig. 4 is to show that the probe pulses that have been used to record the Raman spectra are not sufficiently intense to cause appreciable photodissociation of endogenous ligands. The failure to do this in earlier work [e.g., 19] opens the possibility that the ligand X, postulated by Woodruff et al., is photodissociable in the putative Fe-X photoproduct. Several controls have been carried out for the picosecond experiments in Fig. 4 that show that this pitfall has been avoided [20].

Taken together, the results from the initial work by Findsen et al. [17], the subsequent studies by Woodruff and his co-workers, and those in Fig. 4 and [20], indicate that the scheme in Fig. 5 provides an accurate description of the ligand dynamics that occur on CO photodissociation from carbon-monoxy cytochrome oxidase. Photolysis of the Fe–CO bond leads to rapid (~1 ps) migration of the CO to its Cu$_B$-binding site, where it remains bound into the microsecond time regime. The Fe–His bond stays intact in the photoproduct but occurs initially in a thermally unrelaxed configuration. It only achieves its equilibrium geometry and stretching frequency on the microsecond time scale. Varotsis and Babcock [21] showed that photolysis of the heme $a_3^{2+}$-O$_2$ bond in the initial complex formed during the dioxygen/cytochrome oxidase

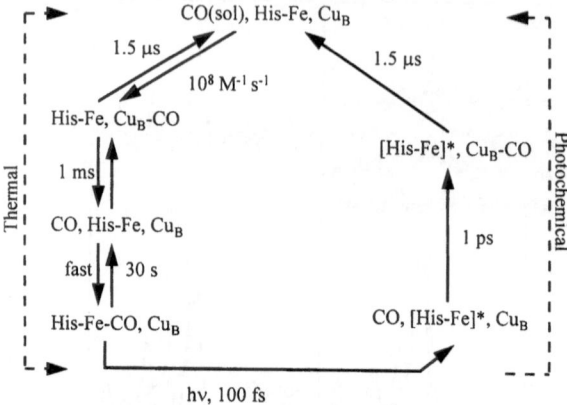

FIG. 5. Overall scheme for photochemical and thermal CO ligation/photodissociation dynamics in cytochrome oxidase

reduction process could also be photolyzed. The vibrational properties of the immediate photoproduct they observed had properties identical to those of the carbonmonoxy photoproduct, which suggests that, in general, ligand photolysis in cytochrome oxidase produces a common transient photolysis product that is characterized by an upshifted Fe-His vibration.

## Conclusions

Spectroscopic studies and biochemical assays on site-directed mutants whose construction has been guided by the crystal structures of cytochrome oxidase have yielded considerable insight into the proton channels that operate during the delivery of protons to the dioxygen-reduction site and to the proton-pumping process. The pumping element that couples the redox chemistry in the enzyme to the pump has been more elusive. Two recent proposals have been tested, as described, and neither appears to be a viable mechanism for the coupling process. The ligand dynamics work, however, has provided considerable insight into events in the binuclear center on exogenous ligand association/dissociation in the enzyme. On ligand dissociation, the proximal Fe–His bond remains intact but is found in a nonequilibrium conformation that persists into the microsecond range. This suggests that there is significant inertia exerted by the protein matrix on the proximal histidine, much like that that occurs in hemoglobin [20]. The latter behavior has been correlated with the R $\leftrightarrow$ T state interconversion in the oxygen transport protein and suggests that similar conformational restraints are operational in cytochrome oxidase.

The role that these may play, and their relevance to the proton-pump mechanism, remain to be elucidated. However, the fact that the relaxation of the Fe–His bond following ligand dissociation occurs on the same time scale as CO dissociates from the $Cu_B$ site suggests that a functional link may exist between the two processes. This opens the possibility that ligand dynamics at the two metal centers, heme $a_3$ and $Cu_B$, are coupled and that this coupling signals the onset of events associated with the initiation of the proton-pumping process in the enzyme.

*Acknowledgments.* This work was supported by National Institutes of Health (NIH) grants GM25480 (GTB) and GM26916 (SFM).

# References

1. Ferguson-Miller S, Babcock GT (1996) Heme/copper terminal oxidases. Chem Rev 96:2889–2907
2. Wikström M, Krab K, Saraste M (1981) Cytochrome oxidase—a synthesis. Academic Press, New York
3. Babcock GT, Wikström M (1992) Oxygen activation and the conservation of energy in cell respiration. Nature 356:301–309
4. Sucheta A, Georgiadis KE, Einarsdóttir Ó (1997) Mechanism of cytochrome *c* oxidase—catalyzed reduction of dioxygen to water: evidence for peroxy and ferryl intermediates at room temperature. Biochemistry 36:554–565
5. Kitagawa T, Ogura T (1996) Oxygen activation mechanism at the binuclear site of heme-copper oxidase superfamily as revealed by time-resolved resonance Raman spectroscopy. Prog Inorg Chem 45:431–480
6. Hallén S, Nilsson T (1992) Proton transfer during the reaction between fully reduced cytochrome *c* oxidase and dioxygen: pH and deuterium isotope effects. Biochemistry 31:11853–11859
7. Tsukihara T, Aoyama H, Yamashita E, et al. (1995) Structures of metal sites of oxidized beef heart cytochrome *c* oxidase at 2.8 Å. Science 269:1069–1074
8. Tsukihara T, Aoyama H, Yamashita E, et al. (1996) The whole structure of the 13-subunit oxidized cytochrome *c* oxidase at 2.8 Å. Science 272:1136–1144
9. Iwata S, Ostermeier C, Ludwig B, et al. (1995) Structure at 2.8 Å resolution of cytochrome *c* oxidase from *Paracoccus denitrificans*. Nature 376:660–669
10. Rich PR (1995) Electron transfer processes in proteins. Aust J Plant Physiol 22:479–484
11. Wikström M, Bogachev A, Finel M, et al. (1994) Mechanism of proton translocation by the respiratory oxidases. The histidine cycle. Biochim Biophys Acta 1187:106–111
12. Hosler JP, Espe MP, Zhen Y, et al. (1995) Analysis of site-directed mutants locates a non-redox-active metal near the active site of cytochrome *c* oxidase of *Rhodobacter sphaeroides*. Biochemistry 34:7586–7592
13. Qian J, Shi W, Pressler M, et al. (1997) Aspartate-407 in *Rhodobacter sphaeroides* cytochrome *c* oxidase is not required for proton pumping or manganese binding. Biochemistry 36:2539–2543
14. Rousseau DL, Ching YC, Wang J (1993) Proton translocation in cytochrome *c* oxidase: redox linkage through proximal ligand exchange on cytochrome $a_3$. J Bioenerg Biomembr 25:165–176
15. Woodruff WH (1993) Coordination dynamics of heme-copper oxidases. The ligand shuttle and the control and coupling of electron transfer and proton translocation. J Bioenerg Biomembr 25:177–188
16. Gelles J, Blair DF, Chan SI (1987) The proton-pumping site of cytochrome *c* oxidase: a model of its structure and mechanism. Biochim Biophys Acta 853:205–236
17. Findsen EW, Centeno JA, Babcock GT, et al. (1987) Cytochrome $a_3$ hemepocket relaxation subsequent to ligand photolysis from cytochrome oxidase. J Am Chem Soc 109:5367–5372
18. Woodruff WH, Einarsdóttir Ó, Dyer RB, et al. (1991) Nature and functional implications of the cytochrome $a_3$ transients after photodissociation of CO-cytochrome oxidase. Proc Natl Acad Sci USA 88:2588–2592

19. Mitchell DM, Adelroth P, Hosler JP, et al. (1996) A ligand-exchange mechanism of proton pumping involving tyrosine-422 of subunit I of cytochrome oxidase is ruled out. Biochemistry 35:824–828
20. Schelvis JPM, Deinum G, Varotsis CA, et al. (1997) Low power ps resonance Raman evidence for histidine ligation to heme $a_3$ after photodissociation of CO from cytochrome $c$ oxidase. J Am Chem Soc (in press)
21. Varotsis CA, Babcock GT (1995) Photolytic activity of early intermediates in dioxygen activation and reduction by cytochrome oxidase. J Am Chem Soc 117:11260–11269

# Time-Resolved Resonance Raman Study of Dioxygen Reduction by Cytochrome c Oxidase

Teizo Kitagawa[1], Takashi Ogura[1], Shun Hirota[1],
Denis A. Proshlyakov[1], Jorg Matysik[1], Evan H. Appelman[2],
Kyoko Shinzawa-Itoh[3], and Shinya Yoshikawa[3]

*Summary.* Six oxygen-associated vibrations were observed for reaction intermediates of bovine cytochrome $c$ oxidase with $O_2$ using time-resolved resonance Raman spectroscopy. The isotope shift of the $Fe-O_2$ stretching frequency for an asymmetrically labeled dioxygen, $^{16}O^{18}O$, has established that the primary intermediate of cytochrome $a_3$ is an end-on type dioxygen adduct of $Fe_{a3}$. The subsequent intermediates at about 0.1–1 ms following the initiation of reaction yielded Raman band pairs at 804/764, 785/751, and 356/342 cm$^{-1}$ for $^{16}O_2/^{18}O_2$ derivatives in $H_2O$ and, while these frequencies were the same between the $H_2O$ and $D_2O$ solutions, the final intermediate appearing at about 3 ms gave the Raman bands at 450/425 cm$^{-1}$ in $H_2O$ and 443/417 cm$^{-1}$ in $D_2O$. The last bands are reasonably assigned to the Fe-OH(D) stretching mode. The extended measurements at lower temperatures and longer delay times have revealed that the 804/764 cm$^{-1}$ pair appears before the 785/751 cm$^{-1}$ pair and that the conversion from the former to the latter species is significantly delayed in $D_2O$ than in $H_2O$, suggesting that this step of electron transfer is tightly coupled with proton pumping. The experiments using $^{16}O^{18}O$ have established that all the 804/764, 785/751, and 356/342 cm$^{-1}$ bands arise from the Fe=O heme but definitely not from the Fe-O⁻-O⁻ heme, whose presence has been long postulated. It is noted that the reaction intermediates of oxidized cytochrome $c$ oxidase with hydrogen peroxide yield the same three sets of oxygen isotope-sensitive bands as those that have been seen for intermediates of dioxygen reduction, indicating the identity of intermediates.

*Key words.* Cytochrome $c$ oxidase—Oxygen activation—Resonance Raman—Peroxy intermediates—Proton pump

[1] Institute for Molecular Science, Okazaki National Research Institutes, 38 Nishigounaka, Myodaiji, Okazaki, Aichi 444, Japan
[2] Chemistry Division, Argonne National Laboratory, 9700, South Case Ave., Argonne, IL 60439, USA
[3] Department of Life Science, Faculty of Science, Himeji Institute of Technology, 1479-1 Kanaji, Kamigori-cho, Akoh-gun, Hyogo 678-12, Japan

## Introduction

The terminal oxidase of the mitochondrial respiratory chain contains two copper redox centers ($Cu_A$ and $Cu_B$) and two heme A groups (Cyt $a$ and Cyt $a_3$) [1] and has been historically called cytochrome $aa_3$ (E.C. 1.9.3.1). A conceptual structure of this enzyme is illustrated in Fig. 1. Heme $a_3$ is of the five-coordinate high-spin type and provides the catalytic site for $O_2$ reduction, while heme $a$ is of the six-coordinate low-spin type and works simply for electron transfer from $Cu_A$ to Cyt $a_3$. The $Cu_A$ center, which is a binuclear copper complex and changes between $Cu^I Cu^{II}$ and $Cu^I Cu^I$, receives electrons from cytochrome $c$ and transfers them to Cyt $a$. The role of $Cu_B$ has not been clarified yet but it has been known that $Cu_B$ is antiferromagnetically coupled with heme $a_3$ in the resting state. Accordingly, $Cu_B$ of the oxidized form is electron paramagnetic resonance (EPR) silent while $Cu_A$ is EPR active.

Full reduction of $O_2$ to $H_2O$ requires four electrons and four protons. In a respiration system the reaction is catalyzed by the heme iron of Cyt $a_3$ ($Fe_{a3}$) in stepwise fashion, that is, repetitions of one electron transfer to oxygen followed by uptake of one proton [2]. In addition to consumption of four protons to yield two water molecules, another four protons are transported across the mitochondrial inner membrane to generate the electrochemical potential to be used for producing ATP from ADP [3,4]. The stepwise reduction of an isolated $O_2$ molecule may imply generation of intermediately reduced oxygen species such as $\cdot O_2^-$, $\cdot OOH$, $O_2^{2-}$, $\cdot OOH^-$, HOOH, $\cdot O^-$,

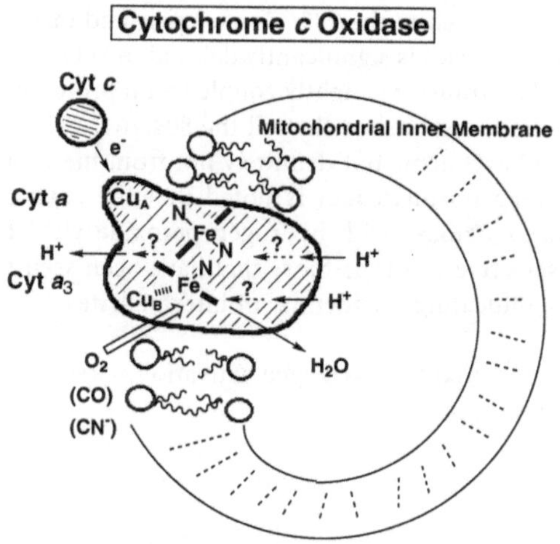

FIG. 1. Conceptual illustration of cytochrome $c$ oxidase. The enzyme is located in the inner membrane of mitochondria and is functionally categorized into two parts, that is, cytochrome $a$ (Cyt $a$) and cytochrome $a_3$ (Cyt $a_3$). Cytochrome $c$ gives electrons to $Cu_A$, which actually consists of a dinuclear copper center. Electrons are transferred from $Cu_A$ to the six-coordinate heme (Cyt $a$) and then to the five-coordinate heme (Cyt $a_3$). $Cu_B$ is antiferromagnetically coupled with the iron of Cyt $a_3$ in the resting state. $O_2$ binds to the $Fe^{II}$ ion of Cyt $a_3$, to which CO and $CN^-$ also can be bound. On reduction of one $O_2$ molecule, four protons are translocated across the membrane from the matrix to cytosol side; in addition, four protons are consumed in the matrix side to generate two water molecules

and ·OH, which are called active oxygen and are very toxic to organisms, but no such dangerous intermediates are normally released in the respiration process. Then, how is the molecular oxygen reduced in the mitochondria and how is the electron transfer coupled with the proton transport? The purpose of our study is to answer these questions.

The three-dimensional structure of bovine cytochrome c oxidase (CcO) with $M_r = 2 \times 10^5$ and 13 subunits has been revealed recently with X-ray crystallography [5]. The reaction mechanism of this enzyme has been investigated with various spectroscopic techniques including time-resolved [6–8], cryo-trapped absorption [9], and EPR [10,11] spectroscopy, but more recently a breakthrough was made with time-resolved resonance Raman (TR$^3$) spectroscopy by Babcock's group [12,13], Rousseau's group [14,15], and our group [16–19]. The course of progress has been summarized in a comprehensive review article [20]; here, some exciting results from recent TR$^3$ experiments [19] are explained.

# Reaction of Reduced CcO with O$_2$

## Time-Resolved Resonance Raman Spectra

Resonance Raman (RR) spectroscopy allows us to observe selectively the vibrational spectra of a chromophore by tuning the excitation wavelength into an absorption band of a molecule. Application of this technique to heme proteins has brought about unique and important information on dynamical as well as static structures [21]. In the time-resolved measurements, the reaction was initiated by photolysis of CO-inhibited CcO in the presence of $O_2$, and after a certain delay time ($\Delta t_d$) RR spectra were determined. In these TR$^3$ experiments, a device for simultaneous measurements of Raman and absorption spectra [22] and a sample circulation system for regenerating the reacted enzyme during one cycle [18] were constructed. Because the Raman intensities of the oxygen-associated vibrations are extremely weak, we measured the RR spectra of intermediates for $^{16}O_2$ and $^{18}O_2$ at the same time and calculated their differences.

Figure 2 shows the TR$^3$ spectra observed with the delay times of 0.1 (A), 0.27 (B), 0.54 (C), 2.7 (D), and 5.4 (E) ms for the H$_2$O solution at 3°C [19]. The visible absorption spectrum simultaneously observed with spectrum (A) is close to that of compound A obtained in cryogenic measurements [9]. Spectrum (A) shows the Fe–O$_2$ stretching ($\nu_{Fe-O2}$) RR bands at 571/544 cm$^{-1}$ for the $^{16}O_2/^{18}O_2$ derivatives. The frequencies and isotope shift are very close to those seen for HbO$_2$ and MbO$_2$ [23]. In the next stage ($\Delta t_d = 0.27$ ms), new Raman bands appear at 804/764, 785/750, and 356/342 cm$^{-1}$. Initially a single band was reported around 800 cm$^{-1}$ [17], but the presence of two bands there was pointed out from higher resolution experiments [18] and is confirmed in Fig. 2, although they were not resolved in studies by the other two groups [13,15]. Spectrum (D), which exhibits only the 785/750 cm$^{-1}$ pair near 800 cm$^{-1}$, strongly suggests that the 804/764 cm$^{-1}$ pair precedes the 785/750 cm$^{-1}$ pair. The 450/425 cm$^{-1}$ pair, which arises from the Fe–OH stretch ($\nu_{Fe-OH}$) of the Fe$_{a3}$$^{III}$–OH heme [13,15,17], appears at $\Delta t_d = 2.7$ ms (spectrum D), but disappears at $\Delta t_d = 5.4$ ms (spectrum E) because of the exchange of the bound $^{18}$OH$^-$ anion with bulk water. This $\nu_{Fe-OH}$ frequency is significantly lower than those of *aquamet*Hb and *aquamet*Mb at

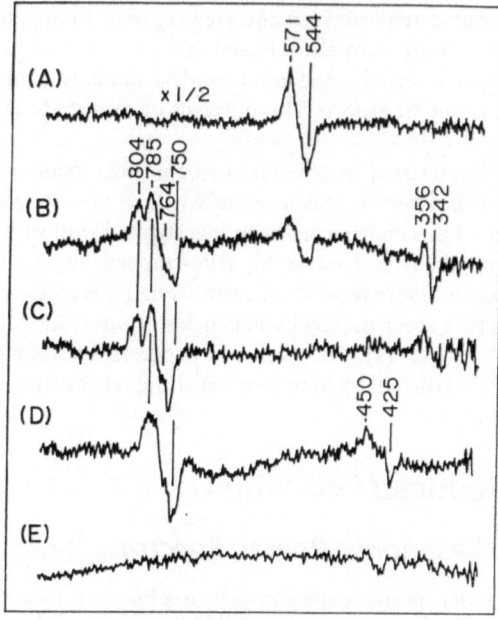

RAMAN SHIFT/cm⁻¹

Fig. 2. Time-resolved resonance Raman (TR³) difference spectra of reaction intermediates of cytochrome $c$ oxidase (CcO) in $H_2O$. The Raman difference spectra obtained by subtracting the spectrum of the corresponding $^{18}O_2$ derivative from the spectrum of the $^{16}O_2$ derivative at each delay time are depicted. Therefore, positive and negative peaks denote the contributions of $^{16}O_2$ and $^{18}O_2$ derivatives, respectively. Delay time after initiation of the reaction is 0.1 (A), 0.27 (B), 0.54 (C), 2.7 (D), and 5.4 (E) ms. Excitation wavelength, 423 nm; temperature, 3°C. (From [19], with permission)

495 and 490 cm⁻¹, respectively [24], presumably as a result of the strong interaction between the bound OH⁻ group and $Cu_B$. The relative intensities of the 804/764 cm⁻¹ bands to the 356/342 cm⁻¹ bands had varied with $\Delta t_d$ in repeated experiments, suggesting that these pairs arise from separate molecular species.

Figure 3 shows similar difference RR spectra observed with $\Delta t_d$ = 0.1 (A), 0.54 (B), 2.7 (C), 6.5 (D) and 11 (E) ms for the $D_2O$ solution at 3°C. The $\nu_{Fe-O_2}$ frequencies in spectrum (A) are the same as those in Fig. 2(A). With the delay times of $\Delta t_d$ = 0.54 (B) and 2.7 ms (C), however, a single band was observed around 800 cm⁻¹ (at 804/764 cm⁻¹), contrary to the case for the $H_2O$ solution. Therefore, it was once misunderstood [18] that the 785/750 cm⁻¹ bands in the $H_2O$ solution were shifted to 796/766 cm⁻¹ in $D_2O$, giving rise to an overlapping band centered around 800 cm⁻¹. However, it was determined from Fig. 3 that the lifetime of the 804/764 cm⁻¹ species is so different between the $H_2O$ and $D_2O$ solutions that the 785/750 cm⁻¹ species was not generated in $D_2O$ at $\Delta t_d$ = 2.7 ms when it was generated in $H_2O$. It is noted that the 356/342 cm⁻¹ bands are clearly seen in spectra (B) and (C) at the same frequencies as those in the $H_2O$ solution.

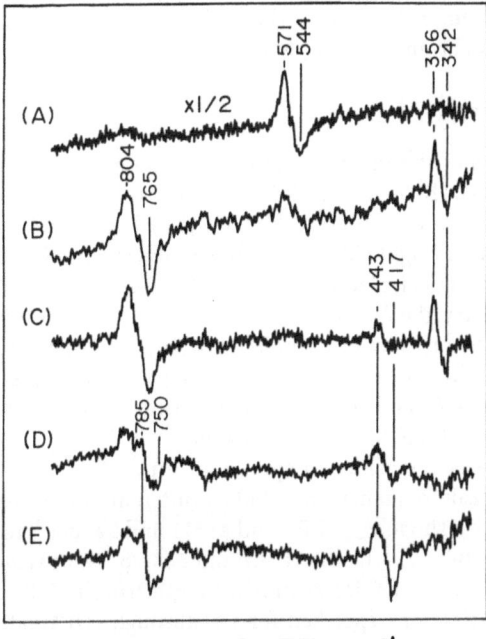

FIG. 3. TR³ difference spectra of reaction intermediates of CcO in $D_2O$. Delay time is 0.1 (*A*), 0.54 (*B*), 2.7 (*C*), 6.5 (*D*), and 11 (*E*) ms. Other conditions are the same as those for Fig. 2. (From [19], with permission)

Spectra (D) and (E) for the $D_2O$ solution demonstrated that the $785/750\,cm^{-1}$ bands did appear at the same frequencies but much later than those in the $H_2O$ solution. It is estimated that the conversion rate from the $804/764\,cm^{-1}$ species to the $785/750\,cm^{-1}$ species in $D_2O$ is approximately one-fifth of that in $H_2O$. The subsequent appearance of the $443/417\,cm^{-1}$ bands, which arise from the Fe–OD stretch, is consistent with the results shown in Fig. 2. The experiments with $\Delta t_d = 0.3$–$0.5\,ms$ for $D_2O$ solutions (data not shown) yielded the 804/764 and $356/342\,cm^{-1}$ bands in a manner similar to spectrum (B).

When the mixed-valence CO-bound enzyme, which had only two electrons in a molecule, was used as a starting compound instead of the fully reduced CO-bound enzyme, which had four electrons, the primary intermediate gave the $O_2$ isotope-sensitive bands at $571/544\,cm^{-1}$ [25,26]. The subsequent intermediate, which is characterized by the difference absorption peak at 607 nm (intermediates-fully oxidized), was found to give RR bands at 804/764 and $356/342\,cm^{-1}$ for $^{16}O_2/^{18}O_2$ [19]. These bands lasted as long as $\Delta t_d = 4.2\,ms$ but the bands at 785/750 and $450/425\,cm^{-1}$ did not appear in this case. These results have proved that the species giving rise to the $804/764\,cm^{-1}$ pair has an oxidation state higher than that of the $785/750\,cm^{-1}$ species. Hirota et al. [27] carried out similar TR³ experiments for the reaction intermediates of fully reduced *E. coli* cytochrome *bo* with $O_2$, and found the $O_2$ isotope-sensitive bands at 568/535, 788/751, and 361/347 cm$^{-1}$. They failed to identify the bands corresponding to the bands of cytochrome $aa_3$ at 804/764 cm$^{-1}$.

## Binding Geometry of $O_2$

To determine whether the dioxygen adduct of CcO adopts the side-on or end-on geometry, Ogura et al. [18] examined RR spectra of $^{16}O^{18}O$-bound CcO. The results are shown in Fig. 4, where the observed and simulated isotope-difference spectra are depicted on the left and right sides, respectively, and the combination of the isotope species in the difference calculations is specified in the middle. If $^{16}O^{18}O$ binds to $Fe_{a3}^{II}$ in the end-on geometry, the $\nu_{Fe-O_2}$ frequencies for Fe–$^{16}O$–$^{18}O$ and Fe–$^{18}O$–$^{16}O$ should be different. As these two species are generated by equal amounts, two $\nu_{Fe-O_2}$ RR bands should appear. On the other hand, if the binding is of the side-on type, the $\nu_{Fe-O_2}$ frequencies for Fe($^{16}O^{18}O$) and Fe($^{18}O^{16}O$) are identical and are located in the middle of the $\nu_{Fe-O_2}$ frequencies of the $^{16}O_2$- and $^{18}O_2$-adducts.

The difference-peak intensities in spectra (B) and (C) are weaker than those in spectrum (A) and peak frequencies are slightly different. If it is assumed that the $^{16}O_2$ and $^{18}O_2$ species give a single $\nu_{Fe-O_2}$ RR band at 571 and 544 cm$^{-1}$, respectively, but the $^{16}O^{18}O$ species gives two $\nu_{Fe-O_2}$ bands at 567 and 548 cm$^{-1}$ with Gaussian band shapes ($\Delta\nu_{1/2} = 12.9$ cm$^{-1}$) and intensities as drawn by spectrum (E), the difference calculations for the combinations specified for (A) through (D) yielded the patterns as depicted on the right side. The residuals in the subtraction of the observed minus the simulated spectra are depicted below each simulated spectrum. The calculated difference spectrum, $^{16}O^{18}O - (^{16}O_2 + {}^{18}O_2)/2$ (spectrum D), gives positive peaks at 563 and 552 cm$^{-1}$ and troughs at 574 and 541 cm$^{-1}$, which are in good agreement with the observed spectra. This means that the Fe-O$_2$ stretching frequencies for the Fe–$^{16}O$–$^{18}O$ and Fe–$^{18}O$–$^{16}O$ species are 567 and 548 cm$^{-1}$, respectively. The magnitude of the isotopic frequency shifts and the simple normal coordinate calculations allowed us to estimate the Fe–O–O bond angle to be close to 120°. Thus, this experiment established that the binding of O$_2$ to $Fe_{a3}$ is of the end-on type in CcO·O$_2$.

## Assignments of Transient RR Bands Around 800cm$^{-1}$

Spectra (B) and (C) in Fig. 2 at $\Delta t_d = 0.27$ and 0.54 ms give two oxygen isotope-sensitive bands around 800 cm$^{-1}$, and the higher frequency component arises from the species with the Fe$^V$ oxidation level (compound I of peroxidase). In this frequency region, two kinds of oxygen-associated bands are expected; one is the peroxy O$^-$–O$^-$ stretch ($\nu_{OO}$) and the other is an ironoxo Fe=O stretch ($\nu_{Fe=O}$). These two modes cannot be distinguished by the $^{16}O_2$ and $^{18}O_2$ isotopic frequency shifts. However, if $^{16}O^{18}O$ is used to produce such intermediates, the distinction would be possible. The peroxy intermediate for $^{16}O^{18}O$ is expected to give an additional $\nu_{OO}$ band in a middle of those for the $^{16}O_2$ and $^{18}O_2$ derivatives. In contrast, the oxo intermediate for $^{16}O^{18}O$ is expected to yield two bands at the same frequencies as those seen for the $^{16}O_2$ and $^{18}O_2$ derivatives but with half their intensity.

Such a distinction was satisfactorily carried out with $^{16}O^{18}O$ on the 804/764 and 785/750 cm$^{-1}$ bands. The results are shown in Fig. 5, where the TR$^3$ difference spectra observed with $\Delta t_d = 1.1$ ms at 5°C for the H$_2$O and D$_2$O solutions are presented on the

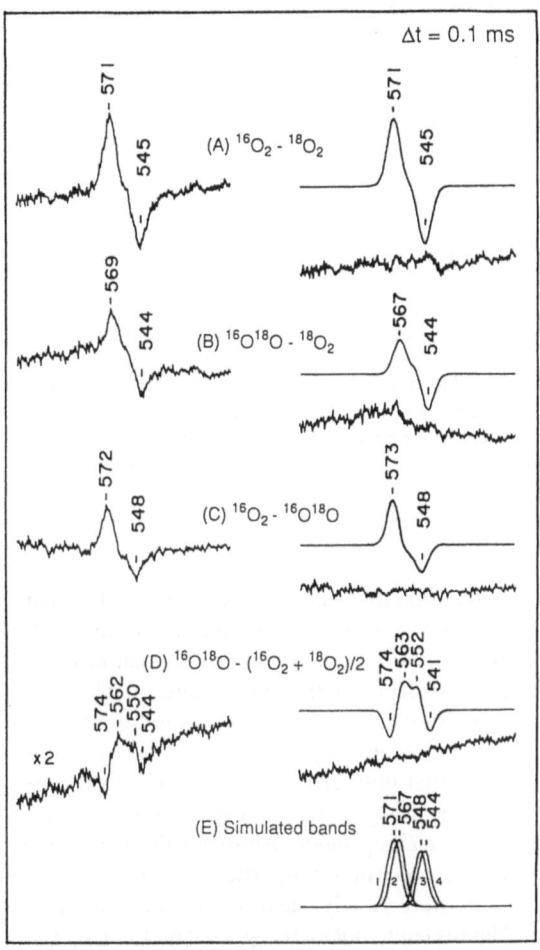

FIG. 4. TR³ difference spectra in the $Fe^{III}$-$O_2^-$ stretching region of CcO at $\Delta t_d = 0.1$ ms: *left side*, observed spectra; *right side*, calculated spectra. (*A*) $^{16}O_2$–$^{18}O_2$; (*B*) $^{16}O^{18}O$–$^{18}O_2$; (*C*) $^{16}O_2$–$^{16}O^{18}O$; (*D*) $^{16}O^{18}O$ – ($^{16}O_2$ + $^{18}O_2$)/2; (*E*) Fe–$^{16}O_2$ (1), Fe–$^{16}O^{18}O$ (2), Fe–$^{18}O^{16}O$ (3), and Fe–$^{18}O_2$ (4) stretching Raman bands assumed in the simulation. The peak intensity ratios are 6:6:5:5, and all have the Gaussian band shape with a FWHM of 12.9 cm⁻¹. In the calculation for the $^{16}O^{18}O$ spectrum, (spectrum 2 + spectrum 3)/2 was used. The differences between the observed and calculated spectra are depicted with the same ordinate scale as that of the observed spectra under the individual calculated spectra. Experimental conditions: probe beam, 423 nm, 4 mW; pump beam, 590 nm, 210 mW; accumulation time, 4800 s. (From [18], with permission)

left and right sides, respectively. The various isotope combinations are defined in the center of the figure. In trace (A) for the $H_2O$ solution, there are two bands at 804 and 785 cm⁻¹ for $^{16}O_2$, and they downshift to 764 and 751 cm⁻¹ for $^{18}O_2$, in agreement with Fig. 2(B), obtained in independent experiments with different batches of enzyme

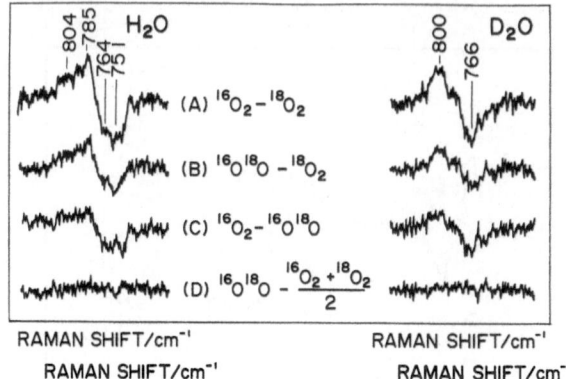

FIG. 5. Higher resolution TR$^3$ difference spectra of CcO reaction intermediates in the ~800-cm$^{-1}$ region for $\Delta t_d = 1.1$ ms at 5°C for $H_2O$ (*left*) and $D_2O$ solutions (*right*). The ordinate scales are common to all spectra. The spectra combined in the difference calculations are specified in the middle of the figure. Resolution, 0.43 cm$^{-1}$/channel. (From [18], with permission)

preparations. It is evident that the peak positions and spectral patterns of the difference spectra (B) and (C) in Fig. 5 for the $H_2O$ solution are alike and similar to those of spectrum (A) but their intensities are approximately half of those in spectrum (A). The same features are also seen for the $D_2O$ solution (right side). The point to be emphasized is that there is no difference peak in the bottom traces for either the $H_2O$ or $D_2O$ solution. This feature definitely differs from that seen in Fig. 4(D).

These results indicate that only one atom of $O_2$ is primarily responsible for the two RR bands. In other words, neither the bands at 804/764 nor 785/750 cm$^{-1}$ are assignable to the O–O stretching mode. Although the two sets of RR bands arise from an Fe=O stretch, the electronic properties of their hemes seem to be distinct. The 804/764 cm$^{-1}$ bands were clearly identified on Raman excitation at 441.6 nm, but the 785/750 cm$^{-1}$ bands could not be recognized on Raman excitation at 441.6 nm. The experiments with mixed-valence enzyme demonstrated that the 804/764 cm$^{-1}$ species has a higher oxidation state than the 785/750 cm$^{-1}$ species. Consequently, the latter is firmly assigned to the $\nu_{Fe=O}$ band of non-ionized porphyrin with an oxo-ferryl iron.

## Assignment of the Transient Band Around 350cm$^{-1}$

Relative intensities of the 804/764 and 356/342 cm$^{-1}$ bands change with the delay time and, on excitation at 441.6 nm, the 356/342 cm$^{-1}$ bands are not enhanced while the 804/764 cm$^{-1}$ bands are clearly observed. Therefore, the two sets of bands are considered to arise from separate molecular species. To clarify the assignment of this band, Ogura et al. [18] examined the $^{16}O^{18}O$ effect on the 356/342 cm$^{-1}$ bands. The results are shown in Fig. 6, where spectra (A–D) were obtained for the indicated combinations of $O_2$ isotopes in the $H_2O$ solution and spectrum (E) was obtained for combination (A) in the $D_2O$ solution. Spectra in Figs. 5 and 6 are drawn on the same wavenumber scale. It is noticed that the 356-cm$^{-1}$ band is very narrow. The inset of Fig. 6 (spectra A′ and

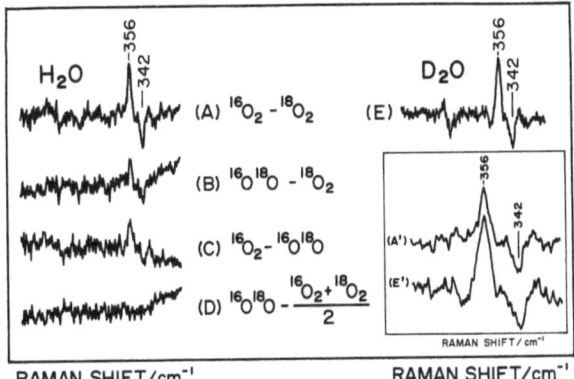

FIG. 6. $TR^3$ difference spectra of CcO reaction intermediates in the ~350-cm$^{-1}$ region for $\Delta t_d$ = 0.5 ms at 5°C for $H_2O$ (*left*) and $D_2O$ solutions (*right*). The ordinate scales are common to all spectra. The spectra combined in the difference calculations are specified in the middle of the figure. The *inset* depicts the plots of spectra $A$ and $E$ expanded by a factor of 2.5 in the wavenumber axis. (From [18], with permission)

$E'$) shows plots of spectra (A) and (E) with the wavenumber axis expanded by 2.5 fold. The 356-cm$^{-1}$ band appears to be somewhat broader in $D_2O$ than in $H_2O$, but both spectra exhibit a flat region between the positive and negative peaks. This means that the separation between the $^{16}O_2$ and $^{18}O_2$ peaks is larger than their bandwidths, and thus the narrowness of the band is not the consequence of the close proximity of the $^{16}O_2$ and $^{18}O_2$ bands. This also means that the peak positions of the difference spectrum correctly represent those of each individual spectrum.

It is evident from spectrum (E) that there are no deuteration effects on the absolute frequencies of the 356/342 cm$^{-1}$ bands. Therefore, this band cannot be an M–OH stretching mode. As was seen for the bands around 800 cm$^{-1}$, spectra (B) and (C) do not differ from each other in shape or position of the peaks, which are also close to those of spectrum (A). If the 356/342 cm$^{-1}$ bands arose from the Fe–OOH stretch, the frequency difference between the Fe–$^{16}O^{18}OH$ and Fe–$^{16}O^{16}OH$ stretches and that between the Fe–$^{18}O^{16}OH$ and Fe–$^{18}O^{18}OH$ stretches should be as large as 2.5 cm$^{-1}$ unless the Fe–O–O bond-angle is close to 90°, and there should be some difference peaks in Fig. 6(D) as observed for the Fe–$O_2$ adduct in Fig. 4(D). Actually, however, there is no difference peak in spectrum (D) in Fig. 6. Accordingly, it became clear again that only one oxygen atom from the $O_2$ molecule is primarily responsible for the 356 cm$^{-1}$ band.

## Reaction of Oxidized CcO with $H_2O_2$

The reaction of ferric heme proteins with $H_2O_2$ usually yields the $Fe^V$-level intermediate first, which is then successively reduced to the $Fe^{IV}$ and $Fe^{III}$ oxidation levels. The reaction of oxidized CcO with $H_2O_2$ has been extensively studied with visible absorption [28], EPR [29], and RR [30–32]. The visible absorption spectra of reaction intermediates are displayed in Fig. 7, where the differences between the intermediates and

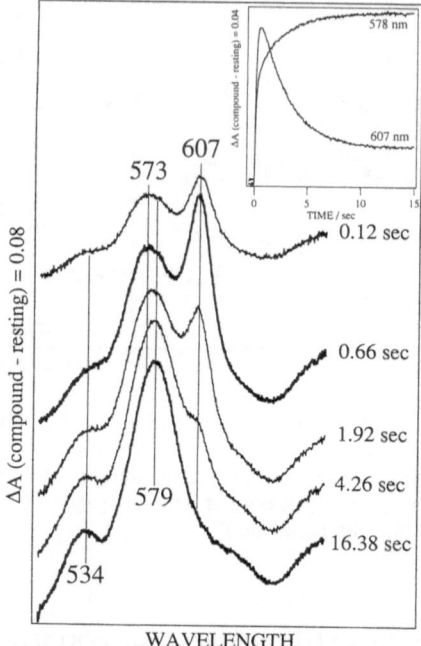

FIG. 7. Transient absorption spectra for reaction intermediates of oxidized CcO with $H_2O_2$. The spectra are represented as difference spectra with regard to the spectrum of fully oxidized enzyme, and delay times after initiation of the reaction are specified at the *right* side of each spectrum. The zero line for each spectrum is shifted for clear presentation. The *inset* depicts the behavior of absorbances at 607 and 578 nm against delay times. (Full scale of $\Delta A$ is 0.04 in terms of the difference, intermediate minus fully oxidized.) (From [31], with permission)

fully oxidized enzyme for several delay times are depicted. The so-called "607-nm" form is generated first and then it is replaced by the "580-nm" form. The rate of formation of the 607-nm form was found to be proportional to the concentration of $H_2O_2$ and was considered to be the primary intermediate in this reaction. This fact means that the 607-nm form has the $Fe^V$ oxidation level of the heme $a_3$. When the concentration of $H_2O_2$ is increased, the 580-nm form is developed rapidly. This is interpreted in the following way. Under high concentrations of $H_2O_2$, the 607-nm form developes much faster and the extra $H_2O_2$ acts as a reductant to the 607-nm form to yield the 580-nm form, which has the $Fe^{IV}$ oxidation level of the heme $a_3$. Accordingly, relative populations of the 607-nm and 580-nm forms can be regulated by the concentration of $H_2O_2$ and pH [32].

Figure 8 shows the absorption (right) and Soret-excited RR (left) spectra of intermediates that were measured simultaneously with an improved Raman/absorption simultaneous measurement device [30]. In this experiment the reaction intermediates in question are retained in a steady state for a certain period of time by adding $H_2O_2$ at a constant rate to the circulating solution. The upper spectra were obtained under

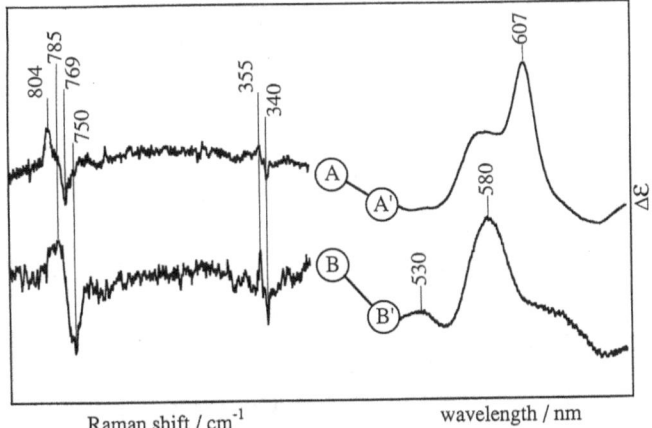

FIG. 8. The steady-state Raman/absorption simultaneously observed spectra of the "607-nm" and "580-nm" forms of cytochrome $c$ oxidase at pH 8.5. Raman spectra $A$ and $B$ are shown as a difference of $H_2^{16}O_2$ compound minus $H_2^{18}O_2$ compound. Spectra $A/A'$: laser, 427 nm, 2.5 mW; accumulation time, $3 \times 2400$ s for each isotope; cytochrome $c$ oxidase, 50 μM; initial concentration of $H_2O_2$, 1 mM. Spectra $B/B'$: laser power, 7.5 mW; accumulation time, $3 \times 800$ s for each isotope; cytochrome $c$ oxidase, 10 μM; initial concentration of $H_2O_2$, 5 mM. Absorption spectra are represented as a difference with regard to the spectrum of the resting enzyme; ordinate full scale $\Delta\varepsilon = 11.7\,mM^{-1}\,cm^{-1}$; pathlength, 0.6 mm. (From [30], with permission)

the conditions where the 607-nm form is dominant, while the lower spectra were obtained under the conditions where the 580-nm form is dominant. Both RR spectra are represented as the difference between the derivatives from $H_2^{16}O_2$ and $H_2^{18}O_2$, and accordingly positive and negative peaks correspond to the vibrations associated with $^{16}O$ and $^{18}O$, respectively. It is noted that when the 607-nm form is dominant, the 804/769 $cm^{-1}$ bands are clearly observed and when the 580-nm form is dominant, broad bands centered around 785/750 $cm^{-1}$ and sharp bands at 355/340 $cm^{-1}$ are intensified. Examination of the intensity of the 355/340 $cm^{-1}$ bands relative to those of the 804/769 and 785/750 $cm^{-1}$ bands indicated that the species giving rise to the 355/340 $cm^{-1}$ bands is different from the two species giving rise to the 804/769 and 785/750 $cm^{-1}$ bands. However, the 355/340 $cm^{-1}$ species seems to have an absorption spectrum similar to that of the 580-nm form. This implies that the 580-nm form consists of multiple intermediate species [32]. It is stressed that all these oxygen isotope-sensitive bands are identical with those observed in the dioxygen reaction.

Hitherto the "607-nm" form has been believed to be a peroxo species, namely Fe–O–O–X (X=H or $Cu_B$). Because both the $O^--O^-$ and Fe=O stretching frequencies are located around 800 $cm^{-1}$, it is impossible to determine which type of vibrations was observed for the 607-nm form in Fig. 8. To sort out the two possible modes, experiments using $H_2^{16}O^{18}O$ have been performed. The results are shown in Fig. 9, where RR spectra of the 607-nm form excited at 607 nm are displayed for intermediates derived with $H_2^{16}O_2$ (A), $H_2^{18}O_2$ (B), and $H_2^{16}O^{18}O$ (C) [31]. The band of the $H_2^{16}O_2$ derivative at

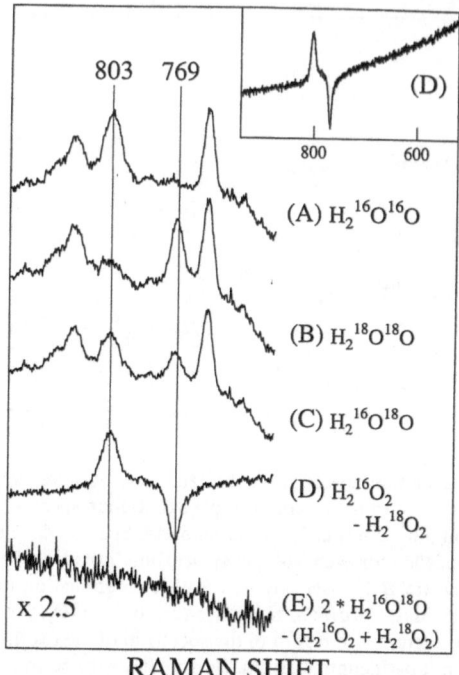

803   769

(D)

800        600

(A) $H_2{}^{16}O{}^{16}O$

(B) $H_2{}^{18}O{}^{18}O$

(C) $H_2{}^{16}O{}^{18}O$

(D) $H_2{}^{16}O_2$
    - $H_2{}^{18}O_2$

x 2.5

(E) 2 * $H_2{}^{16}O{}^{18}O$
    - ($H_2{}^{16}O_2$ + $H_2{}^{18}O_2$)

RAMAN SHIFT

FIG. 9. The 607-nm excited steady-state resonance Raman (RR) spectra in the 800-cm$^{-1}$ region of the "607-nm" form of CcO formed in the reaction of the oxidized enzyme with hydrogen peroxide. Hydrogen peroxide forms used are $H_2{}^{16}O_2$ (*A*), $H_2{}^{18}O_2$ (*B*), and $H_2{}^{16}O{}^{18}O$ (*C*). Spectra *D* and *E* show difference spectra: spectrum D = spectrum A − spectrum B; spectrum E = spectrum C − (spectrum A + spectrum B)/2. The *inset* (spectrum *D*) is the same as spectrum D in the main frame but shows the full frequency range measured. Experimental conditions: cross section of flow cell, $0.6 \times 0.6$ mm$^2$; slit width, 4.2 cm$^{-1}$; laser, 607 nm, 100 mW at the sample; total accumulation time, 78, 78, and 156 min for spectra A, B, and C, respectively, and 160 min for the resting enzyme; cytochrome *c* oxidase, 50 µM, pH 7.45. (From [31], with permission)

803 cm$^{-1}$ (A) is shifted to 769 cm$^{-1}$ for the $H_2{}^{18}O_2$ derivative (B), while other bands arising from the porphyrin macrocycle remain unshifted. This is most clearly seen in the difference spectrum (D) [= spectrum (A) − spectrum (B)]. The inset shows the same difference spectrum in a wider spectral range (930–550 cm$^{-1}$). It is evident that there is no other oxygen-associated band in this frequency region.

In spectrum (C) for the $H_2{}^{16}O{}^{18}O$ derivative, there are two bands at 803 and 769 cm$^{-1}$, and their intensities relative to the porphyrin bands are reduced to half of those in spectra (A) and (B). This is most clearly seen by the difference spectrum (E) [= spectrum (C) − (spectrum (A) + spectrum (B))/2], which exhibits no band in the 700–900 cm$^{-1}$ region. This means that the 803/769 cm$^{-1}$ bands arise from a species that has a single oxygen atom on the heme iron. Consequently, the peroxide structure (Fe–O–O–X), postulated for the 607-nm form, is no longer valid.

# Conclusion

The reaction intermediates of CcO characterized by RR spectroscopy and other spectroscopic techniques are interrelated in Fig. 10. Two kinds of reactions, that is, dioxygen cycle (right) and peroxide cycle (left), were treated in this chapter. The first intermediate in the reaction of reduced CcO with $O_2$, which has been called Compound A, has now been demonstrated to be a dioxygen adduct of cytochrome $a_3$ with $v_{Fe-O_2}$ at 571 cm$^{-1}$. The binding of $O_2$ is of the end-on type. The subsequent intermediates, called Compound B, might be mixtures of intermediates and it is difficult to correlate them with a species with specific oxygen isotope-sensitive bands. The intermediate that is generated in the reaction of mixed-valence enzyme with $O_2$ gives the oxygen isotope-sensitive bands at 804 and 356 cm$^{-1}$. The last intermediate in the dioxygen cycle has an $Fe_{a_3}^{III}$-OH$^-$ heme with $v_{Fe-OH}$ RR band at 450 cm$^{-1}$. This OH$^-$ group is exchangeable with bulk water.

In the reaction of oxidized enzyme with $H_2O_2$, the first intermediate, the 607-nm form, has long been thought to have the $Fe_{a3}$-O-O-X structure, but RR experiments demonstrated that it contains an oxo-iron heme. The subsequent intermediates give a broad difference peak at 580 nm and seem to have multiple components. RR spectroscopy can sort out two species within the 580-nm form. One gives the oxygen

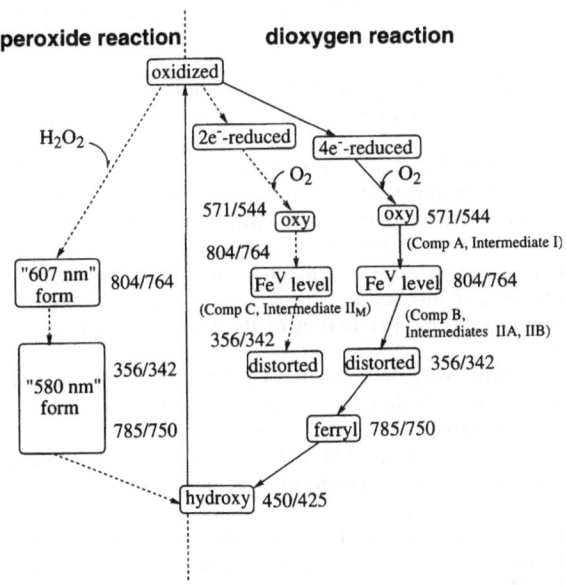

FIG. 10. Reaction mechanism of cytochrome $c$ oxidase, interrelations of intermediates obtained with different techniques, and the specific oxygen-associated vibrational frequencies ($^{16}O_2$ derivatives/$^{18}O_2$ derivatives in cm$^{-1}$ units) of individual intermediates. In the dioxygen reaction the fully reduced (4e$^-$-reduced) and the mixed-valence (2e$^-$-reduced) species are contained. The peroxide reaction gives rise to two absorption forms, but the "580-nm" form is of multiple species and is further sorted out by vibrational frequencies. The "peroxy" and "ferryl" forms in the reversed reaction are presumably the same as the "607-nm" and "580-nm" forms of the peroxide reaction, respectively

isotope-sensitive bands at 356 and approximately $800\,cm^{-1}$ and the other gives the band at $785\,cm^{-1}$. The latter is unquestionably assigned to the $Fe^{IV}=O$ stretching mode.

# References

1. Wikstrom M, Krab K, Saraste M (1981) Cytochrome oxidase; a synthesis. Academic, New York
2. Babcock GT, Wikstrom M (1992) Oxygen activation and conservation of energy in cell respiration. Nature 356:301–309
3. Wikstrom M (1981) Energy-dependent reversal of the cytochrome oxidase reaction. Proc Natl Acad Sci USA 78:4051–4055
4. Sone N, Hinkle PC (1982) Proton transport by cytochrome $c$ oxidase from the thermophilic bacterium PS3 reconstituted in liposomes. J Biol Chem 257:12600–12604
5. Tsukihara T, Aoyama H, Yamashita E, et al. (1995) The whole structure of the 13-subunit cytochrome $c$ oxidase at 2.8 Å. Science 269:1069–1074
6. Orii Y (1988) Intermediates in the reaction of reduced cytochrome oxidase with dioxygen. Ann NY Acad Sci 550:105–117
7. Blackmore RS, Greenwood C, Gibson QH (1991) Studies of the primary oxygen intermediates in the reaction of fully reduced cytochrome oxidase. J Biol Chem 266:19245–19249
8. Oliveberg M, Malmstrom BG (1992) Reaction of dioxygen with cytochrome $c$ oxidase reduced to different degrees: indication of a transient dioxygen complex with copper B. Biochemistry 31:3560
9. Chance B, Saronio C, Leigh JS Jr (1975) Functional intermediates in the reaction of membrane-bound cytochrome oxidase with oxygen. J Biol Chem 250:9226–9237
10. Clore GM, Andreasson L-E, Karlsson B, Aasa R, Malmstrom BG (1980) Characterization of the low-temperature intermediates of the reaction of fully reduced soluble cytochrome oxidase with oxygen by electron paramagnetic resonance and optical spectroscopy. Biochem J 185:139–154
11. Blair DF, Witt SN, Chan SI (1985) Mechanism of cytochrome $c$ oxidase-catalyzed dioxygen reduction at low temperatures. Evidence for two intermediates at three electron level and entropic promotion of the bond-breaking step. J Am Chem Soc 107:7389–7399
12. Varotsis C, Woodruff WH, Babcock GT (1989) Time-resolved Raman detection of $v(Fe-O)$ in an early intermediate in the reduction of $O_2$ by cytochrome oxidase. J Am Chem Soc 111:6439; (1990) 112:1297
13. Varotsis C, Zhang Y, Appelman EH, Babcock GT (1993) Resolution of the reaction sequence during the reduction of $O_2$ by cytochrome $c$ oxidase. Proc Natl Acad Sci USA 90:237
14. Han S, Ching Y-C, Rousseau DL (1990) Cytochrome $c$ oxidase: decay of the primary intermediate involves direct electron transfer from cytochrome a. Proc Natl Acad Sci USA 87:2491–2495
15. Han S, Ching Y-C, Rousseau DL (1990) Ferryl and hydroxy intermediates in the reaction of oxygen with reduced cytochrome $c$ oxidase. Nature 348:89–90
16. Ogura T, Takahashi S, Shinzawa-Itoh K, et al. (1990) Observation of the $Fe^{II}\text{-}O_2$ stretching Raman band for compound A of cytochrome oxidase at room temperature. J Am Chem Soc 112:5630–5631
17. Ogura T, Takahashi S, Shinzawa-Itoh K, et al. (1991) Time-resolved resonance Raman investigation of cytochrome oxidase catalysis: observation of a new oxygen-isotope sensitive Raman band. Bull Chem Soc Jpn 64:2901–2907

18. Ogura T, Takahashi S, Hirota S, et al. (1993) Time-resolved resonance Raman elucidation of the pathway for dioxygen reduction by cytochrome c oxidase. J Am Chem Soc 115:8527–8536

19. Ogura T, Hirota S, Proshlyakov DA, et al. (1996) Time-resolved resonance Raman evidence for tight coupling between electron transfer and proton pumping of cytochrome c oxidase upon the change from the $Fe^V$ oxidation level to the $Fe^{IV}$ oxidation level. J Am Chem Soc 118:5443–5449

20. Kitagawa T, Ogura T (1997) Oxygen activation mechanism at binuclear site of heme-copper oxidase superfamily as revealed by time-resolved resonance Raman spectroscopy. Prog Inorg Chem 45:431–479

21. Kitagawa T, Ogura T (1993) Time-resolved resonance Raman spectroscopy of heme proteins. Adv Spectrosc 21:139–188

22. Ogura T, Kitagawa T (1988) A novel optical device for simultaneous measurements of Raman and absorption spectra: application to photolabile reaction intermediates of heme proteins. Rev Sci Instrum 59:1316–1320

23. Nagai K, Kitagawa T, Morimoto H (1980) Quaternary structures and low frequency molecular vibrations of haems of deoxy and oxyhaemoglobin studied by resonance Raman scattering. J Mol Biol 136:271–289

24. Asher SA, Schuster TM (1979) Resonance Raman examination of axial ligand bonding and spin-state equilibria in metmyoglobin hydroxide and other hemederivatives. Biochemistry 18:5377–5387

25. Han S, Ching Y-C, Rousseau DL (1990) Time evolution of the intermediates formed in the reaction of oxygen with mixed valence cytochrome c oxidase. J Am Chem Soc 112:9445–9450

26. Varotosis C, Woodruff WH, Babcock GT (1990) Direct detection of a dioxygen adduct of cytochrome $a_3$ in the mixed valence cytochrome oxidase/dioxygen reaction. J Biol Chem 265:11131–11136

27. Hirota S, Mogi T, Ogura T, et al. (1994) Observation of the Fe-$O_2$ and $Fe^I$=O stretching Raman bands for dioxygen reduction intermediates of cytochrome bo isolated from Escherichia coli. FEBS Lett 352:67–70

28. Vygodina TV, Konstantinov AA (1988) $H_2O_2$-induced conversion of cytochrome c oxidase peroxy complex to oxoferryl state. Ann NY Acad Sci 550:124–138

29. Fabian M, Palmer G (1995) The interaction of cytochrome oxidase with hydrogen peroxide: the relationship of compounds P and F. Biochemistry 34:13802–13810

30. Proshlyakov DA, Ogura T, Shinzawa-Itoh K, et al. (1996) Microcirculating system for simultaneous determination of Raman and absorption spectra of enzymatic reaction intermediates and its application to the reaction of cytochrome c oxidase with hydorogen peroxide. Biochemistry 35:76–82

31. Proshlyakov DA, Ogura T, Shinzawa-Itoh K, et al. (1994) Selective resonance Raman observation of the "607 nm" form generated in the reaction of oxidized cytochrome c oxidase with hydrogen peroxide. J Biol Chem 269:29385–29388

32. Proshlyakov DA, Ogura T, Shinzawa-Itoh K, et al. (1996) Resonance Raman/absorption characterization of the oxo-intermediates of cytochrome c oxidase generated in its reaction with hydrogen peroxide: pH and $H_2O_2$ concentration dependence. Biochemistry 35:8580–8586

# Human Cytochrome c Oxidase Analyzed with Cytoplasts

Yasuo Kagawa

*Summary.* The reaction of cytochrome *c* oxidase (COX) is the final step in human respiration. Although COX has been studied extensively in microorganisms, no tissue-specific functions or bioenergetic disease, such as mitochondrial encephalomyopathy and aging, occur in these organisms. The physiological roles of human COX were analyzed by using cytoplasts (cells without nucleus), $\rho^\circ$ cells (cells without mtDNA), COX mutants, and their cybrids. Recent developments of the Human Genome Project will become an important factor in the study of human bioenergetics. The roles of the COX are not only the generation and regulation of the electrochemical potential ($\Delta\mu H^+$) across the inner mitochondrial membrane, but also the regeneration of NAD in mammalian cells. The major subunits I–III ($H^+$ pump) of COX are encoded by mtDNA and the remaining ten minor subunits by nDNA. The cause of the age-dependent decline of COX has been attributed to the accumulation of mutations in mtDNA. However, the involvement of nDNA in the decline is also important because of telomere shortening in somatic cells, and, thus, age-dependent mtDNA expression was analyzed with the cytoplasts. The roles of the control regions of the COX genes are to coordinate both mtDNA and nDNA, depending on the energy demand of the cells.

*Key words.* Human mitochondria—Cytochrome *c* oxidase—Heteroplasmy—Cytoplast—Cybrid

## Introduction

The major pathway by which human cells produce ATP is oxidative phosphorylation (OXPHOS). The reaction of cytochrome *c* oxidase (COX) is the final step in the electron transport of OXPHOS [1] as well as the regulating step of energy metabolism [2]. Although COX has been studied extensively in bacteria and yeast [2], the study of human COX is needed to understand tissue specificity (isoforms L and H in the subunits VIa and VIIa), development, aging, and diseases such as mitochondrial encephalomyopathy that are not found in microorganisms. In contrast to microor-

Department of Biochemistry, Jichi Medical School, 3311-1, Yakushiji, Minamikawachi, Kawachi-gun, Tochigi 329-04, Japan

ganisms capable of alcoholic fermentation, COX-deficient human cells cannot survive in glucose media unless pyruvate is added: this is also important in the therapy of mitochondrial diseases [3]. OXPHOS varies greatly depending on the activity of the tissues, especially in the muscle and the brain, and hence there are many encephalomyopathies caused by defective mitochondria. COX of beef heart has also been used extensively because it is available in large quantity and is crystallizable [4].

However, recent success in the genetic mapping of human DNA [5] will facilitate the future study of human COX. The detailed data on human mtDNA [6], mutations, diseases, and so on are available via internet (http://www.gen.emory.edu/mitomap.html). Human COX can easily be studied biologically because of the abundant human cell lines established and the many mutations available from patients [1,7]. Human COX contain four redox active (heme a, $a_3$, $Cu_A$, and $Cu_B$) and inactive (Zn, etc.) metal centers [4]. The three largest subunits (I = 57 kDa, II = 26 kDa, and III = 30 kDa), which are encoded by mtDNA [6], constitute a catalytic center [4]. The subunit III is larger, but its electrophoretic velocity is faster than the subunit II. In yeast COX, there are only six minor subunits, while in mammalian COX there is a minimum of ten minor subunits (IV, 17 kDa; Va, 12 kDa; Vb, 11 kDa; VIa, 9.4 kDa; VIb, 10 kDa; VIc, 8.5 kDa; VIIa, 6.2 kDa; VIIb, 5.9 kDa; VIIc, 5.5 kDa; VIII, 4.9 kDa) [8,9]. These are encoded by nDNA.

There are tissue-specific isoforms among two minor subunits (VIa and VIIa) in human cells and an additional one (VIII) in other mammals [9,10]. An L (liver) isoform is ubiquitously expressed, and an H (heart) isoform is expressed primarily in adult heart and skeletal muscles [10]; for example, they are written as VIa-H and VIa-L. These isoforms are not found in the largest subunits. The minor subunits may be needed for the subunit assembly, stabilization, and regulation [2,11]. For example, the mammalian subunit Vb contains a stabilizing Zn center [4], VIb is a modulator, and VIa-H is the ATP sensor [11].

The expression of the subunits of COX in human cells was analyzed by transcription factors and their *cis*-elements of COX genes [8,12–14] and mtDNA [15,16]. The cell-level studies became possible by isolating mtDNA in the cytoplasts and nDNA in the $\rho^\circ$ cells [17], as described in the following sections.

## Mitochondrial Disease Mutations Affecting Cytochrome *c* Oxidase

The COX activity in each mitochondrion can be shown by staining with 3,3-diaminobenzidine tetrachloride and copper sulfate under electron microscopy (Fig. 1A). The cells with a mutation affecting mtDNA, such as maternally inherited cardiomyopathy (CM), contain unstained mitochondria (Fig. 1B). This clearly shows that the major subunits are essential for COX activity. Because mtDNA has much higher information density (no exon) than nDNA, and is exposed to reactive oxygen species produced in mitochondria, it would seem to be an excellent target for mutations giving rise to human diseases. Mutations causing mtDNA diseases fall into four groups: (1) protein synthesis mutations, (2) insertion-deletion mutations, (3) missense mutations, and (4) copy number mutations [1]. COX activity is affected in all cases of group 1, in most cases of group 2, and in some cases of groups 3 and 4.

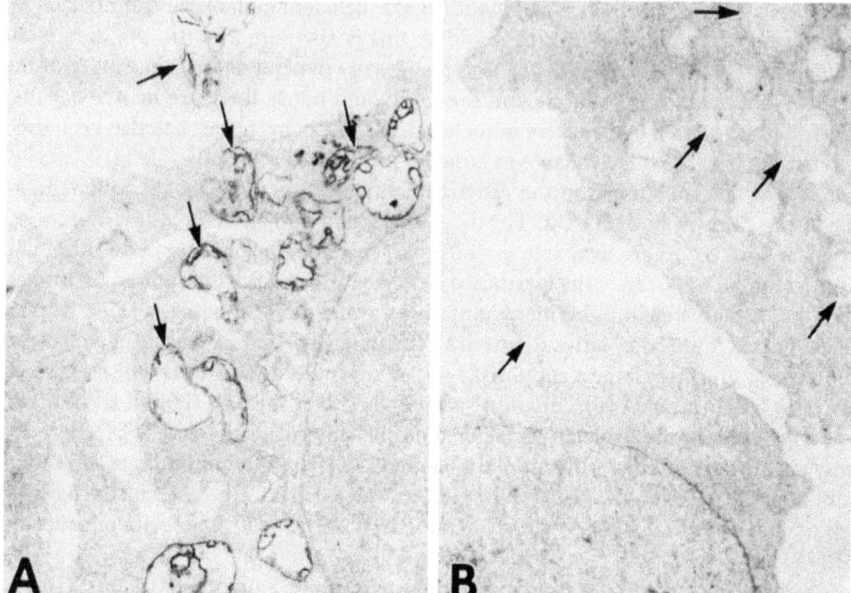

FIG. 1A,B. Idenficication of cytochrome $c$ oxidase (COX) activities in individual mitochondria by electron microscopy. **A** Human cells with predominantly normal mtDNA. **B** Human cells from a cardiomyopathy patient with mostly mutant CM4269 mtDNA. *Arrows* indicate mitochondria with stained COX activity. *Bar*: 1 μm

1. All protein synthetic mutations (syn⁻) known to date have been tRNA mutations associated with mitochondrial myopathy and other systemic phenotypes. They are mitochondrial encephalomyopathy, lactic acidosis, and strokelike symptoms (MELAS) [7], myoclonic epilepsy and ragged-red fiber disease (MERRF) [18], and CM [17]. The major mutations found in MERRF, MELAS, and CM are localized at np 8344 (A to G; tRNA$^{Lys}$) [18], np 3243 (A to G; tRNA$^{Leu(UUR)}$) [7], and np 4269 (A to G; tNRA$^{Ile}$) [17], respectively. Owing to the global mitochondrial translation failure caused by loss of the tRNA, COX activity is lost.

2. Deletions in mtDNA have been found to cause the majority of cases of chronic progressive external ophthalmoplegia (CPEO) and Kearn–Sayre syndrome (KSS) [1]. Most of the deletions occur between $O_H$ and $O_L$. The COX subunit IV is still detectable, but the subunits I, II, Va, Vb, and VIc are greatly reduced in the muscle fiber of patients with CPEO [19]. Some cases of ocular myopathies are associated with autosomal dominant mutations. A microdeletion (15 bp, Phe–Ala–Gly–Phe–Phe) in a hightly conserved region of the gene for the subunit III of COX causes recurrent myoglobinuria because of the destruction of the muscle [20]. Affected individuals in these pedigrees harbor multiple mtDNA deletions, and all the deletions are flanked by direct repeats, suggesting an increased frequency of slip-replication. There are many other mutations including deletions of mtDNA [21], diabetes mellitus [22] and other degenerative diseases [1].

3. Examples of missense mutation are those in Leber's hereditary optic neuropathy (LHON) [1]. A number of mutations are associated with LHON, but about 50% of LHON cases are caused by mutation in ND4 of NADH dehydrogenase (np 11778 R to H). However, COX subunit III mutations are also pathogenic: missense mutations at np9438A to G (78 Gly to Ser) and 9804G to A (200 Ala to Thr) that cause LHON [23].

4. A marked decrease in the number of mtDNA is associated with a familial mitochondrial myopathy [24].

## Heteroplasmy and Homoplasmy

Each human cell has hundreds of mitochondria and thousands of mtDNAs. In most diseases, the cells of patients carry a mixture of both mutant and wild-type (normal) mtDNA and thus the COX-positive and -negative mitochondria are found in one cell [25]. This heterogeneous state of the cells is called heteroplasmy (Fig. 2). On the other hand, the homogeneous state of cells containing pure mutant or normal mtDNA is called homoplasmy. During mitotic or meiotic cytokinesis of a heteroplasmic cell, both mutated and normal mtDNA are divided unevenly into daughter cells. Consequently, during repeated cell divisions, a heteroplasmic cell line is separated into two kinds of homoplasmic cell lines containing either normal or mutated mtDNA. This phenomenon of replicative segregation of a cell line is called stochastic segregation (Fig. 2, right bottom). The same MELAS mutation in mtDNA may cause encephalomyopathy [7], diabetes mellitus [22], or other diseases [1]. This is attributed to the stochastic segregation of the mutant mtDNA, which is unevenly distributed among different tissues of the patient during development [1].

## Cells Without nDNA (Cytoplasts) and Those Without mtDNA ($\rho^\circ$ Cells)

Cells without mtDNA, which are called $\rho^\circ$ cells, are very useful to study in mitochondrial genetics (Fig. 2, left) as they are pure models of COX-less cells with mutant mtDNA [17,26]. Although $\rho^\circ$ yeast cells have been used for a long time, human $\rho^\circ$ cells have been obtained only recently [17,26]. Because it lacks histone, mtDNA is easily removed without damaging nDNA by adding ethidium bromide in a culture medium. Theoretically, $\rho^\circ$ cells lack OXPHOS and can survive in a glucose medium. However, in contrast to yeast cells, adding pyruvate to a glucose medium is essential to support human $\rho^\circ$ cells because most NAD is reduced in the absence of COX. Human $\rho^\circ$ cells or cells with mutant mtDNA lacking OXPHOS cannot survive in the galactose pyruvate medium called DM170 [17]. Thus, we used glucose-rich media with pyruvate (0.1 mg/ml) to prevent an extreme reductive state, and we selected both cells with mutant mtDNA and human $\rho^\circ$ cells from wild-type cells using the glucose pyruvate medium [17]. The $\rho^\circ$ cells synthesize minor subunits that are detected by the monoclonal antibodies [19].

Cells without a nucleus that contain mitochondria are called cytoplasts. Cells are enucleated by centrifugation in the presence of cytochalasin B (10 µg/ml). By fusing cytoplasts with the $\rho^\circ$ cells, the properties of mutant mtDNA and nDNA can be analyzed (Fig. 2, left).

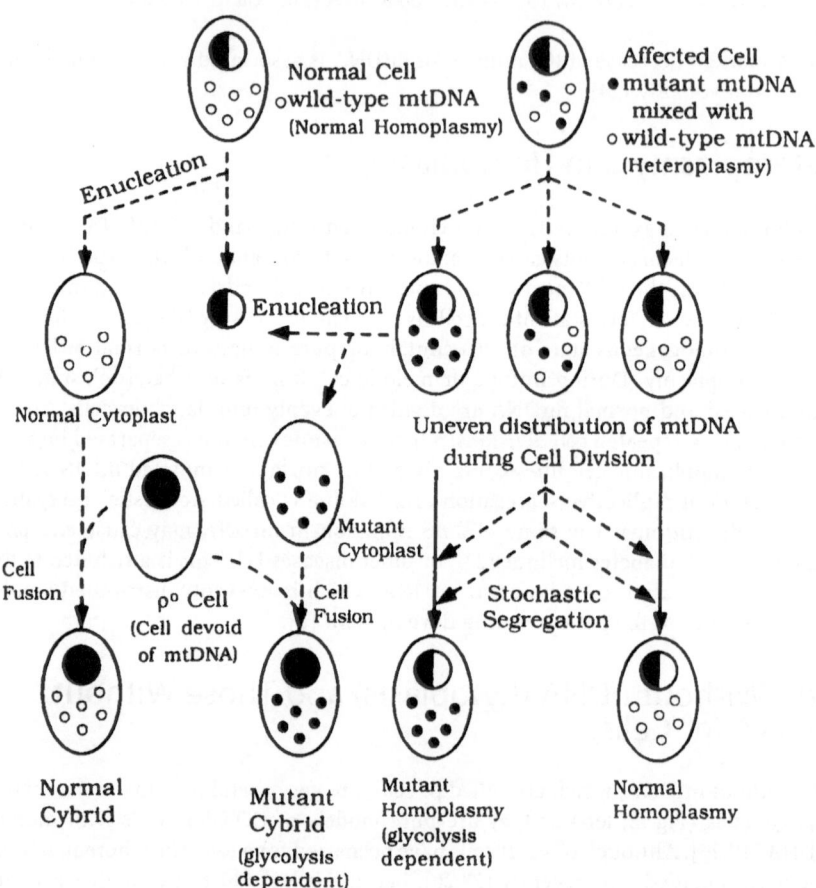

FIG. 2. Model illustrating heteroplasmy, homoplasmy, cytoplasts, and cybrids. *Small open circles* represent wild-type mtDNA encoding the largest subunits I, II, and III of COX; *small closed circles* indicate mutant mtDNA; *large circles* indicate nDNA encoding ten small subunits of COX and many proteins for translation and transcription of mtDNA; and *ovals* indicate the cell surface

## Cytochrome *c* Oxidase Tested in Cybrids

To analyze COX, we fused the cytoplasts with the $\rho^{\circ}$ cells to form cybrids [3,17,26]. We restored COX of the $\rho^{\circ}$ cells by introducing normal mtDNA in the cybrids (Fig. 3, left, HeEB). Human skin fibroblasts isolated from a MELAS patient were grown in glucose- and pyruvate-rich media [25,26]. During the growth of these heteroplasmic fibro-

**CONTROL**        **MELAS3243**        **CM4269**

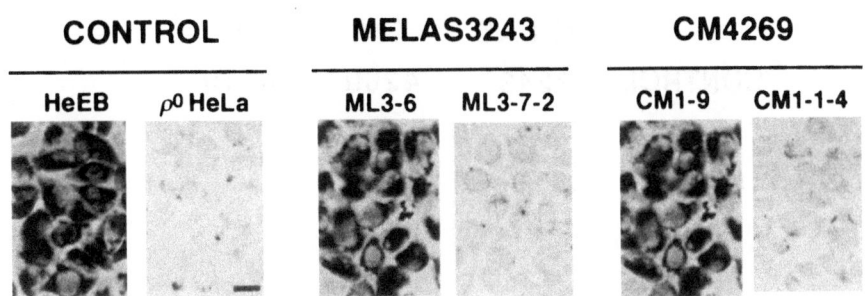

HeEB    $\rho^0$ HeLa        ML3-6    ML3-7-2        CM1-9    CM1-1-4

FIG. 3. Cytochrome oxidase activity of Mitochondria of cybrids obtained by fusing $\rho^0$ HeLa cells and cytoplasts containing normal and defective mtDNA. Cytochrome c oxidase activity was stained cytochemically [26]. *HeEB*, cybrids of $\rho^0$ cells and cytoplasts from normal HeLa cells; $\rho^0$ *HeLa*, HeLa cells without mtDNA; *ML3-6* and *ML3-7-2* are cybrids rich, respectively, in normal mtDNA and mutant MELAS mtDNA derived from mutant at np 3243; and *CM1-9* and *CM1-1-4* are cybrids rich in normal mtDNA and mutant CM mtDNA derived from mutant at np. *Bar:* 50 μm

blasts, stochastic segregation occurred (Fig. 2, right). On fusion with $\rho^0$ cells, the heteroplasmic cytoplasts from the mutant cell gave cybrid lines both with and without COX as a result of the stochastic segregation as shown by staining COX activity (Fig. 3, middle and right). The cell line with 95% wild-type mtDNA was called ML3-6 and that with concentrated MELAS mtDNA was called ML3-7-2.

Similarly, we could obtain cell lines from the patient with cardiomyopathy that segregated into the line rich in the wild-type mtDNA (CM1-9) and that rich in mutant-type mtDNA (CM1-1-4). These cells were enucleated, and the resulting cytoplasts were fused with HeLa $\rho^0$ cells. Thus, COX activity in the HeLa $\rho^0$ cells was restored in the cybrids from cytoplasts of original HeLa cells, the ML3-6 line and CM1-9 line (Fig. 3, HeEB, ML-3-6, and CM1-9, respectively). All the mtDNA-dependent subunits in the cybrid of the HeLa $\rho^0$ cells were now synthesized by introducing either wild-type mtDNA (Fig. 4). However, when the mutant mtDNAs from ML3-7-2 or CM1-1-4 lines were introduced into the HeLa $\rho^0$ cells, the activity was not restored (Figs. 3 and 4). Other activities dependent on mtDNA, such as NADH dehydrogenase, were also restored only when the wild-type mtDNA introduced [17].

In the cells with deleted mtDNA, fusion protein of a deleted fusion mtRNA is translated in the heteroplasmic cells. This indicated an intermitochondrial communication, because the deletion resulted in the lack of several tRNAs. This should cause global translation failure in homoplasmic cells. However, in heteroplasmic cells containing some amounts of normal mtDNA, there should be a sufficient supply of tRNAs from the coexisting normal mtDNA.

Under fluorescence microscopy using rhodamine 123, mitochondria lacking COX are swollen (Fig. 5). The morphology of defective mitochondria is normalized by the introduction of normal mtDNA and the resulting restoration of $\Delta\mu H^+$ (Fig. 5). The fluorescence of the COX-less mitochondria indicates that there still remains enough $\Delta\mu H^+$ to import matrix proteins or accumulate rhodamine 123, even in $\rho^0$ cells [17].

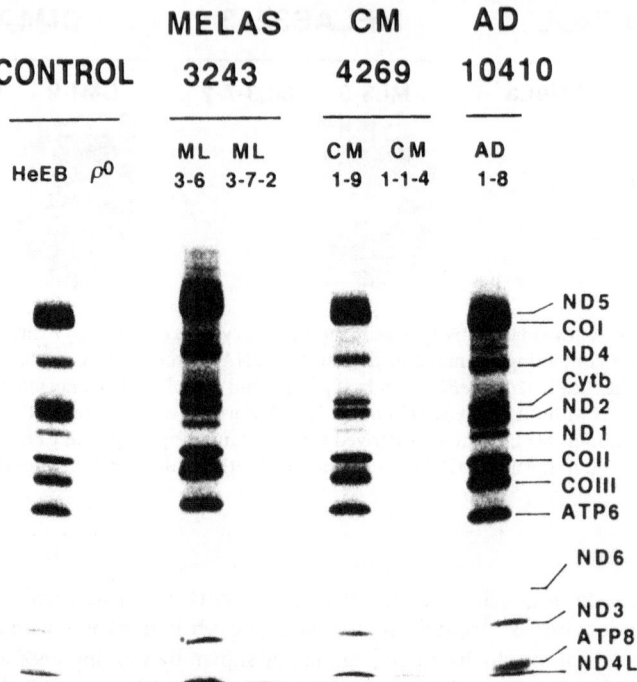

FIG. 4. Autoradiogram of polyacrylamide gel electrophoresis of translation products of mtDNA in cybrids obtained by fusing ρ° HeLa cells and cytoplasts containing normal and defective mtDNA. The 13 bands in the lanes are the translation products of mtDNA and are labeled with [$^{35}$S]methionine in the presence of emetine (0.2 mg/ml) to inhibit formation of nuclear gene products in the cybrids. The names of the bands (subunits) are indicated on the right-hand side, from *top* (*ND5*) to *bottom* (*ND4L*); subunits I, II, and III of COX are indicated as *COI*, *COII*, and *COIII*, respectively. The names of the cells or cybrids produced the [$^{35}$S]proteins are indicated on the lanes from *left* to *right*. CONTROL: HeEB, original HeLa cells; ρ°, ρ° HeLa cells; *MELAS 3243*: ML3-6, cybrids of HeLa ρ° cells and ML3-6 cells or ML3-7-2 cells; *CM 4269*: cybrids of HeLa ρ° cells and cardiomyopathy cells CM 1-9 or CM 1-1-4, as detailed in Fig. 3; and *AD 10410*: cybrids of HeLa ρ° cells and cytoplasts obtained from a cell line of Alpers disease in which mtDNA is not affected

# Decrease of Cytochrome *c* Oxidase in Aged Tissues

The decrease in COX is one of the greatest changes during human aging. Numerous studies have shown a continuous decline in the bioenergetic capacity of both physical and mental activity [1,27,28]. This is caused by the decrease in both the number of cells and their activities. The decrease in the numbers of the cells in aged tissue (or weight loss of the organs) is attributed mainly to telomere shortening in the somatic cells. The decrease in cell activities is characterized by lowered tissue respiration of aged subjects [27,28]. We found in vivo age-related reductions in the activities of COX

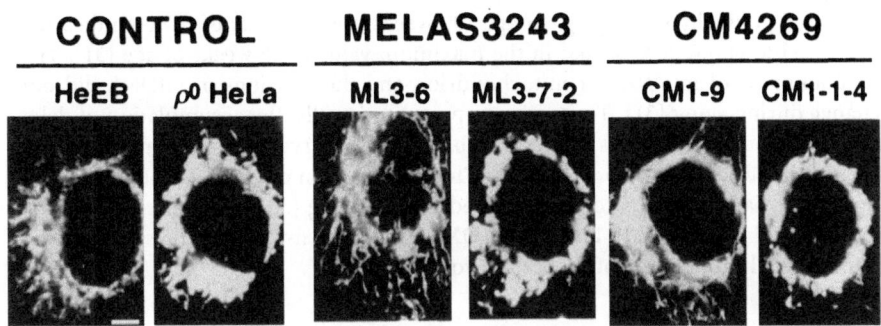

FIG. 5. Morphology of mitochondria observed under fluorescence photomicrography. The abnomrality was exclusively observed in $\rho^\circ$ HeLa cells and cybrids without COX. Cells grown in a living cell observation chamber were treated with Rhodamin R123 (10 μg/ml) for 2 min and washed several times with culture medium. The cybrids were observed with a fluorescence microscope (TMD-EF, Nikon, Tokyo Japan) with excitation light at 495 nm. *Left to right: HeEB,* cybrids of $\rho^\circ$ cells and cytoplasts from normal HeLa cells; $\rho^\circ$ *HeLa,* HeLa cells devoid of mtDNA; *ML3-6* and *ML3-7-2* are, respectively, cybrids with normal mtDNA and mutant MELAs mtDNA; and *CM1-9* and CM1-1-4 are as described under Fig. 3. *Bar:* 10 μm

in human fibroblasts from donors of various ages (0–97 years) [26]. The protein synthetic activity of mtDNA and COX in fibroblasts from an aged donor of 97 years was only about 15% of that in fetal fibroblasts.

During aging in *Drosophila*, decreases in COX activity and its major subunit, I, are correlated with the life-span curve without affecting the glucose phosphate isomerase [29]. There are also reports on the decrease of COX activity with age in human liver obtained directly by biopsy [30]. Although the loss of COX is attributed to the accumulation of the mutated mtDNA during aging [27,28,30], the roles of nDNA in aging should also be considered [29].

To test both possibilities, cytoplasts were prepared from the fibroblasts obtained from the aged donors and were fused with the $\rho^\circ$ HeLa cells described in the previous section [26]. The decreased COX activity (15% of the fetus) and the amounts of the subunits I, II, and III in aged cells (more than 80 years old) were restored by fusing with the $\rho^\circ$ cells [26]. The COX activity of the resulting cybrids was nearly the same as that in the cybrids from fetal cells and the $\rho^\circ$ HeLa cells, the cybrids of the original HeLa cells and the $\rho^\circ$ HeLa cells, and the hybrids of both the aged fibroblasts and the $\rho^\circ$ HeLa cells [26]. The restoration of the COX of these cybrids and hybrids was also confirmed in the restored translation of all 13 subunits encoded by mtDNA by the same method as described in Fig. 4. Thus, the aged nDNA is the major factor in the age-dependent decrease in human COX activity.

The telomere shortening theory of cellular senescence is not applicable to the postmitotic cells of the brain and skeletal muscles. Because fresh human cells from these tissues are difficult to obtain, we used C57BL/6 mouse brains and muscles to test for mitochondrial aging [31]. The mitochondrial translational activity in both the brain and skeletal muscle increased progressively up to 21 weeks after birth, then decreased gradually with aging. At age 100 weeks, the amounts of COX synthesized decreased to about 20% of that in the maximum value at 25 weeks of age [31]. The total amounts of mtDNA per mitochondrion and the deletions of mtDNA did not change during aging [31]. Thus, even in postmitotic cells, the accumulation of deletion mutations in mtDNA is not responsible for the observed age-associated decrease in mitochondrial translational activity. The transcription of mtDNA in the rat brain mitochondria also dramatically decreased during aging [32]. The poly A+ mRNAs, including all subunits I, II, and III of COX, encoded by mtDNA in the senescent rat brain are only about 20% of that in the young rats [32].

## Regulation of COX Subunits

Human COX activity is changed by many factors including activity and aging. The regulation of human COX is also very different from that of microorganisms. In yeast, carbon source and oxygen are the major regulators of COX [2], while in human cells these are constantly supplied from the circulating blood and thus the major regulators for human COX are tissue activity and hormone levels. Of all the enzymes in mammals, only four enzyme complexes of OXPHOS (FoF1, complexes I and III, and COX) are encoded by both mtDNA [6] and nDNA [1,2,8–10]. Among these, COX is the key enzyme in the overall regulation of energy metabolism [2]; because it is the only irreversible step, it is limiting in amount with respect to other complexes in human tissues, and the ATP:ADP ratio inhibits the activity of COX [11].

In principle, two general types of regulation are possible: short term and long term. Short-term regulation is immediate, could be affected by an allosteric effector, and does not require protein synthesis. The tissue activity decreases the ATP:ADP ratio, and the ratio changes $H^+/e^-$ stoichiometry of COX by a muscle-specific subunit VIa-H [11], on which ATP is specifically bound at the cholate-binding site (Arg 14 and Arg 17 of VIa-H) [4]. The roles for other minor subunits in the regulation are not clear yet [1,2,8–11,33]. From the biological standpoint, the $H^+/e^-$ stoichiometry of the $H^+$ pump is flexible depending on the ATP:ADP ratio [11], and the elastic energy transfer in the $H^+$-translocation has been discussed in detail previously [34].

Long-term regulation involves protein synthesis of COX. Because the $V_{max}$ for respiration is limited by the amounts of COX, their synthesis increases in response to continuous energy demand, a response that must be achieved by coordinating the expressions of subunits encoded by both nDNA and mtDNA. There is a detailed analysis of the synthesis of COX during brain activity [35]. This response must be dominated by nDNA because the expression of mtDNA is dependent on the DNA and RNA polymerases and the transcription factors encoded by nDNA [15,16,36].

There are several important upstream cis-elements that control the expression of COX [12–14]. The major subunits of COX are expressed by the mitochondrial transcription factor TFA (or TF1) [15,16,36]. There are common transcription factors for the COX subunits: nuclear respiratory factor 1 (NRF-1) [12] for positive activation of genes of COX Vb [37] and VIc-2 [12], cytochrome c and MRP RNA (for COX I, II, and III via mtDNA activation), and nuclear respiratory factor 2 (NRF-2) [8,13] for activation of genes of COX IV and Vb.

# Conclusion

Our cell-level research on human COX using cytoplasts, cybrids, and mutants revealed the functions of mtDNA and roles played by nDNA in diseases, development, and aging. However, more detailed analyses are needed to elucidate the development of COX [9-11] and the tissue-specific phenotypes of diseases related to COX [38,39].

# References

1. Wallace DC (1992) Diseases of the mitochondrial DNA. Annu Rev Biochem 61:1175–1212
2. Poyton RO, McEwen JE (1996) Crosstalk between nuclear and mitochondrial genomes. Annu Rev Biochem 65:563–607
3. Kagawa Y, Hayashi J-I (1997) Gene therapy of mitochondrial diseases using human cytoplasts. Gene Ther 4:6–10
4. Tsukihara Y, Aoyama H, Yamashita E, et al. (1996) The whole structure of the 13-subunit oxidized cytochrome *c* oxidase at 2.8 Å. Science 272:1136–1144
5. Dib C, Faure S, Fizames C, et al. (1996) A comprehensive genetic map of the human genome based on 5264 microsatellites. Nature 380:152–154
6. Anderson S, Bankier AT, Barrell BG, et al. (1981) Sequence and organization of the human mitochondrial genome. Nature 290:457–465
7. Kobayashi Y, Momoi MY, Tominaga K, et al. (1990) A point mutation in the mitochondrial tRNA$^{\text{Leu (UUR)}}$ gene in MELAS (mitochondrial myopathy, encephalopathy, lactic acidosis and strokelike episodes). Biochem Biophys Res Commun 173:816–822
8. Seelan RS, Gopalakrishnan L, Scarpulla RC, et al. (1996) Cytochrome *c* oxidase subunit VIIa live isoform. J Biol Chem 271:2112–2120
9. Capaldi RA, Marusich MF, Taanman JW (1995) Mammalian cytochrome *c* oxidase: characterization of enzyme and immunological detection of subunits in tissue extracts and whole cells. Methods Enzymol 260:117–132
10. Linder D, Freund R, Kadenbach B (1995) Species-specific expression of cytochrome *c* oxidase isozymes. Comp Biochem Physiol B Comp Biochem Mol Biol 112:461–469
11. Frank V, Kadenbach B (1996) Regulation of the $H^+/e^-$ stoichiometry of cytochrome *c* oxidase from bovine heart by intramitochondral ATP/ADP ratios. FEBS Lett 382:121–124
12. Chau CM, Evans MJ, Scarpulla RC (1992) Nuclear respiratory factor 1 activation sites in genes encoding the g subunit of ATP synthase, eukaryotic initiation factor 2a, and tyrosine aminotransferase. J Biol Chem 267:6999–7006
13. Carter RS, Avadhani NG (1994) Cooperative binding of GA-binding protein transcription factors to duplicated transcription initiation region repeats of the cytochrome *c* oxidase subunit IV gene. J Biol Chem 269:4381–4387
14. Tomura H, Endo H, Kagawa Y, et al. (1990) Novel regulatory enhancer in the gene of the human mitochondrial ATP synthase β subunit. J Biol Chem 265:6525–6527
15. Clayton DA (1991) Replication and transcription of vertebrate mitochondrial DNA. Annu Rev Cell Biol 7:453–478
16. Tominaga K, Akiyama S, Kagawa Y, et al. (1992) Upstream region of a genomic gene for human mitochondrial transcription factor 1. Biochem Biophys Acta 1131:217–219

17. Hayashi J, Ohta S, Kagawa Y, et al. (1994) Functional and morphological abnormalities of mitochondria in human cells containing mitochondrial DNA with pathogenic point mutations in tRNA genes. J Biol Chem 269:19060–19066

18. Shoffner JM, Lott MT, Lezza AM, et al. (1990) Myoclonic epilepsy and ragged-red fiber disease (MERRF) is associated with the DNA$^{tRNA\ Lys}$ mutation. Cell 61:931–937

19. Taanman JW, Burton MD, Marusich MF, et al. (1996) Subunit specific monoclonal antibodies show different steady-state levels of various cytochrome-c oxidase subunits in chronic progressive external ophthalmoplegia. Biochim Biophys Acta 1315:199–207

20. Keightley JA, Hoftbuhr KC, Burton MD (1996) A microdeletion in cytochrome $c$ oxidase (COX) subunit III associated with COX deficiency and recurrent myoglobinuria. Nature Genet 12:410–416

21. Kagawa Y, Yuzaki M, Ohta S, et al. (1991) Multiple deletions of mitochondrial DNA in patients with familial mitochondrial myopathy. Prog Neuropathol 7:129–139

22. Remes AM, Majamaa K, Herva R (1992) Adult-onset diabetes mellitus and neurosensory hearing loss in maternal relatives of MELAS patients in a family with the tRNA Leu(UUR) mutation. Neurology 43:1015–1020

23. Johns DR, Neufeld MJ (1993) Cytochrome $c$ oxidase mutations in Leber hereditary optic neuropathy. Biochem Biophys Res Commun 196:810–815

24. Otsuka M, Niijima K, Mizuno Y, et al. (1990) Marked decrease of mitochondrial DNA with multiple deletions in patient with familial mitochondrial myopathy. Biochem Biophys Res Commun 167:680–685

25. Shimoizumi H, Momoi M, Ohta S, et al. (1989) Cytochrome oxidase-deficient myogenic cell lines in mitochondrial myopathy. Ann Neurol 25:615–621

26. Hayashi J, Ohta S, Kagawa Y, et al. (1994) Nuclear but not mitochondrial genome involvement in human age-related mitochondrial dysfunction: function integrity of mitochondrial DNA from aged subjects. J Biol Chem 269:6878–6883

27. Linnane AW, Marzuki S, Ozawa T, et al. (1989) Mitochondrial DNA mutations as an important contributor to aging and degenerative diseases. Lancet i:642–645

28. Ozawa T (1995) Mechanism of somatic mitochondrial DNA mutations associated with age and diseases. Biochim Biophys Acta 1271:177–189

29. Calleja M, Pena P, Ugalde C, et al. (1993) Mitochondrial DNA remains intact during *Drosophila* aging, but the levels of mitochondria transcripts are significantly reduced. J Biol Chem 268:18891–18897

30. Yen TC, Chen YS, King KL, et al. (1989) Liver mitochondrial respiratory functions decline with age. Biochem Biophys Res Commun 178:124–131

31. Takai D, Inoue K, Shisa H, et al. (1995) Age-associated changes of mitochondrial translation and respiratory function in mouse brain. Biochem Biophys Res Commun 217:668–674

32. Fernandez-Silva P, Petruzzella V, Facasso F, et al. (1991) Reduced synthesis of mtRNA in isolated mitochondria of senescent rat brain. Biochem Biophys Res Commun 176:645–653

33. Otsuka M, Mizuno Y, Yoshida M, et al. (1988) Nucleotide sequence of cDNA encoding human cytochrome $c$ oxidase subunit VIc. Nucleic Acids Res 16:10916

34. Kagawa Y, Hamamoto T (1996) The energy transmission in ATP synthase: from the γ-c rotor to the α3β3 oligomer fixed by OSCP-b-stator via the βDELSEED sequence. J Bioenerg Biomembr 28:421–431

35. Hevner RF, Wong-Riley MTT (1993) Mitochondrial and nuclear gene expression for cytochrome oxidase subunits are disproportionately regulated by functional activity in neurons. J Neurosci 13:1805–1819

36. Tominaga K, Hayashi J-I, Kagawa Y, et al. (1993) Smaller isoform of human mitochondrial transcription factor 1: its wide distribution and production by alternative splicing. Biochem Biophys Res Commun 194:544–551

37. Hoshinaga H, Amuro N, Goto Y (1994) Molecular cloning and characterization of rat cytochrome *c* oxidase subunit Vb gene. J Biochem (Tokyo) 115:194–201
38. Shiraiwa N, Ishii A, Iwamoto H, et al. (1993) Content of mutant mitochondrial DNA and organ-dysfunction in a patient with a MELAS-subgroup of mitochondrial encephalomyopathies. J Neurol Sci 120:174–179
39. Kawashima S, Ohta S, Kagawa Y, et al. (1994) Widespread tissue distribution of multiple mitochondrial DNA deletions in familial mitochondrial myopathy. Muscle Nerve 17:741–746

# Redox Behavior of Copper A in Cytochrome Oxidase in the Brain In Vivo: Its Clinical Significance

Yoko Hoshi[1,2], Hideo Eda[3], Osamu Hazeki[1], Yasutomo Nomura[1], Yasuyuki Kakihana[1,4], Satoshi Kuroda[5], and Mamoru Tamura[1]

*Summary.* The accuracy of near-infared measurement of the redox state of cyto-chrome oxidase in situ remains controversial. Our new approach to the measurement of the redox state of cytochrome oxidase resolves the most difficult problem, that the in vivo absorption coefficient of cytochrome oxidase is unknown, in addition to other problems such as the light-scattering effects and marked overlap of absorbance changes attributed to hemoglobin. We applied this method to both animal and clinical investigations. Based on the results obtained from these investigations, we discuss the redox behavior of cerebral cytochrome oxidase in vivo and the significance of the measurement of cytochrome oxidase in clinical medicine. Our conclusion is that cerebral cytochrome oxidase in vivo is fully oxidized under normal physiological conditions and that its oxygen-dependent redox change, which precedes a decline of brain function, occurs only when the oxygen supply is extremely impaired. Thus, the start of the reduction of cytochrome oxidase can be used as an alarm indicating that the brain condition is critical metabolically and functionally.

*Key words.* Cytochrome oxidase—Near-infrared photometry—Hypoxia—Cerebral oxygenation—Mitochondrial energy state

## Introduction

Near-infrared spectroscopy (NIRS), a relatively new technique, enables us to monitor changes in hemoglobin (Hb) and myoglobin (Mb) oxygenation states, blood volume, and the redox state of cytochrome oxidase noninvasively [1]. This technique now finds wide clinical application [2–4] and its usefulness to the measurement of Hb has

[1] Biophysics Group, Research Institue for Electronic Science, Hokkaido University, Kita 12, Nishi 6, Kita-ku, Sapporo 060, Japan
[2] New Energy and Industrial Technology Development Organization, Tokyo 170, Japan
[3] Shimadzu Corporation Technology Research Laboratory, Hadano, Kanagawa 259-13, Japan
[4] Department of Anestheology, School of Medicine, Kagoshima University, Kagoshima 890, Japan
[5] Department of Neurosurgery, School of Medicine, Hokkaido University, Kita 15, Nishi 7, Kita-ku, Sapporo 060, Japan

been confirmed. However, the specificity and accuracy of the measurement of the redox state of cytochrome oxidase are still controversial. This is mainly attributed to the lack of valid absorption spectra for cytochrome oxidase in the near-infrared region in vivo, although many investigators have reported different spectra [5-8]. Therefore, the reported absorption coefficient of copper A, which accounts for more than 85% of cytochrome oxidase absorption in the near-infrared spectrum [9], is ambiguous.

Furthermore, using isolated mitochondria, we found that the apparent absorption coefficient of copper A varied with the mitochondrial energy state [10]. In other words, the apparent absorption coefficient of copper A, which has been believed to be a constant, is actually not a constant. These in vitro mitochondrial data have been confirmed by experiments with the isolated perfused rat head [11]. This questions the validity of any algorithms that contain a constant value for the absorption coefficient of cytochrome oxidase in simultaneous equations.

In this chapter, we describe the following: (1) the problem of near-infrared measurement of the redox state of cytochrome oxidase in situ; (2) the redox behavior of cerebral cytochrome oxidase in vivo; and (3) the significance of the measurement of cytochrome oxidase in clinical medicine.

## Problem of Measurement of the Redox State of Copper A in Cytochrome Oxidase

Figure 1 shows absorbance changes in copper A in isolated mitochondria caused by the aerobic–anaerobic transition measured at 830–760 nm in different energy states. The magnitude of absorbance change varied with the mitochondrial energy state: in state 4 it was about 1.3 and 2.6 times larger than in state 3 and the uncoupled state, respectively. This means that the apparent absorption coefficient of copper A is not a constant. Thus, we have developed a simple and novel algorithm that does not contain the absorption coefficient of copper A for cytochrome oxidase measurement (Hoshi et al., in manuscript).

In the near-infrared region (700–900 nm), the absorbance changes in the range shorter than 780 nm are mainly attributed to Hb, whereas those at wavelengths longer than 780 nm are attributed to both Hb and oxidized cytochrome oxidase (Fig. 2) [12]. Using two pairs of dual wavelengths shorter than 780 nm, we first calculated relative changes in concentrations of oxygenated ([oxy-Hb]) and deoxygenated hemoglobin ([deoxy-Hb]). Based on these values, we can estimate the absorbance change attributed to Hb at wavelengths longer than 780 nm. Because in this range absorbance changes are attributed to both Hb and cytochrome oxidase, subtraction of the estimated value from the measured one provides relative changes in concentration of oxidized cytochrome oxidase ([cyt. ox.]).

The NIRS instrument that was used (Unisoku, Hirakata, Japan) measures Hb oxygenation state and the redox state of cytochrome oxidase according to the following algorithm:

$$\Delta[\text{oxy-Hb}] = -0.912\Delta A_{700-750} + 2.128\Delta A_{730-750}$$

$$\Delta[\text{deoxy-Hb}] = 0.744\Delta A_{700-750} - 1.613\Delta A_{730-750}$$

$$a''_3 \Delta[\text{cyt. ox.}] = 1.53\Delta A_{700-750} - 0.768\Delta A_{730-750} + \Delta A_{805-750} \tag{1}$$

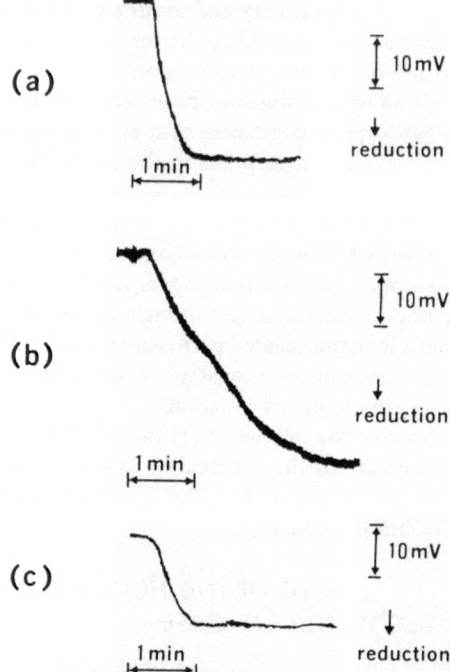

FIG. 1a–c. Absorbance change of copper A in cytochrome oxidase caused by aerobic–anaerobic transition [10]. Rabbit heart mitochondria, 0.8 mg protein/ml, was suspended in a reaction medium containg 0.25 M sucrose, 10 mM Tris-HCl, pH 7.2, 0.2 mM EDTA, and 10 mM glutamate. **a** State 3 (energy level low, respiratory rate high). **b** State 4 (energy level high, respiratory rate low). **c** Uncoupled state. The wavelength pair was 830–760 nm

where $a''_3$ is the proportionality factor (normalized absorption coefficient) for copper A. Because scattering effects prevent determination of the optical pathlength, the results are expressed in relative absorbance units rather than absolute concentration. In contrast to other published algorithms, classical dual wavelength analysis is used in our algorithm. Dual wavelength analysis provides adequate compensation for the light-scattering change of the tissue itself and for the instability of the photomultiplier of the light source.

It must be noted here that the coefficients of Eq. 1 contain instrumentation factors such as the half-width of the optical filters used in addition to the optical pathlength. When a new instrument is assembled, we therefore have to reestimate these factors. Matcher et al. have reported that they found differences when applying four published NIRS algorithms (by University College London, Duke University, Keel University, and our group [12]) to the same in vivo data set [13]. However, the results from our algorithm in their report were different from those that we obtained by the use of our NIRS instrument. This discrepancy might have resulted because they did not take instrumentation factors into account when applying our algorithm.

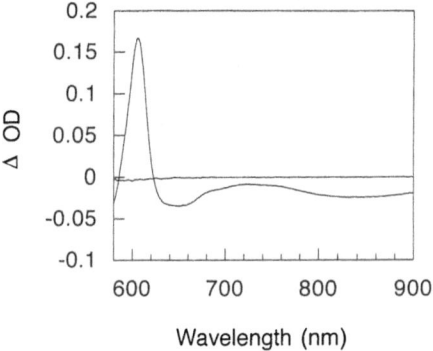

FIG. 2. Anoxic minus aerobic spectrum of fluorocarbon-perfused rat head. Spectrum at 100% $O_2$ was recorded as a flat baseline and the difference spectrum was obtained by changing the perfusate of fluorocarbon solution at 100% $N_2$

## Redox Behavior of Copper A in Cytochrome Oxidase In Vivo in the Rat Brain

Since the pioneering works of Jöbsis et al. [1,14], redox behavior of cerebral cytochrome oxidase has been a big issue. Observations that alterations in arterial oxygen saturation ($SaO_2$) or $PaCO_2$ were positively related to changes both in [cyt. ox.] and in [oxy-Hb] in humans and animals led to the conclusion that cytochrome oxidase was partially reduced under normoxic conditions [1,2,14,15]. By contrast, in blood-free animals, which eliminates the problem of cytochrome oxidase signals being contaminated by the more abundant Hb signals, reduction of cytochrome oxidase was not seen until oxygen delivery was severely impaired [16,17].

Using our new method, we examined the redox behavior of cytochrome oxidase in the rat brain under various conditions. Male Wistar rats (180–250 g) were anesthetized by intraperitoneal injection of urethane (ethyl carbamate, 180 mg/100 g body wt.). They were tracheotomized and mechanically ventilated after an intravenous injection of panchronium bromide (0.02 mg/100 g body wt.). The tidal volume and respiratory rate were adjusted to given arterial $PCO_2$ values of 37–42 mmHg when animals were ventilated with air. For the electroencephalogram (EEG), skin and muscle overlying the calvaria were reflected, two bis electrodes were placed symmetrically on the occipital bone, and a reflectance electrode was placed on the nasal bone. For spectrophotometric measurement, the rat's head was illuminated 5 mm in front of an ear through a light guide, and transmitted light through the cranial bone and cerebral tissue was guided again with another light guide.

Figure 3 shows the relationship between cerebral oxygenation and the EEG in graded hypoxia. Changes in [oxy-Hb] and [cyt. ox.] are expressed as a percentage of the full-scale value. Maximum changes caused by the transition from aerobic (under respiration with 21% oxygen gas) to anaerobic (under respiration with 100% nitrogen gas) are designated 100% (full scale). Each EEG was measured at the points marked on the trace of cytochrome oxidase. When $FiO_2$ was reduced from 21% to 16%, [oxy-Hb]

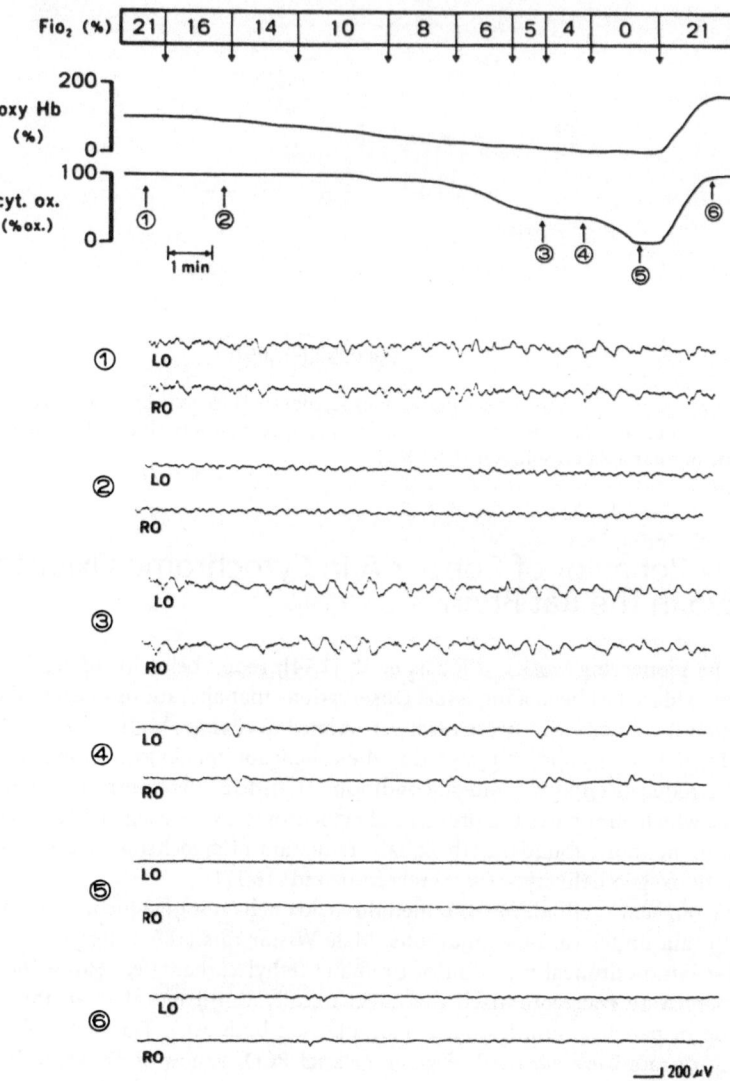

FIG. 3. Relationship between cerebral oxygenation and function under hypoxic conditions in a rat [18]. EEG samples were recorded at the points marked on the trace of cytochrome oxidase

decreased slightly and desynchronization occurred on the EEG (no. 2), while cytochrome oxidase was still in the fully oxidized state. When FiO$_2$ was decreased to 10%, [oxy-Hb] decreased to about 40% and cytochrome oxidase began to be reduced. Decreasing FiO$_2$ further, when [cyt. ox.] decreased to about 35%, resulted in high-voltage slow waves appearing on the EEG (no. 3). Flattening of the EEG (no. 5)

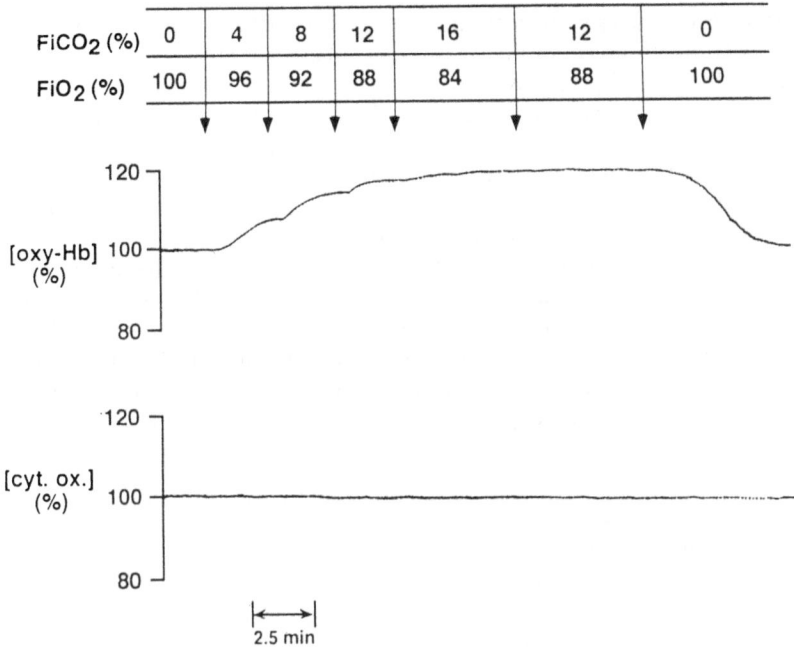

| FiCO$_2$ (%) | 0 | 4 | 8 | 12 | 16 | 12 | 0 |
|---|---|---|---|---|---|---|---|
| FiO$_2$ (%) | 100 | 96 | 92 | 88 | 84 | 88 | 100 |

FIG. 4. Changes in concentrations of oxygenated hemoglobin ([*oxy-Hb*]) and oxidized cytochrome oxidase ([*cyt. ox.*]) caused by hypercapnea (Hoshi et al, in manuscript). FiCO$_2$ was increased stepwise from 0% to 16% under hyperoxic conditions

occurred with a few seconds delay after cytochrome oxidase was fully reduced under anoxic conditions.

Figure 4 shows changes in cerebral oxygenation caused by hypercapnea under hyperoxic conditions. In this case, the values of [oxy-Hb] and [cyt. ox.] under respiration with 100% oxygen were taken as 100%. Stepwise increases in FiCO$_2$ were accompanied by decreases in FiO$_2$ to 88% at most, at which point PaO$_2$ and PaCO$_2$ were 353 and 117mmHg, respectively. Increases in FiCO$_2$ caused increases in total hemoglobin ([t-Hb]) (data not shown). As a result, [oxy-Hb] increased, while further oxidation of cytochrome oxidase was not observed.

Figure 5a shows changes in cerebral oxygenation, cerebral blood volume (CBV), and blood pressure (BP) in epileptic seizures induced by pentylentetrazol (PTZ) administration. Figure 5b shows the EEG monitored simultaneously. Each EEG was monitored at the points marked on the trace of cytochrome oxidase. In this case, changes in CBV were monitored by measuring the absorbance change at 805nm, the isosbestic point between oxy- and deoxyhemoglobin [12]. Intravenous administration of PTZ (5mg/100g body wt.) first caused the transient reduction of cytochrome oxidase. Desynchronization almost simultaneously occurred on the EEG (nos. 2 and 3). Slightly later, BP began to elevate concomitantly with an increase in CBV. When BP reached a maximum, bursts of spikes appeared on the EEG (no. 4). In this period, [oxy-Hb] was increasing and reoxidation of cytochrome oxidase was beginning.

While [oxy-Hb] increased above the preseizure level, further oxidation of cytochrome oxidase was not observed. In the late postictal phase, although BP remained higher and [oxy-Hb] started gradually to decrease (although it was still higher than the preseizure level), cytochrome oxidase was partially reduced. This reduction might have been caused by a lasting arteriovenous shunt that opened during seizures, while the transient reduction of cytochrome oxidase in the preictal period might have been a trigger for an increase in CBF. It should be noted that such cerebral hypoxia could be first detected by measurement of cytochrome oxidase.

From these results, it is concluded that the reduction of cytochrome oxidase occurs only when oxygen supply is extremely compromised. To obtain quantitative information from near-infared signals, we previously determined oxygen dependence of the redox states of copper A and heme $aa_3$ in the cytochrome oxidase of isolated mitochondria (Fig. 6). The apparent oxygen affinity of copper A is much higher than that of Hb, by approximately three orders of magnitude: the oxygen

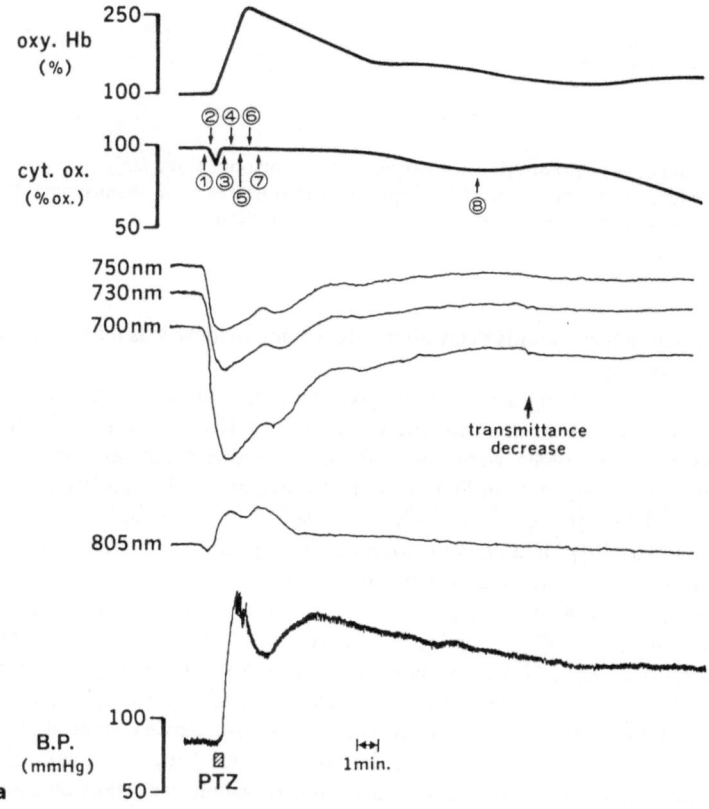

FIG. 5. a Changes in cerebral oxygenation, blood volume (CBV), and blood pressure (BP) caused by pentylentetrazol (*PTZ*); 5 mg/100 g body wt. of PTZ was injected during the time indicated by the *shaded square*. b Changes in the EEG caused by PTZ administration. EEG samples were recorded at the points marked on the trace of *cyt. ox.* [18]

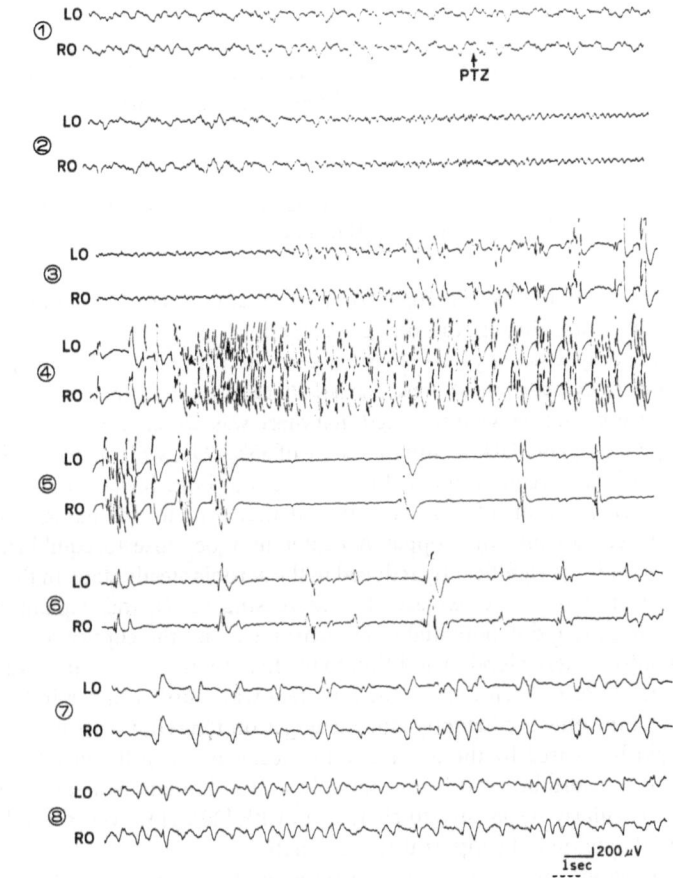

FIG. 5. *Continued*

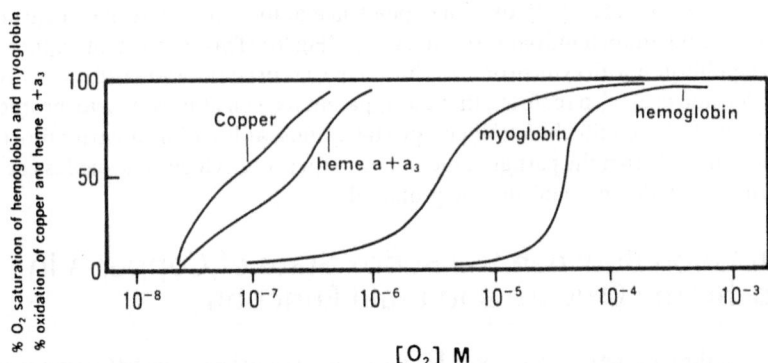

FIG. 6. Effects of oxygen concentration on oxygenation states of hemoglobin (Hb) and myoglobin (Mb) and the redox states of heme $aa_3$ and copper A in cytochrome oxidase [10]. Oxygenation states of Hb and Mb were measured at 37°C. Redox states of heme $aa_3$ and copper A were measured in state 3

concentration required for the half-maximal reduction ($P_{50}$) of copper A is $7.5 \times 10^{-8}$ M. The $P_{50}$ of Hb is $4.2 \times 10^{-5}$ M. Judging from this calibration, our conclusion is acceptable.

## Redox State of Copper A in Cytochrome Oxidase Under Normal Physiological Conditions

A general conclusion that has emerged from recent studies [16,17,19,20] seems to be consistent with ours; that is, the oxygen-dependent redox change of cytochrome oxidase occurs only when oxygen delivery is extremely impaired. However, the issue of whether or not cytochrome oxidase is partially reduced in normoxia remains to be solved. Edwards et al. [21], who reported that there was no relation between changes in cerebral [cyt. ox.] and $SaO_2$ within the range of 85%–99% while increases in $PaCO_2$ from 4.3 to 9.6 kPa were accompanied by increases in both [cyt. ox.] and [t-Hb] in human newborn preterm infants, have argued that cytochrome oxidase might be partially reduced, because the copper A center may be close to equilibrium with cytochrome $c$, which is significantly reduced in the aerobic steady state. In their paper, the oxidation of cytochrome oxidase after increasing $PaCO_2$ was explained by the assumption that the redox potential of cytochrome $c$ and the copper A varied with the mitochondrial energy level in addition to the hemodynamic characteristics of the human neonatal brain. Their data are inconsistent with ours as shown in Fig. 4. This discrepancy of the effect of hemodynamic changes on the redox state of cytochrome oxidase might be caused by the difference in species or the difference between the adult and the newborn brain. It is now thought that in human neonates cytochrome oxidase is particularly responsive to changes in CBF [20]. However, several in vitro mitochondrial data do not support their explanation.

Figure 7 shows effects of oxygen concentration on the redox state of heme $aa_3$ and copper A of isolated mitochondria. Oxygen dependence of the redox state of heme $aa_3$ depends on the mitochondrial energy state as well as the respiratory rate. In contrast, copper A is independent of both the energy state and respiratory rate. According to the data of Sugano et al. [22], oxygen dependence of the redox state of cytochrome $c$ depends on the mitochondrial respiratory rate (Fig. 8). This means that copper A is not in equilibrium with cytochrome $c$, while heme $a$ is in equilibrium with cytochrome $c$ [22]. Sugano et al. also reported that the apparent oxygen affinity of cytochrome $c$ is independent of the mitochondrial energy state. Thus, at least for in vitro mitochondria, it is unlikely that the partial reduction of copper A in cytochrome oxidase occurs in normoxia by the mechanism they proposed.

## Correlation Between the Redox State of Copper A in Cytochrome Oxidase and Brain Function

Studies on the correlation of the cytochrome oxidase signal with the brain energy state of piglets [20] and dogs [23] have shown that the reduction of cytochrome oxidase is highly correlated with a decreased brain energy state. Thus, the reduction of cytochrome oxidase indicates a decrease not only in tissue oxygenation but also in the energy level. As is shown in Fig. 2, however, the reduction of cytochrome oxidase

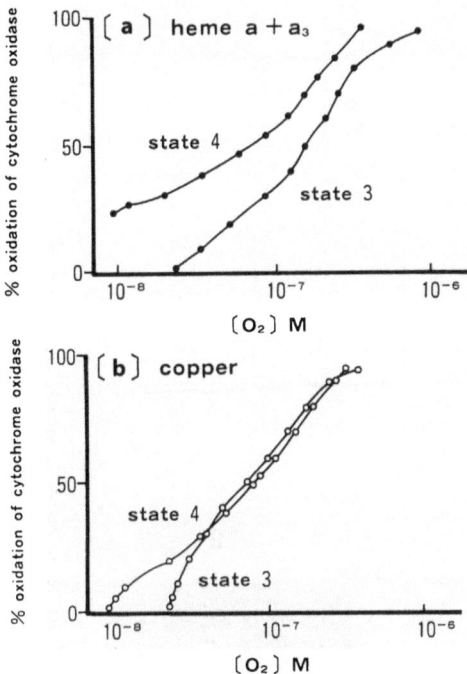

FIG. 7a,b. Effects of oxygen concentration on the redox states of heme $aa_3$ and copper A in state 3 and state 4. Experimental conditions were as described in Fig. 1, except that the medium contained leg hemoglobin as an oxygen indicator

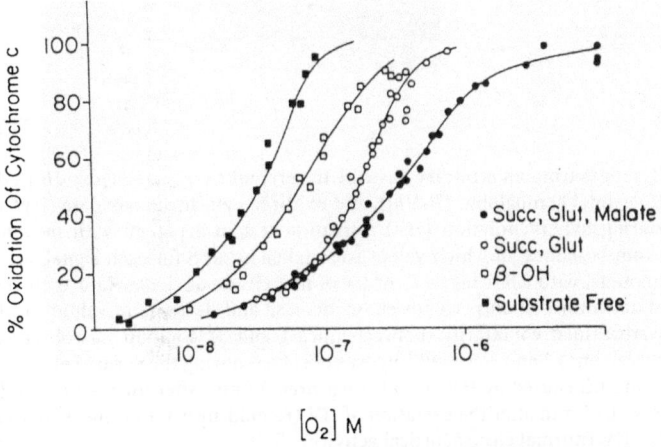

FIG. 8. Effects of substrates on the relation between redox state of cytochrome $c$ and oxygen concentration [21]. Pig heart mitochondria (15 mg protein) and 50 mg wet wt. of *Photobacterium phosphoreum* were suspended in 30 ml of a reaction medium containing 0.5 M mannitol, 0.02 M KCl, 0.1 M $K_2PO_4$ (pH 7.2), and 0.2 mM EDTA

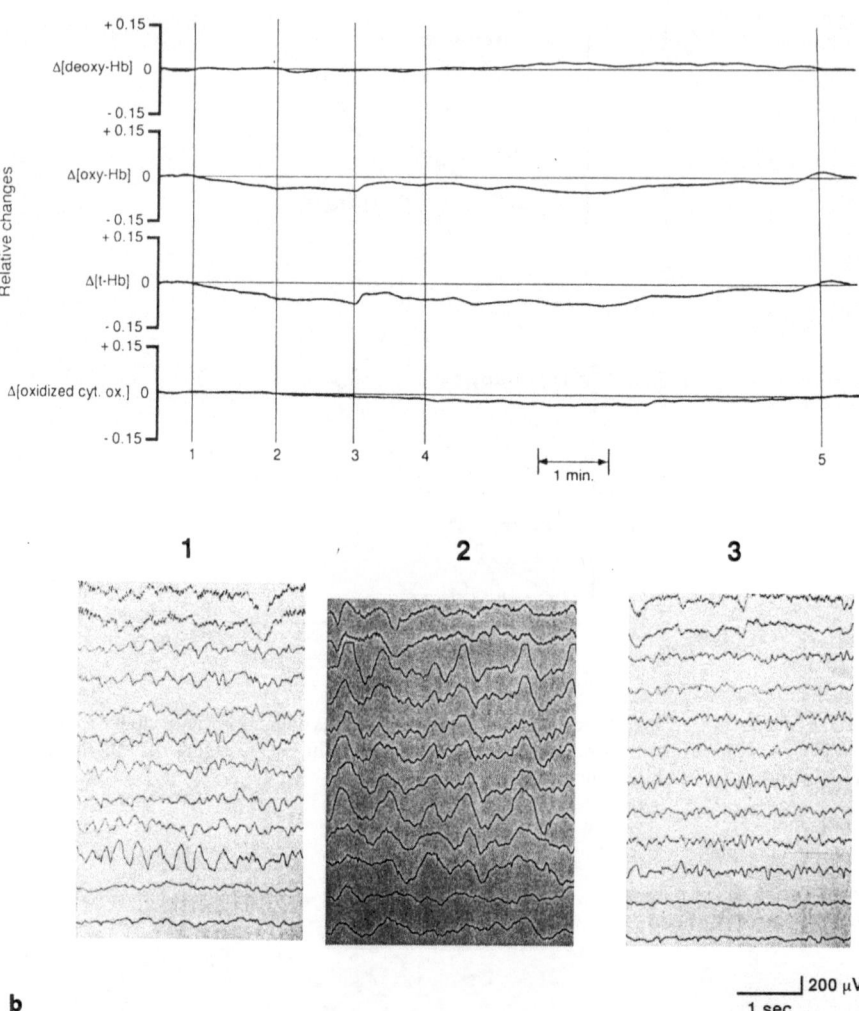

b

FIG. 9. a Changes from an arbitrary baseline in cerebral oxygenated ([*oxy-Hb*]), deoxygenated ([*deoxy-Hb*]), total hemoglobin ([*t-Hb*]), and oxidized cytochrome oxidase ([*cyt. ox.*]) concentrations during hyperventilation (*HV*) activation test in a patient with moyamoya disease. Changes from baselines, of which values were taken as zero for each signal, are presented as relative amounts, with 0.001 taken as order of magnitude of changes for each signal. Upward (*plus*) and downward (*minus*) trends show increase and decrease in values, respectively. The patient overbreathed voluntarily (between lines 1 and 3). Buildup was observed during the period between lines 2 and 3. Rebuild-up was observed during the period between lines 4 and 5. b Changes in EEG caused by HV. 1, EEG measured 2.5 min after the start of HV (build-up); 2, EEG measured 1.5 min after the cessation of HV (rebuild-up); 3, EEG measured 7 min after the cessation of HV (normal electrocortical activity) preceded the appearance of high-voltage slow waves, which indicates a decline of brain function under hypoxic hypoxia [24].

Figure 9a shows NIRS traces from a 9-year-old boy with moyamoya disease during the hyperventilation (HV) activation test. HV first caused decreases in [t-Hb] and [oxy-Hb] and then mild build-up on the EEG at line 2 (EEG-1 in Fig. 9b), while [cyt. ox.] was not changed. He complained of headache at line 3 and ceased HV. After the cessation of HV, build-up disappeared immediately, [t-Hb] and [oxy-Hb] once increased but did not return to the original levels, and cytochrome oxidase started to be reduced. About 1 min after the cessation of HV, diffuse high-voltage slow waves suddenly appeared on the EEG at line 4 (rebuild-up; EEG-2), in which [t-Hb] and [oxy-Hb] decreased further and [deoxy-Hb] increased.

Throughout rebuild-up appearing on the EEG, the patient complained of severe headache and a numb sensation. Reduction of cytochrome oxidase began about 3 min after the cessation of HV, and the high-voltage slow waves also started to disappear. When all the NIRS parameters had returned almost to each baseline level, normal electrocortical activities were observed on the EEG at line 5 (EEG-3). After a while the patient no longer complained of either a headache or a numb sensation. Rebuild-up, which is thought to be caused by both ischemia from hypocapnea and hypoxia from suppression of breathing after HV, reflects the deterioration of brain function. These clinical data also showed that the reduction of cytochrome oxidase preceded a decline of brain function. In addition, complete recovery of both NIRS parameters and the EEG means that irreversible brain damage did not occur during cytochrome oxidase reduction lasting about 6 min.

## Significance of Measurement of the Redox State of Copper A in Cytochrome Oxidase in Clinical Medicine

Several investigators have proposed that [oxy-Hb] may be the best indicator of impending brain hypoxia, while the reduction of cytochrome oxidase may be a prognosticator of irreversible brain damage because it occurs only under extreme hypoxia [16,19]. At the moment, however, NIRS does not provide quantitative information. Thus, it is impossible to judge the degree of cerebral hypoxia by near-infrared measurement of the Hb oxygenation state alone. When Hb oxygenation decreases in sick patients requiring intensive care, for example, we cannot decide whether we should treat them immediately without other monitoring systems.

By contrast, the reduction of cytochrome oxidase is correlated with decreases in cerebral energy level and precedes a decline in brain function. This suggests that the brain funtions so long as cytochrome oxidase is maintained in the fully oxidized state. It therefore appears that the start of cytochrome oxidase reduction can be used as an alarm that the brain condition is critical metabolically and functionally, even though absolute values are lacking.

As to quantitation, it is thought that a differential pathlength factor (DPF) [25] enables quantitation. In fact, mean values and standard deviations are often evaluated for the NIRS data assuming that the DPF is constant. However, DPF varies with several factors such as the kind and age of tissue. Values for DPF are in a relatively wide range; for instance, the values in the infant head are approximately 3–5 [25]. This means that

NIRS data cannot be compared across subjects unless the DPF is quantified for each of the subjects.

Before the advent of NIRS, we evaluated tissue oxygenation from several indirect variables such as $PaO_2$. These methods, however, sometimes do not reflect tissue oxygenation correctly. As is shown in Fig. 5a, even though oxygen supply was apparently sufficient, cerebral hypoxia actually occurred in the preictal period and the late postictal phase. Thus, direct measurement of tissue oxygen concentration is essential for evaluating tissue oxygen sufficiency. Among the various clinical monitoring systems, only near-infrared measurement of cytochrome oxidase can meet this demand.

In conclusion, we believe that near-infrared measurement of cytochrome oxidase is very useful in clinical medicine. As there are still several methodological problems in NIRS, however, when inexplicable data are obtained, reevaluation of the algorithm must be considered. Observations such as motor task-induced oxidation of cytochrome oxidase in the adult human brain [26] and the oxidation of cerebral cytochrome oxidase with $FiO_2$ of 12% and 8% in hypoxic piglets [20] seem to be a case in point.

# References

1. Jöbsis FF (1977) Noninvasive infrared monitoring of cerebral and myocardial oxygen sufficiency and circulatory parameters. Science 198:1264–1267
2. Brazy JE, Lewis DV, Mitnick MH, Jöbsis FF (1985) Noninvasive monitoring of cerebral oxygenation in preterm infants: preliminary observations. Pediatrics 75:217–225
3. Skov L, Pryds O (1992) Capillary recruitment for presentation of cerebral glucose influx in hypoglycemic, preterm newborn: evidence for glucose sensor? Pediatrics 90:193–195
4. Wyatt JS, Cope M, Delpy DT, et al. (1986) Quantitation of cerebral oxygenation and haemodynamics in sick newborn infants by near infrared spectrophotometry. Lancet 2:1063–1066
5. Chance B (1966) Spectrophotometric observations of absorbance changes in the infrared region in suspensions of mitochondria and in submitochondrial particles. In: Peisach G, Aisen P, Blumberg WE (eds) The biochemistry of copper. Academic, New York, pp 293–301
6. Ferrari M, Hanley DF, Wilson DA, Traystman RJ (1990) Redox changes in cat brain cytochrome-*c* oxidase after blood-fluorocarbon exchange. Am J Physiol 258:H1706–H1713
7. Jöbsis FF (1979) Oxidative metabolic effects of cerebral hypoxia. Adv Neurol 26:299–318
8. Wray S, Cope M, Delpy DT, et al. (1988) Characterization of the near-infared absorption spectra of cytochrome *aa*$_3$ and haemoglobin for the non-invasive monitoring of cerebral oxygenation. Biochim Biophys Acta 933:184–192
9. Beinert HR, Shaw RW, Hansen RE, Hartzell CR (1980) Studies on the origin of near infrared (800–900 nm) absorption of cytochrome *c* oxidase. Biochim Biophys Acta 591:458–470
10. Hoshi Y, Hazeki O, Tamura M (1993) Oxygen dependence of redox state of copper in cytochrome oxidase in vitro. J Appl Physiol 74:1622–1627
11. Inagaki M, Tamura M (1993) Preparation and optical characteristics of hemoglobin-free isolated perfused rat head in situ. J Biochem (Tokyo) 113:650–657
12. Hazeki O, Tamura M (1988) Quantitative analysis of hemoglobin oxygenation state of rat brain in situ by near-infrared spectrophotometry. J Appl Physiol 64:796–802

13. Matcher SJ, Elwell CE, Cooper CE, et al. (1995) Performance comparison of several published tissue near-infrared spectroscopy algorithms. Anal Biochem 227:54–68
14. Jöbsis FF, Keizer JH, LaManna JC, Rothenthal M (1977) Reflectance spectrophotometry of cytochrome $aa_3$ in vivo. J Appl Physiol 43:858–872
15. Hempel FG, Kariman K, Saltzman HA (1980) Redox transitions in mitochondria of cat cerebral cortex with seizures and hemorrhagic hypotension. Am J Physiol 238:H249–H256
16. Ferrari M, Williams MA, Wilson DA, et al. (1995) Cat brain cytichrome-c oxidase redox changes induced by hypoxia after blood-fluorocarbon exchange transfusion. Am J Physiol 269:H417–H424
17. Sylvia AL, Piantadosi CA (1988) $O_2$ dependece of in vivo brain cytochrome redox responses and energy metabolism in bloodless rats. J Cereb Blood Flow Metab 8:163–172
18. Hoshi Y, Tamura M (1993) Dynamic changes in cerebral oxygenation in chemically induced seizures in rat: study by near-infrared spectrophotometry. Brain Res 603:215–221
19. Cooper CE, Matcher SJ, Wyatt JS, et al. (1994) Near-infrared spectroscopy of the brain: relevance to cytochrome oxidase bioenergetics. Biochim Soc Trans 22:974–980
20. Tsuji M, Naruse H, Volpe J, Holtzman D (1995) Reduction of cytochrome $aa_3$ measured by near-infred spectroscopy predicts cerebral energy loss in hypoxic piglets. Pediatr Res 37:253–259
21. Edwards AD, Brown GC, Cope M, et al. (1991) Quantitation of concentration changes in neonatal human cerebral oxidized cytochrome oxidase. J Appl Physiol 71:1907–1911
22. Sugano T, Oshino N, Chance B (1974) Mitochodrial functions under hypoxic conditions. The steady states of cytochrome $c$ reduction and of energy metabolism. Biochim Biophys Acta 347:340–356
23. Tamura M, Hazeki O, Nioka S, et al. (1988) The simultaneous measurements of tissue oxygen concentration and energy state by near-infrared and nuclear magnetic resonance spectroscopy. In: Lahiri S (ed) Chemoreceptors and reflexes in breathing. Oxford, New York, pp 91–93
24. Ginsburg DA, Pasternak EB, Gurvitch AM (1977) Correlation analysis of delta activity generated in cerebral hypoxia. Electroencephalogr Clin Neurophysiol 42:445–455
25. van der Zee P, Cope M, Arridge SR, et al. (1992) Experimentally measured optical pathlengths for the adult head, calf and forearm and the head of the newborn infant as a function of inter optode spacing. Adv Exp Med Biol 16:143–153
26. Obrig H, Hirth C, Döge C, et al. (1995) Assesment of localized changes in hemoglobin and cytochrome-oxidase oxidation during performance of a motor task. J Cereb Blood Flow Metab 15(suppl 1):S78

# Crystallization of Bovine Heart Mitochondrial Cytochrome c Oxidase for X-Ray Diffraction at Atomic Resolution (2.8 Å)

Kyoko Shinzawa-Itoh[1], Rieko Yaono[1], Ryousuke Nakashima[1], Hiroshi Aoyama[2], Eiki Yamashita[2], Takashi Tomizaki[2], Tomitake Tsukihara[2], and Shinya Yoshikawa[1]

*Summary.* Four types of crystals were obtained from bovine heart cytochrome c oxidase (ferrocytochrome c: oxygen oxidoreductase, EC 1.9.3.1.), a large multicomponent membrane protein. Three of these crystals (hexagonal bipyramidal, tetragonal plate, and tetragonal column) were obtained from an enzyme preparation stabilized with alkyl polyethelene glycol monoether-type detergents. The tetragonal column crystals diffracted X-rays up to 5 Å resolution, but this is far lower than the atomic resolution. The orthorhombic crystals have been obtained from an enzyme preparation stabilized with decyl β-D-maltoside, which diffracted X-rays up to 2.6 Å resolution. Crystals sufficient for X-ray diffraction experiments had not been obtained from enzyme preparations using any other alkyl-sugar-type detergent commercially available. These results suggest that crystallization of many membrane proteins to achieve the atomic resolution level is possible if a detergent of appropriate structure is available.

*Key words.* Cytochrome c oxidase—Crystallization—Membrane protein—Nonionic detergent

## Introduction

Cytochrome c oxidase is the terminal enzyme of mitochondrial respiration that reduces molecular oxygen to water, coupled with proton pumping across the mitochondrial inner membrane. The physiological importance of this oxidase has stimulated many investigations on its structure and function [1,2]. For elucidation of the reaction mechanism, the three-dimensional structure at atomic resolution is indispensable. However, a method for crystallization of large membrane proteins has not yet been generally established [3].

We have crystallized this enzyme, purified from beef heart muscle, as the first example of membrane protein crystals of higher animals that diffract X-rays [4].

---

[1] Department of Life Science, Faculty of Science, Himeji Institute of Technology, 1479-1, Kanaji, Kamigori-cho, Akoh-gun, Hyogo 678-12, Japan
[2] Institute for Protein Research, Osaka University, 3-2 Yamada-oka, Suita, Osaka 565, Japan

However, resolution of the X-ray diffraction (8 Å) was far lower than the atomic resolution. Thus, we have been trying to optimize the crystallization conditions [5–7]. Through this research, the structure of detergents to stabilize the membrane protein molecules homogeneously in aqueous solution has turned out to be one of the most critical factors for crystallization of the membrane protein. The highest resolution obtained from the tetragonal crystals containing an alkyl polyethyleneglycol monoether was 5.0 Å [8]. From the enzyme stabilized with decyl β-D-maltoside, however, orthorhombic crystals that gave a resolution adequate for structural analysis were obtained [9].

## Materials and Methods

Alkyl oxyethelene-type detergents were obtained from Nikko Chemicals (Tokyo, Japan), and decyl β-D-maltoside was supplied by Sigma (St Louis, MO, USA) or Anatrace (Maumee, OH, USA). Cytochrome *c* oxidase was prepared with the method previously described [10] with slight modifications. Various nonionic detergents were used instead of Brij-35 as in the previous method [4]. Ammonium sulfate concentrations for the fractionations in the presence of each replaced detergent were adjusted by detergent replacement. All the reagents used were of the highest grade commercially available.

## Results and Discussion

Tetragonal plate crystals appeared readily from the enzyme using alkyloxyethelene-type detergents when the ionic strength of the medium was lower (30 mM sodium phosphate buffer pH 7.4 or less) and the concentration of detergent or protein was higher. These crystals appeared without any amorphous protein precipitate in the salting-in region. This crystallization significantly improved the purity of the enzyme preparation. Thus, the crystalline sample was used for surveying the crystallization conditions. Tetragonal plate crystals diffracted X-rays at rather low resolution (>15–20 Å).

Hexagonal bipyramidal crystals were obtained only from enzyme preparations using Brij-35 (polydispersed) as the detergent at low ionic strength (1–3 mM sodium phosphate buffer, pH 7.4). Hexagonal crystals diffracted X-rays up to 6.0 Å resolution. The space groups and cell dimensions are $P6_2$ or $P6_4$ and $a = b = 208.7$ Å, $c = 282.3$ Å.

On the other hand, in the presence of sodium phosphate buffer pH 7.4 at 50 mM or higher, tetragonal column crystals were obtained using ammonium sulfate as the precipitant and hexamethylene glycol as an additive. Tetragonal column crystals diffracted X-rays up to 5.0 Å resolution. The space group and the cell dimension of the crystal are $I4_1$ or $I4_3$ and $a = b = 253$ Å, $c = 507$ Å.

The effects of the length of alkyl chain and ethyleneglycol unit on the crystal conditions are summarized in Table 1. The tetragonal plate crystals were obtained from the preparations stabilized with decyl and dodecyl polyethylene glycol monoether; the ethylene glycol unit was 6–23. The requirement of detergent structure for these crystals were quite low, and the resolution of X-ray diffraction was also quite low (>15 Å). The hexagonal bipyramidal crystals were obtained only using Brij-35, which

TABLE 1. Crystallization of bovine heart cytochrome $c$ oxidase preparations each stabilized with a different nonionic detergent

| Detergent | Crystal system (crystal habit) | | | CMC (mM) |
|---|---|---|---|---|
| | Hexagonal (bipyramid) | Tetragonal (column) | Tetragonal (plate) | |
| C12E23 | + | − | + | $9.0 \times 10^{-2}$ |
| C12E8 | − | + | + | $7.0 \times 10^{-2}$ |
| C12E7 | − | + | + | $6.9 \times 10^{-2}$ |
| C12E6 | − | − | + | $6.9 \times 10^{-2}$ |
| C12E5 | − | − | − | $6.8 \times 10^{-2}$ |
| C10E8 | − | + | + | $9.0 \times 10^{-1}$ |
| C10E7 | − | + | + | $9.0 \times 10^{-1}$ |

CMC, critical micellar concentration.
The symbols + and − denote crystallizable and noncrystallizable, respectively.

is polydispersed, as the detergent. The resolution of the hexagonal bipiramidal crystal was 6.0 Å. The tetragonal column crystals were obtained from the preparation stabilized with decyl and dodecyl polyethelene glycol monoether; the ethylene glycol units were 7–8. The critical micell concentration of alkyl polyethylene glycol monoether depends on the length of the saturated hydrocarbon tail. A change in alkyl chain length, C-10 to C-12, involving a large difference in critical micellar concentration (CMC), hardly affects the crystallization condition. On the other hand, neither C12E23 nor C12E6 has provided the tetragonal column crystal.

Michel proposed that the compactness of the detergent is the most important factor for crystallization of membrane proteins [11]. However, the hexagonal crystal obtained from the enzyme preparation stabilized with C12E23 was not obtained when any other dodecyl polyethylene glycol monoethers (see Table 1), which have a much shorter ethylene glycol chain than C12E23, were used. The result is consistent with the finding that larger detergents were more effective for crystallization of large membrane protein complexes. On the other hand, neither C12E23 nor C12E6 has provided the tetragonal column crystal that was obtained when C12E7 or C12E8 was used. Thus, all these results indicate that an optimum size of detergent is critical for crystallization of large membrane proteins in addition to the compactness.

The orthorhombic crystals (the space group and cell dimensions are $P2_12_12_1$ and $a = 189.1$ Å, $b = 210.5$ Å, $c = 178.6$ Å) were obtained from the preparation stabilized with decyl β-D-maltoside and diffracted X-rays up to 2.6 Å resolution. The crystallization condition for this orthorhombic crystal is extremely sensitive to the structure of the detergent molecules attached to the protein. When dodecyl β-D-maltoside or sucrose monocaprate were used instead of decyl maltoside, the crystal exhibited the resolution of the X-ray diffraction, far lower (~10 Å) than the atomic resolution.

The flexible polyethylene glycol chain gives significantly low specificity in the size of the alkyl polyethylene glycol monoether for crystallization. The flexibility of the polyethylene glycol chain of alkylpolyethylene glycol monoether attached to the enzyme in the crystals has been shown with solid nuclear magnetic resonance (NMR) technique [12]. On the other hand, the rigid maltoside portion of the decyl maltoside seems responsible for the strict specificity of the detergent structure for the crystallization conditions and thus also for the high resolution of the crystals, giving the atomic resolution of the X-ray diffraction.

Our current results show how critical is the selection of detergent for crystallization of membrane proteins, and that crystallization of many membrane proteins to achieve the atomic resolution level is possible if a detergent of appropriate structure is available.

# References

1. Malmstrom BG (1990) Cytochrome *c* oxidase as a redox-linked proton pump. Chem Rev 90:1247–1260
2. Babcock GT, Wikstrom M (1992) Oxygen activation and the conservation of energy in cell respiration. Nature 356:301–309
3. Garavito RM, Picot D (1990) The art of crystallizing membrane proteins. Methods (Orlando) 1:57–69
4. Yoshikawa S, Tera S, Takahashi Y, et al. (1988) Crystalline cytochrome *c* oxidase of bovine heart mitochondrial membrane: composition and x-ray diffraction studies. Proc Natl Acad Sci USA 85:1354–1358
5. Yoshikawa S, Shinzawa K, Tsukihara T, et al. (1991) Crystallization of beef heart cytochrome *c* oxidase. J Crystal Growth 110:247–251
6. Yoshikawa S, Shinzawa-Itoh K, Ueda H, et al. (1992) Strategies for crystallization of large membrane protein complexes. J Crystal Growth 122:298–302
7. Shinzawa-Itoh K, Yamashita H, Yoshikawa S, et al. (1992) Single crystals of bovine heart cytochrome *c* oxidase at fully oxidized resting, fully reduced and CO-bound fully reduced states are isomorphous with each other. J Mol Biol 226:987–990
8. Shinzawa-Itoh K, Ueda H, Yoshikawa S, et al. (1995) Effects of ethyleneglycol chain length of dodecyl polyethyleneglycol monoether on the crystallization of bovine heart cytochrome *c* oxidase. J Mol Biol 246:572–575
9. Tsukihara T, Aoyama H, Yamashita E, et al. (1995) Structure of metal sites of oxidized bovine heart cytochrome *c* oxidase at 2.8 Å. Science 268:1069–1074
10. Yoshikawa S, Choc MG, O'Toole MC, Caughey WS (1977) An infrared study of CO binding to heart cytochrome c oxidase and hemoglobin A. J Biol Chem 252:5498–5508
11. Michel H (1983) Crystallization of membrane proteins. Trends Biochem Sci 8:56–59
12. Tuji S, Shinzawa-Itoh K, Erata T, et al. (1992) A high resolution solid-state $^{13}$C-NMR study on crystalline bovine heart cytochrome *c* oxidase and lysozyme. Eur J Biochem 208:713–720

# Mechanism of Dioxygen Reduction by Cytochrome c Oxidase as Studied by Time-Resolved Resonance Raman Spectroscopy

Takashi Ogura[1], Denis A. Proshlyakov[1], Jörg Matysik[1], Evan H. Appelman[2], Kyoko Shinzawa-Itoh[3], Shinya Yoshikawa[3], and Teizo Kitagawa[1]

*Summary.* Time-resolved resonance Raman (TR$^3$) spectroscopy has been applied to cytochrome c oxidase (CcO) to elucidate the mechanism of dioxygen reduction. Six oxygen isotope-sensitive Raman bands have been identified in the TR$^3$ spectra. The "607-nm species" defined by difference absorption spectrum, which is referenced against the oxidized enzyme, is demonstrated to have an Fe=O heme, although it has long been believed to have an Fe–O–O–X (X=H or Cu$_B$) heme. The one-electron reduction of this Fe=O intermediate, which yields the oxoferryl intermediate, is demonstrated to be coupled with proton transfer in the protein. The mechanism of dioxygen reduction by CcO is discussed on the basis of the structures of the reaction intermediates.

*Key words.* Cytochrome c oxidase—Dioxygen reduction—Resonance Raman—Peroxy intermediate—Bioenergetics

## Introduction

Bovine heart cytochrome c oxidase (CcO) is the terminal enzyme in the mitochondrial respiratory chain and is a 13-subunit membrane protein with a molecular weight of 200 kDa. CcO catalyzes dioxygen reduction to water, and this electron-transfer reaction is coupled with vectorial proton translocation across the membrane. The electrochemical gradient thus produced is ultimately utilized to synthesize adenosine triphosphate (ATP) from adenosine diphosphate (ADP) and inorganic phosphate (Pi) [1]. In the human body, more than 90% of dioxygen taken up by the lungs is reduced by CcO. During four-electron reduction of dioxygen, one-, two-, and three-electron-reduced oxygen species are involved. These intermediately reduced species of dioxygen are highly reactive and thus toxic to living cells, and must be trapped at the

---

[1] Institute for Molecular Science, Okazaki National Research Institutes, 38 Nishigounaka, Myodaiji, Okazaki, Aichi 444, Japan
[2] Chemistry Division, Argonne National Laboratory, Argonne, IL 60439, USA
[3] Department of Life Science, Faculty of Science, Himeji Institute of Technology, 1479-1, Kanaji, Kamigori-cho, Akoh-gun, Hyogo 678-12, Japan

catalytic site (heme $a_3$-Cu$_B$) during turnover. Actually, those activated oxygen species have rarely been detected in the cells during active respiration.

Our purpose is to elucidate how activated oxygen species are trapped at the heme $a_3$-Cu$_B$ site. To determine the structure of activated oxygen species at the catalytic site, we have adopted time-resolved resonance Raman (TR$^3$) spectroscopy because this technique gives us both structural and kinetic information with regard to the heme and its vicinity. In this study, we have identified six oxygen isotope-sensitive bands for reaction intermediates in a time-resolved fashion with which we could determine the structure of the intermediates.

## Materials and Methods

CcO is isolated from bovine heart according to the method described elsewhere [2]. TR$^3$ spectra are obtained by using a special device developed in this laboratory [3,4]. Simultaneous measurements of Raman and absorption spectra of the same sample volume are performed with a device also developed in this laboratory [5,6]. The reaction of CcO with dioxygen is initiated by photolyzing CO from carbonmon-oxy CcO in the presence of dioxygen by a 590-nm illumination that falls on the absorption maximum of the $a_3$-CO heme. Excitation wavelengths employed to excite resonance Raman scattering are 406.7, 416.0, 420.0, 423.0, 425.0, 430.0, 441.6, 580.0, and 607.0 nm.

## Results

At a delay time ($\Delta t$) of 0.1 ms after initiation of the reaction of CcO with dioxygen, Raman band pairs at 571/544 cm$^{-1}$ and 435/415 cm$^{-1}$ for the $^{16}O_2/^{18}O_2$ pair are observed. The $^{16}O^{18}O$ experiments give two bands at 567 and 548 cm$^{-1}$, which establish that the band at 571/544 cm$^{-1}$ for the $^{16}O_2/^{18}O_2$ pair results from the Fe–O$_2$ stretching vibration of an end-on dioxygen complex, and the Fe–O–O bond angle is estimated to be about 120°. The 435/415 cm$^{-1}$ pair is assigned to the Fe–O–O bending vibration of the same molecular species.

At $\Delta t = 0.1$–3 ms, four oxygen isotope-sensitive Raman bands are observed at 804/764, 356/342, 785/750, and 450/425 cm$^{-1}$ for the $^{16}O_2/^{18}O_2$ pair in this order at 3°C. The frequency of the band at 450/425 cm$^{-1}$ shows a downshift to 443/417 cm$^{-1}$ in D$_2$O, while those of the other three bands do not. The band at 450/425 cm$^{-1}$ is accordingly assigned to the Fe$^{3+}$-OH$^-$ stretching vibration. This band loses its intensity at $\Delta t = 5.4$ ms, which is interpreted as a result of the exchange of the OH$^-$ group with bulk water. The $^{16}O^{18}O$ experiments reveal that all the frequencies of bands at 804/764, 356/342, and 785/750 cm$^{-1}$ are sensitive to the mass of only one oxygen atom of dioxygen. These results establish that the species giving the bands at 804/764, 356/342, and 785/750 cm$^{-1}$ have iron-oxo heme. The band at 785/750 cm$^{-1}$ is reasonably assigned to the Fe=O stretching vibration of an oxo-ferryl intermediate (Fe$^{4+}$=O).

The band at 804/764 cm$^{-1}$ is observed before the band at 785/750 cm$^{-1}$. The band at 804/764 cm$^{-1}$ is resonance enhanced on excitations at 441.6 and 607.0 nm in addition to at 423.0 and 430.0 nm, while the band at 785/750 cm$^{-1}$ is enhanced at 580.0 nm as well as at 423.0 and 430.0 nm. This difference in excitation profile implies that the electronic structures and thus the oxidation states of these two species are different. If the reaction is initiated with the mixed-valence (two-electron reduced) and CO-bound form as the starting material, we see the band at $804/764\,\mathrm{cm}^{-1}$ but not at $785/750\,\mathrm{cm}^{-1}$. These observations lead us to conclude that, although both species have the Fe=O heme, the species that gives the $804/764\,\mathrm{cm}^{-1}$ band has one oxidative equivalent higher than the species that gives the $785/750\,\mathrm{cm}^{-1}$ band. In other words, the $804/764\,\mathrm{cm}^{-1}$ and $785/750\,\mathrm{cm}^{-1}$ species belong to the peroxy and ferryl oxidation levels, respectively. The time profile of the band at $356/342\,\mathrm{cm}^{-1}$ seems to coincide with that at $804/764\,\mathrm{cm}^{-1}$. The conversion rate of the $804/764\,\mathrm{cm}^{-1}$ to the $785/750\,\mathrm{cm}^{-1}$ species becomes approximately one-fifth in $D_2O$ compared to that in $H_2O$. This fact suggests that the electron-transfer step to the $804/764\,\mathrm{cm}^{-1}$ species that yields the $785/750\,\mathrm{cm}^{-1}$ species is coupled to proton transfer in the protein.

In the reaction of oxidized CcO with $H_2O_2$, two spectrally distinct forms are observed, which exhibit 607- and 580-nm absorption peaks in the difference spectra obtained by subtracting the spectrum of oxidized CcO (compound minus oxidized CcO). These "607-nm" and "580-nm" species give the Raman bands at 804/769 and $785/750\,\mathrm{cm}^{-1}$ for the $H_2{}^{16}O_2/H_2{}^{18}O_2$ pair on excitation at 607 and 580 nm, respectively. These results suggest that the three intermediate species seen in the dioxygen and hydrogen peroxide reactions are common.

## Discussion

In the four-electron reduction reaction of dioxygen by CcO, existence of the following four reaction intermediates is established by $TR^3$ spectroscopy. The first one is an end-on type oxygenated intermediate with an Fe–O–O bond angle of approximately $120°$, which is characterized by the Fe-O stretching and Fe-O-O bending Raman bands at 571 and $435\,\mathrm{cm}^{-1}$, respectively. The Fe-O-O geometry of this intermediate is virtually identical with those of oxyhemoglobin and oxymyoglobin. The second intermediate has an Fe=O heme, which is characterized by the Fe=O stretching Raman band at $804\,\mathrm{cm}^{-1}$ and has one oxidative equivalent higher than the $Fe^{4+}$=O intermediate does, although the second intermediate belongs to the peroxy oxidation level. Accordingly, we propose to call this intermediate the "perferryl" intermediate instead of the "peroxy" intermediate. The most likely location of the oxidative equivalent is iron; that is, this intermediate has an $Fe^{5+}$=O heme [7]. The so-called 607-nm absorption intermediate observed in the hydrogen peroxide reaction is demonstrated to be identical with the perferryl intermediate. The third intermediate, characterized by the Fe=O stretching Raman band at $785\,\mathrm{cm}^{-1}$, has an $Fe^{4+}$=O heme. Judging from the vibrational frequency and the amplitude of the shift on $^{18}O_2$ substitution, this intermediate resembles compound II of peroxidases [8]. The fourth intermediate, characterized by the Fe-O stretching Raman band at $450\,\mathrm{cm}^{-1}$, has an $Fe^{3+}$-$OH^-$ heme. The $OH^-$ group is exchangeable with bulk water. The Fe-O stretching frequency is notably lower than those of aquamethemoglobin and aquametmyoglobin, and this difference must be caused by a significant interaction of the $OH^-$ group with some distal residue in the case of CcO. The species that gives the $356$-$\mathrm{cm}^{-1}$ band seems to appear coincidently with the perferryl intermediate in the dioxygen reaction but with ferryl intermediate in the hydrogen peroxide reaction. Thus, it is considered to be located between the perferryl and ferryl intermediates. This fact is tentatively interpreted to indicate that the 356-cm$^{-1}$ band results from the His-Fe=O bending vibration of a deformed Fe=O heme [7].

The reaction of the one-electron reduction of the perferryl intermediate to the ferryl intermediate becomes significantly slower in $D_2O$ than in $H_2O$. This fact strongly suggests that this electron-transfer step is coupled with proton transfer in the protein. In conclusion, this study establishes that the "607-nm" species has an Fe=O heme and demonstrates an electron-transfer step which is coupled with proton transfer in the protein.

# References

1. Wikström M, Krab K, Saraste M (1981) Cytochrome oxidase; a synthesis. Academic Press, New York
2. Yoshikawa S, Choc MG, O'Toole MC, Caughey WS (1977) An infrared study of CO binding to heart cytochrome *c* oxidase and hemoglobin A. J Biol Chem 252:5498–5508
3. Ogura T, Yoshikawa S, Kitagawa T (1989) Raman/absorption simultaneous measurements for cytochrome oxidase compound A at room temperature with a novel flow apparatus. Biochemistry 28:8022–8027
4. Ogura T, Takahashi S, Hirota S, Shinzawa-Itoh K, Yoshikawa S, Appelman EH, Kitagawa T (1993) Time-resolved resonance Raman elucidation of the pathway for dioxygen reduction by cytochrome *c* oxidase. J Am Chem Soc 115:8527–8536
5. Ogura T, Kitagawa T (1988) Novel optical device for simultaneous measurements of Raman and absorption spectra: application to photolabile reaction intermediates of hemoproteins. Rev Sci Instrum 59:1316–1320
6. Proshlyakov DA, Ogura T, Shinzawa-Itoh K, Yoshikawa S, Kitagawa T (1996) Microcirculating system for simultaneous determination of Raman and absorption spectra of enzymatic reaction intermediates and its application to the reaction of cytochrome *c* oxidase with hydrogen peroxide. Biochemistry 35:76–82
7. Kitagawa T, Ogura T (1997) Oxygen activation mechanism at the binuclear site of heme-copper oxidase superfamily as revealed by time-resolved resonance Raman spectroscopy. Prog Inorg Chem 45:431–479
8. Kitagawa T, Mizutani Y (1994) Resonance Raman spectra of highly oxidized metalloporphyrins and heme proteins. Coord Chem Rev 135/136:685–735

# Coupling of Proton Transfer to Oxygen Chemistry in Cytochrome Oxidase: The Roles of Residues I67 and E243

Brigitte Meunier and Peter R. Rich

*Summary.* We are studying mutant forms of cytochrome oxidase to investigate residues of importance for proton movement and for sites of redox-linked protonation. In this chapter, we describe the effects of a mutation in the residue I67. The mutation alters the redox properties of heme $a$, probably by perturbing the p$K$ of E243, a conserved residue that we propose to be a protonation site that is redox-linked to heme $a$. This mutation has little effect on the other redox centers or on the ligand reactions of the binuclear center. The effects are compared with those of the mutation K362M, a change that has no effect on the redox properties of heme $a$ but instead alters the reducibility of the binuclear center, probably by preventing protonations that are required for more than one charge to be accumulated on the heme $a_3$/Cu$_B$ system.

*Key words.* Cytochrome oxidase—Protonation sites—Coupling

## Introduction

We have proposed a model for proton transfer in cytochrome oxidase, based on charge-balancing requirements (see the chapter by P.R. Rich et al, this volume), and have highlighted a conserved glutamate residue (E243 in yeast or E242 in the bovine enzyme) and its surroundings as the most likely region for the charge-balancing protonation changes that are central to the coupling process. The model is being investigated by examination of mutant enzymes from various sources. We present here the preliminary characterisation of a mutation I67N in subunit I of the yeast enzyme. The residue I67 (166 in the bovine enzyme) is close both to heme $a$ and to the conserved glutamate E243 (see Fig. 1 in the chapter by P.R. Rich et al, this volume). The effects of the I67N mutation are compared with those of the mutation K362M in the *Rhodobacter sphaeroides* enzyme (equivalent to K319 in the yeast and bovine enzymes), a residue located in a possible channel that has been proposed to provide a route for the "substrate" protons required for oxygen reduction to water.

---

Department of Biology, University College London, Gower Street, London WC1E 6BT, UK

# Materials and Methods

## Mutant Forms of the Enzymes

The mutant I67N has been produced in yeast [1]. The yeast cytochrome $c$ oxidase has been prepared by Dr. C. Ortwein, Frankfurt, as described in [2]. His-tagged forms of the wild-type and K362M mutant forms of cytochrome oxidase from *R. sphaeroides* were generously provided by Prof. R.B. Gennis.

## Spectrophotometric and Kinetic Measurements

Spectra and transient kinetics at individual wavelengths were monitored with a single-beam scanning/kinetic instrument built in house.

## Kinetics of Reaction of Ferrocytochrome c with Fully Oxidized Cytochrome c Oxidase Monitored by the Flash-Induced Chemical Photoreduction (FIRE) Method

Purified enzyme was dissolved aerobically to about $0.5\mu M$ in $0.5\,ml$ of $50\,mM$ 2(n-morpholino)ethane sulfonic acid (MES), 0.05% lauryl maltoside, $200\,U/ml$ catalase, $50\,U/ml$ superoxide dismutase, $8\mu M$ oxidized cytochrome $c$, and $150\mu M$ 5-methyl phenazinium methosulphate (PMS), at pH 6. Photoreduction of cytochrome $c$ was initiated with a xenon flash filtered with short-pass filters to produce a blue flash, which is actinic for the PMS photoreduction system. In these conditions, the ferrocytochrome $c$ generated per flash was about $0.28\mu M$. The rapid reduction of ferricytochrome $c$ by reduced PMS was followed by a reaction between ferrocytochrome $c$ and oxidase. The same sample could be used repeatedly at many individual wavelengths by allowing a dark adaptation of at least $20\,s$ between flashes, during which time the system reverted to the fully oxidized state by reaction with oxygen. The redox kinetics of cytochrome $c$ and cytochrome $c$ oxidase were monitored at $550 - (544 + 556)/2\,nm$ ($\varepsilon = 14.7\,mM^{-1}cm^{-1}$) and $603 - (593 + 613)/2\,nm$ ($\varepsilon = 14\,mM^{-1}cm^{-1}$), respectively.

## Steady-State Turnover Behavior of Hemes on Slow Time Scales

Enzymes were dissolved to about $0.7\mu M$ in $0.5\,ml$ $50\,mM$ potassium phosphate, 0.1% lauryl maltoside, and $0.1\mu M$ cytochrome $c$ at pH 7, with a positive pressure of argon above the liquid surface. The reaction was initiated by addition of $10\,mM$ ascorbate and $0.4\mu M$ PMS. Steady-state redox changes of the hemes were monitored at $445-435\,nm$, a wavelength pair that is isosbestic for oxygen intermediates.

# Results

## Effect of the Mutations on the Catalytic Activity

Oxygen consumption activities of the wild-type and mutant enzymes were monitored with $80\mu M$ cytochrome $c$. At pH 7, the turnover numbers of both the yeast and *R. sphaeroides* wild-type enzymes were approximately 700 electrons $s^{-1}$. At this pH, the activities of both the I67N and K362M mutants were only a few percent of the activity of the wild-type enzymes.

## Major Effects of the I67N Mutation

The kinetics of reaction of ferrocytochrome $c$ with the oxidized forms of the wild-type and I67N mutant forms of yeast oxidase were monitored by the FIRE method [3]. Less than one electron per flash per oxidase was generated, so that mainly the one-electron-reduced E state was generated.

As shown in Table 1, the number of electrons donated to both enzymes was roughly equivalent. In the wild-type enzyme, the electron distribution between heme $a$ and the binuclear center was consistent with their relative midpoint potentials. However, the level of reduction of heme $a$ was significantly lower in the I67N mutant, indicating a lower midpoint potential relative to the binuclear center. A lower midpoint potential of heme $a$ was demonstrated by direct redox titration of heme $a$ in the cyanide-ligated form of the enzyme [4].

The FIRE data also showed that electron donation to the binuclear center could still occur (Fig. 1). Furthermore, the reactions of the binuclear center with ligands and oxygen was unchanged, and the steady-state spectrum in the presence of reduced cytochrome $c$ indicated a relatively normal mixture of oxygen intermediates (data not shown).

## Major Effects of the K362M Mutation

Figure 2 shows the optical changes in the Soret region during a series of substrate pulses given to the wild-type and K362M forms of the *R. sphaeroides* enzyme. The reaction was initiated by addition of PMS, ascorbate, and cytochrome $c$ (1). In the wild type, a steady state composed of a mixture of oxygen intermediates was rapidly attained (panel b). After 300 s, the mixture became anaerobic and both hemes became reduced. Oxygen was rapidly mixed into the sample (2). This caused a reoxidation back to the steady-state level, which was followed by heme rereduction. In (3), dithionite was added to the sample. The behavior of the K362M was dramatically different. Addition of PMS, ascorbate, and cytochrome $c$ caused a rapid reduction of heme $a$. The signal observed was about half the signal of the fully reduced wild-type enzyme, but no further reduction occurred even after 2000 s or after addition of dithionite (3).

TABLE 1. Electron distribution in one-electron reduction of cytochrome $c$ oxidase using the flash-induced chemical photo reduction (FIRE) method

| Electron (μM):<br>total of electrons produced per flash | Wild type: 0.28 (100%) | I67N: 0.275 (100%) |
|---|---|---|
| Electrons remaining on cytochrome $c$ | 0.05 (18%) | 0.067 (24%) |
| Electrons remaining on heme $a$ | 0.071 (25%) | 0.027 (10%) |
| Electrons transferred to the binuclear center | 0.159 (57%) | 0.181 (66%) |

The estimation of electrons on the redox centers has been done, as described in Fig. 1, 0.14 s after the flash, which corresponded to the peak of heme $a$ reduction.

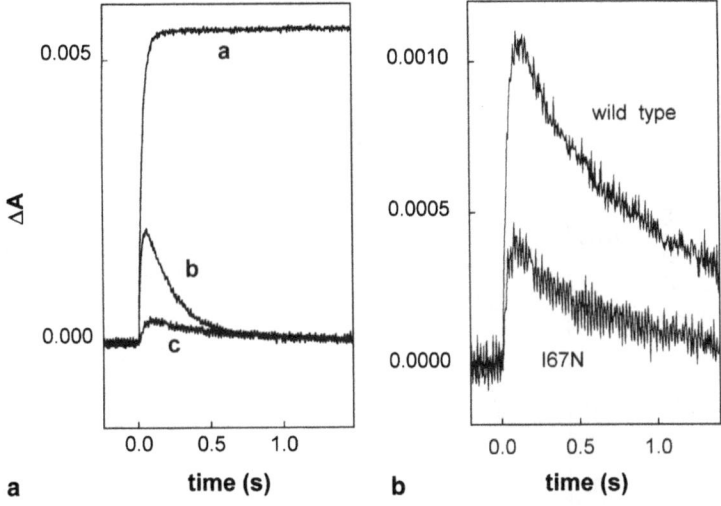

FIG. 1a,b. One-electron reduction of cytochrome $c$ oxidase by ferrocytochrome $c$, using the flash-induced chemical photo reduction (FIRE) method: comparison between I67N and the wild-type yeast enzymes. **a** Kinetics traces obtained with the wild-type enzyme (see Materials and Methods). The yield of ferrocytochrome $c$ per flash was first estimated by monitoring the reduction of cytochrome $c$ (*trace a*). After this measurement, cytochrome $c$ oxidase was added to the same sample. The photoreduction was repeated, and the reductions of cytochrome $c$ (*trace b*) and of cytochrome oxidase (*trace c*) were monitored. Ferrocytochrome $c$ rapidly reduces oxidase, and initially a mixture of ferro/ferricytochrome $c$ and singly reduced oxidase is formed. This is followed by a redistribution of electrons between oxidases, ultimately to be reoxidized by oxygen. Thus, cytochrome $c$ and the oxidaze return to the fully oxidized state in the dark and can be photoactivated again. From the difference between the cytochrome $c$ reduction level in the absence of oxidase (*trace a*) and its reduction level in presence of oxidase (*trace b*), we can determine how many electrons have entered the oxidase at any given time and, from *trace c*, estimate how many of them have appeared on heme $a$. The difference between the total number of electrons donated by cytochrome $c$ and the number of electrons appearing on heme $a$ gives the number that have been transferred to the binuclear center. The data obtained for wild-type and I67N enzymes are presented in Table 1. **b** shows the level of reduction of heme $a$ of the wild type and I67N; the wild-type trace was an expansion of trace c in **a**

## Discussion

Our data show that the major effect of the mutation I67N in yeast is the lowering of the midpoint potential of heme $a$. Heme $a$ therefore becomes more difficult to reduce, leading to a decrease in the catalytic activity of the enzyme. In the published structure of the beef heart enzyme, I67 (I66 in beef heart) is located between heme $a$ and the conserved glutamate residue E243 (E242 in beef heart) [5]. E243 is likely to be one of the groups in which the charge-balancing protonation changes occur, associated with redox changes. From our preliminary results, we propose that substitution of an asparagine for I67 modulates the E243, interfering with its redox-linked protonation and in turn lowering the midpoint potential of heme $a$.

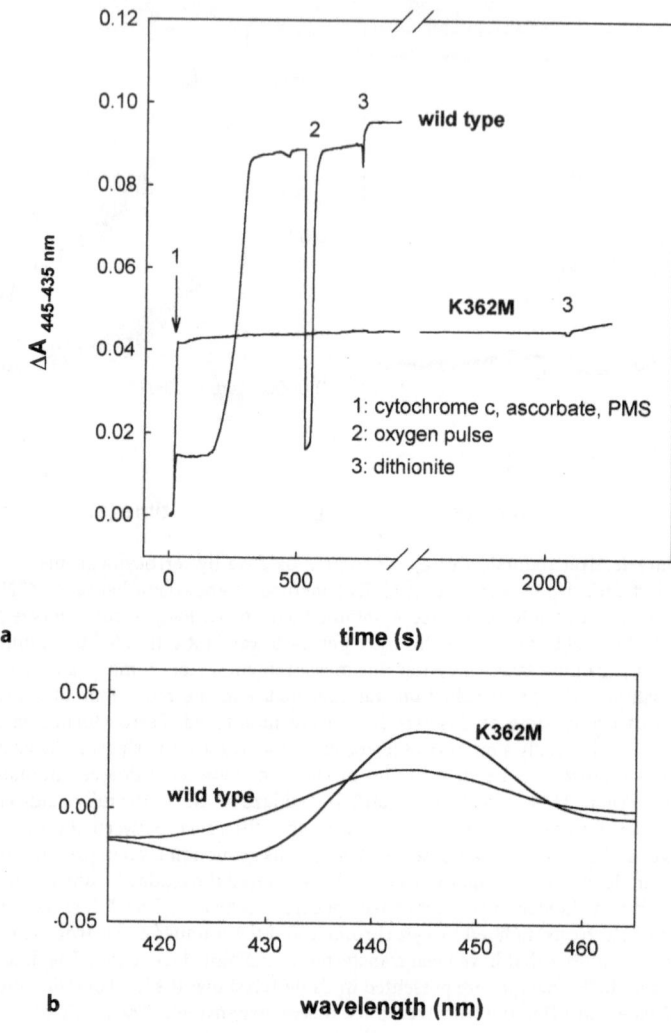

FIG. 2a,b. Steady-state turnover behavior of K362M and wild-type *Rhodobacter sphaeroides* enzymes. The measurements were performed as described in Materials and Methods. **a** Steady-state redox changes of the hemes monitored at 445–435 nm, a wavelength pair isosbestic for oxygen intermediates. (*1*) The reaction was initiated by 10 mM ascorbate, 0.2 μM cytochrome *c*, and 0.4 μM PMS. (*2*) Oxygen was mixed into the sample. (*3*) Dithionite was added to the sample. **b** Aerobic steady-state difference spectra taken after the addition of ascorbate, cytochrome *c*, and 5-methyl phenazinium methosulphate (PMS)

In contrast, the redox properties and reducibility of heme *a* were not affected in the K362M mutant. Instead, further electron transfer to the binuclear center was perturbed so that generation of the oxygen intermediates was no longer possible. Initial experiments (not shown) suggest that one reducing equivalent may be transferred, but the transfer of two is not possible. Thus, the enzyme is unable to function catalytically because it is unable to produce the intermediate that can react with oxygen to produce the peroxy form.

*Acknowledgments.* This work was supported by an EPSRC (GR/J28148) award to P.R. and an EC Fellowship award to B.M. (BIO2-CT-94-8197). Parts of this work are being undertaken collaboratively with Prof. R. Gennis (Urbana, IL, USA) and Prof. U. Brandt and Dr. C. Ortwein (Frankfurt, Germany).

# References

1. Meunier B, Colson A-M (1994) Random deficiency mutations and reversions in the cytochrome *c* oxidase subunits I, II and III of *Saccharomyces cerevisiae*. Biochim Biophys Acta 1187:112–115
2. Geier BM, Schägger H, Ortwein C, et al. (1995) Kinetic properties and ligand binding of the eleven subunit cytochrome *c* oxidase from *Saccharomyces cerevisiae* isolated with a novel large scale purification method. Eur J Biochem 227:296–302
3. Moody AJ, Brandt U, Rich PR (1991) Single electron reduction of "slow" and "fast" cytochrome *c* oxidase. FEBS Lett 293:101–105
4. Ortwein C, Link TA, Meunier B, et al. (1997) Structural and functional analysis of deficient mutants in subunit I of cytochrome *c* oxidase from *Saccharomyces cerevisiae*. Biochim Biophys Acta (in press)
5. Tsukihara T, Aoyama H, Yamashita E, et al. (1996) The whole structure of the 13-subunit oxidized cytochrome *c* oxidase at 2.8 Å. Science 272:1136–1144

# Respiration of *Helicobacter pylori*, *cb*-Type Cytochrome *c* Oxidase, and Inhibition of NADH Oxidation by $O_2$

NOBUHITO SONE[1], SAKURA TSUKITA[1], KUMIKO NAGATA[2], and TOSHIHIDE TAMURA[2]

*Summary.* *Helicobacter pylori* is a microaerophilic gram-negative spiral bacterium residing in human stomachs. A *cb*-type cytochrome *c* oxidase terminating the respiratory chain was purified almost to homogeneity after solubilizing the membranes with Triton X-100 and succeeding anion exchange, Cu-chelating, and gel-filtrating chromatography. The enzyme was composed of two or three subunits (58, 30, and 23 kDa), probably bearing three C hemes and two protohemes. One protoheme reacted with CO, and seemed to be high spin and to form a binuclear center with $Cu_B$. The enzyme actively oxidized soluble cytochrome *c* from this bacterium with a $K_m$ of $0.9\,\mu M$ ($TN_{max}$ of about 250). Yeast cytochrome *c* and *N,N,N′,N′*-tetramethyl *p*-phenylene diamine (TMPD) also were oxidized at similar maximal velocities with larger $K_m$s. The effect of $O_2$ on inactivation of NADH oxidase activity of membranes during incubation was also examined.

*Key words.* Cytochrome *c* oxidase—Cytochrome *bc*—*Helicobacter pylori*—*c*-type cytochrome—Heme-copper oxidase

## Introduction

*Helicobacter pylori* is a gram-negative spiral bacterium residing in human stomachs. This bacterium is known to be a microaerophilic (5%–7% $O_2$ preferable) but obligate aerobe [1,2]. Oxidase activity measurement of the membrane fraction showed the presence of strong (about $0.3\,\mu mol\,min^{-1}\,mg^{-1}$ protein) cytochrome *c* and *N,N,N′,N′*-tetramethyl *p*-phenylene diamine (TMPD) oxidase activities [3]. Redox and CO-difference spectra indicated that the responsible terminal oxidase is not an $aa_3$-type but a *cb*-type cytochrome *c* oxidase, which showed high $O_2$ affinity (about $0.4\,\mu M$) and was very susceptible to cyanide ($Ki = 2.5\,\mu M$). A *cb*-type cytochrome *c* oxidase was found in bacteroids fixing nitrogen in root nodula as the terminal oxidase with a very low $K_m$ for $O_2$ [4,5]. This type of oxidase has also been found in purple bacteria such as *Rhodobacter capsulata* [6], *Rhodobacter sphaeroides* [7], and *Paracoccus denitrifi-*

[1] Kyushu Institute of Technology, Kawazu 680, Iizuka, Fukuoka 820, Japan
[2] Hyogo College of Medicine, Mukogawa-cho 1-1, Nishinomiya, Hyogo 663, Japan

*cans* [8]. These enzymes have a high-spin heme-$Cu_B$ binuclear center for dioxygen reduction to water in the largest subunit (subunit I), and thus belong to the heme-copper oxidase superfamily with well-known cytochrome $aa_3$-type cytochrome *c* oxidases [9,10]. Multiple alignment of amino acid sequences also indicate that those of *cb*-type enzymes so far as is known are closely homologous with subunit I of cytochrome $aa_3$ [4,10,11]. We report here trials to purify the terminal oxidase from *H. pylori* and to clone the genes for this enzyme. In addition, we also found it interesting to analyze why parts of the respiratory system of this bacterium are susceptible to oxygen.

# Materials and Methods

## Materials

*Helicobacker pylori* NCTC11637 was cultured and its membrane fraction was prepared as described previously [3]. TMPD and *o*-tolidine were purchased from Wako Pure Chemical (Osaka, Japan), and equine cytochrome *c* (type VI) and yeast cytochrome *c* (type VIII) were provided by Sigma (St. Louis, MO, USA). Diethylammoethanyl cellulose (DEAE-cellulose) (DE52) was a product of Whatman (Maidstone, Kent), and Q-sepharose FF, chelating sepharose FF, and a Superdex gel filtration column for high pressure liquid chromatography (HPLC) (2000 HR10/30) were purchased from Pharmacia (Uppsala, Sweden). Other chemicals, inhibitors, and detergents were obtained as described previously [12].

## Purification Procedure

The membrane fraction from *H. pylori* was washed once with 50 mM HEPES hydroxylethylpiperazine ethanesulfonic buffer (pH 7.0) containing 0.5% sodium cholate, 0.5 M NaCl, and 15% (w/v) glycerol by centrifugation. The resulting residues were extracted with 5 ml of a mixture of 3% Triton X-100, 50 mM HEPES buffer, 0.1 M NaCl, and 1 mM phenylmethylsulfonyl fluoride (PMSF) for 1 h with stirring. The soluble fraction was diluted with an equal volume of $H_2O$, and applied on a DEAE-cellulose column (1 × 3 cm), and the pass-through fraction was applied on a Q-sepharose column (0.8 × 4 cm) equilibrated with water. The column was washed with 20 mM NaCl containing 1% Triton X-100, HEPES buffer, and glycerol, then with 40 mM, and then with 50 mM, so that the red band moved slowly. The red band was eluted by raising the NaCl concentration to 80 mM, and the eluate was fractionated and measured for TMPD oxidase activity. The active fractions were pooled and applied on a chelating Sepharose column (0.4 × 2 cm) loaded with Cu(II) and the buffer with 0.5 M NaCl. The column was first washed by the buffer containing 1% sucrose monolaurate, 5 mM imidazol, and 0.5 M NaCl, and the oxidase was eluted by raising the concentration of imidazol to 10 mM. The red-colored eluate (0.4 ml) was fractionated through a sepharose column (1 × 30 cm) equilibrated with 20 mM HEPES buffer (pH 7.0) containing 0.1 M NaCl, 0.2% sucrose monolaurate, and 10% glycerol.

## Analytical Procedures

Oxygen uptake of the membrane fraction was followed with a No. 4005 oxygen electrode (Yellow Springs Instrument, Yellow Springs, OH, USA) in a semiclosed vessel

(2.3 ml) containing the reaction medium of 50 mM sodium phosphate buffer, pH 7.1. The contents of these vessels were all thermostatically controlled (35°C) and mixed with a magnetic stirrer.

Cytochrome $c$ oxidase and TMPD oxidase activities of the purified enzyme were measured by following the pH change with ascorbate as a final electron donor according to the following equation [13]: ascorbate $H \cdot Na + H^+ + 1/2\ O_2 =$ dehydroascorbate $+ H_2O + Na^+$. The net alkali formation was back titrated with aliquots of 5 mM HCl. Absorption spectra were measured on a DU-70 spectrophotometer (Beckmen Instruments, Fullerton, CA, USA). Heme contents were determined according to the method of Berry and Trumpower [14]. Polyacrylamide gel electrophoresis with sodium dodecyl sulfate (SDS-PAGE), protein determination, and heme staining were carried out as previously [12].

## Cloning and Sequencing of the Gene

For cloning of the largest subunit of the *H. pylori* *cb*-type oxidase gene, two sets of primers were designed for polymerase chain reaction (PCR) targeting the very conserved region in the heme-copper oxidase superfamily including *cb*-type: 5′-CA(A/G)TGGTGGTA(T/G)GGNGAT/CAA for QWWYGHN in helix VI as a sense primer, and 5′-ATNGTCCA(A/G)TCNGT(A/C)TA(G/T)AG for HYTDWTI in helix X as an antisense primer. The PCR product with *H. pylori* genomic DNA as a template, almost 400 bp, was cloned into pT7Blue T-vector and sequenced. The adjacent sequences were also obtained by the cassette PCR method. The plasmid vector and enzymes for this work were the products of Takara Shuzo (Kyoto). General gene manipulations followed those of Sambrook et al. [15].

# Results and Discussion

## Purification

The membrane fraction was first washed with cholate to remove peripheral membrane protein and then solubilized with Triton X-100. The media for chromatography contained glycerol (10%–15%) for stabilization of the enzyme. A typical result of

TABLE 1. Summary of the purification of cytochrome $c$ oxidase

| Step | Total protein (mg) | Total activity (μmol/min) | Yield (%) | Specific activity (μmol min$^{-1}$ mg$^{-1}$) |
|---|---|---|---|---|
| 1. Washed membrane | 49.8 | 17.8 | (100) | 0.38 |
| 2. Triton X-100-solubilized | 17.0 | 9.87 | (56) | 0.58 |
| 3. Q-sepharose | 1.12 | 4.75 | (27) | 4.2 |
| 4. Chelating sepharose | 0.56 | 1.10 | (6.2) | 2.0 |
| 5. Gel filtration | 0.10 | 0.67 | (3.4) | 6.7 |

$N,N,N',N'$-tetramethyl *p*-phenylene diamine (TMPD) oxidase activity was measured spectrophotometrically at 562 nm with a single-beam spectrophotometer using 0.1 mM TMPD as described previously [12].

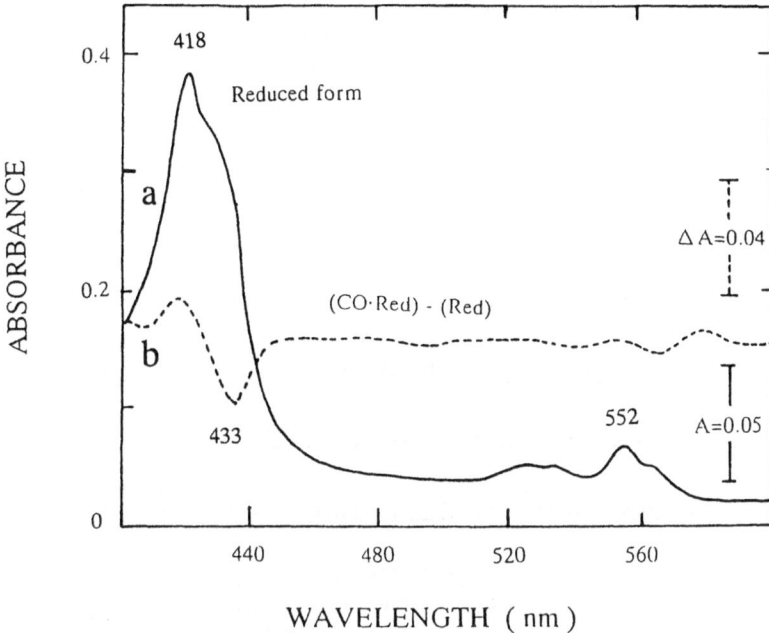

FIG. 1. Spectra of *Helicobacter pylori* cytochrome *c* oxidase. *a*, Na$_2$S$_2$O$_4$-reduced spectrum; *b*, CO-reduced *minus* reduced difference spectrum. Reduction was carried out in sodium ascorbate with a very tiny amount of Na$_2$S$_2$O$_4$. *A*, absorbance

purification is summarized in Table 1. The last step (gel filtration) showed that the enzyme was about 210 kDa with the detergent.

## Chromophores and Subunit Structure

A reduced form spectrum of the purified cytochrome oxidase (Fig. 1a) shows that only *c*-type cytochromes are dominant, and a shoulder around 560 nm indicates the presence of *b*-type cytochromes in the enzyme. A CO difference spectrum (Fig. 1b) suggests that *b*-type cytochrome is reacting with CO as in the case of cytochrome $a_3$. Analysis of heme content by pyridine hemechrome showed that content of C-heme and protoheme were about 18 and 10 nmol/mg protein, respectively. Subunit structure was examined by SDS-PAGE. Two main (58- and 23-kDa) and one faint (30-kDa) bands were stained for protein, and at least 58- and 23-kDa bands showed peroxidase activity from covalently bound heme (not shown). No hitherto known largest subunit with the heme-copper binuclear center was reported to bear heme C as well [4–7].

## Catalytic Properties

Kinetic constants of the purified enzyme with cytochromes *c* from different sources and TMPD are summarized in Table 2. The *H. pylori* enzyme showed high molecular activity with these electron donors except equine cytochrome *c* as in the membrane

TABLE 2. Oxidase activities and inhibition by KCN

| | Substrate | | | |
|---|---|---|---|---|
| | *Helicobacter pylori* cyt. *c*-553 | Yeast cyt. *c* | Equine cyt. *c* | TMPD |
| $V_{max}$ (s$^{-1}$) | 252 | 250 | 72.6 | 247 |
| $K_m$ (μM) | 0.9 | 15.2 | 1.1 | 108 |
| 150 (μM) for KCN | — | — | — | 2.6 |

```
         VI                                                      VII
          *
1. GIQDAMFQAWYGHNAVGFFLTAGFIAIMYYFIPKRAEKPIYSYRLSIIHFWALIFLYIW
2. :V :V    :::::::: :  : :L   ::: ::D V:: :: :V V::     :::
3. :V :  T W::::::: :  : : LGM ::: ::Q R:: ::K:    ::     :::
4. GSNDALIQWWYGHNAVAFVFTSGVIGTIYYFLPKESGQPIFSYKLTLFSFWSLMFVYIW
        **                                    VIII
1. AGPHHLHYTALPDWTQTLGMIFSIMLWMPSWGGMINGLMTLSGAWDKLRTDPVLRMLVV
2. ::::::: :   :::A ::   ::  :  :::: :: : :  :: :    :II  M:
3. ::::::: :   :::AS:: V:: I:     :: : :  :  :  :    :II  M:
4. AGGHHLIYSTVPDWVQTLSSVFSVVLILPSWGTAINMLLTMRGQWHQLKDSPLIKFLVL
   IX                                   * *    X
1. SVAFYGMSTFEGPMMSIKVVNSISHYTDWTIGHVHSGALGWVGFVSFGALYCLVPWAWN
2. A  ::  A: :     V:S::   :::E:G::::: : : AIY   I:  I L :
3. A  ::  A: :     :A::   :::T: ::::: : :N:MIT    Y:  RL GG:
4. ASTFYMLSTLEQSINAIKSVNALAHYTDWTIGHVHDGVLGWVGFTLIASMYHMTPRLFN
```

FIG. 2. Multiple alignment of segment VI–X region of subunit I of *cb*-type oxidases. The hydrophobic segments are indicated by *overlines* and *underlines* and are *numbered*. The conserved His residues liganding metal centers are indicated by *asterisks*. *1, Bradyrhizobium japonicum* [4]; *2, Azorhizobium caulinodans* [16]; *3, Paracoccus denitrificans* [8]; *4, H. pylori* (current study)

[3]. The very low $K_m$ of *H. pylori* cytochrome *c*-552 suggests that this is the physiological substrate. This cytochrome has been partially purified from the soluble fraction of *H. pylori* on disruption by sonic oscillation, and is probably present in the periplasma.

## Gene for Subunit I of the H. pylori Enzyme

Cloning of a main portion of subunit I of *H. pylori* oxidase has been attained as to PCR-amplified DNA. The deduced amino acid sequences are compared with those of *cb*-type terminal oxidases such as that of *Paracoccus denitrificans* [8], *Azorhizobium caulinodans* [16], and *Bradyrhizobium japonicum* [4] (Fig. 2). These three bacteria belong to the alpha-division group of proteobacteria, and the sequences are very similar to each other. On the other hand, *H. pylori* is classified in the gamma division, and its sequence is somewhat different from the others. Membrane-spanning alpha helices VI–X surround the heme-copper binuclear center where dioxygen is to be reduced [17,18]. Five His (*) residues, which are liganding two metal atoms, are all conserved, and residues in their vicinities in helices VI, VII, and X are also well conserved in four *cb*-type oxidases.

The Glu and Tyr in helix VI, which is supposed to be very important for electron and proton transport, are not found in the alignment of *cb*-type oxidase. Also Lys in helix VIII, which is supposed to confer a proton pathway, is not found in the present alignment although two or three hydroxyl residues are conserved. Two His residues in helix X are separated by Val instead of Phe, which is conserved in the heme-copper oxidase superfamily except the *cb*-type, indicating the aromatic side chain is not necessary for electron transfer from low-spin heme to the high-spin heme of the binuclear center. Among the heme-copper oxidase superfamily, the *cb*-type cytochrome *c* oxidases seem to be a primitive group, because (1) they are expressed under microaerophilic conditions or found in microaerophilic bacteria, (2) their smaller subunit is *c*-type cytochrome(s), totally different from the subunits II and III of the rest of the superfamily, and (3) their (subunit I) sequences seem closely related to that of NO reductase, which may be the origin of the oxidase superfamily when dioxygen was available [10,11].

As reported previously, the rate of oxygen uptake of the membrane fraction of *H. pylori* with NADH was very low in comparison with that of cytochrome *c* oxidation

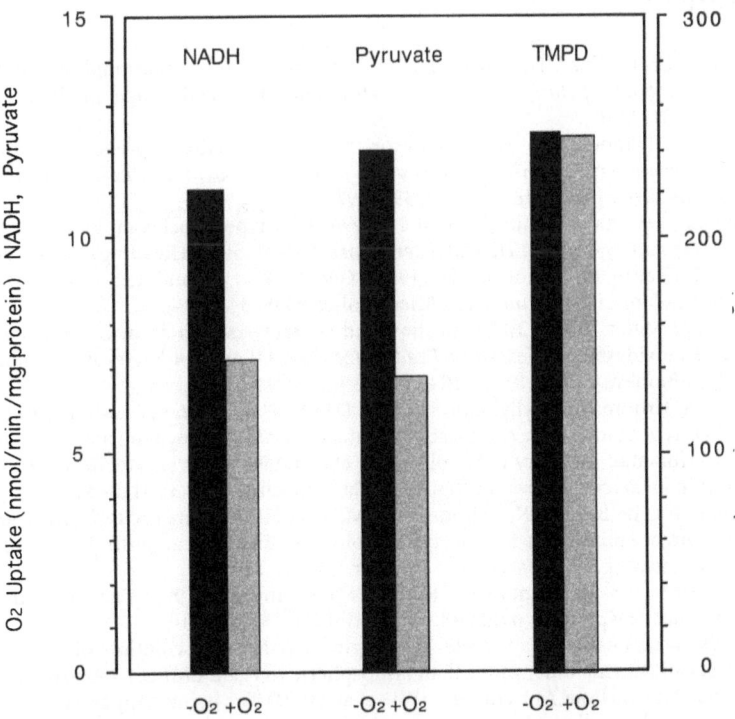

Fɪɢ. 3. Inactivation by oxygen of NADH oxidase and pyruvate oxidase activity during incubation. The *H. pylori* membrane fractions prepared by sonication under an atmosphere of $N_2$ were incubated for 2 h in air $(+O_2)$ or in $N_2$ $(-O_2)$

[3], although they are comparable in most bacterial preparations. We suggested that this may be the result of detachment of the type II NADH dehydrogenase from the membrane during the preparation. In additon, instability of the dehydrogenase under oxygen may be partly responsible, because oxygen uptakes with NADH and pyruvate severely decreased after 2-h exposure under air even when kept on ice (Fig. 3).

## Conclusion

We have succeeded in purifying the terminal oxidase from a very small amount of membrane fraction of *H. pylori*. Our data showed that the terminal oxidase in *H. pylori* is a *cb*-type cytochrome *c* oxidase. *H. pylori* is a strict microaerobe; the most suitable gaseous conditions for its growth are 5%–7% oxygen and 7%–10% carbon dioxide in nitrogen. *H. pylori* lives in the mucous layer of the human stomach, most frequently sited in the "grooves" at the junction of the cells as the microaerophilic niche. This type of the enzyme is known to be essential for nitrogen-fixing bacteroids [4,5]. We would like to suggest that this type of oxidase may serve as the terminal oxidase in many microaerophilic bacteria.

## References

1. Goodwin CS, Collins MD, Blincow E (1986) The absence of the thermoplasmaquinones in *Campylobacter pylori*, and its temperature and pH growth range. FEMS Microbiol Lett 32:1137–1140
2. Moss CW, Lambert-Fair MA, Nicholson MA, et al. (1990) Isoprenoid quinones of *Campylobacter cryaerophila*, *C. cinaedi*, *C. fenneliae*, *C. byointestinalis*, *C. pylori*, and "*C. upsaliensis*". J Clin Microbiol 28:395–397
3. Nagata K, Tsukita S, Tamura T, Sone N (1996) A *cb*-type cytochrome *c* oxidase terminates respiratory chain in *Helicobacter pylori*. Microbiology (Reading) 142:1757–1763
4. Bott M, Preisig O, Hennecke H (1992) Genes for a second terminal oxidase in *Bradyrhizobium japonicum*. Arch Microbiol 158:335–343
5. Keefe RG, Maier RJ (1993) Purification and characterization of an O$_2$-utilizing cytochrome *c*-oxidase complex from *Bradyrhizobium japonicum* bacteroid membranes. Biochim Biophys Acta 1183:91–104
6. Gray KA, Grooms M, Myllykallio H, et al. (1994) *Rhodobacter capsulatus* contains a novel *cb*-type cytochrome *c* oxidase without a CuA center. Biochemistry 33:3120–3127
7. Garcia-Horsman JA, Berry E, Shapleigh JP, et al. (1994) A novel cytochrome *c* oxidase from *Rhodobacter sphaeroides* that lacks CuA. Biochemistry 33:3113–3119
8. de Gier JWL, de Boer APN, Reijnders WNM, et al. (1995) *Paracoccus denitrificans cco* locus. Direct submission to European Molecular Biology Laboratory (EMBL) Data Bank, Heidelberg, under accession number U34353
9. Garcia-Horsman JA, Barquera B, Rumley MJ, Gennis RB (1995) The superfamily of heme-copper respiratory oxidases. J Bacteriol 176:5587–5600
10. Castresana J, Lueben M, Saraste M, Higgins DG (1994) Evolution of cytochrome oxidase, an enzyme older older than atmospheric oxygen. EMBO J 13:2516–2525
11. Van der Ost J, de Boer APN, de Gier JWL, et al. (1994) The heme-copper oxidase family consists of three distinct types of terminal oxidases and is related to nitric oxide reductase. Microbiol Lett 121:1–10
12. Tashiro H, Sone N (1995) Preparation and characterization of the hydrophilic CuA-cytochrome *c* domain of subunit II of cytochrome *c* oxidase from thermophilic bacillus PS3. J Biochem 11:521–526

13. Nicholl P, Sone N (1984) Kinetics of cytochrome *c* and TMPD oxidation by cytochrome *c* oxidase from the thermophilic bacterium PS3. Biochim Biophys Acta 767:240–247
14. Berry EA, Trumpower BL (1987) Simultaneous determination of hemes a, b, and c from pyridine hemechrome spectra. Anal Biochem 161:1–15
15. Sambrook E, Fritsch F, Maniatis T (1989) Molecular cloning: a laboratory manual, 2nd edn. Cold Spring Habor Laboratory Press, Cold Spring Habor, NY
16. Mandon K, Alexandre K, Elmerich C (1994) Functional analysis of the *fixNOPQ* region of *Azorhizobium caulinodans*. J Bacteriol 176:2560–2568
17. Iwata S, Ostermeier C, Ludwig B, Michel H (1995) Structure at 2.8 Å resolution of cytochrome *c* oxidase from *Paracoccus denitrificans*. Nature 376:660–669
18. Tsukihara T, Aoyama H, Yamashita E, et al. (1995) Structure of metal sites of oxidized bovine heart cytochrome *c* oxidase at 2.8 Å. Science 269:1069–1074

# The Regulation by ATP of the Catalytic Activity and Molecular State of *Thiobacillus novellus* Cytochrome *c* Oxidase

KAZUO SHOJI, KIYOAKI HORI, MINORU TANIGAWA, and TATEO YAMANAKA

*Summary.* *Thiobacillus novellus* cytochrome *c* oxidase has one heme *a* molecule and one copper atom in the minimal structural unit consisting of one molecule each of two subunits (32 and 23 kDa). The oxidase occurs as a monomer of the unit in the presence of 0.5% *n*-octyl-β-D-thioglucoside and as a dimer of the unit in the presence of 0.5% Tween 20. The heme molecule in the monomer is completely reactive with CO, that is, the monomeric oxidase appears to be cytochrome $a_3$, while one of the two heme molecules in the dimer reacts with CO, that is, the dimeric oxidase appears to be cytochrome $aa_3$. The [s]-v curve in the oxidation of ferrocytochrome *c* catalyzed by the dimeric oxidase is sigmoidal. On addition of ATP, the molecular mass of the dimeric oxidase becomes half and the [s]-v curve changes to a hyperbola from a sigmoid. Thus, ATP regulates the molecular states and catalytic properties of the oxidase.

*Key words.* Cytochrome *c* oxidase—ATP—Cytochrome $a_3$—Cytochrome $aa_3$—*Thiobacillus novellus*

## Introduction

Cytochrome *c* oxidase [1] isolated from the sulfur-oxidizing bacterium *Thiobacillus novellus* has a unique property [2]: it contains one heme *a* molecule and one copper atom in the minimal structural unit consisting of two subunits. The oxidase occurs as a monomer of the unit in the presence of 0.5% *n*-octyl-β-D-thioglucoside (OTG), while it occurs as a dimer of the unit in the presence of 0.5% Tween 20. The heme *a* molecule in the monomeric oxidase is completely reactive with CO, while the two heme molecules in the dimeric oxidase combine with CO by 50%. Thus, the monomeric oxidase appears to be cytochrome $a_3$, while the dimeric oxidase appears to be cytochrome $aa_3$. Further, the $E_{m,7.0}$ (midpoint redox potential at pH 7.0) value of the heme *a* molecule in the monomeric oxidase is +0.26 V, whereas the values of two heme

Department of Industrial Chemistry, College of Science and Technology, Nihon University, Kanda-Surugadai 1-5, Chiyoda-ku, Tokyo 101, Japan

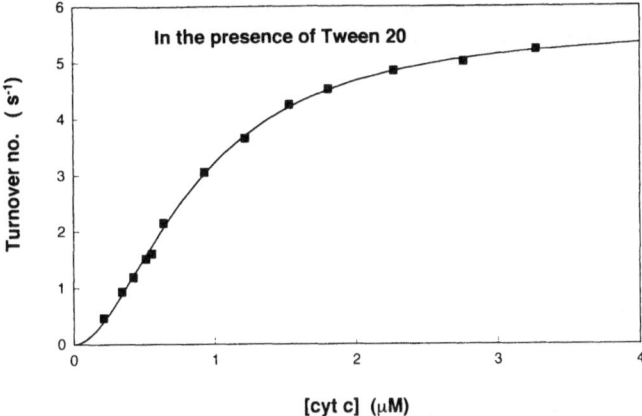

FIG. 1. The [s]-v curve in the oxidation of ferrocytochrome *c* catalyzed by *Thiobacillus novellus* cytochrome *c* oxidase. The reactions were performed in 10 mM HEPES-KOH buffer (pH 7.3) containing 0.5% Tween 20. The reactions of the enzymatic oxidation of ferrocytochrome *c* were started by addition of the oxidase

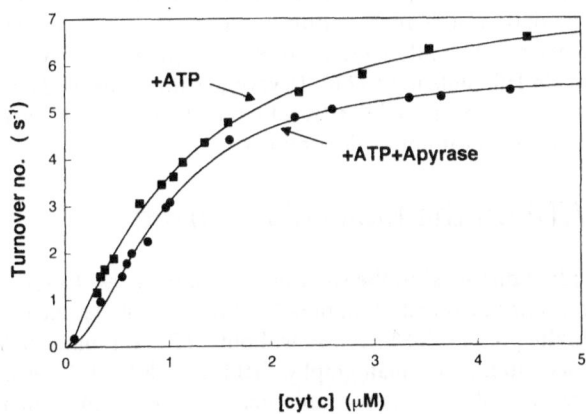

FIG. 2. The [s]-v curve in the oxidation of ferrocytochrome *c* catalyzed by *T. novellus* cytochrome *c* oxidase. The reactions were performed in 10 mM HEPES-KOH buffer (pH 7.3) containing 0.5% Tween 20. +*ATP*, 700 μM ATP was added to the oxidase solution; +*ATP* +*apyrase*, 2 unit/ml apyrase was added 5 min after 700 μM ATP had been added to the oxidase solution. The reactions of the enzymatic oxidation of ferrocytochrome *c* were started by addition of the oxidase

*a* molecules in the dimeric oxidase are +0.18 and +0.36 V, respectively. The monomeric oxidase as well as the dimeric oxidase catalyzes the oxidation of ferrocytochrome *c* with concomitant reduction of molecular oxygen to water. The [s]-v curve in the oxidation of ferrocytochrome *c* catalyzed by the monomeric oxidase is hyperbolic,

TABLE 1. Molecular mass of *T. novellus* cytochrome *c* oxidase in the presence and absence of ATP

|  | Molecular mass (kDa) |
| --- | --- |
| Summation of molecular masses of subunits | 55 (35 + 23) |
| In the presence of Tween 20 (−ATP) | 160 |
| In the presence of Tween 20 (+ATP) | 70 |

while that in the oxidation of ferrocytochrome *c* catalyzed by the dimeric oxidase is sigmoidal.

## Effects of ATP on the Catalytic Properties

As described in the Introduction, the [s]-v curve in the oxidation of ferrocytochrome *c* catalyzed by *T. novellus* cytochrome *c* oxidase is sigmoidal in the presence of 0.5% Tween 20 (Fig. 1). The addition of 700 μM ATP to the reaction mixture makes the curve hyperbolic. When added ATP is decomposed by apyrase, the curve is restored to the original sigmoidal curve (Fig. 2). As the dimeric oxidase shows a sigmoidal [s]-v curve in the oxidation of ferrocytochrome *c*, it appears that the oxidase has two cytochrome *c*-reacting sites and that the two sites interact intimately with each other. Thus, the *n* value in Hill plots is 1.8–2.0. However, as the Eadie–Hofstee plots for the reaction do not show two linear lines (results not shown), the $K_m$ values for cytochrome *c* of the two sites cannot be determined.

## Effects of ATP on the Molecular Aspects

The dimeric oxidase dissolved in the solution containing 0.5% Tween 20 dissociates to the monomeric enzyme on addition of 600 μM ATP (Table 1); the molecular mass of the oxidase in the presence of Tween 20 without ATP is estimated to be 160 kDa in high performance liquid chromatography (HPLC), while it is estimated to be 70 kDa in the presence of ATP. As the molecular mass of the monomeric enzyme is estimated to be 55 kDa as the summation of the masses of two subunits, 32 and 23 kDa, the masses of 160 and 70 kDa seem to be those of the dimeric and monomeric enzymes, respectively, considering the mass of the detergent bound to the enzyme molecules.

Cytochrome $aa_3$ shows an absorption trough around 451 nm in the second-derivative absorption spectrum of its CO-bound form. The trough is characteristic of the cytochrome *a* component [3]. The dimeric oxidase of *T. novellus* shows the trough at 452 nm, while the trough disappears on addition of ATP (Fig. 3). This seems to mean that cytochrome $aa_3$ dissociates to cytochrome $a_3$ on addition of ATP, resulting in disappearance of the cytochrome *a* component. Further, the difference absorption spectrum, reduced *minus* reduced + ATP in the solution containing Tween 20, shows two peaks at 444 and 600 nm, respectively (Fig. 4). This seems also to show that cytochrome *a* component in the dimeric oxidase disappears on addition of ATP.

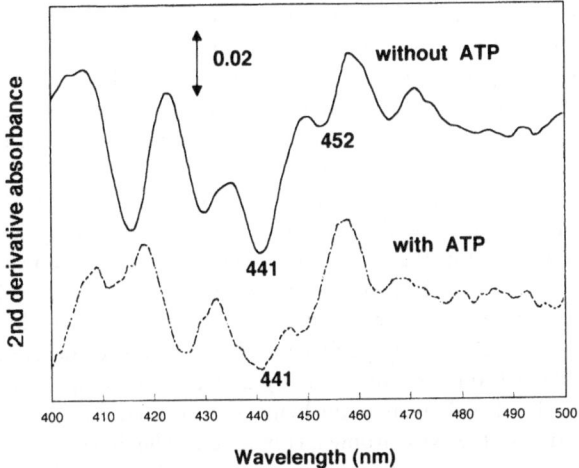

FIG. 3. Second-derivative absorption spectra of CO-bound form of *T. novellus* cytochrome *c* oxidase. Concentrations of Tween 20 and ATP were 0.5% and 700 μM, respectively

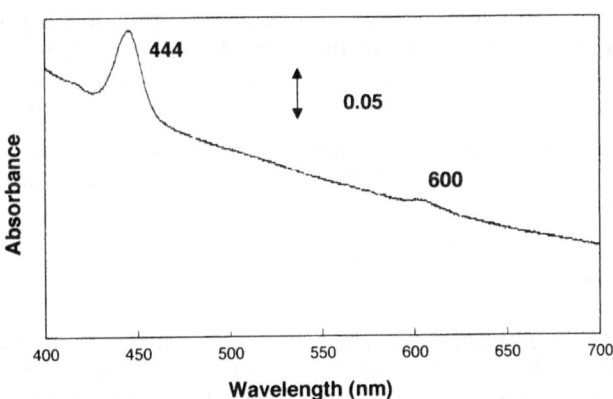

FIG. 4. Difference absorption spectrum, [reduced] − [reduced + ATP] of *T. novellus* cytochrome *c* oxidase. The enzyme was dissolved in 10 mM phosphate buffer in the presence of 0.5% Tween 20. Concentration of ATP was 700 μM

## Conclusion

*Thiobacillus novellus* cytochrome *c* oxidase appears to be an allosteric enzyme; the [s]-v curve in the oxidation of ferrocytochrome *c* catalyzed by the dimeric enzyme is sigmoidal with the *n* value in Hill plots of about 2 in the solution containing Tween 20, and the curve becomes hyperbolic by the addition of ATP. The effector, ATP, makes the dimeric oxidase, cytochrome $aa_3$, dissociate to the monomeric enzyme, cytochrome $a_3$. Thus, the changes in the absorption spectra of the dimeric oxidase caused by ATP show the disappearance of the cytochrome $a$ component from cytochrome $aa_3$. A preliminary experiment has shown that the dimeric oxidase shows proton pumping activity while the monomeric enzyme does not (unpublished results). As proton pumping by cytochrome $c$ oxidase is related to the biosynthesis of ATP [4], the dissociation of the dimeric enzyme to the monomeric enzyme by ATP seems to be of physiological significance; the presence of a sufficient amount of ATP makes the enzyme change to the state that does not have the proton pumping activity.

It seems very interesting that the interconversion of cytochrome $aa_3$ and cytochrome $a_3$ occurs easily with *T. novellus* cytochrome $c$ oxidase. In cytochromes $aa_3$ of bovine heart [5] and *Paracoccus denitrificans* [6], the heme of cytochrome $a$ has six ligands and that of cytochrome $a_3$ has five ligands. When one cytochrome $a_3$ molecule is changed to a cytochrome $aa_3$ molecule by dimerization, a ligand from the other cytochrome $a_3$ molecule will ligate to the heme of the cytochrome $a_3$ molecule, resulting in the formation of the cytochrome $a$ component. The determination of the sixth ligand to the heme of the cytochrome $a$ component is now being undertaken by resonance Raman spectroscopy. Further, the base sequence of DNA encoding the enzyme is under study in our laboratory.

The copper atom in the monomeric oxidase shows a signal at $g = 2.0$ in the electron spin resonance (ESR) spectrum, and the atoms in the dimeric enzyme do not show the signal [2]. Therefore, in the *T. novellus* enzyme, only cytochrome $a_3$ and $Cu_A$ appear sufficient for the oxidation of ferrocytochrome $c$ and reduction of oxygen to water. This may give us a clue to elucidating the mechanism in the oxidation of ferrocytochrome $c$ and the reduction of molecular oxygen by cytochrome $c$ oxidase.

*Acknowledgment.* This research was supported in part by grant-in-aid for Scientific Research on Priority Areas. No. 280, the Ministry of Education, Science, Sports and Culture, Japan.

# References

1. Yamanaka T, Fujii K (1980) Cytochrome $a$-type terminal oxidase derived from *Thiobacillus novellus*. Molecular and enzymatic properties. Biochim Biophys Acta 591:53–62
2. Shoji K, Yamazaki T, Nagano T, et al. (1992) *Thiobacillus novellus* cytochrome $c$ oxidase contains one heme $a$ molecule and one copper atom per catalytic unit. J Biochem (Tokyo) 111:46–53
3. Felsch JS, Horvath MP, Gursky S, et al. (1994) Probing protein-cofactor interactions in the terminal oxidases by second derivative spectroscopy: study of bacterial enzymes with cofactor substitutions and heme A model compounds. Protein Sci 3:2097–2103
4. Wikström M (1989) Identification of the electron transfers in cytochrome oxidase that are coupled to proton-pumping. Nature 338:776–778
5. Tsukihara T, Aoyama H, Yamashita E, et al. (1995) Structure of metal sites of oxidized bovine heart cytochrome $c$ oxidase at 2.8 Å. Science 269:1069–1074
6. Iwata S, Ostermeier C, Ludwig B, et al. (1995) Structure at 2.8 Å resolution of cytochrome $c$ oxidase frome *Paracoccus denitrificans*. Nature 376:660–669

# Part 2
# Cytochrome P-450 Monooxygenases

# Structure–Function Studies of P-450BM-3

Sandra E. Graham-Lorence and Julian A. Peterson

*Summary.* P-450BM-3 is essentially a three-domain enzyme that contains a substrate-binding heme domain termed P-450BMP and an NADPH-P-450 reductase domain termed BMR, which itself is comprised of a flavin adenine dinucleolide (FAD)-binding and a flavin mononucleolide (FMN) -binding domain, all of which we have expressed, purified, and characterized. We have crystallized and determined the structure of P-450BMP which has a structure similar to the other three P-450 structures, P-450$_{cam}$, P-450$_{terp}$, and P-450$_{eryF}$, thus implying a common P-450 structural fold. However, there are differences in the substrate- and redox-partner-binding regions. P-450BM-3 oxidizes long-chain saturated and unsaturated fatty acids. Generally, the more rigid the fatty acid, the more regio- and stereoselective the monooxygenation. We have been able to change the regioselectivity of monooxygenation by mutating phe-87, which lays over pyrrole ring C, to val. Additionally, we have mutated arg-47, which is at the mouth of the access channel, and found that although it does not affect the selectivity of monooxygenation, it does affect substrate binding. Finally, we have reconstituted P-450BM-3 using various domain combinations to characterize electron transfer in P-450BM-3. Subtle changes in the product profile of palmitic acid metabolites indicate that domain–domain interactions may alter the dynamics of substrate binding to the enzyme resulting in a different product profile.

*Key words.* P-450—Monooxygenation—Reductase—Protein–protein interaction—Mutagenesis

## Introduction

Cytochromes P-450 (P-450) are members of a gene superfamily of hemoproteins in which the fifth coordinating ligand of the heme iron is a thiolate side chain of a cysteinyl residue. More than 70 different families of P-450s [1] are found both in prokaryotes and in eukaryotes where they detoxify exogenous organic compounds

Department of Biochemistry, The University of Texas Southwestern Medical Center at Dallas, 5323 Harry Hines Boulevard, Dallas, TX 75235–9038, USA

such as insecticides, herbicides, and drugs, or synthesize required regulatory compounds such as growth hormones in plants, and steroids, eicosanoids, and prostaglandins in mammals. There are more than 500 sequenced P-450s at the present time; however, current information can be found at David Nelson's Webpage http://drnelson.utmem.edu/nelsonhomepage.html. The substrates for P-450s are organic, hydrophobic compounds. Thus, it was not surprising to find that the heme pocket, the site of catalysis in P-450s, is buried within the protein in a hydrophobic environment. In this microenvironment, P-450s usually catalyze a monooxygenation reaction forming an alcohol or an epoxide. With few exceptions, P-450s require molecular oxygen and an exogenous source of electrons (which are obtained from NAD(P)H via their redox partners) and catalyze the reaction:

$$R - CH_3 + O_2 + NAD(P)H + H^+ \rightarrow R - CH_2OH + NAD(P)^+ + H_2O$$

The P-450 family of proteins can be classified on the basis of their redox partner requirement. Class I P-450s require a flavin-containing NAD(P)H-iron sulfur protein reductase and an iron sulfur protein that shuttles electrons one at a time from the reductase to the P-450. This class of proteins is generally found in prokaryotes, where they oxidize hydrophobic natural products and enable them to be used as sources of carbon and energy for growth. Other members of this class are found in the mitochondria of eukaryotes, where they monooxygenate certain steroids or cleave the C-17–C-20 bond in cholesterol to form pregnenolone. Class II P-450s utilize a flavin adenine dinucleolide/flavin mononucleolide (FAD/FMN) -containing NADPH-P-450 reductase to transfer electrons from NADPH to the P-450. These P-450s are, with only one exception to date, eukaryotic, endoplasmic reticulum-bound (microsomal) proteins found most abundantly in the liver and in steroidogenic organs (e.g., adrenals and gonads). The only exception is the soluble prokaryotic protein P-450BM-3 from *Bacillus megaterium* in which the reductase is fused to the P-450 domain via a peptide linker, thus forming a self-sufficient holoenzyme. Finally, class III P-450s do not require an external source of electrons because they do not catalyze a monooxygenation reaction, but rather they rearrange endoperoxides and hydroperoxides. Only a few of these proteins are known, for example, prostacyclin synthase [2], thromboxane synthase [3], and allene oxide synthase [4].

Of the crystallized P-450s whose structures have been released for general distribution [5–8], only one is a class II P-450, P-450BMP [6], the P-450 domain of P-450BM-3, while the other three are class I P-450s: P-450$_{cam}$ [5], P-450$_{terp}$ [7], and P-450$_{eryF}$ [8]. As shown in the ribbon diagram of the four structures (Fig. 1), they appear remarkably alike. On close inspection of the structures of these proteins on a graphics workstation, one can discern the differences, which are located largely in the regions associated with substrate binding (i.e., helices A, B', F, G), and those regions associated with redox-partner binding (i.e., a new helix, J', and a difference in charge on helices C and L). The similarity between P-450BM-3 and eukaryotic fatty acid monooxygenases has been noted [9]. Thus, P-450BM-3 appears to be a good model for eukaryotic microsomal P-450s. Additionally, P-450BM-3 regioselectrively and stereoselectively monooxygenates long-chain fatty acids, especially arachidonic acid [10], as do several eukaryotic P-450s involved in fatty acid metabolism. Thus, we have chosen to study P-450BM-3 as a model for eukaryotic P-450 monooxygenation, electron

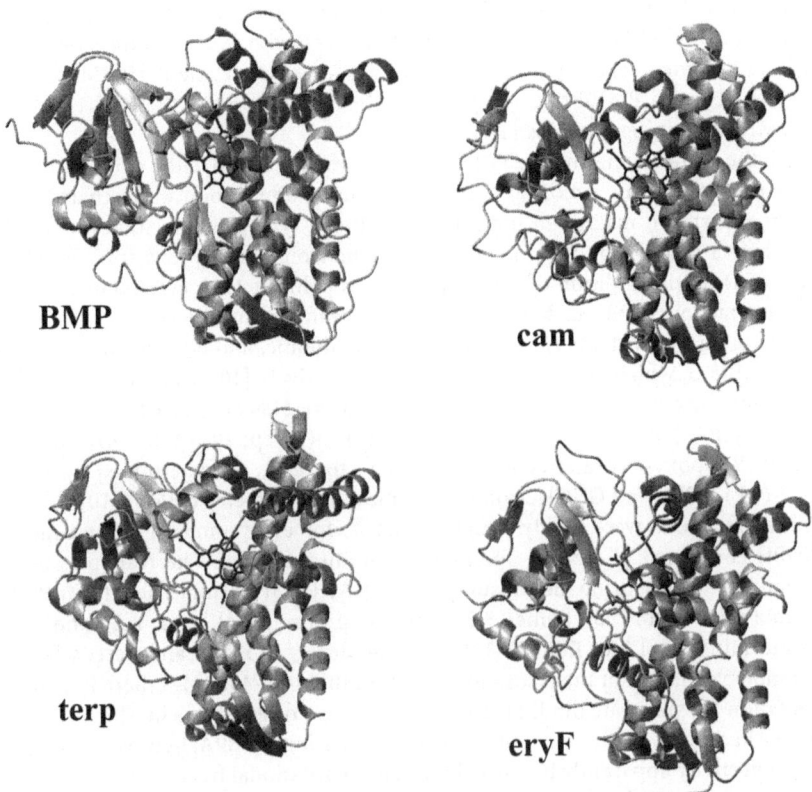

FIG. 1. Ribbon diagrams of the secondary structural elements of P-450s. The atomic coordinates of P-450BMP (*BMP*) [6], P-450$_{cam}$ (*cam*) [5], P-450$_{terp}$ (*terp*) [7], and P-450$_{eryF}$ (*eryF*) [8] were used to construct this figure. The pyrrole carbon atoms of the heme rings were overlaid in three-dimensional space for each of these molecules using the program InsightII from Biosym Corporation (Molecular Simulations, San Diego, CA, USA). The program MolMol [8a] was then used to create the ribbon diagrams. α-Helices are shown by the *helical coil*, strands of β-sheets are shown by *flat arrows*, and random coils and loops are shown by the *lines* between the other secondary structural elements. The heme ring is shown in the center of each of the molecules by the *dark lines*

transfer from the NADPH-P-450 reductase to the P-450, protein–protein interaction, and fatty acid metabolism.

## P-450BM-3 Structure and Function

The initial evidence for the presence of two domains in P-450BM-3 came from Fulco's laboratory [11,12]. Fulco's group first purified, cloned, sequenced, and expressed this fatty acid monooxygenase from *Bacillus megaterium* [11–13]. Using trypsinolysis, in the presence of substrate, they found that they obtained two domains: a 55-kDa hemoprotein domain found to be a P-450 (P-450BMP) and a 66-kDa flavoprotein domain (P-450BMR) [12]. Because of our interest in studying soluble bacterial P-450s, we obtained the plasmid P-450BM-3-2A from Dr. Fulco, and have expressed the holoenzyme P-450BM-3 in a different strain of E. coli with extremely high yields [14]. We have designed expression vectors encoding the holoenzyme P-450BM-3 [10], P-450BMP [15], and P-450BMR [16] and, as is be described next, the FAD-containing and FMN-containing domains [17].

Both the holoenzyme and the P-450 domain, when reduced with sodium dithionite in the presence of an atmosphere of carbon monoxide, produce the characteristic 450 nm absorption maximum indicative of a P-450-type protein. Additionally, they both show a shift of the Soret band of the heme from 418 nm (low spin) to 398 nm (high spin) on the addition of substrate, such as palmitic [14], arachidonic (AA), or eicosapentaenoic (EPA) acid [10]. This spectral shift readily allowed us to determine spectral binding constants ($Ks$) of palmitic, arachidonic, and eicosapentaenoic acids for P-450BM-3, which are 2, 1.2, and 1.6 µM, respectively [10,11,14]. The products of limited monooxygenation of palmitic acid (PA) were 21% ω-1, 44% ω-2, and 35% ω-3 hydroxy-PA [18], 80% 18-OH-AA and 14($S$),15($R$)-epoxy-AA for AA, and 97% 17($S$),18($R$)-epoxy-eicosatetraenoic acid (EET) for EPA, as determined by reverse phase-(RP) HPLC and GC/electron impact mass spectrometry (EIMS) [10].

These reactions were highly coupled; that is, for every mole of $O_2$ and mole of NADPH consumed, there was one mole of product formed. The reaction rate was determined by measuring the consumption of NADPH at 340 nm. The rate for PA was 1.6 µmol min$^{-1}$nmol$^{-1}$ of heme, 3.2 ± 0.4 µmol min$^{-1}$nmol$^{-1}$ for AA, and 1.4 ± 0.2 µmol min$^{-1}$nmol$^{-1}$ for EPA [10,14]. The high degree of regio- and stereoselectivity of monooxygenation of the eicosanoids implies that in P-450BM-3 there is a specific orientation for substrate binding. There must be two orientations in which AA binds over the heme iron in the active site and one orientation for EPA; however, in PA there is less selectivity apparently because there is more rotational freedom of the substrate in the access channel and active site.

As mentioned, with the high yields of the P-450BMP obtainable, the P-450 domain was crystallized and the structure determined by X-ray diffraction. This was the second P-450 whose structure had been determined, and shortly after we solved the third structure, P-450$_{terp}$. Therefore, we were able to determine what regions comprised the conserved "structural fold" among P-450s, as well as those regions that are more variable [19,20]. The structurally conserved regions include the four-α-helix bundle composed of α-helices D, E, I, and L, α-helices C, J, and K, and β-sheets 1 and 2. Also conserved is the heme-binding region and a region designated the "meander." Less conserved in three dimensions are α-helices A, B, and K' and β-sheets 3 and 4, with the least conserved regions being α-helices F and G and loops.

One of the more striking features of P-450BMP (Fig. 2) is the 20-Å-long access channel that leads from the surface opening or mouth to the active site at the heme iron. On the protein surface, around the mouth, is a hydrophobic patch with Arg-47 and Lys-41 just at the beginning of the access channel. As discussed later, we believe that the carboxylate group of the fatty acid is oriented by electrostatic interaction with Arg-47. Following the initial interaction with the hydrophobic surface region, the aliphatic chain of the fatty acid is driven by hydrophobic interactions down the long access channel to the active site where it interacts with the heme. The length of the access channel is determined by the length and orientation of α-helices F and G and the F-G loop. Thus, it is not surprising that P-450$_{cam}$ and P-450$_{terp}$, which have monot-

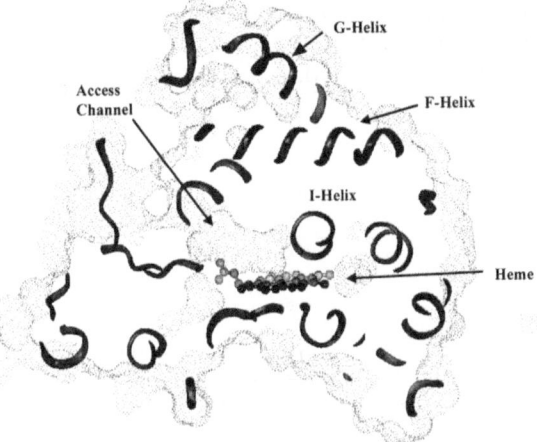

FIG. 2. Connolly surface of P-450BMP illustrating the substrate access channel. The atomic coordinates of molecule A of P-450BMP were used by the program InsightII to visualize the substrate access channel of this molecule. The dotted surface was constructed in the program by "rolling" a 1.4-Å sphere on the surface of the molecule. This surface represents those parts of the molecule that are solvent exposed. Indicated in this figure are the heme ring in the *center*, the substrate access channel, and the coil of helices I, F, and G. This representation is an 8-Å-thick slab taken vertically through the BMP molecule; the β-sheet-rich region would be to the left in this figure

erpenes as their substrates, have much shorter F and G helices. The presence of two molecules of P-450BMP in the asymmetric unit of crystals of the hemoprotein domain of P-450BM-3 (BMP) was noted in our examination of the structure of this protein [6]. These molecules differ in the region of the substrate access channel and led us to propose that there was dynamic motion associated with substrate binding to P-450BM-3 [6]. The presence of this motion was illustrated by the dynamic simulations of P-450BMP with the F-G loop and β-sheet 1 [21]. These areas form the mouth of the access channel and interact with hydrophobic substrates.

## P-450BM-3 Mutations: Understanding Selective Monooxygenation of Fatty Acids

We initially proposed that Arg-47 and Phe-87 were involved in substrate binding and orientation in the substrate-binding site [6]. Thus, we have focused our studies of substrate-binding reactions on these residues, with Arg-47 located at the mouth of the access channel and Phe-87 lying directly above the C-ring of the heme (crystallographic C-ring). We hypothesized that Arg-47 binds the carboxylate group of fatty acids, anchoring the fatty acid. This electrostatic interaction should facilitate passage of the fatty acid down the access channel. Therefore, we changed the arginine to an alanine and a glutamate. To determine if the Phe-87 assists in the selectivity of monooxygenation in the active site, we changed the phenylalanine to a valine and a tyrosine.

TABLE 1. Stoichiometry of NADPH and oxygen utilization during the metabolism of arachidonic acid by different isoforms of P450BM-3

| P450 | NADPH (µM) | $O_2$ (µM) | $H_2O_2$ (µM) | NADPH/$O_2$ ratio |
|------|-----------|-----------|--------------|-------------------|
| Wild type | 125 | 126 | N.D. | 1.0 |
|  | 189 | 190 | N.D. | 1.0 |
| R47A mutant | 125 | 140 | 5.8 | 0.9 |
|  | 189 | 190 | 8.6 | 1.0 |
| R47E mutant | 145 | N.D. | N.D. | — |
| E87V mutant | 125 | 124 | 5.8 | 1.0 |
|  | 189 | 189 | 8.6 | 1.0 |
| F87Y | 138 | 65 | N.D. | 2.1 |
|  | 145 | 75 | N.D. | 1.9 |
|  | 290 | 137 | N.D. | 2.1 |

The experiment was performed as described previously [22]. N.D., not detectable.

Of the Arg-47 mutants, the addition of AA to R47E showed a small shift in the Soret band of the heme indicative of a low- to high-spin conversion of the heme iron. This is presumed to reflect substrate binding in the active site. However, R47E did not show any turnover with AA as substrate when assayed for oxygen consumption, and as expected, no product was detected [22]. R47A showed approximately the same amount of spectral shift from low- to high spin as did the wild-type (WT) enzyme; however, unlike R47E, it was catalytically active [22]. As shown in Table 1, R47A metabolized AA slower than WT and EPA faster than WT. However, R47A was not as tightly coupled as WT, producing small yet detectable amounts of $H_2O_2$. Binding constants for AA and EPA were the same for both mutants, 11 µM for AA and 33 µM for EPA. Thus, it appears that while both mutants are capable of binding substrate (although at reduced affinities), the negative charge at the mouth of the access channel appears to prevent proper binding of substrate, inhibiting catalysis; a neutral charge however still permits the enzyme to function.

The other mutants were made at residue Phe-87 directly over the heme. F87Y did show a small amount of oxygen consumption, approximately equal to one turnover; however, rather than a stoichiometry of 1 mole of NADPH per 1 mole of $O_2$ consumed, there were 2 moles of NADPH per 1 mole of $O_2$ [22]. F87Y also did not show any product formation. On the other hand, F87V appeared to be as catalytically active as the wild type, and although it was not as tightly coupled as WT, it did form product. Interestingly, the product formed from AA was 100% 14,15-epoxy-AA. Thus, this mutation causes the P-450-BM-3 protein to become more regioselective in its monooxygenation of AA. We believe that this increased selectivity results from the larger active site pocket above the heme in which the ω-end of AA can curl back on itself. In contrast, with EPA as a substrate in which there is an additional double bond between C17 and C18, F87V becomes less regioselective than WT, forming 28% 14,15- and 72% 17,18-epoxy-ETA, while WT forms only 17,18-epoxy-ETA. While there remains additional room in the active site in F87V, EPA is less flexible and therefore cannot readily curl back on itself as does AA [22].

# Protein–Protein Interaction and Electron Transport

As discussed, P-450BM-3 is a self-sufficient monooxygenase with both hemoprotein and reductase domains on the same polypeptide chain [11,12]. During our initial studies of the electron-accepting properties of this protein, we observed that reduction of the holoenzyme under an atmosphere of carbon monoxide did not result in the accumulation of reduced semiquinone forms of the flavins [23]. This observation was in direct contrast to the results expected from prior studies with microsomal P-450s and their reductase [24,25]. To identify the cause of this difference, we have expressed the reductase domain in *E. coli* and found it to contain stoichiometric amounts of both FAD and FMN [16]. The electron-accepting properties of the purified reductase domain were similar to those of the domain in the holoenzyme. We concluded, from these results, that in contrast to the microsomal P-450 reductase there is no air-stable semiquinone form of the reductase of P-450BM-3. Rather, the two-electron reduced form of FMN is thermodynamically favored over the one-electron reduced state, and at equilibrium the two-electron reduced form predominates [26].

Our further studies of electron transfer between the domains and the protein–protein interactions that facilitate this electron transfer have included the expression

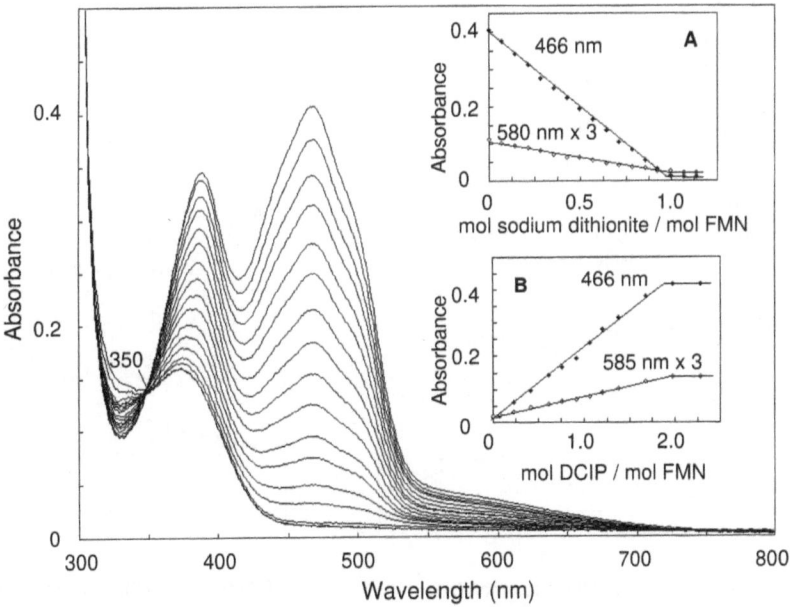

FIG. 3. Reduction of the FMN-binding domain of P-450BM-3. The FMN-binding domain was titrated anaerobically with a standardized solution of sodium dithionite as described previously [17]. *Inset A*, a replot of the absorbance changes that occurred at two wavelengths during the titration, shows that the flavin accepts two electrons without the accumulation of intermediates during this reduction process. *Inset B*, a replot of the absorbance changes that occurred during the reoxidation of the reduced form of the FMN-binding domain shows that the electrons that were input can be removed. DCIP, 2,6-dichloroindophenol

of each of the domain combinations possible without shuffling the coding sequences for these domains. Thus, we have expressed the heme domain, reductase domain, FMN-binding domain, FAD-binding domain, and heme FMN-binding domain [15–17,27,28]. We have examined the electron-accepting properties of each of these domain combinations and, as can be seen in Fig. 3, reduction of the FMN-binding domain with sodium dithionite, an obligate one-electron donor, does not lead to an accumulation of one-electron reduced forms [17]. Rather, the FMN is smoothly reduced from the fully oxidized state to the two-electron reduced state with no accumulation of one-electron reduced forms. The lack of accumulation of reduced forms must result from interdomain electron transfer where 2 one-electron reduced forms react to give 1 two-electron reduced molecule and one molecule that is fully oxidized. Titration of the FAD-binding domain with sodium dithionite (Fig. 4) results first in the accumulation of the neutral, blue semiquinone form of FAD, as evidenced by the increase in absorbance at 585 nm. This was followed by the complete reduction of the FAD to the two-electron reduced form [17].

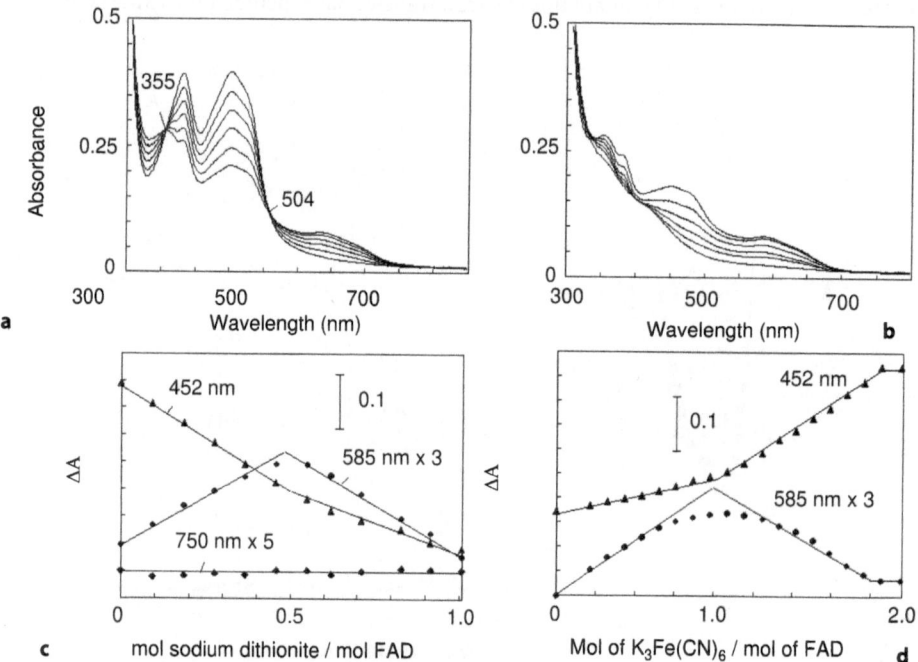

Fig. 4a–d. Reduction of the FAD-binding domain of P-450BM-3. The FAD-binding domain was titrated anaerobically with sodium dithionite as described previously [17]. a Absorbance changes that occurred during the addition of the first electron equivalent with the accumulation of the neutral, blue semiquinone form of the flavin. b Absorbance changes that occurred during the addition of the second electron equivalent, resulting in the complete reduction of the flavin. c Replot of the data from a and b. d Plot of the reoxidation of the reduced FAD-binding domain with potassium ferricyanide (absorbance spectra not shown) to show that the reduction was completely reversible

In the past, we have tried to reconstitute the monooxygenase activity of P-450BM-3 by employing the reductase and heme domains [15]. Because of the autooxidizibility of the reductase domain, we used limiting concentrations of this domain and excess of the heme domain. As was expected, the specific activity of the excess component was not optimal; however, the reductase domain was found to be competent for fatty acid monooxygenation [15]. Additional reconstitution experiments with the variety of expressed domain combinations have permitted us to firmly establish that the two-electron reduced form of the FMN-binding domain is thermodynamically incapable of reducing the heme domain [26]. Additionally, we have observed that the product distribution from fatty acid oxidation by various domain combinations is subtly different [28]. In the case of the holoenzyme, palmitic acid is oxidized multiple times to give a variety of mono-, di-, and tri-oxidized products [18] as seen on the thin-layer chromatogram in Fig. 5.

Under the conditions of these experiments with excess reductase domain, there were essentially no products formed because the FMN domain had the opportunity to

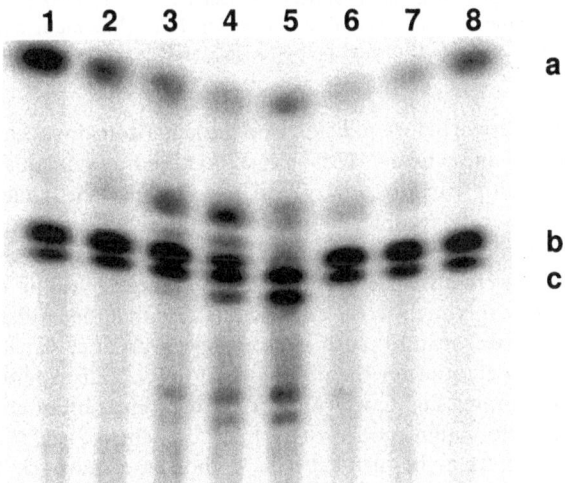

FIG. 5. Reconstitution of the palmitic acid hydroxylase activity of P-450BM-3. A printout of a PhosphorImager scan of palmitic acid metabolites produced by P-450BM-3 and the reconstitution systems. The reactions were conducted as described [28]. Aliquots from the BMP/FMN:30FAD system were taken at 2 and 7 min after addition of NADPH (*lanes 1* and *2*, respectively). At that time, all the initial NADPH was consumed and a new portion of NADPH was added (200 μM, final concentration). An aliquot from that reaction was taken after an additional 7 min (*lane 3*). *Lane 4* represents the metabolites produced by 60 nM P-450BM-3 within 3.5 min, when all NADPH was consumed. After that, a new aliquot of NADPH was added (200 μM, final concentration), and an aliquot was taken 10 min later (*lane 5*). *Lane 8* represents the metabolites produced by BMP:10FMN:30FAD within 2 min of the reaction after utilizing 200 μM of NADPH; *lane 7*, the metabolites produced 2 min after addition of a second aliquot of NADPH; and *lane 6*, the metabolites produced 6 min after adding a third aliquot of NADPH. *Band a* corresponds to palmitic acid; *bands b* and *c* refer to ω-3/ω-2 and ω-1 monohydroxy palmitic acid

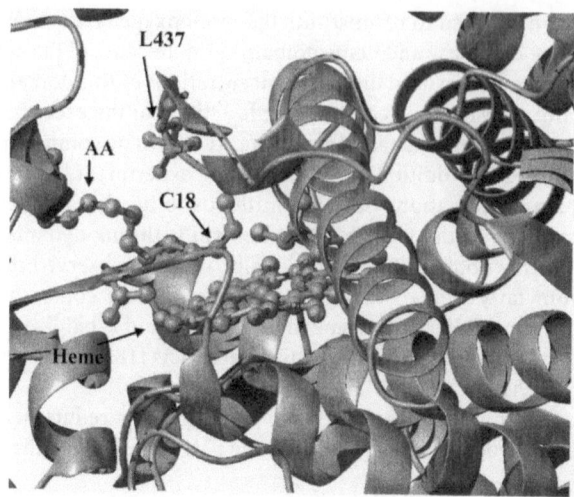

FIG. 6. Model of arachidonic acid in the substrate-binding site of P-450BMP. Arachidonic acid was modeled into the substrate access channel of molecule A of P-450BMP [10]. The secondary structural elements of the molecule are as described in Fig. 1. Arachidonic acid (*AA*) and the side chain of leucine-437 (*L437*) are shown as ball-and-stick models

be fully reduced (inactive) before collision with the heme domain. Reconstitution with the heme/FMN domain or heme-, FMN-, and FAD-binding domains gave different combinations of multiply oxidized fatty acid. For example, the heme/FMN-binding domain gave some secondary metabolites during the time of the experiment while reconstitution with each of the domains resulted in essentially no secondary metabolism. These results were obtained even when the reduced pyridine nucleotide was added to the reaction mixture several times to establish that the availability of reductant was not the limiting feature causing the change in product distribution [28].

Our hypothesis to account for the difference in product distribution can be best understood by examining Fig. 6, which shows a representation of arachidonic acid bound in the substrate-binding pocket of the heme domain of P-450BM-3. Seen in the upper part of this figure is the turn between strands $\beta4$-1 and $\beta4$-2. The side chain from Leu-437 in this turn forms a part of the substrate-binding pocket for the fatty acid [7,19]. Continuing from this $\beta$-sheet to the carboxy terminus of the heme domain, there are only 14 amino acids and the only secondary structural elements in this sequence are random coil and strand $\beta3$-2.

Thus, we have proposed that the presence of the FMN domain (a 22-kDa mass) or the reductase domain (a 66-kDa mass) can result in distortion of the secondary structure of this portion of the molecule, thereby increasing the "volume" of the active site as the reductase mass bound to the heme domain is increased. This increased volume would permit hydroxylated fatty acids to fit into the substrate-binding site and permit them to be further oxidized [28].

*Acknowledgments.* The research reported here was supported in part by research grants from the National Institutes of Health (NIH) R01 GM43479 and R01 GM50858.

# References

1. Nelson DR, Kamataki T, Waxman DJ, et al. (1993) The P-450 superfamily: update on new sequences, gene mapping, accession numbers, early trivial names of enzymes, and nomenclature (Review). DNA Cell Biol 12:1–51
2. Pereira B, Wu KK, Wang LH (1994) Molecular cloning and characterization of bovine prostacyclin synthase. Biochem Biophys Res Commun 203:59–66
3. Haurand M, Ullrich V (1985) Isolation and characterization of thromboxane synthase from human platelets as a cytochrome P-450 enzyme. J Biol Chem 260:15059–15067
4. Song WC, Funk CD, Brash AR (1993) Molecular cloning of an allene oxide synthase: a cytochrome P-450 specialized for the metabolism of fatty acid hydroperoxides. Proc Natl Acad Sci USA 90:8519–8523
5. Poulos TL, Finzel BC, Howard AJ (1987) High-resolution crystal structure of cytochrome P-450$_{cam}$. J Mol Biol 195:687–700
6. Ravichandran KG, Boddupalli SS, Hasemann CA, et al. (1993) Crystal structure of hemoprotein domain of P-450BM-3, a prototype for microsomal P-450s. Science 261:731–736
7. Hasemann CA, Ravichandran KG, Peterson JA, et al. (1994) Crystal structure and refinement of cytochrome P-450$_{terp}$ at 2.3 Å resolution. J Mol Biol 236:1169–1185
8. Cupp-Vickery JR, Poulos TL (1995) Structure of cytochrome P-450$_{eryF}$ involved in erythromycin biosynthesis. Nat Struct Biol 2:144–153
8a. Koradi R, Billeter M, Wüthrick K (1996) Molmol: a program for display and analysis of macromolecular structures. J Mol Graphics 14:51–55
9. Ruettinger RT, Wen LP, Fulco AJ (1989) Coding nucleotide, 5′ regulatory, and deduced amino acid sequences of P-450BM-3, a single peptide cytochrome P-450:NADPH-P-450 reductase from Bacillus megaterium. J Biol Chem 264:10987–10995
10. Capdevila JH, Wei S, Helvig C, et al. (1996) The highly stereoselective oxidation of polyunsaturated fatty acids by P-450BM-3. J Biol Chem 271:22663–22671
11. Narhi LO, Fulco AJ (1986) Characterization of a catalytically self-sufficient 119,000-dalton cytochrome P-450 monooxygenase induced by barbiturates in Bacillus megaterium. J Biol Chem 261:7160–7169
12. Narhi LO, Fulco AJ (1987) Identification and characterization of two functional domains in cytochrome P-450BM-3, a catalytically self-sufficient monooxygenase induced by barbiturates in Bacillus megaterium. J Biol Chem 262:6683–6690
13. Wen LP, Fulco AJ (1987) Cloning of the gene encoding a catalytically self-sufficient cytochrome P-450 fatty acid monooxygenase induced by barbiturates in Bacillus megaterium and its functional expression and regulation in heterologous (Escherichia coli) and homologous (Bacillus megaterium) hosts. J Biol Chem 262:6676–6682
14. Boddupalli SS, Estabrook RW, Peterson JA (1990) Fatty acid monooxygenation by cytochrome P-450BM-3. J Biol Chem 265:4233–4239
15. Boddupalli SS, Oster T, Estabrook RW, et al. (1992) Reconstitution of the fatty acid hydroxylation function of cytochrome P-450BM-3 utilizing its individual recombinant hemo-and flavoprotein domains. J Biol Chem 267:10375–10380
16. Oster T, Boddupalli SS, Peterson JA (1991) Expression, purification, and properties of the flavoprotein domain of cytochrome P-450BM-3. Evidence for the importance of the amino-terminal region for FMN binding. J Biol Chem 266:22718–22725
17. Sevrioukova IF, Truan G, Peterson JA (1996) The flavoprotein domain of P-450BM-3: expression, purification, and properties of the FAD-and FMN-binding sub-domains. Biochemistry 35:7528–7535
18. Boddupalli SS, Pramanik BC, Slaughter CA, et al. (1992) Fatty acid monooxygenation by P-450BM-3: product identification and proposed mechanisms for the sequential hydroxylation reactions. Arch Biochem Biophys 292:20–28

19. Hasemann CA, Kurumbail RG, Boddupalli SS, et al. (1995) Structure and function of cytochromes P-450: a comparative analysis of three crystal structures. Structure (Lond) 3:41–62
20. Graham-Lorence SE, Peterson JA (1996) P-450s—structural similarities and functional differences. FASEB J 10:206–214
21. Paulsen MD, Ornstein RL (1995) Dramatic differences in the motions of the mouth of open and closed cytochrome P-450BM-3 by molecular dynamics simulations. Proteins 21:237–243
22. Graham-Lorence SE, Falck JR, Wei S, et al. (1997) An active site mutation, F87V, converts P-450BM-3 into a highly stereoselective arachidonic acid epoxygenase. J Biol Chem 272:1127–1135
23. Peterson JA, Boddupalli SS (1992) P-450BM-3: reduction by NADPH and sodium dithionite. Arch Biochem Biophys 294:654–661
24. Iyanagi T, Makino N, Mason HS (1974) Redox properties of the reduced nicotinamide adenine dinucleotide phosphate-cytochrome P-450 and reduced nicotinamide adenine dinucleotide-cytochrome b5 reductases. Biochemistry 13:1701–1710
25. Iyanagi T, Anan FK, Imai Y, et al. (1978) Studies on the microsomal mixed function oxidase system: redox properties of detergent-solubilized NADPH-cytochrome P-450 reductase. Biochemistry 17:2224–2230
26. Sevrioukova IF, Shaffer C, Ballou DP, et al. (1996) Equilibrium and transient state spectrophotometric studies of the mechanism of reduction of the flavoprotein domain of P-450BM-3. Biochemistry 35:7058–7068
27. Sevrioukova IF, Peterson JA (1996) Domain-domain interaction in cytochrome P-450BM-3. Biochimie (Paris) 78:744–751
28. Sevrioukova IF, Truan G, Peterson JA (1997) Reconstitution of the fatty acid hydroxylase activity of cytochrome P-450BM-3 utilizing its functional domains. Arch Biochem Biophys 340:231–238

# Oxygen Binding to P-450$_{cam}$ Induces Conformational Changes of Putidaredoxin in the Ferrous P-450$_{cam}$– Reduced Putidaredoxin Complex

Hideo Shimada[1], Masashi Unno[1], Yoko Kimata[1], Ryu Makino[2], Futoshi Masuya[3], Takashi Obata[3], Hiroshi Hori[3], and Yuzuru Ishimura[1]

*Summary.* P-450$_{cam}$ catalyzes the conversion of *d*-camphor to 5-*exo*-hydroxy-camphor at the expense of 1 mole each of NADH and oxygen. Two reducing equivalents from NADH are transferred to P-450$_{cam}$ via two redox-linked proteins, putidaredoxin reductase (PdR) and putidaredoxin (Pd). Pd serves as a one-electron shuttle between PdR and P-450$_{cam}$. Reduced Pd has been known to form a tight complex with ferric or ferrous P-450$_{cam}$. The former complex yields oxidized Pd and ferrous P-450$_{cam}$. Binding of oxygen to the latter complex affords the product formation, degrading into 5-*exo*-hydroxy-camphor, water, oxidized Pd, and ferric P-450$_{cam}$. In this study, we show evidence that in the ferrous P-450$_{cam}$-reduced Pd complex, binding of a ligand such as $O_2$, CO, and NO to P-450$_{cam}$ induces the conformational changes of reduced Pd at least at the redox center. The significance of this finding is discussed.

*Key words.* Cytochrome    P-450—Iron-sulfur    protein—Putidaredoxin—EPR— Oxygen

## Introduction

Cytochrome P-450 (P-450) is a group of heme proteins that participates in the monooxygenation reactions of a wide variety of hydrophobic substances including steroids, fatty acids, and hydrocarbons [1]. More than 400 species of P-450 have been isolated from animals, plants, insects, and microorganisms and have been recognized to form a gene superfamily [2]. Among many P-450 species, P-450$_{cam}$ inducible in a soil bacterium *Pseudomonas putida* on growth on *d*-camphor has been the focus of P-450 research because of the ease of isolation of the enzyme in large quantities and its ability catalyze the oxidation of an inactivated carbon [3,4]. P-450$_{cam}$ hydroxylates *d*-camphor at the 5-*exo* position with nearly 100% stereo- and regiospecificity at the

[1] Department of Biochemistry, School of Medicine, Keio University, 35 Shinanomachi, Shinjuku-ku, Tokyo 160, Japan
[2] Department of Biochemistry, Faculty of Science, Rikkyo University, Nishi-ikebukuro, Toshima-ku, Tokyo 171, Japan
[3] Division of Biophysical Engineering, Department of Systems and Human Science, Graduate School of Engineering Science, Osaka University, Toyonaka, Osaka 560, Japan

expense of 1 mole each of NADH and molecular oxygen. Two electrons from NADH are transferred to P-450$_{cam}$ via an electron-transfer chain composed of putidaredoxin reductase (PdR), a flavine adenine dinucleotide- (FAD-) containing protein, and putidaredoxin (Pd), a 2Fe-2S iron-sulfur protein. Pd serves as a one-electron carrier from PdR to P-450$_{cam}$. In the reaction, reduced Pd donates an electron to ferric camphor-bound P-450$_{cam}$, yielding the ferrous form of the enzyme, which subsequently combines with molecular oxygen to form an oxy-ferrous intermediate. Subsequent reduction of the oxy-intermediate by reduced Pd affords oxidized Pd, ferric P-450$_{cam}$, water, and 5-*exo*-hydroxycamphor.

At the step at which Pd donates an electron to ferric P-450$_{cam}$, Pd can be replaced by a variety of reductants including spinach ferredoxin, bovine adrenodoxin, and methyl viologen. But at the step of reduction of the oxy-intermediate, such reductants as just shown cannot replace Pd. Based on these and other results, Gunsalus and his co-workers considered that Pd has a role more than that of a electron donor to P-450$_{cam}$ and proposed that Pd serves as an effector in the monooxygenation catalyzed by P-450$_{cam}$ [5]. We have infrared evidence supporting this proposal: Pd binding to P-450$_{cam}$ induces changes in the active site structure of P-450$_{cam}$ as seen by a shift in C-O stretch frequency of CO-ferrous P-450$_{cam}$ from 1940 to 1932 cm$^{-1}$ [6]. The effector action exerted by Pd, however, is poorly understood. Therefore, for elucidation of the mechanisms of the effector action and also of the monooxygenation reaction, studies on the Pd–P-450$_{cam}$ complex are required.

In this chapter, we show for the first time evidence that in the ferrous P-450$_{cam}$–reduced Pd complex, oxygen binding to P-450$_{cam}$ induces the conformational changes of reduced Pd by employing electron paramagnetic resonance (EPR) spectroscopy. The current finding suggests a possible regulation of the structure and function of Pd by oxygen.

# Materials and Methods

## Enzyme Preparations

The wild-type P-450$_{cam}$ and its mutants were expressed in *Escherichia coli* strain JM109 and purified with the procedures described previously [7]. Purified preparations with the RZ value (A392/A280) greater than 1. 5 were employed in this study. Pd and PdR were expressed also in *Escherichia coli* strain JM109 and were purified according to the methods described by Gunsalus and Wagner [8]. Pd and PdR were found to be homogeneous on sodium dodecyl sulfate-polyacrylamide gel electrophoresis (SDS-PAGE). Pd was stored in 0.1 M Tris-HCl, pH 8.0, containing 10 mM 2-mercaptoethanol. For the EPR spectral measurement of Pd described next, 2-mercaptoethanol was removed from Pd by using centricon 10 (Amicon, Beverly, MA, USA) before the measurement. The concentrations of P-450$_{cam}$, PdR, and Pd were determined spectrophotometrically by using extinction coefficients [8].

## EPR Spectroscopy

For spectroscopy, 150–200 μl of oxidized Pd or the mixture of oxidized Pd and ferric P-450$_{cam}$ in 50 mM potassium phosphate, pH 7.4, containing 50 mM KCl and 1 mM *d-*

camphor was transferred into a screw-capped EPR tube. After putting on a screw-cap with a rubber septum, the protein solution was degassed with several cycles of evacuation and subsequent flushing with oxygen-free N$_2$ gas. Subsequently, Pd and P-450$_{cam}$ were reduced by a trace amount of solid sodium dithionite, which had been placed inside the tube. The mixture was allowed to stand for 5 min on ice to complete the reduction. When necessary, CO or NO was anaerobically introduced to the tube. The tube was slowly immersed in liquid nitrogen: the protein solution was usually frozen within 15–30 s. EPR measurements were carried out on a Varian E-12 EPR spectrometer (San Fernado, CA) at the X-band (9.363-GHz) microwave frequency.

## Results and Discussion

### Effect of Heme Ligand on the P-450$_{cam}$–Putidaredoxin Complex

The EPR spectrum of reduced Pd measured at liquid nitrogen temperature, shown at the top of Fig. 1, has $g = 2.02$ and $g = 1.94$ signals near 330 and 345 millitesla (mT), respectively, as reported [9]. When P-450$_{cam}$ was added to reduced Pd, the derivative type $g = 1.94$ signal near 345 mT changed slightly; this change became maximum when the amount of P-450$_{cam}$ was equal to that of Pd. The middle spectrum in Fig. 1 is of reduced Pd with the same amount of P-450$_{cam}$. Further addition of P-450$_{cam}$ did not alter the spectrum (data not shown), indicating that the middle spectrum of Fig. 1 was that of reduced Pd in a 1:1 complex with P-450$_{cam}$.

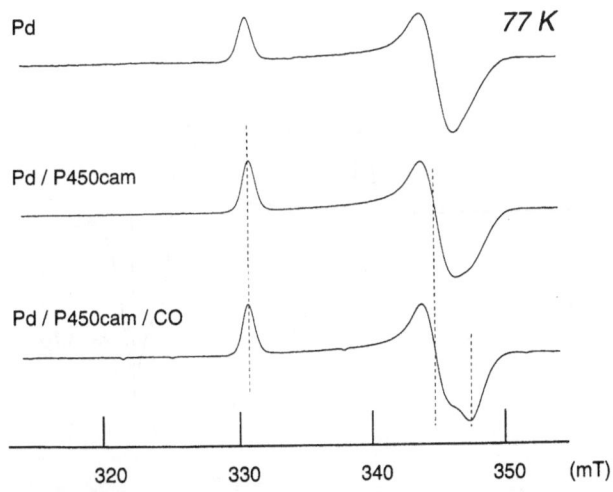

FIG. 1. Electron paramagnetic resonance (EPR) spectra of reduced putidaredoxin (Pd) measured at 77 K under various conditions: 200 μM each of Pd and/or P-450$_{cam}$ in 50 mM potassium phosphate, pH 7.4, containing 50 mM KCl and 1 mM d-camphor were reduced by sodium dithionite in an anaerobic EPR tube. *Top spectrum*, reduced Pd; *middle spectrum*, reduced Pd with an equal amount of P-450$_{cam}$; *bottom spectrum*, obtained by exposing a 1:1 mixture of ferrous P-450$_{cam}$ and reduced Pd to CO

On the addition of CO to the 1:1 complex of Pd and P-450$_{cam}$, the EPR spectrum of Pd changed evidently as shown by a shift of the trough of the derivative type $g = 1.94$ signal near 345 mT to a higher magnetic field. A minor change was also observed as a slight shift of the $g = 2.02$ signal to a lower magnetic field. In the absence of P-450$_{cam}$, CO did not affect the spectrum of reduced Pd. Hence, it can be concluded that binding of CO to the heme of P-450$_{cam}$ alters the EPR spectrum of Pd via P-450$_{cam}$.

EPR spectral measurements were also carried out by employing *Pseudomonas putida* cells containing PdR, Pd, and P-450$_{cam}$ in a 1:2:2 ratio [8]. The cells in the same buffer as used for Pd and P-450$_{cam}$ were reduced with sodium dithinite. On exposure to CO, reduced Pd in the cells showed an almost identical spectral change with that just described, supporting the present in vitro finding.

CO induces spectral changes of Pd when Pd forms the complex with P-450$_{cam}$. Requirements for the complex formation in the present finding were also demonstrated by the following experiments. The complex of Pd and P-450$_{cam}$ is stabilized by iron pairs formed between acidic amino acid residues of Pd and basic residues of P-450$_{cam}$ [10,11,14]. Therefore, high ionic strength can destabilize ionic interaction and thus favor the dissociation of the two proteins. When KCl (1 M) was added to a 1:1 mixture of Pd and P-450$_{cam}$, CO induced only ≈40% of the spectral changes observed under standard conditions. Dissociation of two proteins from the complex is also favored by the addition of PdR, which competively binds to Pd. Addition of a 5-molar excess of PdR to the 1:1 mixture of Pd and P-450$_{cam}$ again decreased the CO-induced EPR spectral changes to ≈40% of that obtained under standard conditions.

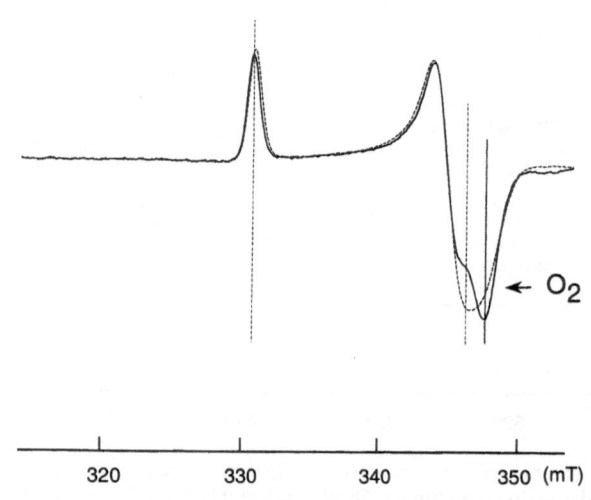

320        330        340        350 (mT)

FIG. 2. Effects of oxygen on the EPR spectrum of reduced Pd in the presence of the Asp251Asn mutant P-450$_{cam}$ in a ferrous state. *Dotted line*, 200 µM reduced Pd with the same amount of ferrous mutant of P-450$_{cam}$; *solid line*, effects of oxygen. (Other details of the experiment as in Fig. 1)

Nitric oxide added in place of CO was also found to induce an EPR spectral change of reduced Pd in the ferrous P-450$_{cam}$-reduced Pd complex. The observed EPR signal of Pd near 345 mT was almost identical to the CO-induced signal (data not shown). In the P-450$_{cam}$ system, oxygen, a ligand to ferrous heme, is the substrate for the monooxygenation reaction. On addition of oxygen to the ferrous P-450$_{cam}$-reduced Pd complex, Pd is readily oxidized, giving an electron to the oxy-ferrous P-450$_{cam}$ and thus allowing product formation. One way to see the effects of oxygen on the complex of ferrous P-450$_{cam}$ and reduced Pd is to employ a mutant P-450$_{cam}$ at the 251-position such as Asp251Asn, Asp251Gly, or Asp251Ala [12]. The mutation at the 251-position has been known to retard the rate of electron transfer from reduced Pd to the oxy-intermediate by 500 to 1000 fold, but this retardation does not affect the subsequent $d$-camphor monooxygenation [12,13].

The mutant P-450$_{cam}$ at the 251-position was found to transmit the signal of CO or NO binding to the heme to reduced Pd as the wild-type P-450$_{cam}$ did. Thus, the mutant P-450$_{cam}$ was used, in place of the wild-type enzyme, to observe the effects of oxygen on reduced Pd. Oxygen was introduced into a 1:1 mixture of reduced Pd and a ferrous form of the mutant P-450$_{cam}$ in an anaerobic EPR tube that had been placed on ice. After quick mixing, the tube was immediately immersed into liquid nitrogen. When the mixture of the two protein with oxygen was allowed to stand on ice for a while after mixing, the EPR signal of reduced Pd disappeared, indicating oxidation of reduced Pd. Thus, these procedures should be done quickly. Figure 2 shows the effects of oxygen on the EPR spectrum of reduced Pd in the presence of an equal amount of the Asp251Asn-P-450$_{cam}$. Oxygen changed the spectrum of reduced Pd as shown by a shift of the trough of the $g = 1.94$ signal to a higher magnetic field. As seen, the EPR spectrum was almost identical to that observed in the presence of CO or NO. Other 251-mutants, Asp251Gly and Asp251Ala, also produced the same results (data not shown).

# Implication for the Mechanisms of Signal Transmission from P-450$_{cam}$ to Putidaredoxin

When $O_2$, CO, or NO binds to the ferrous high spin heme of P-450$_{cam}$, the active site structure of P-450$_{cam}$ possibly changes, including movement of the axial ligand toward the heme as observed in heme proteins such as hemoglobin, myoglobin, and horseradish peroxidase. The current finding is that Pd changes its conformation at least at the metal center on binding of $O_2$, CO, or NO to P-450$_{cam}$. This observation suggests that ligand binding to P-450$_{cam}$ evidently alters the active site structure as well as the Pd-binding site structure of P-450$_{cam}$. This is the first evidence that ligand binding to P-450$_{cam}$ alters its conformation.

With regard to the transmission of the conformational changes of P-450$_{cam}$ to reduced Pd, the interaction of Pd and P-450$_{cam}$ at the binding site plays an essential role. Arg-112, a surface residue of P-450$_{cam}$, has been proposed to form the Pd-binding site of P-450$_{cam}$ in ferric and ferrous states [14]. Thus, an attempt was made to understand the mechanisms of the transmission of conformational changes from P-450$_{cam}$ to Pd by employing a mutant P-450$_{cam}$ at Arg-112. Although the mutant P-450$_{cam}$ in its ferric state forms the complex with reduced Pd, its affinity toward reduced Pd is

TABLE 1. Effects of mutation at the 112-position of P-450$_{cam}$ on the electron-transfer reaction from reduced putidaredoxin (Pd) to ferric P-450$_{cam}$ and redox potential of P-450$_{cam}$

$$P\text{-}450_{cam}^{3+} + Pd\,(Fe^{2+}) \underset{k_{-1}}{\overset{k_1}{\rightleftharpoons}} P\text{-}450_{cam}^{3+}/Pd\,(Fe^{2+}) \overset{k_2}{\longrightarrow} P\text{-}450_{cam}^{2+} + Pd\,(Fe^{3+})$$

| Mutant | $k_1$ (M$^{-1}$s$^{-1}$) | $k_{-1}$ (s$^{-1}$) | $K_d$ (μM) | $k_2$ (s$^{-1}$) | Redox potential (mV) |
|---|---|---|---|---|---|
| Arg-112 | $1.8 \times 10^7$ | 3.5 | 0.19 | 42 | −132 |
| Lys-112 | $5.7 \times 10^6$ | 25 | 4.4 | 18 | −162 |
| Cys-112 | — | — | 480 | 4.0 | −182 |
| Met-112 | — | — | 470 | 1.3 | −200 |
| Tyr-112 | — | — | 110 | 0.16 | −195 |

The data are from Unno et al. [14]. The values of the dissociation constant were calculated according to the equation $K_d = k_{-1}/k_1$.

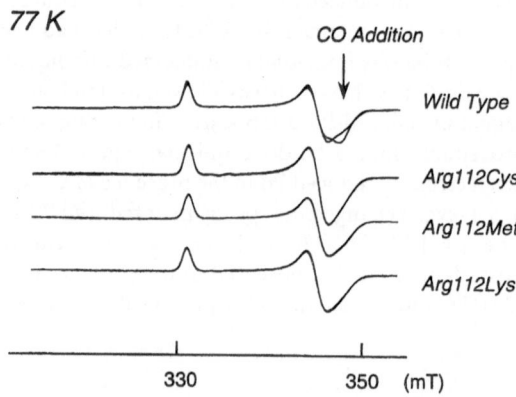

FIG. 3. Effects of mutation at 112-position of P-450$_{cam}$ on the ligand-induced EPR spectral changes of reduced Pd in its complex with P-450$_{cam}$. 200 μM of putidaredoxin and 200 μM–2 mM of the wild-type or the mutant P-450$_{cam}$ in 50 mM potassium phosphate, pH 7.4, containing 50 mM KCl and 1 mM d-camphor were mixed. (Other details as in Fig. 1)

decreased by the mutation (Table 1). Therefore, the mutant P-450$_{cam}$ in its ferrous state has possibly a lower affinity for reduced Pd, although we have not yet determined this. As the wild-type P-450$_{cam}$ has nearly the same affinity toward reduced Pd in either its ferric or ferrous state [14–16], we tentatively employed the dissociation constants listed in Table 1 to estimate the amount of the mutant P-450$_{cam}$ required for an almost 100% complex formation 200 μM Pd. As shown in Fig. 3, we could not detect a discernible spectral change of reduced Pd on addition of either the mutant P-450$_{cam}$ alone or the mutant P-450$_{cam}$ with CO or NO except for the Lys-112 mutant, which showed a slight broadening of the 1.94 signal at the higher magnetic region. In each experiment with the mutant P-450$_{cam}$, further addition of the mutant P-450$_{cam}$ by fivefold did not show any changes in addition to those seen in Fig. 3.

Arg-112 of P-450$_{cam}$ is located at the surface of the protein, but its guanidino group is pointing toward the inside of the protein and forms a hydrogen bond with the propionate side chain of the heme located in the interior of the enzyme [17]. The same propionate side chain forms a hydrogen bond with His-355, which is the second residue from the axial ligand to the heme Cys-357. As reported previously, the mutation at the 112-position lowers the redox potential of the heme (Table 1). These facts suggest Arg-112 is structurally and functionally related to the heme. Therefore, the binding of the ligand to the heme is possibly sensed by the Arg-112 residue. As the data obtained by employing the mutant P-450$_{cam}$ suggest, the structure of the amino acid residue at the 112-position plays an essential role in the transmission of the conformational change that occurred at the heme to reduced Pd.

The current finding is that oxygen binding to the P-450$_{cam}$ heme induces a conformational change of reduced Pd when Pd is associated with P-450$_{cam}$. This finding suggests a new type of regulation in the camphor monooxygenation system. We are now trying to find the functional significance of this ligand-induced conformational change in reduced Pd.

*Acknowledgments.* This work was supported in part by grants-in-aid for scientific research on priority areas from the Ministry of Education, Science, Sports and Culture of Japan, by the Special Coordination Funds of the Science and Technology Agency of Japan, and by grants from Keio University.

# References

1. Oritiz de Montellano PR (1995) Cytochrome P-450: structure, mechanism and biochemistry, 2nd edn. Plenum, New York
2. Nelson DR, Kamataki T, Waxman DJ, et al. (1993) The P-450 superfamily: update on new sequences, gene mapping, accession numbers, early trivial names of enzymes, and nomenclature. DNA Cell Biol 12:1–51
3. Gunsalus IC, Meeks JR, Lipscomb JD, et al. (1974) Bacterial monooxygenases—the P-450 cytochrome system. In: Hayaishi O (ed) Molecular mechanisms of oxygen activation. Academic, New York, pp 559–613
4. Shimada H, Sligar SG, Yeom H, et al. (1997) Heme monooxygenase—a chemical mechanism for cytochrome P-450 oxygen activation. In: Fuanabiki T (ed) Oxygenases and model system, vol 19. Catalysis by metal complexes. Kluwer, Dordrecht, pp 195–221
5. Lipscomb JD, Sligar SG, Namtvedt MJ, et al. (1976) Autooxidation and hydroxylation reaction of oxygenated cytochrome P-450$_{cam}$. J Biol Chem 251:1116–1124
6. Ishimura Y, Makino R, Iizuka T, et al. (1987) Resonance Raman and infrared spectral studies on carbon monoxide and dioxygen complexes of cytochrome P-450. In: Sato R, Omura T, Imai Y, Fujii-Kuriyama Y (eds) Cytochrome P-450: new trend. Yamada Science Foundation, pp 151–153
7. Imai M, Shimada H, Watanabe Y, et al. (1989) Uncoupling of the cytochrome P-450$_{cam}$ monooxygenase reaction by a single mutation, threonine-252 to alanine or valine: a possible role of the hydroxy amino acid in oxygen activation. Proc Natl Acad Sci USA 86:7823–7827
8. Gunsalus IC, Wagner GC (1978) Bacterial P-450$_{cam}$ methylene monooxygenase components: cytochrome m, putidaredoxin, and putidaredoxin reductase. Methods Enzymol 52:166–188

9. Tsibris JCM, Tsai RL, Gunsalus IC, et al. (1968) The number of iron atoms in the paramagnetic center ($g = 1.94$) of reduced putidaredoxin, a nonheme iron protein. Proc Natl Acad Sci USA 59:959–965

10. Stayton PS, Sligar SG (1990) The cytochrome P-450$_{cam}$ binding surface as defined by site-directed mutagenesis and electrostatic modeling. Biochemistry 29:7381–7386

11. Pochapsky TL, Lyons T, Ratnaswamy G, et al. (1996) A structure-based model for cytochrome P-450$_{cam}$–putidaredoxin interactions. Biochimie 78:723–733

12. Shimada H, Makino R, Imai M, et al. (1990) Mechanism of oxygen activation by cytochrome P-450$_{cam}$. In: Yamamoto S, Nozaki M, Ishimura Y (eds) International symposium on oxygenases and oxygen activation. Yamada Science Foundation, pp 133–136

13. Gerber NC, Sligar SG (1994) A role of Asp-251 in cytochrome P-450$_{cam}$ oxygen activation. J Biol Chem 269:4260–4266

14. Unno M, Shimada H, Toba Y, et al. (1996) Role of Arg112 of cytochrome P-450$_{cam}$ in the electron transfer from reduced putidaredoxin. J Biol Chem 271:17869–17874

15. Sligar SG, Gunsalus IC (1976) A thermodynamic model of regulation: modulation of redox equilibrium in camphor monooxygenase. Proc Natl Acad Sci USA 73:1078–1082

16. Hintz MJ, Peterson JA (1981) The kinetics of reduction of cytochrome P-450$_{cam}$ by reduced putidaredoxin. J Biol Chem 256:6721–6728

17. Poulos TL, Finzel BC, Howard AJ, et al. (1987) High-resolution crystal structure of cytochrome P-450$_{cam}$. J Mol Biol 195:687–700

# Crystal Structure of Nitric Oxide Reductase Cytochrome P-450$_{nor}$ from *Fusarium oxysporum*

SAM-YONG PARK[1], HIDEAKI SHIMIZU[2], SHIN-ICHI ADACHI[1],
YOSHITSUGU SHIRO[1], TETSUTARO IIZUKA[1], and HIROFUMI SHOUN[3]

*Summary.* Nitric oxide reductase cytochrome P-450$_{nor}$ (P-450$_{nor}$) isolated from *Fusarium oxysporum* is the heme enzyme catalyzing NO reduction in the fungal denitrification process. The three-dimensional structures of this P-450$_{nor}$ in the ferric resting and the ferrous-CO states were determined by X-ray diffraction and refined at 2.0 Å to an R-factor of 19.7% and 19.9%, respectively. Although the overall structure of P-450$_{nor}$ is basically similar to those of monooxygenase cytochrome P-450s, the pocket in the heme distal side is widely opened, implicating the binding of NADH in this site. On binding of CO to the heme iron, the water molecules located in the heme distal pocket are rearranged in their positions, resulting in formation of the hydrogen bound network from the water molecule adjacent to the iron ligand to the protein surface of the distal pocket through the hydroxyl group of Ser-286 and the carboxyl group of Asp-393. Thus, this hydrogen-bonding network possibly provides a pathway of proton delivery in the NO reduction reaction.

*Key words.* Nitric oxide reductase—Cytochrome P-450—Crystal structure—Denitrifying fungus

## Introduction

Nitric oxide (NO) is now recognized as an important molecule in numerous biological processes, serving as a mediator in cellular signal transduction pathways [1,2] and as an intermediate in biological denitrification processes as well [3]. The NO in mammalian cells is produced by nitric oxide synthases (NOSs), which catalyze the stepwise oxidation of L-arginine to NO and L-citrulline using molecular oxygen (O$_2$), the so-called monooxygenation reaction [4–6]. The NOS is a heme enzyme having a reductase domain at its C-terminal region. Optical absorption and resonance

[1] Institute of Physical and Chemical Research (RIKEN), 2-1 Hirosawa, Wako, Saitama 351-01, Japan
[2] Facutly of Science, Gakushuin University, Toshima-ku, Tokyo 170, Japan
[3] Institute of Applied Biochemistry, University of Tsukuba, 1-1-1 Tennodai, Tsukuba, Ibaraki 305, Japan

Raman spectral data show that the heme iron is coordinated, as in cytochrome P-450 (P-450), to a cysteine throlate [7–10]. Because of these structural and functional similarities, the heme domain of NOS is currently thought to be a P-450-like monooxygenase.

The P-450$_{nor}$ isolated from the denitrifying fungus *Fusarium oxysporum* is a water-soluble heme enzyme consisting of 403 amino acid residues and having a molecular weight of approximately 46 000. The enzyme reduces NO to $N_2O$ in the following sequence: $2NO + NADH + H^+ \rightarrow N_2O + NAD^+ + H_2O$. The turnover number of this reaction was estimated to be in excess of 30 000 min$^{-1}$, indicating a rapid and effective NO-diminishing system. Unlike monooxygenation reactions by general P-450s, the NO reduction reaction occurs without any aid from other proteinaceous redox mediators such as a flavoprotein or an iorn-sulfur protein [11–13], suggesting that the electrons utilized in the NO reduction reaction can be directly transferred from NADH to the heme site. On the basis of several kinetic and spectroscopic observations in the reaction of the P-450$_{nor}$ with NO and NADH, we have proposed a mechanism for the NO reduction reaction as follows [14]: (i) NO binds to the ferric resting enzyme; (ii) the resultant ferric NO complex is reduced with two electrons from NADH to yield the characteristic intermediate; and (iii) the NO bound to the intermediate reacts with another NO to generate $N_2O$ and $H_2O$. In terms of NO and protons, this catalytic reaction can be represented as follows: $(NO)^{2-} + NO + 2H^+ \rightarrow N_2O + H_2O$. On the basis of the crystal structure of P-450$_{nor}$, the reaction mechanism of the NO reduction, particularly the precise pathway for the protons required in this reaction, can be discussed.

## Materials and Methods

P-450$_{nor}$ was crystallized with vapor diffusion using the sitting-drop technique. Crystals were grown in 100 mM MES (2-(N-morpholimo)ethanesulfonic acid) buffer at pH 5.6 using polyethylene glycol (PEG) 4000 as a precipitant. The initial drops were composed of 2 μl of the protein solution (50 mg ml$^{-1}$) and 2 μl of the precipitant solution and were equilibrated with the precipitant solution in a 1-ml reservoir.

The crystal belongs to the orthorhombic space group of $P2_12_12_1$ with unit cell dimensions $a = 54.99$ Å, $b = 82.66$ Å, and $c = 87.21$ Å, containing one molecule in an asymmetric unit. The X-ray data were collected at stations BL-6A and BL-18B of Photon Factory (Tsukuba, Japan). Intensity data were obtained using a Weissenberg camera for the macromolecular crystallography and imaging plates as the detector [15]. The crystallographic structure of P-450$_{nor}$ was solved by combination of the phase information obtained from MAD (multiwavelength anomalous diffraction), MIR (multiple isomorphous replacement), and MIRAS (multiple isomorphous replacement with anomalous scattering), using the program MLPHARE [16] in the CCP4 suite. Phases were calculated to 2.5 Å resolution, and the overall figure of merit was 0.525. A density modification program, Solomon [17], was used for phase improvement. Structural evaluation with PROCHECK [18] indicated that the refined structure has good geometric parameters. Almost all backbone dihedral angles fall within allowed regions in the Ramachandran plot. The result of data reduc-

TABLE 1. Statistics of crystallographic analysis

| Data set | | Resolution (Å) | Reflections measured/ unique | Completeness (%), overall/ outer shell | R$_{merge}$ (%)[a], overall/ outer shell |
|---|---|---|---|---|---|
| Native | | 2.0 | 84152/24733 | 90.1/79.5 | 5.4/10.7 |
| CO complex | | 2.0 | 77274/22706 | 80.2/63.3 | 4.4/10.7 |
| MIR analysis (wavelength) | | | | | |
| Fe | (1.738Å) | 2.5 | 20268/9796 | 68.9/54.7 | 4.4/8.0 |
| Fe | (1.741Å) | 2.5 | 19120/9518 | 68.6/53.9 | 4.6/9.8 |
| Fe | (1.700Å) | 2.5 | 58998/13476 | 97.9/94.6 | 4.7/6.9 |
| ErCl3 | (1.00Å) | 2.5 | 38638/11999 | 79.4/62.1 | 5.8/26.6 |
| K2PtCl4 | (1.00Å) | 2.5 | 35784/10834 | 73.3/53.3 | 4.6/16.5 |
| KAu(CN)2 | (1.00Å) | 2.5 | 37748/11850 | 79.7/61.9 | 2.8/8.1 |
| Thimersal-1 | (1.00Å) | 2.5 | 35910/11300 | 75.9/58.2 | 7.6/24.0 |
| Thimersal-2 | (1.00Å) | 3.0 | 23929/7146 | 83.2/63.1 | 7.0/12.7 |
| Overall figure of merit | | | 0.525 (20.0–2.5Å) | | |

| Refinement statistics | | |
|---|---|---|
| | Ferric-P450$_{nor}$ | Ferrous-CO P450$_{nor}$ |
| Refinement resolution (Å) | 10.0–2.0 | 10.0–2.0 |
| R cryst/R free (%, all data) | 19.7/26.6 | 19.9/26.9 |
| rmsd Bond lengths (Å) | 0.006 | 0.007 |
| rmsd Bond angles (Å) | 1.324 | 1.305 |
| Water atoms | 159 | 139 |

MIR, multiple isomorphous replacement.
Outer shell, the highest resolution shell (final resolution 2 Å was 2.24–2.00 Å; 2.5 Å was 2.79–2.50 Å; 3.0 Å was 3.26–3.00 Å).
[a] $R_{merge} = \Sigma\Sigma_i \mid \langle I(h) \rangle \text{-} I(h)_i \mid / \Sigma\Sigma_i \langle I(h) \rangle$, where $\langle I(h) \rangle$ is the mean intensity after rejections.

tion is summarized in Table 1. Coordinates have been deposited at Brookhaven Databank, ID code 1ROM.

Crystals of the carbon monoxide (CO) complex of ferrous P-450$_{nor}$ were prepared by soaking the crystal of the ferric resting enzyme in the CO-saturated mother liquid in the presence of dithionite for about 1 day. For the X-day diffraction measurements, the crystals were installed into glass capillaries under N$_2$ atmosphere, and pure CO was flushed into the capillary just before sealing. The crystal of the ferrous CO complex of the fungal NOR has the space group of $P2_12_12_1$ with cell dimensions of $a = 55.14$ Å, $b = 82.82$ Å, and $c = 87.17$ Å. The structural refinement was carried out after constructing the starting model from the coordinates of the ferric resting enzyme. The root mean square (rms) deviation of the bond lengths and bond angles from standard values were 0.006 Å and 1.305°, respectively. The comparatively small temperature factor of bound CO molecules (C = 17 Å$^2$, O = 22 Å$^2$) suggests that they had an occupancy close to 100%. The result of data reduction is summarized in Table 1. Coordinates have been deposited at Brookhaven Databank, ID code 2ROM.

# Results and Discussion

## Crystal Structure of P-450*nor* in the Ferric Resting State

### Overall Structure

The overall fold of P-450$_{nor}$ is shown in Fig. 1 as a ribbon diagram. The molecule has a triangular prism shape with an edge length of approximately 60 Å and a thickness of approximately 30 Å. The fungal nitric oxide reductase (NOR) is a mixed α- and β-structure, which can be described as consisting of two domains: the helix-rich α-domain, dominated by a four-helix bundle involving the D, E, I, and L α-helices and the β-sheet-rich β-domain. The fungal NOR contains 17 α-helices, 5 β$_{10}$-helices, and 12 β-strands in 4 β-sheets.

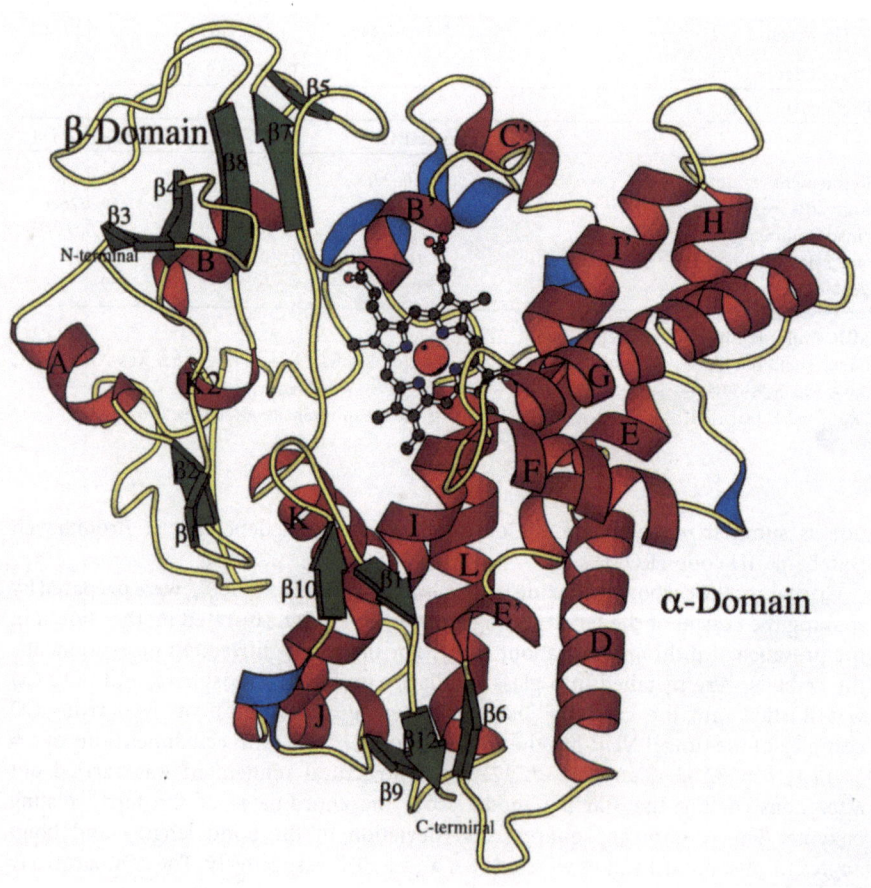

FIG. 1. The tertiary structure of P-450$_{nor}$ in the ferric resting state represented in a ribbon diagram: α-helices, *red coils*; β$_{10}$ helices, *sky-blue coils*; β-strands, *green arrows*; loops, *yellow tubes*. The helices, β-strands, and N- and C-termini are labeled; the heme is represented by a ball-and-stick model

## Heme Distal Side Structure

In comparison of the overall protein folding of the P-450$_{nor}$ with those of monooxygenase P-450s, one of the most striking differences is observed in the heme distal side structure, including the F- and G-helices. In superposition of the crystal structures between the P-450$_{nor}$ and P-450$_{cam}$, the rms differences for the helical C$_\alpha$ atoms are 2.80 Å and 1.99 Å for with and without the F/G region, respectively. In a side view of the molecules (Fig. 2), we find that the F- and G-helices in the P-450$_{nor}$ are "flipped up," compared with the position of these helices in P-450$_{cam}$, producing a large cavity at the heme distal pocket. In a top view, the F- and G-helices in the P-450$_{nor}$ significantly slide downward (in the orientation of Fig. 1), relative to those in monooxygenase P-450s, resulting in the distal pocket being widely open toward the solvent region.

The open heme distal pocket in P-450$_{nor}$ is more clearly seen in its molecular surface model, in which the heme (red-colored) can be seen from outside the molecule, while the heme pocket of P-450$_{cam}$ is closed in its crystal structure (Fig. 3). The heme cavity of the P-450$_{nor}$ is more widely opened than that of the substrate-free P-450$_{BM3}$ [19]. This open heme distal pocket agrees well with the kinetic data for ligand binding to the heme iron; CO binding to the ferrous iron and NO binding to the ferric iron are much faster in the fungal NOR than the corresponding ones in P-450$_{cam}$, most probably because there is less steric constraint. However, it should be noted that the heme environment is basically hydrophobic and solvent water cannot freely enter the iorn site.

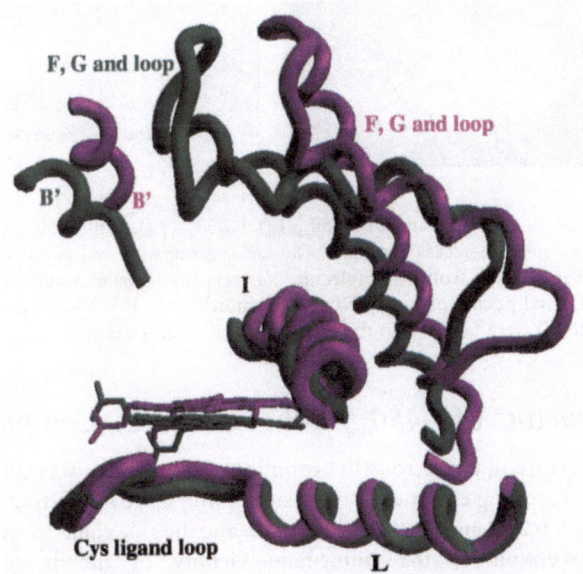

Fig. 2. Comparison of the distal (B'-, F-, G-helices) and proximal sites (Cys ligand loop, L-helix) of the P-450$_{nor}$ (*violet*) and P-450$_{cam}$ (*green*); the superposition in clued C$_\alpha$ atoms from the B'- (NOR 47–57, cam 67–77), E- (NOR 139–153, cam 149–163), and L- (NOR 343–368, cam 348–373) helices, with a fitting error of 0.883 Å. NOR, nitric acid reductase

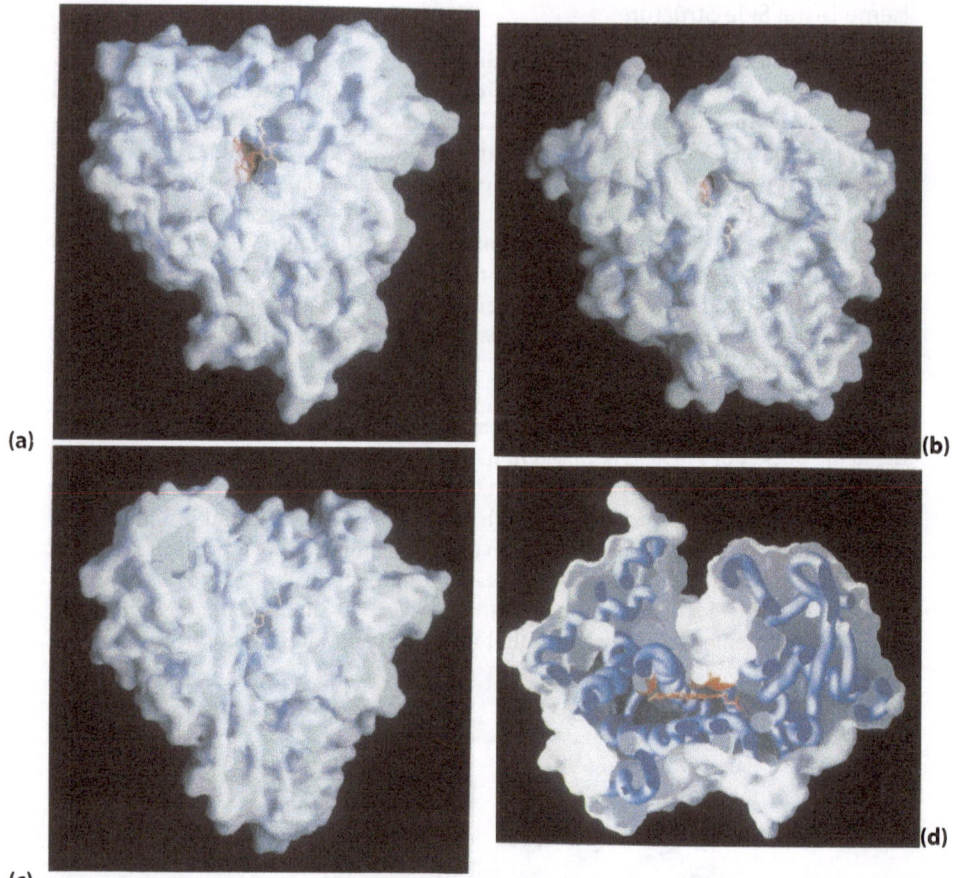

**(a)**

**(b)**

**(c)**

**(d)**

FIG. 3. Molecular surface diagrams of P-450$_{nor}$ (*a*), P-450$_{BM3}$ (*b*), and P-450$_{cam}$ (*c*) viewing along their respective substrate-access channels. The heme groups are *red bonds* in which the *red-colored* heme can be seen from the molecular surface. (*d*) Molecular surface model to show widely opened distal pocket of fungal NOR. The coordinates of P-450$_{cam}$ (code, 1phc) and P-450$_{BM3}$ (code, 2hpd) were taken from the Brookhaven Protein Data Bank

## Crystal Structure of P-450$_{nor}$ in the Ferrous CO Complex

When the structure of the ferrous CO complex of the P-450$_{nor}$ was superimposed on that of the ferric resting enzyme (5–403 residues with 0.248 Å of fitting error), the rms differences were 0.24 Å and 0.83 Å) for the main and the side chain atoms, respectively. With the exception of the immediate vicinity of the heme, the overall protein structures of the P-450$_{nor}$ in these two states are essentially identical to one another.

On CO binding to the heme iron of the P-450$_{nor}$, the water molecules in the heme distal pocket change their position, as shown in Fig. 4. Wat-69 (Wat-63 in the ferric resting form) remains hydrogen bonded to the hydroxyl group of Thr-243 in the ferrous CO form, but Wat-72 and Wat-113 are removed on CO binding, so that Wat-

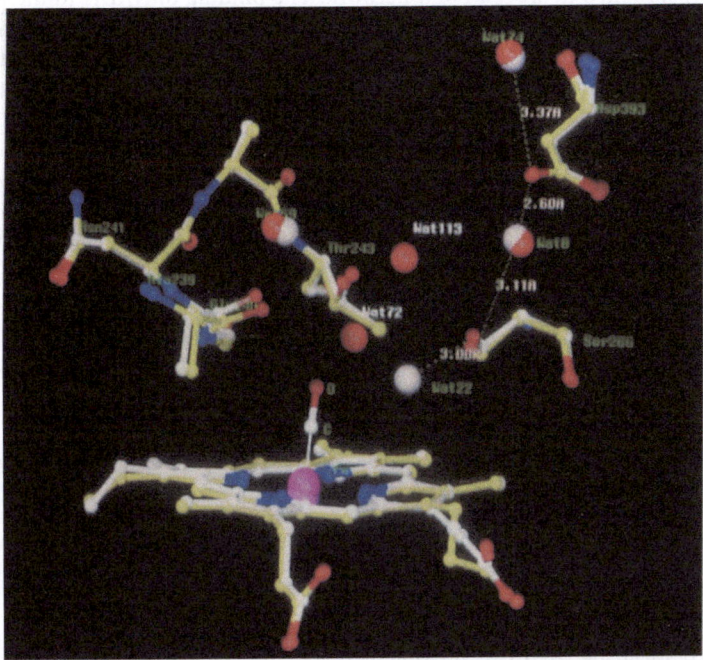

FIG. 4. Rearrangement in location of the water molecules in the heme distal pocket comparing the ferric resting (*yellow bonds*) and the ferrous CO complex (*white bonds*). Molecules have been superimposed in all C$_\alpha$ atoms, with a fitting error of 0.248 Å. Water molecules in the ferric resting enzyme are indicated by *red balls*; those from ferrous-CO are indicated with *white balls*. Wat-72 and Wat-113 are located only in the ferric resting enzyme, while Wat-22 is only seen in the ferrous CO enzyme. The hydrogen bonds in the ferrous CO form are represented as *green broken lines* with bond distances shown in *white numbers*

69 is isolated from the solvent water in the ferrous CO form. Instead, a new water molecule (Wat-22) is located just adjacent to the iron-bound CO and then interacts with the Ser-286 hydroxyl group through a hydrogen bond. The distance between CO and Wat-22 is 3.04 Å. Wat-22 shows a relatively high temperature factor (42 Å$^2$), indicative of its high mobility. The hydroxyl group of Ser-286 is further hydrogen bonded with another water molecule (Wat-8, 21 Å$^2$) which is also hydrogen bonded with a carbonyl group of Asp-393. As the Asp-393 is a residue involved in the hydrophilic surface of the distal pocket, the hydrogen-bonding network from Wat-22 to Asp-393 through Ser-286 and Wat-8 connects the iron-bound ligand with the solvent region in the distal heme pocket.

## Conclusion

The sructural difference corresponding to the functional difference of P-450$_{nor}$ from the monooxygenase P-450s are the presence of an NADH-binding site near the active site and the readily availability of solvent protons to the ligand-binding site through the correctly designed hydrogen bond network. The binding of NADH to the heme distal pocket is presumed to be convenient for the direct two-electron transfer to the ferric-NO moiety for formation of the reaction intermediate. This is sharply contrasted to the monooxygenase P-450s, of which the protein surface in the heme proximal side is thought to be the interaction site of its reductase (flavoprotein or iron-sulfur protein). The reductase modulates the two-electron transfer step from NADH to the one-electron transfer step to the heme site of P-450. In addition, in the monooxygenation reaction, the readily availability of protons undesirably destabilizes the ferrous-$O_2$ complex, as was suggested for the case of the T252A mutant of P-450$_{cam}$ [20]. In contrast, the ferric-NO complex of the thiolate-bound heme, which is the first stage in the NO reduction, is sufficiently stable even in a protic environment, and as a result the location of the water molecule (Wat-69) adjacent to the iron-bound NO serves as a convenient and effective proton donor for intermediate formation, or N–N bond formation and N–O bond cleavage in $N_2O$ and $H_2O$ formation.

*Acknowledgments.* We thank Prof. N. Sakabe, Dr. N. Watanabe, and Dr. M. Suzuki for their advice in the data collection at PF. Data collection at PF was performed under approval of the PF Program Advisory Committe (Proposanl No. 94G253).

# References

1. Culotta E, Koshland DE Jr (1992) NO news is good news. Science 258:1862–1865
2. Stamler JS, Singel DS, Loscalzo J (1992) Biochemistry of nitric oxide and its redoxy-activated forms. Science 258:1898–1902
3. Coyne MS, Arunakumari AA, Averill RA, Tiedje JM (1989) Immunological identification and distribution of dissimilatory heme cd1 and nonheme copper nitrite reductases in denitrifying bacteria. Appl Environ Microbiol 55:2924–2931
4. Stuehr DJ, Kwon NS, Nathan CF, et al. (1991) $N^\omega$-Hydroxy-L-arginine is an intermediate in the biosynthesis of nitric oxide from L-arginine. J Biol Chem 266:6259–6263
5. Stuehr DJ, Ikeda-Saito M (1992) Spectral characterization of brain and macrophage nitic oxide synthases. J Biol Chem 267:20547–20546
6. Klatt P, Schmidt K, Uray G, Mayer B (1993) Multiple catalytic functions of brain nitric oxide synthase. J Biol Chem 268:14781–14787
7. White KA, Marletta MA (1992) Nitric oxide synthase is a cytochrome P-450 type hemoprotein. Biochemistry 31:6627–6661
8. McMillan K, Bredt DS, Hirsch DJ, et al. (1992) Cloned expressed rat cerebellar nitric oxide synthase contains stoicometric amounts of heme, which binds carbon monoxide. Proc Natl Acad Sci USA 89:11141–11145
9. Wang J, Stuehr DJ, Ikeda-Saito M, Rousseau DL (1993) Heme coordination and structure of the catalytic site in nitric oxide synthase. J Biol Chem 268:22255–22258
10. Wang J, Rousseau DI, Abu-Soud HM, Stuehr DJ (1994) Heme coordination of NO in NO synthase. Proc Natl Acad Sci USA 91:10512–10516
11. Shoun H, Suyama W, Yasui T (1989) Soluble, nitrate/nitrite-inducible cytochrome P-450 of the fungus *Fusarium oxysporum*. FEBS Lett 244:11–14
12. Shoun H, Tanimoto T (1991) Denitrification by the fungus *Fusarium oxysporum* and involvement of cytochrome P-450 in the respiratory nitrite reduction. J Biol Chem 266:11078–11082
13. Nakahara K, Tanimoto T, Hatano K, et al. (1993) Cytochrome P-450 55A1 (P-450dNIR) acts as nitric oxide reductase empolying NADH as the direct electron donor. J Biol Chem 268:8350–8355

14. Shiro Y, Fujii M, Iizuka T, et al. (1995) Spectroscopic and kinetic studies on reaction of cytochrome P-450$_{nor}$ with nitric oxide. J Biol Chem 270:1617–1623
15. Sakabe N (1983) A focusing Weissenberg camera with multi-layer-line screens for macromolecular crystallography. J Appl Crystallogr 16:542–547
16. The CCP4 (1994) CCP4 collaborative computational project 4. Acta Crystallogr D50:760–763
17. Abrahams JP, Leliew AGW, Lutter R, Walker JE (1994) Structure at 2.8 Å resolution of F1-ATPase from bovine heart mitochondria. Nature 370:621–628
18. Laskowski RA, MacArthur MW, Moss DS, Thronton JM (1993) PROCHECK—a program to check the stereochemical quality of protein stuctures. J Appl Crystallogr 26:283–291
19. Ravichandran KG, Boddupalli SS, Hasemann CA, et al. (1993) Crystal structure of hemoprotein domain of P-450$_{BM3}$, a prototype for microsomal P-450s. Science 261:731–736
20. Raag R, Martinis SA, Sligar SG, Poulos TL (1991) Crystal structure of the cytochrome P-450$_{cam}$ active site mutant Thr252Ala. Biochemistry 30:11420–11429

# Analysis of NAD(P)H Binding Site of Cytochrome P-450$_{nor}$

Naoki Takaya, Takashi Kudo, and Hirofumi Shoun

*Summary.* Cytochrome P-450$_{nor}$ (P-450$_{nor}$) is an enzyme that catalyzes reduction of nitric oxide (NO) to generate nitrous oxide (N$_2$O) using NADH as a direct electron donor. From recent studies, it was suggested that the N-terminal region consisting of 160 amino acid residues is critical for NAD(P)H recognition (Kudo et al, Biochimie 78:792–799, 1996). We found a sequence (74 A–X–G–X–X–A 79) similar to the pyridine nucleotide-binding motif (G–X–G–X–X–G/A) in the region that corresponds to the B′-helix in P-450$_{cam}$. We constructed mutant proteins in the yeast expression system at the Ala-74 and Gly-76 residues. One of the mutant proteins, whose Gly-76 was substituted with Val, showed marked reduction of the catalytic activity. Little difference was observed between wild-type and mutant proteins in the absorption spectra as well as the NO-binding state measured by the laser photolysis method. These results suggest that the 76th Gly is involved in NAD(P)H binding.

*Key words.* Cytochrome P-450$_{nor}$—NAD(P)H—Site-directed mutagenesis—Denitrification—*Fusarium oxysporum*

## Introduction

Cytochrome P-450$_{nor}$ (P-450$_{nor}$) is an enzyme that is involved in denitrification by the fungus *Fusarium oxysporum* [1]. In contrast to other P-450 monooxygenases, it catalyzes reduction of nitric oxide (NO) to generate nitrous oxide (N$_2$O) as follows [2,3]:

$$2NO + NADH + H^+ \rightarrow N_2O + NAD + H_2O$$

P-450$_{nor}$ catalyzes this reaction without any electron-donating proteins such as P-450 reductases and receives electrons directly from reduced nicotinamide adenine nucleotide (NADH). It is suggested that the catalytic turnover is near diffusion limit [4], so that it seems that P-450$_{nor}$ would recognize and bind NADH efficiently. There are no reports of direct binding of nucleotide cofactors to cytochrome P-450 other than P-450$_{nor}$, and the mechanism of NADH binding to the enzyme is interesting.

Institute of Applied Biochemistry, University of Tsukuba, 1-1-1 Tennodai, Tsukuba, Ibaraki 305, Japan

156

To study the mechanism of NADH recognition of P-450$_{nor}$, we constructed the chimeric enzymes between P-450$_{nor}$ and a related enzyme P-450$_{nor2}$ of *Cylindrocarpon tonkinense* [5], which can utilize NADPH as an electron donor as well as NADH. Analyzing NADH/NADPH specificity of the chimeric proteins, it was suggested that the N-terminal region consisting of 160 amino acid residues was critical for NAD(P)H recognition [6]. A sequence (74 A–X–G–X–X–A 79) similar to the pyridine nucleotide-binding motif (G–X–G–X–X–G/A) is found in the region that corresponds to the B'-helix in P-450$_{cam}$.

In this study, we mutagenized the conserved amino acid residues in the motif and demonstrated that Gly-76 is important for the nitric oxide reductase (Nor) activity of P-450$_{nor}$. It is possible that the residue is critical for the binding of NADH to the enzyme.

## Materials and Methods

### Strains, Culture, and Media

*Escherichia coli* K-12 strain HB101 was used for the DNA manipulations. The strains were grown in LB medium (1% tryptone, 0.5% yeast extract, 0.5% NaCl). *Saccharomyces cerevisiae* INVSC2 (*MATa, Δhis200, ura3-167*) was cultured as described [6].

### Site-Directed Mutagenesis

The P-450$_{nor}$ cDNA was mutagenized by two successive polymerase chain reactions (PCR). In the first PCR, pFP-450-118 [6], which contains the 1374-bp fragment of P-450$_{nor}$ cDNA, was used as a template. Thirty cycles at 94°C for 1 min for denaturation, at 55°C for 2 min for anealing, and at 72°C for 1 min for extension were performed with two sets of primers in each mutagenesis as follow: for the A74V mutant, 5'-GAGCTTAGCGTCAGTGGAAAGC-3' and M13-47 (Takara Shuzo, Otsu, Japan), 5'-GCTTTCCACTGACGCTAAGCTC-3' and M13-RV (Takara); for the G76V mutant, 5'-GCGCCAGTGTCAAGCAGC-3' and M13-47, 5'-GCTTGCTTGACACTGGCGC-3' and M13-RV; for the A74V, G76V mutant, 5'-GCTTAGCGTCAGTGTCAAGCAAGC-3' and M13-47, 5'-GCTTGCTTGACACTGACGCTAAGC-3' and M13-RV, respectively. In the second PCR, both the first PCR products were purified with polyacrylamide gel electrophoresis and mixed, then used for the templates of the second PCR with the same conditions with M13-47 and M13-RV as primers.

### Construction of Expression Plasmids

For the A74V and the A74V, G76V mutants, the *Hind*III-*Bam*HI fragments of the second PCR products were cloned once into the site of pBluescript II KS(+). This was then digested with *Bam*HI and ligated with the 1-kb *Bam*HI fragment of pfp-450-118. For the G76V mutant, the *Hind*III-*Bam*HI fragment of the second PCR product was cloned once into the site of pfp-450-20, which had been constructed by inserting the P-450$_{nor}$ cDNA into the *Eco*RI site of pUC18. Then, it was digested with *Bam*HI and ligated with the 1-kb *Bam*HI fragment of pfp-450-118.

Each resultant plasmid was digested with *Kpn*I and *Xba*I; the 1.4-kb fragments were purified and inserted into the *Kpn*I-*Xba*I site of pYES2 to give rise to pYESFO-74V, pYESFO-76V, and pYESFO-74V76V, respectively.

## Expression and Purification of the Mutant Proteins

The plasmid pYESFO1 [6] and its derivatives were introduced into *S. cerevisiae* by the method of Ito et al. [7]. Recombinant enzymes were produced as described [7]. Preparation of the soluble fraction of the yeast was essentially done as described by Kudo et al. [6]. The G76V protein was purified by the method of Kudo et al. [6] using DEAE-cellulose column chromatography and fast protein liquid chromatography (FPLC) equipped with a Mono Q HR5/5 column (Pharmacia Biotech, Tokyo, Japan).

## Analytical and Assay Methods

Spectrophotometric measurements were done with a Shimadzu UV2100 spectrophotometer (Shimadzu, Kyoto, Japan). The P-450$_{nor}$ content was determined according to Omura and Sato [8] using the extinction coefficient of $86.9 \text{mM}^{-1}\text{cm}^{-1}$ for the absorbance difference between 448 and 490 nm. Nor activity was assayed using the soluble fraction containing 20 pmol of P-450$_{nor}$ as described [1].

# Results and Discussion

## Expression of the A74V, G76V, and A74V, G76V Mutant Proteins

Soluble fractions were prepared from the yeast transformants harboring the plasmids pYESFO1 [6], pYESFO-74V, pYESFO-76V, and pYESFO-74V76V cultured in YEPG medium and assayed for Nor activity (Fig. 1). The soluble fraction of the transformant with pYESFO-74V remained approximately 60% of the activity of pYESFO1, indicating that Ala-74 is not essential for Nor activity. In contrast, the yeast strain expressing G76V decreased Nor activity to less than 10%. Double mutation in the 74th and the 76th residues to Val caused a similar extent of decrease in Nor activity as the G76V mutant. These results indicated that Gly-76 was important for the enzyme activity. As the 76th residue corresponds to the highly conserved Gly of the NADH-binding motif [9], it seemed that Gly-76 would contribute to NADH recognition or binding.

## Characteristics of the G76V Mutant Proteins

To investigate the role of Gly-76, the G76V mutant was purified and analyzed. The absorption peaks of the mutants were observed at 413 nm, 418 nm, and 448 nm for native, dithionite-reduced (Fig. 2), and CO-liganded forms (data not shown), respectively, which were quite well matched to the wild-type P-450$_{nor}$ [2,5]. These results suggest that the heme was properly incorporated into the mutant as the wild-type P-450$_{nor}$. The NO-binding form of the G76V mutant showed an absorption peak at 431 nm (Fig. 2), which exactly matches that of the wild type [4], suggesting that the

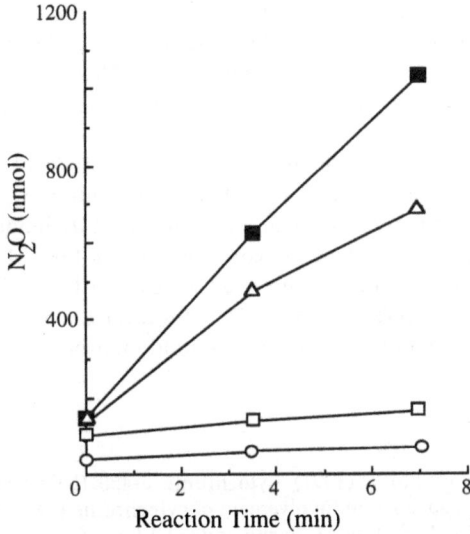

FIG. 1. Nitric oxide reductase activity of the soluble fractions of the cells producing wild-type and mutant enzymes. *Closed squares*, wild type; *triangles*, the A74V mutant; *circles*, the G76V mutant; *open squares*, the A74V, G76V mutant

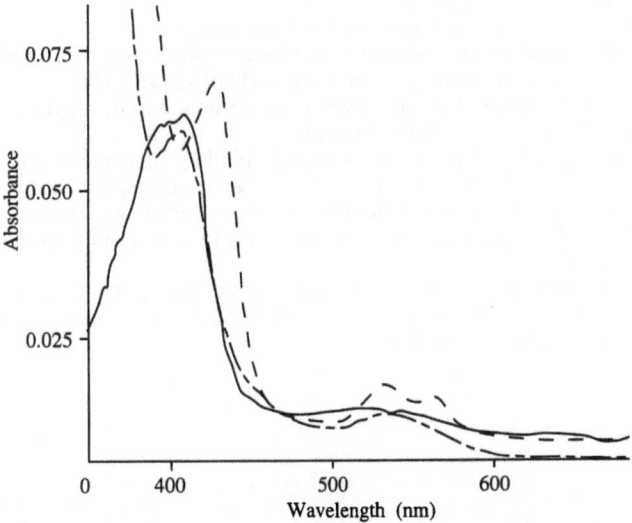

FIG. 2. Absorption spectra of the G76V mutants. *Solid line*, ferric (native); *broken line*, ferrous (dithionite reduced); *dash-dot line*, ferrous plus NO

mutation does not perturb NO to bind to the heme iron in the G76V mutant. The binding constant of NO to the G76V mutant was indistinguishable from the wild type by the laser photolysis method (data not shown). These results suggest that Gly-76 is not involved in NO binding to the heme. It is possible that Gly-76 may be important for binding of NADH.

Although the primary structure of Ala-74 to Gly-76 of P-450$_{nor}$ resembles the NADH-binding motif, our recent crystallographic studies suggest that the region does not form the $\beta\alpha\beta$ structure shown in common NAD(P)H-binding domains [9]; rather, it constitutes an $\alpha$-helical structure corresponding to the B'-helix of P-450$_{cam}$ (manuscript in preparation), which is supposed to be a substrate recognition site for P-450 superfamily [10]. Therefore, in the P-450$_{nor}$, the mechanism to recognize NADH would be different from that of other nucleotide-binding proteins.

# References

1. Shoun H, Tanimoto T (1991) Cytochrome P-450 (NOR) of the denitrifier fungus *Fusarium oxysporum* and involvement of cytochrome P-450 in the respiratory nitrate respiration. In: Archakov AI, Bachmanova GI (eds) Cytochrome P-450: biochemistry and biophysics. Joint Stock Company, Moscow, pp 626–628
2. Nakahara K, Tanimoto T, Hatano K, et al. (1993) Cytochrome P-450 55a1 (P-450dNIR) acts as nitric oxide reductase employing NADH as the direct electron donor. J Biol Chem 268:8350–8355
3. Shiro Y, Kato M, Iizuka T, et al. (1994) Kinetics and thermodynamics of CO binding to cytochrome P-450$_{nor}$. Biochemistry 33:8673–8677
4. Shiro Y, Fujii M, Iizuka T, et al. (1995) Spectroscopic and kinetic studies on reaction of cytochrome P-450$_{nor}$ with nitric oxide. J Biol Chem 270:1617–1623
5. Usuda K, Toritsuka N, Matsuo Y, et al. (1995) Denitrification by the fungus *Cylindrocarpon tonkinense*: anaerobic cell growth and two isozyme forms of cytochrome P-450$_{nor}$. Appl Environ Microbiol 61:883–889
6. Kudo T, Tomura D, Dai XQ, Shoun H (1996) Two isozymes of P-450$_{nor}$ of *Cylindrocarpon tonkinense*: molecular cloning of the cDNA and genes, expression in the yeast, and the putative NAD(P)H binding site. Biochimie (Paris) 78:792–799
7. Ito H, Fukuda Y, Murata K, et al. (1983) Transformation of intact yeast cells treated with alkali cations. J Bacteriol 153:163–168
8. Omura T, Sato R (1964) The carbon monooxide-binding pigment of liver microsomes. II. Solublization, purification and properties. J Bacteriol 239:2379–2385
9. Wirrenga RK, Terpstra P, Hol WGJ (1986) Prediction of the occurrence of the ADP-binding $\beta\alpha\beta$-fold in proteins, using an amino acid sequence fingerprint. J Mol Biol 187:101–107
10. Gotoh O (1992) Substrate recognition site in cytochrome P-450 family 2 (CYP2) proteins inferred from comparative analyses of amino acid and coding nucleotide sequences. J Biol Chem 267:83–90

# Substitutions of Artificial Amino Acids O-Methyl-Thr, O-Methyl-Asp, S-Methyl-Cys, and 3-Amino-Ala for Thr-252 of Cytochrome P-450$_{cam}$: Probing the Importance of the Hydroxyl Group of Thr-252 for Oxygen Activation

Yoko Kimata[1], Hideo Shimada[1], Tada-aki Hirose[2], and Yuzuru Ishimura[1]

*Summary.* In the monooxygenation reaction catalyzed by cytochrome P-450$_{cam}$ (P-450$_{cam}$), Thr-252, a hydroxyl amino acid at the active site, is essential for the reductive cleavage of the O–O bond of oxygen: the hydroxyl (OH) group of Thr has been proposed to serve as an acid catalyst for the O–O bond scission. In this study, four different artificial amino acids were incorporated into the 252-position to verify the role of the OH group. The catalytic activities of the mutant enzymes suggest that the OH group does not function as the catalyst. Then, we propose that the OH group serves as an anchor of an acid catalyst and facilitates its catalytic action to cleave the O–O bond; water is possibly the catalyst.

*Key words.* Artificial amino acid—Mutagenesis—Cytochrome P-450—Catalytic mechanism—Hydrogen bond

## Introduction

Cytochrome P-450$_{cam}$ (P-450$_{cam}$) in *Pseudomonas putida* is a heme-containing monooxygenase that catalyzes the reaction: $d$-camphor + NADH + H$^+$ + O$_2$ ⇒ 5-$exo$-hydroxycamphor + NAD$^+$ + H$_2$O. In this reaction, the O–O bond of oxygen is cleaved; one of the oxygen atoms is incorporated into $d$-camphor, and the other atom is converted to H$_2$O. The mechanism of dioxygen scission to form an active oxygen species that monooxygenates substrate is one of the major questions in the reaction catalyzed by P-450$_{cam}$ as well as other heme-containing monooxygenases. At the oxygen scission step, two protons are required. Thr-252 has been proposed to participate in a proton delivery system (Fig. 1) through a hydrogen-bonding network (Thr-252–H$_2$O-Asp-251-Lys–176/Arg-186) extending from the interior active site to the outside of the protein [1–5]. In this system, the OH group of Thr-252 is an acid catalyst for the oxygen scission that donates a proton to the heme-bound oxygen: the OH group can act as a donor and acceptor of the hydrogen bond, which property is required for the

---

[1] Department of Biochemistry, [2] Pharmaceutical Institute, Keio University, 35 Shinanomachi, Shinjuku-ku, Tokyo 160, Japan

**a**

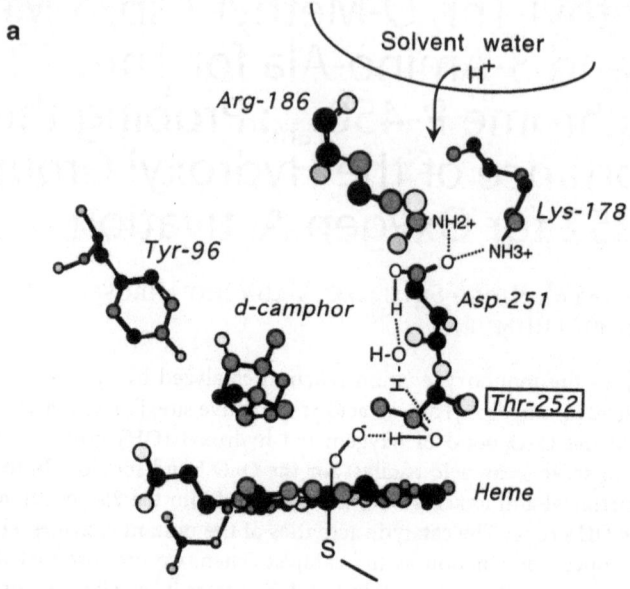

**b**

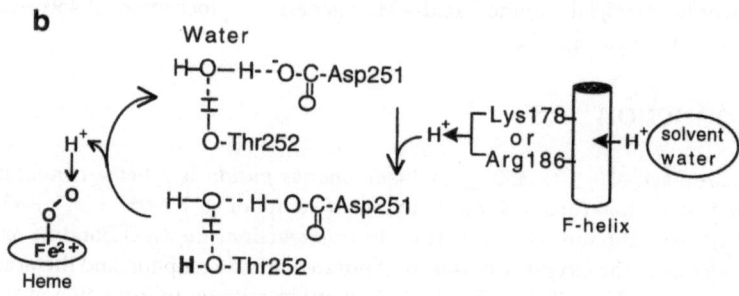

**c**

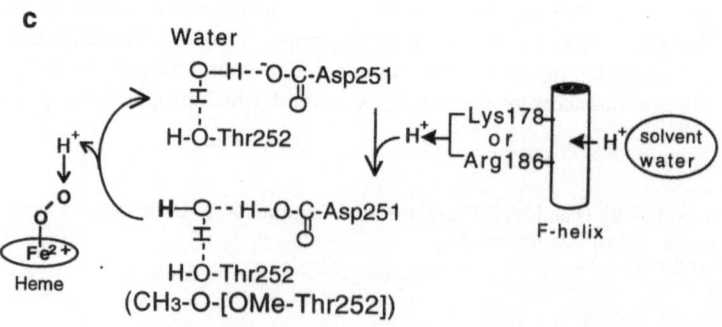

TABLE 1. Enzyme activity of the wild-type and mutant P-450$_{cam}$ carrying an artificial or natural amino acid at the 252 or 251 position

| P-450$_{cam}$ species | 5-*exo*-Hydroxycamphor formation per O$_2$ consumed (%) | Rate of oxygen consumption (µM/min per µM P-450$_{cam}$) |
|---|---|---|
| 252-Thr (wild-type) | 100 | 1304 |
| 252-*O*-Me-Thr[a] | 100 | 410 |
| 252-Ser | 85 | 830 |
| 252-*O*-Me-Asp[a] | 76 | 140 |
| 252-Asn | 57 | 420 |
| 252-Ile | 44 | 277 |
| 252-Val | 19 | 360 |
| 252-Leu | 17 | 7 |
| 252-Cys | 7 | 690 |
| 252-*S*-Me-Cys[a] | <5[b] | 102 |
| 252-3-Amino-Ala[a] | <5[b] | 73 |
| 252-Ala | 5 | 1150 |
| 251-Asp (wild-type) | 100 | 1304 |
| 251-Ala | 89 | 3 |

[a] Mutants containing artificial amino acid in this study.
[b] Less than detection limit.

proposed role of Thr-252. However, because a mutant P-450$_{cam}$ with *O*-methyl-Thr at the 252 position retained a considerable monooxygenase activity (see Table 1), the role of Thr-252 as a proton donor was questioned [6]. Thus, to further investigate the role of the OH group, three other artificial amino acids were site-specifically incorporated into the 252 position of P-450$_{cam}$. These artificial amino acids can be categorized to serve as (1) a proton donor but not an acceptor of a hydrogen bond or (2) an acceptor of a hydrogen bond but not a proton donor. The effects of these mutations on catalytic activities were evaluated by comparison with those of other site-directed mutants prepared in our laboratory.

## Materials and Methods

Construction of mutant P-450$_{cam}$ expression plasmid and biosynthetic incorporation of artificial amino acids into P-450$_{cam}$ have been previously described [6]. The mutant P-450$_{cam}$ synthesized in vitro was partially purified with DEAE column chromatography. The content of P-450$_{cam}$ was estimated from the absorption difference at 446 and

---

FIG. 1a–c. Proposed proton relay systems of cytochrome P-450$_{cam}$ in previous papers. a Schematic of proposed proton transfer pathway in the active site of cytochrome P-450$_{cam}$. Thr-252 is located close to the oxygen-binding site. b In the originally proposed proton relay system [4,5], the direct proton donor to the heme-bound oxygen is the hydroxyl group of Thr-252. c In the alternative proton relay system [6], the direct proton donor is the water molecule, whose existence is not yet certain. The system in part c is presented to explain the result that the *O*-methyl-Thr mutant also works well with a methoxy group (R–O–CH$_3$) in place of the hydroxyl group (R–OH) of Thr-252 [6]

500 nm in the CO-difference spectra ($\varepsilon_{mM}$ = 93), as described [6]. The catalytic activity of P-450$_{cam}$ was assessed by measuring the oxygen consumption and 5-*exo*-hydroxycamphor formation in a reconstituted system as previously detailed [1].

## Results and Discussion

Catalytic activities of *O*-methyl-Thr, *O*-methyl-Asp, *S*-methyl-Cys, and 3-amino-Ala mutants are presented in Table 1, together with other site-directed mutants at the 252 or 251 position [1,3] prepared in our laboratory. Characteristic of several mutants at the 252-position is the reduced efficiency for the product (5-*exo*-hydroxycamphor) formation, which results from the uncoupling of oxygen consumption from hydroxylation of the substrate [1]. For example, the Ala-252 mutant, which lacks the OH group of Thr-252, showed greatly reduced product formation per oxygen consumed (coupling ratio, 5%), although oxygen-consuming activity was as high as that of the wild-type enzyme [1]. On the other hand, mutants at the 251 position, which retains Thr-252, exhibited a low oxygen-consuming activity but a considerably high (about 90%) coupling ratio [3], as exemplified by Ala-251 in Table 1. Hence, we conclude that the major function of Thr-252 can be assessed by determining the coupling ratio.

The mutant P-450$_{cam}$ with *O*-methyl-Thr and *O*-methyl-Asp retained high coupling ratios (100% and 76%, respectively), although the rates of oxygen consumption were reduced to about one-third and one-ninth that of the wild-type enzyme, respectively. Because *O*-methyl-Thr and *O*-methyl-Asp cannot donate a proton, these results indicate that a proton donor at this 252 position is not required for efficient coupling. Also, a mere proton donor did not fulfill the function at the 252 position; the mutant P-450$_{cam}$ with 3-amino-Ala ($-NH^{3+}$), which can serve as a proton donor, showed severe uncoupling (<5% coupling). Thus, the previously proposed role of the OH group of Thr-252 as an acid catalyst for the O–O bond scission, which recruits protons through a hydrogen-bonding network consisting of Thr-252-H$_2$O-Asp-251-Lys-176/Arg-186 (Fig. 1a), is inadequate.

We have recently proposed [6] (Fig. 1c) that the water located between Thr-252 and Asp-251 in this system can be a terminal proton donor and serve as an acid catalyst. In this proposal, the OH group of Thr-252 acts as an acceptor of a hydrogen bond formed between the Thr and the water. Water molecule(s) found at the active site of the Ala-252 mutant in a X-ray analysis were suggested to be a cause of the uncoupling [7]. However, in the wild-type enzyme, efficient coupling can be obtained on behalf of the function of Thr-252 that serves as an anchor of the water in the proton delivery system by forming a hydrogen bond with it: this bond formation between Thr-252 and water can facilitate the catalytic function of the water. High coupling ratios of those P-450$_{cam}$ enzymes that carry an oxygen atom in the side chain of the 252-residue (Thr, Ser, Asn, *O*-methyl-Thr, and *O*-methyl-Asp) can also be explained by the ability of the oxygen atom to accept a hydrogen bond. In contrast, the activity of the 3-amino-Ala mutant can be reduced because the amino group that probably protonated as-NH$^{3+}$ in the protein does not act as a hydrogen bond acceptor.

The *S*-methyl-Cys and Cys mutants exhibited a low ratio of coupling (see Table 1), indicating that the sulfur atom cannot replace the oxygen atom in the O–O bond scission. The sulfur atom is known to form a less stable hydrogen bond than does the oxygen atom, as is shown by an equilibrium constant that is more than 100 fold smaller [8]. Therefore, an unstable hydrogen bond between the sulfur atom and water may not allow the sulfur atom to act as an anchor of water in the proton delivery system, which causes uncoupling of the reaction. The results shown by Val and Leu, both of which showed a low coupling ratio, support this idea because the carbon atom cannot form a hydrogen bond with a donor atom. Although the relatively higher catalytic activity of the Ile mutant (see Table 1) is unexpected, the spatial arrangement of the Ile residue within the active site may hold the water molecule at a favorable position for proton transfer without forming a hydrogen bond.

# References

1. Imai M, Shimada H, Watanabe Y, et al. (1989) Uncoupling of the cytochrome P-450$_{cam}$ reaction by a single mutation, threonine-252 to alanine or valine: A possible role of the hydroxy amino acid in oxygen activation. Proc Natl Acad Sci USA 86:7823–7827
2. Martinis SA, Atkins WM, Stayton PS, et al. (1989) A conserved residue of cytochrome P-450$_{cam}$ is involved in heme-oxygen stability and activation. J Am Chem Soc 111:9252–9253
3. Shimada H, Makino R, Imai M, et al. (1990) Mechanism of oxygen activation by cytochrome P-450$_{cam}$. In: Nozaki M, Yamamoto S, Ishimura Y (eds) International symposium on oxygenases and oxygen activation. Yamada Science Foundation, Japan, pp 133–136
4. Shimada H, Makino R, Unno M, et al. (1993) Proton and electron transfer mechanism in dioxygen activation by cytochrome P-450$_{cam}$. In: Lechner MC (ed) Proceedings of 8th international symposium on cytochrome P-450. Libbey, Paris, pp 299–306
5. Gerber NC, Sligar SG (1994) A role for Asp-251 in cytochrome P-450$_{cam}$ oxygen activation. J Biol Chem 269:4260–4266
6. Kimata Y, Shimada H, Hirose T, et al. (1995) Role of Thr-252 in cytochrome P-450$_{cam}$: a study with unnatural amino acid mutagenesis. Biochem Biophys Res Comman 208:96–102
7. Raag R, Martinis SA, Sligar SG, et al. (1991) Crystal structure of the cytochrome P-450$_{cam}$ active site mutant Thr252Ala. Biochemistry 30:11420–11429
8. Fersht AR (1987) The hydrogen bond in molecular recognition. Trends Biochem Sci 8:301–304

# From a Monooxygenase to an Oxidase: The P-450 BM3 Mutant T268A

GILLES TRUAN and JULIAN A. PETERSON

*Summary.* P-450s are heme monooxygenases that are widely distributed among living organisms. The P-450 catalytic cycle requires an input of two electrons and a molecule of oxygen. The generic reaction is the insertion of an oxygen atom in a substrate and the release of a water molecule. If the catalytic cycle does not complete successfully, the available electrons can generate reduced oxygen species like peroxide, superoxide, or water. These reactions are defined as uncoupled. Understanding how uncoupling can occur and the role of the side-chain amino acids around the heme is of great importance for describing the catalytic cycle. P-450 BM3 is a soluble P-450 isolated from *Bacillus megaterium*. It catalyzes hydroxylation of various fatty acids, as well as epoxidations of double bonds. We have constructed the mutant T268A and analyzed the effect on arachidonic acid (AA) and palmitic acid (PA) binding and metabolism. Data indicate that the mutation changes the binding and the coupling for both AA and PA metabolism. Cumene hydroperoxide-driven reactions are unaffected in the mutant. These data support the hypothesis of a role of T-268 in maintaining substrate position so that the reaction can be fully coupled. The role of substrate movement in generating active oxygen species is discussed.

*Key words.* B. *megaterium*—Fatty acid—Monooxygenase—Oxygen activation—P-450

## Introduction

The formation of a highly oxidizing species, the iron–oxo complex, is a key point of the P-450 catalytic cycle. Of great interest are the residues that are involved in molecular oxygen cleavage. To this end, the T-252 of P-450$_{cam}$ and its homologues in other P-450s have been extensively studied. In P-450$_{cam}$, mutation of this residue to an alanine leads to uncoupling NADPH and $O_2$ consumption [1,2]. According to the crystal structure of wild-type and T252A mutant of P-450$_{cam}$, T-252 could stabilize the

Department of Biochemistry, The University of Texas Southwestern Medical Center in Dallas, 5323 Harry Hines Boulevard, Dallas, TX 75235-9038, USA

dioxygen complex, preventing its conversion into hydrogen peroxide [3]. Nevertheless, the data of Kimata et al. showed that the hydrogen of the hydroxyl group in T-252 was not necessary for catalytic function [4].

P-450 BM3 is isolated from *Bacillus megaterium* and can oxidize long-chain fatty acids. P-450 BM3 contains two distinct functional domains, a heme domain and a reductase domain, and is a fully self-sufficient P-450 [5,6]. The tertiary structure of the heme domain of P-450 BM3 has been solved at 2.0 Å [7]. Sligar and co-workers have reported the construction and the analysis of lauric acid metabolism by the T268A mutant of P-450 BM3 [8].

We also have constructed the T268A mutant (T268A) and extensively analyzed the binding as well as the metabolism of arachidonic acid (AA) and palmitic acid (PA), both substrates of P-450 BM3 with significantly higher rates of turnover than lauric acid [9]. Here we present these data along with the degree of NADPH/O$_2$ coupling for AA and PA, as well as cumene hydroperoxide-driven reactions for PA metabolism.

## Material and Methods

All chemicals were obtained from Sigma Chemical (St. Louis, MO, USA) and of the purest grade available. In vitro site-directed mutagenesis was achieved with polymerase chain reaction (PCR) with a degenerate oligonucleotide containing several mutations as one of the primers. The PCR fragment was purified and served as a primer for a second amplification. The mutated fragment was inserted in plasmid pIBI-BM3/T268A and the entire introduced region was sequenced to establish that there were no undesired mutations.

Spectra were recorded on a Hewlett-Packard model 8452A diode array spectrophotometer (San Fernando, CA, USA). Substrate-binding experiments were performed at 20°C. Spectra were recorded after each addition of AA or PA until the absorbance at 392 nm remained constant. All kinetic experiments were carried out at 20°C in 50 mM (N-2-morpholino) propane sulfonic acid (MOPS)/Tris buffer, pH 7.4. NADPH consumption was followed spectrophotometrically at 340 nm. Oxygen consumption was followed with a Gilson model 5/6 oxygraph (Gilson, Middleton, WI, USA) with a Clark-type oxygen electrode (YSI) (Yellow Springs, OH, USA).

Radioactive substrates ($^{14}$C-PA, Amersham, Arlington, IL, USA, and $^{14}$C-AA, NEN, Boston, MA, USA) were mixed with cold AA and PA. NADPH (4 mM) or cumene hydroperoxide (2 mM) was added at time zero, and the reaction was followed for 10 min (NADPH) or 30 min (cumene hydroperoxide). Reactions were acidified with citric acid; products were extracted with ethyl acetate and separated on a silica gel 60A alumina plate. Quantification of product formation utilized a Phosphorimager and the Image Quant software (Molecular Dynamics, Sunnyvale, CA, USA).

## Results and Discussion

Conversion of the low-spin to the high-spin form of the iron is a good way of measuring substrate interactions with the heme of P-450s. Binding experiments for both AA and PA revealed that substrate binding is affected by the mutation T268A. As seen in

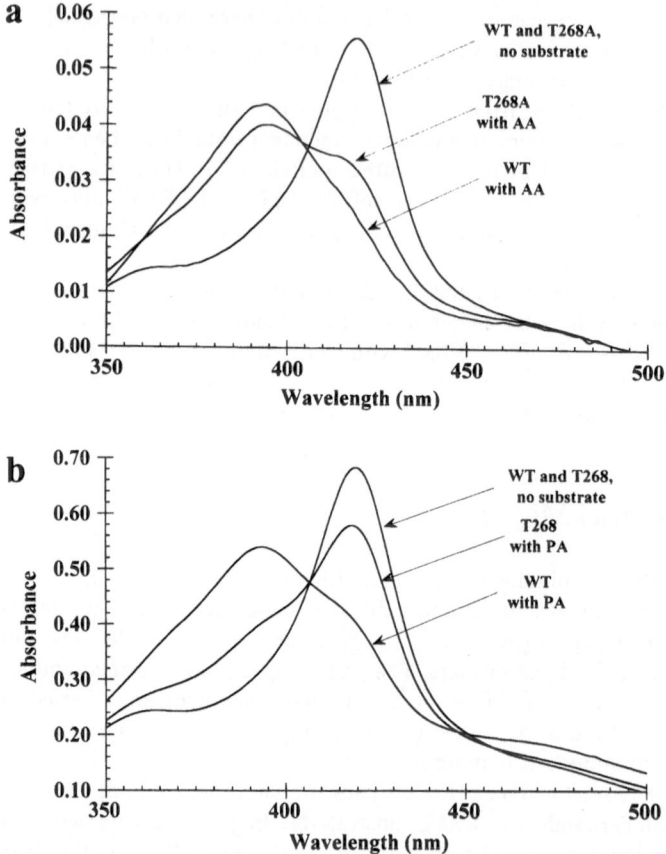

FIG. 1a,b. Spectral changes on addition of arachidonic acid (*AA*) and palmitic acid (*PA*) to oxidized P-450 BM3 wild-type (*WT*) and mutant T268A. Spectra were recorded between 350 and 500 nm. **a** Spectra with AA. P-450 concentration (WT and T268A) was 0.4 μM. **b** Spectra with PA. P-450 concentration (WT and T268A) was 4.8 μM

Fig. 1, the amount of high-spin present with a saturating amount of substrate differs between the wild-type (WT) and T268A enzyme. Fig. 1a shows that a saturating amount of AA cannot convert T268A to 100% high-spin as in the WT. This effect is more pronounced with PA binding (almost 100% conversion in the WT and very low in T268A). Titrations of the formation of the high-spin species allowed us to calculate the dissociation constants for AA for the WT and T268A ($K_s$ = 1.2 μM for WT and 5.5 μM for T268A). Thus, binding of fatty acids is impaired in T268A. Furthermore, the extent of high-spin conversion, a measure of the ability of the substrate to remove the sixth axial water molecule, is decreased. These data suggest that the mutation affects the positioning of the substrate.

Table 1 presents a summary of the coupling data, that is, the stoichiometry of NADPH oxidation versus oxygen consumption and product formation. The addition

TABLE 1. NADPH and $O_2$ consumption for P-450 BM3 wild-type enzyme and mutant T268A with arachidonic acid (AA) and palmitic acid (PA) metabolism

|  | Wild type | T268A |
|---|---|---|
| AA metabolism: |  |  |
| NADPH consumption |  |  |
| (nmol min$^{-1}$nmol$^{-1}$ of P-450 | 4310 | 1400 |
| $O_2$ consumption |  |  |
| (nmol min$^{-1}$nmol$^{-1}$ of P-450) |  |  |
| With catalase | 3360 | 600 |
| Without catalase | 3360 | 1320 |
| NADPH oxidation/$O_2$ consumption[a] | 1.09 | 0.97 |
| NADPH consumed/$O_2$ consumed[b] | 1 | 0.99 |
| $H_2O_2$ produced/NADPH consumed[c] | 0 | 0.91 |
| PA metabolism: |  |  |
| NADPH consumed/$O_2$ consumed[b] | 1 | 1.02 |
| $H_2O_2$ produced/NADPH consumed[c] | 0 | 0.94 |

[a] The coupling ratio was determined as the ratio between the rate of NADPH oxidation and the rate of $O_2$ consumption.
[b] The ratio was determined in a reaction in which NADPH was limiting compared to the AA concentration ($[AA] = 10 \times [NADPH]$) to ensure complete consumption of NADPH. The amount of oxygen consumed was deduced from the endpoint of the curve before catalase was added.
[c] The amount of $H_2O_2$ produced was calculated as twice the difference between the total amount of $O_2$ consumed and the amount of $O_2$ produced after catalase addition.

of catalase does not affect NADPH or oxygen consumption in the WT, suggesting a very tight coupling. On the contrary, catalase affects NADPH and oxygen consumption in the T268A. The coupling ratio shows that almost 90% (for AA) and 95% (for PA) of the electrons serve to form hydrogen peroxide. There is also a decrease in NAPDH and oxygen consumption in the T268A, which could result from the augmentation of the dissociation constant in the mutant T268A.

AA metabolites formation was analyzed (Fig. 2a) by thin-layer chromatography (TLC). In the T268A mutant, the amount of AA metabolites formed is about 1% of the amount formed by the WT. Rates of product formation were calculated for both WT and T268A (3360 and 37.6 min$^{-1}$, respectively). Addition of catalase to the reaction with the mutant did not change the accumulation or rate of formation of AA metabolites. This result demonstrates that the reaction is still driven by electrons coming from NADPH rather than utilizing hydrogen peroxide as an oxygen donor.

P-450s can function with an alternate oxygen donor and in the absence of pyridine nucleotides. Oxygen activation occurs by release of cumene alcohol, directly generating the high-valence iron-oxo species. We used CuOOH as a source of oxygen; both WT and T268A can function with CuOOH, as shown in Fig. 2b. The $k_{cat}$ of the reaction is slow compared to that driven by NADPH (0.2 compared to 4440 mol min$^{-1}$ mol$^{-1}$). The WT and T268A have the same turnover number, suggesting that the mutant is affected in a step before the formation of the iron-oxo species.

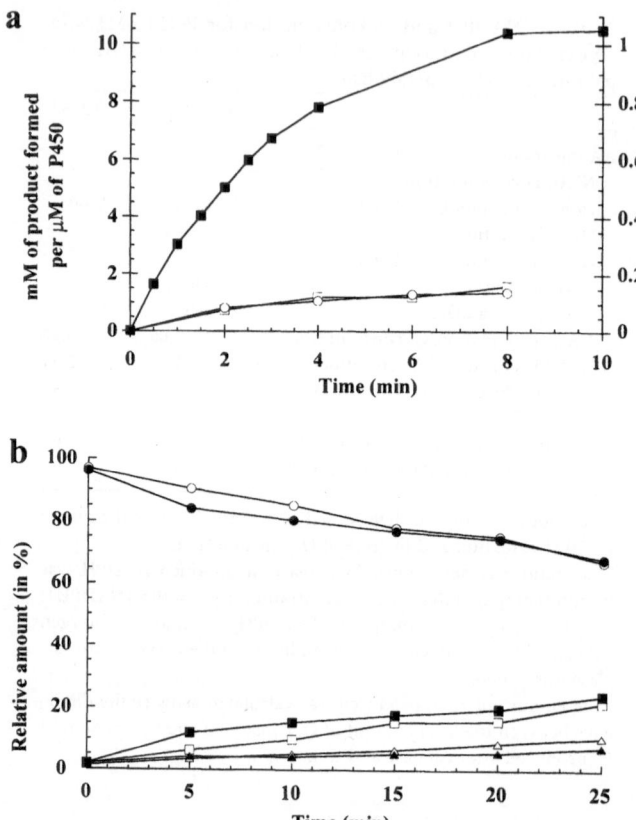

FIG. 2a,b. Comparison of WT and T268A activities. **a** AA product formation was followed after NADPH addition for 10 min. AA concentration was 200 μM. The vertical axis on the *left* is for WT AA product formation (*closed squares*); the vertical axis on the *right* is for the mutant (*open squares*, minus catalase; *open circles*, with catalase). **b** PA product formation driven by cumene hydroperoxide. PA concentration was 100 μM. The vertical axis displays the relative amounts produced as the percentage of total PA introduced at the beginning of the experiment. Circles show PA consumption with WT (*open circles*) and T268A (*closed circles*). Squares show formation of 13- and 14-hydroxy-PA with WT (*open squares*) and T268A (*closed squares*). Triangles show formation of 15-hydroxy-PA with WT (*open triangles*) and T268A (*closed triangles*)

Studies on the role of the conserved T-252 in P-450$_{cam}$ have led to the conclusion that this residue is important for oxygen activation [1–3], although controversy exists as to which atom is required for this function. In this chapter we have demonstrated that the Thr-268 of P-450 BM3 plays a role in substrate binding as well. The positioning of substrate is clearly different in T268A, affecting the displacement of the water axial ligand and the affinity for substrate. Our results also suggest that the mutation T268A affects a step preceding the formation of the iron-oxo species.

The results presented herein emphasize a role of the threonine in the catalytic activation of molecular oxygen as well as in the binding of substrate. Our hypothesis is that these two steps are simultaneous and that the threonine residue could help to hold the substrate during catalysis.

## References

1. Imai M, Shimada H, Watanabe Y, et al. (1989) Uncoupling of the cytochrome P-450$_{cam}$ monooxygenase reaction by a single.mutation, threonine-252 to alanine or valine: possible role of the hydroxy amino acid in oxygen activation. Proc Natl Acad Sci USA 86:7823–7827
2. Martinis SA, Atkins WM, Stayton PS, et al. (1989) A conserved residue of cytochrome P-450 is involved in heme-oxygen stability and activation. J Am Chem Soc 111:9252–9253
3. Raag R, Martinis SA, Sligar SG, et al. (1991) Crystal structure of the cytochrome P-450$_{CAM}$ active site mutant Thr252Ala. Biochemistry 30:11420–11429
4. Kimata Y, Shimada H, Hirose T, et al. (1995) Role of Thr-252 in cytochrome P-450$_{CAM}$: a study with unnatural amino acid mutagenesis. Biochem Biophys Res Commun 208:96–102
5. Narhi LO, Fulco AJ (1987) Identification and characterization of two functional domains in cytochrome P-450 BM-3, a catalytically self-sufficient monooxygenase induced by barbiturates in *Bacillus megaterium*. J Biol Chem 262:6683–6690
6. Narhi LO, Fulco AJ (1986) Characterization of a catalytically self-sufficient 119,000-dalton cytochrome P-450 monooxygenase induced by barbiturates in *Bacillus megaterium*. J Biol Chem 261:7160–7169
7. Ravichandran KG, Boddupalli SS, Hasemann CA, et al. (1993) Crystal structure of hemoprotein domain of P-450 BM3, a prototype for microsomal P-450s. Science 261:731–736
8. Yeom H, Sligar SG, Li H, et al. (1995) The role of Thr268 in oxygen activation of cytochrome P-450 BM3. Biochemistry 34:14733–14740
9. Boddupalli SS, Estabrook RW, Peterson JA (1990) Fatty acid monooxygenation by cytochrome P-450 BM-3. J Biol Chem 265:4233–4239

# Thiolate Adducts of the Myoglobin Cavity Mutant H93G as Models for Cytochrome P-450

Mark P. Roach[1], Stefan Franzen[2], Phillip S.H. Pang[3], William H. Woodruff[2], Steven G. Boxer[3], and John H. Dawson[1,4]

*Summary.* A recent development in the field of heme proteins has been the engineering of cavity mutants in which the axial coordinating residue is replaced by a smaller, noncoordinating residue, leaving a cavity that can then be filled by exogenous ligands. In this manner, potential models for the cysteinate-ligated cytochrome P-450 monooxygenases have been prepared using the H93G cavity mutant of sperm whale myoglobin in which the coordinating histidine has been replaced by glycine. Magnetic circular dichroism (MCD) spectroscopy has been used for structural characterization of several ferric and ferrous thiolate-H93G adducts. Ferric mixed-ligand complexes can be prepared with neutral sixth ligands. The model breaks down in two cases: (i) when anionic ligands are added to the ferric-thiolate complex and (ii) on reduction of the ferric-thiolate complex. Results are discussed in the context of the H93C mutants of myoglobin and the stabilizing influences on the cytochrome P-450 heme–cysteinate complex.

*Key words.* Cytochrome P-450—Cavity mutant—Myoglobin—Magnetic circular dichroism (MCD) spectroscopy—Heme–thiolate complex

## Introduction

The cytochrome P-450 enzymes utilize a thiolate-heme active site to activate molecular oxygen for ambient temperature insertion into C–H bonds through a proposed thiolate "push" mechanism involving a highly reactive oxo-ferryl intermediate [1–4]. Structural studies of alkyl- and aryl-porphyrin model systems have yielded a wealth of information on many different heme coordination arrangements, but facile oxidation of free thiols in organic solvents has limited the feasibility of using simple iron

[1] Department of Chemistry and Biochemistry, University of South Carolina, 730 South Main St., Columbia, SC 29208, USA
[2] Biosciences and Biotechnology Group CST-4 MS J586, Los Alamos National Laboratories, Los Alamos, NM 87545, USA
[3] Department of Chemistry, Stanford University, Stanford, CA 94305, USA
[4] School of Medicine, University of South Carolina, 730 South Main St., Columbia, SC 29208, USA

porphyrin systems as mimics of the cytochrome P-450 coordination sphere. The thiol oxidation problem has only recently been overcome by the elegant synthesis of an iron porphyrin with a tethered thiol that retains its structure during catalysis [5].

Construction of mutant heme proteins with thiolate ligation by substitution of the endogenous proximal heme-iron ligand with cysteine has also proven to be a promising approach to modeling the cytochrome P-450 active site and has met with success in human myoglobins (Mbs) H93C and H93C/H64V [6–8] and in horse heart myoglobin H93C/H64V [9]. In addition, the human mutant Mbs exhibit enhanced reactivity to heterolytic cleavage of bound peroxides and increased monooxygenase activity [7,8] in accordance with the proposed thiolate "push" mechanism for oxygen activation by cytochrome P-450.

Yet another recent development has been the engineering of cavity mutant heme proteins where the proximal ligating residue is replaced with a smaller, noncoordinating residue. This leaves a proximal cavity that can be filled by exogenous ligands. Cavity mutants have been prepared for cytochrome $c$ peroxidase [10], heme oxygenase [11], horseradish peroxidase [12], and sperm whale Mb [13–15]. In the current study, we employed the H93G cavity mutant of sperm whale Mb [13–15] in the first effort to mimic the coordination structure of cytochrome P-450 with thiolate complexes of a heme protein cavity mutant.

The high degree of fine structure exhibited in magnetic circular dichroism (MCD) spectra makes it an electronic "fingerprinting" technique well suited for investigations of axial ligation in structurally undefined iron porphyrins and heme proteins [16]. Ligation assignments are made through comparisons of spectra with those of defined heme iron centers. A series of ferric and ferrous adducts of H93G with ethanethiol (EtSH) and benzenethiol (BzSH) have been prepared and characterized in context of native Mb and cytochrome P-450$_{CAM}$ through electronic absorption and MCD spectroscopy to investigate the potential of the H93G-thiolate ensemble as a model for cytochrome P-450.

# Experimental Methods

## Protein Preparation and Purification

The sperm whale Mb mutant was expressed and purified as previously described [13]. As purified, the protein contains imidazole (Im) in the proximal pocket and exists as a mixture of oxidation states: ferric and oxyferrous. Pure samples are oxidized completely by addition of solid potassium ferricyanide followed by desalting on a P6DG (Bio-Rad, (Richmond, CA, USA)) gel filtration column. Because this process also removes exogenous ligands, the resulting ligand-free sample is termed "H93G(X)" to indicate the present uncertainty of the coordination structure of this species.

## Preparation of Ligand Adducts

Imidazole, potassium cyanide, and sodium dithionite were purchased from Aldrich. All protein samples were handled at 4°C at concentrations of approximately 60 μM in

100 mM potassium phosphate buffer pH 7.0. The ferric ethanethiol (EtSH) and ferric benzenethiol (BzSH) adducts were prepared from ferric H93G(X) through addition of minimal volumes of 1 M stock solutions of EtSH and BzSH (as monitored by spectrophotometric titrations). The ferric H93G ethanethiol/imidazole adduct was prepared by addition of minimal volumes of a 6.8 mM Im stock solution to the ferric H93G(EtSH) sample. The cyanoferric adduct was prepared from both ferric H93G(EtSH) and ferric H93G(X) through addition of minimal volumes of a 1 M KCN stock solution. The deoxyferrous samples were prepared from ferric H93G(EtSH) and ferric H93G(X) by exchanging the atmosphere of the cuvette with nitrogen followed by addition of solid sodium dithionite.

## Spectroscopic Techniques

Electronic absorption spectra were obtained on a Cary 210 spectrophotometer (Varian Instruments, Sunnyvale, CA, USA) interfaced to an IBM PC. MCD spectra were recorded at a magnetic field strength of 1.41 T with a JASCO J500A spectropolarimeter (Japan Spectroscopic, Tokyo) equipped with a JASCO MCD 1B electromagnet and interfaced to a Gateway 2000 4DX2-66V PC through a JASCO IF-500-2 interface unit (same as above). Spectroscopic data handling has been described elsewhere [17].

## Results and Discussion

The coordination structure of the ferric exogenous ligand-free species termed "H93G(X)" is the subject of current investigations in our laboratories and is not discussed further here. Titration of ferric H93G(X) with EtSH results in a species with an absorption maximum at 392 nm (Fig. 1). Further characterization of this adduct by MCD spectroscopy shows that, like the substrate-free state of ferric cytochrome P-$450_{CAM}$, the heme iron is a five-coordinate high-spin thiolate complex [2,4]. This assignment is made on the basis of the similarities between the shape and position of the MCD Soret trough (whose minimum is located at 346 nm) with the MCD Soret feature of substrate-free cytochrome P-$450_{CAM}$ (Fig. 1). Ligation of cysteine in H93C human Mb [7] and H93C/H64V horse heart Mb [9] also results in five-coordinate high-spin heme–thiolate complexes.

Wild-type human, horse heart, and sperm whale Mbs are known to be six-coordinate high-spin complexes with histidine and water as the axial ligands [18]. The finding that water is not bound to the myoglobin–thiolate complexes in the two ferric H93C mutants and ferric H93G(EtSH) indicates that binding of thiolates (in place of the wild-type His-93) to ferric myoglobin disfavors binding of water in the position trans to the thiolate. This observation stands in contrast to ferric cytochrome P-$450_{CAM}$, which has a thiolate/aquo structure in the absence of substrate [2,4].

Addition of Im to ferric H93G(EtSH) yields an adduct whose MCD and absorption spectra are displayed in Fig 2. The similarities between the spectra of this derivative and those of the Im complex of ferric cytochrome P-$450_{CAM}$ provide evidence for ligation of both EtSH and Im in this H93G derivative. Unfortunately, the myoglobin cavity mutant model for cytochrome P-450 breaks down when potassium cyanide is added to ferric H93G(EtSH). The resulting species exhibit spectra that are quite

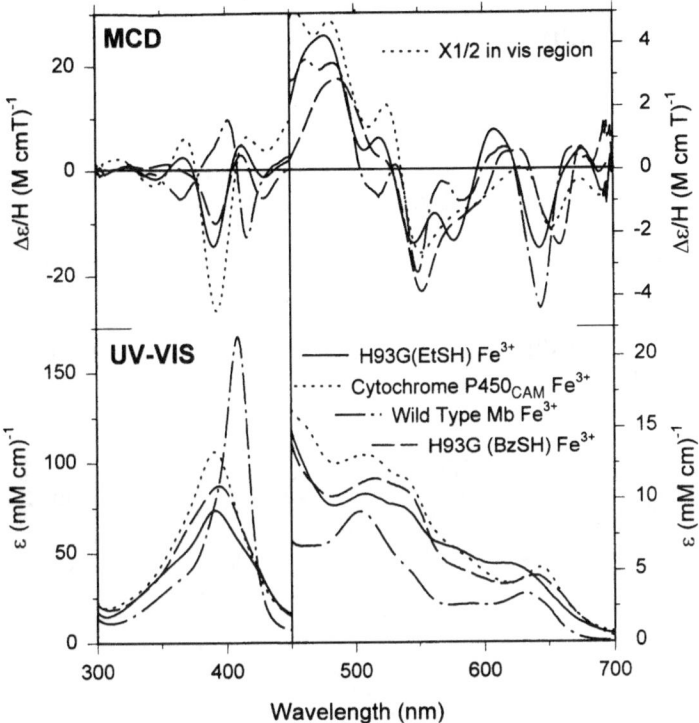

FIG. 1. Magnetic circular dichroism (*MCD*) and electronic absorption (*UV-VIS*) spectra of ferric ethanethiolate-bound H93G myoglobin [*H93G(EtSH)*] pH 7.0, 30 mM EtSH (*solid line*); ferric substrate-free cytochrome *P-450*~CAM~ pH 7.0 (*dotted line*) (replotted using data from [20]); *ferric wild-type* myoglobin (*Mb*) (*dot-dashed line*) (replotted using data from [21]); and ferric benzenethiol-bound H93G myoglobin [*H93G(BzSH)*] pH 7.0 (*long dashed line*)

distinct from those of cyanoferric cytochrome P-450 (Fig. 3) and instead resemble the spectra of wild-type cyanoferric Mb (Fig. 3). We conclude that the bound EtSH is displaced on addition of cyanide, followed by ligation of an internal ligand such as the distal His-64. This ligation arrangement seems particularly likely because addition of cyanide to ferric H93G(X) yields a spectroscopically identical species (data not shown).

The ferric H93G(EtSH) heme–thiolate unit is also disrupted on reduction with sodium dithionite. The resulting derivative exhibits absorbance and MCD spectra that are different from those of the deoxyferrous state of cytochrome P-450~CAM~ (Fig. 4) and similar to the spectra obtained on similar reduction of ferric H93G(X) (data not shown). The coordination structure of this species is currently under further investigation because its MCD spectrum (Fig. 4) has the same general band shape but is much less intense and is slightly blue-shifted relative to that of wild-type deoxy Mb (Fig. 4). These results are in accord with the previous observations that displacement of Cys-93 occurs on reduction of human Mb H93C [7,8] and horse heart Mb H93C/H64V [9].

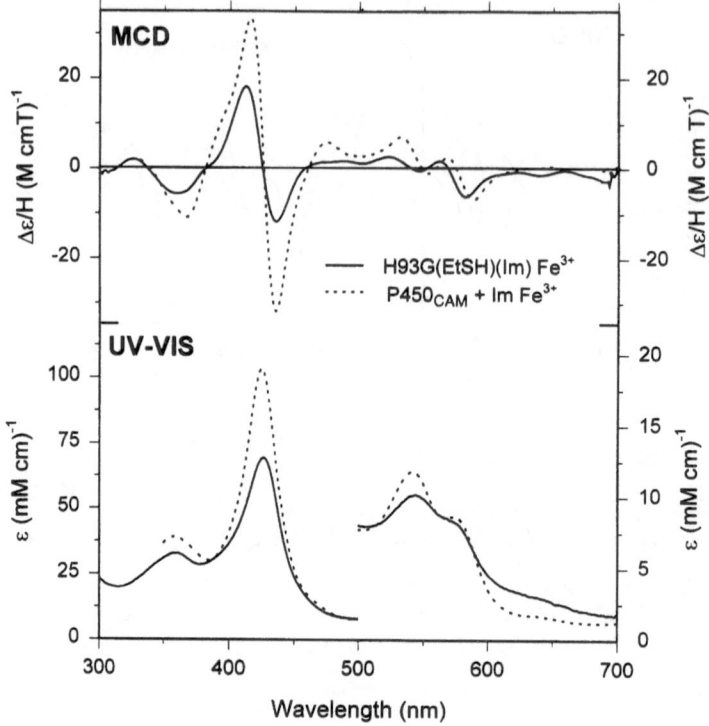

FIG. 2. MCD and UV-VIS spectra of ferric EtSH and imidazole-(Im-) bound H93G myoglobin [*H93G(EtSH)(Im)*] pH 7.0, 30 mM EtSH, 7 mM Im (*solid line*); and ferric Im-bound cytochrome P-450$_{CAM}$, pH 7.0 (*dotted line*) (replotted using data from [22])

Some general conclusions may be drawn from the results of the manipulations of the system. Ferric H93G(EtSH) can bind a neutral ligand such as Im and remain thiolate ligated. However, binding of an anionic ligand such as cyanide or a one-electron reduction results in displacement of the thiolate. These results indicate that the heme–iron complex of H93G cannot tolerate an overall −1 charge. In this behavior, H93G(EtSH) is different from cytochrome P-450, which retains its cysteinate ligand in deoxyferrous and cyanoferric states. Poulos has suggested that heme–thiolate complexes require stabilization through hydrogen bonding and favorable locations near the positive terminus of the proximal helix dipole, as have been observed for cytochrome P-450$_{CAM}$ and chloroperoxidase from *Caldariomyces fumago* [19]. The absence of these stabilizing influences in Mb are most likely responsible for the breakdown of the H93G cytochrome P-450 model when the net charge of the heme–iron complex is −1.

Having established the limitations of our model, we have begun to refine it using different thiols. Ferric complexes prepared with aliphatic thiols, including examples of cyclic and benzylic thiols, have essentially identical spectra. However, the absorbance and MCD spectra of the ferric H93G(BzSH) complex (see Fig. 1) are a

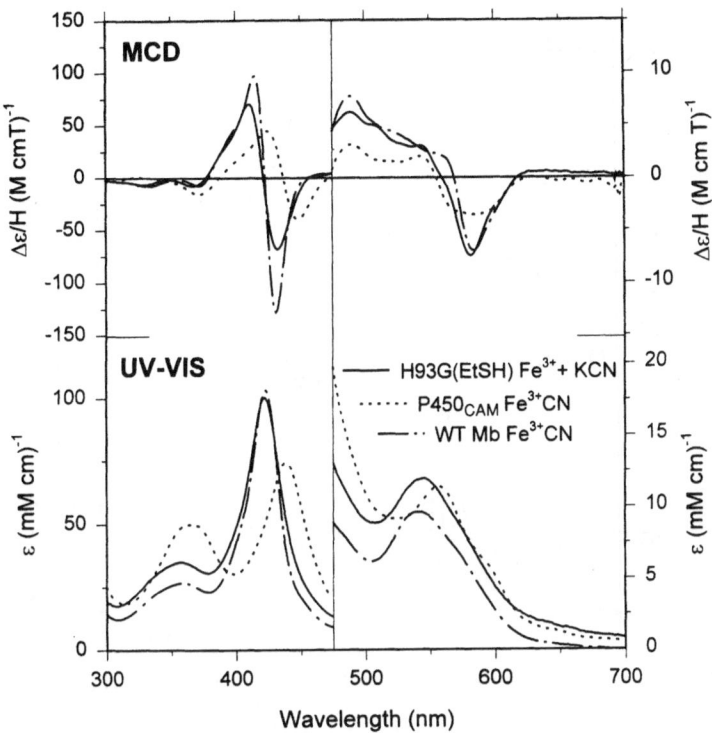

FIG. 3. MCD and UV-VIS spectra of the reaction product of ferric H93G(EtSH) + potassium cyanide, pH 7.0, 30 mM EtSH, 30 mM potassium cyanide (*solid line*); cyanoferric cytochrome P-450$_{CAM}$, pH 7.0 (*dotted line*) (replotted using data from [23]); and cyanoferric myoglobin, pH 7.0 (*dot-dashed line*) (replotted using data from [21])

much closer match to those of ferric substrate-free cytochrome P-450$_{CAM}$ (Fig. 1), especially in the visible region. This is encouraging from the standpoint of future mechanistic studies on alkylhydroperoxide generated oxo-ferryl H93G-thiolate species because the electron-donating properties of aromatic thiolates as ligands are highly influenced by their phenyl substituents. This ability to alter the electronic properties of the thiolate ligand is an advantage that utilization of the H93G cavity mutant system has over conventional amino acid ligand substitution or synthesis of thiol-tethered model systems in which the nature of the ligand cannot be conveniently altered. The versatility of this simple cavity mutant model system will make it a valuable tool with which to systematically probe the requirements for oxygen activation by heme systems.

*Acknowledgments.* This work was supported by National Institutes of Health grant GM26730 (to J.H.D). The JASCO J-500 spectropolarimeter was purchased with NIH grant RR-03960 and the electromagnet was obtained with a grant from Research Corporation. We wish to thank Alycen E. Pond and Drs. Eric D. Coulter and Masanori Sono for numerous helpful discussions and Drs. Edmund W. Svastits and John J. Rux

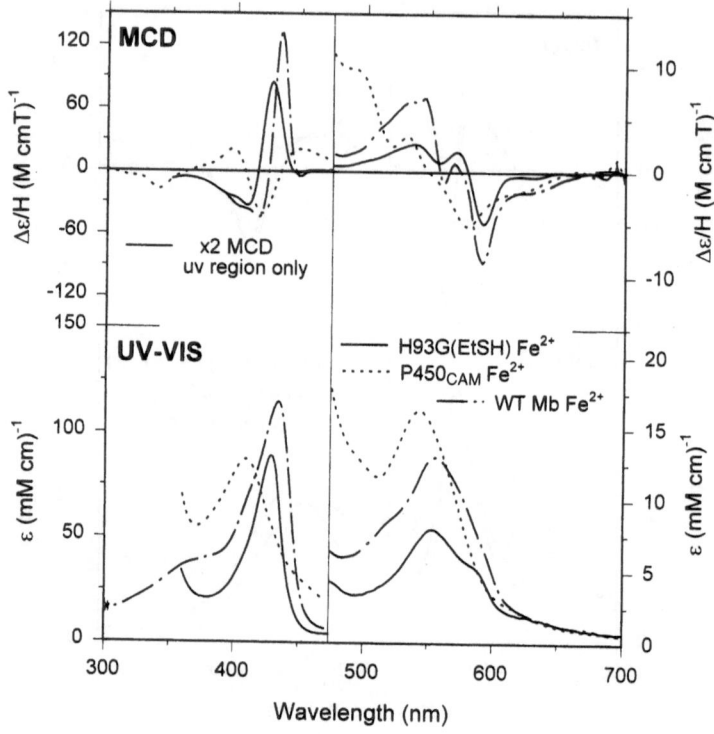

FIG. 4. MCD and UV-VIS spectra of the reaction product of ferric H93G(EtSH) + sodium dithionite, pH 7.0 (*solid line*); deoxyferrous cytochrome P-450$_{CAM}$, pH 7.0 (*dotted line*) (replotted using data from [20]); deoxyferrous myoglobin, pH 7.0 (*dot-dashed line*) (replotted using data from [21])

for assembling the custom MCD data analysis software. J.H.D and M.P.R also wish to thank Prof. Yuzuru Ishimura and all members of the symposium organizing committee for the warm hospitality extended to them during the Keio University Medical School Symposium "Oxygen Homeostasis and Its Dynamics" where the results of this research have been presented.

# References

1. Dawson JH, Holm RH, Trudell JR, et al. (1976) Oxidized cytochrome P-450. Magnetic circular dichroism evidence for thiolate ligation in the substrate-bound form. Implications for the catalytic mechanism. J Am Chem Soc 98:3707–3709
2. Dawson JH (1988) Probing structure-function relations in heme-containing oxygenases and peroxidases. Science 240:433–439
3. Liu HI, Sono M, Kadkhodayan S, et al. (1995) X-ray absorption near edge studies of cytochrome P-450$_{CAM}$, chloroperoxidase, and myoglobin. Direct evidence for the electron releasing character of a cysteine thiolate proximal ligand. J Biol Chem 270:10544–10550

4. Sono M, Roach MP, Coulter ED, et al. (1996) Heme-containing oxygenases. Chem Rev 96:2841–2887
5. Higuchi T, Shimada K, Maruyama N, et al. (1993) Heterolytic O–O bond cleavage of peroxy acid and effective alkane hydroxylation in hydrophobic solvent mediated by an iron porphyrin coordinated by thiolate anion as a model for cytochrome P-450. J Am Chem Soc 115:7551–7552
6. Adachi S, Nagano S, Watanabe Y, et al. (1991) Alteration of human myoglobin proximal histidine to cysteine or tyrosine by site-directed mutagenesis: characterization and their catalytic activities. Biochem Biophys Res Commun 180:138–144
7. Adachi S, Nagano S, Ishimori K, et al. (1993) Roles of proximal ligand in heme proteins: replacement of proximal histidine of human myoglobin with cysteine and tyrosine by site-directed mutagenesis as models for P-450, chloroperoxidase, and catalase. Biochemistry 32:241–252
8. Matsui T, Shingo N, Ishimori K, et al. (1996) Preparation and reactions of myoglobin mutants bearing both proximal cysteine ligand and hydrophobic distal cavity: protein models for the active site of cytochrome P-450. Biochemistry 35:13118–13124
9. Hildebrand DP, Ferrer JC, Tang H, et al. (1995) Trans effects on cysteine ligation in the proximal His93Cys variant of horse heart myoglobin. Biochemistry 34:11598–11605
10. McRee DE, Jensen GM, Fitzgerald MM, et al. (1994) Construction of a bisaquo heme enzyme and binding by exogenous ligands. Proc Natl Acad Sci USA 91:12847–12851
11. Sun J, Loehr TM, Wilks A, et al. (1994) Identification of histidine 25 as the heme ligand in human liver heme oxygenase. Biochemistry 33:13734–13740
12. Newmyer SL, Sun J, Loehr TM, et al. (1996) Rescue of the horseradish peroxidase his170→ala mutant activity by imidazole: importance of ligand tethering. Biochemistry 35:12788–12795
13. Barrick D (1994) Replacement of the proximal ligand of sperm whale myoglobin with free imidazole in the mutant His-93→Gly. Biochemistry 33:6546–6554
14. DePillis GD, Decatur SM, Barrick D, et al. (1994) Functional cavities in proteins: a general method for proximal ligand substitution in myoglobin. J Am Chem Soc 116:6981–6982
15. Decatur SM, Boxer SG (1995) $^1$H-NMR characterizations of myoglobins where exogenous ligands replace the proximal histidine. Biochemistry 34:2122–2129
16. Dawson JH, Dooley DM (1989) Magnetic circular dichroism spectroscopy of iron porphyrins and heme proteins. In: Lever ABP, Gray HB (eds) Iron porphyrins, part III. VCH, New York, pp 1–135
17. Huff AM, Chang CK, Cooper DK, et al. (1993) Imidazole- and alkylamine-ligated iron (II,III) chlorin complexes as models for histidine and lysine coordination to the iron in dihydroporphyrin-containing proteins: characterization with magnetic circular dichroism spectroscopy. Inorg Chem 23:1460–1466
18. Antonini E, Brunori M (1971) Hemoglobin and myoglobin in their reactions with ligands. Elsevier, Amsterdam, pp 43–46
19. Poulos TL (1996) The role of the proximal ligand in heme proteins. J Biol Inorg Chem 1:356–359
20. Andersson LA (1982) The active site environments of heme mono-oxygenases: spectroscopic investigations of cytochrome P-450 and secondary amine mono-oxygenase. Ph.D. thesis, University of South Carolina, Columbia, SC, USA.
21. Dawson JH, Kadkhodayan S, Zhuang C, et al. (1992) On the use of iron octa-alkylporphyrins as models for protoporphyrin IX-containing heme systems in studies employing magnetic circular dichroism spectroscopy. J Inorg Biochem 45:179–192
22. Sono M, Dawson JH, Hall K, et al. (1986) Ligand and halide binding properties of chloroperoxidase: peroxidase-type active site heme environment with cytochrome P-

450 type endogenous axial ligand and spectroscopic properties. Biochemistry 25:347–356

23. Dawson JH, Andersson LA, Sono MS, et al. (1992) Systematic trends in the spectroscopic properties of low-spin ferric ligand adducts of cytochrome P-450 and chloroperoxidase: the transition from normal to hyper spectra. New J Chem 16:557–582

# Pronounced Effects of Axial Thiolate Ligand on Oxygen Activation by Iron Porphyrin

Tsunehiko Higuchi, Yasuteru Urano, Masaaki Hirobe, and Tetsuo Nagano

*Summary.* A distinctive structural feature of P-450 is the unusual thiolate coordination to heme. We have succeeded in the preparation of the first synthetic thiolato-iron porphyrin (SR complex) that retains its structure during catalytic oxidation. Experiments using the SR complex have revealed that the thiolate ligand greatly accelerates the rate of the O–O bond cleavage and its heterolysis, even in highly hydrophobic media. The results of kinetic isotope effects in the oxidative demethylation of *p*-dimethoxybenzene unambiguously showed that the formed active intermediates of heme-thiolates are different from those of hemes coordinated by imidazole or chloride.

*Key words.* Cytochrome P-450—Thiolate—Porphyrin—Axial ligand—O–O bond

## Introduction

Cytochrome P-450s play important roles in diverse in vivo processes. The enzymes participate in the synthesis of physiologically significant biomolecules, including various steroids and prostaglandins, by utilizing molecular oxygen and are mainly responsible for the metabolism of xenobiotics. The mechanism of their catalytic activities and their structure–function relationships have been the subject of extensive investigation in the field of biomimetic chemistry (for reviews, see [1–3]).

The distinctive structural features of P-450 as a hemoprotein are the unusual thiolate coordination to heme iron and also the high hydrophobicity of its active site [4]. Among many heme enzymes, only P-450 can hydroxylate unactivated alkanes and arenes. Our interest has been focused on the relative effect of an axial thiolate ligand on the reactivity of heme in hydrophobic media as a model of the P-450 pocket. It was thought necessary, for investigation of the axial ligand effect of thiolate, to prepare an iron porphyrin coordinated by thiolate that can retain its complexation during catalytic oxidation using itself. Several synthetic iron porphyrins having thiolate anion as an intramolecular ligand have been reported as P-450 model complexes. However, the

Faculty of Pharmaceutical Sciences, University of Tokyo, 7-3-1 Hongo, Bunkyo-ku, Tokyo 113, Japan reactivities of those model complexes as catalysts on oxidations have never been described [5].

## Design, Synthesis, and Spectroscopy of a New Iron Porphyrin Coordinated by a Thiolate Anion (SR Complex)

For the synthesis of a stable heme-thiolate, it was important to determine how the complex is designed to have tight Fe–S binding and how its sulfur atom is protected from oxidants or active species. We expected that the stability of the thiolato-iron porphyrin complex toward oxidation increases largely by protecting the sulfur atom with steric hindrance of bulky groups introduced on the proximal site of the porphyrin molecule. Further, the spacer for the thiolate ligand was designed to be an appropriate length, to be directed toward the iron center by intramolecular hydrogen bonding in the spacer part, and to provide CH-$\pi$ interaction with the introduced bulky molecules.

We have succeeded in the synthesis of a unique iron porphyrin ligated by an alkylthiolate anion (named the "swan resting" form of porphyrin because the shape topologically suggested a swan resting, burying its head in its feathers; the name of the complex is abbreviated as SR complex), which retains its axial thiolate coordination during catalytic oxidation reactions (Fig. 1) [6]. To our surprise, compound SR could be purified by silica gel column chromatography in air without decomposition by dioxygen and was obtained as dark brown microcrystals by recrystallization. The UV-visible (UV-vis) spectra of CO-$Fe^{2+}$ (SR) exhibited a typical hyperporphyrin spectra for heme-thiolate. The crystal field parameter (rhombicity), which was derived from $g$ values of electron paramagnetic resonance (EPR) spectra of SR ($Fe^{3+}$), indicated axial thiolate coordination. Elemental analysis and fast atom bombardment (FAB) mass spectra also supported the structure shown in Fig. 1. Extended X-ray absorption fine structure (EXAFS) analysis of the SR complex ($Fe^{3+}$) indicated an Fe–S bond length of 2.24 Å, which is quite close to that in native ferric P-450 enzymes [4,5]. The SR complex could be stored at room temperature in air for several months.

## Relative Effect of the Axial Ligand on the Rate of O–O Bond Cleavage

The catalytic cycle of P-450 is a multistep reaction; therefore, the effect at each step should be evaluated using the SR complex to elucidate the axial ligand effect on oxygen activation by P-450. First, we examined the catalytic activity of SR on the peroxide shunt reaction of P-450 using alkyl hydroperoxides and compared it with that of Fe(TPP)Cl to investigate the relative effect of a thiolate ligand on O–O bond cleavage. 2,4,6-Tri-*tert*-butylphenol (TBPH) and 1,1-diphenyl-2-picrylhydrazine (DPPH) were chosen as substrates because both are known to trap reactive intermediates very rapidly to produce, almost stoichiometrically, the 2,4,6-tri-*tert*-butylphenoxyl (TBP) or 1,1-diphenyl-2-picrylhydrazyl (DPP) radical[7,8]. Thus, the rate of O–O bond cleavage can be estimated by extrapolation to be the formation rate of TBP· or DPP·. Toluene was used as the solvent, taking into account the highly

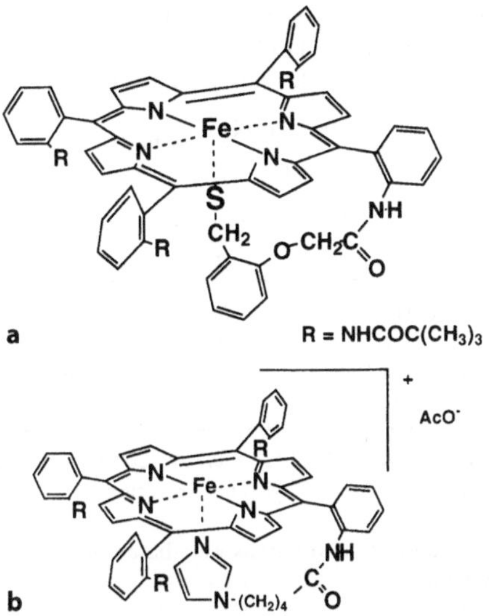

FIG. 1. **a** Complex of iron porphyrin coordinated by a thiolate anion (SR complex) (for "swan resting"). UV-Visible (UV-vis) spectra of SR($Fe^{II}$)-CO complex, $\lambda_{max}$ 383 nm (Soret), 459 nm (Soret); electron paramagnetic resonance (EPR) spectra of SR($Fe^{III}$) at 77 K, $g_x = 1.96$, $g_y = 2.21$, $g_z = 2.32$; rhombicity = 1.02; extended X-ray absorption fine structure (EXAFS) spectra of SR ($Fe^{III}$), Fe–S = 2.20 Å, Fe–N = 2.00 Å; elemental analysis, calculated C = 69.51, H = 5.66, N = 9.40, S = 2.68, and found C = 69.23, H = 5.52, N = 9.38, S = 2.60; fast atom bombardment (FAB) mass spectra $(M + 1)^+ = 1160$. **b** Imidazole (NP complex). UV-visible Spectra of NP ($Fe^{II}$)-CO complex, $\lambda_{max}$ 423 nm (Soret); EPR spectra of SR(FeIII) at 77 K, $g = 6.0, 2.0$; elemental analysis, calculated for $[Fe(Por)Im]^+ \cdot AcO^-$, C = 57.69, H = 5.45, N = 9.34, and found C = 57.87, H = 5.00, N = 9.27; High-resolution Fab mass, observed m/e = 1131.4668 $(M-OAc)^+$, calculated for $(C_{67}H_{67}N_{10}O_4Fe) = 1131.4696$

hydrophobic environment of the active site of P-450. EPR spectra of the reaction solution (15 s after the start of the reaction at 25°C) exhibited low-spin spectra of the thiolate-ligated iron porphyrin in addition to the signal of the DPP· formed, and the high-spin signal ($g = 6$) increased very slightly. All these results supported the conclusion that SR catalyzes the oxidation of the substrates by organic hydroperoxides while the axial thiolate ligand remains essentially unoxidized. Comparison of the rates of the reactions (Table 1) shows SR to have catalytic activity about 60–240 fold higher than that of Fe(TPP)Cl. Thus, the acceleration of the catalytic reaction by thiolate ligation is undoubtedly the result of the enhancement of O–O bond scission because the concentration of the peroxides used in these reactions is so high that the O–O bond cleavage step is rate determining. The cyclic voltammogram of SR in dimethyl formamide (DMF) showed a clear, reversible reduction couple [Fe(III)/Fe(II)] at −0.45 V versus saturated calomel electrode (SCE), which is more negative than that for Fe(TPP)Cl (−0.27 V versus saturated calomel electrode (SCE)). The negativity of the redox potential of SR is probably caused by electron donation from the thiolate to the iron atom.

TABLE 1. Observed initial rates of TBP· or DPP· formation on the oxidation of 2,4,6-tri-*tert*-butylphenol (TBPH) or 1,1-diphenyl-2-picrylhydrazine (DPPH) with alkyl hydroperoxides catalyzed by SR or Fe(TPP)Cl

$$\text{ROOH} + 2\text{ArXH} \xrightarrow[\text{X = O or R'-N}]{\text{Iron Porphyrin}} \text{ROH} + 2\text{ArX·} + \text{H}_2\text{O} \qquad (1)$$

| Substrate | Oxidant | k (turnover number/min)[a] | | $k_{SR}/k_{FeTPPCl}$ |
| --- | --- | --- | --- | --- |
| | | SR | Fe(TPP)Cl | |
| TBPH | PhC(CH$_3$)$_2$OOH | 21 | 0.35 | 58 |
| | *t*-Bu-OOH | 8.5 | 0.080 | 110 |
| DPPH | PhC(CH$_3$)$_2$OOH | 20 | 0.085 | 235 |
| | *t*-Bu-OOH | 7.5 | 0.041 | 182 |

Conditions: Solvent = toluene; [TBPH] = 0.2 M; [DPPH] = 0.1 M; [oxidant] = 5 × 10$^{-2}$ M; [SR] = [FeTPPCl] = 10$^{-4}$ M. These reactions were carried out at 20°C under an argon atmosphere.

[a] k: Observed initial rates of the reactions were based on the catalysts (turnover number of catalysts/min).

# Relative Effect of the Axial Ligand on the Mode of the O–O Bond Cleavage

It remains controversial, in the reaction mechanism of P-450, whether the heme of P-450 can cleave the O–O bond heterolytically in such a highly hydrophobic active site, because the active site does not contain any amino acid residue to be an effective acid-base catalyst for heterolytic O–O bond scission. A report by White et al. proposed a mechanism involving homolysis of the O–O bond for P-450 on the basis of the result of experiments using the enzyme itself and peroxy phenyl acetic acid (PPAA) as a peroxide probe [9]. Therefore, we investigated the effect of thiolate ligand on the mode of the O–O bond cleavage. A new iron porphyrin axially and intramolecularly coordinated by imidazole, termed the NP complex, was prepared to compare the effect of imidazole as an axial ligand with that of thiolate [10]. The structure of NP is a modification of the imidazole-ligated heme prepared by Collman et al. [11]. The UV-vis and EPR spectra, elemental analysis, and high-resolution FAB mass spectra all supported the structure shown in Fig. 1. We compared the catalytic activity of NP with that of SR complex in an oxidation system using PPAA, which is a useful probe for the determination of the mode of the O–O bond cleavage. SR gave PAA quantitatively (run 1), while Fe(TPP)Cl mainly catalyzed the formation of toluene (run 2), benzyl alcohol, and carbon dioxide (Table 2). Complex NP showed moderate reactivity, intermediate between those of SR and Fe(TPP)Cl (run 3). Therefore, we can unambiguously conclude that SR breaks the O–O bond of peroxyacids heterolytically in benzene [10].

This result indicates that the thiolate ligand enhances heterolytic cleavage of the peroxyacid–iron porphyrin complex even in highly hydrophobic media without the assistance of acid or base. In contrast, it was deduced from our data that Fe(TPP)Cl catalyzes the homolysis of peroxyacid in benzene. This conclusion con-

TABLE 2. Oxidation of 2,4,6-tri-*tert*-butylphenol (TBPH) and admantane with peroxyacid catalyzed by iron porphyrins

| Run | Iron porphyrin | Substrate | Peroxyacid | Products (yield %)[a] | | | |
|---|---|---|---|---|---|---|---|
| 1 | SR | TBPH | PPAA | PAA (100) | PhCH$_3$ (0) | PhCH$_2$OH (0) | TBP· (kcat = 12.5)[b] |
| 2 | Fe(TPP)Cl | TBPH | PPAA | PAA (7) | PhCH$_3$ (61) | PhCH$_2$OH (28) | TBP· (kcat = 0.043)[b] |
| 3 | NP | TBPH | PPAA | PAA (58) | PhCH$_3$ (15) | PhCH$_2$OH (0) | TBP· (kcat = 0.33)[b] |
| 4 | SR | TBPH | PPAA + PAA[c] | PAA (98)[d] | PhCH$_3$ (0) | PhCH$_2$OH (0) | TBP· (kcat = 12.1)[b] |
| 5 | Fe(TPP)Cl | TBPH | PPAA + PAA[c] | PAA (85)[d] | PhCH$_3$ (1) | PhCH$_2$OH (14) | TBP· (kcat = 0.083)[b] |
| 6 | SR | TBPH | Perlauric acid | | Lauric acid (89) | | |
| 7 | Fe(TPP)Cl | TBPH | Perlauric acid | | Lauric acid (25) | | |
| 8[e] | SR | Adamantane[f] | mCPBA | | 1-Adamantanol (80) | 2-Adamantanol (8) | |
| 9[e] | Fe(TPP)Cl | Adamantane[f] | mCPBA | | 1-Adamantanol (9) | 2-Adamantanol (2) | |
| 10[e] | NP | Adamantane[f] | mCPBA | | 1-Adamantanol (34) | 2-Adamantanol (5) | |
| 11[e] | SR | Adamantane[f] | PPAA | | 1-Adamantanol (65) | 2-Adamantanol (7) | |
| 12[e] | Fe(TPP)Cl | Adamantane[f] | PPAA | | 1-Adamantanol (11) | 2-Adamantanol (2) | |

PPAA, peroxyphenyl acetic acid; NP (complex), imidazole.

These reactions were carried out in benzene at 25° under argon for 10 min. PPAA: PhCH$_2$CO$_3$H, PAA: PhCH$_2$CO$_2$H. [Fe(Por)] = 0.10 mM; [peracid] = 1.0 mM;

[TBPH] = 10 mM. Benzaldehyde was not detected. Otherwise as noted.

[a] Yields are based on the oxidants used. Products were determined by GLC and/or GCMS.

[b] kcat, the initial rate of TBP· formation (turnover number/s).

[c] One equivalent amount of PAA to PAA was added.

[d] Yield based on two equivalent amounts to the added PPAA.

[e] [Fe(Por)] = [peracid] = 1.0 mM. Ketone product was not detected in any case.

[f] [Adamantane] = 0.50 M.

cerning the O–O bond homolysis by Fe(TPP)Cl in benzene is consistent with the results described by Watanabe and co-workers [12]. It is expected that a more strongly electron-donating axial ligand would increase both the proportion and rate of heterolytic O–O bond scission. Therefore, the order of donative character of the examined ligands can be estimated to be as follows: thiolate > imidazole ≫ chloride anion. These results are supported by a related and independent work that was published in the same year as ours [13]. Our study using the SR complex should contribute to the discussion on the mode of the O–O bond cleavage in the P-450 catalytic cycle.

Further, the reactivity of the active species formed by cleavage of the O–O bond of peroxyacid–iron porphyrin complexes was examined by using unactivated alkanes as substrates in benzene (see Table 2). The SR complex catalyzed the hydroxylation of adamantane so efficiently that the yield of adamantanols based on the used m-chloroperoxybenzoic acid (mCPBA) reached 88% (run 8). In the reaction, 80% of SR was confirmed to remain undecomposed by EPR and UV-vis examinations. On the other hand, only a low or moderate yield of adamantanol was obtained by catalysis with Fe(TPP)Cl or NP, although mCPBA was completely consumed (runs 9, 10). Both Fe(TPP)Cl and NP also retained their structures after completion of the reactions. The degree of activity of iron porphyrins for heterolysis of PPAA in benzene is thought to correlate positively with the degree of hydroxylation activity toward alkanes [10,14].

## Axial Ligand Effect on the Reactivity of the Active Intermediate of Iron Porphyrin

Oxidative O-dealkylation of alkyl aryl ethers is one of the major metabolic reactions catalyzed by cytochrome P-450. So far, there are two generally accepted mechanisms, that is, the H-atom abstraction mechanism and the ipso-substitution mechanism.

H-atom abstraction mechanism ($k_H/k_D > 6$):

ipso-substitution mechanism ($k_H/k_D \approx 1$):

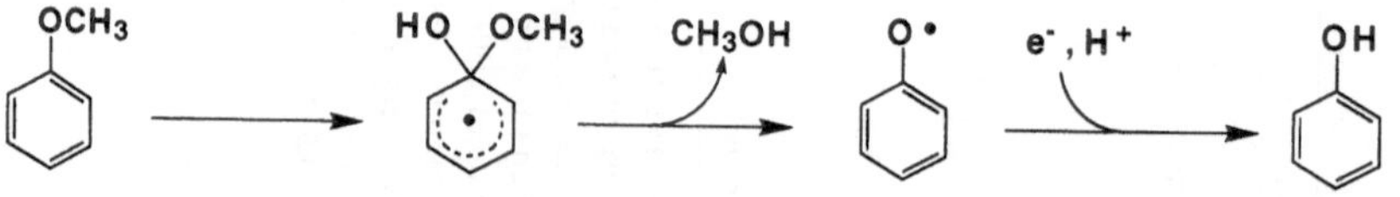

Clear differences between these two mechanisms are observed in the kinetic isotope effects (KIEs) and in the origin of the oxygen atom of the resulting phenolic hydroxy group. It is thought that the mechanism that actually operates depends on the oxidizing system used; namely, in the cytochrome P-450-dependent enzymatic reaction and the iron porphyrin-iodosylbenzene (PhIO) systems, the former mechanism operates, and in the hydroxyl radical-mediated reactions, the latter does. However, to our knowledge no investigation has been carried out for the exploration of an axial ligand effect on the O-dealkylation mechanism.

We have already shown that p-dimethoxybenzene is a useful probe for the evaluation of the reactivity of several active species such as OH radical, the formed intermediate in $O_2$–$Cu^{2+}$-ascorbic acid system, and microsomes [15]. Next, we examined the O–demethylation mechanisms of p-dimethoxybenzene in various iron porphyrin-oxidant systems and the rat liver microsome (Ms)-NADPH/$O_2$ system. We used six kinds of iron porphyrins including the SR complex. First, we investigated the modes of O–O bond cleavage mediated by these iron porphyrins by using PPAA, and the following results were obtained. In every iron porphyrin–PPAA system examined, phenylacetic acid (PAA) was the major product, which indicated the predominance of the heterolytic O–O bond cleavage and compound I formation. It was revealed that, among the iron porphyrin–oxidant systems, the O-demethylation mechanism similar to that which operates in the Ms-NADPH/$O_2$ system occurs only in the SR-PPAA system. In these two systems, p-dimethoxybenzene was O-demethylated with high KIE values (up to 12), which clearly showed that the reaction proceeded by the means of H-atom abstraction mechanism. On the other hand, in other iron porphyrin–oxidant systems low KIE values (nearly equal to 1) were observed, which showed that the reaction proceeded in the ipso-substitution manner. These data provide evidence that the axial thiolate ligand has a great influence on the reactivity of the high valent oxo-iron porphyrin intermediate, although the results are rather preliminary.

## Conclusion

We have developed a unique SR complex having axial thiolate as a structural model of the cytochrome P-450 active site. The use of this complex has led us to find several pronounced axial ligand effects of thiolate on the activation of oxygen by heme. We are now attempting to determine the structure of an active intermediate of SR by direct spectral observation, as the structure of the active species of cytochrome P-450 has not yet been clarified

*Acknowledgments.* We are grateful to Professor Yuzuru Ishimura and the Organizing Committee of Keio University International Symposia for Life Sciences and Medicine for inviting us to the 1996 Conference on Oxygen Homeostasis and Its Dynamics.

## References

1. Ortiz de Montellano P (ed) (1986, 1995) 1st and 2nd edns. Cytochrome P-450: structure, mechanism, and biochemistry, Plenum, New York
2. Meunier B (1992) Metalloporphyrins, a versatile catalyst for oxidation reactions and oxidative DNA cleavage. Chem Rev 92:1411–1456
3. Sono M, Roach MP, Dawson JH, et al. (1996) Heme-containing oxygenases. Chem Rev 96:2841–2887
4. Poulos TL, Finzel BC, Howard AJ (1987) High-resolution crystal structure of cytochrome P-450$_{cam}$ J Mol Biol 195:687–700

5. Dawson JH, Sono M (1987) Cytochrome P-450 and chloroperoxidase: thiolate-ligated heme enzymes spectroscopic determination of their active site structures and mechanistic implications of thiolate ligation. Chem Rev 87:1255–1276
6. Higuchi T, Uzu S, Hirobe M (1990) Synthesis of a highly stable iron porphyrin coordinated by alkylthiolate anion as a model for cytochrome P-450 and its catalytic activity in O–O bond cleavage. J Am Chem Soc 112:7051–7053
7. Traylor TG, Lee WA, Stynes DV (1984) Model compound studies related to peroxidases. Mechanisms of reactions of hemins with peracids. J Am Chem Soc 106:755–764
8. Yuan LC, Bruice TC (1986) Influence of nitrogen base ligation and hydrogen bonding on the rate constants for oxygen transfer from percarboxylic acid and alkyl hydroperoxides to (meso-tetraphenylporphinato)manganese(III) chloride. J Am Chem Soc 108:1643–1650
9. White RE, Sligar SG, Coon HJ (1980) Evidence for a homolytic mechanism of peroxide oxygen-oxygen bond cleavage during substrate hydroxylation by cytochrome P-450. J Biol Chem 255:11108–11111
10. Higuchi T, Shimada K, Hirobe M, et al. (1993) Heterolytic O–O bond cleavage of peroxy acid and effective alkane hydroxylation in hydrophobic solvent mediated by an iron porphyrin coordinated by thiolate anion as a model for cytochrome P-450. J Am Chem Soc 115:7551–7552
11. Collman JP, Brauman JI, Doxsee KM, et al. (1980) Synthesis and characterization of "tailed picket fence" porphyrins. J Am Chem Soc 102:4182–4192
12. Watanabe Y, Yamaguchi K, Morishima I, et al. (1991) Remarkable solvent effect on the shape-selective oxidation of olefins catalyzed by iron(III) porphyrins. Inorg Chem 30:2581–2582
13. Yamguchi K, Watanabe Y, Morishima I (1993) Direct observation of the push effect on the O–O bond cleavage of acylperoxoiron(III) porphyrin complexes. J Am Chem Soc 115:4058–4065
14. Higuchi T, Hirobe M (1996) Four recent studies in cytochrome P-450 modelings: a stable iron porphyrin coordinated by a thiolate ligand; a robust ruthenium porphyrin-pyridine N-oxide derivatives system; polypeptide-bound iron porphyrin; application to drug metabolism studies. J Mol Catal A Chem 113:403–422
15. Urano Y, Higuchi T, Hirobe M (1996) Substrate-dependent changes of the oxidative O-dealkylation mechanism of several chemical and biological oxidizing systems. J Chem Soc Perkin Trans 2:1169–1173

# Superoxide via Peroxynitrite Blocks Prostacyclin Synthesis

Volker Ullrich and Minghui Zou

*Summary.* Prostacyclin and nitric oxide, the major vasodilators, relax smooth muscle by different mechanisms. Superoxide efficiently quenches nitric oxide under formation of peroxynitrite, for which we have reported an inhibitory action on prostacyclin formation. Here we report that concentrations of peroxynitrite less than micromolar inhibit prostacyclin synthase in a purified state and also in endothelial cells. Antioxidants at physiological concentrations could not prevent the inactivation. Peroxynitrite addition did not affect the heme–thiolate bond but resulted in a positive reaction with an anti-nitrotyrosine antibody in Western blots. Because tetranitromethane also causes inhibition, we concluded that nitration of a tyrosine at the active site is a likely mechanism of prostacyclin synthase inhibition by peroxynitrite. Catalysis by the heme–thiolate entity or a suggested tyrosinate structure of the tyrosine could explain the observed high reactivity and sensitivity. Only the active enzyme, but not an inhibited form, can be nitrated.

*Key words.* Nitric oxide—Oxygen radicals—Tyrosine nitration—Vasoconstriction—Heme–thiolate protein

## Introduction

Vascular tone and blood flow are controlled by a complex network of regulatory pathways in which prostacyclin and nitric oxide (NO = EDRF, [endothelium-derived relaxing factor]) together with a postulated "endothelium-derived hyperpolarizing factor" (EDHF) and adenosine form the vasodilating mediators. They are counteracted by thromboxane $A_2$, leukotriene $C_4$, endothelins, or platelet-aggregating factor (PAF), which are usually formed only in response to pathophysiological events. A rapid reaction of NO with superoxide ($O_2^-$) under generation of peroxynitrite has been described [1] that indirectly renders superoxide a vasoconstrictor through scavenging of NO. This pathway seems to gain even greater importance as we found that peroxynitrite inactivates prostacyclin synthase [2]. The endothelial

Faculty of Biology, University of Konstanz, Fach M611, D78457 Konstanz, Germany

189

vasodilation through two independent pathways is thus consecutively blocked by superoxide.

The crucial question remained whether this inhibition observed with the microsomal enzyme has significance in cellular systems in view of the strong antioxidant potential of the intact cell. The high sensitivity in the submicromolar range was in favor of a pathophysiological role, which we have confirmed by investigations with whole cells. Another point of interest was the mechanism involved in this inactivation, because the enzyme only contained one SH group, which was coordinated to the ferric heme iron [3]. Such heme–thiolate structures are however not suitable targets, as judged from our work on other P-450 proteins [4]. The alternate hypothesis of a tyrosine nitration was tested in the current investigations.

# Materials and Methods

## Materials

Prosta-5, 13-dien-1-oic acid, 9,11-epidioxy-15-hydroxy-diethylamine (1-$^{14}$C-PGH$_2$) was synthesized from $^{14}$C-1-arachidonic acid according to [2]. Diethylamide NON-Oate; 9,11-dideoxy-9$\alpha$,11$\alpha$-methanoepoxy-prostaglandin F$_{2\alpha}$ (U46619); spermine NONOate; and 12-hydroxyheptadecatrienoic acid (HHT) were obtained from Cayman SPI (Massy, France). The monoclonal antibody against nitrotyrosine was obtained from Upstate Biotechnology (Lake Placid NY, U.S.A.). Tetranitromethane (TNM), trans-2-phenylcyclopropylamine (tranylcypromine), hypoxanthine, xanthine oxidase, and hydrogen peroxide were purchased from Sigma Biochemicals (Deisenhofen, Germany). Organic solvents were obtained from Merck Darmstadt (Darmstadt, Germany).

## Peroxynitrite Synthesis

Peroxynitrite was prepared using a quenched-flow reaction as described by Reed et al. [4]. The concentration of peroxynitrite was determined spectrally in 0.8 M NaOH ($E_{302} = 1670\,M^{-1}\,cm^{-1}$).

## Purification of PGIS from Bovine Aorta Microsomes

Bovine prostacyclin synthase (PGIS) was purified as previously described [3]. Protein concentrations of purified PGIS were estimated by the method of Bradford [5]. The enzyme was identified as a 52-kDa protein on polyacrylamide gel electrophoresis by Western blots with a polyclonal antibody [6] and by its binding difference spectrum with tranylcypromine (absorbance peak at 435 nm). Spectrophotometric studies were performed using an Aminco DW-2 dual wavelength spectrometer (SOPRA, Büttelborn, Germany).

## Activity Assays

PGIS and thromboxane synthase (TxS) activity in microsomes and cells was assayed by monitoring 6-keto-PGF$_{1\alpha}$ or TxB$_2$ formation on thin layer chromatagraphy (TLC) after incubation with [1-$^{14}$C]-PGH$_2$ as described earlier [3].

## Cell Culture

Human EaHy 926 cells (kindly provided by Dr. M. Edgell, University of North Carolina (Chapel Hill), USA) were routinely cultured in Dulbecco's modified Eagles medium (DMEM) with 10% fetal calf serum (FCS). Two days before use, the confluent cells were transferred to 0.5% serum. Cells were washed several times with phosphate buffer saline (PBS) buffer, pH 7.4. PGIS activity was monitored on TLC by the formation of $^{14}$C-6-keto-PGF$_{1\alpha}$ from $^{14}$C-PGH$_2$.

## Western Blot Analyses

Western blots were performed as described [7]. The membrane was blocked with 5% milk powder in PBS/0.1% Tween 20 for 2 h and incubated overnight at 4°C with the polyclonal antibody directed against PGIS at a dilution of 1 μg/ml. The blot was washed several times with PBS/0.1% Tween 20 and was incubated with a goat anti-rabbit antibody for 45 min at a dilution of 1:7500. Antibody binding was visualized by the enhanced chemiluminescence technique, according to the instructions of the supplier (Amersham Braunschweig, Germany). The blot, stained with a polyclonal antibody against PGIS, was stripped by incubating the membrane in stripping buffer (100 mM 2-mercaptoethanol, 2% sodium dodecyl sulfate [SDS], 62.5 mM Tris-HCl, pH 6.7) at 50°C for 30 min with agitation. The membrane was then washed, blocked, and incubated overnight at 4°C with a monoclonal antibody against nitrotyrosine at a dilution of 1 μg/ml and with a goat anti-mouse antibody for 45 min in a dilution of 1:7500. Antibody binding was visualized as described.

# Results

We first addressed the question of the physiological or pathophysiological significance of our previously described inhibition of PGI$_2$ synthase by peroxynitrite [2]. To test this, we applied peroxynitrite to the EaHy 926 endothelial cell line for which the expression of PGI$_2$ synthase has been described [8]. We preincubated the cells with increasing concentrations of peroxynitrite and followed the formation of 6-keto-PGF$_{1\alpha}$ by addition of 100 μM $^{14}$C-PGH$_2$ and quantitation of the total prostaglandin pattern by TLC after a 5-min incubation period (Table 1).

It was apparent that the very sensitive inhibition of 6-keto-PGF$_{1\alpha}$ formation observed with the microsomal enzyme could be reproduced in whole cells. An IC$_{50}$ value of approximately 100 nM could be calculated, which was close to the value in aortic microsomes [2]. This result was difficult to understand in view of the reported interaction of peroxynitrite with glutathione (GSH) [9], which amounts to about 5 mM in living cells. Therefore, a study of peroxynitrite inhibition of microsomal PGI$_2$ synthase in the presence of varying GSH concentrations was added, and indeed no inhibitory effect was observed (Fig. 1).

Several other antioxidants were also tried with a similar outcome (Table 2). The higher inhibition observed with several antioxidants was first puzzling but found an explanation in a certain background of radioactivity on the TLC plate from an unspecific oxidation of $^{14}$C-PGH$_2$ by decomposing peroxynitrite in control assays. Because antioxidants block this process, the corresponding 6-keto-PGF$_{1\alpha}$ spots contained less

TABLE 1. Inhibition by peroxynitrite of $^{14}$C-PGH$_2$-induced prostaglandin I$_2$ synthesis in human EaHy 926 cells

| Agents | Inhibition (%) | Agents | Inhibition (%) |
|---|---|---|---|
| Peroxynitrite | | Sodium nitrite | |
| 0.01 μM | 23.4 ± 4.7 | 10 μM | 1.7 ± 2.3 |
| 0.1 μM | 56.2 ± 7.9 | 100 μM | 3.5 ± 2.7 |
| 1 μM | 75.1 ± 8.1 | 1 mM | 1.9 ± 0.9 |
| 10 μM | 85.5 ± 6.7 | Sodium nitrate | |
| 100 μM | 92.9 ± 5.6 | 10 μM | 1.7 ± 0.8 |
| | | 100 μM | 3.7 ± 1.7 |
| | | 1 mM | 1.9 ± 3.7 |

PGH$_2$, prostaglandin I$_2$.
We cultured 5 × 10$^6$ cells of the human EaHy 926 cell line routinely in Dulbecco's modified Eagle's medium (DMEM) with 10% serum. Two days before use, the cells were transferred to 0.5% serum and treated with the indicated concentrations of peroxynitrite, sodium nitrite, and sodium nitrate in 1 ml medium supplemented with 100 mM HEPES buffer, pH 7.4, for 15 min. The reaction was started by adding 100 μM $^{14}$C-PGH$_2$ and incubated for 5 min; 6-keto-PGF$_{1\alpha}$ was analyzed as described in Materials and Methods. The data (means ± SEM) represent eight samples from three different experiments.

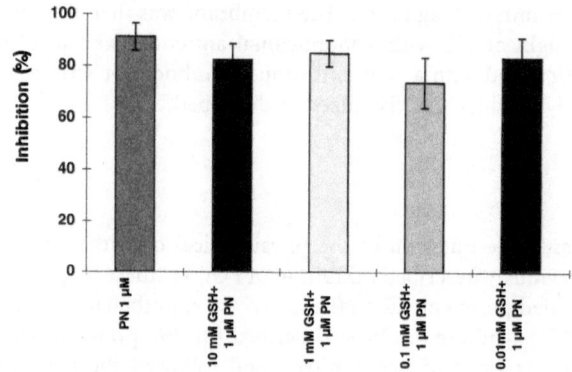

FIG. 1. Effect of glutathione (GSH) on peroxynitrite (PN) inhibition of prostacyclin synthesis in bovine aortic microsomes

background radioactivity, so that the increase in inhibition by these compounds is not real.

We concluded from the results described so far that peroxynitrite reacted extremely rapidly with PGI$_2$ synthase with a low activation energy. This directed our interest to the chemical mechanism involved in the inhibition process. The most plausible explanation was linked to the heme–thiolate (P-450) nature of the enzyme together with the reported reactivity of metal–thiolate structures against peroxynitrite, as evidenced from studies on alcohol dehydrogenase [10] and aconitase [11,12]. A similar oxidation of the thiolate ligand by PGI$_2$ synthase would destroy the typical absorption at 418 nm of the ferric enzyme and the 450 nm absorption band of the reduced enzyme in the presence of carbon monoxide [13]. Corresponding difference spectra with the oxidized enzyme after addition of 1–10 μm peroxynitrite did not show any spectral change (results not shown), and also the 450 nm band, which appears as a shoulder of a large heme chromophore in aortic microsomes, was resistant to the same concentrations of peroxynitrite (Fig. 2). As a control, hypochlorite converted the P-450 chromophore to P-420; i.e., the heme–thiolate linkage was destroyed [14].

One may argue that other essential sulfhydryl groups could be the target of peroxynitrite, but from the primary structure of the bovine enzyme it is clear that no other cysteine except the heme–binding Cys-469 is present [3]. As an alternate target, tyrosine residues have to be considered because the formation of 3-nitrotyrosine by peroxynitrite has been reported [15,16]. A monoclonal antibody, as well as a polyclonal antiserum, is commercially available and both were tried in Western blots of peroxynitrite-treated microsomes. A solubilization followed by DEAE chromatography resulted in a 10% purified preparation that reacted positively with the monoclonal antinitrotyrosine antibody after treatment of the enzyme with 1 μM peroxynitrite (Fig. 3).

Double staining with a polyclonal antiserum raised against PGI$_2$ synthase verified the identity of the nitrated enzyme with PGI$_2$ synthase. Only a second minor band reacted with the antinitrotyrosine antibody but because it also stained positively with the PGI$_2$ synthase antibody it probably was a degradation product of the enzyme. The fact that no other bands appeared in the Western blots of the microsomal fraction, which contained only about 0.5% PGI$_2$ synthase, can be taken as proof for the selective staining of PGI$_2$ synthase in aortic microsomes.

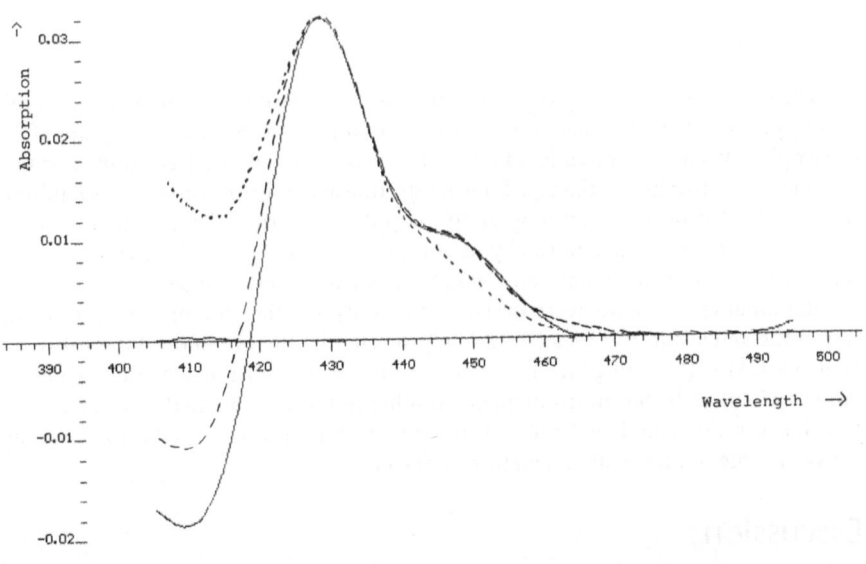

FIG. 2. Carbon monoxide difference spectra of reduced bovine aortic microsomes after peroxynitrite and hypochlorite treatment

TABLE 2. Effects of antioxidants on peroxynitrite-induced inhibition of 6-keto-$PGF_{1\alpha}$ formation in purified prostacyclin synthase (PGIS) assayed with $^{14}C$-$PGH_2$

| Agents | Inhibition (%) |
|---|---|
| Peroxynitrite (1 µM) (control) | 78.9 ± 6.3 |
| +SOD (100 U/ml) | 68.2 ± 5.4 |
| +Catalase (10 000 U/ml) | 68.9 ± 4.2 |
| +SOD + catalase | 64.0 ± 7.8 |
| +Desferal (1 mg/ml) | 94.2 ± 6.2 |
| +Mannitol (1 mM) | 95.2 ± 2.8 |
| +Uric acid (1 mM) | 98.9 ± 0.7 |
| +DMSO (1 mM) | 95.9 ± 10.5 |
| +Tryptophan (1 mM) | 74.6 ± 3.2 |
| +DTPA (1 mM) | 97.2 ± 0.4 |
| +EDTA (1 mM) | 78.4 ± 5.3 |
| +Hematin (10 µM) | 96.7 ± 3.2 |
| +$Fe^{2+}$ (10 µM) | 97.5 ± 3.1 |
| +$Fe^{3+}$ (10 µM) | 95.7 ± 4.2 |
| +α-Tocopherol (1 mg/ml) | 73.7 ± 4.5 |
| +Ascorbic acid | 81.3 ± 7.1 |
| +GSH (10 mM) | 77.4 ± 4.7 |
| +Dithiothreitol (1 mM) | 78.6 ± 5.7 |

DTPA, diethylenetriaminepentaacetic acid, desferal, desferrioxamine; DMSO, dimethyl sulfoxide; EDTA, ethylenediaminetetraacetic acid disodium salt dihydrate; GSH, glutathione; SOD, superoxide dismutase.

Such data would favor a tyrosine nitration as the underlying mechanism of inhibition by peroxynitrite. Because HPLC detection of nitrotyrosine in a hydrolysate of the electrophoretically separated band was not sensitive enough, we treated the purified enzyme with tetranitromethane, a known nitrating agent for tyrosine [17]. An inhibition was indeed observed starting at 10–100 µM tetranitromethane, which is in the lower concentration range normally used (Table 3). The Western blot with the anti-nitrotyrosine monoclonal antibody was also positive (results not shown).

Additional experiments were designed to clarify whether the presumed tyrosine was at the active site and hence could be protected by the presence of the substrate analog U46619 [18]. This protection could be observed against inactivation by peroxynitrite as well as by tetranitromethane. Another important clue to the mechanism of tyrosine nitration could be the observation that nitration was detectable only with an active enzyme but not with a denatured enzyme.

## Discussion

In continuation and extension of our first report on $PGI_2$ synthase inhibition by peroxynitrite, we are now able to verify this inhibition also for intact endothelial cells.

TABLE 3. Effects of the PGH$_2$ analog, U46619, on inhibition of prostacyclin synthase by peroxynitrite and tetranitromethane

| Agent (µM) | Inhibition without U46619 (%) | Inhibition with U46619 (%) | |
|---|---|---|---|
| | | 10 µM | 100 µM |
| Peroxynitrite | | | |
| 1 | 76.5 ± 7.8 | 8.5 ± 10.7 | 1.7 ± 3.4 |
| 10 | 87.4 ± 7.6 | 32.4 ± 11.1 | 16.4 ± 5.9 |
| Tetranitromethane | | | |
| 10 | 22.7 ± 8.1 | 2.5 ± 1.5 | 1.1 ± 0.7 |
| 100 | 67.3 ± 9.7 | 30.2 ± 9.3 | 15.4 ± 9.7 |
| 1000 | 96.8 ± 2.1 | 54.5 ± 9.5 | 27.3 ± 6.7 |

Purified PGIS (113 pmol/ml) was incubated with the given concentrations of
per-oxynitrite and tetranitromethane for 15 min in 0.1 ml of 100 mM KPi
buffer, pH 7.4. U46619 was added 10 min before the incubation. The reaction was started by dilution to 1 ml with 100 mM KPi buffer, pH 7.4, and
addition of 100 µM $^{14}$C-PGH$_2$ and then incubated for 3 min with shaking; 6-
keto-PGF$_1$ was extracted and analyzed as described in Table 1. PGIS activity
was expressed as percent inhibition compared with untreated enzyme. The
data (mean ± SEM) represent 10 samples from three assays.

The fact that a very similar and low IC$_{50}$ value for the EaHy926 cell line and for the
purified enzyme was obtained (~0.1 µM) indicated that the antioxidant potential of
the cells does not seem to interfere with the action of peroxynitrite. This was surprising because a reaction of peroxynitrite with glutathione had been reported [9], but as
much as 10 mM glutathione was found not to interfere with the inactivation of PGI$_2$
synthase. Several conclusions can be drawn from these findings:

First, becaused there is little doubt that micromolar concentrations of peroxynitrite
can build up in vivo, it is now conceivable that PGI$_2$ synthase is a likely target because
it can react with peroxynitrite without interference by cellular oxidants. Second, it is
obvious that PGI$_2$ synthase does not react with its apoprotein but that an intact active
site seems to be involved, leading to an autocatalytic process of inactivation. Third,
the chemical modification introduced by peroxynitrite cannot consist in a sulfhydryl
oxidation because only one cysteine can be found in the primary structure of
PGI$_2$ synthase and this cysteine residue is involved in the heme binding, which
was not found to be influenced by peroxynitrite treatment. We rather presented
evidence for a tyrosine nitration as documented by a positive Western blot with an
antinitrotyrosine monoclonal antibody. The staining was completely blocked in the
presence of 10 mM 3-nitrotyrosine, and, interestingly, at 1 µM peroxynitrite no bands
other than the 52-kDa band of PGI$_2$ synthase were stained. This again proves the
high selectivity of this enzyme for its reaction with peroxynitrite at the given low
concentration.

Support for a tyrosine nitration as the underlying mechanism of inactivation also
came from tetranitromethane treatment, which resulted in an inhibition of PGI$_2$
synthase. Both nitrations, that by peroxynitrite and that by tetranitromethane, could
be blocked by the stable substrate analog U46619. From this, location of the nitrated
tyrosine at the active site of PGI$_2$ synthase would be likely. Considering the previously

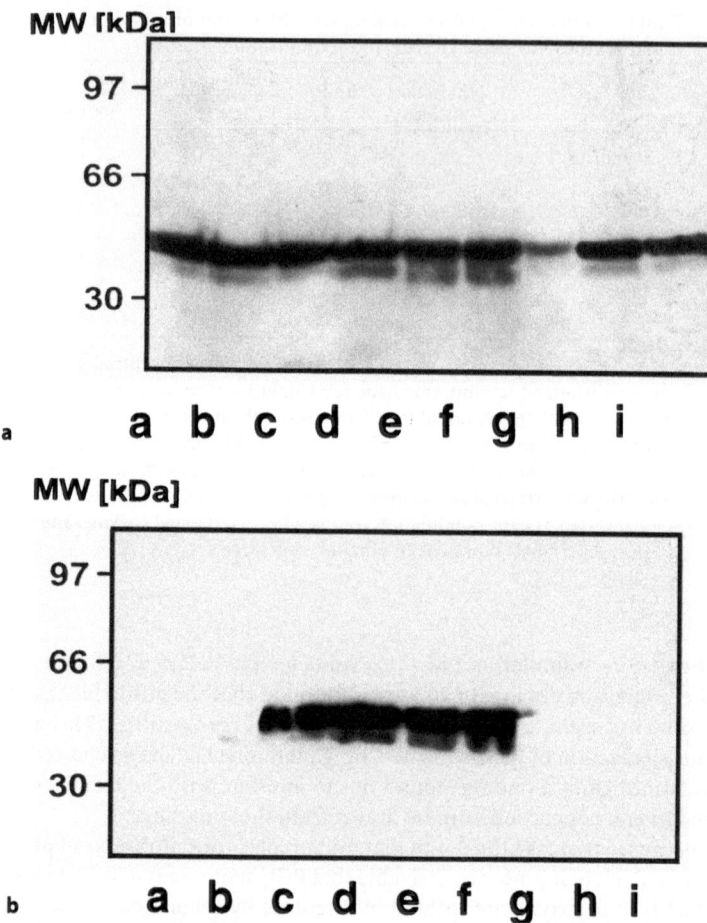

FIG. 3. Immunodetection of nitrotyrosine and prostacyclin synthase (PGIS). Purified PGIS (1 nmol) was treated with: buffer, *lane a*; decomposed peroxynitrite, *lane b*; 1 and 0.1 μM of peroxynitrite, *lanes c* and *d*; 100 mU/ml xanthine oxidase /0.1 mM hypoxanthine plus 1 mM diethylamine NONOate, *lane e*; 1 mM 1,2,3-oxadiazolium, 5-amino-3-(4-morpholinyl)-, chloride (SIN-1), *lane f*; 100 mU/ml xanthine oxidase /0.1 mM hypoxanthine, *lane g*; NO generated from 1 mM diethylamine, *lane h*; and 100 μM hypochlorite, *lane i*; Western blots were performed as described in Methods and Materials. **a** Immunodetection of PGIS (52 kDa band) with polyclonal antiserum against PGIS. **b** Immunodetection of nitrotyrosine with a monoclonal antibody against 3-nitrotyrosine from the same blot

proposed fitness of heme–thiolate proteins with regard to their reaction with peroxides [19], one can postulate a formal homolytic cleavage of the peroxide bond by the heme–sulfur entity, followed either by an immediate electron transfer from the $NO_2$ radical to the ferryl species or by first a nitration and then oxidation:

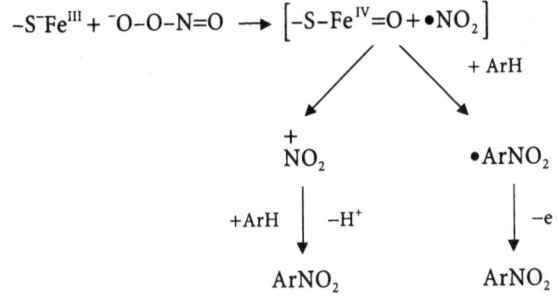

In summary, the proposed mechanism of peroxynitrite inhibition provides a satisfactory explanation for the high sensitivity of PGI$_2$ synthase toward peroxynitrite and also for the insensitivity against cellular antioxidants because these cannot compete effectively with the rapid catalytic process at the heme–thiolate center. Conditions of increased superoxide production in the presence of vascular NO can thus create enough peroxynitrite to block prostacyclin synthesis. The situation would become critical and pathophysiological if the rate of superoxide radical generation approached or equaled that of NO. Then both vasodilators, NO and prostacyclin, would be eliminated with the consequences of severe vasoconstriction and ischemia. This could further enhance superoxide production through conversion of xanthine dehydrogenase to the oxidase form [20], resulting in a vicious cycle with leukocyte adherence and activation [21].

# References

1. Huie RE, Padmaja S (1995) The reaction rate of nitric oxide with superoxide. Free Radical Res Commun 18:195–199
2. Zou MH, Ullrich V (1996) Peroxynitrite formed by simultaneous generation of nitric oxide and superoxide selectively inhibits bovine aortic prostacyclin synthase. FEBS Lett 382:101–104
3. Hara S, Miyata A, Yokoyama C, et al. (1994) Isolation and molecular cloning of prostacyclin synthase from bovine endothelial cells. J Biol Chem 269:19897–19903
4. Reed JW, Ho HH, Jolly WL (1974) Chemical synthesis with a quenched flow reactor: hydroxytrihydroborate and peroxynitrite. J Am Chem Soc 96:1248–1249
5. Bradford MM (1976) A rapid and sensitive method for the quantitation of microgram quantities of protein utilizing the principle of protein-dye binding. Anal Biochem 72:248–254
6. Siegle I, Nüsing R, Brugger R, et al. (1994) Characterization of monoclonal antibodies generated against bovine and porcine prostacyclin synthase and quantitation of bovine prostacyclin synthase. FEBS Lett 347:221–225
7. Klein T, Nüsing RM, Ullrich V (1994) Selective inhibition of cyclooxygenase-2. Biochem J 48:1605–1610
8. Edgell CJS, McDonald CC, Graham IB (1983) Permanent cell line expressing human factor VIII-related antigen established by hybridization. Proc Natl Acad Sci USA 80:3734–3737
9. Radi R, Beckman JS, Bush KM, et al. (1991) Peroxynitrite oxidation of sulfhydryls. J Biol Chem 266:4244–4250
10. Crow JP, Beckman JS, McCord JM (1995) Sensitivity of the essential zinc-thiolate moiety of yeast alcohol dehydrogenase to hypochlorite and peroxynitrite. Biochemistry 34:3544–3552

11. Castro L, Rodriguez M, Radi R (1994) Aconitase is readily inactivated by peroxynitrite, but not by its precursor, nitric oxide. J Biol Chem 269:29409–29415
12. Hausladen A, Fridovich I (1994) Superoxide and peroxynitrite inactivate aconitase, but nitric oxide does not J Biol Chem 269:29405–29408
13. Ullrich V, Castle L, Weber P (1981) Spectral evidence for the cytochrome P-450 nature of prostacyclin synthase. Biochem Pharmacol 30:2033–2036
14. Shimizu T, Hirana K, Takahashi M, et al. (1988) Site-directed mutageneses of rat liver cytochrome P-450$\alpha$: axial ligand and heme incorporation. Biochemistry 27:4138–4141
15. Ischiropoulos HI, Zhu L, Chen J, et al. (1992) Peroxynitrite-mediated tyrosine nitration catalyzed by superoxide dismutase. Arch Biochem Biophys 298:431–437
16. Beckman JS, Ischiropoulos H, Zhu L, et al. (1992) Kinetics of superoxide dismutase and iron catalysed nitration of phenolics by peroxynitrite. Arch Biochem Biophys 298:438–445
17. Sokolovsky M, Riordan JF, Vallee BL (1966) Tetranitromethane, a reagent for the nitration of tyrosyl residues in proteins. Biochemistry 5:3582–3589
18. Hecker M, Ullrich V (1989) On the mechanism of prostacyclin and thromboxane $A_2$ biosynthesis. J Biol Chem 264:141–150
19. Ullrich V, Brugger R (1994) Prostacyclin and thromboxane synthases: new aspects of hemethiolate catalysis. Angew Chem Int Ed Engl 33:1911–1919

# The N=N Bond Cleavage of Angeli's Salt Is Markedly Enhanced by Cytochrome P-450 1A2: Effects of Distal Amino Acid Mutations on the Formation of Nitric Oxide Complexes

Yoshinori Shibata, Hideaki Sato, Ikuko Sagami, and Toru Shimizu

*Summary.* Angeli's salt, $Na_2N_2O_3$, is known to release $NO^-$ or $NO^{\bullet}$. We studied the effect of rat liver cytochrome P-450 1A2 (P-450 1A2) in regard to its catalysis of the N=N bond scission of Angeli's salt with absorption spectroscopy. We also examined the contribution of putative distal amino acids of P-450 1A2 to the reaction with the salt. It was found that wild-type $Fe^{3+}$ P-450 1A2 markedly enhances the N=N scission of the salt up to 100 fold in terms of the optical absorption spectral change of P-450 1A2. A ferric wild-type P-450 1A2–NO complex was formed when the salt was added and the complex was then changed to a six-coordinated ferrous NO complex. Glu318Asp, Glu318Ala, and Thr319Ala mutants at the putative distal site of P-450 1A2 formed a five-coordinated ferrous NO complex that was not formed with the wild type. The Glu318Ala mutant did not form the ferric NO complex with the addition of Angeli's salt. Effects of substrates such as phenanthrene and 1,2:3,4-dibenzanthracene on the P-450 1A2 spectral change were also examined. It was suggested that the N=N bond of Angeli's salt is cleaved with the P-450 1A2 active site and that $NO^-$ or $NO^{\bullet}$ is released. This hypothetical N=N scission mechanism seems relevant to the O=O scission with this enzyme during the monooxidation.

*Key words.* Nitric oxide—Angeli's salt—Cytochrome P-450—Hemoprotein—Distal site—Redox change

Angeli's salt, $Na_2N_2O_3$ or $^-O-N=N^+-(OH)(O^-)$ in aqueous solution, is known to induce relaxation of vascular smooth muscle in vitro and lower blood pressure in vivo [1,2]. This is relevant to functions mediated by nitric oxide [3,4]. Angeli's salt spontaneously decomposes and forms $N_2O$ and $NO_2^-$ at neutral pH with a mechanism as follows [5,6]:

$$^-O-N=N^+ \Big\langle \begin{matrix} O^- \\ OH \end{matrix} \longrightarrow HNO + NO_2^-$$

Institute for Chemical Reaction Science, Tohoku University, 2-1-1 Katahira, Aoba-ku, Sendai, Miyagi 980-77, Japan

$$2HNO \rightarrow HO-N=N-OH \rightarrow N_2O + H_2O$$

$$NO^- \rightarrow NO^{\bullet} + e^-$$

We examined the interaction of Angeli's salt with rat liver microsomal cytochrome P-450 1A2 (P-450 1A2) with optical absorption spectra. Roles of distal amino acids of P-450 1A2 were also examined using putative distal mutants of this enzyme.

The Soret absorption band at 393 nm of the high-spin $Fe^{3+}$ wild-type P-450 1A2 moved to 435 nm by the addition of Angeli's salt. This spectral change rate was linear and approximately $10^{-2} s^{-1}$. This spectral change rate was of the first order with the Angeli's salt concentration. The Soret band moved further, to around 440 nm, concomitant with the disappearance of the 435-nm absorption peaks.

In the presence of large and less flexible substrates such as phenanthrene and 1,2:3,4-dibenzanthracene (dibenzanthracene), the spectral change rates were altered. Namely, the 435-nm peak formation rate was accelerated by 3.2 fold on the addition of phenanthrene, while it was decelerated by 2 fold on the addition of dibenzanthracene.

Well-conserved Glu318 and Thr319 should be located at the distal site of the P-450 1A2 heme and must be very important in $O_2$ molecule activation during the catalytic

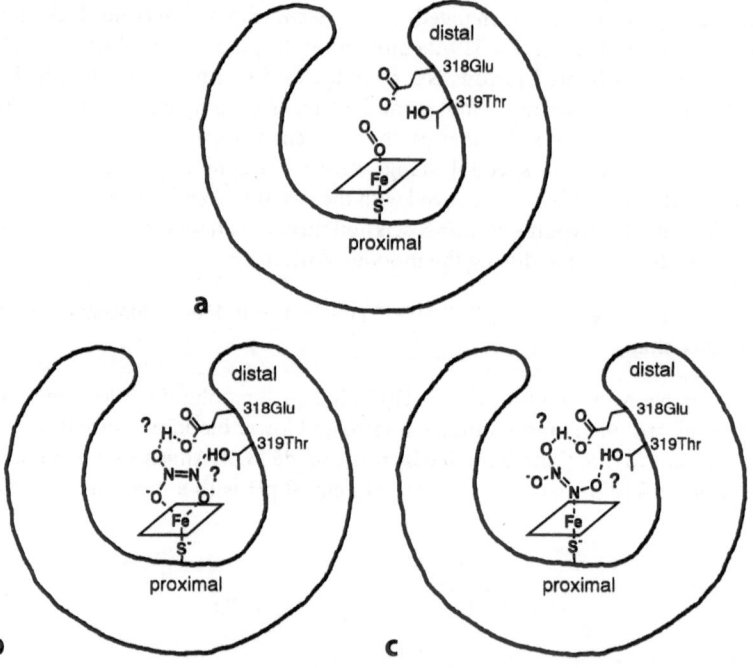

FIG. 1a–c. Hypothetical heme active site structure of cytochrome P-450 1A2 in the presence of $O_2$ (a) or Angeli's salt (b,c). Two geometries of the coordination structure, coplanar (b) and coaxial (c), of Angeli's salt with the heme plane are considered, if Angeli's salt coordination is a reality. In the coplanar case, the N–N axis lies nearly parallel to the heme plane [10], while in the coaxial case, the internuclear axis of N=N is nearly perpendicular [11]. The O-atom coordination to the heme iron is also possible in the coaxial case [11]

reaction [7,8] (Fig. 1a). Soret absorption of the Glu318Asp mutant at 393 nm first moved to 435 nm and then to 440 nm by adding Angeli's salt. A new peak at 400 nm was observed for this mutant concomitantly with the 440-nm-peak decrease. The addition of Angeli's salt to the Glu318Ala mutant formed a new peak around 400 nm. The Soret peak at 393 nm of the Thr319Ala mutant moved to 435 nm, but did not move further to 440 nm with the addition of Angeli's salt. Concomitant with the decrease of the 435-nm band of this mutant, a new peak appeared at 400 nm.

Angeli's salt spontaneously decomposes on the order of $10^{-4}s^{-1}$, giving rise to $N_2O$ and $NO_2^-$ [5,6]. An important observation of our study is that Angeli's salt decomposes at the rate $10^{-2}s^{-1}$, which is 100 fold higher that the autodecomposition rate [5,6]. Thus it is clear that P-450 1A2 facilitates the decomposition of Angeli's salt. Because the rate-determining step of the Angeli's salt decomposition is the N=N bond cleavage [5,6], P-450 1A2 apparently enhances the N=N bond scission. The distal site mutations further enhance the N=N bond cleavage rate in terms of the spectral change, suggesting again that the distal amino acid(s) contributes to the N=N bond scission (Fig. 1b,c). Perhaps $H^+$ of the distal amino acids or $H_2O$ in the distal site assists the N=N bond cleavage; that is relevant to the O=O bond scission in the P-450 functions, as shown in Fig. 1b,c.

It was previously reported that by adding an NO-releasing reagent to the Glu318Ala mutant, a five-coordinated NO–ferrous complex having a peak at 400 nm was formed after forming the NO ferric complex [9]. This is perhaps associated with a $Fe^{3+}\cdots N-O\cdots H^+\cdots$distal carboxylate (of Glu318) hydrogen bond. When the hydrogen bond is broken by the Glu318Ala mutation, accessibility of $OH^-$, $H_2O$, or $NO^-$ to the heme iron increases and facilities the heme iron reduction. Figure 2 summarizes

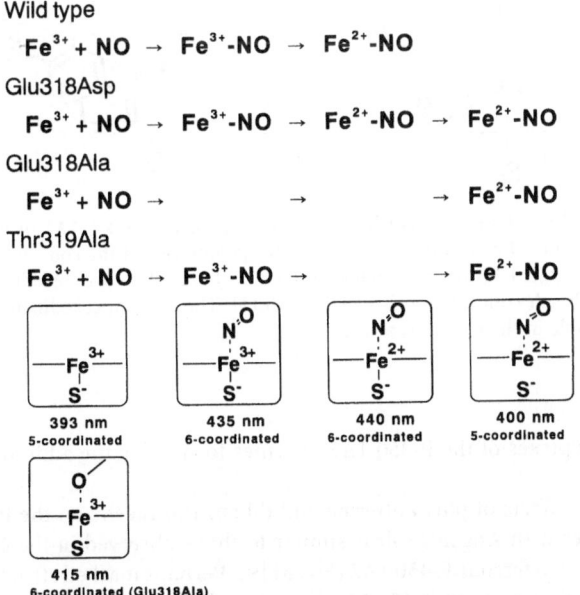

FIG. 2. NO-bound complexes of the P-450 1A2 enzymes formed by Angeli's salt

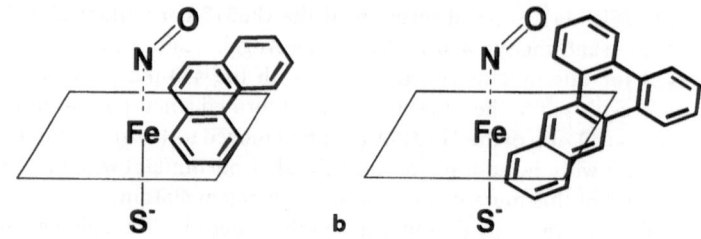

FIG. 3a,b. Hypothetical NO-bound heme structure of P-450 1A2 in the presence of phenanthrene (a) or dibenzanthracene (b)

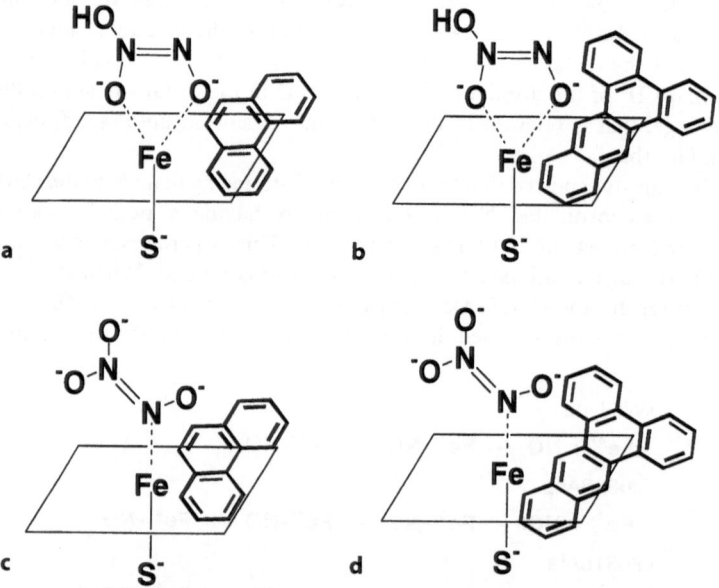

FIG. 4a–d. Hypothetical Angeli's salt-bound heme structure of P-450 1A2 in the presence of phenanthrene (a,c) or dibenzanthracene (b,d). Two geometries of the coordination structure, coplanar (a,b) and coaxial (c,d), of Angeli's salt with the heme plane are considered as in Fig. 1b and 1c, if Angeli's salt coordination is a reality [10,11]. The O-atom coordination to the heme iron is also possible in the coaxial case [11]

NO-bound complexes of the P-450 1A2 enzymes formed by the addition of Angeli's salt.

The opposite effects of phenanthrene and dibenzanthracene on the $P\text{-}450_{1A2}$ spectral change rates with Angeli's salt is similar to those observed in the CO- and NO-binding affinities to ferrous P-450 1A2 (Fig. 3) [9]. Perhaps marked structural changes in the heme environment of P-450 1A2 before or during the N=N bond scission are caused by the addition of those less flexible polycyclic hydrocarbons (Fig. 4).

# Summary

1. The P-450 1A2 active site effectively catalyzes the N=N bond scission of Angeli's salt. The role of $H^+$ of the active site in the N=N cleavage and in the complex stability is implied (Fig. 1b,c).
2. P-450 1A2–NO complex structures are markedly influenced by the distal site mutations (Fig. 2). This is perhaps associated with $H_2O$ or $OH^-$ in the active site or hydrogen bonds such as $Fe^{3+}\cdots N-O\cdots H^+\cdots Glu$ carboxylate of the distal site (Fig. 1b,c).
3. Opposite effects of phenanthrene and dibenzanthracene on the P-450 1A2 spectral change rates with Angeli's salt were observed, suggesting that structural changes in the heme active site were caused by those substrates in different fashions (Fig. 4).

*Acknowledgment.* This work was supported in part by a grant from Takeda Science Foundation and grants-in-aid from the Ministry of Education, Science, Sports and Culture of Japan for General (7680670) and for Priority Area (biometallics) (9235201) to T.S.

# References

1. Fukuto JM, Chiang K, Hszieh, R, et al. (1992) The pharmacological activity of nitroxyl: a potent vasodilator with activity similar to nitric oxide and/or endothelium-derived relaxing factor. J Pharmacol Exp Ther 263:546–551
2. Zamora R, Feelisch M (1994) Bioassay discrimination between nitric oxide ($NO^{\cdot}$) and nitroxyl ($NO^-$) using L-cystein. Biochem Biophys Res Commun 201:54–62
3. Feelisch M, Stamler JS (eds) (1996) Methods in nitric oxide research. Wiley, Chichester
4. Lancaster J JR (ed) (1996) Nitric oxide, principles and actions. Academic Press, San Diego
5. Bonner FT, Ravid B (1975) Thermal decomposition of oxyhyponitrite (sodium trioxodinitrate (II)) in aqueous solution. Inorg Chem 14:558–563
6. Hughes MN, Wimbledon PE (1976) The chemistry of trioxodinitrites. Part I. Decomposition of sodium trioxodinitrite (Angelis's salt) in aqueous solution. J Chem Soc Dalton 703–707
7. Ishigooka M, Shimizu T, Hiroya K, Hatano M (1992) Role of Glu318 at the putative distal site in the catalytic function of cytochrome P-450d. Biochemistry 31:1528–1531
8. Shimizu T, Murakami Y, Hatano M (1994) Glu318 and Thr319 mutations of cytochrome P-450 1A2 remarkably enhance homolytic O–O cleavage of alkyl hydroperoxides. An optical absorption spectral study. J Biol Chem 269:13296–13304
9. Nakano R, Sato H, Watanabe A, et al. (1996) Conserved Glu318 at the cytochrome P-450 1A2 distal site is crucial in the nitric oxide complex stability. J Biol Chem 271:8570–8574
10. Yi GB, Khan MA, Richter-Addo GB (1995) Metalloporphyrins with $X[N_2O_2]^-$ ligands. Novel high-spin (*N*-Phenyl-*N*-nitrosohydroxylaminato)(*meso*-tetraarylporphyrinato)iron(III). Inorg Chem 34:5703–5704
11. Yi GB, Khan MA, Richter-Addo GB (1995) The first metalloporphyrin nitroamine complex: Bis(diethylnitrosoamine)(*meso*-tetraphenylporphyrinato)iron(III) perchlorate. J Am Chem Soc 117:7850–7851

# Langmuir–Blodgett Films of Cytochrome P-450$_{scc}$: Molecular Organization and Thermostability

OLEG L. GURYEV, ALEXANDER V. KRIVOSHEEV, and SERGEY A. USANOV

*Summary.* Langmuir–Blodgett (LB) films of cytochrome P-450$_{scc}$ (CYP11A1; P-450$_{scc}$) and its complex with fluorescein isothiocyanate-labeled adrenodoxin (AD) were deposited on the solid supports, and their conformational characteristics at the level of a secondary structure were analyzed by means of spectroscopic and circular dichroism measurements. The interrelationship between spin-state transitions and appropriate secondary structure of P-450$_{scc}$ incorporated in LB films as well as the temperature influence on the P-450$_{scc}$ secondary structure both in LB films and in solution have been studied. P-450$_{scc}$ in LB films is in a low spin state that is preserved even if the films are prepared from the P-450$_{scc}$:AD complex where P-450$_{scc}$ exists in the high-spin state. The temperature growth from 22°C to 60°C results in decrease of α-helix content: 35%–22% in solubilized high-spin P-450$_{scc}$, 17%–10% in solubilized low-spin P-450$_{scc}$, and 13%–10% in LB film. The P-450$_{scc}$ α-helix structure in the film returned to its initial state after heating and cooling to room temperature. The secondary structure of the film-immobilized heme protein was the most stable relative to temperature effects. Thus, LB films represent a specifically organized molecular structure, providing an essential retention of heme protein secondary structure.

*Key words.* Cytochrome P-450$_{scc}$ (CYP11A1)—Langmuir–Blodgett films—Spin-state equilibrium—Circular dichroism spectroscopy

## Introduction

Cytochrome P-450$_{scc}$ (P-450$_{scc}$, CYP11A1) is an integral membrane protein localized in the inner mitochondrial membrane of the adrenal cortex [1]. It catalyzes the key biosynthetic reaction of the main steroidogenic pathways, conversion of cholesterol to pregnenolone. This heme protein is the terminal component of the electron-transfer system in eukaryotic cells, which also includes [2Fe-2S]-type ferredoxin, adrenodoxin (AD), and a flavoprotein, NADPH-adrenodoxin reductase (ADR). AD

---

Institute of Bioorganic Chemistry, Academy of Sciences of Belarus, Zhodinskaya St. 5/2, 220141 Minsk, Belarus

forms tight complexes both with ADR and P-450$_{scc}$, thus providing an electron transfer between them [2,3].

One of the problems arising when enzymes are used in biosensors, bioreactors, and bioelectronics is an appropriate immobilization of protein molecules on solid supports. The Langmuir–Blodgett (LB) technique has come to be considered as a suitable tool for modification of the surfaces by macromolecules. This approach allows us to obtain ordered mono- and multilayers of protein molecules on a variety of substrates, yielding surfaces with reproducible and controlled properties [4].

Recently, LB films of P-450$_{scc}$ were prepared on the solid supports and their spectral properties were investigated [5]. Being immobilized, heme protein changes its spin state from an initially high to a low spin. This transition is reversible, because after the solubilization of heme protein the spin-state equilibrium tends to be shifted again toward the high-spin state. The conformational changes at the level of the secondary structure taking place in P-450$_{scc}$ LB films are not yet ascertained. This chapter presents analysis of the interrelationship between spin-state transitions and appropriate secondary structure components of P-450$_{scc}$ incorporated in LB film. Circular dichroism (CD) spectroscopy was used to (a) determine the solid-state P-450$_{scc}$ secondary structure; (b) compare the film-bound heme protein secondary structure with that in solution; and (c) analyze the influence of temperature increase on the secondary structure of film-incorporated and soluble P-450$_{scc}$. The effect of P-450$_{scc}$:AD complex formation on P-450$_{scc}$ spin states in LB film has been also studied.

## Materials and Methods

### Materials

P-450$_{scc}$ and AD were affinity purified using specific adsorbents from bovine adrenocortical mitochondria [6,7]. The P-450$_{scc}$:AD complex was formed by dialysis of a heme protein and ferredoxin mixture (1:2 molar ratio) in 20 mM Tris buffer, pH 7.4.

### Deposition of LB Films of the Proteins

LB films of P-450$_{scc}$, P-450$_{scc}$, and AD were prepared in a $55 \times 200 \times 7$ mm LB trough (MDT, Zelenograd, Russia) at 25 mN/m surface pressure; 10 mM sodium phosphate buffer, pH 7.4, was used as a subphase. LB films were transferred onto dimethyldichlorosilane-treated quartz slides. The preformed protein monolayer was transferred from the subphase surface onto the activated supports by "touching" the subphase surface with the silanized support (analogous to the Langmuir–Schaefer method). The subphase excess was removed from the film surface by gentle nitrogen flux. The prepared films were stored in 50 mM sodium phosphate buffer, pH 7.4. The same procedure was used to prepare films from P-450$_{scc}$:AD complex.

### Labeling of AD with Fluorescein Isothiocyanate

Modification of AD with fluorescein isothiocyanate (FITC) was carried out as described elsewhere [8]. Gel filtration on Sephadex G-25 and chromatography on diethyl aminoethyl- (DEAE-) sepharose were used to separate AD–FITC conjugates from the free label. The modification degree of AD was calculated using the molar extinction coefficient for FITC of $\varepsilon_{493} = 62\,\text{mM}^{-1}\text{cm}^{-1}$ in 30 mM Tris buffer, pH 7.4. The ferredoxin concentration was determined using an extinction coefficient of $\varepsilon_{600} = 2.1\,\text{mM}^{-1}\text{cm}^{-1}$ (FITC has no absorbance in this range). We determined the AD molar extinction coefficient at 493 nm as $6.1\,\text{mM}^{-1}\text{cm}^{-1}$. To calculate the FITC net absorbance, the value of ferredoxin absorbance was subtracted from the absorbance of AD-FITC conjugate.

The modification effect on the AD ability to form a complex with P-450$_{scc}$ was studied by optical difference spectroscopy titration of low-spin heme protein using AD as a ligand. Low-spin preparation of P-450$_{scc}$ was obtained by enzymatic conversion of endogenous cholesterol to pregnenolone. P-450$_{scc}$ : AD complex dissociation constant ($K_d$) was determined from the dependence of the absorbance changes on free AD concentration. The concentration of free AD was calculated from this equation:

$$[AD]_{free} = [AD]_{total} - (\Delta A / \Delta A_{max}) \cdot [P450_{scc}] \tag{1}$$

P-450$_{scc}$ : AD complex formation, facilitated by exogenous cholesterol, results in absorbance changes in the Soret region (393–416 nm), thus indicating low- to high-spin transition of the heme iron.

## Analytical Methods

Absorption spectra of protein LB films were recorded using a Shimadzu UV3000 spectrophotometer (Shimadzu, Kyoto, Japan). Circular dichroism measurements were performed using a Jasco J20 spectropolarimeter (Japan Spectroscopic, Tokyo, Japan) with a thermostated cell. The sample was preincubated at the given temperature for 10 min. The $\alpha$-helix content was determined from the residual ellipticity at 222 nm according to Chen and Yang [9]. The presented spectra were averaged from four or five individual scans after subtraction of the baseline for the corresponding buffer or blank quartz stripes.

The absolute amount of the heme protein in the optical path of LB films was calculated using the following assumption. Because 409 nm is an isobestic point at P-450$_{scc}$ spin transitions, the absorbance in this point remains constant. Calculating the absorbance value at 409 nm for the protein solution of a known concentration, we assumed the concentration of the LB film "theoretical solution" to the estimated as $A_{409}$ solution/$A_{409}$ film ratio.

## Results

### Spectral Characterization of P-450$_{scc}$ LB Films

In accordance with conventional purification protocol, isolated P-450$_{scc}$ still contains tightly bound cholesterol molecules. At neutral pH and room temperature, about 70% of P-450$_{scc}$ is in a high-spin state [10]. Thus, taking these data into account, we assumed our P-450$_{scc}$ preparation to be mostly in a high-spin state, and it is hereinafter referred to as "high-spin P-450$_{scc}$." The absorption spectrum of P-450$_{scc}$ in solution is characterized by the peaks at 393, 526, and 645 nm, corresponding to the substrate-

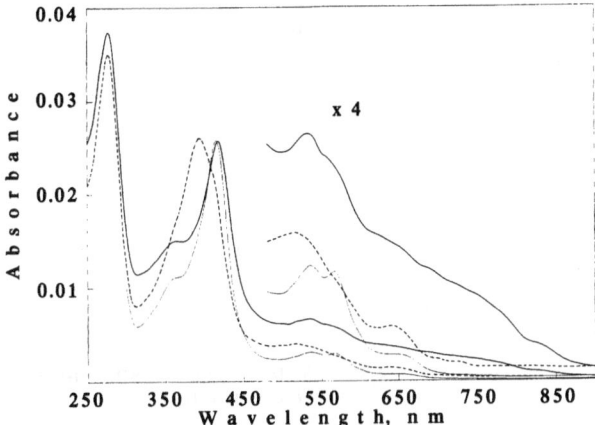

FIG. 1. Absorption spectra of high-spin cytochrome P-450$_{scc}$ (*dashed line*), Tween 20-induced low-spin cytochrome P-450$_{scc}$ (*thin solid line*), and 50 layers of P-450$_{scc}$ containing Langmuir-Blodgett (LB) film (*thick solid line*)

bound high-spin form of the heme protein. The spectra of heme protein in LB films differed from those in solution. The observed peaks at 416, 535, and 570 nm and a shoulder around 360 nm are very similar to the pattern for a Tween 20-induced low-spin heme protein (Fig. 1). These data clearly indicate that P-450$_{scc}$ in LB films exists in a low-spin form (detailed results and discussion of the optical spectral and electron-transfer properties of P-450$_{scc}$ in thin solid films were previously reported [5]).

## Adrenodoxin Modification with FITC

The spin-state equilibrium of P-450$_{scc}$ is highly dependent on the binding of AD or substrate. However, even long-term incubation (2–5 days) of P-450$_{scc}$ LB films with 50 μM cholesterol or 4 μM AD did not cause any evident changes in the Soret region of the heme protein spectrum (data not shown).

The treatment of AD with FITC results in modification of the ferredoxin free amino groups. Figure 2 represents the absorption spectrum for the AD–FITC preparation, which differs significantly from those of the initial reagents. The high absorbance of FITC in the visible region of the spectrum overlaps greatly with the absorbance of ferredoxin. Absorption maximum of the protein-bound FITC was found to be shifted to the long-wave area by approximately 5 nm. The peak location at 498 nm indicates the hydrophobic environment of the FITC molecule.

A two-step gel filtration on Sephadex G-25M and DEAE-sepharose allows complete removal of the noncovalently bound FITC, as judged from the sodium dodecyl sulfate-poly acrylamide gel electrophoresis data. The modified AD migrates as a single band with a molecular weight corresponding to the native ferredoxin and was specifically colored (data not shown). The modification ratio is equal to 1.45 FITC molecules per AD molecule.

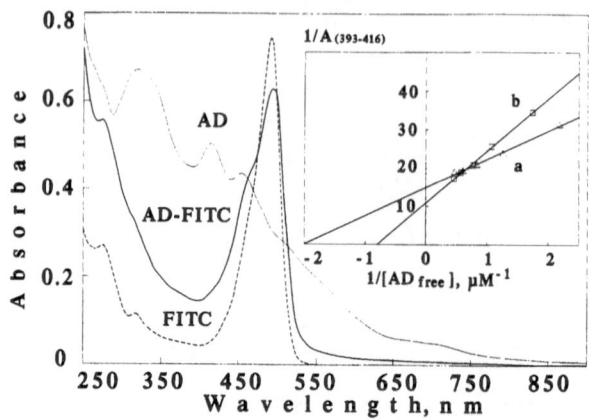

FIG. 2. Absorption spectra of free fluorescein isothiocyanate (*FITC*) (*dashed line*), free adrendo-doxin (*AD*) (*thin solid line*), and FITC-labeled AD (*AD-FITC, thick solid line*). Inset: double-reciprocal plot of low-spin P-450$_{scc}$ titration with AD (*a*) and FITC-labeled AD (*b*)

AD is a well-known high-spin effector for P-450$_{scc}$ [3]. The complex formation between low-spin heme protein and ferredoxin is characterized by the absorption maximum shift in the Soret region from 416 to 393 nm. Figure 2 (inset) shows the dependence of P-450$_{scc}$ spectral changes on the free AD concentration plotted in double-reciprocal coordinates. The $K_d$ for modified AD is increased to 0.56 ± 0.06 μM in comparison with 1.35 ± 0.1 μM for the native AD, thus showing that modified AD is characterized by a lower affinity to P-450$_{scc}$. It was previously shown that electrostatic interactions provided by the free carboxyl groups of ferredoxin play an important role in the process of the complex formation between AD and P-450$_{scc}$ [11]. We believe that the decrease in the complex formation capability of the FITC-modified AD might result from the relatively large spatial size of the reagent that could affect but not totally prevent the process.

## Properties of LB Films Containing P-450$_{scc}$:AD Complex

According to our previous experience, AD was found to form monolayers on the air-water interface. However, these monolayers could not be transferred onto the hydrophobic quartz surface. It has been found that P-450$_{scc}$:AD–FITC complex could be deposited onto the substrate. To evaluate the ratio between P-450$_{scc}$ and AD in the films, two types of film with equal amount of the layers were obtained: (1) P-450$_{scc}$:AD and (2) P-450$_{scc}$:AD-FITC. Because of the high absorbance of P-450$_{scc}$, the spectrum of LB films containing P-450$_{scc}$:AD resembles the spectrum of substrate-free P-450$_{scc}$ (Fig. 3). The AD absorbance in the film is masked by the higher absorbance of heme protein. To visualize the presence of AD in the films, we used FITC-modified AD preparation. In the case of AD–FITC, one can observe the absorption peak at 498 nm that corresponds to FITC absorption maximum in the visible region, thus proving the presence of FITC–AD in the films (to compare, refer to Fig. 2). According to the data

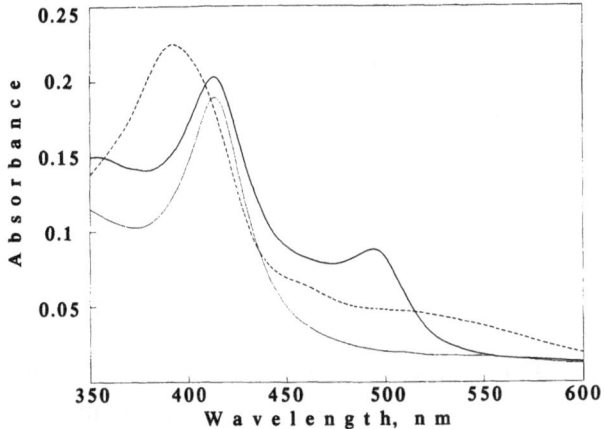

FIG. 3. Absorption spectra of initial 2:1 (mole:mole) mixture of P-450$_{scc}$ and AD (*dashed line*), 28 layers of LB film obtained from P-450$_{scc}$ and AD mixture (*thin solid line*), and 28 layers of LB film obtained from P-450$_{scc}$ and FITC-labeled AD mixture (*thick solid line*)

presented in Fig. 3, the low-spin state of P-450$_{scc}$ in the films is steadily preserved even if the films were prepared from P-450$_{scc}$ : AD complex. Assuming that the spectral properties of AD, P-450$_{scc}$, and FITC in the films are not significantly different from those for the solubilized ones, we have calculated the P-450$_{scc}$/AD molar ratio in the film as 2:1.

## Circular Dichroism Measurements of P-450$_{scc}$ LB films

To determine the α-helix content of high-spin and Tween 20-induced low-spin soluble P-450$_{scc}$ and P-450$_{scc}$-containing LB films, the far-ultraviolet CD spectra were recorded (Fig. 4). At 22°C, two bands with negative ellipticity at 208 nm and 220 nm were observed for all samples. The α-helix content for high-spin heme protein was calculated to be 35%. This value is slightly lower than that known from the literature (about 40%) [12] but is comparable with that for cytochrome P-450$_{LM2}$ (CYP2B4) (33%) [13]. However, the CD spectra of Tween 20-induced low-spin P-450$_{scc}$ and P-450$_{scc}$ LB films were similar to one another but differed from the high-spin heme protein spectrum. It is necessary to point out that the far-ultraviolet CD spectra (Fig. 4) of low-spin P-450$_{scc}$ in solution and in the film have a similar shape near 220 nm but vary at 208 nm. A small ellipticity minimum shift from 208 to 205 nm is observed in the film versus the solution. The temperature increase of high- and low-spin forms of P-450$_{scc}$ in solution up to 60°C results in an ellipticity decrease (Fig. 4). The negative double maxima tended to decrease and be transformed into a lower intensity single negative maximum. Temperature treatment of the high-spin heme protein significantly decreases α-helix content in the protein molecule from 35% to 22% (Fig. 5). In contrast, temperature-induced helical decrease of low-spin P-450$_{scc}$ in solution (17%–10%) and in the film (13%–10%) was less dramatic (Fig. 5). The secondary structure

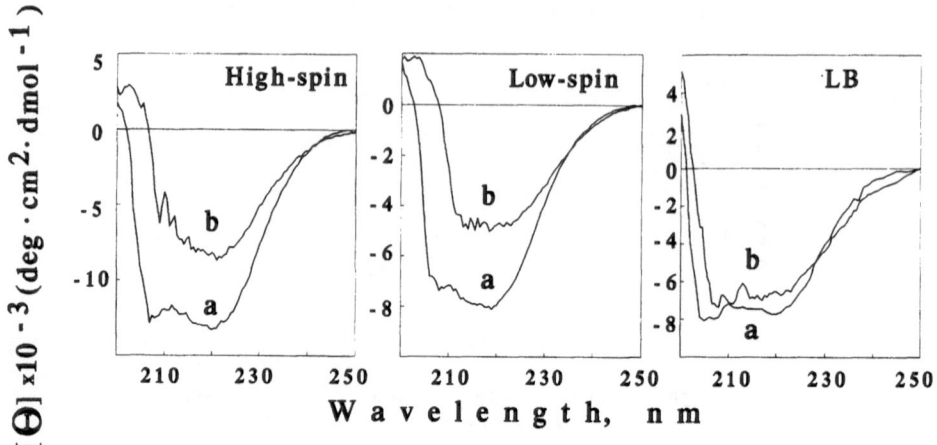

FIG. 4. Circular dichroism spectra of soluble high-spin P-450$_{scc}$ (*High-spin*), Tween 20-induced low-spin P-450$_{scc}$ (*Low-spin*), and 70 layers of P-450$_{scc}$-containing LB film (*LB*) registered at 22°C (*a*) and at 60°C (*b*)

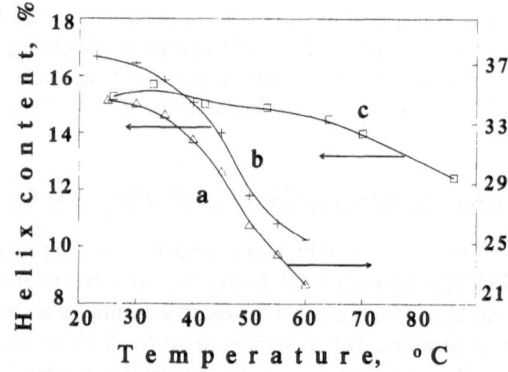

FIG. 5. Temperature dependence of the helix content in soluble high-spin P-450$_{scc}$ (*a*), Tween 20-induced low-spin P-450$_{scc}$ (*b*), and 70 layers of P-450$_{scc}$-containing LB film (*c*)

of the protein immobilized in the film was more stable to temperature increase than that of Tween 20-induced low-spin heme protein in solution.

It should be noted that the temperature increase results in solution transparency changes for both high- and low-spin P-450$_{scc}$. The opaque solutions indicate irreversible thermal denaturing of the protein. In contrast, the helical component of the secondary structure of P-450$_{scc}$ in the film reverted from 10% after heating to 86°C to 13% after cooling to room temperature.

# Discussion

Among cytochrome P-450 species, P-450$_{scc}$ is a unique object for spin-state investigation. The spin-state equilibrium of this heme protein is affected by temperature [14,15] and pH [16,17] and depends on the binding of substrates [18] and AD [19]. It was discussed earlier that spin-state transitions of the P-450$_{scc}$ molecule could be connected with its conformational changes, and a model comprising two conformations was proposed [17]. Under certain pH and temperature conditions the same P-450$_{scc}$ spin-state form may have different conformations; conversely, a given conformation may give rise to the different spin states. Our results on optical absorbance and CD measurements confirm this model.

We found, additionally, that the spin-state equilibrium of P-450$_{scc}$ is shifted from a high- to a low-spin state during the process of LB film preparation. Our data clearly demonstrate the existence of two stable conformers of P-450$_{scc}$ characterized by two distinct spin states. At 22°C, high-spin conformation contains 35% of α-helix, and the low-spin conformation is characterized by 17% of α-helix content. In the solid state, P-450$_{scc}$ maintains a helical pattern similar to that observed in solution for low-spin heme protein. However, heme protein in the film is more stable to heating. Thus, LB films represent a specifically organized molecular structure providing efficient preservation of heme protein secondary structure. The P-450$_{scc}$ organization stipulated by LB film formation is considered to be the main stabilizing factor.

It was shown [20] that at the air–water interface protein molecules are able to change their conformation, which results in surface unfolding of the protein. The unfolding also could lead to changes in the globule spatial organization, exposing its inner hydrophobic parts to the air–water interface. The data reported by Tronin et al. [20] suggest that the proteins organized in LB films preserve their secondary and tertiary structures. Our results indicate that the increasing hydrophobic interactions and loss of water during LB film preparation and high-temperature treatment could be considered as the factors responsible for the P-450$_{scc}$ conformational and spin-state transitions in the film.

# Conclusions

1. The preparation of P-450$_{scc}$ LB film leads to changes in its spin state from an initially high- to a low-spin state. AD, high-spin effector of P-450$_{scc}$, being associated in the complex with the heme protein, does not maintain the high-spin state of the heme protein in the solid films.
2. Spin-state transitions of P-450$_{scc}$ in LB films are associated with changes in the secondary structure of the protein molecule.
3. The secondary structure of P-450$_{scc}$ is preserved in LB films after heating to 86°C, while in solution it denatures irreversibly. The helical part of P-450$_{scc}$ secondary structure in the film reverts to its initial state after heating and then cooling to room temperature.

# References

1. Usanov SA, Chashchin VL, Akhrem AA (1990) Molecular mechanisms of adrenal steroidogenesis and aspects of regulation and application. In: Ruckpaul K, Rein H (eds) Frontiers in biotransformation, vol 3. Akademie-Verlag, Berlin, pp 1–57
2. Chu J-W, Kimura T (1973) Studies on adrenal steroid hydroxylases. Complex formation of the hydroxylase components. J Biol Chem 248:5183–5187
3. Katagiri M, Takikawa O, Sato H, et al. (1977) Formation of a cytochrome P-450$_{scc}$-adrenodoxin complex. Biochem Biophys Res Commun 77:804–809
4. Lvov YuM, Erokhin VV, Zaitsev SYu (1991) Langmuir-Blodgett protein films. Biol Membr (Lond) 9:1477–1513
5. Guryev OL, Erokhin VV, Usanov SA, et al. (1996) Cytochrome P-450$_{scc}$ spin state transitions in the thin solid films. Biochem Mol Biol Int 39:205–214
6. Usanov SA, Pikuleva IA, Chashchin VL, et al. (1984) Chemical modification of adrenocortical cytochrome P-450$_{scc}$ with tetranitromethane. Biochim Biophys Acta 790:259–267
7. Akhrem AA, Shkumatov VM, Chashchin VL (1977) Isolation of individual components of the steroid hydroxylating system from bovine adrenocortical mitochondria. Bioorg Khim (Moscow) 3:780–786
8. Harlow E, Lane D (1988) Antibodies. A laboratory manual. Cold Spring Harbor Lab, Cold Spring Harbor, NY, p 354
9. Chen YH, Yang JT (1971) A new approach to the calculation of scondary structure of globular proteins by optical rotary dispersion and circular dichroism. Biochem Biophys Res Commun 44:1285
10. Lange R, Pantaloni A, Saldana JL (1992) The cholesterol-side-chain-cleaving cytochrome P-450 spin-state equilibrium. 2. Conformational analysis. Eur J Biochem 207:75–79
11. Lambeth JD, Geren LM, Millett F (1984) Adrenodoxin interaction with adrenodoxin reductase and cytochrome P-450$_{scc}$. J Biol Chem 259:10025–10029
12. Akhrem AA, Adamovich TB, Lapko VN, et al. (1985) A model of molecular organization of the cholesterol side chain cleavage cytochrome P-450 from adrenal cortex mitochondria. In: Vereczkey L, Magyar K (eds) Cytochrome P-450. Biochemistry, biophysics and induction. Akademiai Kiado, Budapest, pp 113–120
13. Chiang YL, Coon MJ (1979) Comparative study of two highly purified forms of liver microsomal cytochrome P-450: circular dichroism and other properties. Arch Biochem Biophys 195:178–187
14. Akhrem AA, Lapko VN, Lapko AG, et al. (1979) Isolation, structural organization and mechanism of action of mitochondrial steroid hydroxylating systems. Acta Biol Med Ger 38:257–273
15. Lambeth JD, Kitchen SE, Farooqui AA, et al. (1982) Cytochrome P-450$_{scc}$-substrate interactions. Studies of binding and catalytic activity using hydroxycholesterols. J Biol Chem 257:1876–1884
16. Jefcoate CR, Orme-Johnson WH, Beinert H (1976) Cytochrome P-450 of bovine adrenocorticl mitochondria. Ligand binding to two forms resolved by EPR spectroscopy. J Biol Chem 251:3706–3715
17. Lange R, Larroque C, Anzenbacher P (1992) The cholesterol-side-chain-cleaving cytochrome P-450 spin-state equilibrium. 1. Thermodynamic analysis. Eur J Biochem 207:69–73
18. Kido T, Arakawa M, Kimura T (1979) Adrenal cortex mitochondrial cytochrome P-450 specific to cholesterol side chain cleavage reactions. Spectral changes induced by detergents, alcohols, amines, phospholipids, steroid hydroxylase inhibitors and steroid substrates and conditions for adrenodoxin binding to the cytochrome. J Biol Chem 254:8377–8385

19. Kido T, Kimura T (1979) The formation of binary and ternary complexes of cytochrome P-450$_{scc}$ with adrenodoxin and adrenodoxin reductase: adrenodoxin complex. The implication in ACTH function. J Biol Chem 254:11806–11815
20. Tronin A, Dubrovsky T, Dubrovskaya S, et al. (1996) Role of protein unfolding in monolayer formation on air-water interface. Langmuir 12:3272–3275

# Kinetic Analysis of Successive Reactions Catalyzed by Cytochromes P-450$_{17\alpha,\text{lyase}}$ and P-450$_{11\beta}$

Takeshi Yamazaki, Hiroko Tagashira-Ikushiro, Takashi Ohno, Tadashi Imai, and Shiro Kominami

*Summary.* Several kinds of steroidogenic cytochrome P-450 (P-450) catalyze two- or three-step successive monooxygenase reactions in which the product of the preceding reaction is metabolized by the same enzyme without leaving the catalytic site. Adrenal androgen formation by P-450$_{17\alpha,\text{lyase}}$ and aldosterone production by P-450$_{11\beta}$ consist of two and three successive monooxygenation reactions, respectively. We analyze here kinetic parameters of these reactions using the rapid quenching method. The analysis revealed the reaction pathways and the regulatory steps of the successive reactions. (I) Guinea pig P-450$_{17\alpha,\text{lyase}}$ catalyzes androstenedione production from progesterone via 17α-hydroxyprogesterone as the intermediate. (II) Bovine P-450$_{17\alpha,\text{lyase}}$ catalyzes a successive reaction for androgen formation from pregnenolone but not from progesterone, because the enzyme releases 17α-hydroxyprogesterone 50 times faster than 17α-hydroxypregnenolone. (III) Bovine P-450$_{11\beta}$ catalyzes a successive reaction for aldosterone production from 11-deoxycorticosterone via corticosterone and 18-hydroxycorticosterone as the intermediates, but not via 18-hydroxy-11-deoxycorticosterone. (IV) The protein–protein interaction of P-450$_{11\beta}$ and P-450$_{\text{scc}}$ stimulates the release of the intermediate metabolite, corticosterone, from P-450$_{11\beta}$. This quick release of the intermediate prevents aldosterone formation. It is demonstrated that release rate of intermediate metabolite from the enzyme regulates successive reactions of steroidogenic P-450s.

*Key words.* Steroid hormone—Rapid quenching method—Successive reaction—P-450$_{17\alpha,\text{lyase}}$—P-450$_{11\beta}$

## Introduction

Most steroidogenic P-450s, such as P-450$_{\text{scc}}$ (product of the CYP11A gene), P-450$_{17\alpha,\text{lyase}}$ (product of the CYP17 gene), P-450$_{11\beta}$ (product of the CYP11B gene), and P-450$_{\text{arom}}$ (product of the CYP19 gene), catalyze two- or three-step successive monooxygenase reactions in which the products of the preceding reactions are metabolized by the

Faculty of Integrated Art & Sciences, Hiroshima University, 1-7-1 Kagamiyama, Higashihiroshima, Hiroshima 739, Japan same enzyme without leaving the catalytic site [1–5]. It is widely accepted that P-450$_{scc}$ and P-450$_{arom}$ catalyze three-step monooxygenations and do not dissociate the reaction intermediates [1].

P-450$_{17\alpha,lyase}$ catalyzes the 17$\alpha$-hydroxylation of pregnenolone or progesterone and the subsequent $C_{17}$–$C_{20}$ bond cleavage reaction to form androgens [2]. The reaction of androgen formation by the P-450 has been analyzed by double-substrate and double-label experiments. When $^{14}$C-labeled pregnenolone and $^3$H-labeled 17$\alpha$-hydroxypregnenolone were incubated with adrenal cultured cells, most of the dehydroepiandrosterone was produced from [$^{14}$C]pregnenolone [6]. A similar result was obtained in progesterone metabolism by rat ovarian microsomes [7]. We concluded from these experiments that androgens are synthesized from pregnenolone or progesterone without the intermediate metabolites leaving the P-450s.

In adrenal glands, significant amounts of 17$\alpha$-hydroxylated steroids, which are the precursors of glucocorticoid, are released in the reactions of P-450$_{17\alpha,lyase}$ [8]. The reaction of P-450$_{17\alpha,lyase}$ is located at the branching point of the biosynthesis of androgen and corticoid and might be controlled in an organ-specific manner [1–4,7]. To clarify the control mechanism of adrenal steroidogenesis, we analyzed reactions of guinea pig and bovine P-450$_{17\alpha,lyase}$ by the rapid quenching method.

P-450$_{11\beta}$ catalyzes a three-step successive reaction to form aldosterone from 11-deoxycorticosterone as well as a one-step monooxygenation to form glucocorticoid in bovine adrenal glomerulosa cells [5,9]. There are two possible reaction pathways for aldosterone formation [10,11]. Because the reactions proceed without intermediates leaving the P-450, actual intermediate metabolites cannot be identified in the steady-state reaction condition.

P-450$_{11\beta}$ catalyzes only the formation of glucocorticoids in bovine adrenal fasciculata-reticularis cells. The reactions of P-450$_{11\beta}$ are reported to be modified by protein–protein interaction with P-450$_{scc}$ [9,12,13]. Coexistence of P-450$_{scc}$ prevents aldosterone formation by P-450$_{11\beta}$ but stimulates 11$\beta$-hydroxylation. The rapid quenching experiments revealed the regulatory step of the reactions, which was modified by the interaction with P-450$_{scc}$.

## Materials and Methods

Guinea pig adrenal P-450$_{17\alpha,lyase}$, and bovine adrenal P-450$_{11\beta}$ and P-450$_{scc}$, were purified as previously described [12,14]. Bovine P-450$_{17\alpha,lyase}$ was expressed in *E. coli* and purified as reported by Waterman with some modifications [15]. Proteoliposomes containing guinea pig P-450$_{17\alpha,lyase}$, bovine P-450$_{17\alpha,lyase}$, or bovine P-450$_{11\beta}$ were prepared as described elsewhere [13,16].

A rapid quenching device was constructed as previously described [16] (Fig. 1). Three micropipettes were placed in the reaction vessel, which contained P-450-proteoliposomes, the appropriate electron-transfer system, and $^3$H-labeled steroid. The micropipettes were driven by $N_2$ gas pressure, which was controlled by a personal computer via electrovalves. Reaction was started by addition of NADPH from the pipette. Immediately after the initiation of the reaction, an excess amount of nonlabeled substrate was added as a chaser to prevent rebinding of the released [$^3$H]steroids to the P-450. The reaction was quenched by the addition of HCl. This apparatus can

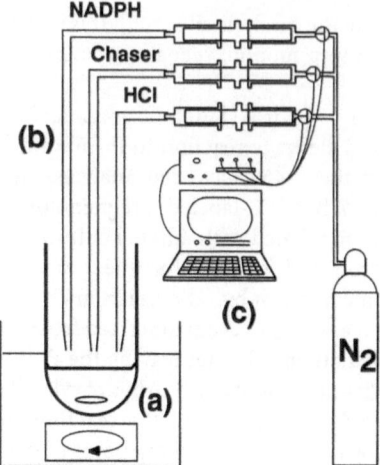

Fig. 1. Schematic view of the rapid quenching device. A reaction vessel (*a*) contains ³H-labeled steroids, P-450-proteoliposomes, and the electron-transfer system in 0.5 ml of potassium phosphate buffer. NADPH, non-labeled steroids as the chaser, and HCl were added from micropipettes (*b*) at the appropriate times. The solutions in the micropipettes were pushed out by $N_2$ gas pressure, which was controlled by a microcomputer (*c*)

follow the reaction at intervals of less than 0.1 s. The steroid metabolites were extracted and analyzed by HPLC as previously described [6,13,16].

## Results and Discussion

The multistep reactions of P-450$_{17\alpha,lyase}$ and P-450$_{11\beta}$ were barely analyzed at the steady-state condition because intermediate steroids were trapped in the P-450s and were present in extremely small quantities during the reactions [16]. In the system of rapid quenching experiments, most of the [³H]-labeled substrate bound to P-450 before the start of reaction, where P-450 was present in excess to the substrate [16].

### Guinea Pig P-450$_{17\alpha,lyase}$

Metabolism of [³H]progesterone by guinea pig P-450$_{17\alpha,lyase}$ was analyzed by the rapid quenching method. Figure 2a shows that the production of 17α-hydroxyprogesterone increased until 10 s after the initiation of reaction and then gradually decreased. The decrease of 17α-hydroxyprogesterone indicated that it was further metabolized to androstenedione by the second monooxygenation before the dissociation from the active site of P-450, because released [³H]steroid cannot be metabolized by the P-450 in the presence of excess amount of nonlabeled steroid as the chaser. It is concluded from the data shown in Fig. 2a that 17α-hydroxyprogesterone is the actual intermediate for androstenedione formation from progesterone. The rapid quenching data were analyzed kinetically by the method described elsewhere [16]. The rate constants

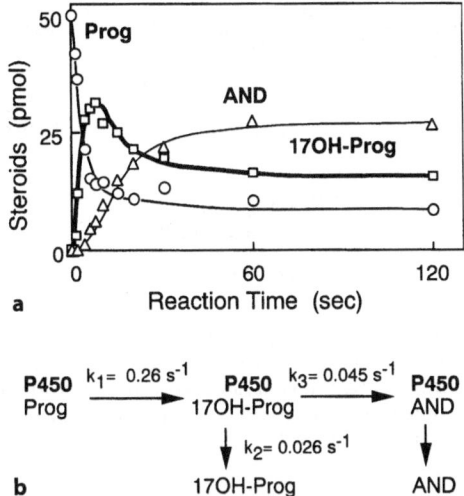

**a**

$P450$    $k_1= 0.26 \, s^{-1}$    $P450$    $k_3= 0.045 \, s^{-1}$    $P450$
Prog                    17OH-Prog                    AND

$\downarrow$ $k_2= 0.026 \, s^{-1}$    $\downarrow$

**b**    17OH-Prog                    AND

Fig. 2a,b. **a** Time course of progesterone metabolism catalyzed by guinea pig P-450$_{17\alpha,\text{lyase}}$ at 10°C. *Prog, 17OH-Prog* and *AND* represent progesterone, 17α-hydroxyprogesterone, and androstenedione, respectively. The *solid lines* represent the theoretical curves obtained using rate constants, which are shown in **b**. **b** The scheme of the successive reaction of androstenedione production from progesterone. *P-450* represents P-450$_{17\alpha,\text{lyase}}$. The values of $k_1$–$k_3$ were obtained by curve fitting to the data shown in **a** [16]

$k_1$, $k_2$, and $k_3$ were estimated by fitting the data to the reaction scheme of Fig. 2b. Solid lines in Fig. 2a are the theoretical curves derived from the scheme of Fig. 2b. Good fitting of the theoretical lines to the observed data indicates validity of the analysis.

## Bovine P-450$_{17\alpha,lyase}$

The reactivity of bovine P-450$_{17\alpha,\text{lyase}}$ is different from the guinea pig enzyme. This P-450 catalyzes the formation of dehydroepiandrosterone from pregnenolone but not androstenedione formation from progesterone, although both pregnenolone and progesterone are good substrates for 17α-hydroxylation [17]. These reactions were analyzed kinetically by the rapid quenching method in the same manner as that of the guinea pig P-450$_{17\alpha,\text{lyase}}$ (Fig. 3). The analysis revealed that the release rate of 17α-hydroxyprogesterone was 10 times higher than that of 17α-hydroxypregnenolone ($k_2$ in Fig. 3). The quick release of the intermediate prevents further metabolism to form androstenedione.

## Bovine P-450$_{11\beta}$

There are two possible reaction pathways for aldosterone formation from 11-deoxycorticosterone catalyzed by bovine P-450$_{11\beta}$; the first intermediate might be 18-hydroxy-11-deoxycorticosterone or corticosterone [10,11] (Fig. 4a). Reactions

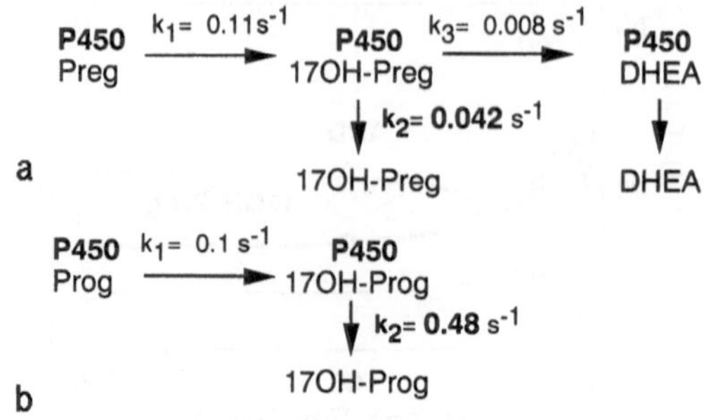

FIG. 3a,b. The reaction scheme of dehydroepiandrosterone production from pregnenolone (a) and metabolism of progesterone (b) by bovine P-450$_{17\alpha,lyase}$. *Preg, 17OH-Preg,* and *DHEA* represent pregnenolone, 17α-hydroxypregnenolone, and dehydroepiandrosterone, respectively. The rate constants $k_1$–$k_3$ were obtained by theoretical curve fitting to data of rapid quenching experiments performed at 10°C (data not shown)

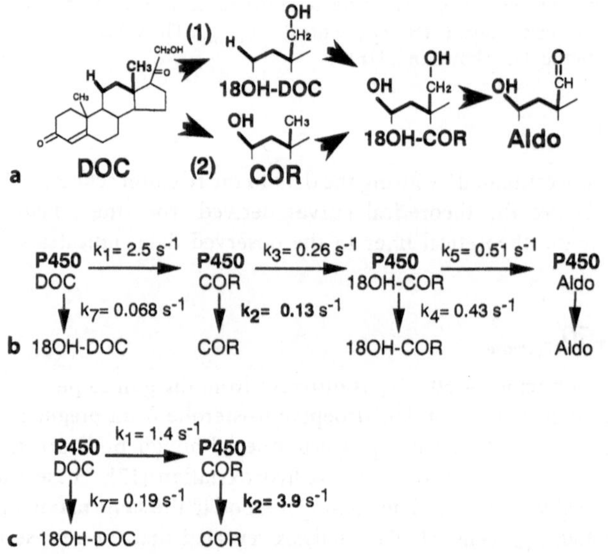

FIG. 4a–c. **a** Two possible reaction pathways for aldosterone formation from 11-deoxycorticosterone by bovine P-450$_{11\beta}$. *DOC, 18OH-DOC, COR, 18OH-COR,* and *Aldo* represent 11-deoxycorticosterone, 18-hydroxy-11-deoxycorticosterone, corticosterone, 18-hydroxycorticosterone, and aldosterone, respectively. **b** The reaction scheme of aldosterone production from 11-deoxycorticosterone by the pathway *(2)* in **a**. *P-450* represents bovine P-450$_{11\beta}$. Rate constants for each step were obtained by theoretical curve fitting to data of rapid quenching experiments at 37°C (data not shown). **c** The reaction scheme of 11-deoxycorticosterone metabolism by bovine P-450$_{11\beta}$ in the presence of P-450$_{scc}$. Rate constants for each step were obtained by curve fitting to data of rapid quenching experiments at 37°C (data not shown)

of the P-450 were analyzed by the rapid quenching method using 11-[$^3$H]deoxycorticosterone as the substrate. The experimental data fit well to the reaction scheme shown in Fig. 4b but do not fit to the other scheme that involves 18-hydroxy-11-deoxycorticosterone as a intermediate. The analysis revealed that corticosterone is the first intermediate for aldosterone formation by bovine P-450$_{11\beta}$.

Coexistence of P-450$_{scc}$ prevents aldosterone and 18-hydroxycorticosterone formation by P-450$_{11\beta}$ but stimulates 11$\beta$-hydroxylation [9,12,13]. Kinetic parameters were calculated from the results of rapid quenching experiments for P-450$_{11\beta}$ in the presence of P-450$_{scc}$ (Fig. 4c). As can be seen in Fig. 4b and 4c, $k_2$, a release rate of corticosterone from P-450$_{11\beta}$, was increased 30 fold by the addition of P-450$_{scc}$. These data indicate that protein–protein interaction between P-450$_{11\beta}$ and P-450$_{scc}$ stimulates the release of corticosterone from P-450$_{11\beta}$, and the quick release of the intermediate prevents aldosterone formation.

## Conclusion

P-450$_{17\alpha,lyase}$ and P-450$_{11\beta}$ catalyze multistep successive monooxygenase reactions without intermediates leaving the P-450s. The actual intermediates of the successive reactions were identified by the rapid quenching experiments as 17$\alpha$-hydroxylated steroids for androgen formation by P-450$_{17\alpha,lyase}$ and corticosterone and 18-hydroxycorticosterone for aldosterone synthesis by P-450$_{11\beta}$. Kinetic analysis of the successive reactions with the rapid quenching device revealed that the release rate of intermediate steroids from the P-450s regulates reaction pathways. Quick release of intermediates prevents the successive reactions.

## References

1. Takemori S, Yamazaki T, Ikushiro S (1993) Cytochrome P-450-linked electron transport system in monooxygenase reaction. In: Omura T, Ishizuka Y, Fujii-Kuriyama Y (eds) Cytochrome P-450. Kodansha, Tokyo, pp 44–63
2. Shinzawa K, Kominami S, Takemori S (1985) Studies on cytochrome P-450 (P-450$_{17\alpha,lyase}$) from guinea pig adrenal microsomes. Dual function of a single enzyme and effect of cytochrome $b_5$. Biochim Biophys Acta 833:151–160
3. Kominami S, Higuchi A, Takemori S (1988) Interaction of steroids with adrenal cytochrome P-450 (P-450$_{17\alpha,lyase}$) in liposome membranes. Biochim Biophys Acta 937:177–183
4. Kominami S, Inoue S, Higuchi A, et al. (1989) Steroidogenesis in liposomal system containing adrenal microsomal cytochrome P-450 electron transfer components. Biochim Biophys Acta 985:293–299
5. Ikushiro S, Kominami S, Takemori S (1989) Adrenal P-450$_{11\beta}$-proteoliposomes catalyzing aldosterone synthesis. Biochim Biophys Acta 984:50–56
6. Yamazaki T, Nawa K, Kominami S, et al. (1992) Cytochrome P-450$_{17\alpha,lyase}$ mediating pathway of androgen synthesis in bovine adrenocortical cultured cells. Biochim Biophys Acta 1134:143–148
7. Yamazaki T, Marumoto T, Kominami S, et al. (1992) Kinetic studies on androstenedione production in ovarian microsomes from immature rats. Biochim Biophys Acta 1125:335–340

8. Takemori S, Kominami S (1989) The role of cytochromes P-450 in adrenal steroidogenesis. Trends Biochem Sci 9:393–396
9. Takemori S, Kominami S, Yamazaki T, et al. (1995) Molecular mechanism of cytochrome P-450 dependent aldosterone biosynthesis in the adrenal cortex. Trends Endocrinol Metab 6:267–273
10. Kim CY, Sugiyama T, Okamoto M, et al. (1983) Regulation of 18-hydroxycorticosterone formation in bovine adrenocortical mitochondria. J Steroid Biochem 18:593–599
11. Vinson GP, Laird SM, Whitehouse BJ, et al. (1991) The biosynthesis of aldosterone. J Steroid Biochem 39:851–858
12. Ikushiro S, Kominami S, Takemori S (1992) Adrenal P-450$_{scc}$ modulates activity of P-450$_{11\beta}$ in liposomal and mitochondrial membranes. J Biol Chem 267:1464–1469
13. Kominami S, Harada D, Takemori S (1994) Regulation mechanism of the catalytic activity of bovine adrenal cytochrome P-450$_{11\beta}$. Biochim Biophys Acta 1192:234–240
14. Kominami S, Shinzawa K, Takemori S (1982) Purification and some properties of cytochrome P-450 specific for steroid 17α-hydroxylation and $C_{17}$–$C_{20}$ bond cleavage from guinea pig adrenal microsomes. Biochem Biophys Res Commun 109:916–921
15. Barnes HJ, Arlotto MP, Waterman MR (1991) Expression and enzymatic activity of recombinant cytochrome P-450 17α-hydroxylase in *Escherichia coli*. Proc Natl Acad Sci USA 88:5597–5601
16. Tagashira H, Kominami S, Takemori S (1995) Kinetic studies of cytochrome P-450$_{17\alpha,lyase}$ dependent androstenedione formation from progesterone. Biochemistry 34:10939–10945
17. Sakaki T, Shibata M, Yabusaki Y, et al. (1989) Expression of bovine cytochrome P-450$_{c17}$ cDNA in *Saccharomyces cerevisiae*. DNA (NY) 8:409–418

# Changing Substrate Specificity and Product Pattern in Adrenal Cytochrome P-450-Dependent Steroid Hydroxylases

Benjamin Böttner[1], Peirang Cao[2], and Rita Bernhardt[2]

*Summary.* In the human adrenal cortex, the mitochondrial cytochrome P-450-dependent 11β-hydroxylase (CYP11B1) is responsible for the formation of cortisol, whereas the aldosterone synthase (CYP11B2), being 93% identical in its primary structure to CYP11B1, catalyzes the biosynthesis of aldosterone. Both proteins receive the necessary electrons for oxygen activation via an electron-supporting system consisting of adrenodoxin and adrenodoxin reductase. We have cloned the cDNA of human CYP11B1 and CYP11B2 using reverse transcriptase-polymerase chain reaction (RT-PCR) methodology. Site-directed mutagenesis and computer modeling were used to investigate the molecular basis for the regioselectivity of steroid hydroxylation. Replacement of three amino acids of CYP11B2 by the corresponding residues of CYP11B1 were sufficient to increase cortisol formation from about 5% to 85% of the level obtained with CYP11B1. The aldosterone synthase activities of the mutant CYP11B2 proteins were suppressed to 10% of the CYP11B2 activity. When replacing these three residues of CYP11B1 by the amino acid found exclusively in CYP11B2, the mutant is able to form aldosterone. The capacity amounts to about 20% of that of CYP11B2. Taking into account that in human adrenals, CYP11B1 is considerably more strongly expressed than CYP11B2, a potential role of point mutations of CYP11B1 as a cause of hyperaldosteronism can be envisioned. As data from the literature suggest also that factors other than the primary structures of CYP11B1 and CYP11B2 could affect the amount of aldosterone formed, we investigated whether the supply of electrons to bovine CYP11B0 would change the amount of aldosterone production. For this purpose, mutant variants of adrenodoxin obtained by site-directed mutagenesis and possessing variable abilities to transfer electrons were studied. It could be demonstrated that mutants with an increased rate constant for the transfer of the first electron were able to increase the amount of aldosterone produced by bovine CYP11B0.

---

[1] Max-Delbrück-Centrum für Molekulare Medizin, Robert-Rössle-Straße-10, D-13122 Berlin, Germany
[2] Universität des Saarlandes, Fachrichtung 12.4-Biochemie, Postfach 15 11 50, D-66041 Saarbrücken, Germany

*Key words.* Steroid hydroxylases—CYP11B1—CYP11B2—Substrate specificity—Electron transfer

## Introduction

In the adrenal gland, essential steroid hormones such as glucocorticoids, mineralo-corticoids, and androgens are produced. Cortisol, the major glucocorticoid in humans, is synthesized in the zona fasciculata/reticularis under control of pituitary-derived adrenocorticotropic hormone (ACTH), whereas the most potent mineralo-corticoid, aldosterone, is secreted from zona glomerulosa cells primarily in response to angiotensin II and potassium [1–3]. The synthesis of cortisol and aldosterone is catalyzed by a series of monooxygenases called P-450 enzymes. The initial steps, the side-chain cleavage of cholesterol and the successive dehydrogenation of preg-nenenolone at the 3β-position to form progesterone, are identical in both pathways. For cortisol biosynthesis, progesterone is 17α-hydroxylated to 17α-hydroxyprogest-erone, which contrasts with the aldosterone synthesis pathway, where no 17α-hydrox-ylation occurs because of the lack of 17α-hydroxylase/17-20-lyase expression in the zona glomerulosa. Both pathways then proceed with 21-hydroxylation to 11-deoxy-cortisol or 11-deoxycorticosterone, respectively. 11-Deoxycortisol in glucocorticoid synthesis is 11β-hydroxylated to yield cortisol, whereas 11-deoxycorticosterone in mineralocorticoid synthesis, in addition, is 18-hydroxylated and 18-oxidized to form aldosterone, with corticosterone and 18-hydroxycorticosterone as further metabolic intermediates [4].

In humans the final steps in cortisol and aldosterone production are performed by two distinct enzymes, namely 11β-hydroxylase (P-450$_{11\beta}$) and aldosterone synthase (P-450$_{aldo}$). The genes encoding these enzymes, CYP11B1 (P-450$_{11\beta}$) and CYP11B2 (P-450$_{aldo}$), have been isolated from a genomic library [5] and were shown by sequence comparison to share an overall homology of 93%. Overproduction of aldosterone and 18-hydroxy-11-deoxycorticosterone can lead to hypertension. In most cases, aldoster-one-producing adenomas are the cause of hyperaldosteronism. Another reason is the rare occurrence of a chimeric gene consisting of the promotor region of the CYP11B1 and important regions (C-terminal of His-256) [6] of the CYP11B2 gene, which insti-gates glucocorticoid-remediable hyperaldosteronism. P-450$_{11\beta}$ enzymes of other spe-cies have been extensively studied, indicating that in bovine [7], porcine [8], and frog [9] adrenal cortex synthesis of gluco- and mineralocorticoids is catalyzed by a single enzyme. Conversely, synthesis of rat [10] and mouse [11] gluco- and mineralocorti-coids have been separated in evolution, similar to the human system, and both reactions are carried out by distinct enzymes. For their functional activity, CYP11B1 and CYP11B2 need electrons to be able to activate molecular oxygen and then perform steroid hydroxylation. The electrons are supplied by an NADPH-dependent redox system consisting of a flavoprotein, adrenodoxin reductase, and an iron-sulfur pro-tein, adrenodoxin.

We intended to gain insight into the principles underlying the different regioselec-tivities involved in 11β-hydroxylation and 18-hydroxylation/oxidation in the human enzymes. Because these proteins are 93 % identical, yet carry out separate reactions to yield different steroid hormones, it remained elusive on which structure–function relationships these diversities could be based. We carried out a computer-based sequence alignment with three of the four P-450 proteins that have been crystallized so far, namely, P-450$_{cam}$ from *Pseudomonas putida* [12], P-450$_{BM3}$ from *Bacillus megaterium* [13], and P-450$_{terp}$ from another *Pseudomonas* species [14]. Based on this, we performed site-directed mutagenesis on a region supposedly analogous to the P-450$_{cam}$ I-helix and subsequently analyses of the mutants by transient transfection experiments using COS-1 cells. In addition, we were interested in investigating whether factors others than the primary structure of the P-450, especially the efficiency of electron transfer, could affect the amount of aldosterone formed. For this purpose the effect of using mutants of adrenodoxin with different abilities for the transfer of the first electron on the amount of aldosterone formed by bovine CYP11B0 has been studied.

# Materials and Methods

## Materials

Oligonucleotides were synthesized at BioTez (Berlin, Germany). Restriction enzymes, Klenow fragment of DNA polymerase I, T4 polynucleotide kinase, bovine alkaline phosphatase, T4 DNA ligase, and DHFR cells were purchased from New England Biolabs (Beverly, MA, USA) or Boehringer Mannheim (Taq polymerase was obtained from Perkin Elmer-Cetus Branchburg, NJ, USA) and the pALTER in vitro mutagenesis system from Promega (Madison, WI, USA). [$^3$H]Deoxycorticosterone and [$^3$H]deoxycortisol were obtained from Dupont-New England Nuclear (Bad Homburg, Germany) Deoxycorticosterone (DOC), corticosterone, deoxycortisol, cortisol, chloroquine, cell culture-tested HEPES, dimethyl sulfoxide, L-$\alpha$-phosphatidylcholine (type II E), L-$\alpha$-phosphatidylethanolamine (type III) and cardiolipin were purchased from Sigma (St. Louis, MO, USA). Radioimmunoassays were performed with active coated radioimmunoassay kits from Diagnostic System Laboratories (Webster, TX, USA).

## Methods

### Insertion of Mutations into the CYP11B1 and CYP11B2 cDNA by Site-Directed Mutagenesis

Mutations were produced using the pALTER mutagenesis kit as previously described [15]. All standard procedures were carried out as previously described [16].

### Cell Culture and Transfection

COS-1 cells were maintained at 37°C and 8% $CO_2$ in Dulbecco's modified Eagle's medium (DMEM) supplemented with 10% fetal bovine serum (FCS), 100 units penicillin/ml, and 0.1 mg streptomycin/ml, as previously described [15]. Transfections were done as described previously [17] except for some minor modifications.

## Hydroxylase Assays

Transfected cells were incubated for 48 h with either [1,2-$^3$H] deoxycortisol or [$^3$H]DOC. For extraction of steroids the medium was combined with 4 vol of an ethanol/acetone mixture (1:1) and incubated at room temperature [18]. After pelleting of the debris, the supernatant was transferred to a fresh tube and evaporated to the original volume. Steroids were extracted with 2 vol of methylene chloride and the organic phase was evaporated. The residue was dissolved in 0.5 ml of hexane and centrifuged at maximal speed, giving a small pellet. The supernatant was then rotated in a speed vac to dryness and the extraction products were redissolved in 150 µl of 10 % (v/v) isopropanol in hexane. After addition of a combination of internal steroid standards the samples were subjected to normal-phase HPLC using a Lichrosorb Diol column (Merck, Darmstadt, Germany). A gradient solvent system was run starting with 15% (v/v) isopropanol in $n$-hexane at a flow rate of 1.3 ml/min. The standards were monitored by UV detection at 254 nm, and the radioactivity was assayed with a Betascan 386 radiation detector (Kubisiak GmbH, Dobel, Germany) fitted to the HPLC. Alternatively, steroids were measured using an active aldosterone or active cortisol radioimmunoassay (Diagnostic System Laboratories, Webster, USA), respectively. Moreover, high performance thin-layer chromatography (HPTLC) (Merck) has been used to analyze the reaction products in several sets of experiments. The HPTLC was developed once in $n$-hexane/isopropanol (8:2) and once in methylene chloride/methanol/$H_2O$ (300:100:1). The reaction products were analyzed by autoradiography or exposed to Fuji image plates and quantitated on a Fuji bioimaging analyzer (BAS2000, Fuji Photo Film, Kanagawa, Japan).

## Alignment of P-450 Sequences

Alignment was performed as previously described [15]. Briefly, an initial alignment of the P-450$_{aldo}$ and P-450$_{11\beta}$ sequences was performed using HOMOL, a database created by D. Nelson in which he related 155 different P-450 proteins to P-450$_{cam}$ (personal communication). Updating of the sequences and further optimization was carried out by hand. In addition, we included the sequences of P-450$_{scc}$, P-450$_{arom}$, and P-450$_{aldo/11\beta}$ from other species by using CLUSTAL. The first amino acid of the different steroidogenic proteins to be considered was the first residue of the mature sequences.

## Protein Purification

Bovine CYP11B0, adrenodoxin reductase, and mutants of adrenodoxin were isolated as previously described [19, 20]. The concentrations were determined using extinction coefficient $\varepsilon_{414} = 9.8\,mM^{-1}\,cm^{-1}$ [21] and $\varepsilon_{450} = 10.9\,mM^{-1}\,cm^{-1}$ [22] for adrenodoxin and adrenodoxin reductase, respectively. The concentration of CYP11B0 was determined from a CO-dithionite-reduced difference spectrum using $\Delta\varepsilon_{450-490} = 91.0\,mM^{-1}\,cm^{-1}$ [23].

## Preparation of Liposomes

The liposome preparation was done as described previously [23] with minor modifications: a phospholipid mixture composed of L-α-phosphatidylcholine, L-α-phosphatidylethanolamine, and cardiolipin at a weight ratio of 2:2:1 (5 mg) was dispersed by means of a vortex mixer at 4°C in 50 mM potassium phosphate buffer (pH 7.4) containing 500 mM NaCl, 0.1 mM EDTA, 0.1 mM dithiothreitol (DTT), 100 μM DOC, 20% glycerol, and 1% sodium cholate. The mixtures were then ultrasonicated in a waterbath at 0°C until the suspension became clear. The purified bovine CYP11B0 (2.5 nmol) was added to the suspension and incubated for 12 h at 4°C. The mixtures were dialyzed at 4°C for 36 h agaist 50 mM potassium phosphate buffer (pH 7.4) containing 500 mM NaCl, 0.1 mM EDTA, 0.1 mM DTT, and 20% glycerol. The dialyzed suspension was centrifuged at 15 300 × g for 1 h. The supernatant was collected and stored at −70°C.

## Enzyme Assays

CYP11B0 assays were performed using DOC as substrate according to [24] except for some modifications. The volume of the assay mixture was 0.5 ml and contained 17.5 pmol CYP11B0, 350 pmol adrenodoxin reductase, and various amounts of adrenodoxin mutants 4–128 and 4–108. The concentrations of the substrate of DOC for adrenodoxin mutants 4–128 and 4–108 were 3.0 and 5.0 μM, respectively, with trace amounts of [$^{14}$C]DOC in 50 mM potassium phosphate buffer (pH 7.4). The reaction was started by addition of 50 nmol NADPH and performed at 37°C for 10 min. Chloroform was used to stop the reaction and to extract products. Product analysis was performed by HPTLC as decribed earlier.

# Results and Discussion

The first aim of our study was to understand the structural basis for the regioselectivity of the steroid hydroxylations catalyzed by CYP11B1 and CYP11B2. This concern is related to the question whether point mutations of CYP11B1 and CYP11B2 would be able to convert a glucocorticoid-producing enzyme to a mineralocorticoid-producing one and vice versa.

The three-dimensional structure of the mitochondrial steroid hydroxylases has not yet been determined. Therefore, we performed computer modeling to obtain a model of the structures of CYP11B1 and CYP11B2, which was necessary for doing a rational design of these enzymes. On aligning the human CYP11B1 and CYP11B2 peptide sequences to the P-450$_{cam}$, P-450$_{terp}$, and P-450$_{BM3}$ sequences and structures, we postulated a domain encompassing amino acid residues 299–338 as a region homologous to the bacterial I-helix to be of possible importance for the regiospecificity of steroid hydroxylation. Subsequent residue-swapping experiments clearly showed that, after expression of the mutant proteins in COS-1 cells, a CYP11B2 protein previously having a comparatively weak 11β-hydroxylase activity for 11-deoxycortisol could be converted to an efficient 11β-hydroxylase derivative. Replacing only single amino acid residues of CYP11B2 at positions 301, 302, and 320 for the respective ones of the CYP11B1 protein gave rise to only slightly increased hydroxylation potentials.

However, multiple replacement mutants with combined substitutions at positions 301/320 and 301/302/320 acquired 60% and 85% of the CYP11B1 wild-type 11β-hydroxylase activity, respectively (Fig. 1). Assaying the aldosterone synthase activities elucidated that single mutations at these positions already had significant effects on

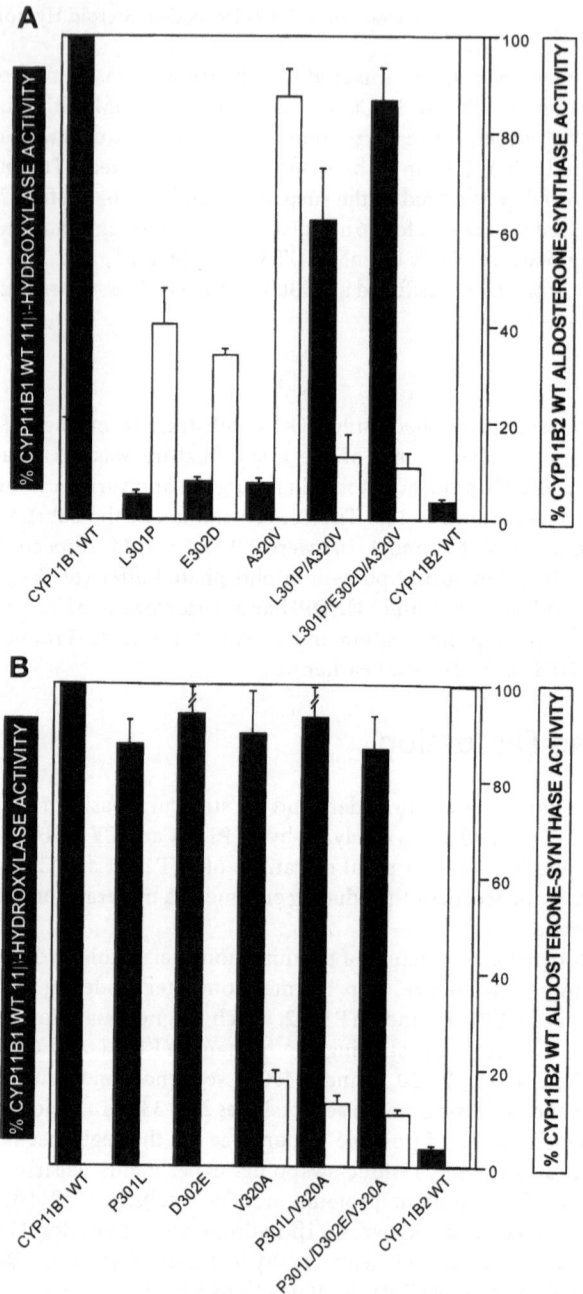

Fig. 1A,B. 11β-Hydroxylase and aldosterone synthase activities of CYP11B1 and CYP11B2 mutants. **A** 11β-Hydroxylase activities (*black bars*) and aldosterone synthase activities (*white bars*) for CYP11B2 mutants in comparison with the wild-type CYP11B1 and CYP11B2 proteins. **B** Activities of CYP11B1 mutants. COS-1 cells were transiently transfected with the respective cDNA variants and subsequently incubated with [$^{3}$H]deoxycortisol or [$^{14}$C]deoxycorticosterone for 48 h. Steroids were extracted from the cell culture medium and assayed for reaction products by thin-layer chromatography. Radioactivity was quantitated using a phosphoimager BAS2000 system and plotted against the wild-type values set equal to 100%

aldosterone production; the effects were even more severe in combinations (Fig. 1). Taken together, these observations suggest that from an evolutionary perspective the structure-to-function relationships in the CYP11B2 protein have developed into a highly effective mineralocorticoid-synthesizing molecule already sensitive to minor perturbations. From a physiological point of view, mutations altering one of the critical I-helix residues in the human CYP11B2 allele could be a possible cause for certain forms of "aldosterone synthase deficiencies" (hypoaldosteronism).

Based on these results on the human CYP11B2 we further asked whether the human CYP11B1 protein could be endowed with a novel mineralocorticoid-synthesizing function by introduction of CYP11B2 residues mapping to the same putative I-helix. Successive in vitro analysis revealed that a single replacement in this region in vitro can lead to the production of about 20% of the CYP11B2 wild-type aldosterone level (see Fig. 1). Thus, a new specificity, aldosterone synthase activity, has been rationally designed into the CYP11B1 enzyme. Interestingly, the double and the triple mutants did not form significantly more aldosterone as compared to the single-replacement mutants. Given the theoretical possibility that such a mutation would arise in a human individual, and further considering the amount of CYP11B1 protein present in the adrenal which because of the much higher promoter strength is 100–1000 fold more abundant than the CYP11B2 protein, one would obtain an aldosterone concentration elevated by a factor of 20–200. This conclusion means that apart from the described type of glucocorticoid-remediable aldosteronism (GRA), a hereditary form of hypertension arising from chimeric gene duplications and ectopic expression of a CYP11B1/CYP11B2 hybrid gene [25], point mutations in the CYP11B1 gene could constitute a second cause of GRA.

To investigate whether changes in the amount of aldosterone formed could also have other causes in addition to point mutations of the CYP11B1 and CYP11B2 gene, bovine CYP11B0 has been investigated in an reconstitution assay. Bovine CYP11B0 is expressed in zona glomerulosa as well as in zona fasciculata/reticularis [26]. It is able to synthesize both cortisol and aldosterone. Aldosterone production requires three consecutive hydroxylation steps of DOC with corticosterone and 18-OH corticosterone as intermediates. As we have produced a mutant of adrenodoxin that could increase the efficiency of corticosterone formation by a factor of about four to five [20], it was tempting to investigate whether this increase would lead to an acceleration of the further hydroxylation steps and thereby to an accumulation of aldosterone. The mutant 4–108 is a deletion mutant of adrenodoxin, missing residues 109–128 and the first three amino acids. As a control, mutant 4–128 was used, which shows behavior identical to wild-type adrenodoxin [20]. It can be deduced from Fig. 2 that, in fact, the $V_{max}$ value for the conversion of DOC to 18-OH corticosterone as well as aldosterone was increased. The increase is about 5 fold for the conversion of DOC to aldosterone and 1.2 fold for the conversion of DOC to 18-OH corticosterone (Table 1). The ratios of the activities of aldosterone versus corticosterone and 18-OH corticosterone formation, respectively, are about 3 fold higher with the deletion mutants (Table 2). These results clearly demonstrate that the efficiency of electron transfer can modulate the relative amount of aldosterone formed by bovine CYP11B0. Whether a similar regulation is true for human CYP11B2 is under investigation.

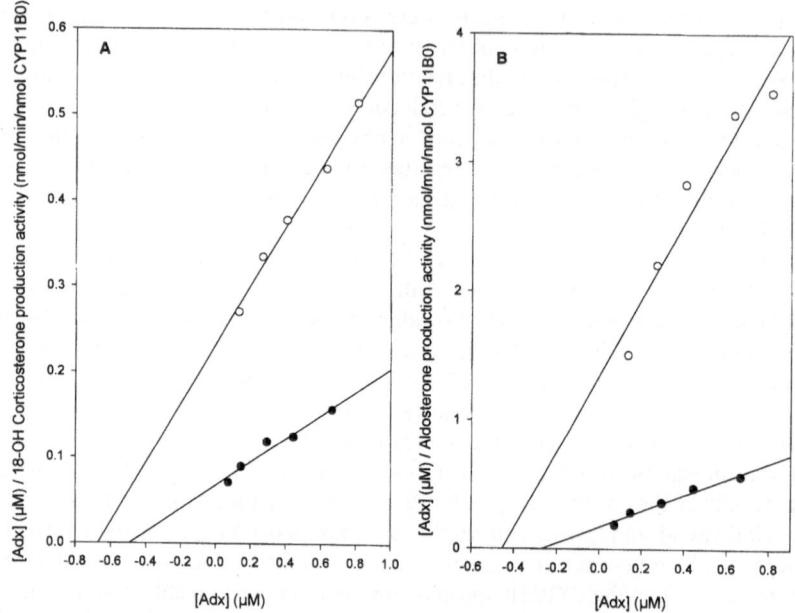

FIG. 2A,B. CYP11B0-dependent conversion of deoxycorticosterone to 18OH corticosterone (**A**) and to aldosterone (**B**) with adrenodoxin mutant 4-108. Reaction mixtures consisted of 17.5 pmol bovine CYP11B0, 350 pmol adrenodoxin reductase, and specified amounts of adrenodoxin in 500 µl reaction buffer. The reactions were started by addition of NADPH and carried out at 37°C for 10 min. The formation of 18-hydroxycorticosterone and aldosterone from deoxycorticosterone were analyzed using high performance thin-layer chromatography. *Open and closed circles* represent adrenodoxins [*Adx*] 4–128 and 4–108, respectively

TABLE 1. Kinetic parameters for different mutants of adrenodoxin

| | 4-128 | | 4-108 | |
| Reaction | $K_m$ (µM) | $V_{max}$ (nmol product formed per min/nmol CYP11B0) | $K_m$ (µM) | $V_{max}$ (nmol product formed per min/nmol CYP11B0) |
| --- | --- | --- | --- | --- |
| DOC → aldosterone | 0.420 ± 0.03 | 0.322 ± 0.02 | 0.267 ± 0.01 | 1.541 ± 0.06 |
| DOC → 18-OH corticosterone | 0.729 ± 0.06 | 3.028 ± 0.18 | 0.252 ± 0.02 | 3.882 ± 0.31 |
| DOC → Corticosterone[a] | 1.30 ± 0.1 | 5.6 ± 0.4 | 0.2 ± 0.02 | 19.7 ± 1.8 |

Reactions were performed in a system consisting of 17.5 pmol bovine CYP11B0 in liposomes, 350 pmol adrenodoxin reductase, and different amounts of adrenodoxin in 500 µl 50 mM phosphate buffer (pH 7.4). The reactions were started by addition of NADPH and carried out at 37°C for 10 min, stopped, and extracted with 3 volumes of chloroform. The products from deoxycorticosterone (DOC) with trace amounts of [14C]DOC were analyzed by high performance thin layer chromatography (HPTLC). DOC concentrations for 3 and 5 µM were used.

[a] Data from [20].

TABLE 2. Ratios of aldosterone and 18-hxdroxycorticosterone formation activities of different mutants of adrenodoxin

| Activity ratio | 4-128 | 4-108 |
|---|---|---|
| Aldosterone/corticosterone formation activity | 0.091 ± 0.006 | 0.260 ± 0.016 |
| Aldosterone/18-0H corticosterone formation activity | 0.128 ± 0.012 | 0.444 ± 0.037 |

The individual product formation activity (nmol product per min/nmol CYP11B0) was determined by HPTLC as described under Materials and Methold. DOC with trace amounts of [$^{14}$C]DOC was used as substrate. The system contains 17.5 pmol bovine CYP11B0 in liposomes. 350 pmol adrenodoxin reductase, and 310 pmol adrenodoxin.

*Acknowledgments.* This work was supported by a grant Be 1343/2–4 from the Deutsche Forschungsgemeinschaft and a grant from the Stiftungsfonds Schering AG im Stifterverband für die Deutsche Wissenschaft to C.P.

# References

1. Parker KL, Schimmer BP (1993) Transcriptional regulation of the adrenal steroidogenic enzymes. Trends Endocrinol Metab 4:46–50
2. Aguilera G (1993) Factors controlling steroid biosynthesis in the zona glomerulosa of the adrenal. J Steroid Biochem Mol Biol 45:147–151
3. Kawamoto T, Mitsuuchi Y, Toda K, et al. (1992) Role of steroid 11 beta-hydroxylase and steroid 18-hydroxylase in the biosynthesis of glucocorticoids and mineralocorticoids in humans. Proc Natl Acad Sci USA 89:1458–1462
4. Miller WL, Tyrell JB (1995) The adrenal cortex. In: Felig P, Baxter J, Frohman L (eds) Endocrinology and metabolism. McGraw-Hill, New York, pp 555–711
5. Mornet E, Dupont J, Vitek A, et al. (1989). Characterization of two genes encoding human steroid 11 beta-hydroxylase (P-450(11) beta). J Biol Chem 264:20961–20967
6. Pascoe L, Curnow KM, Slutsker L, et al. (1992) Glucocorticoid-suppressible hyperaldosteronism results from hybrid genes created by unequal crossovers between CYP11B1 and CYP11B2. Proc Natl Acad Sci USA 89:8327–8331
7. Wada A, Ohnishi T, Nonaka Y, et al. (1985) Synthesis of aldosterone by a reconstituted system of cytochrome P-45011 beta from bovine adrenocortical mitochondria. J Biochem (Tokyo) 98:245–256
8. Yanagibashi K, Shackleton CH, Hall PF (1988) Conversion of 11-deoxycorticosterone and corticosterone to aldosterone by cytochrome P-450 11 beta-/18-hydroxylase from porcine adrenal. J Steroid Biochem 29:665–675
9. Nonaka Y, Takemori H, Halder SK, et al. (1995) Frog cytochrome P-450 (11 beta, aldo), a single enzyme involved in the final steps of glucocorticoid and mineralocorticoid biosynthesis. Eur J Biochem 229:249–256
10. Matsukawa N, Nonaka Y, Ying Z, et al. (1990) Molecular cloning and expression of cDNAS encoding rat aldosterone synthase: variants of cytochrome P-450(11 beta). Biochem Biophys Res Commun 169:245–252
11. Domalik LJ, Chaplin DD, Kirkman MS, et al. (1991) Different isozymes of mouse 11 beta-hydroxylase produce mineralocorticoids and glucocorticoids. Mol Endocrinol 5:1853–1861

12. Poulos TL, Finzel BC, Howard AJ (1987) High-resolution crystal structure of cyto-chrome P-450cam. J Mol Biol 195:687–700
13. Ravichandran KG, Boddupalli SS, Hasermann CA, et al. (1993) Crystal structure of hemoprotein domain of P-450BM-3, a prototype for microsomal P-450s. Science 261:731–736
14. Hasemann CA, Ravichandran KG, Peterson JA, et al. (1994) Crystal structure and refinement of cytochrome P-450terp at 2.3 Å resolution. J Mol Biol 236:1169–1185
15. Böttner B, Schrauber H, Bernhardt R (1996) Engineering a mineralocorticoid- to a glucocorticoid-synthesizing cytochrome P-450. J Biol Chem 271:8028–8033
16. Sambrook J, Fritsch EF, Maniatis T (1989) Molecular cloning: a laboratory manual, 2nd edn. Cold Spring Harbor Laboratory, Cold Spring Harbor, NY
17. Zuber MX, Mason JI, Simpson ER, et al. (1988) Simultaneous transfection of COS-1 cells with mitochondrial and microsomal steroid hydroxylases: incorporation of a steroidogenic pathway into nonsteroidogenic cells. Proc Natl Acad Sci USA 85:699–703
18. Kahlil MW, Walton JS (1985) Identification and measurement of 4-oestren-3,17-dione (19-norandrostenedione) in porcine ovarian follicular fluid using high performance liquid chromatography and capillary gas chromatography-mass spectrometry. J Endo-crinol 107:375–381
19. Akhrem AA, Lapko VN, Lakpo AG, et al. (1979) Isolation, structual organization and mechanism of action of mitochondrial steroid hydroxlyation systems. Acta Biol Med Ger 38:257–274
20. Uhlman H, Kraft R, Bernhardt R (1994) C-terminal region of adrenodoxin affects its integrity and determines differences in its electron transfer function to cytochrome P-450s. J Biol Chem 269: 22557–22564
21. Huang JJ, Kimura T (1973) Studies on adrenal steroid hydroxylases. oxidation-reduc-tion properties of adrenal iron-sulfur protein (adrenodoxin). Biochemistry 12:406–409
22. Chu JW, Kimura T (1973) Molecular catalytic properties of adrenodoxin reductase (a flavoprotein). J Biol Chem 248:2089–2094
23. Ikushiro S, Kominami S, Takemori S (1989) Adrenal cytochrome P-450(11) beta-proteoliposomes catalyzing aldosterone synthesis: preparation and characterization. Biochim Biophys Acta 984:50–56
24. Lombardo A, Defaye G, Guidicelli C, et al. (1982) Integration of purified adrenocortical cytochrome P-450(11) beta into phospholipid vesicles. Biochem Biophys Res Commun 104:1638–1645
25. Lifton RP, Dluhy RG, Powers M, et al. (1992) A chimaeric 11 beta-hydroxylase/aldos-terone synthase gene causes glucocorticoid-remediable aldosteronism and human hypertension. Nature 355:262–265
26. Mitani F, Shimizu T, Ueno R, et al. (1982) Cytochrome P-450(11) beta and P-450scc in adrenal cortex: zonal distribution and intramitochondrial localization by the horse-radish peroxidase-labeled antibody method. J Histochem Cytochem 30:1066–1074

# Inhibition Studies of Steroid Conversions Mediated by Human CYP11B1 and CYP11B2 Expressed in Cell Cultures

Karsten Denner[1] and Rita Bernhardt[2]

*Summary.* Nonsteroidogenic lung fibroblast-derived V79 Chinese hamster cells were genetically engineered to express human mitochondrial cytochrome P-450 as an analytical tool to study adrenal steroidogenesis. Two V79-derived cell lines were established expressing the enzymatically active human 11β-hydroxylase and aldosterone synthase in a stable and constitutive manner. These cell lines were used for kinetic measurements of the human isoenzymes and for studying side effects of drugs on gluco- and mineralocorticoid genesis.

*Key words.* Steroid hydroxylases—Stable expression—Adrenal—V79 fibroblasts—Cytochrome P-450

## Introduction

Adrenal cortex produces several kinds of steroid hormones, such as mineralocorticoids, glucocorticoids, and androgens, through multistep reactions catalyzed by a group of monooxygenases called P-450 enzymes. Aldosterone, the most potent mineralocorticoid in humans, and cortisol, the major glucocorticoid, are synthesized from 11-deoxycorticosterone (DOC) and 11-deoxycortisol, respectively, by the action of two different cytochrome P-450 isoenzymes. The 11β-hydroxylase (P-450$_{11\beta}$), encoded by the gene CYP11B1, is found in the zona fasciculata/reticularis and catalyzes predominately the formation of glucocorticoids. The aldosterone synthase (P-450$_{aldo}$), encoded by the gene CYP11B2, is found in the zona glomerulosa and catalyzes the hydroxylation of DOC to form corticosterone, 18-hydroxycorticosterone, and aldosterone. From the clinical aspect it is interesting to characterize the properties of aldosterone synthase and 11β-hydroxylase because acquired and inborn errors in the synthesis or action of mineralocorticoids and glucocorticoids have been reported. Heterologous expression of cytochromes P-450 has been repeatedly shown to facili-

[1] Fachbereich Chemie, Institut für Biochemie, Freie Universität Berlin, Thiellalee 63, 14195 Berlin, Germany
[2] Universität des Saarlandes, Fachrichtung 12.4-Biochemie, Postfach 151150, D-66041 Saarbrücken, Germany

tate studies at the molecular and cellular level without the need for tedious purification and reconstitution of cytochromes P-450 [1,2]. We used stably transfected cell lines that express constant amounts of 11β-hydroxylase and aldosterone synthase to study these human steroidogenic enzymes regarding their steroid metabolism and inhibition profiles.

## Methods

The CYP11B1 and CYP11B2 cDNAs were obtained by polymerase chain reaction (PCR) cloning using specific oligonucleotide primers according to sequences published by Mornet et al. [3] and Kawamoto et al. [4]. COS-1 cells were transfected by the diethylaminoethyl (DEAE) dextran method. Stable transfections of V79MZ cells were carried out by the calcium–phosphate coprecipitation technique as cotransfection of plasmids pMTneo, pSVLh11B1, and pSVLh11B2, respectively. For enzymatic assays cells were grown in 24-well plates and incubated with 0.5 ml medium containing either [³H]-11-deoxycortisol or [³H]-11-deoxycorticosterone together with the corresponding unlabeled steroid. Steroids were extracted with methylene chloride. The extract was redissolved in 150 µl of 10% (v/v) isopropyl alcohol in n-hexane.

After the addition of a combination of internal steroid standards, the samples were subjected to normal-phase HPLC using a Lichrosorb-Diol (Merck, Darmstadt, Germany) column. A gradient solvent system was run starting with 10% isopropyl alcohol in n-hexane at a flow rate of 1.3 ml/min. The standards were monitored by UV detection at 254 nm; the radioactivity was assayed online with a Betascan radiation detector (Beta Ray Kubisiak GmbH, Dobel, Germany) fitted to the HPLC. Alternatively, steroids were analyzed by thin-layer chromatography. The extracts were redissolved in 20 µl methylene chloride and spotted onto a glass-backed silica-coated thin-layer chromatography (TLC) plate. Chromatography was performed in methylene chloride/methanol: $H_2O$ at a ratio of 300:20:1. A phosphor imager (BAS 2000, Fujifilm/Raytest, Straubenhardt, Germany) was used for quantification. Adrenaline, noradrenaline, dopamine, serotonine, synephrine, and histamine were obtained from Sigma (St. Louis, MO, USA).

## Results and Discussion

The full-length cDNAs, including the presequences, were obtained from total RNA of a surgically removed normal human adrenal gland by PCR cloning using specific oligonucleotide primers for CYP11B1 and CYP11B2, respectively. The cloned cDNAs, inserted into expression vectors (pSVLh11B1 and pSVLh11B2), were transfected into COS-1 cells, and steroid hydroxylase assays were performed to proof the enzymatic activity of the constructs. The V79MZ cells, which do not express endogenous cytochrome P-450 and are therefore defined for heterologous cDNA-encoded cytochrome P-450 expression, were used as host cells for stable transfection. Transfection of V79MZ Chinese hamster cells was carried out as cotransfection of plasmid pMTneo342-2, and pSVLh11B1 and pSVLh11B2, respectively. G418-resistant colonies were propagated.

Enzymatic activity in transfected V79MZ cells was tested by following the hydroxylation of 11-deoxycorticosterone. No activity was detected in parental V79MZ cells. Several clones that exhibited the highest conversion rates were selected for subcloning. Cells were subcloned by limited dilution and checked by Northern and Southern blotting as well as enzymatic activity as described previously [5]. The cell lines V79MZh11B1, stably expressing CYP11B1, and V79MZh11B2, stably expressing CYP11B2, were used for further steroid metabolism studies as well as inhibition studies with substances of pharmacological interest. Metyrapone is used as a diagnostic inhibitor of the 11β-hydroxylation reaction. Spironolactone is known to function as a competitive antagonist of mineralocorticoids. Fluconazole, ketoconazole, and miconazole are azole derivatives that are applied as antimycotic drugs. The steroid derivative 18-ethynylprogesterone is known as a potent inhibitor of aldosterone biosynthesis in bovines [6,7]. The 11β-hydroxylation in the V79MZh11B1 as well as in V79MZh11B2 cells was strongly inhibited by ketoconazole, miconazole, and

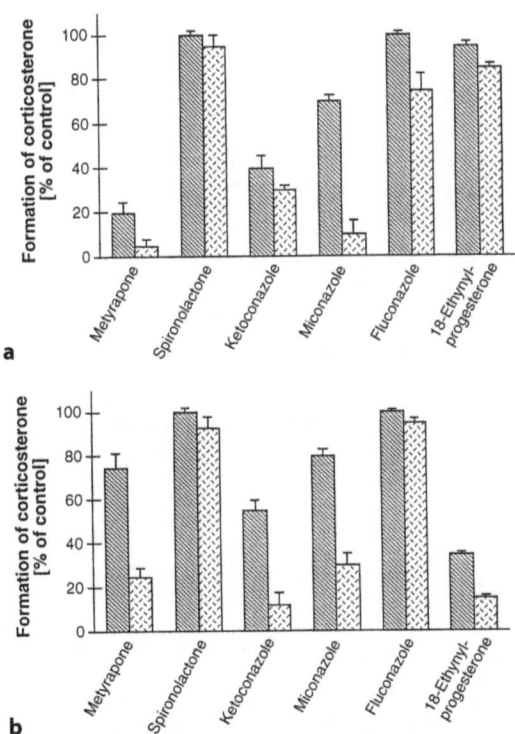

Fig. 1a,b. The influence of potential inhibitors on the formation of corticosterone catalyzed from V79MZ cells stably transfected with the cDNA of CYP11B1 (a) or CYP11B2 (b). Cells were incubated with 0.5 ml of medium containing 11-deoxycorticosterone (DOC) as substrate and the indicated substance (*darker striped bars*, 0.5 μM; *lighter hatched bars*, 5 μM) for 6 h. After extraction the steroids were analyzed by HPTLC and quantitated using a phosphor imager. Each value is the mean of three or more independent determinations

metyrapone in a dose-dependent manner. The inhibitory potency of spironolactone on steroid hydroxalase activities was minor.

As shown in Fig. 1, metyrapone inhibits 11β-hydroxylation more effectively in V79MZh11B1 cells compared with V79MZh11B2 cells. The inhibition of fluconazole on 11β-hydroxylase and aldosterone synthase was minor compared with the other azole derivatives. The 18-ethynylprogesterone was a more potent inhibitor of aldosterone synthase than of 11β-hydroxylase. The inhibition of hydroxylation activities of CYP11B1- and CYP11B2-transfected V79 cells by 18-ethynylprogesterone was characterized in more detail (Figs. 2, 3). As shown in Fig. 2, the 11β-hydroxylase activity of V79MZh11B1 cells was inhibited in a competitive manner by 18-ethynylprogesterone. The maximal rates for the conversion of 11-deoxycorticosterone to corticosterone was about 2.8 nmol corticosterone formed per 6h from $4 \times 10^5$ cells. The inhibition of aldosterone synthase expressed in V79MZh11B2 cells by 18-ethynylprogesterone was characterized by a drastic decrease of $V_{max}$ (Fig. 2b). Thus, this compound inhibits the 11β-hydroxylation mediated by aldosterone synthase more effectively than the same reaction catalyzed by 11β-hydroxylase.

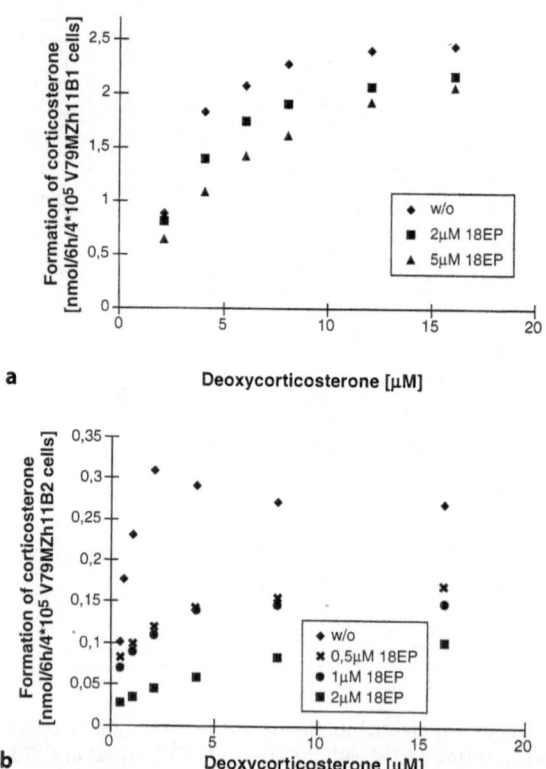

Fig. 2a,b. Inhibition of hydroxylation activities of CYP11B1- (a) or CYP11B2- (b) transfected V79 cells by 18-ethynylprogesterone. Deoxycorticosterone was added to the cells, and the formation of corticosterone was measured in the presence of 18-ethynylprogesterone (*18EP*). Each point is the mean of three or more independent determinations

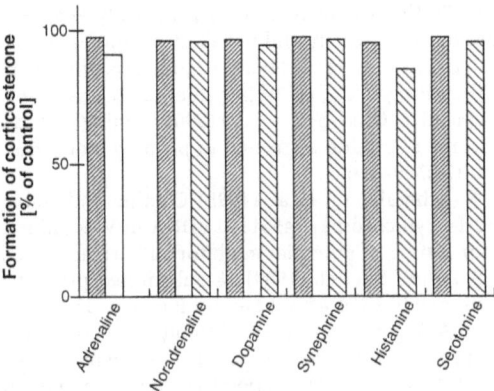

FIG. 3. The influence of biogenic amines on the conversion of 11-deoxycorticosterone to corticosterone by CYP11B1-transfected V79MZ cells. Cells were incubated with 0.5 ml of medium containing 4 µM DOC and the given biogenic amines (*darker stripped bars*, 5 µM; *unpatterned bars*, 50 µM; *lighter stripped bars*, 200 µM) for 6 h. Each point is the mean of three independent determinations

An interaction of biogenic amines with adrenal steroidogenic enzymes, namely an inhibition of sheep 11β-hydroxylase and bovine 21-hydroxylase by noradrenaline and synephrine, respectively, has been described [8]. To study the effect of synephrine and the catecholamines adrenaline, noradrenaline, and dopamine on human enzymes, we followed the conversion of 11-deoxycorticosterone in V79MZh11B1 and V79MZh11B2 cells. Histamine and serotonine were included in these measurements. However, in our systems no effect of these biogenic amines on 11β-hydroxylase (Fig. 3) or aldosterone synthase activities (data not shown) could be observed.

Taken together, our results demonstrate that the established cell lines V79MZh11B1 and V79MZh11B2 may be applied in drug design to check newly developed drugs for interference with 11β-hydroxylase and aldosterone synthase activity, respectively. Azoles were used as model substances showing inhibition of 11β-hydroxylase as well as aldosterone synthase. The steroid derivative 18-ethynylprogesterone inhibits aldosterone synthase more effectively than 11β-hydroxylase. Thus, the V79MZ-based cell lines expressing cytochromes P-450 are valuable tools for the investigation of these enzymes in a highly defined cellular system.

*Acknowledgments.* We are grateful to Dr. A. Piffeteau at the Lab. de Chimie Organique Biologique, Université Pierre at Marie Curie, Paris, France, for providing the steroid derivative 18-ethynylprogesterone, and to Dr. P. Swart at Departement of Biochemistry, University of Stellenbosch, South Africa, for stimulating discussions. This work was supported by a grant Be1343/2-4 from the Deutsche Forschungsgemeinschaft.

# References

1. Zuber MX, Simpson ER, Waterman MR (1988) Expression of bovine 17α-hydroxylase cytochrome P-450 cDNA in nonsteroidogenic (COS1) cells. Science 234:1258–1261

2. Doehmer J (1993) V79 Chinese hamster cells genetically engineered for cytochrome P-450 and their use in mutagenicity and metabolism studies. Toxicology 82:105–118
3. Mornet E, Dupont J, Vitek A, et al. (1989) Characterization of two genes encoding human steroid 11β-hydroxylase (P-45011β). J Biol Chem 264:20961–20967
4. Kawamoto T, Mitsuuchi Y, Ohnishi T, et al. (1990) Cloning and expression of a cDNA for human cytochrome P-450aldo as related to primary aldosteronism. Biochem Biophys Res Commun 173:309–316
5. Denner K, Vogel R, Schmalix W, et al. (1995) Cloning and stable expression of the human mitochondrial cytochrome P-45011B1 cDNA in V79 Chinese hamster cells and their application for testing of potential inhibitors. Pharmacogenetics 5:89–96
6. Viger A, Coustal S, Perard S, et al. (1989) 18-Substituted progesterone derivatives as inhibitors of aldosterone biosynthesis. J Steroid Biochem 33:119–124
7. Delorme C, Piffeteau A, Viger A, et al. (1995) Inhibition of bovine cytochrome P-45011β by 18-unsaturated progesterone derivatives. Eur J Biochem 232:247–256
8. Swart P, De Villiers EP, Swart AC, et al. (1993) The interaction of biogenic amines with adrenal cytochrome P-450-dependent enzymes. Biochem Soc Trans 21:413

# Effects of ACTH and Angiotensin II on the Novel Cell Layer Without Corticosteroid-Synthesizing Activity in Rat Adrenal Cortex

Hirokuni Miyamoto, Fumiko Mitani, Kuniaki Mukai, and Yuzuru Ishimura

*Summary.* We previously reported the presence of a novel cell layer containing neither cytochrome $P\text{-}450_{11\beta}$ nor cytochrome $P\text{-}450_{aldo}$ in the rat adrenal cortex. As these cytochrome P-450s are terminal enzymes in biosynthetic pathways for gluco- and mineralocorticoids such as corticosterone and aldosterone, the cell layer can be regarded as inert in producing corticosteroids. This study was designed to examine the chronic effects of adrenocorticotropin (ACTH) or angiotensin II on the novel layer employing histochemical techniques. The novel layer in a normal rat was present between the zona glomerulosa (zG) and the zona fasciculata (zF) as a four- to six-cell stratum, where replicating cells positive to anti-bromodeoxyuridine (anti-BrdU) and anti-proliferating cell nuclear antigen (anti-PCNA) were abundant. When the plasma ACTH level was raised for 20 days, zF formed a thicker zone, while the width of the novel layer attenuated to one to two cells thick. Under this condition, replicating cells were concentrated around the interface of the increased zF and the attenuated layer. A high level of plasma angiotensin II concentration for 20 days caused proliferation of zG together with a decrease of the novel layer to two to three cells thick. In this case, replicating cells were found around the interface of zG and the novel layer. These results thus support our previous hypothesis that the novel layer without steroidogenic activity is a stem- or progenitor-cell zone of the adrenal cortex.

*Key words.* ACTH—Angiotensin II—Adrenal cortex—Cytochrome $P\text{-}450_{aldo}$—Cytochrome $P\text{-}450_{11\beta}$

## Introduction

The adrenal cortex is composed of four zones that differ in their functional and morphological properties as follows. (1) In the outer zone, the zona glomerulosa (zG), cells express cytochrome $P\text{-}450_{aldo}$ ($P\text{-}450_{aldo}$), which is responsible for the synthesis of aldosterone, the strongest mineralocorticoids in mammals, from deoxycorticosterone

Department of Biochemistry, School of Medicine, Keio University, 35 Shinanomachi, Shinjuku-ku, Tokyo 160, Japan

[1,2]. (2) In the next zone, the zona fasciculata (zF), cells contain glucocorticoid-synthesizing enzyme cytochrome $P-450_{11\beta}$ ($P-450_{11\beta}$), which is an enzyme responsible for the synthesis of glucocorticoids, corticosterone in rodents [3,4]. (3) In the inner-most zone adjacent to the medulla, the zona reticularis, cells secrete adrenal andro-gens in some animals [5]. (4) In addition to these classical three zones, we previously reported the presence of a new cell layer as the fourth zone [6,7]. The new cell layer was present between zG and zF, where the cells did not express either $P-450_{aldo}$ or $P-450_{11\beta}$ [6,7]. As $P-450_{aldo}$ and $P-450_{11\beta}$ were expressed specifically in zG and zF cells, respectively, it was suggested that the cells of the new layer lacking these enzymes were functionally undifferentiated [6,7]. Furthermore, replicating cells were abundant in and around the new layer [6,7]. On the basis of these observations, we have proposed that the new layer contains some stem cells or progenitor cells of the adrenal cortex.

We report here further characteristics of the new layer under stimulation of angio-tensin II and adrenocorticotropin (ACTH) secretion, which accelerate the expression of $P-450_{aldo}$ and $P-450_{11\beta}$, respectively. The results showed that these hormonal stimuli caused the hypertrophy of zG or zF around the new layer and the atrophy of the new layer. Localization of the decreased portions in the new layer closely paralleled that of cell replication in and around the layer. Thus, observations in this study were not inconsistent with our hypothesis that the new cell layer is a stem- or progenitor-cell layer in the adrenal cortex [6,7].

## Materials and Methods

### Animals and Treatments

Animals used in this study were female Wistar rats (Sankyo Labo Service, Tokyo, Japan) weighing 140–160g. ACTH secretion was stimulated by administration of metyrapone (Nippon Ciba Geigy, Tokyo, Japan), an inhibitor of the $P-450_{11\beta}$-dependent 11β-hydroxylase reaction, and was suppressed by administering dexam-ethasone (Dekadoron, Banyu Pharmaceutical, Tokyo, Japan), a synthetic glucocorticoid. The dosages of these reagents and the timing of the administration were determined as previously described [8].

Animals were fed with a normal- or low-sodium diet, of which the latter activates the renin-angiotensin system [6,7,9], and were maintained in accordance with the institutional animal care guidelines in Keio University School of Medicine. Especially, animals were handled every day to collect plasma ACTH in their resting condition. The blood samples were obtained at about 11:00 A.M., about the time that the level of plasma ACTH was consistently low in its circadian rhythm.

### Immunohistochemical Detection

$P-450_{aldo}$, $P-450_{11\beta}$, 5-bromo-2'-deoxyuridine (BrdU), and proliferating-cell nuclear antigen (PCNA) were stained as previously described [6,7]. To detect BrdU-incorporated cells, animals were administered BrdU (Sigma) 1h before being killed.

## Biochemical Determinations

The plasma ACTH concentration and the plasma renin activity (PRA), an index of angiotensin II-producing activity, were assayed by using a commercially available RIA kit as previously described [9]. Catalytic activities of P-450$_{aldo}$ and P-450$_{11\beta}$ in term of aldosterone and corticosterone formation from deoxycorticosterone were determined in the mitochondrial lysate of the capsular portion (mainly zG) and the decapsulated portion zonae fasciculata-reticularis [zFR] and medulla) of the adrenal gland, respectively, as described previously [10]. All data were compared using Student's $t$-test for unpaired data and Mann–Whitney's U test.

# Results and Discussion

Figure 1A,B shows the zonal distribution of P-450$_{aldo}$ and P-450$_{11\beta}$ in the adrenal cortex of female Wistar rats fed with a normal- and a low-sodium diet, respectively. The new layer, which stained negative to antibodies for P-450$_{aldo}$ and P-450$_{11\beta}$, was present mainly between zG and zF in the adrenal cortex. The results were in agreement with previous data using male Sprague–Dawley rats [6,7]. To further characterize the new layer, especially the effects of ACTH and angiotensin II on the proliferation of cells in and around the layer, concentrations of these hormones in rat plasma were regulated by feedback control mechanism(s) as described in Materials and Methods. With these manipulations, we were able to examine chronic effects of endogenously secreted ACTH and angiotensin II.

In a normal rat (Fig. 1A), part of the zG had P-450$_{aldo}$-positive cells. All the zF and a part of the zona reticularis (zR) had P-450$_{11\beta}$-positive cells. The new layer with cells

FIG. 1A,B. Localization of cytochromes P-450$_{aldo}$ and P-450$_{11\beta}$, and a novel cell layer in the adrenal cortex of female Wistar rats. **A** Tissue section of the adrenal cortex of a rat maintained under normal feeding. **B** Adrenal cortex after a low-sodium diet for 20 days. Violet, presence of P-450$_{aldo}$ in zona glomerulosa (zG); brown, P-450$_{11\beta}$ in zonae Fasciculata-Reticularis layer (zFR). A unstained by P-450s between zG and zona fasciculata (zF) shows the novel cell layer. ×25

negative to both anti-P-450$_{aldo}$ and anti-P-450$_{11\beta}$ was a four- to six-cell stratum that was composed of two parts as follows: one was a P-450$_{aldo}$-unstained cell stratum in zG and the other a P-450$_{aldo}$- and P-450$_{11\beta}$-unstained cell layer between zG and zF. As shown in Figs. 1B and 2A, when angiotensin II production was stimulated (about 1.5 fold higher than the normal level), the zG increased to five to six cells and that of the new layer diminished slightly to two to three cells in thickness. Anti-BrdU- and anti-PCNA-positive cells, replicating cells, were abundant in and around the boundary between the increased zG and the decreased new layer (data not shown). The thickness of the P-450$_{11\beta}$-positive zF did not seem to be different from that in the normal cortex. The result of the staining was not contradictory to the fact that the plasma ACTH level in sodium-restricted rats was similar to that in normal rats [8]. In a previous experiment under such a condition [2], the specific activities of P-450$_{aldo}$ and P-450$_{11\beta}$ paralelled the immunostaining of the P-450s. It was suggested, therefore, that angiotensin II simultaneously affected the thickness of the new layer and that of the zG with the activity of P-450$_{aldo}$, although it did not affect functions of zF cells.

When the ACTH level in plasma was raised to about threefold that of the normal level, angiotensin II production was suppressed by one-eighth of the normal level. In this case, P-450$_{aldo}$-positive zG cells almost disappeared (Fig. 2B), but a few cells were found just below the capsule. The specific activity of P-450$_{aldo}$ was significantly lower than activity in the normal level; $121.2 \pm 35.6$ pmol min$^{-1}$ mg$^{-1}$ of mitochondrial protein in the normal group vs. $20.2 \pm 31.0$ in the metyrapone-treated group ($n = 4$) ($P < .05$). In contrast, a P-450$_{11\beta}$-positive zone in the metyrapone-treated rat became thicker but the specific activity of P-450$_{11\beta}$ was similar to the activity in the normal rat adrenal cortex ($n = 4$; $14.9 \pm 5.5$ nmol min$^{-1}$ mg$^{-1}$ of mitochondrial protein in the metyrapone-

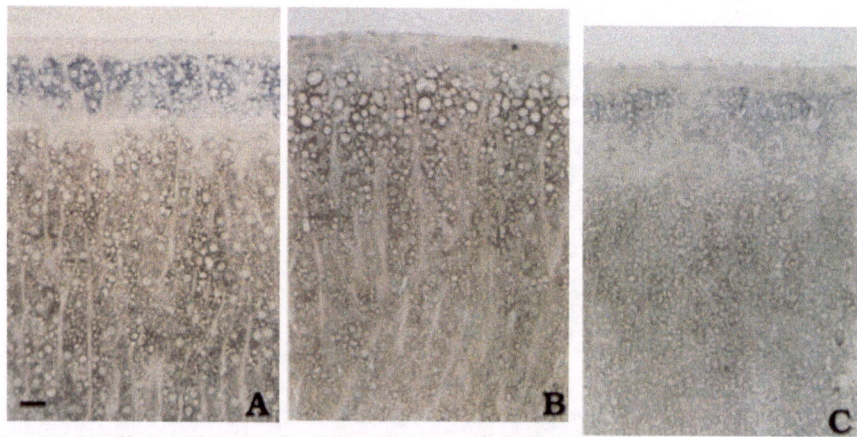

FIG. 2A–C. Effects of the administration of metyrapone/dexamethasone with a normal or low-sodium diet on the localization and thickness of the novel layer. A Adrenal tissue section of a rat after a low-sodium diet for 20 days. B Adrenal tissue with metyrapone treatment and a normal diet for 20 days. C Adrenal tissue section with dexamethasone treatment and a normal diet for 15 days. *Bar*, 50 μm

treated group and $13.4 \pm 5.4$ in the normal group). The new layer decreased to 1–2 cells thick (Fig. 2B), and replicating cells were concentrated in and around the boundary between the hypertrophied zF and the atrophied new layer (data not shown). Furthermore, when ACTH secretion was suppressed by one-fifth of the normal level (Fig. 2C), the new layer increased to 7–14 cells thick. The thickness of the zG also increased (Fig. 2C). Although plasma renin activity under this condition was about threefold higher than the normal level, the specific activity of P-450$_{aldo}$ was markedly low (<10 pmol min$^{-1}$ mg$^{-1}$ protein). In contrast, the thickness of the zF decreased and the intensity of the P-450$_{11\beta}$ staining was weak (Fig. 2C). Concomitantly, the specific

TABLE 1. Thickness of each zone in the adrenal cortex under adrenocorticotropin (ACTH) or angiotensin II stimulation

| Zone | ACTH + | ACTH − | Angiotensin II + | Angiotensin II − | ACTH and angiotensin II + |
|---|---|---|---|---|---|
| zG | A | U or H | H | U | H |
| uz | A | H | A | H | A (almost disappeared) |
| zF | H | A | U | U | H |

zG, zona glomerulosa; uz, a zone unstained by P-450s, i.e., the novel cell layer; zF, zona fasciculata; A, atrophy of the zone; H, hypertrophy of the zone; U, thickness of zone was unchanged.

+, a high concentration of the hormone in plasma; −, a low concentration of the hormone.

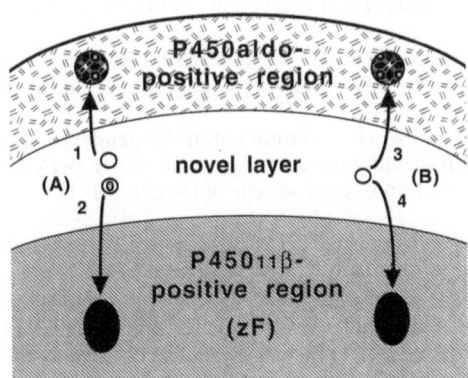

FIG. 3. Putative models for the role of the novel cell layer as a stem- or progenitor-cell layer in the adrenal cortex. The novel layer means a cell layer without expression of either P-450$_{aldo}$ or P-450$_{11\beta}$. The novel layer in this scheme shows a cell layer in a normal rat that is composed of P-450$_{aldo}$ and P-450$_{11\beta}$ nonexpressing cells in zG and a two- to three-cell layer between zG and zF as shown in Fig. 1A. In "model (A)", *arrows 1* and *2* show that the P-450$_{aldo}$- and the P-450$_{11\beta}$- positive cells arise from two kinds of progenitors, respectively, in the novel cell layer. In "model (B)", *arrows 3* and *4* show that the positive cells are generated from one kind of stem cells in the novel layer

activity of P-450$_{11\beta}$ was significantly lower ($7.0 \pm 2.1$ nmol min$^{-1}$ mg$^{-1}$ protein) than that in the normal rat adrenal cortex ($P < .05$). Thus, changes in the ACTH level in plasma with treatment with metyrapone or dexamethasone affected the angiotensin II level in plasma, and these conditions seemed to affect simultaneously the thickness of the new layer, the zG, and the zF as well as the activities of both P-450$_{aldo}$ and P-450$_{11\beta}$ in the adrenal cortex.

Taken together with previous reports [6–8] (Table 1), the new layer seemed to be present when the steroidogenic activity of either zG or zF cells was normal or was reduced. These observations indicate that the suppression of ACTH or angiotension II production triggers an increase of cell numbers in the new layer, resulting in the atrophy of zF or zG with low steroidogenic activities in the adrenal cortex. Additionally, the stimuli of these hormones caused the attenuation of the new layer and the hypertrophy of zF or zG.

We conclude here that ACTH and angiotensin II regulate the number of undifferentiated cells in the new layer as well as differentiated cells in the adrenal cortex. On the basis of these observations and previous reports [6–8], we propose two models for differentiation of adrenocortical cells (Fig. 3). Further studies are currently under way to determine which model is suitable for a physiological role of the new layer or to examine other physiological roles of the layer.

*Acknowledgments.* We thank Prof. Jun-ichi Hata and Dr. H. Suzuki in the Department of Pathology for their advice concerning immunohistochemistry and Mr. Takayoshi Ogawa for excellent technical assistance.

# References

1. Quinn SJ, Wiliams GH (1988) Regulation of aldosterone secretion. Annu Rev Physiol 50:409–426
2. Shibata H, Ogishima T, Mitani F, et al. (1991) Regulation of aldosterone synthase cytochrome-P-450 in rat adrenals by angiotensin II and potassium. Endocrinology 128:2534 –2539
3. Salmenper M, Kahri AI (1977) Studies on the dependence of mitochondrial 11β- and 18-hydroxylation on the nuclear and mitochondrial DNA synthesis during ACTH-induced differentiation of cortical cells of rat adrenals in tissue culture. Exp Cell Res 104:223–232
4. Arola J, Heikkilä P, Kahri AI (1993) Biphasic effect of ACTH on growth of rat adrenocortical cells in primary culture. Cell Tissue Res 271:169–176
5. O'Hara MJ, Nice EC, Neville AM (1980) Regulation of androgen secretion and sulphoconjugation in adult human adrenal cortex: studies with primary monolayer cell cultures. In: Genazzani AR, Thijssen JHH, Siiteri PK (eds) Adrenal androgens. Raven Press, New York, pp 7–25
6. Mitani F, Suzuki H, Hata J, et al. (1994) A novel cell layer without corticosteroid-synthesizing enzyme in rat adrenal cortex: histochemical detection and possible physiological role. Endocrinology 135:431–438
7. Mitani F, Ogishima T, Miyamoto H, et al. (1995) Localization of P-450$_{aldo}$ and P-450$_{11\beta}$ in normal and regenerating rat adrenal cortex. Endocr Res 21:413–423
8. Mitani F, Miyamoto H, Mukai K, et al. (1996) Effects of long-term stimulation of ACTH- and angiotensin II-secretions on the rat adrenal cortex. Endocr Res 22:421–431

9. Ogishima T, Suzuki H, Hata J, et al. (1992) Zone-specific expression of aldosterone synthase cytochrome P-450 and cytochrome P-450$_{11\beta}$ in rat adrenal cortex: histochemical basis for the functional zonation. Endocrinology 130:2971–2977

10. Ogishima T, Mitani F, Ishimura Y (1989) Isolation of aldosterone synthase cytochrome P-450 from zona glomerulosa mitochondria of rat adrenal cortex. J Biol Chem 264:10935–10938

# Cell-Specific and Hormonally Regulated Gene Expression Directed by the *CYP11B1* Gene Promoter in Rat Adrenal Cortex

Kuniaki Mukai, Fumiko Mitani, and Yuzuru Ishimura

*Summary.* The *CYP11B1* gene encodes steroid 11β-monooxygenase P-450, which is responsible for the last step in the biosynthesis of the major glucocorticoids, corticosterone and cortisol. This gene is expressed in a specific cell zone, the zona fasciculata, of the adrenal cortex. Its expression is stimulated at a transcription level by adrenocorticotropic hormone, the predominant regulator of glucocorticoid synthesis. We describe here that AP-1 transcription factors play crucial roles in both cell-specific and hormonal regulation of the *CYP11B1* gene expression in the adrenal cortex.

*Key words.* Adrenal cortex—*CYP11B1* gene—Glucocorticoid—Steroidogenesis—Transcriptional regulation

## Introduction

Glucocorticoids are produced in a specific zone, the zona fasciculata, of the adrenal cortex in mammals [1]. In the rat, the zone specificity in glucocorticoid synthesis is attributable to the spatially restricted expression of the *CYP11B1* gene, which encodes steroid 11β-monooxygenase P-450 (CYP11B1) [2,3]. CYP11B1 is responsible for the last step in the biosynthesis of the major glucocorticoids, corticosterone and cortisol. The same zone specificity in glucocorticoid synthesis is observed in humans [4].

Adrenocorticotropic hormone (ACTH), a peptide hormone secreted from the anterior pituitary, primarily controls glucocorticoid synthesis [1]. Previous studies have revealed that ACTH stimulates glucocorticoid synthesis, at least by activating expression of the steroidogenic enzyme genes [5,6], including the *CYP11B1* gene [7].

We describe here the mechanisms for the cell-specific and the hormonal regulation of *CYP11B1* gene expression in adrenocortical cells.

Department of Biochemistry, School of Medicine, Keio University, 35 Shinanomachi, Shinjuku-ku, Tokyo 160, Japan

# Transcriptional Activation of the *CYP11B1* Gene by Binding of an AP-1 Factor(s) to the Promoter

The 0.5-kb 5′-flanking region of the *CYP11B1* gene functions as a promoter in adrenocortical Y1 cells [7]. There are at least two adjacent sites in the promoter for the binding of Y1-cell nuclear proteins [8]: the binding site for AP-1 transcription factors [9] and the site for Ad4-binding protein (Ad4BP) [10]. AP-1 factors have been known to be involved in common processes linked to cell growth and differentiation, while Ad4BP, also called steroidogenic factor 1 [11], has been shown to activate various steroidogenic genes. Mutational analyses have revealed that binding of AP-1 factors, but not the Ad4BP, is required for activation of the *CYP11B1* promoter in unstimulated Y1 cells (Fig. 1). Furthermore, several lines of evidence indicate that binding of the two proteins to their own sites is competitive to each other and that binding of AP-1 has a suppressive effect on that of Ad4BP in Y1-cell nuclear extracts.

AP-1 factors are known to be dimers of the Jun family proteins or heterodimers between the Jun and Fos family proteins [9]. When the composition of the AP-1 factor(s) in the nuclear extracts from Y1 cells was analyzed with specific antibodies to the members of the two families, we found that the AP-1 factor(s) is composed of JunD and a Fos-related protein.

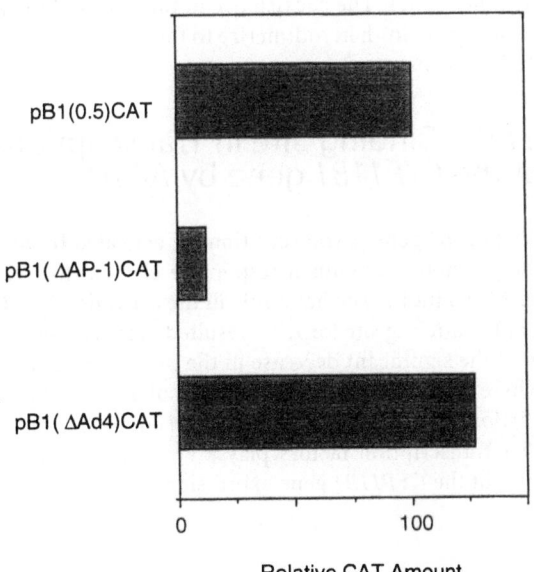

Relative CAT Amount

FIG. 1. Transcriptional activation of the *CYP11B1* gene is dependent on the AP-1-binding site. Adrenocortical Y1 cells were transiently transfected with one of the following chloramphenicol acetyltransferase *(CAT)* reporter plasmids; *pB1(0.5)CAT*, which contains the wild-type promoter sequence, *pB1(ΔAP-1)CAT*, which carries a mutation at the AP-1 binding site in the promoter, and *pB1(ΔAd4)CAT*, which carries a mutation at the Ad4 site [8]. CAT expressed in the transfected cells were determined and shown as relative amounts

TABLE 1. Presence of AP-1 factor(s) in the *CYP11B1*-expressing zone in the rat adrenal cortex

| | | AP-1 | |
|---|---|---|---|
| Adrenocortical zone | *CYP11B1* | Fos-related protein | Jun D |
| Zona glomerulosa | | | + |
| Zona fasciculata | + | + | + |
| Zona reticularis | (+) | + | |

## Presence of the AP-1 Factor(s) in Nuclei of the Cells of the Glucocorticoid-Producing Zone

The adrenal cortex consists of three major zones, the zona glomerulosa, zona fasciculata, and zona reticularis [1]. Table 1 summarizes the results from immunohistochemistry of CYP11B1, JunD, and a Fos-related protein with rat adrenal sections [8]. As reported previously [3], CYP11B1 is expressed in the cells of the zona fasciculata but is not detectable in the zona glomerulosa; the immunoreactivity in the zona reticularis is faint. The distributions of JunD and a Fos-related protein indicate that they coexist in nuclei of the cells in the zona fasciculata that express CYP11B1 but not in the nuclei of the other zones. The coexistence in the nuclei of the zona fasciculata cells suggests that these proteins heterodimerize to form an AP-1 factor that is able to activate the *CYP11B1* gene expression.

## Roles of the AP-1 Binding Site in Transcriptional Activation of the *CYP11B1* gene by ACTH

Expression of the *CYP11B1* gene is transcriptionally activated by ACTH stimuli [7], suggesting that the promoter contains a responsive element for ACTH. There is a possibility that the AP-1-binding site has a role in the activation by ACTH. We found that the mutation at the binding site for AP-1 resulted in loss of the responsiveness to ACTH in addition to the significant decrease in the promoter activity in the unstimulated Y1 cells. These results suggest that binding of AP-1 factor(s) to the site is necessary for activation of the *CYP11B1* gene expression by ACTH.

Collectively, AP-1 transcription factors play a key role in both cell-specific and hormonal regulation of the *CYP11B1* gene expression.

## References

1. Orth DN, Kovacs WJ, DeBold CR (1992) The adrenal cortex. In: Wilson JD, Foster DW (eds) Williams textbook of endocrinology. Saunders, Philadelphia, pp 489–619
2. Ogishima T, Mitani F, Ishimura Y (1989) Isolation of aldosterone synthase cytochrome P-450 from zona glomerulosa mitochondria of rat adrenal cortex. J Biol Chem 264:10935–10938

3. Ogishima T, Suzuki H, Hata J, et al. (1992) Zone-specific expression of aldosterone synthase cytochrome P-450 and cytochrome P-45011β in rat adrenal cortex: histochemical basis for the functional zonation. Endocrinology 130:2971–2977
4. Ogishima T, Shibata H, Shimada H, et al. (1991) Aldosterone synthase cytochrome P-450 expressed in the adrenals of patients with primary aldosteronism. J Biol Chem 266:10731–10734
5. Simpson ER, Waterman MR (1988) Regulation of the synthesis of steroidogenic enzymes in adrenal cortical cells by ACTH. Annu Rev Physiol 50:427–440
6. Waterman MR (1994) Biochemical diversity of cAMP-dependent transcription of steroid hydroxylase genes in the adrenal cortex. J Biol Chem 269:27783–27786
7. Mukai K, Imai M, Shimada H, et al. (1993) Isolation and characterization of rat CYP11B genes involved in late steps of mineralo- and glucocorticoid syntheses. J Biol Chem 268:9130–9137
8. Mukai K, Mitani F, Shimada H, et al. (1995) Involvement of an AP-1 complex in zone-specific expression of the *CYP11B1* gene in the rat adrenal cortex. Mol Cell Biol 15:6003–6012
9. Angel P, Karin M (1991) The role of Jun, Fos and the AP-1 complex in cell-proliferation and transformation. Biochim Biophys Acta 1072:129–157
10. Morohashi K, Honda S, Inomata Y, et al. (1992) A common *trans*-acting factor, Ad4-binding protein, to the promoters of steroidogenic P-450s. J Biol Chem 267:17913–17919
11. Lala DS, Rice DA, Parker KL (1992) Steroidogenic factor 1, a key regulator of steroidogenic enzyme expression, is the mouse homolog of *fushi tarazu*-factor 1. Mol Endocrinol 6:1249–1258

# Gene Organization and Genetic Defects of Bilirubin UDP-Glucuronosyltransferase

Yoshikazu Emi, Shin-ichi Ikushiro, and Takashi Iyanagi

*Summary.* Bilirubin, which is the final product of the oxidative degradation of heme, is conjugated with glucuronic acid in the liver and excreted by the biliary system. We have analyzed the novel bilirubin/phenol uridine diphosphate- (UDP-) glucuronosyltransferase UGT1 gene complex, which encodes a set of first exons encoding a variable amino-terminal domain and four downstream exons encoding the identical carboxyl-terminal domain. UGT1 isozymes are the result of alternate splicing of specific first exons and common exons, and UGT1*1 is the major enzyme involved in glucuronidation of bilirubin in rat liver. The Gunn rat is a mutant strain with unconjugated hyperbilirubinemia that is caused by a single mutation which results in the formation of a common truncated carboxyl terminus for all members of the UGT1 gene family.

*Key words.* Bilirubin—UDP-glucuronosyltransferase—Unconjugated hyperbilirubinemia—Gunn rats—Gene complex

## Introduction

Bilirubin is the oxidative product of the protoporphyrin part of the heme group of proteins such as hemoglobin and cytochromes. In the liver, bilirubin is conjugated with glucuronic acid, and the resulting water-soluble bilirubin glucuronides are excreted into bile with aid of an ATP-dependent anion transporter. Glucuronidation of bilirubin is catalyzed by hepatic enzyme bilirubin uridine diphosphate- (UDP-) glucuronosyltransferase (UGT) (Fig. 1).

In 1938, Gunn described a mutant strain of Wistar rats (Gunn rats) that had hereditary hyperbilirubinemia [1]. The Gunn rat is deficient in hepatic UGT activity toward bilirubin and also in a 3-methylchalanthrene- (3MC-) inducible phenolic substrate, although activity to steroid substrate, chloramphenicol, and morphine is normal. These facts suggest that UGT isozymes comprise at least two families. Isolation of these UGT genes from normal rats should enable elucidation of the molecular processes underlying their absence is Gunn rats. The purpose of this chapter is to

Department of Life Science, Himeji Institute of Technology, Harima Science Park City, 1479-1 Kanaji, Kamigori-cho, Akoh-gun, Hyogo 678-12, Japan

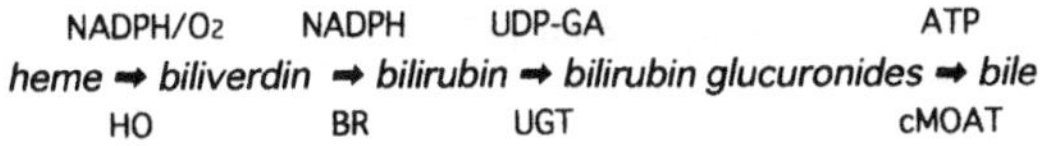

NADPH/O₂    NADPH    UDP-GA    ATP
heme → biliverdin → bilirubin → bilirubin glucuronides → bile
HO    BR    UGT    cMOAT

FIG. 1. Oxidative heme degradation and bilirubin transport pathway in the liver. *UDP-GA*, uridine diphosphate-glucuronic acid; *HO*, heme oxygenase; *BR*, biliverdin reductase; *UGT*, UDP-glucuronosyltransferase; *cMOAT*, canalicular multispecific organic anion transporter

summarize current knowledge on the molecular basis of multiple UGT isozyme deficiencies in Gunn rats.

## The Hyperbilirubinemic Rat (Gunn Rat)

We isolated and sequenced a cDNA designated 4-nitrophenol (4-NP) UGT from normal and homozygous Gunn rats, and sequencing analysis of Gunn rat cDNA clones revealed a single base deletion in the coding region [2,3]. Furthermore, we isolated and sequenced cDNAs from a Gunn rat liver library using mutant 4NP-UGT cDNA as a probe. Three novel cDNAs were identified that had identical 3'-regions of 1362 base pairs containing a single base deletion in the same position as that of the mutant 4NP-cDNA. However, their 5'-regions showed no more than 40% homology with that of 4NP-UGT.

These diverse amino-terminal domains can provide the substrate-binding site, while the conserved carboxyl-terminal domain provides the binding site for the common cosubstrate, UDP-glucuronic acid (UDP-GA). These isozymes constitute a new family, UGT1. On the basis of these results, we proposed that defective UGT isozymes in the Gunn rat are caused by a mutation in a single locus that encodes their carboxyl-terminal domains, which results in the formation of a common truncated carboxyl terminus [4].

## Bilirubin- and Phenol-Metabolizing UGT1 Locus

To elucidate the gene organization of the bilirubin- and phenol-UGT family, we have constructed a physical map of the 120-kbp DNA fragment that contains nine variable region (V) gene sequences (Fig. 2). The organization of the V locus showed several features that indicate phenol (*6–*9) and bilirubin (*1–*5) clusters [5]. The sequenced nine V sequence includes two pseudogenes, *4 and *9. The conserved region (C) is composed of four exons (exons 2,3,4, and 5). In the Gunn rat, a single base deletion, which is at the same position as that found in the Wistar rats, was identified in exon 4. These observations strongly suggest that each unique exon 1 (patterned boxes in Fig. 2) encodes the amino-terminal half of the isozymes, and exons 2–5 (black boxes in Fig. 2) specify the carboxyl-terminal region of all isozymes encoded in this gene complex.

These isozymes are the result of alternate splicing of the unique first exons and the conserved exons 2–5. This conclusion is also supported by the fact that the beginning of the conserved region corresponds exactly to a splice junction and that a single base deletion is found in each of the cDNAs. This introduces an in-frame stop codon in the

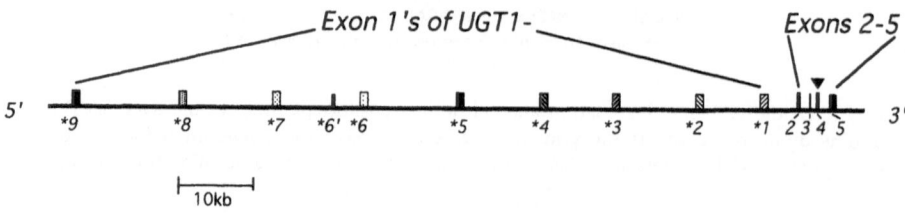

FIG. 2. The UGT1 gene complex that encodes rat bilirubin and phenol UDP-glucuronosyltrans-ferase. *Black boxes* indicate C-terminal exons (2–5); *patterned boxes*, N-terminal exons (*1–*9); *black bar*, introns. *Solid arrowhead* indicate the position where a guanosine (G) residue is deleted in the Gunn rats

shared 3'-terminals of RNAs encoding UGT1 family proteins. Therefore, these all UGT1 isozymes are lacking in the Gunn rats.

## Expression and Regulation of UGT1 Family

We prepared antipeptide antibodies raised against a conserved carboxyl-terminal portion of the isozymes and the variable amino-terminal portions of each isozyme of the phenol cluster (UGT1*6) and the bilirubin cluster (UGT1*1) [6]. Among the isozymes expressed in rat hepatic microsomes, UGT1*1 (54kDa) of the bilirubin cluster was found to be a major form, and minor forms were identified as UGT1*6 (53kDa), UGT1*2 (56kDa), and UGT1*5 (57kDa). Using a combination of two-dimensional sodium dodecyl sulfate gel electrophoresis and immunoblotting, all the isozymes were found to be simultaneously lacking in Gunn rat hepatic microsomes [6].

The effects of various drugs as an inducer on the expression of each UGT1 isozyme were then analyzed. The UGT1*6 and UGT1*7 of the phenol cluster isozymes were significantly induced in 3MC-treated rats. The expression of UGT1*1 and the glucuronidation activity toward bilirubin in rat hepatic microsomes were induced two- to threefold by clofibrate and dexamethasone administration (Table 1). This isozyme is a major component in the hepatic microsomes of untreated rats [6]. UGT1*1 cDNA expression vector was then transfected into COS cells. The expressed protein, localized in the endoplasmic reticulum, was active toward bilirubin ($0.28 \, \text{nmol} \, \text{min}^{-1} \, \text{mg}^{-1}$). This result indicates that the UGT1*1 isozyme is the major enzyme involved in the glucuronidation of bilirubin in rat liver. UGT1*6 (UGT1A1), which can catalyze glu-curonidation of 4NP, is a major 3-MC-inducible form, and the level of its mRNAs is markedly elevated in the liver of rats treated with 3-MC [5,7]. We analyzed the 5'-flanking region of the first exon *6 and identified a xenobiotic responsive element (TGCGTG) in the upstream *6' exon (A1*) [5,7]. These accumulated pieces of evidence suggest that each UGT1 transcriptional unit is under the control of its own promoter, so that each leader exon 1 is differentially spliced to the conserved exons (2–5) to generate nine different mRNAs, thus allowing independent regulation of each isozyme at the level of transcription.

TABLE 1. UDP-glucuronosyltransferase activity toward bilirubin and 4-nitrophenol in drug-treated rat hepatic microsomes

| Drug treatments | UDP-glucuronosyltransferase activities (nmol min$^{-1}$ mg$^{-1}$) | |
| --- | --- | --- |
| | Bilirubin | 4-Nitrophenol |
| Wistar rat | | |
| Untreated | 1.23 ± 0.02 | 84.5 ± 3.2 |
| 3-Methylcholanthrene | 1.28 ± 0.04 | 174.0 ± 4.7 |
| Phenobarbital | 1.47 ± 0.06 | 85.8 ± 2.2 |
| Clofibrate | 2.32 ± 0.13 | 54.9 ± 3.7 |
| Dexamethasone | 2.21 ± 0.16 | 65.0 ± 2.8 |
| Gunn rat | | |
| Untreated | ND | 24.3 ± 2.8 |

UDP, uridine diphosphate; ND, no detectable activity.

Finally, the Gunn rat provides a convenient experimental model for developing therapeutic strategies in the treatment of congenic hyperbilirubinemia in human Crigler–Najjar syndrome [8].

*Acknowledgment.* This work was supported in part by grants-in-aid for scientific research from the Ministry of Education, Science, Sports and Culture of Japan, and Hyogo Science and Technology Association.

# References

1. Gunn CH (1938) Hereditary acholuric jaundice in a new mutant strain of rats. J Hered 29:137–139
2. Iyanagi T, Haniu M, Sogawa K, et al. (1986) Cloning and characterization of cDNA encoding 3-methylcholanthrene inducible rat mRNA for UDP-glucuronosyltransferase. J Biol Chem 261:15607–15614
3. Iyanagi T, Watanabe T, Uchiyama Y (1989) The 3-methylcholanthrene-inducible UDP-glucuronosyltransferase deficiency in the hyperbilirubinemic rat (Gunn rat) is caused by a −1 frameshift mutation. J Biol Chem 264:21302–21307
4. Iyanagi T (1991) Molecular basis of multiple UDP-glucuronosyltransferase isoenzyme deficiencies in the hyperbilirubinemic rat (Gunn rat). J Biol Chem 266:24048–24052
5. Emi Y, Ikushiro S, Iyanagi T (1995) Drug-responsible and tissue-sepecific alternative expression of multiple first exones in rat UDP-glucuronosyltransferase family 1 (UGT1) gene complex. J Biochemi (Tokyo) 117:392–399
6. Ikushiro S, Emi Y, Iyanagi T (1995) Identification and analysis of drug-responsive expression of UDP-glucuronosyltransferase family 1 (UGT1) isozyme in rat hepatic microsomes using anti-peptide antibodies. Arch Biochem Biophys 324:267–272
7. Emi Y, Ikushiro S, Iyanagi T (1996) Xenobiotic responsive element-mediated transcriptional activation in the UDP-glucuronosyltransferase family 1 gene complex. J Biol Chem 271:3952–3958
8. Takahashi M, Ilan Y, Roy Chowdhury N, et al. (1996) Long-term correction of bilirubin-UGT-glucuronosyltransferase deficiency in Gunn rats by administration of a recombinant adenovirus during the neonatal period. J Biol Chem 271:26536–26542

# Imaging of Calcium Oscillations and Activity of Cytochrome P-450$_{scc}$ in Adrenocortical Cells

T. Kimoto, H. Mukai, R. Homma, T. Bettou, D. Nishimura, Y. Ohta, and S. Kawato

*Summary.* We have investigated the molecular mechanisms of signal transduction in steroidogenesis of adrenal fasciculata cells using noninvasive methods. With fluorescence video-microscopy, we discovered that the Ca$^{2+}$ oscillations superimposed on step Ca$^{2+}$ increase occurred in individual Calcium Green-1-loaded cells on stimulation with adrenocorticotropic hormone (ACTH) at 1 pM while the cAMP level was not increased. The Ca$^{2+}$ oscillation lasted for several minutes with a frequency around 0.04 Hz. When the Ca$^{2+}$ signaling was inhibited by the addition of ethyleneglycotetraacetic acid (EGTA), the corticoid production was also considerably suppressed. The results suggest that Ca$^{2+}$ signaling is a second messenger for ACTH-induced steroid hormone synthesis in fasciculata cells. Imaging of cholesterol side-chain cleavage by cytochrome P-450$_{scc}$ in intact cells was investigated with 3β-hydroxy-22,23-bisnor-5-cholenyl ether (cholesterol-resorufin) incorporated into adrenocortical cells using a confocal microscope. On conversion of cholesterol-resorufin by P-450$_{scc}$ to pregnenolone and resorufin, the P-450$_{scc}$ activity was localized by observing the large increase in fluorescence (strong fluorescence spots and weak fluorescence patches) of the free resorufin. A good signal-to-noise ratio for the image was obtained by stimulation of cells with ACTH and NADPH when low density lipoprotein (LDL) was used for delivery of cholesterol-resorufin. Cholesterol traffic was investigated by measuring the movement of LDL in individual cells with video and confocal microscopy. We found a rapid directed motion and very slow motion of endosomes containing fluorescent-labeled LDL molecules in the cytoplasm of the cell, implying that LDL-containing endosomes are actively transported toward lysosomes.

*Key words.* Fluorescence imaging—Microscopic imaging—Cytochrome P-450—Calcium signal—LDL

Department of Biophysics and Life Sciences, Graduate School of Arts and Sciences, University of Tokyo at Komaba, Meguro-ku, Tokyo 153, Japan

# Introduction

On stimulation by adrenocorticotropic hormone (ACTH), signal transduction may occur sequentially through hormone receptor in plasma membrane → movement of second messengers in cytoplasm → steroid hydroxylation by cytochrome P-450s in mitochondria and endoplasmic reticulum (ER), resulting in corticoid production. Cholesterol, the substrate of steroidogenesis, is supplied to adrenocortical cells with low-density lipoprotein (LDL) via LDL receptor-mediated endocytosis. The mechanisms of ACTH action have remained a subject of controversy, especially regarding the movement of second messengers. cAMP is a candidate for the intracellular messenger of ACTH action. $Ca^{2+}$ should also play an important role in steroidogenesis because its removal or the addition of calcium channel blockers reduces or abolishes the corticoid response. Cytochrome P-450$_{scc}$ catalyzes the side-chain cleavage reaction of cholesterol, which is an essential and rate-limiting step of steroid hormone synthesis. With conventional radioactivity assays, we cannot investigate in intact cells the time course of P-450$_{scc}$ activity, which may be affected by hormonal stimulation. The real-time imaging of the activity of enzymes that are located inside living cells is generally very difficult as compared with cell-surface receptors because of the difficulty of delivering signal substrates to the enzymes. Therefore, the achievement of time-dependent activity measurements of cytochrome P-450$_{scc}$ has made an essential contribution to the understanding of molecular mechanisms of steroidogenesis. By a combination of $Ca^{2+}$ signals and P-450$_{scc}$ activity, measurements, we now can investigate how $Ca^{2+}$ signals stimulate or regulate the essential steps of steroid hormone synthesis.

# Materials and Methods

## Preparation of Fasciculata Cell Cultures

Adrenocortical zona fasciculata cells were aseptically isolated by collagenase-DNase digestion from bovine adrenal glands [1]. The isolated cells were cultured in Ham's F-10 medium. Cells were plated onto glass-bottomed dishes previously coated with collagen. The cells were cultured for 1–3 days at a cell density of $1 \times 10^4$ cells/cm$^2$ for imaging and $1 \times 10^5$ cells/cm$^2$ for cell suspension measurements, respectively.

## LDL Isolation and Fluorescence Labeling

LDL was isolated from fresh bovine blood using sequential differential flotation in sodium chloride and sodium bromide solutions. The LDL sample was stored at 4°C under nitrogen and was used within 7 days to avoid denaturation. For P-450$_{scc}$ activity measurements, loading cholesterol-resorufin into LDL was carried out by incubation of 2 ml LDL (4 mg) with 1 ml of liposomes containing 1 mg egg phosphatidlcholine (PC) and 0.1 mg cholesterol-resorufin at room temperature for 24 h. For investigations of LDL traffic, LDL was labeled with a fluorescent lipid analogue, dioctadecyl tetramethylindocarbocyamine perchlorate (DiI), at a molar ratio of 1:20.

## Fluorescence Calcium Imaging

Fluorescence calcium imaging was performed with a video-enhanced fluorescence microscope that consisted of an inverted microscope (Nikon TMD-300, Nikon, Tokyo, Japan) equipped with a xenon lamp for excitation and a SIT camera (Hamamatsu Photonics C-1145, Hamamatsu, Japan). The glass-bottomed dish was mounted on the microscope equipped with a temperature chamber that maintained the air atmosphere at 37°C with high humidity. Fluorescence of Calcium Green-1 was measured above 520 nm with excitation at 450–490 nm; fluorescence of DiI was measured above 590 nm with excitation at 510–560 nm. For image analysis with the ARGUS-50 system (Hamamatsu Photonics), the video output was digitized and the images were stored in frame memory.

## Laser Scanning Confocal Microscope

This microscope (Bio-Rad MRC-600UV) consisted of a Nikon inverted microscope and an argon ion laser. This confocal microscope shared the same Nikon inverted microscope as the video-enhanced microscope. The fluorescence of resorufin anion was measured with excitation at 514 nm and fluorescence above 550 nm. Because of the very low fluorescence intensity of produced resorufin, the fast photon-counting mode was employed to get a good signal-to-noise ratio.

# Results and Discussion

## ACTH-Induced $Ca^{2+}$ Signaling

We discovered that the application of 1 pM ACTH (physiological concentration) induces oscillations or sustained elevation in $[Ca^{2+}]_i$ for more than 80% of adrenal fasciculata cells loaded with Calcium Green-1 using single-cell imaging (Fig. 1) [1]. Cell loading with Calcium Green-1 was performed with 3 μM Calcium Green-1/AM (300 μM stock solution in dimethylsulfoxide [DMSO]) for 10 min at 37°C in the presence of 0.01% Triton X-100. More than 95% of cells were loaded with Calcium Green-1. It should be noted that without this trace amount of Triton X-100 only about 5%–10% of cells were loaded with Calcium Green-1/AM, resulting in a very poor $Ca^{2+}$ response. For about 33% of $Ca^{2+}$ signaling cells, ACTH induced $Ca^{2+}$ oscillations that consist of repetitive $Ca^{2+}$ spikes with a frequency around 0.04 Hz. We observed a step increase in $[Ca^{2+}]_i$ for about 10% of cells and $Ca^{2+}$ oscillations superimposed on a step increase in $Ca^{2+}$ for about 57% of cells. Typical $Ca^{2+}$ oscillations lasted for several minutes. The $Ca^{2+}$ oscillations induced by ACTH were suppressed by the addition of ethyleneglycoltetraacetic acid (EGTA) and thapsigargin. On addition of thapsigargin to inhibit the activity of the calcium pump in the ER, the $Ca^{2+}$ transient appeared because of the blocking of $Ca^{2+}$ uptake into the ER, and thereafter $Ca^{2+}$ oscillations did not appear again. The addition of EGTA and thapsigargin before ACTH stimulation also suppressed $Ca^{2+}$ signaling considerably.

We have achieved nearly a complete loading of Calcium Green-1/AM for more than 90% of fasciculata cells in the presence of 0.01% of Triton X-100 in HEPES buffer. Without this trace amount of Triton X-100, only a small population (5%–10%) of

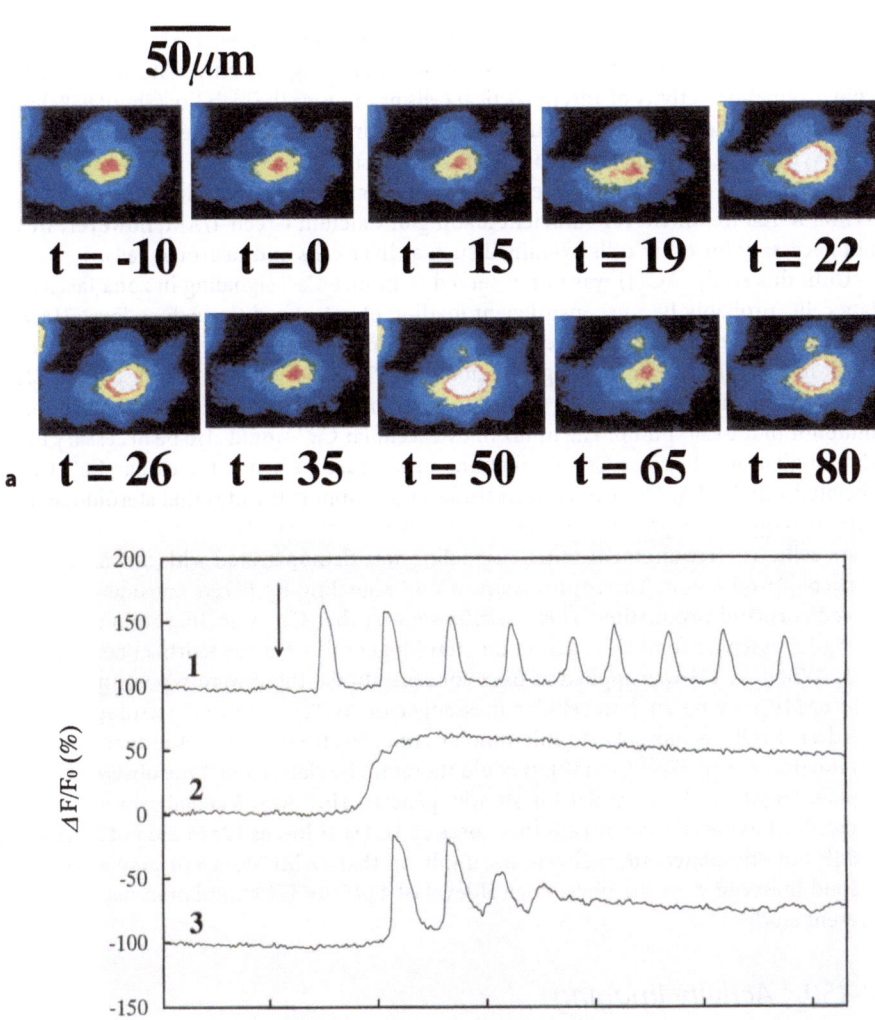

FIG. 1. **a** Pseudo-colored images demonstrate that adrenal fasciculata cells show oscillations of intracellular $[Ca^{2+}]_i$ on 1 pM adrenocorticotropic hormone (ACTH) stimulation at t = 0 s. The fluorescence intensity of Calcium Green-1 is indicated with a *color bar* from blue (low intensity) to red/white (high intensity). **b** The time course of typical $Ca^{2+}$ signaling induced by 1 pM ACTH. *Curve 1*, $Ca^{2+}$ oscillations; *curve 2*, step increase in $Ca^{2+}$; *curve 3*, $Ca^{2+}$ oscillations superimposed on step increase in $Ca^{2+}$. The *vertical scale* ($\Delta F/F_0$) is the ratio of the fluorescence intensity change ($F - F_0$) to the basal fluorescence $F_0$ for Calcium Green-1. The *arrow* indicates addition of ACTH

adrenocortical cells were successfully loaded with Calcium Green-1/AM. It should be noted that the presence of 0.01% Triton X-100 did not significantly disturb the corticoid production activity of adrenocortical cells nor reduce the viability (about 95%) of the cells examined with trypan blue staining. The difficulty of sufficient loading of the acetoxymethyl ester type of calcium dyes was a particular problem for adrenocortical fasciculata cells whose plasma membrane is resistant to incorporation of these dyes. Triton X-100 treatment for sufficient loading of Calcium Green-1/AM, however, was not necessary for other cells examined such as liver cells and neuronal cells.

Until this study, ACTH was not observed to induce $Ca^{2+}$ signaling in zona fasciculata cells, probably because insufficient loading of calcium dyes such as Fura-2/AM and Calcium Green-1/AM into cells without Triton X-100 treatment resulted in a very small fluorescence change. For generation of $Ca^{2+}$ oscillations, $Ca^{2+}$ release from ER would be involved, because $Ca^{2+}$ oscillations were abolished with thapsigargin, an inhibitor of the $Ca^{2+}$ pump. The influx of extracellular $Ca^{2+}$ would also be necessary for $Ca^{2+}$ oscillations, because $Ca^{2+}$ oscillations disappeared when extracellular $Ca^{2+}$ was chelated with EGTA. There have been studies that support the idea that steroidogenic activity requires the presence of extracellular calcium for ACTH stimulation of fasciculata cells. The requirement of $Ca^{2+}$ signaling was demonstrated with our assay for corticoid production. The suppression of $Ca^{2+}$ signaling by EGTA considerably reduced corticoid production. These results suggest that $Ca^{2+}$ signaling should be the second messenger for ACTH action on steroidogenesis of adrenocortical fasciculata cells. When ACTH was applied at high concentrations, there were reports implying that cAMP may be an intracellular messenger of ACTH action on steroidogenesis in adrenal cells. Nanomolar to micromolar concentrations of ACTH induced cAMP production during 10–30 min [2]. It could therefore be claimed that the observed $Ca^{2+}$ signals might not be essential for steroidogenesis. However, Yanagibashi et al. [3] showed in bovine fasciculata cells that doses of ACTH as low as 10 pM did not increase cAMP but stimulated steroidogenesis, implying that cAMP does not play a role of second messenger on the physiological level of 1 pM ACTH stimulation used in the present study.

## P-450~scc~ Activity Imaging

To visualize cytochrome P-450~scc~ activity, we employed cholesterol-resorufin for the following reasons. (a) From cholesterol-resorufin, P-450~scc~ catalyzes the production of not only resorufin but also pregnenolone, which is a physiological precursor of steroid hormones [4]. (b) The fluorescence intensity increases considerably, by about 500 fold, because of the production of resorufin anion. When the time course of side-chain cleavage activity of P-450~scc~ was measured in cell suspensions, we obtained a rapid linear increase of fluorescence intensity on adding 500 μM NADPH, resulting in the production of 11 pmol pregnenolone per 1h per $10^6$ cells (Fig. 2). When we did not add NADPH, the fluorescence increase was small even when the stimulation with 1 nM ACTH was carried out for more than 60 min.

Time-dependent growth of the fluorescence spectrum of resorufin was also examined over 24 h. As resorufin anion is amphiphilic and slowly leaked out from cultured cells into the supernatant solution, we measured fluorescence spectra of the supernatant solution. The specific fluorescence spectrum with a peak at 583 nm was obtained

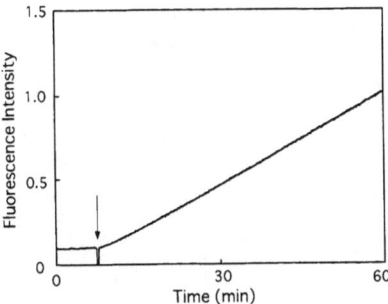

FIG. 2. Time course of activity of P-450$_{scc}$ in cell suspensions. On NADPH addition (*arrow*), a rapid linear increase of resorufin fluorescence was measured using a Hitachi F-3000 spectrofluorometer. Before the measurement, $10\,\mu$M cholesterol-resorufin dissolved in dimethylsulfoxide (DMSO) was incorporated by incubating at 37°C for 40 min into cells suspended in 10 mM Hepes buffer after gentle removal from the culture dish with a spatula

after 3-h incubation. Resorufin production was completely suppressed by the presence of $100\,\mu$M aminoglutethimide, which is a specific inhibitor of P-450$_{scc}$, indicating that this fluorescence increase is solely the result of the P-450$_{scc}$ side-chain cleavage reaction. The rate of conversion from cholesterol-resorufin to pregnenolone plus resorufin was determined to be about 2 nmol pregnenolone per 3 h per $10^6$ cells in the presence of NADPH and about 0.2 nmol pregnenolone per 3 h per $10^6$ cells in the absence of NADPH. This figure of pregnenolone production is in a reasonable agreement with 0.3 nmol pregnenolone per 3 h per $10^6$ cells measured by radioimmunoassay [5].

NADPH was found to be very effective in stimulating P-450$_{scc}$ activity in cells. Without NADPH stimulation, we needed 24-h incubation before a considerable resorufin production could be observed by spectral analysis. The ratio of production of resorufin in the presence and the absence of NADPH was about 10:1 after 3 h and 4/3:1 after 24 h, respectively. NADPH stimulation was most pronounced during the early 3 h of incubation. Because NADPH triggered an immediate increase of resorufin production, there might be a NADPH receptor present in plasma membranes, and this might be a new pathway to stimulate P-450$_{scc}$ activity. In fact, we observed that NADPH induced Ca$^{2+}$ oscillations without a step increase in Ca$^{2+}$ using microscopic single-cell imaging with Calcium Green-1, suggesting the possibility of NADPH receptor-mediated pathway to stimulate P-450$_{scc}$ activity. Because NADPH did not activate the P-450$_{scc}$ reactions in isolated mitochondria, the cellular response of resorufin production induced by NADPH cannot be the result of a direct stimulation of mitochondria by accidentally incorporated NADPH in the cytoplasm via slow pinocytosis or endocytosis. In spectral assay, ACTH was found to be effective to slowly stimulate P-450$_{scc}$ activity. When the effect of ACTH was examined in the presence of NADPH, we observed an almost 1.5-fold increase of resorufin production on application of 100 pM ACTH after 3-h incubation. When 100 pM ACTH was added in the absence of NADPH, the resorufin production was increased by about 4-fold of control cells for 3-h incubation.

A

B

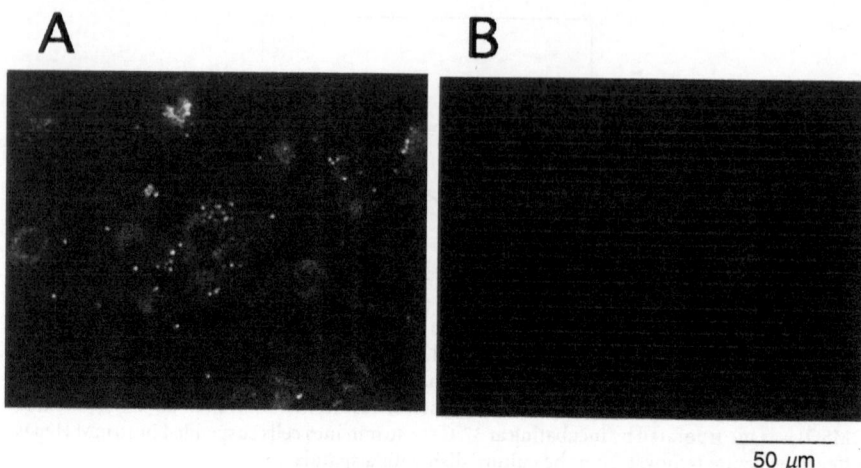

50 μm

FIG. 3. **A** Confocal fluorescence image of resorufin produced in individual cells by cleavage of cholesterol-resorufin by P-450$_{scc}$. The fluorescence image was obtained by accumulating 100 frames with a fast photon-counting mode and subtracting background autofluorescence. **B** No fluorescence was imaged for control cells treated with the same procedures except for the addition of cholesterol-resorufin-loaded low density lipoprotein (LDL)

Because NADPH and ACTH stimulated P-450$_{scc}$ activity considerably for intact cells, we applied these two reagents for intracellular activity imaging to improve the signal-to-noise ratio as much as possible. However, even in the presence of 500 μM NADPH and 1 nM ACTH, the resultant single-cell images of resorufin fluorescence were poor and not significantly stronger than the background autofluorescence of cells when we applied cholesterol-resorufin dissolved in DMSO directly to cells. To obtain further increase of P-450$_{scc}$ activity, we employed facilitated delivery of cholesterol-resorufin to mitochondria using LDL-mediated endocytosis. As shown in Fig. 3, we then achieved appearance of significant fluorescence of the resorufin anion in a single cell. After incubation of cells with cholesterol-resorufin-loaded LDL in a $CO_2$ incubator for 8.5 h at 37°C, strongly fluorescent spots and a broad weak fluorescent region were observed. Strongly fluorescent spots may correspond to mitochondria where the resorufin anion was produced by P-450$_{scc}$, and the broad fluorescent patches may result from leakage of resorufin from mitochondria to cytoplasm. Resorufin fluorescence was not observed in nuclei.

## LDL Traffic Imaging

Cholesterol traffic in the cytoplasm of cells was investigated by imaging the movement of LDL in individual cells with video microscopy [6]. Endosomes containing LDL were observed as a fluorescent particles of about 1–2 μm in diameter. Between 1 and 3 h after the addition of LDL to cells, we discovered that a population of endosomes showed a rapid directed motion (with a period of retrograde motion) with a velocity of 0.5–2 μm/s, and the rest of the endosomes were almost immobile (limited diffusion) within the time scale of 60 s. These results imply that endosomes may slide on microtubules rather than diffusing randomly in the cytoplasm. This finding requires renewing the previous idea that endosomes would undergo free diffusion in cytoplasic liquid without interacting with the cytoskeleton. Three hours after the addition of LDL a clear localization of endosomes around the nuclei was observed, implying that endosomes are slowly moving with a certain direction toward the circumference of nuclei where lysosomes may be enriched.

# References

1. Kimoto T, Ohta Y, Kawato S (1996) Adrenocorticotropin induces calcium oscillations in adrenal fasciculata cells: single cell imaging. Biochem Biophys Res Commun 221:25–30
2. Tremblay E, Payet MD, Gallo-Payet N (1991) Effects of ACTH and angiotensin II on cytosolic calcium in cultured adrenal glomerulosa cells. Role of cAMP production in the ACTH effect. Cell Calcium 12:655–673
3. Yanagibashi K, Papadopoulos V, Masaki E, et al. (1989) Forskolin activates voltage-dependent $Ca^{2+}$ channels in bovine but not in rat fasciculata cells. Endocrinology 124:2383–2391
4. Marrone BL, Simpson DJ, Yoshida TM, et al. (1991) Single cell endocrinology: analysis of P-450$_{scc}$ activity by fluorescence detection methods. Endocrinology 128:2654–2656
5. Yamazaki T, Higuchi K, Kominami S, et al. (1996) 15-lipoxygenase metabolite(s) of arachidonic acid mediates adrenocorticotropin action in bovine adrenal steroidogenesis. Endcrinology 137:2670–2675
6. Kimoto T, Asou H, Ohta Y, et al. (1996) Digital fluorescence imaging of elementary steps of neurosteroid synthesis in rat brain glial cells. J Pharm Biomed Anal 15:1231–1240

# Part 3
# Various Types of Oxidases and Oxygenases

# Fundamentally Divergent Strategies for Oxygen Activation by $Fe^{2+}$ and $Fe^{3+}$ Catecholic Dioxygenases

John D. Lipscomb, Allen M. Orville, Richard W. Frazee, Marcia A. Miller, and Douglas H. Ohlendorf

*Summary.* We propose that the redox state of the active site iron of ring-cleaving catecholic dioxygenases determines both the site of ring cleavage and the mechanism of $O_2$ activation. Spectroscopic and crystallographic studies show that substrates bind to the $Fe^{2+}$ of extradiol dioxygenases as asymmetric chelates in which only one hydroxyl becomes ionized. Charge transfer onto the iron increases the affinity for small molecules such as NO and $O_2$, which bind in another metal coordination site. Oxygen is activated by accepting electron density from the catechol via the iron, promoting nucleophilic attack of the resulting superoxide on the now electron-deficient catechol. In contrast, studies of intradiol $Fe^{3+}$ ring-cleaving dioxygenases show that catechols bind as dianions. This provides a site for electrophilic attack by $O_2$ at a hydroxyl-bearing carbon, which has been shown via model studies to lead to intradiol ring opening. Both dioxygenase classes shift from five-to six-coordinate iron sites during catalysis to allow oxygen binding. In the $Fe^{2+}$ enzymes, this site is occupied before attack on the catechol, while in the $Fe^{3+}$ case, it follows initial attack. In both cases, the expansion of the iron coordination allows the second oxygen required for dioxygenase stoichiometry to be retained for incorporation into the product during the final step.

*Key words.* Dioxygenase—Oxygen—Crystallography—Spectroscopy—Mechanism

## Introduction

Bacterial dioxygenases that catalyze cleavage of the aromatic ring of catecholic substrates as the key step in their biodegradation usually employ an essential mononuclear iron bound in the active site [1,2]. As illustrated in Fig. 1, it has been shown that the dioxygenases can be divided into two large subclasses depending on the redox state of the iron in the enzyme as isolated [1]. Intradiol dioxygenases contain $Fe^{3+}$ and catalyze cleavage of the aromatic ring between the vicinal hydroxyl functions of the catechol with incorporation of both atoms from $O_2$ to yield muconic acid derivatives. In

Department of Biochemistry Medical School and the Center for Metals in Biocatalysis, 4-225 Millard Hall, University of Minnesota, Minneapolis, MN 55455-0347, USA

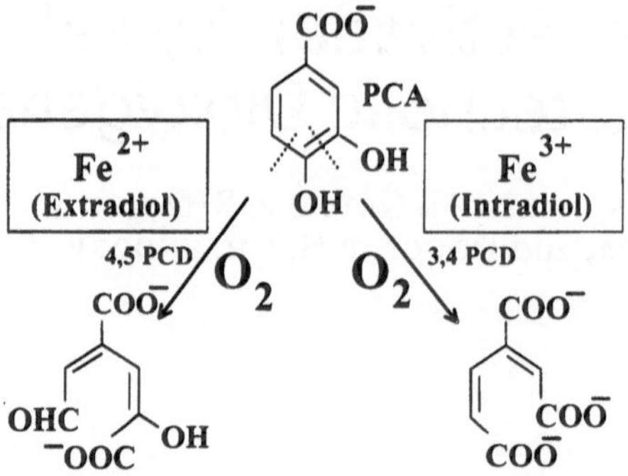

FIG. 1. Aromatic ring cleavage reactions catalyzed by typical $Fe^{2+}$ and $Fe^{3+}$ catecholic dioxygenases. *3,4-PCD*, protocatechuate 3,4-dioxygenase; *4,5-PCD*, protocatechuate 4,5-dioxygenase, *PCA*, protocatechuate

contrast, the extradiol enzymes contain $Fe^{2+}$ and cleave the ring adjacent to, but outside of, the vicinal hydroxyl functions to yield muconic semialdehyde products. The fidelity of ring cleavage position and the correlation with iron redox state is nearly perfect in the many members of this diverse enzyme family that have been investigated.

Although mechanisms have been advanced by some research groups that propose redox cycling of the iron during catalysis [3], no experimental evidence for any change in redox state of the iron has been obtained through spectroscopic studies [1,2]. Consequently, we have developed mechanistic schemes that do not require formal changes in redox state of the metal [1,4–6]. In recent years, sensitive spectroscopic methods to probe the ligation of the metal centers of both $Fe^{2+}$ and $Fe^{3+}$ dioxygenases have been developed. Moreover, crystal structures of representative enzymes in both families are now available for the first time [7–10]. These advances allow a reexamination of mechanistic proposals for the enzymes. Here the experimental evidence and rationale for the current mechanisms are summarized.

## Structure and Mechanism of $Fe^{3+}$ Aromatic Ring-Cleaving Dioxygenases

The structure and mechanism of the $Fe^{3+}$ aromatic dioxygenase class have been investigated most thoroughly using protocatechuate 3,4-dioxygenase (3,4-PCD) and catechol 1,2-dioxygenase isolated from a variety of species. We have solved the crystal structure of 3,4-PCD isolated from *Pseudomonas putida* (classified originally as *P. aeruginosa*) [7]. The 587-kDa protein is composed of 12 protomers each containing one α- and one β-subunit with a single $Fe^{3+}$ liganded by the β-subunit. The structure of the active site, shown schematically in *a* of Fig. 2 (for exact structural representations, see Fig. 3 and the chapter by A.M. Orville et al, this volume), established that the $Fe^{3+}$ resides in a trigonal bipyramidal coordination environment with Tyr-447 and His-462 as axial ligands and Tyr-408 and His-460 and a solvent as equatorial ligands.

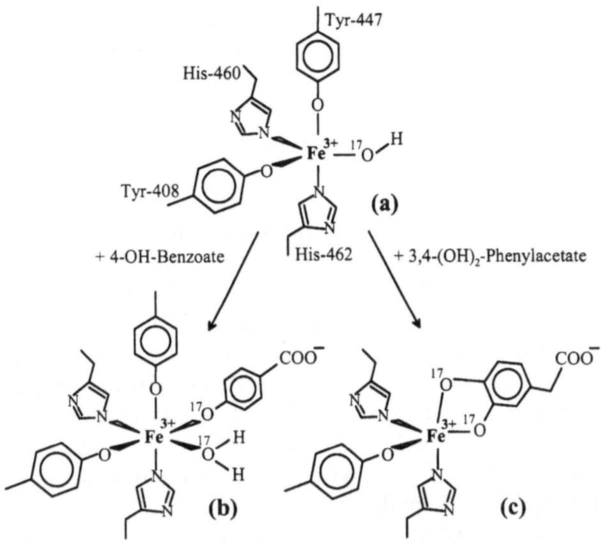

FIG. 2. Schematic structures of the active site $Fe^{3+}$ environment of the intradiol dioxygenase 3,4-PCD determined by spectroscopic and crystallographic procedures. (a) Uncomplexed enzyme; (b) enzyme complex with the inhibitor 4-OH-benzoate; (c) enzyme complex with the alternative substrate 3,4-(OH)$_2$-phenylacetate

A narrow solvent-filled channel at the interface of the α- and β-subunits leads to the solvent-binding site on the iron and comprises the active site of the enzyme.

The intense burgundy color of these enzymes derives from ligand-to-metal charge transfer (LMCT) from the tyrosines to the iron [2]. The $Fe^{3+}$ also gives rise to the sharpest S = 5/2 electron paramagnetic resonance (EPR) spectrum known. Nuclear hyperfine broadening of this signal by $^{17}O$-labeled solvent showed that at least one solvent is present as an iron ligand in the absence of substrates [1]. Binding of substrates resulted in loss of solvent broadening, but when $^{17}O$ was placed specifically in either hydroxyl group of the alternative substrate homoprotocatechuate (HPCA) (c of Fig. 2), broadening from each OH group was observed, showing that they both coordinate the iron [1]. Interestingly, when the monohydroxy-inhibitor 4-$^{17}OH$-benzoate was added (b of Fig. 2), hyperfine broadening was observed in the EPR spectrum from both the inhibitor OH group and solvent, showing that both can bind simultaneously. Extended X-ray absorption fine structure (EXAFS) studies of 3,4-PCD and its HPCA complex showed that the iron is five coordinate in each case. The fact that the coordination number did not change when a chelating ligand was added to an iron coordination sphere with only a single displaceable solvent suggested that an endogenous ligand was also displaced.

One particularly informative family of inhibitors for 3,4-PCD has been the nicotinic acid N-oxides. 2-OH-isonicotinic acid N-oxide (INO) binds to the 3,4-PCD essentially irreversibly in several steps [11]. The early steps are reversible and show no broadening from $^{17}O$-solvent in their EPR spectra, suggesting that solvent is displaced.

The later, tight-binding complex shows hyperfine broadening from $^{17}O$-solvent in the EPR spectrum, indicating that both the inhibitor and solvent bind to the iron. Fluorescence spectroscopy of the bound INO shows that it assumes the ketonized conformation. The optical spectrum of this complex is bleached, similar to a transient intermediate in the 3,4-PCD turnover of protocatechuate (PCA) immediately after the addition of $O_2$, suggesting that PCA also assumes a ketonized conformation in this intermediate.

Through the use of computer modeling techniques and the structure of the uncomplexed enzyme, the binding orientation of PCA in the active site can be predicted. This analysis suggests that PCA would bind with the plane of its ring approximately aligned with the axial axis of the $Fe^{3+}$ coordination and in a position sufficiently far from the $Fe^{3+}$ that only the $PCA^{OH4}$ group would coordinate [7]. The $PCA^{OH3}$ group would point into a pocket near the $Fe^{3+}$ that contains groups capable of H-bonding. This is contradictory to the spectroscopic results indicating a chelated substrate binding, and thus suggests that the substrate binding may proceed through several steps before the chelated structure is achieved as observed for INO binding. Several experiments, described next, support this hypothesis.

We have solved the crystal structures of several aromatic inhibitor complexes of 3,4-PCD [12] (Fig. 3) (described in more detail in the chapter by A.M. Orville et al, this volume). A range of binding positions relative to the iron were found depending on the aromatic ring substituents. When a bulky substituent such as iodine is in the 3-position and an OH is in the 4-position (IHB), a binding orientation much like that predicted on the basis of the hypothetical docking calculation is found. The $IHB^{O4}$

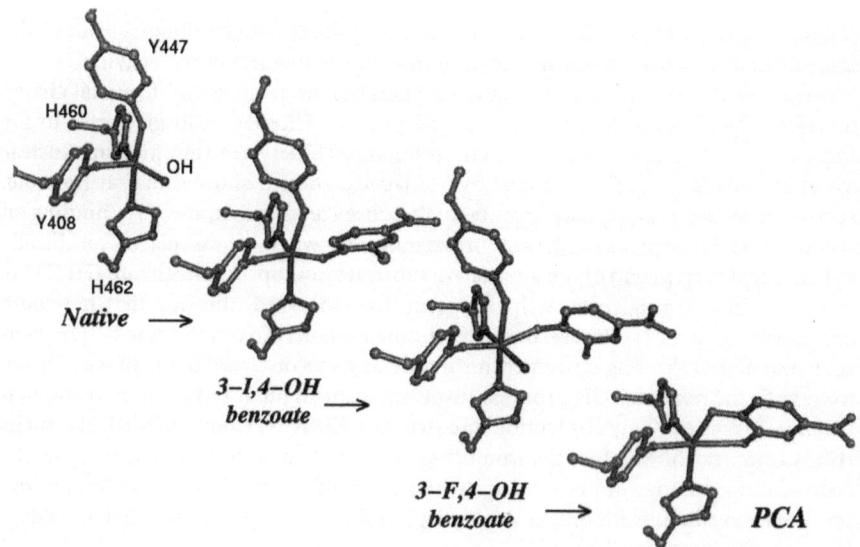

FIG. 3. Crystal structures of the active site $Fe^{3+}$ environment of 3,4-PCD. From *left* to *right*: uncomplexed enzyme; enzyme complex with 3-I,4-OH-benzoate; enzyme complex with 3-F,4-OH-benzoate; enzyme complex with protocatechuate (PCA). The structures are described in detail in [12] and [13]

group binds to the iron in place of the solvent, and $IHB^I$ is positioned away from the iron. When the size of the substituent in the 3-position is reduced, as in the 3-F analog (FHB), the inhibitor moves closer to the iron and the halide binds in a pocket where it can form H-bonds with Arg-457 and Gln-477. In the uncomplexed structure [7], there is a significant distortion of the iron coordination sphere toward octahedral geometry derived primarily from the 93° $Tyr-408^{O\eta}-Fe^{3+}-His-460^{N\epsilon2}$ equatorial bond angle. This yields a fairly open iron coordination face as the result of the 140° $His-460^{N\epsilon2}-Fe^{3+}-Wat-827^{OH}$ equatorial bond angle. When FHB or 4-OH-benzoate (HBA) binds deep in the active site pocket, the $His-460^{N\epsilon2}-Fe^{3+}-phenol^O$ bond angle is close to 90° and a solvent molecule binds in the adjacent equatorial site. Consequently, the trigonal bipyramidal coordination geometry of the uncomplexed enzyme is converted to six-coordinate octahedral. Inhibitors that bind in this way exhibit three to four orders of magnitude higher $K_i$ (and $K_d$) values than those that retain the trigonal bipyramidal geometry and displace solvent.

The association of substrates (PCA or HPCA) and chelating inhibitors (INO) initiates another major change in the $Fe^{3+}$ coordination as illustrated in Fig. 4 [13,14]. The C3–OH group shifts up to bind in the equatorial plane where the single OH groups of the monohydroxy adducts were bound. The C4–OH shifts to bind in the position of the axial Tyr-447, resulting in a displacement of this ligand. The side chain of Tyr-447 rotates about 110° and H bonds to Tyr-16 and Asp-413. This is the best-characterized instance of the release of an endogenous ligand in nonheme systems. One result of the rotation of Tyr-447 is that a pocket is formed in the protein encompassing the site in the equatorial plane adjacent to the site of substrate C3–OH binding. In the INO complex, this iron coordination site is occupied by solvent, while in the PCA and HPCA complexes the site is unoccupied in accord with the spectroscopic studies [1].

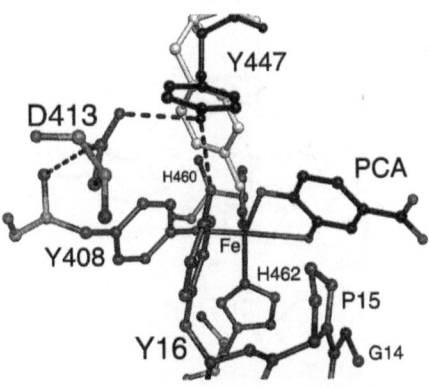

FIG. 4. Crystal structure of the active site environment of the substrate (PCA) complex of 3,4-PCD. The PCA and Tyr-447 in their position in the substrate complex are shown with *black bonds*. The position of Tyr-447 before PCA is added is shown superimposed with *white bonds*. Note the vacant iron coordination site and the H-bonding interactions with Tyr-447, which stabilize it in its position off the $Fe^{3+}$. Tyr-447, PCA, and Tyr-16 form the walls of a pocket proposed to accept $O_2$ during catalysis

One attractive scenario invokes the use of this pocket created near the iron as the $O_2$-binding site during catalysis. The $O_2$ is not known in any system to bind directly to the $Fe^{3+}$, but it could occupy the pocket, which would place it immediately adjacent to the site of ring cleavage on PCA. Because the site is created as a result of PCA binding, this would also account for the observed ordered mechanism of the enzyme in which PCA must bind first.

Spectroscopic studies indicate that the substrate binds to the iron as a dianion. The dissociation of anionic tyrosinate and $OH^-$ ligands on PCA binding would provide two strong bases immediately available to deprotonate the incoming PCA. Oxygen could then initiate electrophilic attack on the strongly negative substrate. Following this attack, the vacant site in the iron coordination might be used to bind the distal oxygen of the substrate-peroxy intermediate and retain it for later incorporation into the substrate as required by the reaction stoichiometry. One test of this proposal was made by examining the complex between 3,4-PCD, INO, and $CN^-$ [13]. It was found that $CN^-$ binds in the new pocket in the position proposed for $O_2$ but is coordinated to the iron. Interestingly, it is coordinated with an unfavorable bent $Fe-C\equiv N$ bond angle, which would weaken the $Fe-CN$ bond but would be quite favorable for the postulated $Fe-O-O-PCA$ bond of the peroxy intermediate (Fig. 5).

The current mechanism, based on the spectroscopic and crystallographic studies, is shown in Fig. 6. We propose that the substrate binds in at least three sequential steps, which may be as follow: (1) weak association with binding through the $PCA^{OH4}$ and displacement of the $OH^-$ ligand; (2) strong association with conversion of the $Fe^{3+}$ coordination to octahedral and return of solvent as neutral $H_2O$; and (3) rotation into a chelate structure with displacement of the axial Tyr-447 and solvent as well as the creation of a pocket for $O_2$ binding. Electrophilic attack of $O_2$ on the substrate must occur in a specific position to yield the intradiol cleavage product. We propose that this position is selected by the position of the putative $O_2$-binding pocket relative to

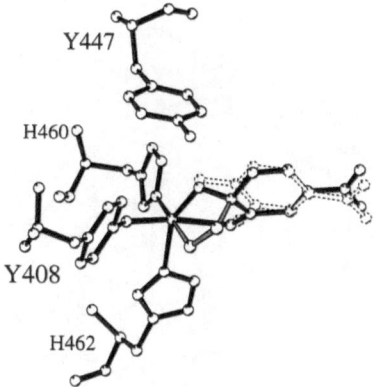

FIG. 5. Hypothetical structure of the peroxy intermediate of the 3,4-PCD reaction cycle. This structure was energy minimized using the structure shown in Fig. 4 and the standard bond lengths for peroxide. The structure predicts nearly ideal Fe-peroxide and peroxide-PCA bond angles. The *dashed line* structure shows the position of PCA in the anaerobic enzyme-substrate complex

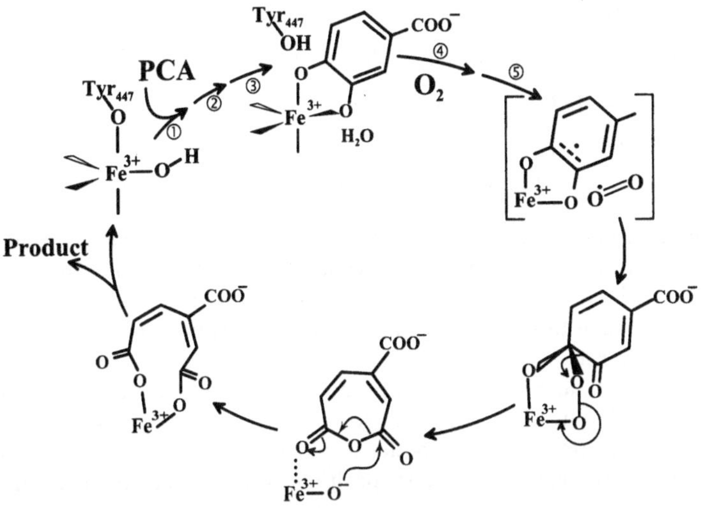

FIG. 6. Proposed reaction cycle for 3,4-PCD based on the structural and spectroscopic data available for $Fe^{3+}$ dioxygenases. Steps 1 through 3 (circled numbers) represent the stages in substrate binding modeled in Fig. 3 with the addition of a Michaelis complex of $O_2$ with the enzyme before binding to the iron. Steps 4 and 5 represent $O_2$ binding in the pocket created as a consequence of substrate binding and initial reaction of $O_2$ with the PCA to yield the hypothetical radical species shown. (See chapter by A.M. Orville et al, this volume, for additional details)

the substrate and by asymmetry in the iron–PCA complex. This asymmetry is apparently created by trans ligand effects. Because the $PCA^{OH3}$ group binds opposite Tyr-408, repulsion by this negative ligand causes the $Fe–PCA^{OH3}$ bond to lengthen relative to the $Fe–O–PCA^{OH4}$ bond, which has His-462 as the trans ligand. Thus, ketonization of $PCA^{OH3}$ following the attack of $O_2$ at $PCA^{OH4}$ will be favored. Model studies indicate that attack of the substrate in this position leads to intradiol ring opening [2].

One test of the proposed mechanism is to replace the critical axial Tyr-447 ligand by site-directed mutagenesis. This would be expected to remove the base needed to deprotonate the substrate and retard this portion of the reaction. However, once the substrate chelates the iron the chemical steps of the reaction would be expected to be less affected because Tyr-447 is displaced from the iron during this part of the cycle. The Y447H mutant 3,4-PCD has a significantly decreased optical spectrum in the visible region as the result of the loss of one of the ligands capable of LMCT [15]. It is found to give the same ring cleavage product as the wild-type enzyme, but its turnover number is decreased 600 fold. When the anaerobic substrate complex is formed, the optical spectrum is nearly identical to that of the wild-type enzyme substrate complex as expected. Transient kinetic analysis of the substrate-binding process shows that it occurs in four discrete steps. We propose that these steps are those modeled by the inhibitor and substrate complexes summarized in Fig. 3 with the addition of an initial Michaelis complex of the enzyme and substrate that precedes PCA binding to the iron. The rates of the steps are as follows, where the $O_2$ concentration is $250\,\mu M$ [1,15]:

*Y477H*

$$E + S \xrightarrow{\text{Fast}} ES_1 \xrightarrow{3.5\,s^{-1}} ES_2 \xrightarrow{0.2\,s^{-1}} ES_3 \xrightarrow{0.09\,s^{-1}} ES_4 \xrightarrow[O_2]{\text{Fast}} ESO_2$$

$$\xrightarrow{\text{Fast}} ESO_2 * \xrightarrow{2.7\,s^{-1}} ESO_2 ** \xrightarrow{0.04\,s^{-1}} E + P$$

*Wild type*

$$E + S \xrightarrow{2.5 \times 10^6\,s^{-1}M^{-1}} ES_4 \xrightarrow[O_2]{125\,s^{-1}} ESO_2 \xrightarrow{450\,s^{-1}} ESO_2 * \xrightarrow{36\,s^{-1}} E + P$$

The rate-limiting step appears to be the final step in which the dianionic chelate with the $Fe^{3+}$ is formed. This is the step in which the Tyr-447 would be displaced to act as a base to facilitate PCA binding in the dianionic form. Histidine would not have a $pK_a$ sufficiently high to efficiently catalyze this process, perhaps accounting for the great decrease in rate.

Following PCA binding, the steps of product formation in the active site of Y447H mutant are rapid, but the product release is slowed 100 fold. This may be because release of the strongly anionic product would be facilitated by the return of the negative tyrosinate ligand to the axial position, which cannot occur in the Y447H mutant.

# Structure and Mechanism of $Fe^{2+}$ Aromatic Ring-Cleaving Dioxygenases

Spectroscopic studies have revealed a quite different coordination environment for the iron of extradiol catecholic ring-cleaving dioxygenases [1,2]. As summarized in *a* of Fig. 7, there is no spectroscopic evidence for charge-donating ligands such as found in the $Fe^{3+}$ enzymes. This has the effect of giving the iron a high redox potential and stabilizing it in the $Fe^{2+}$ state even in the presence of $O_2$. These enzymes are essentially colorless in the visible and have no EPR spectra. Magnetic circular dichroism/Circular dichroism (MCD/CD) [16] and X-ray absorption studies [6] have shown that the iron is five coordinate and assumes square pyramidal coordination geometry. Nitric oxide has been a useful probe of the iron center because it binds with high affinity and the resultant complex exhibits a visible spectrum and an EPR active $S = 3/2$ spin state (*b* of Fig. 7). In the nitrosyl complex, hyperfine broadening detected in the EPR spectrum from $^{17}O$-solvent is observed showing that at least one solvent and the NO occupy $Fe^{2+}$ coordination sites [1]. Addition of substrate (*d* in Fig. 7) causes loss of the hyperfine broadening from solvent, but broadening from either $^{17}OH$-labeled PCA is observed suggesting that the substrate can occupy two $Fe^{2+}$ coordination sites while NO occupies a third. Titration of the complex with NO shows that it is bound three to five orders of magnitude tighter when PCA is also bound, indicating that the binding of substrate and the $O_2$ analog NO are energetically coupled. This shows that NO and substrates do not compete for the same iron sites.

It is reasonable to hypothesize that the increase in NO affinity is caused by a decrease in the redox potential of the iron caused by binding the PCA as an catecholic anion. Electron density from the PCA would be delocalized onto the NO, strengthening the bond to iron. A similar iron-mediated redistribution of electron density in the case of $O_2$ binding would explain why the enzymes have little or no affinity for $O_2$ in the absence of substrates. MCD/CD and X-ray absorption studies have shown that the $Fe^{2+}$ remains five coordinate in a slightly more distorted square pyramidal coordina-

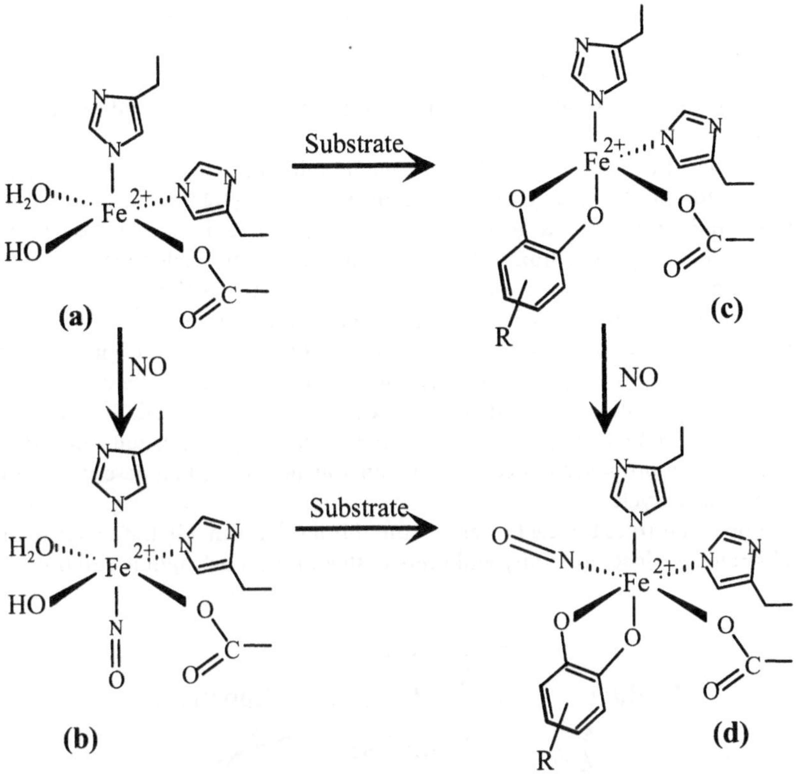

FIG. 7. Schematic structures of the active site $Fe^{2+}$ environment of extradiol dioxygenases determined by spectroscopic procedures. (*a*) Uncomplexed; (*b*) complex with NO; (*c*) complex with the aromatic substrate; (*d*) complex with the aromatic substrate and NO

tion geometry when substrate adds to the enzyme in the absence of NO (*b* in Fig. 7) [6,16], suggesting that the substrate chelates by displacing solvent. Significantly, the X-ray absorption results show that the Fe–PCA$^{OH}$ bonds are different in length. We propose that this is caused by ionization of only one of the OH groups so that the overall charge on the iron center remains neutral. Analysis of the MCD/CD data suggested that the substrate occupies the axial and one base position in the square pyramidal coordination geometry [16]. EXAFS studies show that the iron coordination number increases to six when both NO and substrates are bound, consistent with a chelated substrate in two sites and NO binding in a unique site (*d* of Fig. 7) [6].

The first crystal structure of an $Fe^{2+}$ dioxygenase, 2,3-dihydroxybiphenyl 1,2-dioxygenase, has been solved independently by two research groups (see the chapter by Senda et al, this volume, for detailed structural data) [8–10]. The structure observed is quite consistent with the structure revealed by the spectroscopic studies. The iron coordination is square pyramidal with His at the apical position and His, Glu, and two solvent molecules in the base positions. Studies of the substrate complex show that both solvents are displaced and the iron remains five coordinate [10]. The site for addition of NO and $O_2$ appears to be opposite the Glu in the $Fe^{2+}$ coordination when the substrates are bound. The substrate binds asymmetrically such that the Fe-substrate$^{OH}$ bond furthest removed from the site of ring cleavage is longer than that adjacent to the site of cleavage, in accord with the EXAFS studies described previously.

Considering the spectroscopic and crystallographic data together, the proposed mechanism for the $Fe^{2+}$ dioxygenases is shown in Fig. 8 [1,2,5,6]. In the first step, substrate binds to the $Fe^{2+}$ to form an asymmetric, monoanionic, chelate complex. This results in charge donation to the iron so that it is better able to share electron density with $O_2$ as it binds in another site. In effect, the $Fe^{2+}$ serves to facilitate delocalization of electron density from the catecholic substrate to the $O_2$, to yield a partial positive charge on the ring and a partial negative charge on the bound $O_2$. This nascent superoxide is proposed to initiate nucleophilic attack on the electron-deficient ring in the position of lowest electron density, *meta* to the weak Fe-substrate$^{OH}$ bond and *ortho* to the strong Fe-substrate$^{OH}$ bond. The oxygen insertion and hydrolysis steps are then proposed to occur in a manner analogous to that described for the $Fe^{3+}$ dioxygenase cycle.

Two important roles for the $Fe^{2+}$ arise from this mechanism. First, it serves to bind the substrates in close proximity and prepare them for a nucleophilic reaction. Sec-

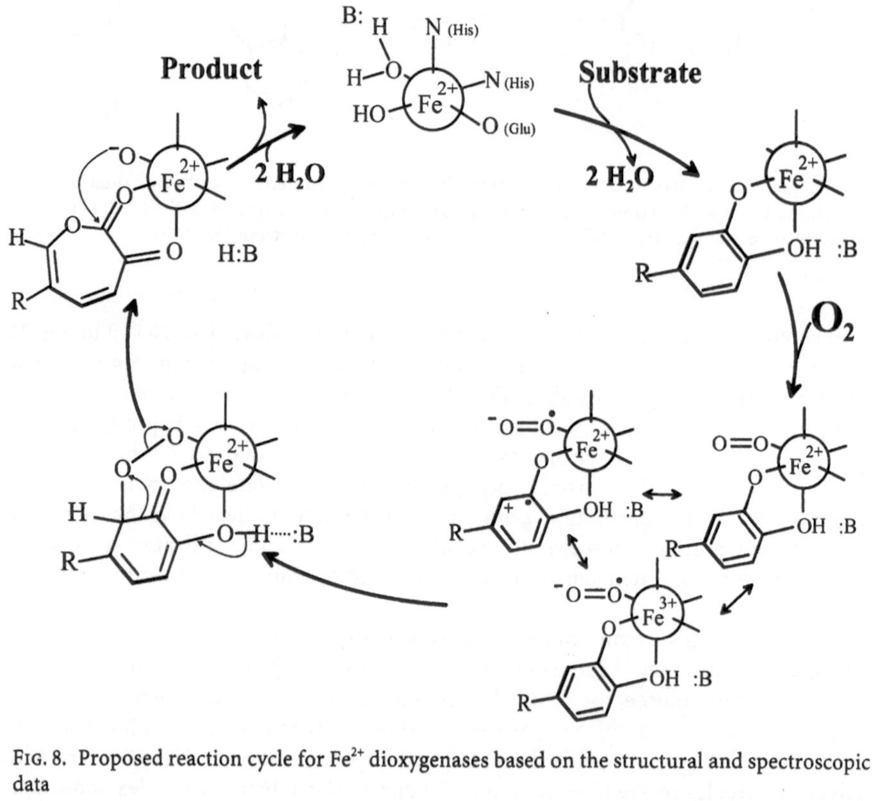

FIG. 8. Proposed reaction cycle for $Fe^{2+}$ dioxygenases based on the structural and spectroscopic data

ond, it promotes single rather than double deprotonation of the aromatic substrate, which lessens the likelihood of direct electrophilic attack by $O_2$ and establishes the lowest electron density on the ring in a position where attack by nucleophilic oxygen must result in extradiol cleavage.

We have recently, discovered a new extradiol dioxygenase that has allowed a test of the proposed mechanism [17]. Homoprotocatechuate 2,3-dioxygenase (2,3-HPCD) from the gram-positive bacterium *Brevibacterium fuscum* is unique among extradiol Fe$^{2+}$ dioxygenases in that it is resistant to inactivation by $H_2O_2$. It was determined that this resistance to peroxide derives from an endogenous catalase activity in which 2 $H_2O_2$ are converted to $O_2$ plus 2 $H_2O$. The double reciprocal plot of initial velocity of the catalase activity versus $H_2O_2$ concentration was found to be parabolic for 2,3-HPCD whereas it is linear for all other catalases. This shows that both $H_2O_2$ molecules must bind in the active site before a reaction occurs.

A proposed mechanism for this reaction is shown in Fig. 9. One $H_2O_2$ is proposed to bind in the normal $O_2$-binding site on the iron in an end-on fashion while the second $H_2O_2$ binds in a side-on fashion in the vicinal binding sites normally occupied by the substrate hydroxyls. From this conformation, two electrons can be transferred through the iron from the side-on to the end-on peroxide to yield the catalase products in a single step. The transfer of electron density in this way is analogous to the usual role proposed for iron in the dioxygenase reaction.

Interestingly, when catecholic substrates or inhibitors are added with $H_2O_2$, the catalase reaction still occurs, but the double reciprocal initial velocity plot is linear, showing that the mechanism has changed. In this case, one $H_2O_2$ reacts at a time with

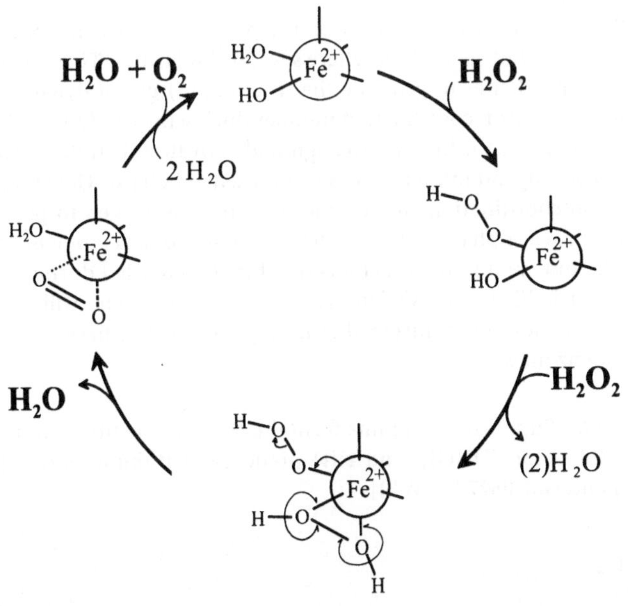

FIG. 9. Proposed reaction cycle for homoprotocatechuate 2,3-dioxygenase catalyzing a catalase reaction. (Adapted from [17], with permission)

the $Fe^{2+}$ center. We believe that the catecholic molecule binds in its usual site of the $Fe^{2+}$ coordination while the first $H_2O_2$ binds in the $O_2$ site. Transfer of electron density from the catechol to the $H_2O_2$ promotes heterolytic bond cleavage to yield $H_2O$ and an iron-bound, electron-deficient oxygen atom. Outer-sphere electron transfer from a second $H_2O_2$ would then complete the reaction in a second step, consistent with the linear kinetic plot. Again, the iron serves to connect the two sites of catecholic molecule and $H_2O_2$ binding to promote the reaction in a manner analogous to that suggested for the dioxygenase reaction.

## Fundamental Differences and Some Similarities in the Mechanisms of $Fe^{3+}$ and $Fe^{2+}$ Dioxygenases

The mechanisms of the $Fe^{3+}$ and $Fe^{2+}$ dioxygenases appear to be fundamentally different in that the former is based on electrophilic attack of $O_2$ on an activated substrate whereas the latter is based on nucleophilic attack by an activated oxygen. In the $Fe^{3+}$ case, the substrate is activated by establishing a ligand set that facilitates binding of the substrate as a dianion. In contrast, the substrate appears to bind as a monoanion to the $Fe^{2+}$ dioxygenases because only a single negative charge from the substrate is necessary to maintain the neutral iron center characteristic of the resting enzyme. Both types of dioxygenase bind $O_2$ after the organic substrate. However, in the case of the $Fe^{3+}$ dioxygenases, this is the result of creation of a site for $O_2$ in the protein through a conformational change caused by substrate binding. In contrast, in the $Fe^{2+}$ dioxygenases, the $O_2$-binding site is created on the iron as a result of anionic substrate binding, which decreases the redox potential of the iron so that a stronger bond with $O_2$ can be established.

Although they are different in the chemical steps that initiate ring cleavage, the two types of ring cleavage dioxygenases do share several similarities. These include (1) use of multiple iron ligand sites to promote the reaction, (2) use of dynamic expansion and contraction of the iron coordination number during the catalytic cycle, (3) use of vicinal sites in the iron coordination to align and activate substrates through chelation, (4) use of *trans* ligand influences to promote asymmetry or O–O bond cleavage, (5) use of an iron coordination site to bind one atom of oxygen to prevent its loss during the ring cleavage phase of the mechanism, and (6) development of a strategy to bind $O_2$ at the time and in the exact location that it is needed during the reaction. We believe that it is likely that variations of these shared traits will be commonly observed in the reaction mechanisms of other types of dioxygenases as well as other types of metalloenzymes.

*Acknowledgments.* Supported by grants from the National Institutes of Health [GM-24689 (JDL), GM-46436 (DHO)], and NIH predoctoral training grants [GM-07323 (AMO, MAM) and GM-08277 (RWF)].

# References

1. Lipscomb JD, Orville AM (1992) Mechanistic aspects of dihydroxybenzoate dioxygenases. In: Sigel H, Sigel A (eds) Metal ions in biological systems, vol 28. Dekker, New York, pp 243–298

2. Que L Jr, Ho RYN (1996) Dioxygen activation by enzymes with mononuclear non-heme iron active sites. Chem Rev 96:2607–2624
3. Howard JB, Rees DC (1991) Perspectives on non-heme iron protein chemistry. Adv Protein Chem 42:199–280
4. Que L Jr, Lipscomb JD, Münck E, et al. (1977) Protocatechuate 3,4-dioxygenase inhibition studies and mechanistic implications. Biochim Biophys Acta 485:60–74
5. Arciero DM, Lipscomb JD (1986) Binding of $^{17}O$-labeled substrate and inhibitors to protocatechuate 4,5-dioxygenase nitrosyl complex: evidence for direct substrate binding to the active site $Fe^{2+}$ of extradiol dioxygenases. J Biol Chem 261:2170–2178
6. Shu L, Chiou YM, Orville AM, et al. (1995) X-ray absorption spectroscopic studies of the Fe(II) active site of catechol-2,3-dioxygenase. Implications for the extradiol cleavage mechanism. Biochemistry 34:6649–6659
7. Ohlendorf DH, Orville AM, Lipscomb JD (1994) Structure of protocatechuate 3,4-dioxygenase from *Pseudomonas aeruginosa* at 2.15 Å resolution. J Mol Biol 244:586–608
8. Sugiyama K, Senda T, Narita H, et al. (1995) Three-dimensional structure of 2,3-dihydroxybiphenyl dioxygenase (BphC enzyme) from *Pseudomonas* sp. strain KKS102 having polychlorinated biphenyl (PCB)-degrading activity. Proc Jpn Acad 71B:32–35
9. Han S, Eltis LD, Timmis KN, et al. (1995) Crystal structure of the biphenyl-cleaving extradiol dioxygenase from a PCB-degrading pseudomonad. Science 270:976–980
10. Senda T, Sugiyama K, Narita H, et al. (1996) Three-dimensional structures of free form and two substrate complexes of an extradiol ring-cleavage type dioxygenase, the BphC enzyme from *Pseudomonas* sp. strain KKS102. J Mol Biol 255:735–752
11. Whittaker JW, Lipscomb JD (1984) Transition state analogs for protocatechuate 3,4-dioxygenase: spectroscopic and kinetic studies of the binding reactions of ketonized substrate analogs. J Biol Chem 259:4476–4486
12. Orville AM, Elango N, Lipscomb JD, et al. (1997) Structures of competitive inhibitor complexes of protocatechuate 3,4-dioxygenase: multiple exogenous ligand binding orientations within the active site. Biochemistry 36:10039–10051
13. Orville AM, Lipscomb JD, Ohlendorf DH (1997) Crystal structures of substrate and substrate analog complexes of protocatechuate 3,4-dioxygenase: endogenous $Fe^{3+}$ ligand displacement in response to substrate binding. Biochemistry 36:10052–10066
14. Elgren TE, Orville AM, Kelley KA, et al. (1997) Crystal structure and resonance Raman studies of protocatechuate 3,4-dioxygenase complexed with 3,4-dihydroxyphenylacetate. Biochemistry 36:(in press)
15. Frazee RW (1994) Cloning, sequencing, expression, and site-directed mutagenesis of *Pseudomonas putida* protocatechuate 3,4-dioxygenase genes. Ph.D. thesis. University of Minnesota, Minneapolis
16. Mabrouk PA, Orville AM, Lipscomb JD, et al. (1991) Variable temperature variable field magnetic circular dichroism studies of the Fe(II) active site in metapyrocatechase: implications for the molecular mechanism of extradiol dioxygenases. J Am Chem Soc 113:4053–4061
17. Miller MA, Lipscomb JD (1996) Homoprotocatechuate 2,3-dioxygenase from *Brevibacterium fuscum*: a dioxygenase with catalase activity. J Biol Chem 271:5524–5535

# Three-Dimensional Structure of an Extradiol-Type Catechol Ring Cleavage Dioxygenase BphC derived from *Pseudomonas* sp. strain KKS102: Structural Features Pertinent to Substrate Specificity and Reaction Mechanisms

Toshiya Senda, Keisuke Sugimoto, Tomoko Nishizaki, Mitsuyo Okano, Takahiro Yamada, Eiji Masai, Masao Fukuda, and Yukio Mitsui

*Summary.* The three-dimensional structure of the BphC enzyme, which is a member of the extradiol-type catechol ring cleavage dioxygenases, was determined using a crystallographic technique at 1.8 Å resolution. The crystal structure of the BphC enzyme complexed with its substrate, 2,3-dihydroxybiphenyl (2,3-DHBP), was also determined. It was revealed that (1) the active site is located inside the barrel-like structure of domain 2 of each subunit, and (2) the Fe ion in the active site coordinates His-145, His-209, and Glu-260. To gain further insight into the reaction mechanism of this enzyme, we have prepared more than 30 mutant proteins and carried out their biochemical and crystallographic analyses. These studies have revealed that (1) the shape of the substrate-binding pocket and its hydrophobic character are important for determining substrate specificity, (2) the shape and location of the putative "oxygen-binding cavity" are consistent with the reaction mechanism previously proposed based on spectroscopic studies on closely related enzymes, and (3) His-194 is indispensable for the enzymatic reaction, most probably playing the role of a proton acceptor whose existence has been suggested by previous spectroscopic studies.

*Key words.* Extradiol-type dioxygenase—Crystal structure—Enzyme mechanism—PCB degradation

---

Department of BioEngineering, Nagaoka University of Technology, 1603-1 Kamitomioka-cho, Nagaoka, Niigata 940-21, Japan

# Introduction

Extradiol-type catechol ring cleavage dioxygenases (hereafter called extradiol-type dioxygenase) catalyze the addition of two atomic oxygens to the catechol ring of a substrate, resulting in an extradiol-type cleavage of the catechol ring. These enzymes contain one nonheme iron (Fe(II)) in their active sites. To date, various types of extradiol-type dioxygenases have been cloned from microorganisms such as pseudomonads [1]. They are the key enzymes in the degradation pathways for aromatic compounds such as benzene, toluene, and polychlorinated biphenyls (PCBs). The details of the catalytic mechanism of the extradiol-type dioxygenases, however, have not been well understood because of a total lack of three-dimensional structural information for this type of enzyme. To gain insights into the catalytic mechanism of the extradiol-type dioxygenases, we have determined the crystal structures of one such dioxygenase, the BphC enzyme derived from *Pseudomonas* sp. strain KKS102 [2], in substrate-free form and in a complex with its substrate, 2,3-dihydroxybiphenyl (2,3-DHBP) [3,4].

The KKS102 BphC enzyme (hereafter the BphC enzyme) is composed of eight identical subunits, each of which is related by 422-point group symmetry. The subunit of the BphC enzyme consists of two domains, domain 1 and domain 2 (Fig. 1). Each domain consists of two repetitions of a βαββ motif. Thus one subunit of this enzyme is composed of four repetitions of the βαββ motifs [3,4]. The Fe ion, which is essential for the enzymatic reaction, is located inside the barrel-like structure of domain 2 (Fig. 1) [3,4]. Based on our crystal structure of the BphC enzyme, we have prepared various mutant BphC enzymes and started enzyme kinetic analyses of these mutant enzymes to elucidate the detailed catalytic mechanism of the BphC enzyme. In

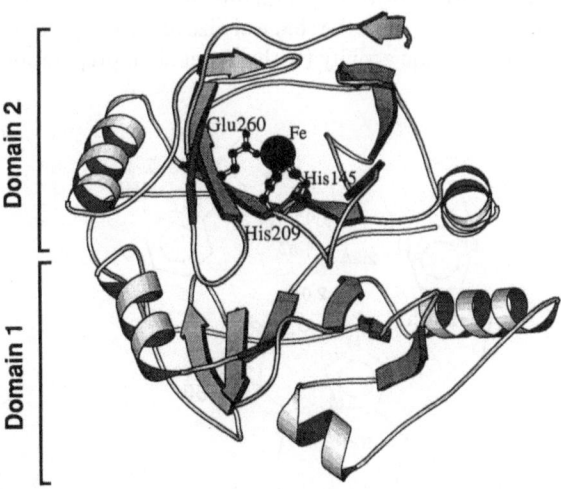

FIG. 1. Schematic ribbon drawing of the subunit of the BphC enzyme shows the side chains of the three Fe ligands *His-145*, *His-209*, and *Glu-260*

addition, the crystal structures of these mutant proteins have also been determined. Here, we briefly summarize the results of our enzyme kinetic and crystallographic studies on the KKS102 BphC enzyme and its mutant enzymes. Details will be published elsewhere.

## Results and Discussion

### Active Site, Coordination Sphere, and Substrate-Binding Pocket

The active site of the BphC enzyme is located deep inside the barrel-like structure of domain 2 (Fig. 1). The Fe ion in the active site coordinates His-145, His-209, and Glu-260, which are completely buried inside the molecule. In addition to these amino acid residues, two water molecules are bound to the Fe ion. The shape of the coordination sphere may be described as a square pyramid [4]. On binding the substrate, 2,3-DHBP, the two water molecules are replaced by the hydroxyl groups of the substrate. The resultant coordination polyhedra may be described as a distorted trigonal bipyramid (Fig. 2) [4].

The substrate binds to a large cavity inside domain 2, which is designated the substrate-binding pocket. In the substrate-free form of the BphC enzyme, a large cluster of electron density, which could not be interpreted clearly, was observed in the substrate-binding pocket [4]. On binding the substrate, the unidentifiable compound in this pocket was replaced by the substrate. The bound substrate is completely buried inside the molecule.

### Changing the Shape of the Substrate-Binding Pocket Drastically Affects Substrate Binding

In our enzyme kinetic analyses of various mutants of the BphC enzyme, the types of amino acid substitutions that change the shape or size of the substrate-binding pocket were found to reduce enzymatic activity (Nishizaki et al, in preparation). When His-

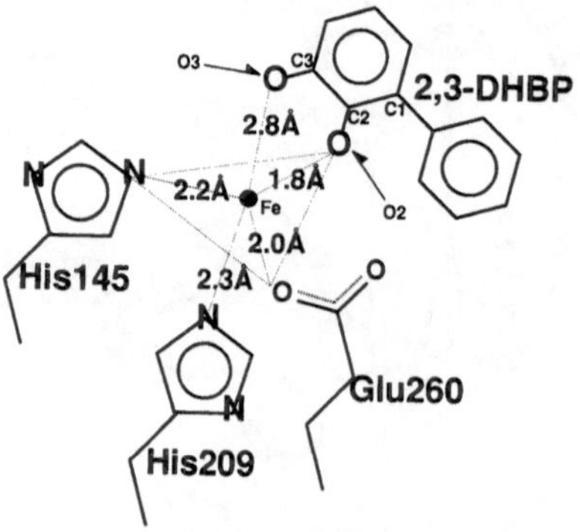

FIG. 2. Schematic drawing of the Fe coordination sphere found in the BphC enzyme complexed with the substrate 2,3-dihydroxybiphenyl (2,3-DHBP). The C-1, C-2, C-3, O-2, and O-3 atoms of the substrate are labeled

240, whose imidazole ring was found to make a stacking interaction with the catechol ring of the substrate (see Fig. 7 in [4]) was replaced by Ala, the $K_m$ value increased drastically (data not shown), indicating a very poor substrate-binding ability of the mutant. Crystal structure of the His240Ala mutant indicated that the size of the substrate-binding pocket was increased because Ala-240 lacked an imidazole ring. Crystals of the mutant (His240Ala) were then soaked in the standard buffer containing substrate 2,3-DHBP at a concentration that would be sufficient for its introduction to the wild enzyme and other mutant enzymes exhibiting similar $K_m$ values as the wild enzyme. The resultant electron density, however, showed no clear indication of the presence of the substrate.

## The Fe Ion Is Not Essential for Substrate Binding

As mentioned, His-145, His-209, and Glu-260 coordinate the Fe ion in the active site. Mutations on these amino acid residues cause a complete loss of enzymatic activity (Nishizaki et al, in preparation). Crystallographic analysis of the His145Ala mutant revealed that the Fe ion is lacking in the active site. However, after soaking this mutant crystal in the standard buffer containing the substrate 2,3-DHBP, a cluster of electron density corresponding to the substrate molecule very clearly appeared. These facts show that (1) the interaction between the hydroxyl groups of the substrate and the enzymatic Fe ion is not essential for substrate binding, and thus (2) the major factor inducing the substrate binding must be the broad-range hydrophobic interaction between the substrate and the substrate-binding pocket, both of which have hydrophobic character.

## Proper Disposition of the Hydrophobic Substrate-Binding Pocket Relative to the Fe Ion Induces the Asymmetric Deprotonation of the Catechol Ring Required for Catalysis

The grossly different distances for the coordinations O-2–Fe (1.8 Å) and O-3–Fe (2.8 Å) (see Fig. 2) strongly indicate that only the O-2 hydroxyl group of the catechol ring is deprotonated. Interestingly this feature of an asymmetric deprotonation is consistent with the previously proposed reaction mechanism [5].

Apparently the enzyme in question is so designed that (1) one of the hydroxyl groups of the catechol ring coordinates to the Fe ion as an equatorial ligand resulting in deprotonation, and (2) the other hydroxyl group coordinates as an axial ligand, which does not require deprotonation because it has a much large coordination distance. Clearly, these features are realized through precisely designed geometrical features of the active site: (a) the inner surface of the substrate-binding pocket having sufficient complementarity with the surface of the substrate, and (b) the proper position and orientation of such a pocket relative to the Fe ion so as to make the two hydroxyl groups properly coordinate to the Fe ion. Needless to say, both these geometrical features are realized through the proper design of the enzyme: proper amino acid sequence, proper conformation of each amino acid residue, and proper polypeptide chain folding as a whole in each subunit of this enzyme.

## Oxygen-Binding Site

For the enzymatic reaction to occur, the participation of a molecular oxygen, $O_2$, is required. Based on the crystal structure of the BphC enzyme complexed with the substrate 2,3-DHBP, we proposed a three-dimensional model for the ternary complex composed of the BphC enzyme, the substrate, and an $O_2$ molecule [4]. Considering the van der Waals radii of the atoms around the Fe ion and the reaction mechanism previously proposed on the basis of spectroscopic studies [5], the cavity around Ala-197 and Val-147 appears to be a highly likely site for the $O_2$ binding. The putative "oxygen-binding cavity" appears to be conveniently located for an $O_2$ molecule to attack the C-1 atom of the substrate [4]

## His-194 is Most Likely a Catalytic Base

According to the previously proposed reaction mechanism of the extradiol-type dioxygenase, the activated oxygen attacks the C-1 (see Fig. 2) atom of the substrate. Concomitantly a proton is transferred from the hydroxyl group attached to the C-3 atom of the substrate to an as yet unidentified base located adjacent to the hydroxyl group [5]. A series of mutants on amino acid residues located around the catechol ring moiety of a bound substrate has been prepared. Enzyme kinetic analyses of these mutants indicated that His-194 (see Fig. 3) is a critical residue for catalysis (Nishizaki et al., in preparation). Most notably, the His 194Phe mutant, harboring a substitution that is sterically almost equivalent, showed no enzymatic activity. The crystal struc-

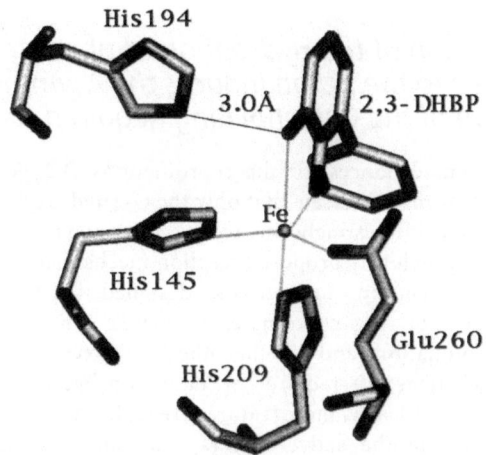

Fig. 3. Amino acid residues apparently directly affecting the electronic nature of the catalysis are shown together with the Fe ion and the bound substrate, 2,3-DHBP. The contact between an imido nitrogen of His-194 and one of the hydroxyl groups of the substrate may be sufficiently close for proton transfer to occur although the connection as shown here is slightly deviated from that of an ideal hydrogen bond

ture of this mutant enzyme showed that little conformational change occurred around the active site on the His194Phe mutation, indicating that the electronically more active nature of the imidazole ring as compared with the phenol ring is essential for catalysis. Thus His-194 in the active site is most likely the "unidentified base" in the previously proposed reaction mechanism [5].

# References

1. Fukuda M (1993) Diversity of chloroaromatic oxygenases. Curr Opin Biotechnol 4:339–343
2. Kimbara K, Hashimoto T, Fukuda M, et al. (1989) Cloning and sequencing of two tandem genes involved in degradation of 2,3-dihydroxybiphenyl to benzoic acid in the polychlorinated biphenyl-degrading soil bacterium *Pseudomonas* sp. strain KKS102. J Bacteriol 171:2740–2747
3. Sugiyama K, Senda T, Narita H, et al. (1995) Three-dimensional structure of 2,3-dihydroxybiphenyl dioxygenase (BphC enzyme) from *Pseudomonas* sp. strain KKS102 having polychlorinated biphenyl (PCB) degrading activity. Proc Jpn Acad 71B:32–35
4. Senda T, Sugiyama K, Narita H, et al. (1996) Three-dimensional structures of free form and two substrate complexes of an extradiol ring-cleavage type dioxygenase, the BphC enzyme from *Pseudomonas* sp. strain KKS102. J Mol Biol 255:735–752
5. Shu L, Chiou YM, Orville AM, et al. (1995) X-ray absorption spectroscopic studies of the Fe(II) active site of catechol 2,3-dioxygenase. Implications for the extradiol cleavage mechanism. Biochemistry 34:6649–6659

# Probing the Reaction Mechanism of Protocatechuate 3,4-Dioxygenase with X-Ray Crystallography

ALLEN M. ORVILLE, JOHN D. LIPSCOMB, and DOUGLAS H. OHLENDORF

*Summary.* Protocatechuate 3,4-dioxygenase (3,4-PCD, E) catalyzes the oxidative ring cleavage of 3,4-dihydroxybenzoate (PCA) to produce $\beta$-carboxy-*cis,cis*-muconate. Previous results suggest that several short-lived intermediates and a conformational change are encountered during formation of the E·substrate complex followed by ternary E·PCA·$O_2$ intermediates and product formation. In this review, the crystal structures of 3,4-PCD as isolated and complexed with 3-I,4-OH-benzoate (IHB), 3-F,4-OH-benzoate (FHB), PCA, or analogs of the ketonized PCA isomers [2-OH-isonicotinate *N*-oxide (INO) and 6-OH-nicotinate *N*-oxide (NNO)] plus $CN^-$ are compared as structural probes of the reaction pathway. The crystal structures of E·IHB and E·FHB appear to mimic the intermediate stages of PCA binding and suggest that the $Fe^{3+}$-OH facilitates C4-OH proton abstraction. In contrast to the monodentate iron complexes of E·IHB and E·FHB, PCA asymmetrically chelates the iron in the anaerobic E·PCA complex. Concurrent dissociation of the axial tyrosinate (Tyr-447) generates an active site base ($pK_a \sim 10$) capable of abstracting the PCA C3-OH proton to yield the dianionic $Fe^{3+}$·PCA complex. After its release from the iron, Tyr-447 forms the top of a small cavity adjacent to the C3–C4 bond of PCA. The ternary E·INO·CN complex mimics an E·PCA·$O_2$ complex, and the structure shows that $CN^-$ binds to the $Fe^{3+}$ in the adjacent cavity. This suggests that 3,4-PCD sequesters PCA and $O_2$ as well as ensuing intermediates during catalysis.

*Key words.* Dioxygenase—Crystal structure—Reaction mechanism—Nonheme iron—Ligand exchange

## Introduction

A common feature of all oxygenases is their absolute requirement for a cofactor or metal center to overcome the reactivity barrier of $O_2$ in a triplet ground state with organic compounds in a singlet ground state. Protocatechuate 3,4-dioxygenase (3,4-

Department of Biochemistry, Medical School, 4-225 Millard Hall, University of Minnesota, Minneapolis, MN 55455-0347, USA

282

PCD, E), an archetypal ring-cleaving dioxygenase, utilizes a nonheme ferric ion to catalyze the ring cleavage of 3,4-dihydroxybenzoate (PCA) and fission of $O_2$ to form β-carboxy-*cis,cis*-muconate (for recent reviews, see [1,2]).

The proposed reaction mechanism of 3,4-PCD based upon previous spectroscopic and kinetic results suggests the dynamic exchange of exogenous and endogenous iron ligands during catalysis [1,2]. To date, however, these deductions have not been corroborated by crystallographic investigations. In this review, the crystal structures of the 3,4-PCD as isolated [3] and complexed with 3-I,4-OH-benzoate (IHB), 3-F,4-OH-benzoate (FHB), PCA, or analogs of ketonized PCA isomers [2-OH-isonicotinate *N*-oxide (INO), or 6-OH-nicotinate *N*-oxide, NNO)] plus CN⁻ are compared to investigate the proposed reaction mechanism. The structures suggest a logical progression of PCA reorientations in the active site that produces a form of the substrate which is susceptible to electrophilic attack by molecular oxygen.

## IHB and FHB as Models for PCA-Binding Intermediates

As isolated, the iron in 3,4-PCD is coordinated by four endogenous and one exogenous ligands in trigonal bipyramidal geometry (Fig. 1a) (see also the chapter by J.D. Lipscomb et al, this volume) [3]. The 3,4-PCD competitive inhibitors, IHB ($K_i > 2\,mM$) and FHB ($K_i \approx 0.9\,\mu M$), are structural analogs of PCA but cannot chelate the $Fe^{3+}$. FHB is a better geometric analog of PCA than IHB because of the similarity of bond distances [C3-F $\cong$ C3-OH (1.35 Å) < C3-I (2.05 Å)] and steric bulk of phenolate and F groups. IHB binds to the iron by displacing the $Fe^{3+}$-OH (Wat-827) (see the chapter by J.D. Lipscomb et al, this volume) [4]. The Gln-477$^{Ne2}$ to IHB$^{C3-I}$ hydrogen bond helps stablilize the complex, and ~40% of the IHB surface area remains solvent exposed [4]. In contrast, FHB binds deeper in the active site crevase (Fig. 1a) (see also the chapter by J.D. Lipscomb et al, this volume) than IHB with only ~20% of the FHB surface area solvent exposed [4]. This results in closer van der Waals interaction of Gly-14, Pro-15, and Ile-491 with the aromatic ring of FHB than IHB. In addition, strong hydrogen bonds are formed between FHB and Tyr-324$^{OH}$, Arg-457$^{NH1}$, and Gln-477$^{Ne2}$, which places the C4-OH of FHB in an $Fe^{3+}$ coordination site opposite Tyr-408$^{OH}$.

## Structure of the Anaerobic E·PCA Complex

Comparison of the E·FHB and E·PCA structures (Fig. 1a) suggests that the orientation of the two aromatic ligands are related to each other by an 18° rotation about C-1 [5]. This rotation allows PCA to form a chelated $Fe^{3+}$ complex concomitant with the dissociation of the axial tyrosinate (Tyr-447). Electrostatic neutrality of the $Fe^{3+}$ center is maintained by Tyr-408 and the dianionic PCA molecule. The $Fe^{3+}$ coordination site opposite His-460$^{Ne2}$ appears to be unoccupied. The chelated PCA·$Fe^{3+}$ complex is rather asymmetric, as indicated by the 2.0 Å PCA$^{OH4}$–$Fe^{3+}$ and ~2.5 Å PCA$^{OH3}$–$Fe^{3+}$ bond distances, and the $Fe^{3+}$ is approximately 0.8 Å out of the PCA plane (Fig. 1b) [5]. The alternative conformation of Tyr-447 is stabilized by hydrogen bonds to Tyr-16$^{OH}$ and Asp-413$^{Oe1}$, where it forms the top of a small, solvent-excluded cavity adjacent to the C3–C4 bond of PCA. This cavity, presumed to be the $O_2$-binding site, is further investigated next.

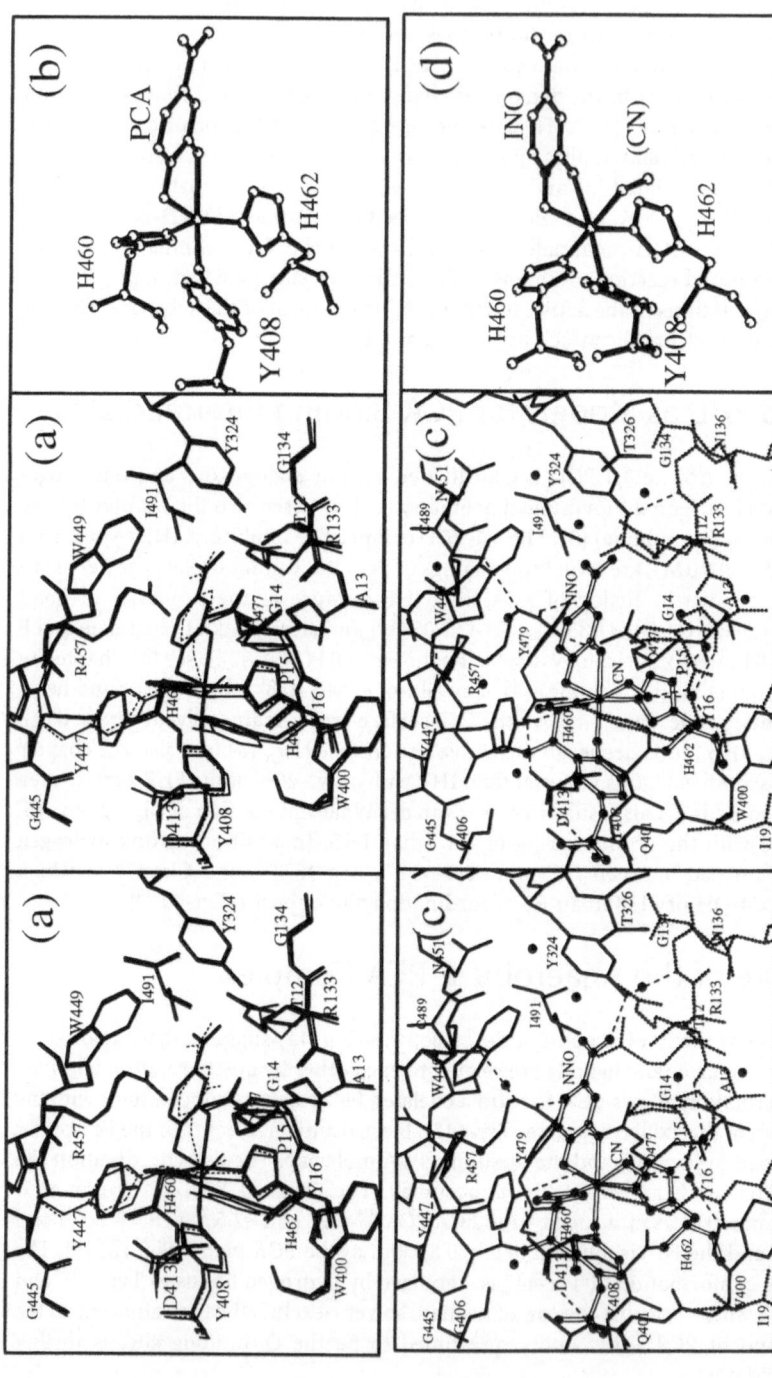

FIG. 1a–d.   Crystal structures of protocatechuate 3,4-dioxygenase (3,4-PCD) as isolated and in several complexes. Divergent stereo views of (a) 3,4-PCD as isolated (*thin solid line*) [3], the E·FHB complex (*thin dashed lines*) [4], and the anaerobic E·PCA complex (*thick solid lines*) [5]. (b) The iron coordination sphere in the 3,4-PCD·PCA (3,4-dihydroxybenzoate) complex [5]. (c) Divergent stereo view of the E·NNO·CN complex [5]. *Dashed lines* indicate hydrogen bonds. (d) The iron coordination sphere in the E·INO·CN complex [5]

# Structures of the E·INO·CN and E·NNO·CN Complexes

Both INO ($K_{D\ overall} = 0.06\,\mu M$) and NNO ($K_{D\ overall} = 0.20\,\mu M$) form high-affinity dead-end complexes with 3,4-PCD [3] that rapidly bleach the burgundy-red color of 3,4-PCD crystals to a pale yellow [5]. Subsequent addition of NaCN to either binary complex causes the yellow crystals to turn purple. The orientation of INO, NNO, and Tyr-447 in the binary (E·INO and E·NNO) or ternary (E·INO·CN, and E·NNO·CN) complexes (Fig. 1c,d) is similar to that of PCA and Tyr-447 in the E·PCA complex (Fig. 1a,b). However, in the E·INO and E·NNO complexes a solvent molecule binds in the $Fe^{3+}$ coordination site opposite His460$^{N\epsilon2}$ [5]. This solvent is displaced by CN$^-$ (an $O_2$ analog) in the ternary complexes [5] in accord with previous spectroscopic results [1].

These structures suggest that ketonization at the PCA$^{C3\text{-}OH}$ (or PCA$^{C4\text{-}OH}$) group may enhance anionic ligand binding in the adjacent $Fe^{3+}$ coordination site to maintain charge neutrality of the metal center. Because nonlinear $Fe^{3+}$–C $\equiv$ N bond angles are observed in the E·NNO·CN (Fig. 1c) and E·INO·CN (Fig. 1d) complexes, steric restraints within the putative $O_2$ cavity appear to force the $Fe^{3+}$–C $\equiv$ N bond into a bent geometry. While this likely disfavors CN$^-$ binding, the active site microenvironment is very well suited to ternary E·PCA·$O_2$ complex formation and sequestration of the ensuing reactive intermediates during catalysis (see the chapter by J.D. Lipscomb et al, this volume).

# Mechanistic Insights from the Crystal Structures

An enhanced reaction mechanism [5] that is consistent with previous spectroscopic and kinetic results [1,2] as well as the new crystal structures is presented in Fig. 2. The binding orientations observed for IHB and FHB are proposed to mimic two intermediate orientations that can be assumed by PCA as it binds to the enzyme. Because the C3-iodo group of IHB prevents the aromatic ring from binding deeper in the active site cavity, the E·IHB complex mimics a weakly bound, early E·PCA intermediate. This intermediate could rapidly shift to a second more tightly bound intermediate, analogous to the E·FHB complex, but before the $Fe^{3+}$–Tyr447$^{OH}$ bond dissociation. The anaerobic E·PCA complex is obtained by rotation of the FHB-like intermediate about PCA$^{C1}$ and concomitant dissociation of Tyr-447. The dissociation of Tyr-447 is proposed to generate the second active site base with a $pK_a$(~10) matching that of the PCA$^{OH3}$ group. This yields the chelated dianionic E·substrate complex that is susceptible to electrophilic attack by $O_2$ at PCA$^{C4}$ as indicated by previous spectroscopic results [1,2].

All the kinetic studies to date indicate that $O_2$ interacts with the enzyme after the substrate is bound [1]. Our structures suggest that 3,4-PCD couples the binding and activation of substrate to the creation of a sequestered $O_2$-binding site immediately adjacent to the C3–C4 bond in PCA where cleavage will occur. CN$^-$ binding in this site in the ternary E·INO·CN and E·NNO·CN complexes strongly supports the binding of $O_2$ in this cavity. The oxygen will take on peroxide character follwing $O_2$ attack on

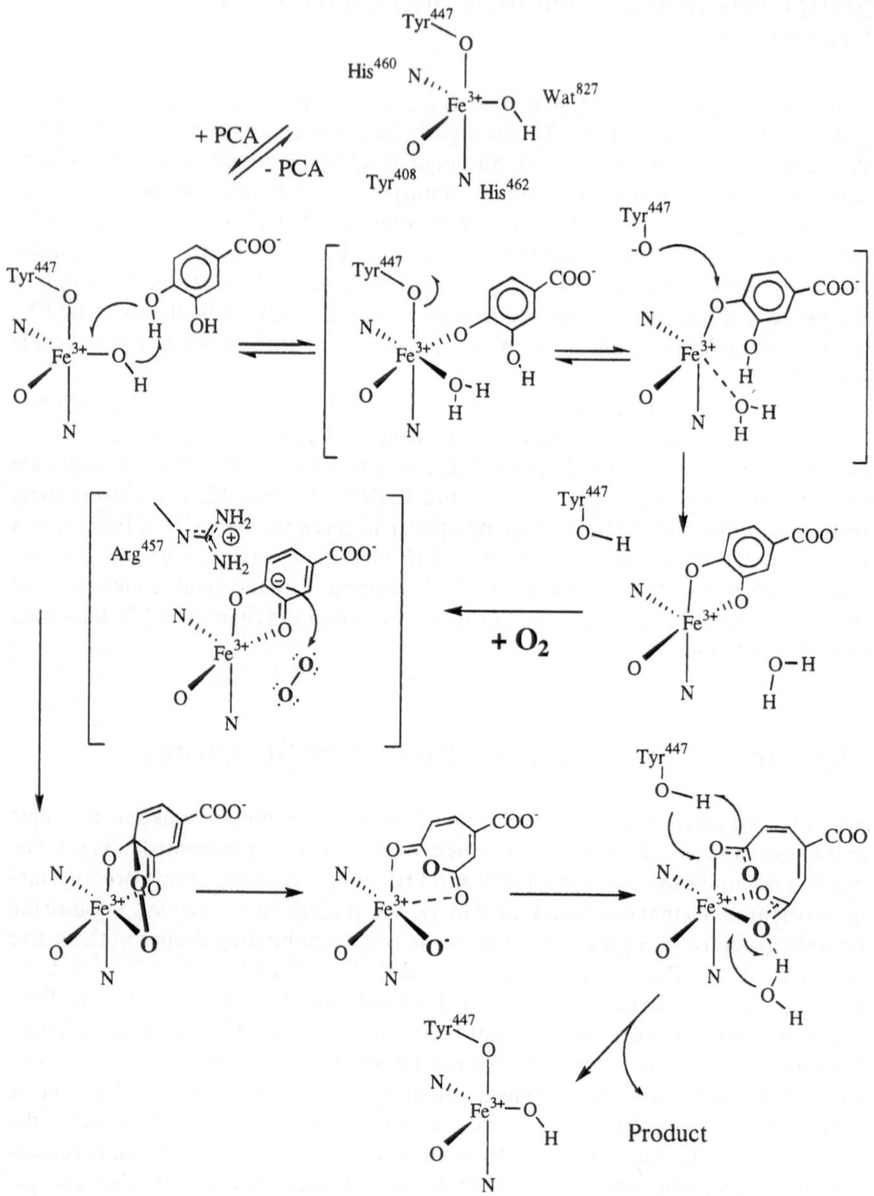

FIG. 2. The proposed reaction mechanism for 3,4-PCD based on the crystal structures (see text for narrative) [5]

the $PCA^{C4}$ and could readily coordinate to the iron in the open binding site. The proposed Fe–O–O–PCA intermediate would necessarily have a nonlinear Fe–O–O bond angle, would maintain a charge neutral iron center, and would be consistent with the size and shape of the cavity.

The endogenous iron ligands are proposed to induce asymmetry in the E·PCA complex and peroxy intermediate that consequently enhances concomitant O–O bond cleavage and insertion into the PCA C3–C4 bond. For example, the $PCA^{OH3}$ and $PCA^{OH4}$ groups are opposite Tyr-408$^{OH}$ and His-462$^{N\epsilon2}$, respectively. *Trans* effects are proposed to increase the $PCA^{OH4}$–$Fe^{3+}$ bond distance as Tyr-408 donates electron density to the iron while the $PCA^{OH3}$–$Fe^{3+}$ bond distance is reduced as His-462 accepts electron density. When $O_2$ binds in the cavity, these distortions enhance both the tendency of $PCA^{OH3}$ to ketonize and the carbanion character at $PCA^{C4}$. The double bond character of the scissile C3–C4 bond in PCA is thereby reduced and electrophilic attack of $O_2$ at $PCA^{C4}$ is favored. Another *trans* effect may induce O–O bond cleavage of the peroxy intermediate because the distal end $O_2$ coordinates to the iron opposite His-460$^{N\epsilon2}$. The anhydride intermediate is therefore obtained by concerted *trans* effects that appear to facilitate both PCA activation and $O_2$ bond cleavage and insertion into the PCA C3–C4 bond.

The second oxygen atom from $O_2$ is proposed to remain in the $O_2$ cavity and coordinate to the iron until it hydrolyzes the anhydride intermediate. Our structures show that the $O_2$ cavity is sequestered so that hydrolysis of the anhydride intermediate occurs before exchange with bulk solvent in accord with the observed reaction stoichiometry. The illustrated structure for the product complex is geometrically reasonable; however, charge repulsion of the product carboxylate groups will likely make it unstable. It seems reasonable to presume that as β-carboxy-*cis,cis*-muconate dissociates from the iron, it deprotonates Tyr-447 and a solvent molecule. The tyrosinate and hydroxide would then coordinate the iron to regenerate the resting state of enzyme. The partially neutralized product molecule would then dissociate from the positive exterior surface of the active site.

These results illustrate several important principles of metalloenzyme catalysis, including (i) the utilization of several iron coordination sites to bind exogenous ligands, (ii) the maintenance of a charge-neutral metal center via ligand exchange, (iii) endogenous iron ligand displacement on exogenous ligand binding, (iv) iron chelation by vicinal hydroxyl groups of aromatic substrates to establish the correct binding orientation for reaction with $O_2$, (v) a conformational change engendered by substrate binding that creates an $O_2$-binding cavity, and (vi) the presence of *trans* ligand effects that perturb the electronic nature of exogenous ligands to promote reaction. One or more of these principles likely apply to a wide variety of metalloenzymes and thus affect an array of biological transformations.

*Acknowledgments.* Supported by grants from the National Institute of Health [GM-46436 (DHO), GM-24689 (JDL)], a NIH predoctoral training grant [GM-07323 (AMO)], and a Doctoral Dissertation Fellowship from the Graduate School, University of Minnesota (AMO).

# References

1. Lipscomb JD, Orville AM (1992) Mechanistic aspects of dihydroxybenzoate dioxygenases. In: Sigel H, Sigel A (eds) Metal ions in biological systems, vol 28. Dekker, New York, pp 243-298
2. Que L Jr, Ho RYN (1996) Dioxygen activation by enzymes with mononuclear non-heme iron active sites. Chem Rev 96:2607-2624
3. Ohlendorf DH, Orville AM, Lipscomb JD (1994) Structure of protocatechuate 3,4-dioxygenase from *Pseudomonas aeruginosa* at 2.15 Å resolution. J Mol Biol 244:586-608
4. Orville AM, Elango E, Lipscomb JD, Ohlendorf DH (1997) Structures of competitive inhibitor complexes of protocatechuate 3,4-dioxygenase: multiple exogenous ligand binding orientations within the active site. Biochemistry 36:10039-10051
5. Orville AM, Lipscomb JD, Ohlendorf DH (1997) Crystal structures of substrate and substrate analog complexes of protocatechuate 3,4-dixygenase: endogenous $Fe^{3+}$ ligand displacement in response to substrate binding. Biochemistry 36:10052-10066

# Nitric Oxide Synthases:
# Structure, Function, and Control

Dawn Harris[1], Susan M.E. Smith[2], Christa Brown[1],
and John C. Salerno[1]

*Summary.* Nitric oxide synthases are large modular enzymes that produce NO and citrulline from arginine at the expense of NADPH and $O_2$. We have modeled the reductase region, which corresponds roughly to the C-terminal half of the molecule, using the available crystal structures of flavodoxin (FMN-binding domain) and ferredoxin NADPH reductase (FAD- and NADPH-binding domains). This has enabled us to identify important sequence regions that interact with cofactors and to show that the N-terminal boundary of the reductase domain extends to within a few residues of the CaM-binding site. A large (40–50 residues) insertion in the FMN-binding domain of cNOS, located ~80 residues downstream from the CaM site, is the major sequence difference between iNOS and cNOS. This insertion is directly adjacent to the CaM-binding site on the three-dimensional structure. Several lines of evidence suggest that it functions as a control element.

*Key words.* Nitric oxide—Nitric oxide synthase—Calmodulin—Control—Reductase

## Introduction

Nitric oxide has been shown to be important in the control of numerous physiological processes. Endothelial nitric oxide synthase (eNOS) regulates vascular tone and smooth muscle tension [1–3]. NO produced by neuronal nitric oxide synthase (nNOS) functions as a diffusible neurotransmitter [4,5]. Both constitutive isoforms (nNOS and eNOS) are regulated by intracellular calcium/calmodulin ($Ca^{2+}$/CaM) and generate NO as an intercellular messenger [6–8]. Inducible nitric oxide synthase (iNOS) is widely expressed in response to cytokines; CaM binds to iNOS even at low levels of intracellular $Ca^{2+}$, leading to the production of cytotoxic levels of NO [9–11].

All NOS isoforms generate NO from L-arginine in a reaction utilizing 2 mol $O_2$ and 1.5 mol of NADPH per NO produced, forming citrulline [12–14]. Nitric oxide synthase monomers are very large (~130–160 kDa per monomer) multidomain modular en-

[1] Biology Department, Rensselaer Polytechnic Institute, 110 8th Street, SCIW 14, Troy, NY 12180, USA
[2] Aeneas Biotechnology, Troy, NY 12180, USA

zymes [15–19]. The active site contains heme with an axial thiolate ligand, spectro-scopically resembling cytochrome P-450 [20–22]. Tetrahydrobiopterin is required for NO synthesis and promotes dimerization, but its role in catalysis is not understood [23–27]. Electron transfer from NADPH is catalyzed by bound flavin mononucleotide (FMN) and flavin adenine dinucleotide (FAD) [18,24,28,29].

## Domain Organization

Bredt and co-workers [18,30] showed that the C-terminal regions of all three NOS isoforms contained sequences homologous to P-450 reductase and proposed loca-tions for the FMN-, FAD-, and NADPH-binding sites. Sheta et al. [31] cleaved nNOS into two fragments during limited trypsinolysis; cleavage occurred within the CaM-binding sequence. The C-terminal (reductase) fragment contained the regions identi-fied by Bredt and Snyder, while the N-terminal (oxygenase) fragment contained the arginine-binding site, tetrahydrobiopterin, and heme. The Masters group pioneered the expression of constructs corresponding to these regions in nNOS, iNOS, and eNOS, as well as nNOS holenzyme, in E. coli [16,17,19] (see also 32–34]).

The oxygenase domains of eNOS and iNOS extend from the N terminal to the CaM-binding site. In nNOS, an additional N-terminal sequence of about 200 amino acid residues is homologous to sequences in the N-terminal region of DLG1 (in *Drosophila melanogaster*) and PSD9 (in rat) [35]. Brenman and co-workers [36] recently reported that this region functions as a protein recognition site.

Cys-415 was identified as the axial heme ligand of bovine nNOS by site-directed mutagenesis [17], and corresponding cysteinyl residues are axial ligands in the other isoforms [37], as previously proposed [20]. The NOS oxygenase domains show no extensive similarity to P-450. Some similarity between pterin-binding proteins and NOS has been reported [19]. Constructs based on regions of apparent similarity can be expressed, but neither these constructs nor constructs including the remainder of the oxygenase domain can be shown to bind tetrahydrobiopterin. The independently expressed construct reported by Nishimura et al. (residues 558–721 in nNOS) does bind *N*-nitroarginine, however, indicating that it contains a significant part of the substrate binding site.

NOS is active only as a dimer [38]. Independently expressed heme protein, but not flavoprotein, is dimeric, indicating that the intermonomer interactions are primarily between the oxygenase regions [17]. In previously suggested "head-to-tail" models, the reductase unit of one monomer reduces the oxygenase unit of the other [39].

## Structure of the Reductase Domains

The FMN-binding modules of all NOS isoforms are homologous to the corresponding region in NADPH P450 redoctase (CPNR) and in turn to flavodoxins, a family of small FMN-binding proteins. Figure 1 contains an alignment of a selected set of NOS and flavodoxin sequences. The iNOS sequence closely corresponds to that of *Desulfovibrio* flavodoxin. A major insertion is present in eNOS and nNOS relative to flavodoxins and CPNR, corresponding to the addition of an approximately 45-residue subdomain. Alignments including numerous additional sequences of similar FMN-binding do-

```
                                 INITIAL FMN BINDING REGION
NOSE_BOVIN   GTLMAKRV.. ...KATILYA SETGRAQSYA QQLGRLFRKA FDPRVLCMD.   556
  NOSB_RAT   GQAMAKRV.. ...KATILYA TETGKSQAYA KTLCEIFKHA FDAKAMSME.   789
NOSM_MOUSE   RKVMASRV.. ...RATVLFA TETGKSEALA RDLATLFSYA FNTKVVCMD.   567
FLAV_ECOLI   .......... ..AITGIFFG SDTGNTENIA KMIQKQL.GK .D.VADVHDI    35
FLAV_DESVH   .......... .MPKALIVYG STTGNTEYTA ETIARELADA .GYEVDSRDA    38

                                 SECOND FMN SITE
NOSE_BOVIN   .EYDVVSL.. ...EHETLVL VVTSTFGNGD PPENGESFAA AL.MEMSGPY   599
  NOSB_RAT   .EYDIVHL.. ...EHEALVL VVTSTFGNGD PPENGEKFGC AL.MEMRHP.   831
NOSM_MOUSE   .QYKASTL.. ...EEEQLLL VVTSTFGNGD CPSNGQTLKK SLFML.....   606
FLAV_ECOLI   AKSSKEDL.. ...EAYDILL LGIPTWYYGE ....AQCDWD DF.FP.....    70
FLAV_DESVH   ASVEAGGLF. ...EGFDLVL LGCSTWGDDS IE..LQDDFI PL.FD.....    76

                     LOCATION OF LOOP INSERT
NOSE_BOVIN   NSSPRPEQHK SYKIRFNSVS CSDPLVSSWR RKRKESSNTD SAGAGTLRFL   649
  NOSB_RAT   NS..VQEERK SYKVRFNSVS SYSDSRKSSG DGPDLRDNFE STGPLANVRF   879
NOSM_MOUSE   .......... .......... .......... .......... .RELNHTFRY   615
FLAV_ECOLI   .......... .......... .......... .........TL EEIDFNGKLV    82
FLAV_DESVH   .......... .......... .......... .........SL EETGAQGRKV    88

                 THIRD FMN SITE
NOSE_BOVIN   CVFGLGSRA. Y.PHFCAFA. AVDTRLEELG GERL   680
  NOSB_RAT   SVFGLGSRA. Y.PHFCAFGH AVDTLLEELG GERI   911
NOSM_MOUSE   AVFGLGSSM. Y.PQFCAFAH DIDQKLSHLG ASQL   647
FLAV_ECOLI   ALFGCGDQED YAEYFCDALG TIRDIIEPRG ATIV   116
FLAV_DESVH   ACFGCGDSS. Y.EYFCGAVD AIEEKLKNLG AEIV   120
```

FIG. 1. Sequence alignment of selected flavodoxins and nitric oxide synthases (NOS) showing regions of homology and major insertion in neuronal NOS (nNOS) and endothelial NOS (eNOS)

mains suggest that only cNOS sequences have such an insertion, which is absent in iNOS. This insertion represents the most important difference between the cNOS and iNOS sequences, and correlates with $Ca^{2+}$/CaM control.

All enzymes that include this module contain three regions involved in FMN binding [40]. Bredt et al. [18] identified a region contributing several residues interacting with FMN, including a shielding aromatic. An additional region near the N terminal of the module forms a number of hydrogen bonds with the terminal phosphate group of FMN, and a third region contributes a second aromatic ring in contact with FMN as well as hydrogen bond partners.

Homology-based models have been constructed using the Homology and Discover modules of the Biosym suite of molecular modeling software. As shown in Fig. 2a, the iNOS FMN domain model structure closely corresponds to *Desulfovibrio vulgaris* flavodoxin. The domain consists of a five-stranded parallel β-sheet, with the FMN-binding site along one edge. The aromatic side chains of two residues (F587 and Y625 in iNOS) are in contact with the isoalloxazine ring system.

Figure 2b shows the structure of the corresponding module of eNOS. Our models for nNOS (not shown) are almost indistinguishable from this. The subdomain corresponding to the insertion is visible on the upper edge of the sheet, opposite the FMN-binding site. The remainder of the module is similar to the corresponding region of iNOS. We lack a solved homologous structure for the insert subdomain; the structure shown is merely intended to convey its position and size.

The FAD- and NADPH-binding domains of NOS can be modeled on the basis of homology with ferredoxin NADPH reductase (FNR). FNR consists of a five-stranded

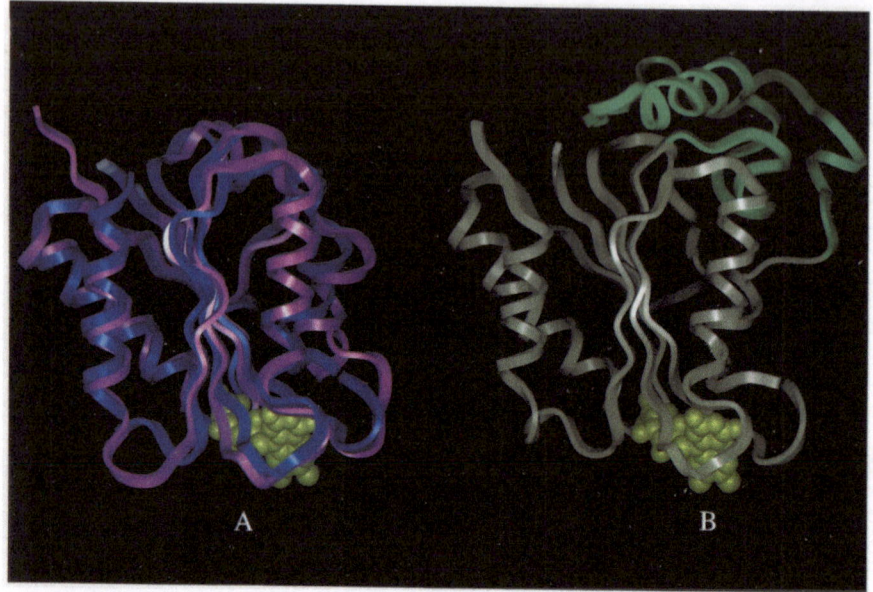

FIG. 2a,b. Structure of NOS flavin mononucleotide-(FMN-) binding domains. Backbone traces of polypeptide are shown as *ribbons*; FMN is shown in *solid symbols* at the bottom of each panel. **a** Overlay of *Desulfovibrio* flavodoxin crystal structure with iNOS FMN-binding domain molecular model. **b** Model of the FMN-binding domain of eNOS with insert shown in green

parallel β-sheet domain and an eight-stranded β-barrel. The structure of CPNR has been solved but is not yet available [41]; this information will extend our understanding of the NOS structure.

Figure 3 shows a homology model of the FAD- and NADPH-binding domains of nNOS. The isoalloxazine ring of FAD lies against the outside of the β-barrel and interacts with a conserved aromatic residue inside the barrel. The adenine binding site is located on a loop connecting two barrel strands. A stacking interaction for the isoalloxazine ring is provided by the last residue in the sheet domain, which is otherwise devoted to NADPH binding.

The NADPH-binding module is a five-stranded antiparallel β-sheet with helical returns. NADPH is a dinucleotide; an adenine-binding site, including a stacking aromatic residue, is located on an $\alpha > \beta$ loop. The nicotinamide moiety was not bound in any solved FNR structure, but can be modeled as stacked with the FAD isoalloxazine ring system.

The requirements of efficient electron transfer imply an orientation of the FMN-binding domain in which the FMN faces the other cofactors. A subdomain is present as an insertion in the FAD-binding domain in all proteins that contain both FNR and flavodoxin modules. It corresponds to an extended loop (with extensive secondary structure) between two strands of the β-barrel and mediates interactions between the FAD- and FMN-binding domains.

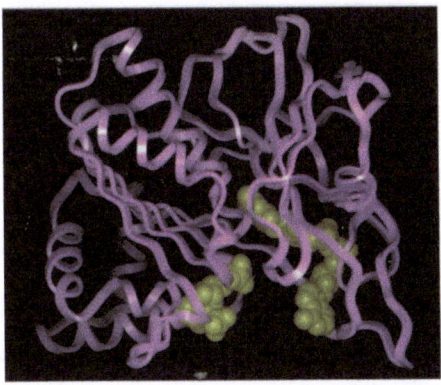

FIG. 3. Molecular model of nNOS flavin adenosine dinucleotide (FAD-) and NADPH-binding domains. FAD is shown as a *solid-symbol* model on the *lower right*; part of the NADPH-binding site (*bottom center*) is occupied by the inhibitor 2′-AMP

## Control of Electron Flow

The primary site of oxygen chemistry in NOS is the heme; the arginine- and tetrahydrobiopterin-binding sites are also located in the N-terminal half of the molecule. The reductase domain function in catalysis is limited to the delivery of reducing equivalents to the catalytic site. It is the delivery of electrons to the catalytic site, however, that is regulated by CaM binding. CaM binding has been reported to affect electron transfer within the reductase domains as well as FMN–heme electron transfer [42].

The organization of the reductase domains suggests that the CaM-binding site is remote from the flavin-binding sites. The C-terminal extension regions of all three NOS isoforms are adjacent to the flavin sites and hence distant from the CaM-binding site. None of these regions, nor the subdomain of the FAD-binding module, is likely to be directly involved in control.

The major insertion in the FMN-binding domain is directly adjacent to the CaM-binding site in our model structure, and the presence of this element correlates perfectly with CaM/$Ca^{2+}$ control. Three groups have noted the difference between cNOS and iNOS in this region. Silvagno et al. [43] identified it as the locus of a tissue-specific alternatively spliced form of nNOS. Lowe et al. [44], in analyzing the tryptic cleavage fragments reported by Sheta et al. [31], pointed out that one early cut site was located in this region and speculated on a possible role in intersubunit interactions. Salerno and co-workers [45], starting from structural considerations, proposed that the insertion was a control element and provided evidence for an autoinhibitory role. In particular, synthetic peptides homologous to the insertions (particularly to the eNOS insertion) are inhibitors of cNOS activity and of CaM binding.

Antibodies raised to these polypeptides bind strongly to both eNOS and nNOS; homology between the two cNOS isoforms leads to cross-reactivity. Figure 4 illustrates the effects of antibody binding on eNOS optical spectra. In Fig. 4a, the effect of antibody binding on the ferriheme optical spectra is shown. The Soret difference spectrum has a peak at 413 nm, indicating a shift toward the low spin state. Figure 4b

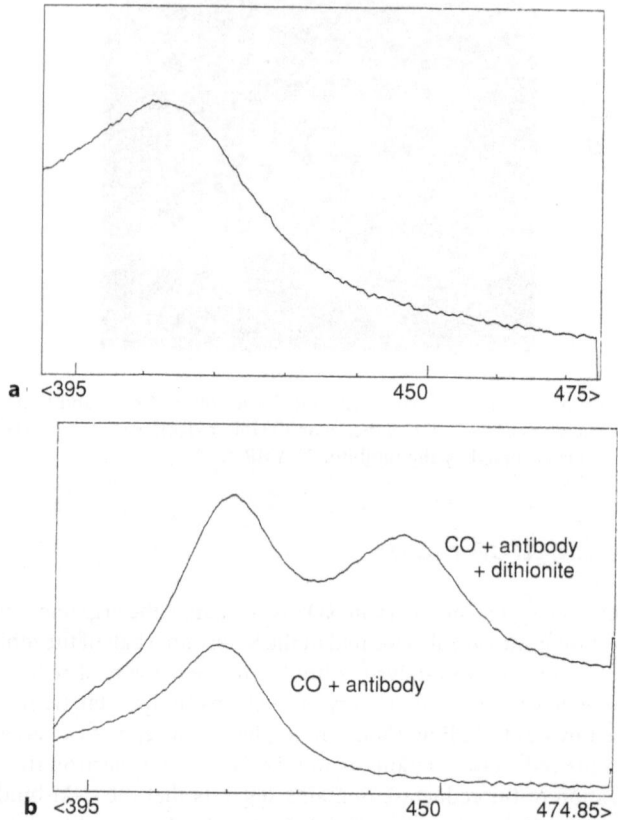

FIG. 4a,b. Effects of antibodies raised to polypeptides based on the FMN domain insert on the heme optical spectrum in the Soret region (395–475 nm shown) of nNOS. **a** Difference spectrum (nNOS as isolated) minus (nNOS plus antibody) showing the effects of antibody binding on ferriheme. **b** *Lower trace* shows difference spectrum of (nNOS plus NADPH plus saturating CO) minus (nNOS plus NADPH plus saturating CO plus antibody); *upper trace* shows difference spectrum of (nNOS plus NADPH plus saturating CO) minus (nNOS plus NADPH plus saturating CO plus antibody plus dithionite)

illustrates the effects of antibody binding on the heme spectrum in the presence of NADPH and CO. Little electron transfer occurs in the absence of either CaM or antibody, but addition of CaM and $Ca^{2+}$ leads to the formation of a 450-nm band (not shown), indicative of flavin to heme electron flow. Addition of antibody in the absence of CaM (lower trace) leads to the formation of a 420-nm species. This suggests that antibody binding has activated electron transfer to the heme, but that it also places strain on the heme site conformation, which results in the formation of the "P-420"-type ferrous CO complex. Addition of dithionite results in the formation of the 450 nm species but does not significantly affect the concentration of the 420-nm CO complex.

The available evidence suggests that the insertion in the FMN-binding domains of cNOSs is placed to interact with bound CaM, and that this insertion functions as a control element. It has already been possible to design novel NOS inhibitors based on this premise, and it seems likely that ligands that activate electron transfer, and hence catalysis, can also be constructed.

# References

1. Ignarro LJ, Buga GM, Wood KS, et al. (1987) Endothelium-derived relaxing factor produced and released from artery and veins is nitric oxide. Proc Natl Acad Sci USA 84:9265–9269
2. Palmer RMJ, Ferrige DS, Moncada S (1987) Nitric oxide release accounts for the biological activity of endothelial relaxing factor, Nature 327:524–526
3. Furchgott RF (1988) Studies on relaxation of rabbit aorta by sodium nitrite: the basis for the proposal that the acid activatable factor from bovine retractor penis is inorganic nitrite and the endothelium-derived relaxing factor is nitric oxide. In: Vanhoutte PM (ed) Vasodilation: vascular smooth muscle, peptides, autonomic nerves and endothelium. Raven, New York, pp 401–404
4. Gally JA, Montague PR, Reeke GN, et al. (1990) The NO hypothesis: possible effects of a short lived rapidly diffusible signal in the development and function of the nervous system. Proc Natl Acad Sci USA 87:3547–3551
5. Garthwaite J, Charles SL, Chess-Williams R (1988) Endothelium derived relaxing factor release on NMDA receptors suggests a role as intercellular messenger in the brain. Nature 336:385–388
6. Abu-Soud HM, Stuehr DJ (1993) Nitric oxide synthases reveal a role for calmodulin in controlling electron transfer. Proc Natl Acad Sci USA 90:10769–10772
7. Bredt DS, Snyder SH (1990) Isolation of nitric oxide synthetase, a calmodulin-requiring enzyme. Proc Natl Acad Sci 87:682–685
8. Bredt DS, Snyder S (1994) Nitric oxide: a physiological messenger molecule. Annu Rev Biochem 63:175–195
9. Knowles RG, Palacios M, Palmer RMJ, et al. (1990) Kinetic characteristics of nitric oxide synthase from rat brain. Biochem J 269:207–210
10. McCall TB, Boughton-Smith NK, Palmer RMJ, et al. (1989) Synthesis of nitric oxide from L-arginine by neutrophils. Release and interaction with superoxide anion. Biochem J 262:293–296
11. Curran RD, Billiar TR, Stuehr DJ, et al. (1989) Hepatocytes produce nitrogen oxides from L-arginine in response to inflammatory products of Kupffer cells. J Exp Med 170:1769–1774
12. Hibbs JB Jr, Taintor RR, Vavrin Z (1987) Macrophage cytotoxicity: role for L-arginine deiminase and imino nitrogen oxidation to nitrite. Science 235:473–476
13. Marletta MA (1993) Nitric oxide synthase structure and mechansim. J Biol Chem 268:12231–12234
14. Stuehr DJ, Kwon NS, Nathan CF, et al. (1991) N-Hydroxy-L-arginine is an intermediate in the biosynthesis of nitric oxide from arginine. J Biol Chem 266:6259–6263
15. Masters BS (1994) Nitric oxide synthases: why so complex? Annu Rev Nutr 14:131–145
16. Masters BS, McMillan K, Sheta EA, et al. (1996) Neuronal nitric oxide synthase, a modular enzyme formed by convergent veolution: stucture studies of a cysteine thiolate-liganded heme protein that hydroxylates L-arginine to produce NO as a cellular signal. FASEB J 10:552–558
17. McMillan K, Masters BS (1995) Prokaryotic expression of the heme- and flavin-binding domains of rat neuronal nitric oxide synthase as distinct polypeptides: inden-

tification of the heme-binding proximal thiolate ligand as cysteine-415. Biochemistry 34:3686–3693

18. Bredt DS, Hwang PM, Glatt CE, et al. (1991) Cloned and expressed nitric oxide synthase structurally resembles cytochrome P-450 reductase. Nature 351:714–718

19. Nishimura JS, Martasek P, McMillan K, et al. (1995) Modular structure of neuronal nitric oxide synthase: localization of the arginine binding site and modulation by pterin. Biochem Biophys Res Commun 210:288–294

20. McMillan K, Bredt DS, Hirsch DJ, et al. (1992) Cloned, expressed rat cerebellar nitric oxide synthase contains stoichiometric amounts of heme, which binds carbon monoxide. Proc Natl Acad Sci USA 89:11141–11145

21. Stuehr DJ, Ikeda-Saito M (1992) Spectral characterization of brain and macrophage nitric oxide synthases. J Biol Chem 267:20547–20550

22. White K, Marletta MA (1992) Nitric oxide synthase is a cytochrome P-450 type heme protein. Biochemistry 31:6627–6631

23. Tayeh MA, Marletta MA (1989) Macrophage oxidation of L-arginine to nitric oxide, nitrate and nitrite. Tetrathydrobiopterin is required as a cofactor, J Biol Chem 264:19654–19658

24. Mayer B, John M, Heinzel B, et al. (1991) Brain nitric oxide synthase is a biopterin and flavin containing multifunctional oxidoreductase. FEBS Lett 288:187–191

25. Kwon NS, Nathan CF, Stuehr DJ (1989) Reduced biopterin as a cofactor in the generation of nitrogen oxides by murine macrophages. J Biol Chem 264:20496–20501

26. Giovanelli J, Campos KL, Kaufman S (1991) Tetrahydrobiopterin, a cofactor for rat cerebellar nitric oxide synthase, does not function as a reactant in the oxygenation of arginine. J Biol Chem 88:7091–7095

27. Tzeng E, Billiar TR, Robbins PD, et al. (1995) Expression of human inducible nitric oxide synthase in a tetrahydrobiopterin (H4b)-Deficient cell line—H4b promotes assembly of enzyme subunits into an active. Proc Natl Acad Sci USA 92:11771–11775

28. Stuehr DJ, Kwon NS, Nathan CF (1990) FAD and GSH participate in macrophage synthesis of nitric oxide. Biochem Biophys Res Commun 168:558–565

29. Hevel JM, White KA, Marletta MA (1991) Purification of the inducible murine macrophage nitric oxide synthase. Identification as a flavoprotein. J Biol Chem 266:22789–22791

30. Bredt DS, Ferris CD, Snyder SH (1992) Cloned and expressed nitric oxide synthase structurally resembles cytochrome P-450 reductase. J Biol Chem 267:10976–10981

31. Sheta EA, McMillan K, Masters BS (1994) Evidence for a bidomain structure of constitutive cerebellar nitric oxide synthase. J Biol Chem 269:15147–15153

32. Matsuoka A, Stuehr DJ, Olson JS, et al. (1994) L-Arginine and calmodulin regulation of the heme iron reactivity in neuronal nitric oxide synthase. J Biol Chem 269:20335

33. Ghosh DK, Stuehr DJ (1995) Macrophage NO synthase—characterization of isolated oxygenase and reductase domains reveals a head-to-head subunit interaction. Biochemistry 34:801

34. Ghosh DK, Abusoud HM, Stuehr DJ (1996) Domains of macrophage NO synthase have divergent roles in forming and stabilizing the active dimeric enzyme. Biochemistry 35:1444–1449

35. Cho K-O, Hunt CA, Kennedy MA (1992) The postsynaptic density fraction contains a homolog of Drosophila discs-large tumor suppressor protein. Neuron 9:929–942

36. Brenman JE, Chao DS, Xia H, et al. (1995) Nitric oxide synthase complesed with dystropyhin and absent from skeletal muscle sarcolemma in Duchenne muscular dystrophy. Cell 82:743–752

37. Chen P-F, Tsai A-L, Wu KK (1994) Cysteine 184 of nitric oxide synthase is involved in heme coordination and catalytic activity. J Biol Chem 269:25062–25066

38. Abu-Soud HM, Loftus M, Stuehr DJ (1995) Subunit dissociation and unfolding of macrophage NO synthase—relationship between enzyme structure, prosthetic group binding, and catalytic function. Biochemistry 34:11167–11175

39. Schmidt HHH, Pollock JS, Nakane M, et al. (1991) Purification of a soluble isoform of guanylyl cyclase-activating-factor synthase. Proc Natl Acad Sci USA 88:365–369

40. Porter TD (1991) An unusual yet strongly conserved flavoprotein reductase in bacteria and mammals. Trends Biochem Sci 16:154–158

41. Kim J-J, Wang M, Roberts DL, et al. (1996) In: Stevenson KJ (ed) Flavins and flavoproteins—1996. University of Calgary Press, Calgary (in press)

42. Abu-Soud HM, Yoho LL, Stuehr DJ (1994) Calmodulin controls neuronal nitric-oxide synthase by a dual mechanism—activation of intra- and interdomain electron transfer. J Biol Chem 269:32050–32054

43. Silvagno F, Xia H, Bredt DS (1996) Neuronal nitric-oxide synthase-mu, an alternatively spliced isoform expressed in differentiated skeletal muscle. J Biol Chem 271:11204–11208

44. Lowe PN, Smith D, Stammers DK, et al. (1996) Identification of the domains of neuronal nitric oxide synthase by limited proteolysis. Biochem J 314:55–62

45. Salerno JC, Harris DE, Irizzary K, et al. (1996) The inhibitory polypeptide of constitutive NO (cNOS) is the missing control element of the inducible isoform (iNOS). In: Ignarro L (ed) Second international conference on the biochemistry and molecular biology of nitric oxide. University of Calrfornia, Los Angeles, P-A20

# Electron Paramagnetic Resonance Studies on Substrate Binding to the NO Complex of Neuronal Nitric Oxide Synthase

Catharina Taiko Migita[1,2], John C. Salerno[3],
Bettie Sue Siler Masters[4], and Masao Ikeda-Saito[1]

*Summary.* The nitric oxide synthases (NOS), a group of flavo-heme enzymes, biosynthesize a physiologically versatile diatomic messenger, nitric oxide (NO), from the substrate L-arginine (L-Arg). We have examined the active site of the neuronal isoform by electron paramagnetic resonance (EPR) spectroscopy of the ferrous NO complex. The substrate-free NO-bound enzyme exhibits a cytochrome P-450-type EPR spectrum of a typical hexa-coordinate NO heme complex with a nonnitrogenous proximal axial heme ligand. The substrate-free NO complex is rather unstable and spontaneously converts to a cytochrome P-420-type penta-coordinate denatured form. Binding of L-Arg enhances the stability of the hexa-coordinate NO form. The EPR spectrum of the NO adduct of the enzyme-arginine complex has an increased $g$-anisotropy and well-resolved hyperfine coupling as the result of the $^{14}N$ of nitric oxide. Significant changes in the NO EPR spectrum were observed on $N^{\omega}$-hydroxy-L-Arg (NHA) binding. These changes are strong indications of a direct interaction between the L-Arg and the bound NO in the distal heme pocket of the enzyme. Electrostatic interactions between the guanidino group of arginine and the bound NO or the steric effect of the substrates appears to affect the Fe-NO geometry, resulting in the observed spectral changes.

*Key words.* Nitric oxide—Nitric oxide synthase—EPR—L-arginine—$N^{\omega}$-hydroxy-L-arginine

## Introduction

Nitric oxide (NO) is a versatile physiological messenger molecule serving as a vasodilator, neurotransmitter, and cytotoxic agent [1–4]. NO is produced by isoenzymes

[1] Department of Physiology and Biophysics, Case Western Reserve University School of Medicine, 2119 Abington Road, Cleveland, OH 44106-4970, USA
[2] School of Allied Health Sciences, Yamaguchi University, 1144 Kogushi, Ube, Yamaguchi 755, Japan
[3] Department of Biology and Center for Biochemistry and Biophysics, Rensselaer Polytechnic Institute, 15th Street, Troy, NY 12180, USA
[4] Department of Biochemistry, University of Texas Health Science Center at San Antonio, 7703 Floyd Curl Drive, San Antonio, TX 78284-7760, USA

termed nitric oxide synthases (NOS), which catalyze the NADPH-dependent conversion of L-arginine (L-Arg) to NO and L-citrulline with $N^{\omega}$-hydroxyl-L-Arg (NHA) being formed as an intermediate [1–4]. Neuronal NOS is a homodimer having heme, flavin adenine dinucleotide (FAD), flavin mononucleotide (FMN), and $H_4$-biopterin as its essential cofactors [1–4]. The axial ligand of the NOS heme iron is a cysteinyl thiolate as in cytochrome P-450 enzymes [5–10], indicating that the cytochrome P-450-type oxygen activation and catalysis [11] are likely to take place at the heme site in NOS [5–10]. The substrate L-Arg, binding to NOS, alters the electronic structure and reactivities of the heme iron [12–14]. The substrate-binding site has been considered to be located close to the sixth coordination position of the heme iron in the distal heme pocket [12,13] as observed in cytochrome P-450 enzymes [15,16].

To delineate substrate–enzyme interactions in NOS, we have measured the electron paramagnetic resonance (EPR) spectra of the ferrous NO complex of neuronal NOS in the presence and absence of the substrates L-Arg and NHA. We have found that the bound substrate directly affects the heme-bound NO and alters the Fe-NO geometry, as manifested by the substrate-dependent changes in the ferrous NO EPR spectrum. In this chapter, we describe the plausible mechanistic interpretation for the perturbations on the Fe-NO geomery induced by the substrate binding.

## Experimental Procedures

Rat neuronal NOS was purified from stably transfected kidney 293 cells as described previously [7]. The NO complex was generated by anaerobic introduction of NO gas into the dithionite-reduced NOS in 0.05 M Tris buffer, pH 7.5, containing 10% glycerol. Formation of the penta-coordinate NO species, the P-420 type NO complex, was achieved by aerobic mixing of NO gas (MG Industries, Malvern, PA, USA) or prolonged incubation with NO gas at room temperature [17,18]. EPR spectra were recorded by a Bruker ESP-300 (Bruker Instruments, Billerica, MA, USA) operating at 9.45 GHz with an Oxford liquid helium flow cryostat (Oxford Instruments North America, Bedford, MA, USA).

## Results and Discussion

Figure 1 illustrates the EPR spectra of the ferrous NO complex for the substrate-free (a), L-Arg-bound (b), and NHA-bound (c) neuronal NOS. The reaction of NO with the substrate-free ferrrous enzyme results in a partial conversion (20%–30%) to the denatured penta-coordinated, P-420-type NO species. The substrate-free NO NOS EPR spectrum (Fig. 1a) is displayed after correction by the spectral subtraction. The substrate-free NOS NO complex exhibits a spectrum that is similar to those of the NO complexes of ferrous cytochrome P-450 enzymes reported by O'Keeffe et al. [17] and Tsubaki et al. [18]. The lack of superhyperfine couplings associated with the $g_2$ component is consistent with the axial cysteine ligation to the heme iron *trans* to the bound NO.

The spectrum of the NO complex in the presence of L-Arg (Fig. 1b) is clearly different from that in its absence (trace a). The spectrum of the L-Arg-bound NO NOS

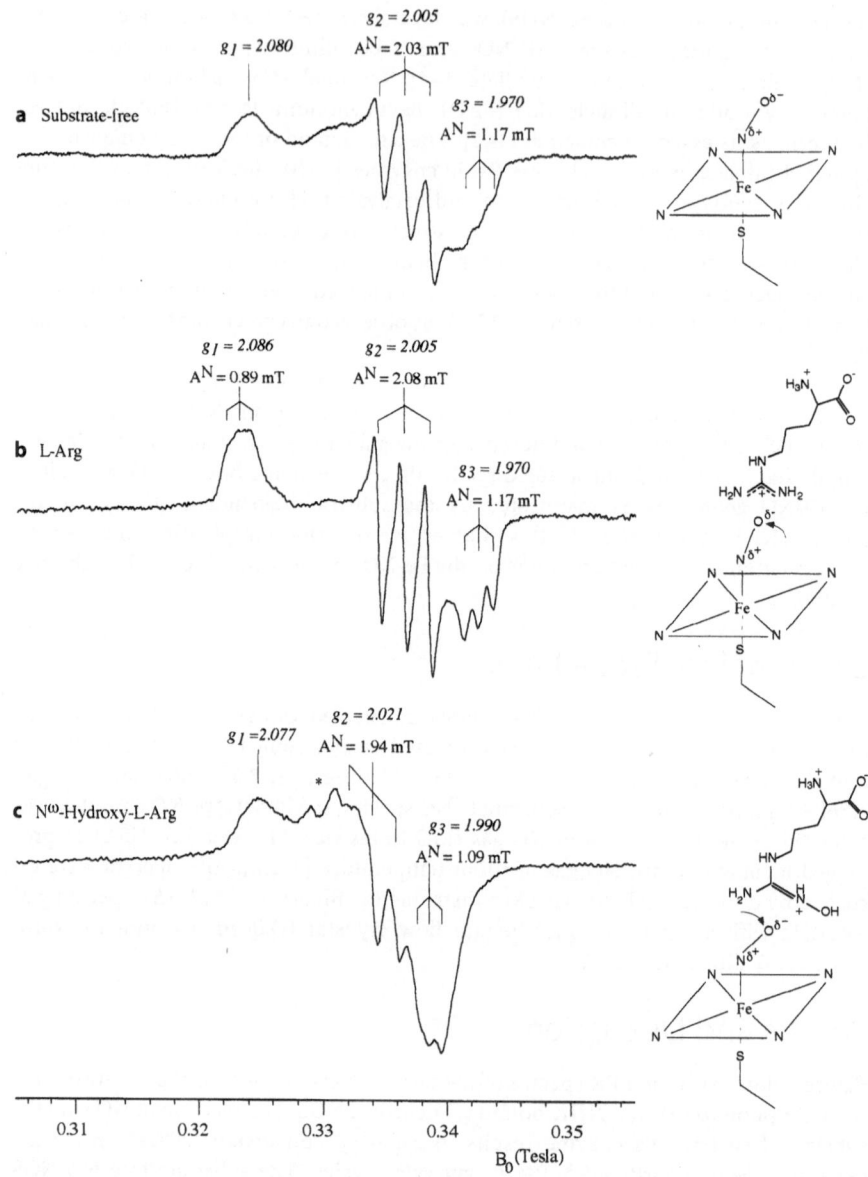

FIG. 1a–c. Electron paramagnetic resonance (EPR) spectra of the ferrous NO complex of substrate-free nitric oxide synthase (NOS) (**a**), that of L-arginine- (L-Arg-) bound NOS (**b**), and $N^{\omega}$-hydroxy-L-Arg-bound NOS (**c**). The proposed Fe–N–O structures are shown to the right of each spectrum. The spectrum for the substrate-free complex (**a**) was displayed after correction for the unavoidable formation of a penta-coordinate denatured species whose spectrum is very similar to spectra of the NO complexes of cytochromes P-420$_{LM}$, P-420$_{CAM}$, and P-420$_{SCC}$, with a sharp triplet splitting at $g_z$ (1.66 mT) [17,18]. Spectra were recorded at 30 K with a field modulation of 0.2 mT at 100 kHz

is more rhombic, and the hyperfine couplings at $g_1$ and $g_3$ components are better resolved: the electronic structure of the Fe-NO system in the NO-bound ferrous NO is affected by L-Arg binding. The L-Arg binding-induced change in the EPR spectrum is an indication that the bound L-Arg directly affects the heme-bound NO and alters the electronic structure of the Fe-NO system. This supports the proposal of the close proximity of the L-Arg and the heme ligand-binding sites deduced from ligand-binding studies by Matsuoka et al. [12], and is also consistent with the L-Arg binding-induced shift in the Fe-NO stretching mode of the NO complex of ferrous NOS [19]. We have noticed that the NOS NO complex formed in the presence of L-Arg is free from the penta-coordinate species as judged from its EPR spectrum. Similar observations have been reported for the cytochrome P-450 enzymes [17,18]. Substrate-enhanced stability of hexa-coordinate NO complexes appears to be a common characteristic of cytochrome P-450-type hemoproteins.

More drastic changes are seen in the EPR spectra of the ferrous NO complexes in the presence of NHA (Fig. 1c). The spectrum is less anisotropic than those of the substrate-free and the L-Arg-bound complexes, with a distinct change in the $g_3$ value from 1.97 to 1.99. Complicated $g_2$ features and an appearance of the pronounced absorption around $g = 2.03$ (signals marked with an asterisk) are indications of the presence of multiple species. This makes assignment of the $g_2$-values and its associated hyperfine coupling difficult.

When NO binds to hemoproteins, spin delocalization from the $2p\pi^*$, $2p\pi$, or $2p\sigma$ orbitals of NO to the iron $dz^2$ orbital occurs, resulting in the admixture of the iron $d$-orbitals with the unpaired electron orbital of the heme–nitrosyl complex. This results in the much larger $g$-anisotropy of nitrosyl heme complexes than that of free radical NO with simultaneous reduction in the contribution of the NO orbitals. This notion is consistent with the inverse correlation between the $g$-anisotropy and the spin density on the nitrogen atom of NO in the cytochrome P-450$_{SCC}$ NO complexes [18]. The changes in the $g$-anisotropy observed on L-Arg and NHA binding to the NOS NO complex could be explained as a result of alterations in the spin delocalization from the NO orbitals to the ferrous iron orbitals, which can be modulated by changes in the Fe-NO geometry.

In ferrous NO complexes of hemoproteins, bound NO assumes a bent, end-on Fe–N–O geometry where the $2p\pi^*$ orbital of NO partly overlaps with the iron $dz^2$ orbital. The characteristic bent Fe-NO geometry is highly likely in the NOS NO complex. Our results show an increase in the $g$-anisotropy of the EPR spectrum of the NOS NO complex on L-Arg binding, indicating that this binding enhances the spin delocalization. Such an enhancement in the spin delocalization is attainable either by increasing or decreasing the Fe–N–O bond angle. An increase in the Fe–N–O bond angle enhances the $2p\sigma$ to $dz^2$ spin delocalization, while a decrease in the bond angle enhances the $2p\pi^*$ to $dz^2$ spin delocalization. However, in the latter case, the $d\pi$ to $2p\pi^*$ back donation is also enhanced, so that the spin density of the NO orbitals increases. Therefore, the L-Arg binding-induced increase in the $g$-anisotropy is likely the result of an increase in the Fe–N–O bond angle. With respect to the interaction between NO and bound L-Arg, an actual attractive force works between the partly anionic oxygen atom of NO and the positive charge on the protonated guanidino group of the bound L-Arg. Here, we propose that the guanidino carbon is placed where it makes the oxygen atom of the bound NO slightly closer to the heme normal than in the substrate-free enzyme, as schematically illustrated in Fig. 1. Because this Fe-NO geometry is favorable for the $2p\sigma$ to $dz^2$ spin delocalization, as discussed earlier, the spin density on the bound NO decreases and results in an increase in the g-anisotropy. .

The hyperfine splitting of the NO EPR spectrum is notably sharpened on binding of L-Arg. In the substrate-free enzyme, a significant range of orientations could be available to the Fe-NO unit. The existence of multiple orientations introduces a distribution in g-anisotropy and hyperfine coupling, leading to line broadening. Binding of L-Arg in the distal pocket of the NOS NO complex will restrict the range of orientations available to NO because of the attractive electrostatic interaction between the oxygen atom of NO and the protonated guanidino group of the bound L-Arg. Steric interactions with bound L-Arg would also help in restricting the geometry of the bound NO. With the restricted range of orientations (possibly with a single orientation) of the Fe-NO unit, the line boadening is reduced, and a well-resolved hyperfine splitting is observed in the L-Arg-bound enzyme.

The effects of the NHA binding on the EPR spectrum of the NOS NO complex can be interpreted in terms of altered Fe-NO geometry. Because NHA binds at the same site as the L-Arg, the position of the guanidino group in L-Arg and NHA relative to the heme are expected to be similar. On the basis of these considerations, we propose the following explanation for the large difference in g-anisotropy observed between the system of Arg-bound NO NOS and of substituted Arg-bound NO NOS: in NHA-bound NOS, NO might bind to ferrous ion so as to decline to the substituted amino group in the guanidino group, because the electron-donating hydroxyl group stabilizes this limiting structure. The latter structure has the localized positive charge not on the nonsubstituted amino nitrogen but on the substituted amino nitrogen, among the alternative resonance structures. As a result, the NO adduct is compelled to have a more bent structure in these systems than in substrate-free NOS, while the NO adduct is less bent in L-Arg-bound NOS. It is also possible that the iron-bound NO experiences steric pressure exerted by the hydroxyl group of the substituted L-Arg, forcing the Fe-NO unit to adopt a further bent Fe–N–O geometry. The more bent structure enhances the $d\pi \rightarrow 2p\pi^*$ back donation as well as $2p\pi^* \rightarrow dz^2$ delocalization, and the total spin on the bound NO increases and causes a drastic reduction in g-anisotropy. The more bent geometry of the NO bound to the NHA NOS is expected to affect the separation among the $d\varepsilon$ orbitals of the heme iron to some extent, but this may not cause a serious change in g-anisotropy because the $d\varepsilon$ orbitals are originally fully occupied in low-spin ferrous heme proteins.

From these results, we infer that the bound oxygen molecule interacts with the substrates L-Arg and NHA in the ferrous heme-$O_2$ adduct in a manner similar to that described for the NO complex in this chapter. We think that such an oxygen interaction is important in that the position of these substrates in the binding site and the orientation of the bound $O_2$ specify the hydroxylation events in the catalytic cycle.

*Acknowledgment.* Supported by NIH (M.I.-S. and B.S.S.M.), the American Heart Association (M.I.-S.), Robert A. Welch Foundation (B.S.S.M.), and the Ministry of Education, Science, Sports and Culture of Japan (C.T.M.).

# References

1. Masters BSS (1994) Nitric oxide synthases: why so complex? Annu Rev Nutr 14:131–145
2. Bredt DS, Snyder SH (1994) Nitric oxide: a physiologic messenger molecule. Annu Rev Biochem 63:175–195
3. Griffith OW, Stuehr DJ (1995) Nitric oxide synthases: properties and catalytic mechanism. Annu Rev Physiol 57:707–736
4. Nathan C, Xie Q (1994) Nitric oxide synthases: roles, tolls, and controls. Cell 78:915–918
5. White KA, Marletta MA (1992) Nitric oxide synthase is a cytochrome P-450 type hemoprotein. Biochemistry 31:6627–6631
6. Stuehr DJ, Ikeda-Saito M (1992) Spectral characterization of brain and macrophage nitric oxide synthases. Cytochrome P-450-like hemeproteins that contain a flavin semiquinone radical. J Biol Chem 267:20547–20550
7. McMillan K, Bredt DS, Hirsch DJ, et al. (1992) Cloned, expressed rat cerebellar nitric oxide synthase contains stoichiometric amounts of heme, which binds carbon monoxide. Proc Natl Acad Sci USA 89:11141–11145
8. Wang J, Stuehr DJ, Ikeda-Saito M, et al. (1993) Heme coordination and structure of the catalytic site in nitric oxide synthase. J Biol Chem 268:22255–22258
9. Sono M, Stuehr DJ, Ikeda-Saito M, et al. (1995) Identification of nitric oxide synthase as a thiolate-ligated heme protein using magnetic circular dichroism spectroscopy—comparison with cytochrome P-450-CAM and chloroperoxidase. J Biol Chem 270:19943–19948
10. McMillan K, Masters BSS (1995) Prokaryotic expression of the heme- and flavin-binding domains of rat neuronal nitric oxide synthase as distinct polypeptides: identification of the heme-binding proximal thiolate ligand as cysteine-415. Biochemistry 34:3686–3693
11. Dawson JH (1988) Probing structure-function relations in heme-containing oxygenases and peroxidases. Science 240:433–439
12. Matsuoka A, Stuehr DJ, Olson JS, et al. (1994) L-Arginine and calmodulin regulation of the heme iron reactivity in neuronal nitric oxide synthase. J Biol Chem 269:20335–20339
13. Salerno JC, Frey C, McMillan K, et al. (1995) Characterization by electron paramagnetic resonance of the interactions of L-arginine and L-thiocitrulline with the heme cofactor region of nitric oxide synthase. J Biol Chem 270:27423–27428
14. McMillan K, Masters BSS (1993) Optical difference spectrophotometry as a probe of rat brain nitric oxide synthase heme-substrate interaction. Biochemistry 32:9875–9880
15. Raag R, Poulos TL (1991) Crystal structures of cytochrome P-450-CAM complexed with camphane, thiocamphor, and adamantane: factors controlling P-450 substrate hydroxylation. Biochemistry 30:2674–2684
16. Sligar SG, Murray RI (1987) Cytochrome P-450 $_{cam}$ and other bacterial P-450 enzymes. In: Ortiz de Montellano PR (ed) Cytochrome P-450. Plenum, New York, pp 429–503
17. O'Keeffe DH, Ebel RE, Peterson JA (1978) Studies of the oxygen binding site of cytochrome P-450. Nitric oxide as a spin-label probe. J Biol Chem 253:3509–3516
18. Tsubaki M, Hiwatashi A, Ichikawa Y, et al. (1987) Electron paramagnetic resonance study of ferrous cytochrome P-450$_{scc}$-nitric oxide complexes: effects of cholesterol and its analogues. Biochemistry 26:4527–4535
19. Wang J, Rousseau DL, Abu-Soud HM, et al. (1994) Heme coordination of NO in NO synthase. Proc Natl Acad Sci USA 91:10512–10516

# Heme Oxygenase: A Central Enzyme of Oxygen-Dependent Heme Catabolism and Carbon Monoxide Synthesis

MASAO IKEDA-SAITO[1], HIROSHI FUJII[2], KATHRYN MANSFIELD MATERA[1], SATOSHI TAKAHASHI[3], CATHARINA TAIKO MIGITA[4], DENIS L. ROUSSEAU[5], and TADASHI YOSHIDA[6]

*Summary.* Heme oxygenase (HO) catalyzes the regiospecific degradation of heme to biliverdin-α by using $O_2$ and electrons donated by NADPH cytochrome P-450 reductase. The enzyme binds one equivalent of heme to form the heme–enzyme complex, and electron donation initiates the three stepwise oxygenase reactions through the two novel heme derivatives, α-hydroxyheme and verdoheme, during which CO and iron-biliverdin-α are produced; heme participates both as a prosthetic group and as a substrate. Electronic states, coordination structures, and reactivities of the HO complexes with heme, α-hydroxyheme, and verdoheme have been studied. The proximal iron ligand has been identified as a neutral imidazole of His-25 and the presence of a distal base has been established. Conversion to α-hydroxyheme is the step responsible for the regiospecificity. α-Hydroxyheme reveals its ferrous neutral radical state as a key property of its reactivity. Requirement of $O_2$ and one reducing equivalent for the conversion of ferric α-hydroxyheme to verdoheme has been determined. The positive charge on the macrocycle causes unique characteristics of the verdoheme complex. Combination of the novel characteristics of the catalytic intermediates and the protein environment appears to be responsible for the unique HO enzyme function.

*Key words.* Verdoheme—α-Hydroxyheme—Biliverdin—Heme—Heme oxygenase—Carbon monoxide—Oxygenase—Hemoprotein

[1] Department of Physiology and Biophysics, Case Western Reserve University School of Medicine, Cleveland, OH 44106-4970, USA
[2] Institute for Life Support Technology, Yamagata Technopolis Foundation, Yamagata 990, Japan
[3] Institute for Physical and Chemical Research (RIKEN), Wako, Saitama 351-01, Japan
[4] School of Allied Health Sciences, Yamaguchi University, 1144 Kogushi, Ube, Yamaguchi 755, Japan
[5] Department of Physiology and Biophysics, Albert Einstein College of Medicine, Bronx, NY 10461, USA
[6] Department of Biochemistry, Yamagata University School of Medicine, Yamagata 990-23, Japan

# Introduction

Heme oxygenase (HO), an amphipathic microsomal protein, catalyzes the regiospecific oxidative degradation of iron protoporphyrin IX (heme hereafter) to biliverdin,
CO, and Fe in the presence of NADPH-cytochrome P450 reductase, which functions as
an electron donor [1–3]. In the catalytic cycle of HO, the enzyme first binds one
equivalent of heme, resulting in the formation of the heme–enzyme complex, which
exhibits optical absorption spectral properties similar to those of myoglobin and
hemoglobin [4,5]. The first electron donated from the reductase reduces the ferric
heme iron to the ferrous state, and a molecule of oxygen binds to form a metastable
oxy form [6]. Following the electron donation to the oxy form, three stepwise oxygenase reactions are initiated in which heme is ultimately converted to the ferric iron-
biliverdin complex through α-hydroxyheme and verdoheme intermediates (Fig. 1).
Finally, the electron donation from the reductase to the iron–biliverdin complex
releases ferrous iron and biliverdin, and the enzyme becomes available for the next
turnover [4]. Heme, therefore, participates both as a prosthetic group of the oxygen
activation and as a substrate of the enzyme catalysis, a property unique to heme
oxygenase [7].

HO has two isozymes, referred to as HO-1 and HO-2 [3]. HO-1, an inducible form,
is mainly distributed in reticuloendothelial cell-rich tissues, such as spleen and liver
[1–3]. HO-1, with a molecular mass of 33 kDa, was first purified from microsomes of
pig spleen and rat liver [4,5]. HO-2, with a molecular mass of 36 kDa, is constitutively

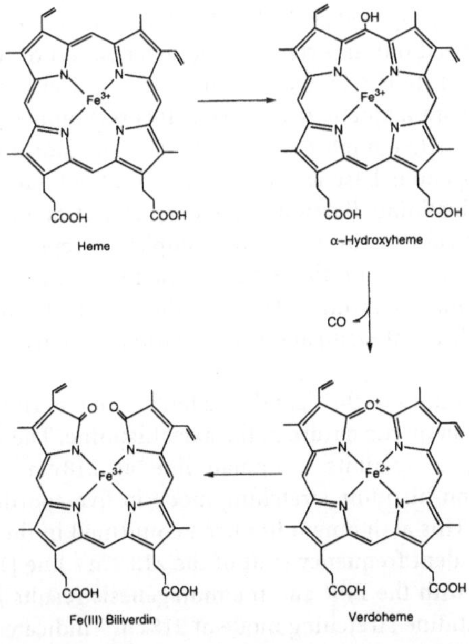

FIG. 1. Intermediates in the reaction of the heme-oxygenase-catalyzed conversion of iron protoporphyrin IX to the biliverdin–iron complex

expressed and is mainly distributed in the brain and testis [3]. The amino acid sequence similarity between HO-1 and HO-2 is about 40%, but there are several stretches of highly conserved sequences including His-25 and His-132 (in HO-1 sequence) [8,9]. Both isoforms display the same enzymatic activity, active site structure, and heme iron coordination and electronic structures; thus, the molecular mechanism of the enzyme action is considered to be analogous between the two isoforms [10].

Understanding of the molecular structure of HO had been severely limited, in part because of the difficulty in obtaining the large amount of the enzyme necessary for structural studies. Recent success in the construction of the bacterial system overexpressing the 30-kDa soluble form of HO-1 [11] has facilitated our structural studies of the enzyme [12–17]. In this chapter, we describe our current knowledge on the active site structures of the heme, α-hydroxyheme, and verdoheme complexes of HO and their implications for the catalytic mechanism of the enzyme action. For details of the HO catalytic mechanism, see the chapter by H. Fujii et al, this volume.

## Active Site Structure of the Heme–HO Complex

Optical absorption spectra of the heme–HO complex were first reported by Yoshida and Kikuchi who have shown that the heme–enzyme complex has spectral properties similar to those of myoglobin and hemoglobin [4–6]. This is an indication that the proximal heme iron ligand is not a cysteine residue as seen with cytochrome P-450-type monooxygenase. Initial site-directed mutagenesis study by Yoshida and his co-workers [11] found that the His-25 to Ala mutation completely abolished the enzyme activity, and His-25 was proposed as the key residue in the HO enzyme function. Our research team has carried out an extensive characterization of the active site of the heme–HO complex and its His-25 to Ala mutant by optical absorption, resonance Raman, and electron paramagnetic resonance (EPR) techniques [12,13]. The ferrous NO form of the heme–HO complex shows a typical six-coordinate heme–NO EPR spectrum with a nitrogenous base as a ligand trans to the bound NO (Fig. 2), being consistent with the histidine ligation. The characteristics of the six-coordinate ferrous NO EPR spectrum of the wild-type complex is replaced by a typical five-coordinate-type EPR spectrum on the His-25 to Ala replacement (Fig. 2). This firmly establishes that the imidazole group of His-25 is the proximal axial heme ligand [15]. Independently, Ortiz de Montellano and his co-workers reached the same conclusion [18].

Resonance Raman studies of the ligand-free ferrous heme–HO complex have provided useful information on the nature of the axial histidine. The ferrous deoxy high-spin heme–HO complex exhibits a Raman line at $218\,cm^{-1}$ [12], a frequency characteristic of an iron-histidine stretching mode in five-coordinate ferrous heme proteins [19] (Fig. 3). This assignment has been confirmed in the study that showed the iron isotope-dependent frequency shift of the $218$-$cm^{-1}$ line [13]. Histidine coordination is consistent with the EPR and the mutagenesis results [11,14,15]. The frequency of the iron-histidine stretching mode at $218\,cm^{-1}$ indicates that the proximal histidine in the heme–HO complex is a neutral imidazole as in myoglobin rather than an imidazolate as in peroxidase enzymes, because the imidazolate-iron stretching

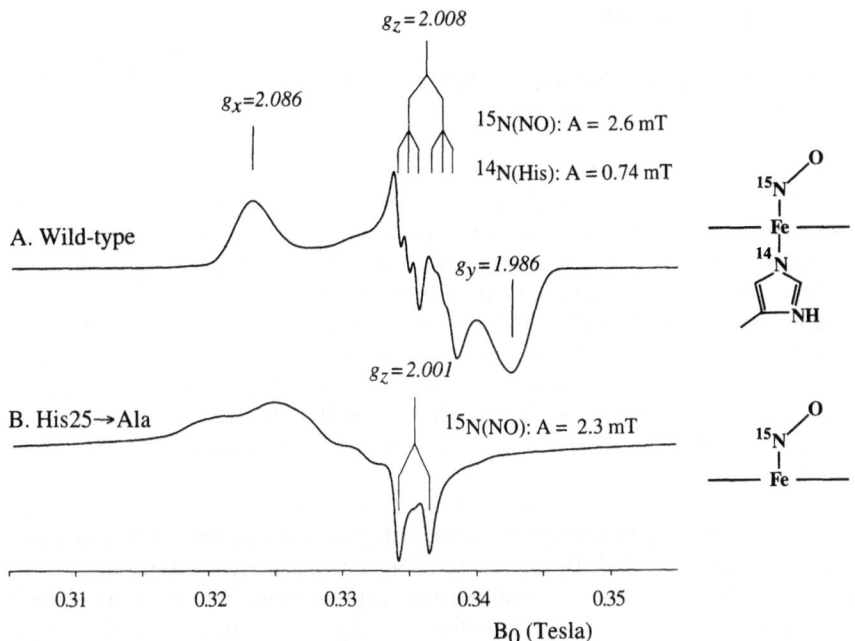

FIG. 2A,B. Electron paramagnetic resonance (EPR) spectra of the ferrous $^{15}NO$ compounds of the heme complexes of (A) wild-type HO-1 (*top*) and (B) His25→Ala (*bottom*). Measurements were carried at 30 K with an incident microwave power of 0.2 mW with a field modulation of 0.2 mT at 100 kHz

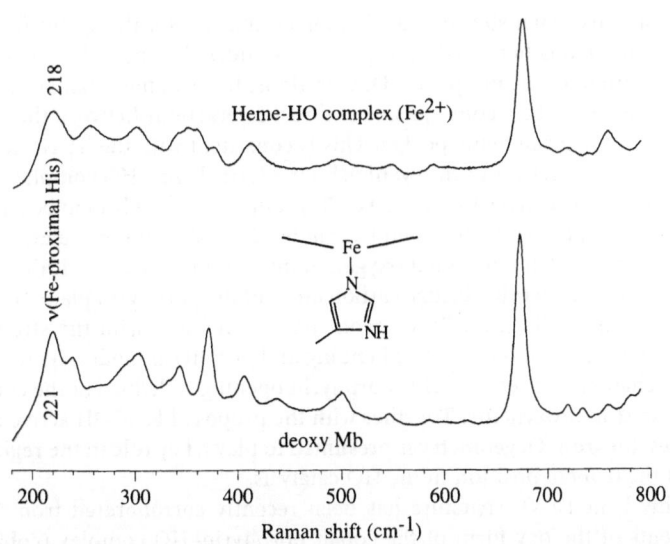

FIG. 3. Low-frequency region of the resonance Raman spectra of the ligand-free ferrous heme–heme oxygenase (HO) complex (*top*) and deoxy Mb (*bottom*). The excitation wavelength was 441.6 nm

mode is in the 240-cm$^{-1}$ region [19] whereas for the neutral imidazole–iron complex the stretching mode is in the 200- to 220-cm$^{-1}$ region, as shown for deoxy Mb in Fig. 3.

The heme pocket structure of the heme–heme oxygenase complex, which has a neutral imidazole as its proximal ligand, is clearly different from that of cytochrome P-450s and peroxidases, where a vectorial polar environment across the heme favors the ferryl–oxo (Fe(IV)=O) intermediates as their activated forms [20]. Thus, the mechanism of oxygen activation in heme oxygenase is expected to be different from these enzymes. Indeed, Wilks and Ortiz de Montellano [21] have ruled out the involvement of a ferryl–oxo intermediate in the first oxygenation step by demonstrating the formation of a stable compound II on reaction of the enzyme with peracids. Instead, ferric peroxide species, Fe(III)–OOH or Fe(III)–OO$^-$, have been proposed as an active intermediate species during the conversion of heme to α-hydroxyheme [21,22].

The resonance Raman measurements of the oxy form of the ferrous heme–HO complex have provided a strong evidence for a highly bent structure of the coordinated dioxygen molecule [15]. The Fe-O$_2$ stretching mode has been detected at 565 cm$^{-1}$, which is the lowest among O$_2$-coordinated heme proteins having neutral histidine as a trans ligand. The Raman spectrum in the low-frequency region exhibits the complicated oxygen isotope-sensitive lines that were interpreted as the result of a strong coupling of the Fe-O-O bending mode (412 cm$^{-1}$) with many porphyrin vibrational modes. The low-frequency Fe-O$_2$ stretching mode and the widespread coupling between the modes of the bound O$_2$ and porphyrin vibrations are unique features only seen in the heme–HO complex but not in the other heme proteins. Normal mode analysis of the isotope shifts indicates a highly bent Fe-O$_2$ angle of approximately 110°.

The neutral proximal histidine coordination established for the ligand-free ferrous and CO-bound ferrous heme–HO complexes precludes the anomalous Fe-proximal linkage as an origin of the unique Fe-O$_2$ vibrations. Instead, the small Fe-O-O angle has been presumed to be a consequence of direct interactions between the bound O$_2$ and residues in the distal heme pocket. This is consistent with the $^1$H-NMR (nuclear magnetic resonance) results of the cyanide bound ferric heme–HO complex where La Mar and his co-workers have found the Fe-CN group is also highly bent because of the steric pressure imposed on the bound cyanide [23]. With the calculated Fe-O-O bending angle of 110°, the terminal oxygen atom could be located within a van der Waals contact with a methine bridge carbon atom of the porphyrin plane (if pointing along that direction). Such an Fe-O$_2$ geometry is consistent with the strong vibrational couplings between the Fe-O-O bending and porphyrin modes. As we describe later in this chapter, the regiospecific porphyrin opening is defined at the conversion from heme to α-hydroxyheme. Together with the proposed Fe-OOH active intermediate species, the iron-O$_2$ geometry is presumed to play a key role in the regiospecific cleavage at the α-*meso* position in the HO catalysis.

The highly bent Fe-O$_2$ structure has been recently corroborated from the EPR measurements of the oxy form of the cobalt porphyrin–HO complex (cobalt–HO). The oxy form of the cobalt-substituted heme protein is different from the its iron counterpart in that it is paramagnetic (3d7, $S = 1/2$), and thus is EPR visible. The high-resolution X-ray crystal structure of oxy cobalt-Mb shows that the metal–O$_2$ structure is very well conserved on the cobalt substitution [24], thus the cobalt–$O_2$ structure assumed in oxy cobalt–HO is expected to be similar to that in the native iron complex. The EPR spectra of oxy cobalt–HO shows the better resolved hyperfine coupling with a larger $g$-value anisotropy than those in the oxy cobalt-Mb spectrum (Fig. 4) (H. Fujii et al, to be published). The larger $g$-anisotropy is an indication of the smaller Co–O–O bond angle in the cobalt–HO complex than in oxy cobalt-Mb, being consistent with the Raman results described earlier. A well-resolved hyperfine coupling observed in the cobalt–HO complex reflects that the range of orientations available to $O_2$ is highly restricted in comparison with cobalt-Mb. Such a restriction might originate from the steric interactions between the bound $O_2$ and the amino acid residues in the distal pocket, and also from the van der Waals interaction between the terminal oxygen atom and the porphyrin α-*meso* carbon.

While His-25 has been identified as the proximal His of the complex, the nature of the distal base has been less clear. Our studies have shown the presence of a distal residue that is responsible for the transition between the high-spin acid and low-spin

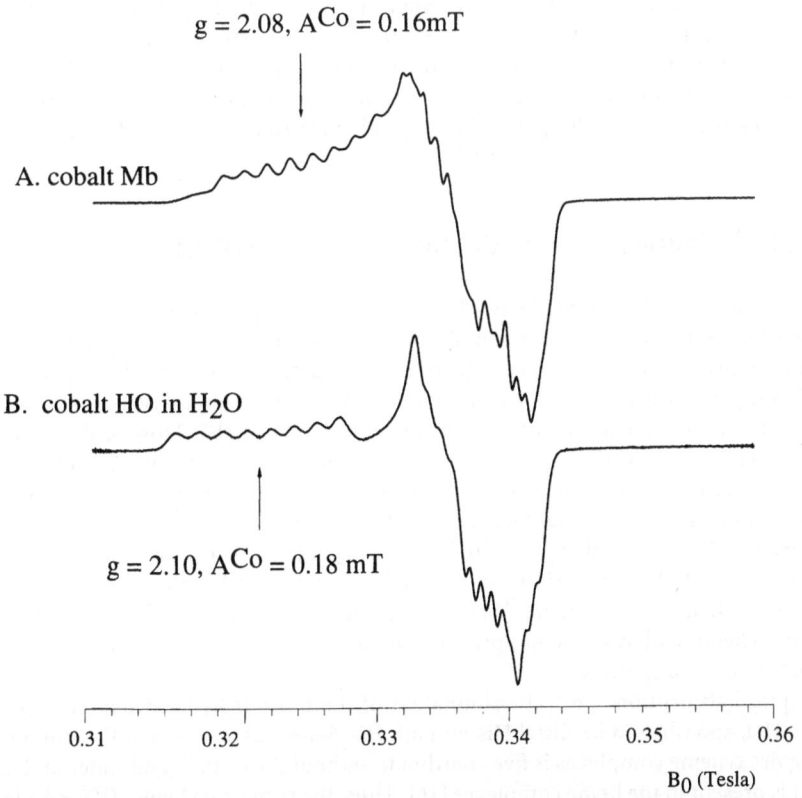

FIG. 4A,B. EPR spectra of the oxy form of (A) cobalt-Mb (*top*) and (B) cobalt porphyrin–HO complex (*bottom*). Measurements were carried at 20 K with an incident microwave power of 0.1 mW with a field modulation of 0.2 mT at 100 kHz

alkaline forms of the hemin complex of HO-1 with a $pK_a$ value of 7.6 [12]. Replacement of the highly conserved His-132 by Ala does not alter the enzymatic activity when ascorbic acid is used as the electron donor [15], leading to our conclusion that His-132 is not an essential residue of the HO catalysis. This conclusion was challenged by Ortiz de Montellano and his co-workers [25], who reported reduced enzyme activities and altered optical absorption spectral properties of the HO complexes with the His-132 to Ser, Ala, and Gly mutations. These observations were used as an evidence for His-132 being present in the distal pocket. Our research team has recently reexamined the same His-132 mutants. We have found that the mutations at position 132 decrease the HO protein stability, resulting in the formation of an inactive, unfolded HO protein that could be removed by gel filtration chromatography.

Using the preparations devoid of unfolded inactive materials, our results (Mansfield Matera et al., submitted for publication) clearly show that the mutations of His-132 by Ser, Gly, and Ala do not alter the spectroscopic properties of the heme–HO complexes and their reactivities, and thus we conclude that His-132 is not the key amino acid residue located at close proximity of the ligand-binding site in the distal pocket but instead His-132 plays an important role in the HO protein stability. The apparent discrepancy between the two results appears to stem from the use of the HO preparation with the unfolded HO protein, which does not form the normal heme complex, in the preparations used by Ortiz de Montellano et al. [25]. Importance of the distal base in controlling the reactivity of heme protein is indisputable; its identification is now in progress in our laboratories.

## α-Hydroxyheme and Verdoheme Intermediates

As described, studies on the heme–HO complexes have provided structural information of the active center, and the molecular mechanism of the first oxygenation step has been proposed. However, knowledge of the structures of intermediate species, the HO complexes with α-hydroxyheme and verdoheme, is also required so as to understand the molecular mechanism of the enzyme action of HO. Most of the previous hydroxyheme and verdoheme studies were carried out on the apomyoglobin complexes or model heme systems [27,28]. During the enzyme turnover, α-hydroxyheme and verdoheme species could not be readily isolated as a single species. The recent interest in CO as a possible physiological messenger [29,30] also warrants the elucidation of the CO biosynthesis reaction, which is associated with the conversion of α-hydroxyheme to verdoheme [31]. To this end, we have chemically synthesized α-hydroxyheme and verdoheme, prepared their HO complexes, and examined their spectroscopic properties.

Optical absorption spectral examination of the ferric α-hydroxyheme complexes with HO, apoMb, and its distal His mutant, His-64Ile, has shown that the iron in the α-hydroxyheme complexes is five coordinate without the sixth ligand water molecule that is present in the heme complexes [16]. Thus, the α-hydroxyheme–HO complexes offer an interesting and unique example in which the altered electronic state of the porphyrin ring changes the coordination structure of the heme iron. The axial heme iron ligand is a neutral imidazole as in the case for the heme–HO complex. More importantly, resonance Raman examinations have revealed redox-dependent changes in the structure of the heme macrocycle [16].

The Raman spectrum of the ferric α-hydroxyheme–HO complex is quite different from that of the heme complex: the presence of two strong lines at about 1220 and 1580 cm$^{-1}$ and the absence of a strong $v_4$ line, which is commonly observed in all ferric hemeproteins with protoheme at about 1370 cm$^{-1}$ (Fig. 5). In contrast, the ferrous form of the complex exhibits a typical heme protein Raman spectrum. Measurements in D$_2$O do not affect the ferric spectrum, while some of the porphyrin vibrational modes exhibit a subtle D$_2$O effect in the ferrous form. We infer from these observations that the α-meso-hydroxy group is deprotonated in the ferric state and it becomes protonated on reduction to the ferrous state. A redox-linked transition between the keto (oxophlorin) and the enol (hydroxyporphyrin) forms of α-

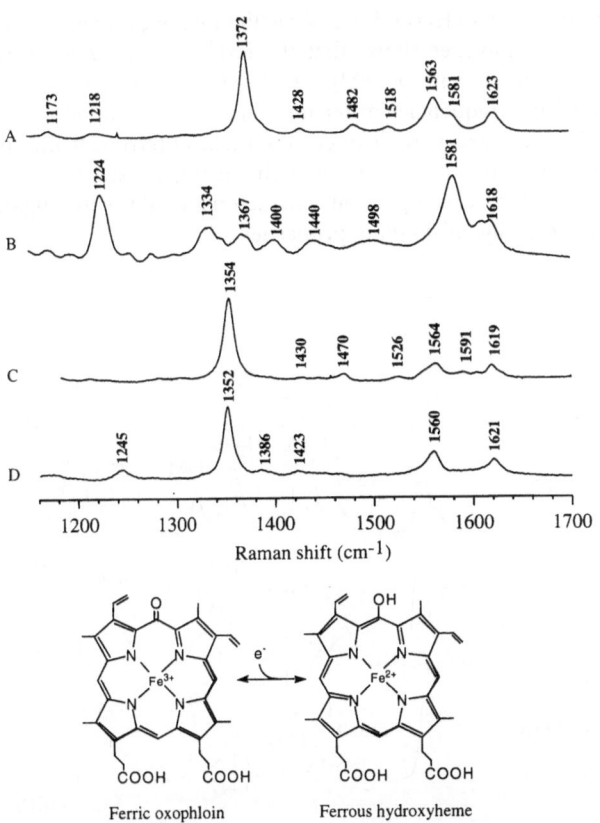

FIG. 5A–D. Resonance Raman spectra of the α-hydroxyheme and heme complexes of HO. From the *top*: ferric heme complex (A), ferric α-hydroxyheme complex (B), ferrous heme complex (C), and ferrous α-hydroxyheme complex (D). Excitation wavelenths were 413.1 nm for the heme complex and ferrous α-hydroxyheme complex, and 406.7 nm for the ferric α-hydroxyheme complex. Schematic diagram for the redox-linked transition between the ferric oxophloin (keto form) and ferrous α-hydroxyheme (enol form) is also shown

hydroxyheme (Fig. 5) takes place. Ferric α-hydroxyheme is present in a resonance structure between oxophlorin, phenolate anion, and a neutral radical forms [26,27]. The resonance structure is essential for oxygen reactivity with the ferric α-hydroxy-heme–HO complex (see the chapter by H. Fujii et al, this volume).

The verdoheme–HO complex has been examined recently by resonance Raman and optical absorption spectroscopy [17]. The high-frequency region of the Raman spectra of the verdoheme–HO and its cyanide and CO complexes are illustrated in Fig. 6. The spectra show an atypical pattern compared with the protoheme complexes, but retain similarity to those of the ferric α-hydroxyheme complexes. In general, two strong lines near 1250 and 1610 cm$^{-1}$ exist, with several weaker lines between them. The Raman spectral features are consistent with the symmetry lowering of the porphyrin by a disruption of the conjugated π-electron system, and indicate that the verdoheme in the HO complex assumes oxohemin structure in the resonance forms proposed by Balch et al. [27]. The comparison of the resonance Raman spectra of the verdoheme complexed with HO and apo-Mb with those of the five- and six-coordinate verdoheme model complexes shows that the ferrous form of the verdoheme–HO complex is six coordinate with possibly a hydroxide as the sixth ligand. The Fe-CO and Fe-CN stretching frequencies of ferrous verdoheme compounds (~470 cm$^{-1}$ for both CO and CN$^-$ complexes) are distinct from those of ferrous heme compounds. It is inferred that the positive charge of the verdoheme ring possesses some of the charge density on the iron atom, causing unique characteristics of the iron ligand-stretching vibrations and altered ligand-binding properties.

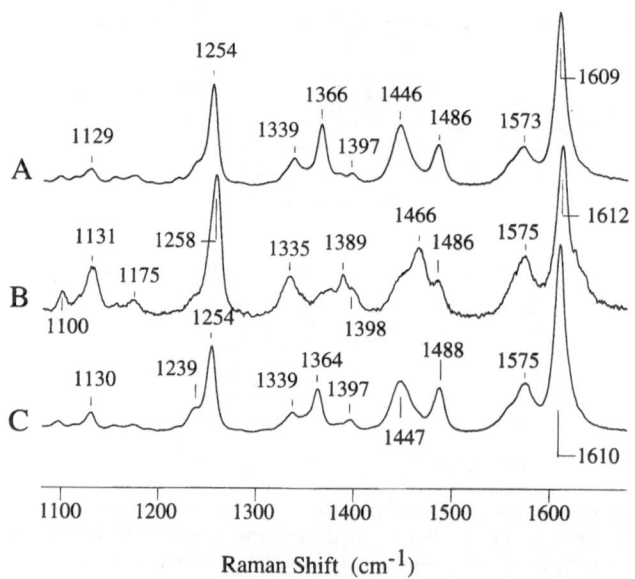

FIG. 6A–C. Resonance Raman spectra of the ferrous verdoheme–HO complex for ligand-free (A), cyande (B), and CO (C) adducts. The excitation wavelength was 441.6 nm

Using β-, γ-, and δ-*meso* isomers of hydroxyheme and verdoheme, the oxygenation step responsible for the regiospecificity has been determined. β-, γ-, and δ-isomers of biliverdin are respectively produced from the β-, γ-, and δ-*meso*-hydroxyheme complexes of HO through the respective verdoheme isomers. β-, γ-, and δ-biliverdin have been obtained from the respective verdoheme isomers in the HO-catalyzed heme catabolism reaction. Therefore, it can be concluded that the initial oxygenase step, the conversion of heme to hydroxyheme, is the step that is responsible for the regiospecificity (to be published).

*Acknowledgments.* We thank Drs. H. Zhou, K. Ishikawa, and N. Takeuchi for their help in the work described in this article. This work was supported by NIH grants (M.I.-S. and D.L.R.), grants in aid from the Ministry of Education, Culture, Sports and Science of Japan (T.Y. and C.T.M.), and a postdoctoral fellowship from RIKEN (S.T.), Japan.

# References

1. Tenhunen R, Marver, HS, Schmid R (1969) Microsomal heme oxygenase. Characterization of the enzyme. J Biol Chem 244:6388–6394
2. Kikuchi G, Yoshida T (1980) Heme degradation by the microsomal heme oxygenase system. Trends Biochem Sci 5:332–325
3. Maines MD (1988) Heme oxygenase: function, multiplicity, regulatory mechanisms, and clinical applications. FASEB J 2:2557–2568
4. Yoshida T, Kikuchi G (1978) Purification and properties of heme oxygenase from pig spleen microsomes. J Biol Chem 253:4224–4229
5. Yoshida T, Kikuchi G (1979) Purification and properties of heme oxygenase from rat liver microsomes. J Biol Chem 254:4487–4491
6. Yoshida T, Noguchi M, Kikuchi G (1980) Oxygenated form of heme-heme oxygenase complex and requirement for second electron to initiate heme degradation from the oxygenated complex. J Biol Chem 255:4418–4420
7. Yoshida T, Kikuchi G (1978) Features of the reaction of heme degradation catalyzed by the reconstituted microsomal heme oxygenase system. J Biol Chem 253:4230–4236
8. Yshida T, Biro P, Cohen T, et al. (1988) Human heme oxygenase cDNA and induction of its mRNA by hemin. Eur J Biochem 171:457–461
9. Rotenberg MO, Maines MD (1991) Characterization of a cDNA-encoding rabbit brain heme oxygenase-2 and identification of a conserved domain among mammalian heme oxygenase isozymes: possible heme-binding site? Arch Biochem Biophys 290:336–344
10. Ishikawa K, Takeuchi N, Takahashi S, et al. (1995) Heme oxygenase-2. Properties of the heme complex of the purified tryptic fragment of recombinant human heme oxygenase-2. J Biol Chem 270:6345–6350
11. Ishikawa K, Sato M, Ito M, et al. (1992) Importance of histidine residue 25 of rat heme oxygenase for its catalytic activity. Biochem Biophys Res Commun 182:981–986
12. Takahashi S, Wang J, Rousseau DL, et al. (1994) Heme-heme oxygenase complex. Structure of the catalytic site and its implication for oxygen activation. J Biol Chem 269:1010–1014
13. Takahashi S, Wang J, Rousseau DL, et al. (1994) Heme-heme oxygenase complex: structure and properties of the catalytic site from resonance Raman scattering. Biochemistry 33:5531–5538

14. Takahashi S, Ishikawa K, Takeuchi N, et al. (1995) Oxygen bound heme-heme oxygenase complex: resonance Raman evidence for a strongly bent structure of the co-ordinated oxygen. J Am Chem Soc 117:6002–6006
15. Ito-Maki M, Ishikawa K, Mansfield Matera K, et al. (1995) Demonstration that histidine 25, but not 132, is the axial heme ligand in rat heme oxygenase-1. Arch Biochem Biophys 317:253–258
16. Mansfield Matera K, Takahashi S, Fujii H, et al. (1996) Oxygen and one reducing equivalent are both required for the conversion of α-hydroxyhemin to verdoheme in heme oxygenase. J Biol Chem 271:6618–6624
17. Takahashi S, Mansfield Matera K, Fujii H, et al. (1997) Resonance Raman spectroscopic characterization of α-hydroxyheme and verdoheme complexes of heme oxygenase. Biochemistry 36:1402–1410
18. Sun J, Loehr TM, Wilks A, et al. (1994) Identification of histidine 25 as the heme ligand in human liver heme oxygenase. Biochemistry 33:13734–11374
19. Kitagawa T (1988) The hemeprotein and the iron-histidine stretching mode. In: Spiro TG (ed) Biological applications of raman spectroscopy, vol 3. Wiley, New York, pp 97–131
20. Dawson JH (1988) Probing structure-function relations in heme-containing oxygenases and peroxidases. Science 240:433–439
21. Wilks A, Ortiz de Montellano PR (1993) Rat liver heme oxygenase. High level expression of a truncated soluble form and nature of the meso-hydroxylating species. J Biol Chem 263:22357–22362
22. Nogichi M, Yoshida T, Kikuchi G (1983) A stoichiometric study of heme degradation catalyzed by the reconstituted heme oxygenase system with special consideration of the production of hydrogen peroxide during the reaction. J Biochem (Tokyo) 93:1027–1036
23. Hernandez G, Wilks A, Paoless R, et al. (1994) Proton NMR investigation of substrate-bound heme oxygenase: evidence for electronic and steric contributions to stereoselective heme cleavage. Biochemistry 33:6631–6641
24. Brucker EA, Olson JS, Phillips GN, et al. (1996) High resolution crystal structures of the deoxy-, oxy- and aquomet-forms of cobalt myoglobin. J Biol Chem 271:25419–25422
25. Wilks A, Ortiz de Montellano PR, Sun J, et al. (1996) Heme oxygenase (HO-1): His-132 stabilizes a distal water ligand and assists catalysis. Biochemistry 35:930–936
26. Fujii H (1990) Studies on the electronic structure and reactivities of catalytic intermediates in heme enzymes. Ph.D. thesis, Kyoto University, Kyoto, Japan
27. Balch AL, Latos-Grazynski L, Noll BC, et al. (1993) Structural characterization of verdoheme analogs. Iron complexes of octaethyloxoporphyrin. J Am Chem Soc 115:1422–1429.
28. Morishima I, Fujii H, Shiro Y, et al. (1995) Studies on iron(II)-meso-oxoporphyrin π-neutral radical as a reaction intermediate in the heme catabolism. Inorg Chem 34:1528–1535
29. Verma A, Hirsch DJ, Glatt CE, et al. (1993) Carbon monoxide: a putative neural messenger. Science 259:381–384
30. Suematsu M, Goda N, Sano T, et al. (1995) Carbon monoxide: an endogenous modulator of sinusoidal tone in the perfused rat liver. J Clin Invest 96:2431–2437
31. Yoshida T, Noguchi M, Kikuchi G (1982) The step of carbon monoxide liberation in the sequence of heme degradation catalyzed by the reconstituted microsomal heme oxygenase system. J Biol Chem 257:9345–9348

# Heme Degradation Mechanism by Heme Oxygenase: Conversion of α-*meso*-Hydroxyheme to Verdoheme IX$_\alpha$

H. Fujii[1], K. Mansfield Matera[2], S. Takahashi[3], C.T. Migita[4], H. Zhou[5], T. Yoshida[5], and M. Ikeda-Saito[2]

*Summary.* Heme oxygenase (HO) is a microsomal enzyme that catalyzes the degradation of iron protoporphyrin IX (heme) to biliverdin IX$_\alpha$ through two novel heme derivatives, α-*meso*-hydroxyheme and verdoheme IX$_\alpha$. Using the recombinant HO protein and chemically synthesized α-hydroxyheme and verdoheme, we have elucidated HO-catalyzed heme degradation mechanisms. Heme in HO is first hydroxylated to form α-*meso*-hydroxyheme via a ferric hydroperoxide intermediate that is produced by the one-electron reduction of ferrous oxy-complex. The hydrogen bond between iron-bound dioxygen and a protein moiety may play an important role in the dioxygen reduction and regiospecific hydroxylation processes. We have shown that one electron as well as one molecular oxygen are required to degrade α-*meso*-hydroxyheme to verdoheme IX$_\alpha$ and that two alternative degradation pathways are probable: ferric and ferrous. We propose here a degradation mechanism whereby the ferrous π-neutral radical formed from ferric α-*meso*-hydroxyheme by an intramolecular electron transfer reacts with superoxide, which is produced from the reduction of dioxygen by ferric or ferrous α-*meso*-hydroxyheme to form verdoheme IX$_\alpha$ and CO. We also propose the degradation mechanism of verdoheme IX$_\alpha$ to biliverdin IX$_\alpha$ via dioxygen activation processes on verdoheme iron.

*Key words.* Heme oxygenase—Reaction mechanism—Heme—α-Hydroxyheme—Verdoheme IX$_\alpha$

[1] Institute for Life Support Technology, Yamagata Technopolis Foundation, Yamagata 990, Japan
[2] Department of Physiology and Biophysics, Case Western Reserve University School of Medicine, Cleveland, OH 44106-4970, USA
[3] Institute of Physical and Chemical Research (RIKEN), Wako, Saitama 351-01, Japan
[4] School of Allied Health Science, Yamaguchi University, 1144 Kagushi, Ube, Yamaguchi 755, Japan
[5] Department of Biochemistry, Yamagata University School of Medicine, Yamagata 990-23, Japan

315

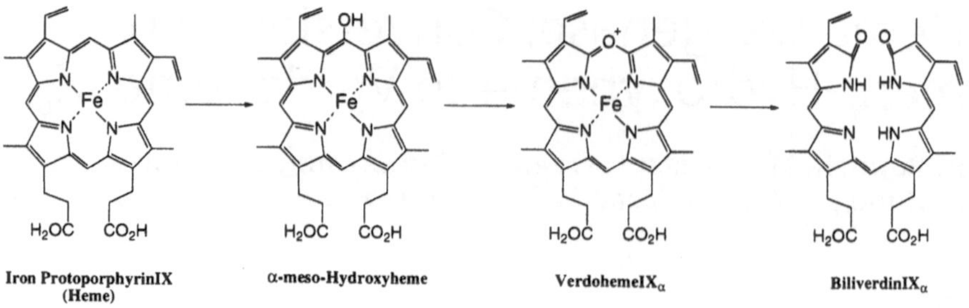

FIG. 1. Reaction sequence of intermediates in the heme degradation reaction by heme oxygenase

## Introduction

Heme oxygenase (HO), an amphipathic microsomal protein, catalyzes the regiospecicfic oxidative degradation of heme (iron protoporphyrin IX) to biliverdin $IX_\alpha$, CO, and iron in the presence of NADPH-cytochrome P-450 reductase, which functions as an electron donor [1–7]. Heme degradation catalyzed by HO is physiologically important because it is the process not only of dissimilation of heme but also of biosynthesis of CO, which serves as a physiological messenger in vivo [8,9]. The HO catalysis is quite complex and is known to contain two novel heme derivatives: $\alpha$-meso-hydroxyheme and verdoheme $IX_\alpha$ (Fig. 1) [6,7]. Recombinant HO preparation has facilitated spectroscopic and mechanistic studies on the HO reactions [10—19]. In this chapter, we discuss the heme degradation mechanism by HO on the basis of our recent studies using the recombinant HO protein and chemically synthesized heme groups for the reaction intermediates. (For the active site structure in heme derivative–HO complexes, see the chapter by Ikeda-Saito, this volume.)

## Conversion of Heme to α-meso-Hydroxyheme

The heme oxygenase reaction starts from the formation of the ferric heme–HO complex. The ferric heme–HO complex is reduced to a ferrous form by an electron donated from NADPH-cytochrome P-450 reductase. The ferrous heme–HO complex binds molecular oxygen to form a ferrous oxy-complex [20]. Further electron donation from the reductase forms $\alpha$-meso-hydroxyheme. The overall reaction is similar to the oxygen activation mechanism proposed for cytochrome P-450s. In cytochrome P-450s, an iron(IV) porphyrin $\pi$-cation radical, formed by the heterolytic cleavage of the O–O bond in the ferric–heme hydroperoxide complex, is thought to be a reactive species of the monooxygenation reaction [21,22]. However, this does not appear to be the case for heme oxygenase catalysis. Ortiz de Montellano and his co-workers showed that heme was converted to verdoheme on the reaction of the heme—HO complex with $H_2O_2$ but that the reaction with m-chloroperoxybenzoic acid yielded a ferryl-oxo porphyrin [17]. On the basis of these observations, they concluded that the ferryl-oxo species is not the active form and proposed that the hydroperoxide or peroxy anion directly reacts at the $\alpha$-meso position to form $\alpha$-meso-hydroxyheme.

Interestingly, the ferrous oxy-form of the heme–HO complex easily receives the second electron from the reductase while oxy-Mb does not, despite the fact that heme—HO complex and Mb have heme irons in similar coordination and electronic structures in their deoxy and ferric forms (for details see the chapter by M. Ikeda-Saito et al, this volume). Differences in the electronic structure of ferrous oxy-forms between HO and Mb could be responsible for the different reactivities toward the second electron. Indeed, resonance Raman spectra of oxy-HO and electron paramagnetic resonance (EPR) spectra of oxy cobaltporphyrin–HO are significantly different from those of Mb [13; see also the chapter by Ikeda-Saito et al, this volume]. Changes in the coordination structure of the oxygen molecule or hydrogen bonding between the bound oxygen and protein might affect the redox potential of the oxy-form of the heme–HO complex, leading to further reduction of coordinated oxygen to hydroperoxide in HO.

A possible mechanism is summarized in Fig. 2. The ferric heme–HO complex is reduced to a ferrous deoxy-complex. Dioxygen then binds to the ferrous deoxy-complex to form a ferrous oxy-complex, whose valence structure can be presented either as the ferrous-O$_2$, 3, or as the ferric-superoxide, 4, complex. Our ESR study indicated that the iron-bound dioxygen forms a hydrogen bond with a proton in the distal side. This hydrogen bond orients the iron-bound dioxygen to the α-*meso*-position and also donates a proton to the ferric superoxide complex (4) to form a ferric hydroperoxide complex (5). As suggested by nuclear magnetic resonance (NMR) and resonance Raman studies, iron-bound dioxygen is placed close to the α-*meso*-position so that a homolytic O–O bond cleavage or an electrophilic attack of the hydroperoxide to the α-*meso*-position forms only α-*meso*-hydroxyheme (7), followed by its keto form (8) with release of H$_2$O.

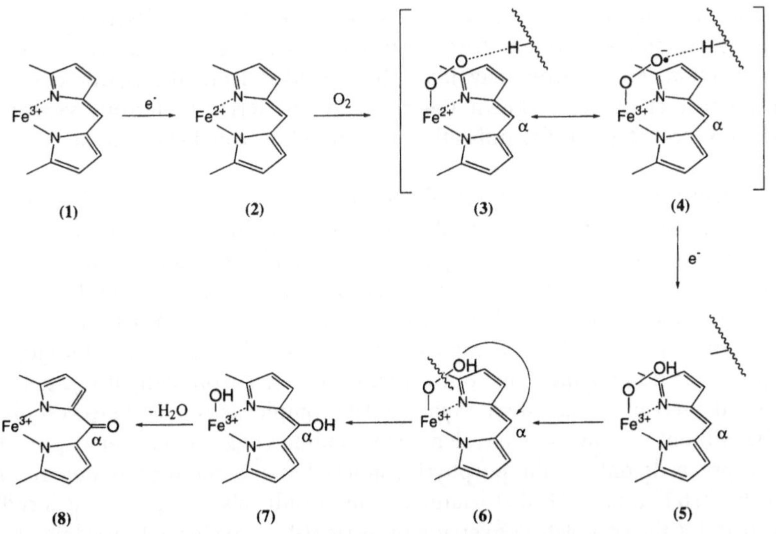

FIG. 2. Proposed mechanism of heme oxygenase-catalyzed conversion of heme to α-*meso*-hydroxyheme. Substituents of the porphyrin ring are omitted for simplicity

# Conversion of α-*meso*-Hydroxyheme to Verdoheme IX$_\alpha$

In the catalytic cycle of HO, α-*meso*-hydroxyheme cannot be isolated because of its short lifetime. Thus, to investigate the degradation mechanism of α-*meso*-hydroxyheme, we have prepared an α-*meso*-hydroxyheme–HO complex from chemically synthesized α-*meso*-hydroxyheme under anaerobic conditions [15]. The α-*meso*-hydroxyheme–HO complex is highly reactive with molecular oxygen. When the ferric complex was exposed to oxygen, the optical absorption spectrum of the α-*meso*-hydroxyheme-HO complex changed to a new spectrum having a broadened Soret band (401 nm) with decreased intensity and two small peaks at 639 and 686 nm. These two peaks correspond to verdoheme IX$_\alpha$–HO and its CO adduct, implying a partial formation (about 30%) of verdoheme IX$_\alpha$ from α-hydroxyheme in the absence of a reducing agent. However, the optical absorption spectrum of the ferrous α-*meso*-hydroxyheme–HO complex was reproduced on addition of sodium dithionite into the solution reacted with oxygen. Thus, the oxygen reaction with the α-hydroxyheme–HO complex in the absence of a reductase only partially converts α-hydroxyheme to verdoheme and mainly produces an oxidized product. The partial formation of verdoheme without reductant may be due to an intermolecular side-reaction of superoxide formed from ferric α-*meso*-hydroxyheme–HO and oxygen. On the other hand, in the presence of a reducing agent, quantitative formation of the verdoheme IX$_\alpha$–HO complex was attained in the reaction of ferric α-*meso*-hydroxyheme–HO with oxygen. These data clearly show that an electron donation from a reductase is required for the complete conversion of ferric α-hydroxyheme to verdoheme IX$_\alpha$ in HO.

Our current result is contrary to previous reports on the model compound studie by Sano et al. [23] and on the α-*meso*-hydroxyheme–HO complex by Ortiz de Montellano et al. [24] asserting the formation of verdoheme IX$_\alpha$ without reducing agents. In model systems, the reaction proceeds via a ferrous π-neutral radical intermediate formed from intramolecular electron transfer. In the model reaction, an electron may likely be provided from the pyridine solvent. Indeed, Balch et al. have shown the reduction of ferric to ferrous verdoheme in the presence of pyridine, indicating that pyridine can behave as a reductant [25].

The ferrous form of the α-*meso*-hydroxyheme–HO complex is also reactive with oxygen. When oxygen was introduced to the solution of the ferrous CO form of the α-*meso*-hydroxyheme–HO complex, the original spectrum of the CO form of α-*meso*-hydroxyheme–HO complex was replaced by the spectrum of the CO form of the verdoheme IX$_\alpha$–HO complex, which has absorption maxima at 408 and 638 nm. The reaction is quite rapid, as it completes within the dead-time (2 ms) of the stopped-flow apparatus. Interestingly, in contrast with the ferric α-*meso*-hydroxyheme–HO complex, the ferrous α-*meso*-hydroxyheme–HO complex is converted to verdoheme IX$_\alpha$–HO without a reducing equivalent. This finding suggests that the degradation reaction proceeds only at the porphyrin moiety because the degradation reaction is much faster than CO dissociation. This result also implies the need of an electron for the complete conversion of ferric α-hydroxyheme to verdoheme IX$_\alpha$ in HO.

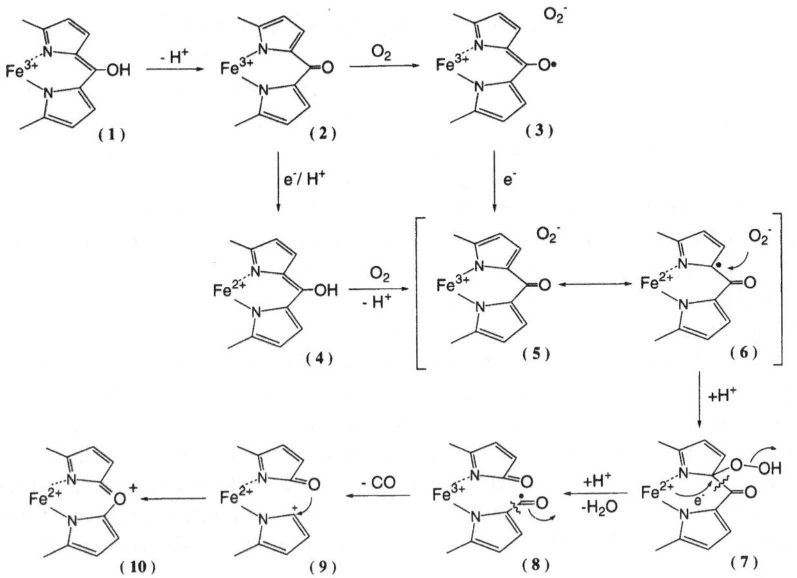

FIG. 3. Proposed mechanism of heme oxygenase-catalyzed conversion of α-*meso*-hydroxyheme to verdoheme IX$_\alpha$. Substituents of the porphyrin ring are omitted for simplicity

Furthermore, the present study shows that the conversion of α-*meso*-hydroxyheme to verdoheme IX$_\alpha$ could proceed in two parallel paths: (1) from ferric α-*meso*-hydroxyheme–HO complex or (2) from ferrous α-*meso*-hydroxyheme–HO complex, as shown in Fig. 3. In the proposed mechanism (Fig. 3), ferric α-*meso*-hydroxyheme–HO complex (2) first reacts with dioxygen to form ferric α-*meso*-hydroxyheme π-cation radical (3) and superoxide. An electron from a reductase rereduces 3 to ferric α-*meso*-hydroxyheme (5). As presented in the model study by Sano et al., ferric α-*meso*-hydroxyheme contains the electronic state of ferrous π-neutral radical, where an electron in the porphyrin moiety transfers to the ferric iron center of α-*meso*-hydroxyheme [23]. Thus, 5 contains the electronic structure of ferrous *meso*-hydroxyheme π-neutral radical (6) as a resonance structure. Then, 6 reacts with superoxide to form a hydroperoxide complex, and bond rearrangement of 7 releases CO and water to form verdoheme. On the other hand, the alternative path starts from the reduction of ferric iron of 2 to its ferrous state (4); 4 reacts with dioxygen to form the common intermediate, 5, following the same path as the first case. In the in vivo heme oxygenase reaction, α-*meso*-hydroxyheme may be degraded to verdoheme by the two alternate paths, depending on the concentration of reductase and dioxygen. The proposed mechanism by Ortiz de Montellano et al. that ferric verdoheme is produced from 2 without a reducing reagent and an electron is consumed by the reduction of ferric verdoheme, is ruled out by our observation that α-*meso*-hydroxyheme is not completely converted to verdoheme without a reductant.

## Conversion of Verdoheme $IX_\alpha$ to Biliverdin $IX_\alpha$

Verdoheme $IX_\alpha$–HO complex was prepared from a chemically synthesized verdoheme and HO protein to investigate the heme-degradation processes after verdoheme $IX_\alpha$. When the solution of this complex was exposed to oxygen in the presence of a reductant, the complete conversion to biliverdin $IX_\alpha$ was attained within 30 min. Interestingly, the reaction was much slower than the steps before verdoheme. Thus, the degradation of verdoheme $IX_\alpha$ to biliverdin $IX_\alpha$ may be the rate-limiting step in the heme oxygenase reaction. The reaction was accelerated as the amount of reductant increased but inhibited if either oxygen or reductant was missing. This finding implies that oxygen activation occurs at the verdoheme iron. Moreover, zinc-substituted verdoheme $IX_\alpha$–HO was not converted to biliverdin $IX_\alpha$ even in the presence of oxygen and a reductase. The participation of a $\pi$–radical intermediate in the reaction was also investigated using a verdoheme $\pi$–radical complex prepared by the reduction of the zinc verdoheme—HO complex. However, no biliverdin $IX_\alpha$ was observed in the reaction of the verdoheme $\pi$–radical complex with oxygen. While further study is needed to define the degradation mechanism of verdoheme $IX_\alpha$, the present results indicate that the degradation occurs via oxygen activation processes.

*Acknowledgment.* This work was supported by an NIH grant (M.I.-S.), grants in aid from the Ministry of Education, Culture, Sports and Science of Japan (T.Y. and C.T.M.), and a postdoctoral fellowship from RIKEN (S.T.).

## References

1. Tenhunen R, Marver HS, Schmid R (1969) Microsomal heme oxygenase. Characterization of the enzyme. J Biol Chem 244:6388–6394
2. Kikuchi G, Yoshida T (1980) Heme degradation by the microsomal heme oxygenase system. TIBS 5:332–325
3. Yoshida T, Kikuchi G (1978) Purification and properties of heme oxygenase from pig spleen microsomes. J Biol Chem 253:4224–4229
4. Yoshida T, Kikuchi G (1979) Purification and properties of heme oxygenase from rat liver microsomes. J Biol Chem 254:4487–4491
5. Maines MD (1988) Heme oxygenase: function, multiplicity, regulatory mechanisms, and clinical applications. FASEB J 2:2557–2568
6. Yoshida T, Noguchi M, Kikuchi G, et al. (1981) Degradation of mesoheme and hydroxymesoheme catalyzed by the heme oxygenase system. Involvement of hydroxyheme in the seqence of heme catabolism. J Biochem 90:125–131
7. Yoshida T, Noguchi M, Kikuchi G, et al. (1980) A new intermediate of heme degradation catalyzed by the heme oxygenase system. J Biochem 88:557–563
8. Yoshida T, Noguchi M, Kikuchi G (1982) The step of carbon monoxide liberation in the sequence of heme degradation catalyzed by the reconstituted microsomal heme oxygenase system. J Biol Chem 257:9345–9348
9. Verma A, Hirsch DJ, Glatt CE, et al. (1993) Carbon monoxide: a putative neural messenger. Science 259:381–384
10. Ishikawa K, Takeuchi N, Takahashi S, et al. (1995) Heme oxygenase-2. Properties of the heme complex of the purified tryptic fragment of recombinant human heme oxygenase-2. J Biol Chem 270:6345–6350

11. Takahashi S, Wang J, Rousseau DL, et al. (1994) Heme-heme oxygenase complex. Structure of the catalytic site and its implication for oxygen activation. J Biol Chem 269:1010–1014
12. Takahashi S, Wang J, Rousseau DL, et al. (1994) Heme-heme oxygenase complex: structure and properties of the catalytic site from resonance Raman scattering. Biochemistry 33:5531–5538
13. Takahashi S, Ishikawa K, Takeuchi N, et al. (1995) Oxygen bound heme-heme oxygenase complex: resonance Raman evidence for a strongly bent structure of the coordinated oxygen. J Am Chem Soc 117:6002–6006
14. Ito-Maki M, Ishikawa K, Mansfield Matera K, et al. (1995) Demonstration that histidine 25, but not 132, is the axial heme ligand in rat heme oxygenase-1. Arch Biochem Biophys 317:253–258
15. Mansfield Matera K, Takahashi S, Fujii H, et al. (1996) Oxygen and one reducing equivalent are both required for the conversion of alpha-hydroxyhemin to verdoheme in heme oxygenase. J Biol Chem 271:6618–6624
16. Takahashi S, Mansfield Matera K, Fujii H, et al. (1997) Resonance Raman spectroscopic characterization of α-hydroxyheme and verdoheme complexes of heme oxygenase. Biochemistry
17. Wilks A, Ortiz de Montellano PR (1993) Rat liver heme oxygenase. High level expression of a truncated soluble form and nature of the *meso*-hydroxylating species. J Biol Chem 263:22357–22362
18. Sun J, Loehr TM, Wilks A, et al. (1994) Identification of histidine 25 as the heme ligand in human liver heme oxygenase. Biochemistry 33:13734–11374
19. Wilks A, Ortiz de Montellano PR, Sun J, et al. (1996) Heme oxygenase (HO-1): His-132 stabilizes a distal water ligand and assists catalysis. Biochemistry 35:930–936
20. Yoshida T, Noguchi M, Kikuchi G (1980) Oxygenated form of heme-heme oxygenase complex and requirement for second electron to initiate heme degradation from the oxygenated complex. J Biol Chem 255:4418–4420
21. Dawson JH (1988) Probing structure-function relations in heme-containing oxygenases and peroxidases. Science 240:433–439
22. Fujii H (1993) Effect of electron-withdrawing power of substituents on the electronic structure and reactivity of oxo iron(IV) porphyrin π-cation radical complexes. J Am Chem Soc 115:4641–4648
23. Morishima I, Fujii H, Shiro Y, et al. (1995) Studies on iron(II)-*meso*-oxoporphyrin π-neutral radical as a reaction intermediate in the heme catabolism. Inorg Chem 34:1528–1535
24. Sono M, Roach MP, Coulter ED, et al. (1996) Heme-containing oxygenases. Chem Rev 96:2841–2887
25. Balch AL, Latos-Grazynski L, Noll BC, et al. (1993) Structural characterization of verdoheme analogs. Iron complexes of octaethyloxoporphyrin. J Am Chem Soc 115:1422–1429

# A Spectroscopic Study on the Intermediates of Heme Degradation by Heme Oxygenase

Yoshiaki Omata and Masato Noguchi

*Summary.* Heme is degraded by heme oxygenase to biliverdin with the aid of NADPH-cytochrome P-450 reductase serving as an electron donor. The process involves three monooxygenase reactions and produces one molecule of carbon monoxide. Several intermediates such as oxygenated heme and verdoheme bound to heme oxygenase are spectrally distinguishable during the degradation of heme to biliverdin. We investigated the kinetics of heme degradation by stopped-flow spectrophotometry; the conversion of heme into its oxygenated form was four times as fast as the formation of verdoheme, and its cleavage to biliverdin was three times slower than the preceding step. These results indicate that the opening of the porphyrin ring should be considered the rate-determining step in the heme degradation reaction. We also measured the redox potential of the heme–heme oxygenase complex. The potential proved to be −76 mV at pH 7.0, which was far lower than those of ascorbate and cytochrome $b_5$. This is consistent with the adoption of NADPH-cytochrome P-450 reductase as the physiological reducing system.

*Key words.* Heme oxygenase—Verdoheme—Biliverdin—Redox potential—Stopped-flow spectrophotometry

## Introduction

Heme oxygenase catabolizes heme to biliverdin with concomitant formation of carbon monoxide (CO) that derives from α-methene carbon [1]. The involvement of CO in a signal transduction system called the heme-CO pathway has been discussed [2]. The heme oxygenase is presently considered the only enzyme producing CO in mammals. The heme oxygenase reaction consists of three monooxygenase reactions in which NADPH-cytochrome P-450 reductase ($f_{P2}$) serves as an electron donor [3]. The heme is assumed to be converted into biliverdin via hydroxyheme and verdoheme [4].

Department of Medical Biochemistry, Kurume University School of Medicine, 67 Asahi-machi, Kurume, Fukuoka 830, Japan

However, the exact chemical properties of the intermediates and the rates of the individual steps have not been fully established.

Heme oxygenase is not a heme protein, but the complex of heme oxygenase and heme exhibits absorption spectra that are characteristic of heme proteins such as myoglobin [5]. In this study, we measured the redox potential of the heme–heme oxygenase complex from a thermodynamic point of view and monitored the absorption spectral changes of the complex during the heme degradation with a stopped-flow spectrophotometer to compare the relative rates of the intermediate steps.

## Materials and Methods

A truncated DNA of heme oxygenase-1 gene encoding the soluble portion of rat heme oxygenase-1, which lacked the membrane-binding domain (22 amino acids at C-terminus), was amplified by polymerase chain reaction (PCR) from rat spleen cDNA library (CLONTECH Laboratories, Palo Alto, CA, USA). The cloned DNA sequence exactly matched the corresponding sequence of DNA for rat heme oxygenase-1 in the database. The obtained DNA was inserted into the expression vector pBAce [6] followed by transfection into *Escherichia coli* JM109. The 10-1 culture of *E. coli* in the induction medium gave 19.8 g of cells, wet weight. The cells were lysed with lysozyme and sonication, and the heme oxygenase expressed was purified by a modified method of Wilks and Ortiz de Montellano [7]. It consisted of ammonium sulfate fractionation, hydroxyapatite column chromatography, and POROS HQ anion exchange chromatography run on a ConSep LC100 (PerSeptive Biosystems, Framingham, MA, USA). The process yielded 170 mg of purified enzyme that was homogenous and which had a molecular mass of 31 kDa as judged by sodiumdodecyl sulfate-polyacrylamide gel electrophoresis (SDS-PAGE). The specific activity of the purified enzyme was comparable to that of the native enzyme purified from rat spleen [5]. Heme oxygenase was mixed with a 1.5-fold excess of heme, and the heme–heme oxygenase complex was obtained by removal of unbound heme by hydroxyapatite column chromatography. The purification of $f_{P2}$ was done according to the method of Yasukochi and Masters [8].

The heme–heme oxygenase complex (0.1 mM) containing 25 μM each of 2,3,5,6-tetramethylphenylenediamine, β-naphthoquinone, gallocyanine, indigo tetrasulfonate, 2-hydroxy-α-naphthoquinone, and anthraquinone-2-sulfonate as redox mediators in 0.1 M potassium phosphate buffer (pH 7.0) was anaerobically reduced with sodium dithionite and titrated by additions of small aliquots of 25 mM potassium ferricyanide. The absorption spectra were recorded with a 2-mm light-pass cuvette, and the corresponding redox potentials were measured at 25°C.

The mixture of the complex and $f_{P2}$ at final concentrations of 14 μM and 2.8 μM, respectively, in 0.1 M potassium phosphate buffer (pH 7.4) was mixed with 150 μM NADPH as a final concentration in the stopped-flow spectrophotometer equipped with a photodiode array detector (Applied Photophysics, Leatherhead, UK), and the absorption spectra were recorded up to 100 s at exponentially increasing time intervals starting 5 ms after mixing.

## Results and Discussion

The absorption spectrum around the Soret region of the heme–heme oxygenase complex was very similar to that of myoglobin, exhibiting the absorbance maximum at 405 nm in the oxidized form and at 430 nm in the reduced form. When the complex was reduced with dithionite and then anaerobically titrated with ferricyanide, the absorbance at 430 nm was decreased and that at 405 nm was increased with an isosbestic point at 415 nm. The relation between the redox potential and the percent reduction determined from the absorbance at 430 nm is shown in Fig. 1. The midpoint potential, −76 mV at pH 7.0, was obtained by the least-squares method (Fig. 1, insert), and the Nernst curve drawn with the midpoint potential and $n$ value of 1 well fitted with the measured potentials as shown in Fig. 1. The redox potential of heme–heme oxygenase complex is far lower than those of ascorbate and cytochrome $b_5$, by 130 and 90 mV, respectively. This appears to explain why $f_{p2}$ was selected as the physiological electron donor for the heme oxygenase system.

The absorption spectrum of the heme–heme oxygenase complex in the longer wavelength region exhibited maxima at 500 and 630 nm, which are typical for ferric

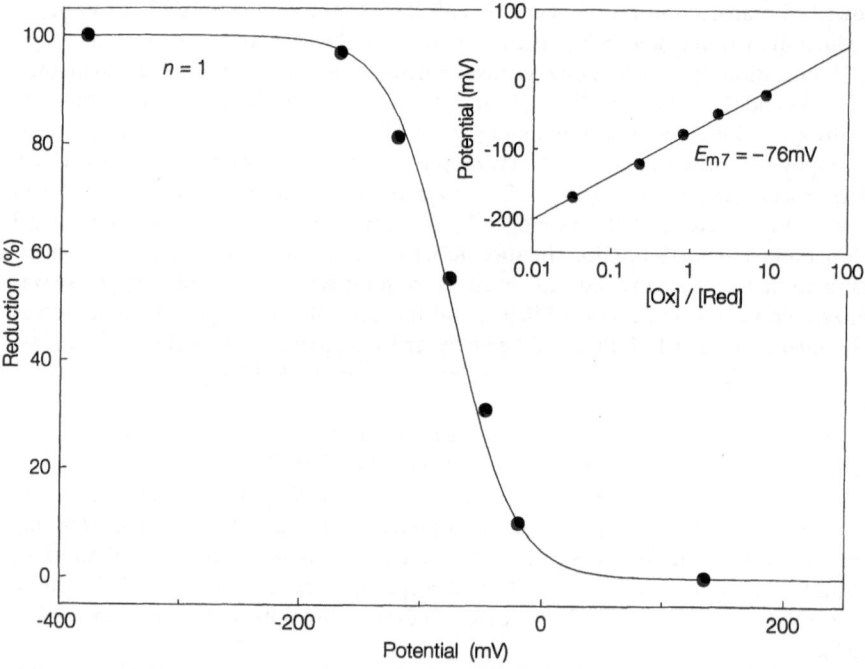

FIG. 1. Oxidative titration of the heme–heme oxygenase complex. Fraction of the reduced form was calculated from the absorbance at 430 nm. The relations between the redox potential and the percent reduction are plotted. The midpoint potential was obtained from the least squares fitting as shown in the *insert*. The estimated Nernst curve ($E_h = E_m + 59/n$ log [Ox]/[Red] mV at 24°C) is drawn with the midpoint potential of −76 mV and $n$ value of 1 for a redox couple accompanying one-electron transfer

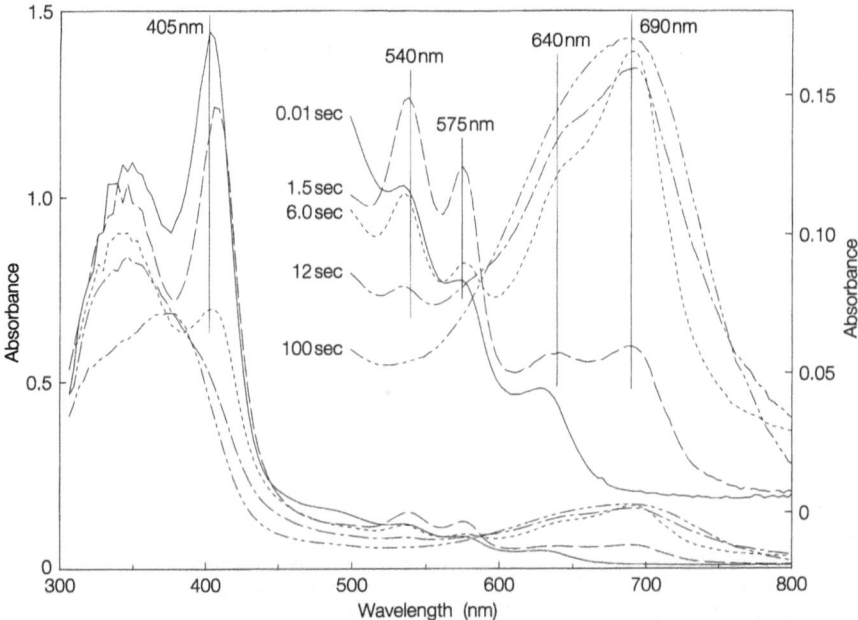

FIG. 2. Stopped-flow spectrophotometry of the spectral changes during heme degradation by the heme–heme oxygenase complex. The reaction conditions are described in the Materials and Methods section. The spectra at 0.01, 1.5, 6, 12, and 100 s after mixing are indicated by the *solid, broken, dotted, dash-dotted,* and *dot-dotted lines,* respectively. *Vertical lines,* except that at 405 nm, denote the wavelengths that were selected for drawing the absorption decay curves in Fig. 3. The *line* at 405 nm indicates the wavelength of the absorption maximum of the complex in the oxidized form

high-spin heme. The mixture of the complex and $f_{P2}$ was mixed with NADPH in the stopped-flow spectrophotometer, and the spectral changes were monitored. The absorption spectra at several time points are shown in Fig. 2. First, the absorbances at 540 and 575 nm, which are attributed to the oxygenated complex [4], increased rapidly and the Soret band red-shifted up to 1.5 s. Then, with the decreasing of these absorbances, the absorbances at 690 nm from verdoheme [9] and at 640 nm from hydroxyheme [10] and the CO-bound form of verdoheme [9] gradually increased up to 6 s. Although the absorbance at 640 nm continued to increase, the absorbance at 690 nm decreased slightly (12 s). The final spectrum at 100 s indicated biliverdin with the absorption maxima at 380 and 685 nm [11].

To compare the relative rates of production and degradation of the intermediates, the absorbance changes at several wavelengths that are characteristic for the possible intermediates were analyzed by curve fitting to the exponential decay model (Fig. 3). The absorbances at 540 and 575 nm rapidly increased at a rate of $1.8 \, s^{-1}$ up to 1.5 s. After that, the absorbance at 575 nm decreased at a rate of $0.4 \, s^{-1}$, but the rate of decrease of the absorbance at 540 nm was $0.2 \, s^{-1}$, which was half as fast as that at

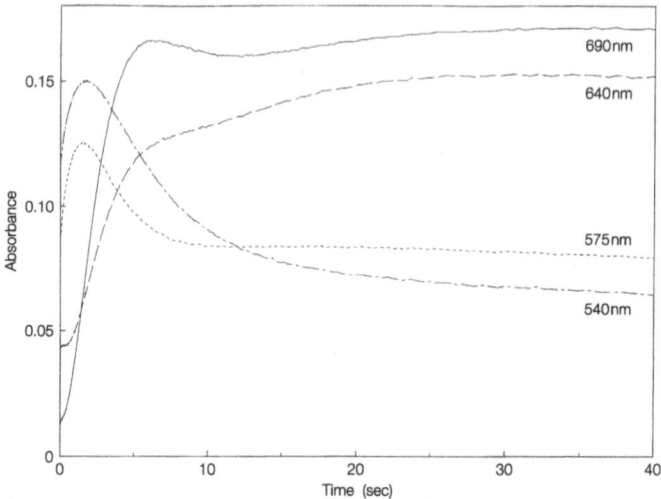

F<small>IG</small>. 3. Stopped-flow kinetics of absorbance changes during heme degradation by the heme–heme oxygenase complex. The absorbance changes at selected wavelengths were extracted from the time-resolved spectra obtained in the experiment shown in Fig. 2. The *solid, broken, dotted,* and *dash-dotted lines* represent the absorption decay curves at 690, 640, 575, and 540 nm, respectively

575 nm. This may be because the oxygenated form and hydroxyheme both have a maximum around 540 nm. The increase of absorbance at 690 nm started with a delay of 0.4 s after mixing and proceeded at a rate of $0.5 s^{-1}$ up to 6 s, followed by a slight decrease. It then increased again at a rate of $0.16 s^{-1}$ after 13 s. The early increase indicates the formation of verdoheme and the latter the production of biliverdin. The decrease between them should correspond to the conversion of a portion of the verdoheme into its CO-bound form. The absorbance change at 640 nm also showed at least two phases: the first increase had a delay of 0.6 s after mixing and proceeded at a rate of $0.3 s^{-1}$ up to 10 s, and the second increase after that proceeded at a slower rate of $0.18 s^{-1}$. As the first increase of the absorbance at 640 nm followed the absorption change at 690 nm, it was attributed to the formation of CO-bound verdoheme. The second increase after 10 s was mainly caused by biliverdin formation, because the rate was similar to the rate of production of biliverdin calculated from the change at 690 nm.

In summary, the redox potential of the heme-heme oxygenase complex was determined for the first time and was found to be −76 mV at pH 7.0. The heme–heme oxygenase complex rapidly binds dioxygen to become the oxygenated form; the rate of this reaction was four times as fast as that of its conversion into verdoheme. The cleavage of the porphyrin ring to biliverdin was three times slower than the formation of verdoheme. Thus, it was concluded that the opening of the porphyrin ring is the rate-determining step in the heme oxygenase reaction.

*Acknowledgment.* Supported by grant-in-aid 08249104 for scientific research on the priority area (Molecular Science on the Specific Roles of Metal Ions in Biological Functions) from the Ministry of Education, Science, Sports and Culture of Japan.

# References

1. Tenhunen R, Marver HS, Pimstone NR, et al. (1972) Enzymatic degradation of heme. Oxygenative cleavage requiring cytochrome P-450. Biochemistry 11:1716–1720
2. Marks GS (1994) Heme oxygenase: the physiological role of one of its metabolites, carbon monoxide and interactions with zinc protoporphyrin, cobalt protoporphyrin and other metalloporphyrins. Cell Mol Biol 40:863–870
3. Yoshida T, Kikuchi G (1978) Features of the reaction of heme degradation catalyzed by the reconstituted microsomal heme oxygenase system. J Biol Chem 253:4230–4236
4. Kikuchi G, Yoshida T (1980) Heme degradation by the microsomal heme oxygenase system. Trends Biochem Sci 5:323–325
5. Yoshida T, Kikuchi G (1978) Purification and properties of heme oxygenase from pig spleen microsomes. J Biol Chem 253:4224–4229
6. Craig SP III, Yuan L, Kuntz DA, et al. (1991) High level expression in *Escherichia coli* of soluble, enzymatically active schistosomal hypoxanthine/guanine phosphoribosyl-transferase and trypanosomal ornithine decarboxylase. Proc Natl Acad Sci USA 88:2500–2504
7. Wilks A, Ortiz de Montellano PR (1993) Rat liver heme oxygenase. High level expression of a truncated soluble form and nature of the *meso*-hydroxylating species. J Biol Chem 268:22357–22362
8. Yasukochi Y, Masters BSS (1976) Some properties of a detergent-solubilized NADPH-cytochrome *c* (cytochrome P-450) reductase purified by biospecific affinity chromatography. J Biol Chem 251:5337–5344
9. Wilks A, Black SM, Miller WL, et al. (1995) Expression and characterization of truncated human heme oxygenase (hHO-1) and a fusion protein of hHO-1 with human cytochrome P-450 reductase. Biochemistry 34:4421–4427
10. Matera KM, Takahashi S, Fujii H, et al. (1996) Oxygen and one reducing equivalent are both required for the conversion of $\alpha$-hydroxyhemin to verdoheme in heme oxygenase. J Biol Chem 271:6618–6624
11. McDonagh AF, Palma LA (1980) Preparation and properties of crystalline biliverdin IXa. Biochem J 189:193–208

# Expression of Heme Oxygenase and Inducible Nitric Oxide Synthase mRNA in a Human Glioblastoma Cell Line

EISHI HARA, KAZUHIRO TAKAHASHI, HIROYOSHI FUJITA, and SHIGEKI SHIBAHARA

*Summary.* Carbon monoxide and nitric oxide (NO) have been shown to function as neural messengers and are synthesized by heme oxygenase and NO synthase, respectively. Expression of each enzyme has been shown to be modulated by heme and NO, raising a possibility of the coordinated regulation of the two enzymes. This study was designed to compare the regulation of expression of the two inducible enzymes, heme oxygenase-1 and inducible NO synthase (iNOS) in a human glioblastoma cell line, A172. Northern blot analysis showed that treatment with cytokines increased the expression of iNOS mRNA but not heme oxygenase-1 mRNA, and treatment with an NO-releasing agent sodium nitroprusside increased the expression of heme oxygenase-1 mRNA but not iNOS mRNA. Thus, the expression levels of iNOS mRNA are not noticeably correlated with those of heme oxygenase-1 mRNA, suggesting that the two enzymes may be induced in a different manner in humans.

*Key words.* Heme oxygenase—Inducible nitric oxide synthase—Human—Glioblastoma—Sodium nitroprusside

## Introduction

Both carbon monoxide (CO) and nitric oxide (NO) are supposed to function as gaseous messenger molecules in various cells and tissues [1–5]. CO is produced during physiological heme degradation catalyzed by heme oxygenase, an essential enzyme in heme catabolism [6,7], and NO is synthesized from L-arginine by NO synthase (NOS) [4,5]. Each enzyme constitutes both inducible and noninducible isozymes. Heme oxygenase-1 and inducible NOS (iNOS) are inducible isozymes of each enzyme [4,5,8,9].

Several lines of evidence suggest a possible correlation in the regulation of expression of heme oxygenase-1 and iNOS. Heme oxygenase-1 is inducible by its own substrate, heme, which is also known as a target molecule for NO [8–10]. The expression of heme oxygenase-1 mRNA was increased by treatment with NO donors in rat

---

[1] Department of Applied Physiology and Molecular Biology, Tohoku University School of Medicine, 2-1 Seiryo-machi, Aoba-ku, Sendai, Miyagi 980-77, Japan

hepatocytes [11] and in T98G human glioblastoma cells [12], suggesting that NO may increase the expression of heme oxygenase-1 mRNA under certain conditions. On the other hand, iNOS transcription is altered (increased or decreased) by cellular iron availability in the murine macrophage cell line J774 [13]. Iron is also one of the heme degradation products catalyzed by heme oxygenase. Furthermore, in porcine vascular endothelial cells, both heme oxygenase and NOS activities were induced in parallel by treatment with cytokines [14]. Recent studies have suggested that upregulation of heme oxygenase-1 is involved in the pathophysiology of certain neurodegenerative diseases such as Alzheimer's disease [15]. It was also suggested that NO plays a role in pathophysiological features of various neurological diseases [16].

This study was therefore designed to explore the possible correlation of the expression of the two inducible enzymes generating CO and NO in humans. As a first step, we examined the expression levels of both heme oxygenase-1 and iNOS mRNA in a glioblastoma cell line.

## Materials and Methods

A human glioblastoma cell line, A172, was cultivated in Dulbecco's modified Eagle's medium supplemented with 10% fetal calf serum (FCS). For induction experiments, A172 cells were exposed for 3–24h to a cytokine mixture containing interferon-$\gamma$ (IFN-$\gamma$) (500 U/ml), interleukin-1$\beta$ (IL-1$\beta$) (5 ng/ml), and tumor necrosis factor-$\alpha$ (TNF-$\alpha$) (500 U/ml). A172 cells were also treated for 5h with sodium nitroprusside (SNP, $Na_2[Fe(CN)_5NO]$). The treated cells were harvested for RNA extraction. Total RNA was subjected to Northern blot analysis as described previously [17]. The hybridization probes for heme oxygenase-1 and heme oxygenase-2 mRNAs were the XhoI/XbaI fragment (−64/923) derived from the human heme oxygenase-1 cDNA, pHHO1 [18], and the HinfI/HinfI fragment derived from the human heme oxygenase-2 cDNA, pHHO2-2 [19], respectively. The cDNA probe for iNOS (0/537) was derived from the human iNOS cDNA, SpHiNOS1 [17]. A $\beta$-actin probe was prepared as described previously [19].

## Results and Discussion

Northern blot analysis showed that the expression levels of iNOS mRNA were not detected in untreated A172 cells, but were noticeably increased at 3h and reached a maximum level at 6h after the addition of the cytokine mixture (Fig. 1) [17]. The expression levels of heme oxygenase-1 (Fig. 1) and heme oxygenase-2 mRNA (data not shown) were not noticeably affected. This observation is consistent in part with a recent report of transient transfection assays showing that the human iNOS gene promoter is able to confer the induction of a reporter gene in response to the mixture of IFN-$\gamma$, IL-1$\beta$, and TNF-$\alpha$ [20]. On the other hand, the lack of induction of heme oxygenase-1 mRNA in cytokine-stimulated cells may represent the cell-type differences in the regulation of heme oxygenase-1.

Heme oxygenase-1 is inducible in a human monocytic leukemia cell line, THP-1, during its differentiation by certain stimulants such as TNF-$\alpha$ or a combination of IFN-$\gamma$ and lipopolysaccharide [21]. Heme oxygenase-1 was established as a stress

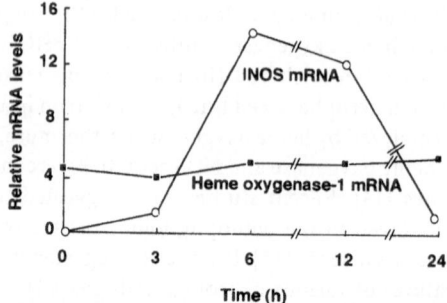

FIG. 1. Effects of cytokine mixture on the expression of heme oxygenase-1 and inducible nitric oxide synthase (*iNOS*) mRNA in A172 cells. Shown is the time course of relative expression levels of heme oxygenase-1 mRNA and iNOS mRNA in the cells treated with a cytokine mixture [17]. The intensity of hybridization signals was quantified with a Bioimage analyzer (BAS 2000, Fuji Film, Tokyo, Japan), and the intensity representing heme oxygenase-1 and iNOS mRNA was normalized with the intensity for β-actin. *Closed squares*, heme oxygenase-1 mRNA; *open circles*, iNOS mRNA

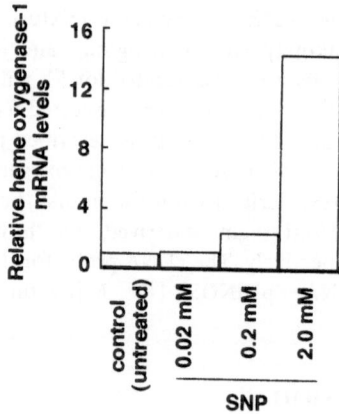

FIG. 2. Effects of sodium nitroprusside (*SNP*) on the expression of heme oxygenase-1 mRNA. Shown are the relative expression levels of heme oxygenase-1 mRNA in A172 cells treated with SNP for 5 h [17]. The intensity of hybridization signals was quantified with a Bioimage analyzer, and the intensity representing heme oxygenase-1 mRNA was normalized with the intensity for β-actin. Expression of iNOS mRNA was undetectable in this series of experiments

protein [9,22], and its human gene promoter contains a putative heat-shock element (HSE) [23]. HSE is responsible for transcriptional activation of the heat-shock protein genes by high temperature and other stresses. However, the activity of human heme oxygenase-1 was not induced by heat shock in several human cell lines, unlike rat heme oxygenase-1 [8,9,18,24], and the lack of heat-mediated inducibility of the heme oxgenase-1 gene was suggested to be caused by the presence of the region located downstream from HSE, which may act as a silencer [9,25]. Therefore there is a need to investigate whether such a putative silencer is also responsible for the lack of induction of heme oxygease-1 mRNA in cytokine-stimulated A172 cells.

To study the possible coordinated regulation of expression of heme oxygenase-1 and iNOS, we examined the effect of SNP, a typical NO-releasing agent, on the expression of heme oxygenase-1 and iNOS mRNA in A172 cells. Northern blot analysis showed that heme oxygenase-1 mRNA expression was increased by the treatment with SNP in a dose-dependent manner, but expression of iNOS mRNA was not detectable under the conditions used (Fig. 2). It has been shown that the first step in the reduction of nitroprusside is the formation of a one-electron reduction product, $[Fe(CN)_5NO]^{3-}$ [26,27]. This radical form of nitroprusside is then processed, leading to NO release. Therefore, we could not exclude the possibility that the effects of SNP on the expression of heme oxygenase-1 mRNA were partly mediated by an Fe-NO complex such as $[Fe(CN)_5NO]^{3-}$. In this context, it is noteworthy that treatment with 1 mM sodium ferrocyanide, 1 mM potassium ferrocyanide, or 1 mM potassium ferricyanide had no noticeable effects on the expression of heme oxygenase-1 mRNA in T98G glioblastoma cells [12]. We are currently investigating the molecular mechanism by which heme oxygenase-1 mRNA expression is increased by SNP.

# References

1. Stevens CF, Wang Y (1993) Reversal of long-term potentiation by inhibitors of haem oxygenase. Nature 364:147–149
2. Verma A, Hirsch J, Glatt CE, et al. (1993) Carbon monoxide: a putative neural messenger. Science 259:381–384
3. Zhuo M, Small SA, Kandel ER, et al. (1993) Nitric oxide and carbon monoxide produce activity-dependent long-term synaptic enhancement in hippocampus. Science 260:1946–1950
4. Lowenstein CJ, Snyder SH (1992) Nitric oxide, a novel biologic messenger. Cell 70:705–707
5. Nathan C (1992) Nitric oxide as a secretary product of mammalian cells. FASEB J 6:3051–3064
6. Tenhunen R, Marver HS, Schmid R (1968) The enzymatic conversion of heme to bilirubin by microsomal heme oxygenase. Proc Natl Acad Sci USA 61:748–755
7. Tenhunen R, Marver HS, Schmid R (1969) Microsomal heme oxygenase. Characterization of the enzyme. J Biol Chem 244:6388–6394
8. Shibahara S (1988) Regulation of heme oxygenase gene expression. Semin Hematol 25:370–376
9. Shibahara S (1994) Heme oxygenase—regulation of and physiological implication in heme catabolism. In: Fujita H (ed) Regulation of heme protein synthesis. AlphaMed Press, Medina, OH, pp 103–116
10. Stamler JS (1994) Redox signaling: nitrosylation and related target interactions of nitric oxide. Cell 78:931–936
11. Kim Y-M, Bergonia HA, Muller C, et al. (1995) Loss and degradation of enzyme-bound heme induced by cellular nitric oxide synthesis. J Biol Chem 270:5710–5713
12. Takahashi K, Hara E, Suzuki H, et al. (1996) Expression of heme oxygenase isozyme mRNAs in the human brain and induction of heme oxygenase-1 by nitric oxide donors. J Neurochem 67:482–489
13. Weiss BG, Werner-Felmayer G, Werner ER, et al. (1994) Iron regulates nitric oxide synthase activity by controlling nuclear transcription. J Exp Med 180:969–976

14. Motterlini R, Foresti R, Intaglietta M, et al. (1996) NO-mediated activation of heme oxygenase: endogenous cytoprotection against oxidative stress to endothelium. Am J Physiol 270:H107–H114
15. Schipper HM, Cisse S, Stopa EG (1995) Expression of heme oxygenase-1 in the senescent and Alzheimer-diseased brain. Ann Neurol 37:758–768
16. Faraci FM, Brian JE (1994) Nitric oxide and the cerebral circulation. Stroke 25:692–703
17. Hara E, Takahashi K, Tominaga T, et al. (1996) Expression of heme oxygenase and inducible nitric oxide synthase mRNA in human brain tumors. Biochem Biophys Res Commun 224:153–158
18. Yoshida T, Biro P, Cohen T, et al. (1988) Human heme oxygenase cDNA and induction of its mRNA by hemin. Eur J Biochem 171:457–461
19. Shibahara S, Yoshizawa M, Suzuki H, et al. (1993) Functional analysis of cDNAs for two types of human heme oxygenase and evidence for their separate regulation. J Biochem (Tokyo) 113:214–218
20. De Vera ME, Shapiro RA, Nussler AK, et al. (1996) Transcriptional regulation of human inducible nitric oxide synthase (NOS 2) gene by cytokines: initial analysis of the human NOS promoter. Proc Natl Acad Sci USA 93:1054–1059
21. Muraosa Y, Shibahara S (1993) Identification of a cis-regulatory element and putative transacting factors responsible for 12-O-tetradecanoylphorbol-13-acetate (TPA)- mediated induction of heme oxygenase expression in myelomonocytic cell lines. Mol Cell Biol 13:7881–7891
22. Shibahara S, Muller RM, Taguchi H (1987) Transcriptional control of rat heme oxygenase by heat shock. J Biol Chem 262:12889–12892
23. Shibahara S, Sato M, Muller RM, et al. (1989) Structual organization of the human heme oxygenase gene and the function of its promoter. Eur J Biochem 179:557–563
24. Okinaga S, Shibahara S (1993) Identification of a nuclear protein that constitutively recognizes the sequence containing a heat-shock element. Its binding properties and possible function of modulating heat-shock induction of the rat heme oxygenase gene. Eur J Biochem 212:167–175
25. Okinaga S, Takahashi K, Takeda K, et al. (1996) Regulation of heme oxygenase-1 gene expression under thermal stress. Blood 87:5074–5084
26. Bates JN, Baker MT, Guerra R Jr, et al. (1991) Nitric oxide generation from nitroprusside by vascular tissue. Evidence that reduction of the nitroprusside anion and cyanide loss are required. Biochem Pharmacol 42:S157–S165
27. Ramakrishna Rao DN, Cederbaum AI (1996) Generation of reactive oxygen species by the redox cycling of nitroprusside. Biochim Biophys Acta 1289:195–202

# The Mechanism of Conversion of Xanthine Dehydrogenase to Xanthine Oxidase

Takeshi Nishino[1], Ken Okamoto[1], Shigeko Nakanishi[1], Hiroyuki Hori[1], and Tomoko Nishino[2]

*Summary.* Xanthine dehydrogenase and xanthine oxidase are complex metalloflavoproteins that represent alternate forms of the same gene product. The cDNAs encoding the enzymes have been cloned from several sources, and structural information is becoming available. Using purified enzyme, comparative analyses between the two forms were attempted by spectroscopic and kinetics methods. The most significant difference between the two forms is the protein conformation around flavin adenine dinucleotide (FAD), which changes the redox potential of the flavin and the reactivity of FAD with the electron acceptors, nicotinamide adenine dinucleotide (NAD) and molecular oxygen. The flavin semiquinone is thermodynamically stable in xanthine dehydrogenase but is unstable in xanthine oxidase. Detailed analyses by stopped-flow techniques suggest that the flavin semiquinone reacts with oxygen to form superoxide anion while the fully reduced flavin reacts to form hydrogen peroxide. Although xanthine dehydrogenase can produce greater amounts of superoxide anion than xanthine oxidase during xanthine oxygen reaction, it seems not to be physiologically significant in the cell, where excess NAD exists under normal conditions.

*Key words.* Xanthine oxidase—Xantine dehydrogenase—Superoxide anion—Oxygen radical—Reperfusion injury

## Introduction

Although mammalian xanthine oxidase exists originally as a dehydrogenase form in freshly prepared samples, it is converted to an oxidase form during purification, either irreversibly by proteolysis or reversibly by sulfhydryl oxidation of the protein molecule [1]. The relationship between xanthine dehydrogenase and oxidase has been the subject of considerable debate, primarily because of its proposed roles in ischemia–reperfusion damage in tissues. The cDNAs encoding the enzyme have been

[1] Department of Biochemistry and Molecular Biology, Nippon Medical School, 1-1-5 Sendagi, Bunkyo-ku, Tokyo 113, Japan
[2] Department of Biochemistry, Yokohama City University School of Medicine, Fukuura 3-9, Yokohama 236, Japan

cloned from several sources, and the expression analyses of cDNA using a baculovirus–insect cell system provide more detailed information about the mechanism of conversion from xanthine dehydrogenase to xanthine oxidase. This chapter presents recent advances of our understanding of biochemistry and molecular biology of these systems.

## Structural Properties of Xanthine-Oxidizing Enzyme

Xanthine dehydrogenases from various sources are proteins of MR 300000 that are composed of two identical subunits. Each subunit contains one molybdopterin, two nonidentical $Fe_2S_2$ centers, and flavin adenine dinucleotide (FAD) [2,3]. The molecular weight of each subunit of the native enzyme prepared without proteolysis is 150000 [4,5]. The full amino acid sequences of the liver enzymes from human [6], rat [5], bovine milk [7], mouse [8], and chicken [9] and of the enzyme from *Drosophila* [10] have been determined from cDNA cloning. The enzymes consist of about 1330 amino acids, and the amino acid sequence is highly homologous among mammalian enzymes, with about 90% identity [5–8]. On the other hand, weaker homology was observed between the mammalian enzyme and the *Drosophila* enzyme with 52% identity [5], and between chicken liver and the mammalian enzyme with 70% identity [9].

By limited proteolysis of the mammalian enzyme with trypsin [5] or the chicken enzyme with subtilisin [9], the enzymes were cleaved into three fragments (20, 40, and 85kDa, respectively). These fragments are only dissociated under denaturation conditions such as in the presence of high concentrations of guanidine hydrochloride [5,9], indicating that the three fragments associate closely with each other. The mammalian enzyme is converted to an oxidase by nicking at these two positions, while the chicken enzyme is not converted to an oxidase. By determination of N-terminal amino acid sequences of each fragment, the 20-KD fragment is assigned to the N-terminal portion, the 85-kDa fragment to the C-terminal portion, and the 40-kDa fragment to an intermediate portion. The sequence comparison [5,9] and chemical modification studies [11] suggested that each redox center is located in different domains, e.g., the two iron–sulfur centers are in the 20-kDa domain, the FAD is in the 40-kDa domain, and the molybdopterin is in the 85-kDa domain [5,9]. By limited proteolysis under particular digestion conditions, the chicken liver enzyme [9] could be isolated as 20-kDa and 85-kDa complexes containing only Fe–S centers and molybdenum centers, supporting the idea that the 40-kDa fragment is in the FAD domain.

## Difference in Structure Between Xanthine Dehydrogenase and Oxidase

Enzyme isolated from rat liver or bovine milk without proteolysis can be purified as a dehydrogenase form by rapid purification [12] or in the presence of sulfhydryl reducing reagents [13]. Alternatively, an oxidase form without proteolysis can be converted to the dehydrogenase form by incubation with sulfhydryl reducing reagents [14]. The dehydrogenase can be converted irreversibly to an oxidase by proteolysis or reversibly by sulfhydryl oxidants[1,5,12]. During conversion from a dehydrogenase to an oxidase by a sulfhydryl oxidant, several cysteine residues were oxidized, but some of them are not involved in the dehydrogenase to oxidase conversion [13–15]. It is likely that two cysteine residues are close enough to form readily a disulfide bridge in the three-dimensional structure of the enzyme [14,15]. On modification with dithio-dipyridine, more than four pairs of cysteine disulfide bridges were formed, but some of them are not considered to be involved in the dehydrogenase to oxidase conversion.

On modification of the protein molecule either by proteolysis or disulfide formation, significant conformational changes occur, particularly around the flavin, resulting in changes in reactivity of the flavin as well as the loss of the nicotinamide adenosine dinucleotide (NAD) binding site [16,17]. It was shown from the spectral perturbation by addition of NAD or its analogue to the enzyme that the dehydrogenase has an NAD-binding site but the oxidase has not [13,15]. The evidence for differences in conformation around the flavin has been provided by active site probe studies using artificial flavins having an ionizable residue at the 6- or 8-position [15–17]. When the normal FAD of the dehydrogenase was replaced by 8- or 6-SH-FAD, the artificial flavin bound the dehydrogenase as a neutral form, whereas it bound the oxidase as an anionic form (Fig. 1). This indicated that the dehydrogenase perturbed the pK of the ionizable substituent by more than 4 units. Such perturbation of the pK suggests the existence of a strong negative charge in the flavin-binding site of the dehydrogenase [15–17]. Evidence for the different conformations between the two types was also provided by another artificial flavin having a reactive residue at the 6-position. The 6-position of FAD is buried in the dehydrogenase, but it is open to the solvent in the oxidase [15].

Although the mammalian enzyme can be easily converted to the oxidase, the chicken enzyme is not converted by either proteolysis or sulfhydryl modification. On proteolysis of the chicken liver enzyme, xanthine NAD activity was decreased without increase of xanthine oxygen activity [18]. The chicken enzyme has an additional 23-amino-acid insertion at the N-terminal region and therefore is a little longer than the mammalian enzyme. This position contains 4 cysteine residues. To know the structural differences for this different property between the chicken and rat enzyme, we analyzed it by constructing the chimeric enzyme. We used a cDNA expression system with a baculovirus–insect cell system. The expressed enzyme, which has all three domains from the rat enzyme, exhibited a dehydrogenase-type character. It had a high activity toward NAD but low activity toward oxygen, and was readily converted to oxidase by treatment with sulfhydryl oxidant, indicating that the expressed enzyme has the same property as the natural enzyme purified from rat liver.

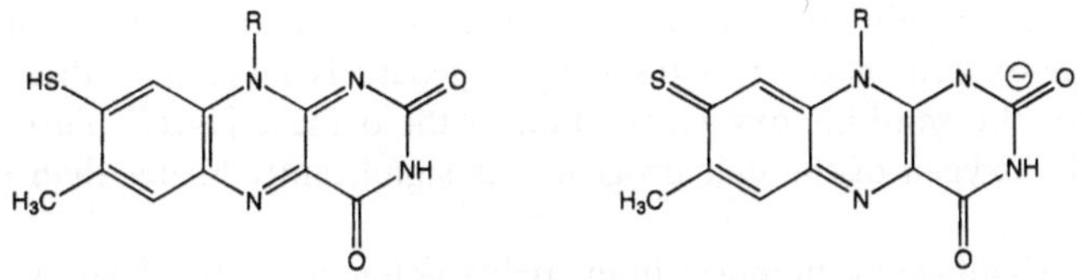

Fig. 1. 8-Mercapto flavin adenine dinucleotide (FAD) bound to xanthine dehydrogenase (*left*; $\lambda_{max}$ = 456 nm) or xanthine oxidase (*right*; $\lambda$ = 579 nm)

However, the chimeric enzyme, which had the 23-amino-acid fragment insertion in the rat enzyme, exhibited the same property as the wild-type rat liver xanthine dehydrogenase. This enzyme was also convertible from xanthine dehydrogenase to xanthine oxidase on modification of a cysteine residue modifier. Another chimeric enzyme, which had the flavin domain of chicken enzyme, was expressed as a typical xanthine dehydrogenase form and could not be converted to xanthine oxidase by modification with sulfhydryl oxidant. Thus, this clearly indicates that the difference in properties between rat and chicken enzymes results from neither the difference in the N-terminal nor C-terminal domain but in an intermediate domain. Comparing the amino acid sequence of the flavin domain of chicken enzyme to that of rat enzyme, generally the sequence exhibits high homology in this region.

However, the region around the cysteine residue was relatively not conserved between the two enzymes. One cysteine residue was modified by fluorodinitrobenzene, resulting in the conversion from xanthine dehydrogenase to oxidase; although this cysteine residue is conserved between the two enzymes, it is followed by a glycine residue in the rat enzyme but by glutamate in the chicken enzyme. Presumably the glutamate residue might disturb the modification of the cysteine residue by the sylfhydryl reagent or, alternatively, this glutamate residue might be a proton-donating residue to stabilize the neutral blue flavin semiquinone. The different amino acid sequence in the 40-kDa domain from the mammalian enzymes might explain why the conversion does not occur in the chicken liver enzyme.

## Difference in Catalytic Properties Between Xanthine Dehydrogenase and Oxidase

Xanthine reacts with the molybdopterin during catalytic turnover. Two electrons are transferred from xanthine to Mo(VI), reducing the metal to Mo(IV). On release of the product of urate from the molybdenum, the electrons are transferred to other centers very rapidly. Pulse radiolysis studies showed that the electron transfer rate was faster than the rate of reduction or reoxidation of the enzyme by the substrates in both xanthine dehydrogenase and oxidase [19,20]. The role of these two centers has been postulated to be as an electron sink [21]. As the reaction site of NAD or oxygen is the flavin cofactor, the difference in properties between the dehydrogenase and the oxidase is mainly caused by the different reactivates of the flavin cofactor with these electron acceptors.

The dehydrogenase reacts rapidly with NAD, but slowly with oxygen. On the other hand, the oxidase reacts rapidly with oxygen and not with NAD. Although the turnover number for the xanthine-NAD reaction of the dehydrogenase is similar to that for the xanthine-oxygen reaction of the oxidase, the turnover number for the xanthine-oxygen reaction of the dehydrogenase is more than threefold lower than that for the xanthine-oxygen reaction of the oxidase [14,15]. Further, the $K_m$ value for the oxygen of the dehydrogenase is significantly higher than that of the oxidase.

There are significant differences in the redox potentials of the flavin between xanthine dehydrogenase and oxidase but not between other centers [14,21]. The redox potential of the $FADH^{\cdot}/FADH_2$ couple is much lower than that of the FAD/

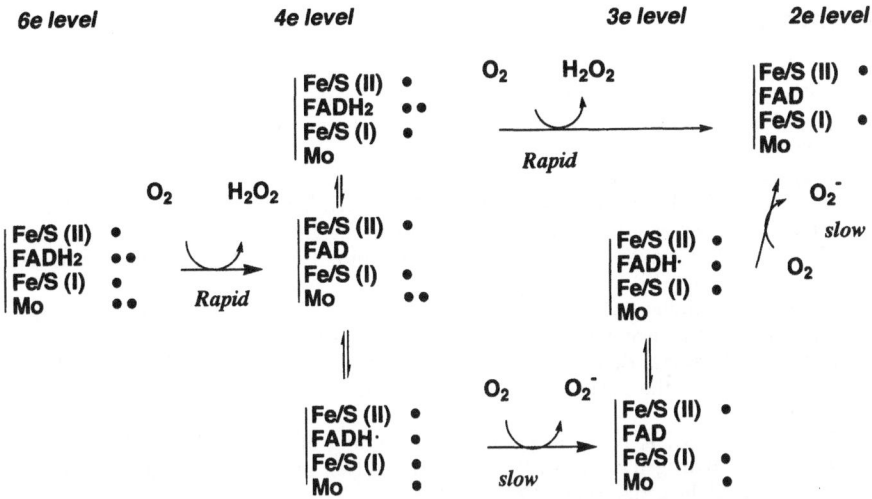

FIG. 2. Reaction of fully reduced xanthine dehydrogenase with $O_2$ to 2-electron reduced level.

FADH· couple in the dehydrogenase, indicating that thermodynamic stabilization of the flavin radicals exists in the dehydrogenase [14,15]. On the other hand, the redox potential of the FADH·/FADH$_2$ couple is higher than that of the FAD/FADH·couple in the oxidase. Such a stabilization of the flavin semiquinone in the dehydrogenase seems to result from the flavin–protein interaction and not the midpotential of flavin, because this property of flavin is maintained even when the flavin has been replaced by artificial flavins having higher redox potentials [22].

Both xanthine dehydrogenase and oxidase produce both $H_2O_2$ and $O_2^-$ when xanthine and molecular oxygen are used as substrates. The dehydrogenase produces more superoxide anion than the oxidase does [14,15] during turnover. Stopped-flow experiments of the oxidative half-reaction with the dehydrogenase [23] have shown that the fully reduced flavin reacts with oxygen to give $H_2O_2$ and the flavin semiquinone reacts to give the superoxide anion (Fig. 2). The greater amount of formation of superoxide anion in the dehydrogenase is well explained by the greater formation of the flavin semiquinone in this form [23]. The slower reaction of the flavin semiquinone with oxygen seems to be one of the reasons why the dehydrogenase reacts slowly with oxygen. In the cell, however, formation of superoxide by the dehydrogenase is almost completely inhibited by NAD, which exists in sufficient amounts under normal conditions [14,23]. The binding of NAD to the dehydrogenase seems to change the redox potential of the FADH/FADH$_2$ couple to a more positive value, resulting in less formation of the flavin semiquinone. On the other hand, the 3-aminopyridine analogue of NAD stabilizes the semiquinone more noticeably [24].

*Acknowledgments.* This work was supported by grants-in-aid (08249104) for science reseach on priority areas from the Ministry of Education, Science, Sports and Culture of Japan.

# References

1. Della Corte E, Stirpe F (1968) The regulation of xanthine oxidase in rat liver: modification of the enzyme activity of rat liver supernatant on storage at −20°C. Biochem J 108:349–351
2. Bray RC (1975) Molybdenum iron-sulfur flavin hydroxylases and related enzymes. In: Boyer PD (ed) The enzymes, vol. XII, part B. Academic Press, New York, pp 299–419
3. Hille R, Massey V (1985) Molybdenum-containing hydroxylase: xanthine oxidase, aldehyde oxidase and, and sulfite oxidase. In: Spiro TG (ed) Molybdenum enzymes, vol 7. Wiley, New York, pp 443–518
4. Nelson CA, Handler P (1968) Preparation of bovine xanthine oxidase and the subunit structure of some iron flavoproteins. J Biol Chem 243:5368–5373
5. Amaya Y, Yamazaki K, Sato M, et al. (1990) Proteolytic conversion of xanthine dehydrogenase from the NAD-dependent type to the $O_2$-dependent type. J Biol Chem 265:14170–14175
6. Ichida K, Amaya Y, Noda K, et al. (1993) Cloning of the cDNA encoding human xanthine dehydrogenase (oxidase): structural analysis of the protein and chromosomal location of the gene. Gene 133:279–284
7. Berglund L, Rasmussen JT, Andersen MD, et al. (1996) Purification of the bovine xanthine oxidoreductase from milk fat globule membrane and cloning of complementary deoxiribonucleic acid. J Dairy Sci 79:198–204
8. Terao M, Cazzaniga G, Ghezzi P, et al. (1992) Molecular cloning of a cDNA coding for mouse liver xanthine dehydrogenase. Biochem J 283:863–870
9. Sato A, Nishino T, Noda K, et al. (1995) The structure of chicken liver xanthine dehydrogenase: cDNA cloning and the domain structure. J Biol Chem 270:2818–2826
10. Keith RP, Riley MJ, Kreitman M, et al. (1987) Sequence of the structural gene for xanthine dehydrogenase (*rosy* locus) in *Drosophilamelanogaster*. Genetics 116:67–73
11. Nishino T, Nishino T (1989) The nicotinamide adenine dinucleotide-binding site of chicken liver xanthine dehydrogenase. J Biol Chem 264:5468–5473
12. Waud WR, Rajagopalan KV (1976) The mechanism of conversion of rat liver xanthine dehydrogenase from an NAD-dependent form (type D) to an $O_2$-dependent type (type O). Arch Biochem Biophys 172:354–364
13. Nakamura M, Yamazaki I (1982) Preparation of bovine milk xanthine oxidase as a dehydrogenase form. J Biochem (Tokyo) 92:1279–1286
14. Saito T, Nishino T (1989) Differences in redox and kinetic properties between NAD-dependent and $O_2$-dependent types of rat liver xanthine dehydrogenase. J Biol Chem 264:10015–10022
15. Hunt J, Massey V (1992) Purification and properties of milk xanthine dehydrogenase. J Biol Chem 267:21479–21485
16. Massey V, Schopfer LM, Nishino T, Nishino T (1989) Differences in protein structure of xanthine dehydrogenase and xanthine oxidase revealed by reconstitution with active site probe study. J Biol Chem 264:10567–10573
17. Saito T, Nishino T, Massey V (1989) Differences in environment of FAD between NAD-dependent and $O_2$-dependent types of rat liver xanthine dehydrogenase shown by active site probe study. J Biol Chem 264:15930–15935
18. Nishino T, Nishino T, Sato A, et al. (1994) Xanthine dehydrogenase: structure and properties. In: Yagi K (ed) Flavins and flavoproteins. de Gruyter, Berlin, pp 699–706
19. Anderson RF, Hille R, Massey V (1986) The radical chemistry of milk xanthine oxidase as studied by radiation chemistry techniques. J Biol Chem 261:15870–15876
20. Kobayashi K, Miki M, Okamoto K, Nishino T (1993) Electron transfer process in milk xanthine dehydrogenase as studied by pulse radiolysis. J Biol Chem 268:24642–24646
21. Hunt J, Massey V, Dunham WR, Sands RH (1993) Redox potentials of milk xanthine dehydrogenase. J Biol Chem 268:18685–18691

22. Nishino T, Nishino T, Schopfer LM, Massey V (1989) Reactivity of chicken liver xanthine dehydrogenase containing modified flavin. J Biol Chem 264:6075–6085
23. Nishino T, Nishino T, Schopfer LM, Massey V (1989) The reactivity of chicken liver xanthine dehydrogenase with molecular oxygen. J Biol Chem 264:2518–2527
24. Schopfer LM, Massey V, Nishino T (1988) Rapid reaction studies on the reduction and oxidation of chicken liver xanthine dehydrogenase by the xanthine/urate and NAD/NADH couples. J Biol Chem 263:13528–13538

# Mechanism-Based Molecular Design of Peroxygenases

YOSHIHITO WATANABE, SHIN-ICHI OZAKI, and TOSHITAKA MATSUI

*Summary.* Comparison of the X-ray structures of cytochrome *c* peroxidase (CcP) and sperm whale myoglobin (Mb) suggests that the Leu-29 → His and His-64 → Leu double mutation of Mb creates a heme crevice similar to the active site structure of CcP and enhance hydrogen peroxide-supported monooxygenation activity. We report here that L29H/H64L Mb significantly increases the rate for the oxidation of both thioanisole and styrene and, more importantly, the enantioselectivity. The 22-fold rate increase compared to wild-type Mb and 97% incorporation of $^{18}O$ from $H_2 {}^{18}O_2$ into the sulfoxide with 97% entantiomeric excess (e.e.) for the R isomer have been observed for thioanisole oxidation by L29H/H64L Mb. The L29H/H64L mutant oxidizes sulfides faster than the wild type with high enantio-selectivity. A great improvement of enantioselectivity, from 9% to 80%, was seen for styrene epoxidation by L29H/H64L Mb, and 92% of the oxygen atom of epoxide formed is derived from peroxide. The present results clearly indicate that His-29 in the mutant directly plays an important role in improving ferryl oxygen transfer activity.

*Key words.* Peroxygenase—Heme—Oxygen activation—Enantioselective oxidation—Myoglobin

## Introduction

Peroxide-dependent heme enzymes such as catalase and peroxidase are known to yield oxo-ferryl porphyrin π-cation radicals, so-called compound I, or their equivalents, by utilizing either hydrogen peroxide or other two-electron-oxidizing reagents [1,2]. As shown in Fig. 1, the oxygen activation process by these heme enzymes requires the heterolytic O–O bond cleavage of a putative hydroperoxo-Fe(III) intermediate, although many reactions of hydrogen peroxide and metal ions are known to afford hydroxyl radical (·OH) because of the homolytic O–O bond cleavage (Fenton reaction). Diffusible hydroxyl radical formation in biological systems is very harmful. To prevent hydroxyl radical formation at the active site of heme enzymes, some

Institute for Molecular Science, Okazaki National Research Institutes, 38 Nishigounaka, Myodaiji, Okazaki, Aichi 444, Japan

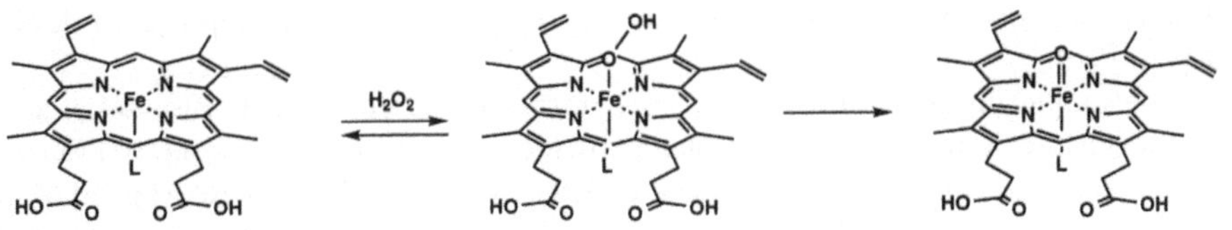

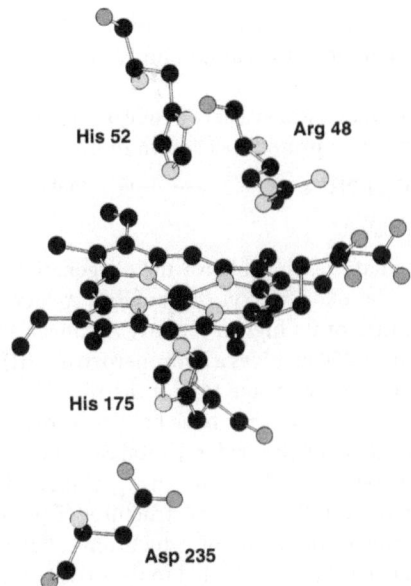

FIG. 1.  Oxygen activation by heme enzymes

FIG. 2.  Active site structure of cytochrome $c$ peroxidase (C$c$P)

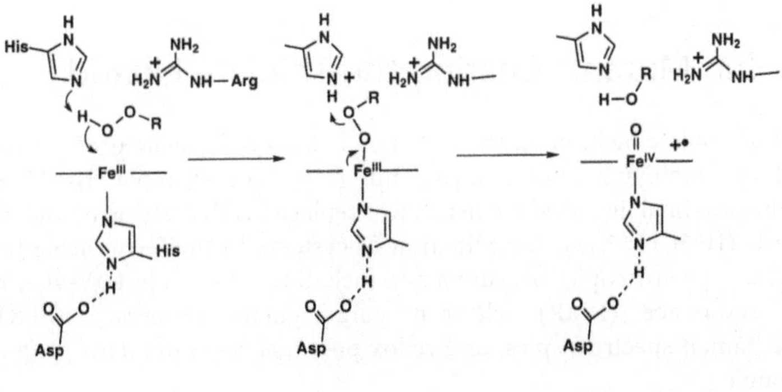

FIG. 3.  Catalytic roles of amino acid residues around heme in the formation of compound I

amino acid residues around the heme are located properly to play important roles (Fig. 2) [3,4].

For example, distal histidine (His-52) of cytochrome $c$ peroxidase (C$c$P) serves as a base to abstract a proton from hydrogen peroxide to assist hydroperoxo-Fe(III) formation. A proton on the distal histidine then interacts with the hydroperoxo ligand to polarize the O–O bond to allow the heterolytic O–O bond cleavage. Protonated arginine (Arg-48) near the heme makes the active site polar to stabilize the polar transition state. At the same time, anionic proximal histidine (His-175) binds to heme and introduces electron density into the $\pi^*$ orbital of the O–O bond to accelerate the heterolysis. These effects are summarized in Fig 3. Among these amino acid residues, the distal histidine is the most crucial amino acid; thus, the replacement of the distal histidine by leucine inhibits the compound I formation rate from $10^7$ $M^{-1}s^{-1}$ to $10^2$ $M^{-1}s^{-1}$ [5].

Cytochrome P-450 catalyzes a wide variety of oxygenation by employing molecular oxygen, two electrons, and two protons, as shown in Eq. 1 [6,7].

$$SH + O_2 + 2H^+ + 2e^- \longrightarrow SOH + H_2O \tag{1}$$

SH : substrate

While the active intermediate responsible for the oxygenation has been a subject of intensive studies, either the oxoperferryl or oxoferryl $\pi$-cation radical is the most attractive candidate because of its high reactivity. A conceivable oxygen activation mechanism of cytochrome P-450 involves a hydroperoxo-Fe(III) intermediate and the following heterolytic O–O bond cleavage to give the reactive intermediate [6,7]. Although very similar reaction mechanisms have been considered for the formation of compound I or its equivalent of peroxidase, catalase, and cytochrome P-450, the absence of any general acid–base catalysts and an arginine residue in the active site of cytochrome P-450 is apparent in their crystal structures (Fig. 4) [8]. Instead, thiolate of a cysteine residue ligates to the heme of cytochrome P-450, and strong electron donation from the thiolate has been attributed to the driving force for the heterolytic O–O bond cleavage [9,10]. In fact, exclusive enhancement of the heterolysis of acylperoxo-Fe(III) porphyrins by the introduction of electron-rich histidines as the sixth ligand [11] is consistent with the proposed role of the thiolate ligand in cytochrome P-450 [9,10].

## Myoglobin Mutants Bearing the Thiolate Ligand

The effect of thiolate ligation on the O–O bond cleavage in heme proteins has been examined by employing human myoglobin (Mb). For example, His-93 in Mb, which is the proximal ligand of the heme, was replaced with cysteine by site-directed mutagenesis (H93C) [12,13]. Coordination of cysteine to the ferric heme iron was confirmed by spectroscopic measurements including UV-visible (UV-vis), nuclear magnetic resonance (NMR), electron paramagnetic resonance (EPR), and resonance Raman spectroscopies, and redox potential measurements of the ferric/ferrous couple.

To examine the role of thiolate ligation in the O–O bond cleavage, a reaction of the ferric form of H93C Mb with cumene hydroperoxide was examined. Heterolytic O–O

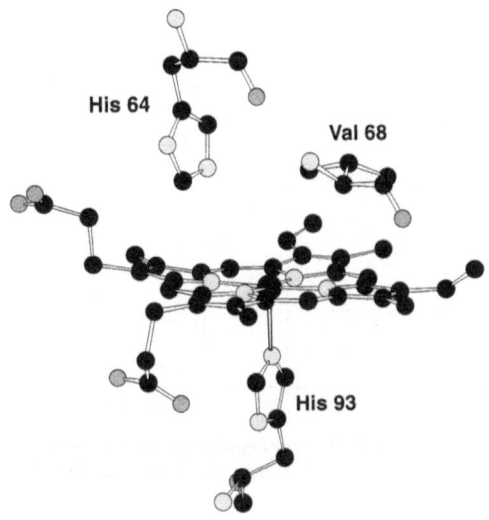

FIG. 4. Active site structure of cytochrome P-450$_{cam}$

bond cleavage of cumene hydroperoxide will afford compound I and cumylalcohol (Eq. 2).

$$Ph\text{-}C(CH_3)_2\text{-}OOH + Fe^{3+} \longrightarrow PhC(CH_3)_2\text{-}O\text{-}O\text{-}Fe^{3+} \qquad (2)$$

$$\longrightarrow PhC(CH_3)_2\text{-}OH + O=Fe\ Por^{+\bullet}$$

On the other hand, homolytic cleavage of the O–O bond affords compound II and Ph(CH$_3$)$_2$O• (Eq. 3) with subsequent elimination of the methyl radical to give acetophenone (Eq. 4).

$$PhC(CH_3)_2\text{-}O\text{-}O\text{-}Fe^{3+} \longrightarrow PhC(CH_3)_2\text{-}O\bullet + O=Fe\ Por \qquad (3)$$

$$PhC(CH_3)_2\text{-}O\bullet \longrightarrow PhC(O)CH_3 + CH_3\bullet \qquad (4)$$

Therefore, the production of cumylalcohol and acetophenone in the reactions of cumene hydroperoxide and H93C Mb as well as wild-type Mb was examined. The time courses of the product formation are shown in Fig. 5. Wild-type Mb reacts with cumene hydroperoxide by both heterolytic and homolytic mechanisms. The ligation of thiolate apparently greatly enhances heterolysis while having very little effect on homolysis.

Although we have attributed enhanced heterolysis of the O–O bond to electron donation from the thiolate ligand, possible participation of the distal histidine (His-64) of H93C Mb in the heterolysis cannot be eliminated (Fig. 6) [14]. In addition, racemic product formation in the epoxidation of styrene catalyzed by H93C Mb

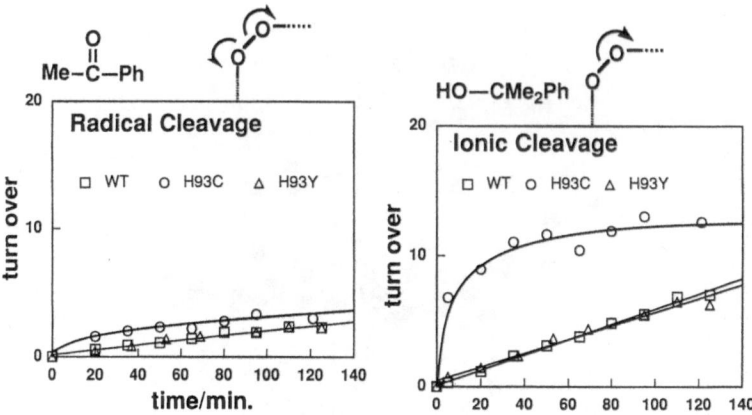

FIG. 5. Time-dependent formation of cumylalcohol and acetophenone in the reactions of cumene hydroperoxide with wild-type human myoglobin (Mb) and H93C and H93Y mutant Mb

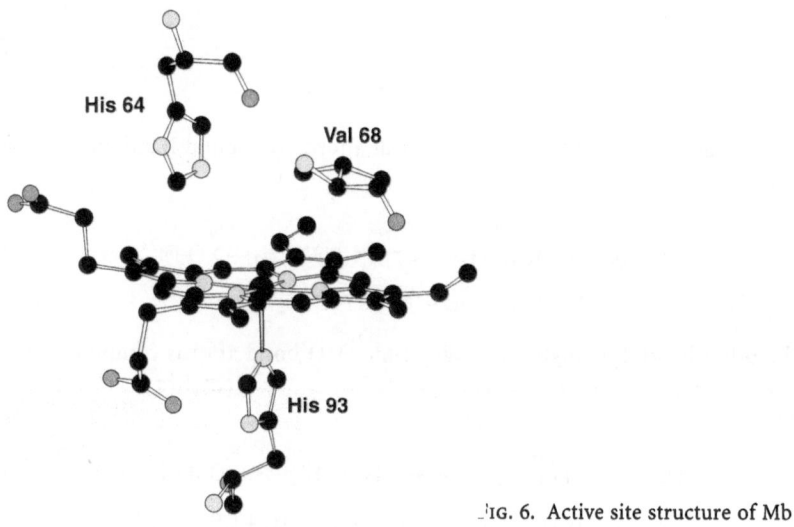

FIG. 6. Active site structure of Mb

implied that its small distal cavity could prevent substrates from accessing the heme and that the reactions proceed other than by the P-450 type mechanism (ferryl oxygen transfer). To clarify whether the distal histidine is involved in the O–O bond cleavage step, the distal histidine of H93C Mb was replaced by smaller and non-polar residues, glycine (H64G/H93C Mb) and valine (H64V/H93C Mb), by site-directed mutagenesis [15]. Various spectroscopic studies on these double-mutated Mbs revealed the ligation of cysteine to the ferric heme as a thiolate form. In the reaction with cumene hydroperoxide, the anionic nature of the proximal cysteine in H64G/H93C and H64V/

H93C Mbs was found to similarly enhance the heterolytic O–O bond cleavage as observed for H93C Mb. The results, which clearly demonstrate that the distal histidine of H93C Mb is not involved in the O–O bond cleavage step, are in good agreement with the proposed role of thiolate ligation for the formation of reactive intermediate, equivalent to compound I, in the catalytic cycle of P-450 reactions.

## Catalytic Roles of the Distal Site Histidine–Asparagine Couple in Peroxidases

The double mutants of Mb, H64G/H93C and H64V/H93C, were good protein models to study the effects of thiolate ligation on oxygen activation by cytochrome P-450; however, the small effect of the distal histidines in H64G/H93C and H64V/H93C on the O–O bond cleavage is contradictory to the role of the distal histidine of peroxidase [4,5]. As pointed out, replacement of the distal histidine of CcP (His-52) with leucine drastically diminishes the formation of compound I from $10^7$ $M^{-1}s^{-1}$ to $10^2$ $M^{-1}s^{-1}$ [5]. The different roles of the distal histidine in Mb and peroxidase seem to be very important in discriminating their biological functions. To understand the reasons for the different roles of the distal histidines, crystal structures of Mb and CcP were compared [3,14]. Figure 7 shows the superimposition of the active sites of Mb and CcP.

There are two crucial differences between Mb and CcP; the distal histidine of Mb is located much closer the the heme iron. Further inspection indicate the formation of a hydrogen bond between the distal histidine (His-82) and aspargine (Asp-82) of CcP. Amino acid sequence alignments [14,16] and X-ray crystal structures [3,16,17–22] show that the hydrogen bond is highly conserved among many plant and fungal peroxidases. Contrary to peroxidases, the Asn–His couple could not be found in the distal site of Mb and hemoglobin [14,23].

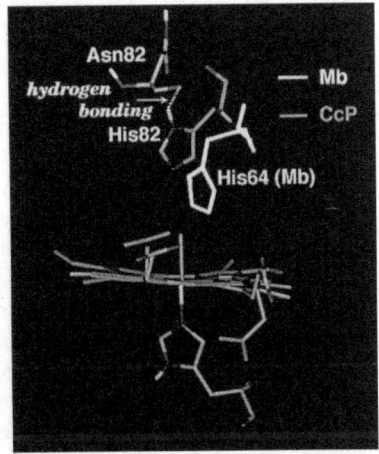

Fig. 7. Superimposition of the active sites of Mb and CcP

TABLE 1. Elementary reaction rate constants $(M^{-1} \cdot s^{-1})$ of native, wild-type, and N70V mutant

| Enzyme | $k_1 \times 10^{-7}$ | $k_2 \times 10^{-6}$ [a] | $k_3 \times 10^{-5}$ [b] |
|---|---|---|---|
| Native HRP | 1.5 | 6.6 | 5.2 |
| Wild-type HRP | 1.4 | 6.2 | 5.7 |
| N70V HRP | 0.12 | 0.12 | 0.20 |

HRP, horseradish peroxidase.
[a] Guaiacol was used as a reducing substrate.
[b] Calculated from the steady-state reaction rate (V) and $k_1$ and $k_2$.

TABLE 2. Kinetic parameters for the oxidation of hydroquinone and ABTS by native, wild-type, and mutant HRPs

| Enzymes | Hydroquinone | | ABTS | |
|---|---|---|---|---|
| | $K_m$ ($\mu$M) | $V_{max}$ ($\mu$M $\cdot$ min$^{-1}$) | $K_m$ ($\mu$M) | $V_{max}$ ($\mu$M $\cdot$ s$^{-1}$) |
| Native HRP | 33 | 281 | 339 | 73 |
| Wild-type HRP | 37 | 283 | 286 | 71 |
| N70V HRP | 19 | 18 | 279 | 217 |

ABTS, 2,2'-azinobis(3-ethylbenzothiazoline-6-sulfonic acid) diazonium salt.

To examine the roles of the highly conserved hydrogen bond in the catalytic activities of peroxidases, the hydrogen bond in horseradish peroxidase (HRP) was eliminated by the site-directed mutation of Asn 70 to valine (N70V) [24–26]. The effect of the mutation on peroxidase activities was examined by kinetic measurements at pH 7 for native, wild-type, and N70V HRPs. The rate of compound I formation by the mutant is reduced to less than 10% of that of the native enzyme (Table 1). As shown in Fig. 3, the distal histidine in peroxidases has been shown to participate in the reaction with hydrogen peroxide; that is, in the first step, the distal histidine serves as a base to accept a proton from hydrogen peroxide to form an Fe(III)-peroxide adduct as a reversible process ($K_1$). The following heterolytic O–O bond cleavage of the peroxide intermediate affords compound I ($k_{hetero}$). Thus, the rate constant for compound I formation ($k_1$) is expressed by using $K_1$ and $k_{hetero}$:

$$k_1 = K_1 k_{hetero}$$

The disruption of the hydrogen bond between Asn-70 and the distal histidine could reduce the basicity of the distal histidine, resulting in a smaller equilibrium constant ($k_1$) to retard the apparent compound I formation (see Table 1). The lower basicity of the distal histidine was confirmed by resonance Raman study [26]. On the basis of the pH titration of the Fe-His (proximal) stretching frequency, the $pK_a$ value for the mutant was estimated to be 5.9. The value is much lower than those for native (7.2) and wild-type (7.2) HRPs [27]. These results indicate that disruption of the distal site Asn–His couple makes the distal histidine less basic. In conclusion, the smaller rate constant for the reaction with hydrogen peroxide is attributed to the less basic distal histidine arising from the disruption of the Asn–His couple.

Peroxidase activities of HRP and N70V HRPs were examined by employing hydroquinone and 2,2'-azinobis(3-ethylbenzothiazoline-6-sulfonic acid) diazonium salt

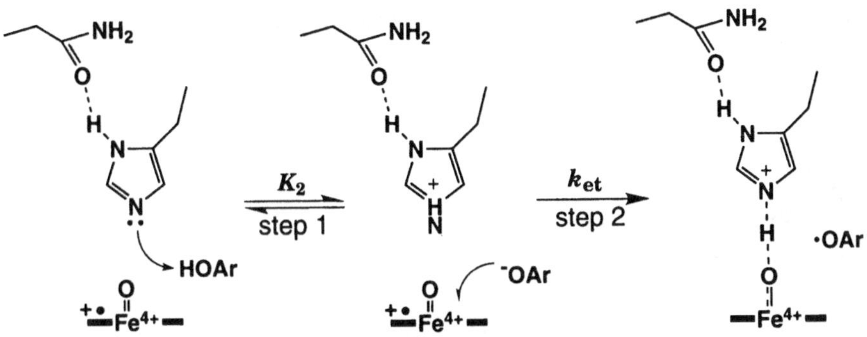

FIG. 8. Plausible mechanism for compound I reduction

(ABTS) as substrates. Although wild-type HRP exhibited hydroquinone oxidation activity comparable to that of the native enzyme, the N70V mutant almost lost its activity (Table 2). In contrast to the oxidation of phenols. ABTS oxidation activity of the mutant was significantly increased. The $V_{max}$ value for the mutant is increased threefold compared to those for native and wild-type HRPs. For both hydroquinone and ABTS oxidation, $K_m$ values of the mutant are almost identical to those for native and wild-type HRPs.

In the reaction of compound I with phenols, the distal histidine abstracts a proton from the substrate (base catalyst) to form a phenoxide anion as a reversible process ($K_2$) [28,29]. The proton transfer is followed by electron transfer from the substrate to the heme ($k_{et}$), giving compound II (Fig. 8). Thus, the rate constant for the reduction of compound I ($k_2$) would be expressed by using $K_2$ and $k_{et}$:

$$k_2 = K_2 k_{et}$$

The involvement of an electron-transfer process in the rate-determining step has been suggested by negative Hammett ρ values for all HRPs [30]. However, the deprotonation process ($K_2$) should show positive Hammett ρ values. Thus, the apparent $k_2$ is the result of these two inverse effects. While the observation of nagative ρ values for the apparent $k_2$ of HRPs indicates that the substituent effect on $k_{et}$ is dominant, the substituent effect on the deprotonation step must be more effective in the reactions of phenols with the mutant, because it is much harder for the mutant to abstract a proton from electron-rich phenols such as methoxyphenol. These considerations suggest that substituent effects on the apparent $k_2$ of phenol oxidation could show smaller negative Hammett ρ values when the mutant is employed. In fact, on the oxidation of methoxyphenol, a less acidic phenol, decreased basicity of the distal histidine substantially depressed the proton abstraction compared to that of native compound I. The resultant smaller $K_2$ values drastically retard the overall reduction rate of the mutant compound I by methoxyphenol to 2%–3% of that for native enzyme.

On the other hand, on the oxidation of hydroxybenzaldehyde, an acidic phenol, the less basic distal histidine of the mutant readily abstracts proton from the substrate, while the following one-electron transfer from the phenolate must be very hard for

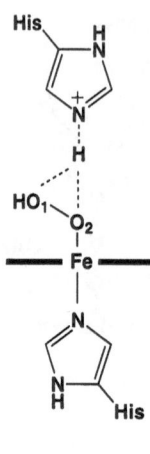

FIG. 9. Expected interaction of the distal histidine of Mb and iron-bound peroxide

both native and mutant HRPs because of the introduction of an electron-withdrawing aldehyde. Therefore, the mutant compound I can oxidize hydroxybenzaldehyde as slowly as native compound I. However, as noted, the large negative substituent effects of phenols ($\rho = -6.9$ to $-3.8$) on the reduction of compounds I are the indication of the involvement of the one-electron-transfer process at the rate-determining step. At the same time, these results also indicate the participation of the deprotonation step in the $k_2$ process.

In the case of ABTS oxidation by wild-type and mutant HRPs, there is no proton abstraction process. Thus, the rate-determining step might be the one-electron-transfer reaction from ABTS to compound II. The higher oxidation rate of ABTS by the H70V mutant (see Table 2) suggests a greater reactivity of the mutant compound II over native and wild-type HRP compounds II. To examine these possibilities, the redox potentials of HRP compound II were studied. The mutant compound II exhibited redox potential approximately 100 mV higher than those of native and wild-type compounds II. A higher redox potential of the mutant compound II could result in the high ABTS oxidation activity of the mutant.

Although the effects of the Asn–His hydrogen bond on the compound I formation and reactivities of peoxidases clearly indicate important roles of the general acid–base catalyst in the peroxidase active site, the peroxidase activities of the H70V mutant are still much higher than those of Mb. Thus, the hydrogen bond between Asn and His is not the sole factor for the discrimination between peroxidase and myoglobin. As noted, we have pointed out the location of the distal histidine of myoglobin as being much closer to the heme iron (Fig. 7). The distal histidine of peroxidase serves as a general acid–base catalyst in the formation of compound I [4,5]. Thus, the hydrogen bond geometry and distance between the distal histidine and the oxygen atom of peroxide are expected to be appropriate to help charge separation of the O–O bond in the transition state of the O–O bond cleavage (Fig. 3). On the basis of the crystal structure of an oxy form of Mb, the distal histidine in Mb lies 4.3 Å above the heme

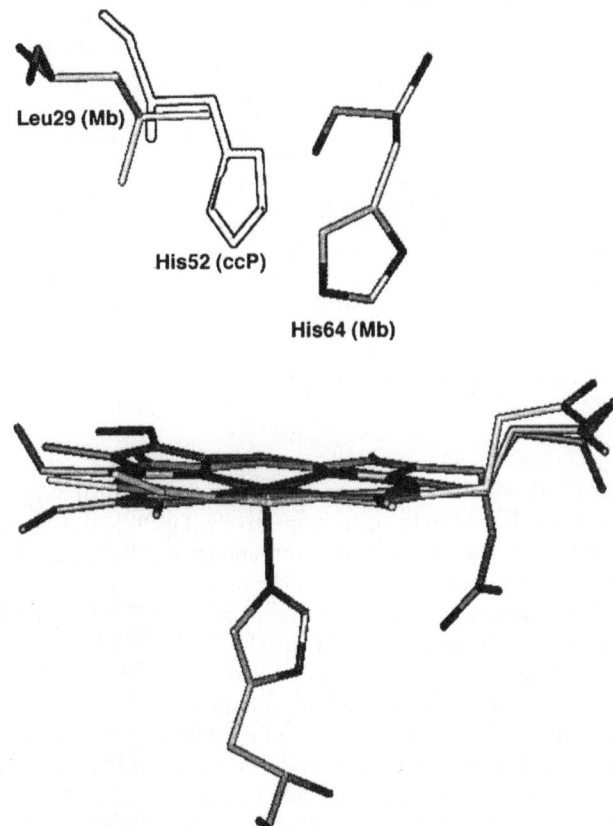

FIG. 10. Superimosition of the distal histidine (His) of CcP in the active site of Mb

TABLE 3. Enantioselective oxidation of thioanisole

| Protein | Turnover (per min) | $H_2{}^{18}O_2$ (%)[a] | e.e. (R) |
|---------|-------------------|------------------------|----------|
| Sperm whale Mb | 0.25 | 92 | 25 |
| His64/Leu Mb | 0.072 | 89 | 27 |
| Leu29/His Mb | 3.9 | 100 | 91 |
| L29H/H64L Mb | 5.5 | 97 | 97 |

Mb, myoglobin; e.e., enantiomeric excess.
[a] An amount of $^{18}O$ in sulfoxide from $H_2{}^{18}O_2$.

TABLE 4. Enantioselective oxidation of styrene

| Protein | Turnover (per min) | $H_2^{18}O_2$ (%)[a] | e.e. (R) |
|---|---|---|---|
| Sperm whale Mb | 0.015 | 20 | 9R |
| H64L Mb | 0.020 | 73 | 34R |
| L29H Mb | 0.093 | 53 | 2S |
| L29H/H64L Mb | 0.14 | 94 | 80R |

[a] An amount of $^{18}O$ in styrene oxide from $H_2^{18}O_2$.

iron, and Nε of His-64 exists at 2.8 and 3.0 Å from the terminal oxygen (O-1) and the oxygen atom (O-2) bound to the iron, respectively (Fig. 9) [14]. If one assumes a very similar structure for the hydroperoxo-iron adduct of Mb, it appears that negative charge cannot be developed on the O-1 atom to cause the heterolysis because O-2 as well as O-1 is positioned to accept a proton from Nε of the distal histidine (Fig. 10). We hypothesize that the distance of the distal histidine would be crucial to differentiate catalytically less active Mb from peroxidase. These considerations suggest that the relocation of the distal histidine of myoglobin to a position similar to that of CcP could allow us to make a myoglobin mutant exhibiting high peroxidase activity. For this purpose, another view of the superimposed structure of CcP and myoglobin is shown in Fig. 9. Similar location of Leu-29 of myoglobin to that of CcP suggests that the Leu-29 → His and His-64 → Leu double mutation of Mb would create a heme crevice similar to the active site structure of CcP, of which the distal histidine lies 5.6 Å above the heme iron [3], and we expect the L29H/H64L mutant would transfer the ferryl oxygen to the substrates efficiently. Thus, we have constructed L29H, H64L, and L29H/H64L mutants [31].

Mutating Leu-29 to a histidine residue improves the rate and enantio-selectivity for the oxidation of thioanisole (Table 3). The rate increases versus wild-type Mb are 15 fold and 22 fold for L29H and L29H/H64L Mb, respectively. As sulfoxidation catalysis, the Leu-29 mutants can compare favorably with native HRP, of which the turnover number is $3.5 \, min^{-1}$ [32]. On the contrary, the elimination of His-64 in the distal pocket causes about 70% decrease in the oxidation rate with respect to the recombinant wild type. The L29H/H64L double mutation enhances the enantiomeric excess from 25% to 97%. The dominant formation of R by the mutation is in contrast wit 97% entantiomeric excess (e.e.) for the S enantiomer given by F41L HRP [32]. The extremely high stereoselectivity with 97% incorporation of $^{18}O$ from $H_2^{18}O_2$ into the sulfoxide for L29H/H64L Mb clearly indicates the ferryl oxygen transfer to thioanisole and rules out the involvement of a protein surface hydroxyl radical and molecular oxygen. Although the enantioselectivity is low for wild-type and H64L Mb, $H_2^{18}O_2$-labeling experiments resulted in approximately 90% incorporation of the labeled oxygen into the sulfoxide.

Styrene epoxidation by H64L Mb was found to proceed at a rate similar to that for wild-type Mb (Table 4) although the turnover number for thioanisole oxidation by the H64L mutant is one-third of that for the wild type. In comparison with wild-type Mb, the L29H mutant oxidizes six times faster, and ninefold enhancement with an improvement of enantio-selectivity from 9% to 80% is seen for L29H/H64L Mb. Interestingly, preferred formation of the $R$ methyl phenyl sulfoxide and $R$ styrene oxide requires the opposite orientation of the phenyl group and the side chain with respect to ferryl oxygen; however, the structural information on the transition state cannot be deduced at this point. The rate of styrene oxidation by L29H/H64L Mb is comparable with those for CcP [33], the Phe-41 and His-42 HRP mutants [32,34] but still one order of magnitude slower than cytochrome P-450 monooxygenase [35].

The oxidation of styrene in the presence of $H_2^{18}O_2$ by wild-type Mb produces epoxide with 20% $^{18}$O-labeled oxygen. The result is consistent with the competition of at least two mechanisms for epoxidation by wild-type Mb: one that incorporates an oxygen atom from hydrogen peroxide employed and another incorporates an atom of molecular oxygen [36–38]. Incubations of styrene and $H_2^{18}O_2$ with L29H/H64L, H64L, and L29H Mb resulted in incorporation of 94%, 73%, and 53% of $^{18}$O in the epoxides, respectively. The value for L29H, bearing two histidines in the active site, is between the incorporation numbers for L29H/H64L and wild-type Mb. The results indicate that the removal of His-64 prevents the incorporation of molecular oxygen and that to place a histidine residue at the 29 position enhances the ferryl oxygen transfer mechanism. Because wild-type and L29H/H64L Mb can form phenyl–iron complex in the presence of phenylhydrazine, the wild type as well as the double mutant has an active site large enough for the substrate to access. Thus, the observed enhancement appears to result from the increase in reactivity of the ferryl species rather than the accessibility of ferryl oxygen for substrates.

Our results clearly indicate that His-29 in the mutants directly plays an important role to improve the hydrogen peroxide-supported monooxygenation activity in terms of the rate and enantio-selectivity. The L29H/H64L Mb mutant, like chloroperoxidase [39–41], could be employed as a practical biocatalyst for the preparation of chiral sulfoxides and epoxides.

## Conclusion

Through the site-directed mutation of Mb and HRP, roles of thiolate ligation in oxygen activation by cytochrome P-450 and factors that discriminate the biological functions of peroxidase from Mb have been studied. More importantly, we have shown the molecular design of myoglobin-based peroxygenase (L29H/H64L Mb), which exhibits high peroxide-dependent enantio-selective oxidation activities.

*Acknowledgments.* This work was supported by grants from the Ministry of Education, Science, Sports and Culture of Japan (grant-in-aid for scientific research no. 07458147 and for Priority Areas, Molecular Biometallics) for Y.W.

# References

1. Dolphin D (1978) The porphyrins, vol VII. Academic Press, New York
2. Everse J, Everse KE, Grisham MB (eds) (1991) Peroxidases in chemistry and biology. CRC, Boca Raton, FL
3. Poulos TL, Freer ST, Alden RA, et al. (1980) The crystal structure of cytochrome c peroxidase. J Biol Chem 255:575–580
4. Poulos TL, Kraut J (1980) The stereochemistry of peroxidase catalysis. J Biol Chem 255:8199–8205
5. Erman JE, Vitello LB, Miller MA, et al. (1993) Histidine 52 is a critical residue for rapid formation of cytochrome c peroxidase compound I. Biochemistry 32:9798–9806
6. White RE, Coon MJ (1980) Oxygen activation by cytochrome P-450. Annu Rev Biochem 49:315–356
7. Ortiz de Montellano PR (ed) (1995) Cytochrome P-450: structure, mechanism, and biochemistry, 2nd edn. Plenum, New York
8. Poulos TL, Finzel BC, Gunsalus IC, et al. (1985) The 2.6Å crystal structure of *Pseudomonas putida* cytochrome P-450. J Biol Chem 260:16122–16130
9. Dawson JH, Sono M (1987) Cytochrome P-450 and chloroperoxidase: thiolate-ligated heme enzymes. Spectroscopic determination of their active site structure and mechanistic implication of thiolate ligator. Chem Rev 87:1255–1276
10. Dawson JH (1988) Probing structure-function relations in heme-containing oxygenases and peroxidases. Science 240:433–439
11. Yamaguchi K, Watanabe Y, Morishima I (1993) Direct observation of the push effect on the O–O bond cleavage of acylperoxoiron (III) porphyrin complexes. J Am Chem Soc 115:4058–4065
12. Adachi S, Nagano S, Ishimori K, et al. (1991) Alteration of human myoglobin proximal histidine to cysteine or tyrosine by site-directed mutagenesis: characterization and their catalytic activities. Biochem Biophys Res Commun 180:138–144
13. Adachi S, Nagano S, Ishimori K, et al. (1993) Roles of proximal ligand in hemoproteins: replacement of human myoglobin with cysteine and tyrosine by site-directed mutagenesis as models for P-450, chloroperoxidase, and catalase. Biochemistry 32:241–252
14. Takano T (1977) Structure of myoglobin refined at 2.0Å resolution. J Mol Biol 110:537–568
15. Matsui T, Nagano S, Ishimori K, et al. (1996) Preparation and reactions of myoglobin mutants bearing both proximal cysteine and hydrophobic distal cavity: protein models for the active site of P-450. Biochemistry 35:13118–13124
16. Baunsgaard L, Dalbøge H, Houen G, et al. (1993) Amino acid sequence of *Coprinus macrorhizus* peroxidase and cDNA sequence encoding *Coprinus cinereus* peroxidase. A new family of fungal peroxidases. Eur J Biochem 213:605–611
17. Sundaramoorthy M, Kishi K, Gold MH, et al. (1994) The crystal structure of manganese peroxidase from *Phanerochaete chrysosporium* at 2.06-Å resolution. J Biol Chem 269:32759–32767
18. Poulos TL, Edwards SL, Wariishi H, et al. (1993) Crystallographic refinement of lignin peroxidase at 2 Å. J Biol Chem 268:4429–4440
19. Edwards SL, Raag R, Wariishi H, et al. (1993) Crystal structure of lignin peroxidase. Proc Natl Acad Sci USA 90:750–754
20. Patterson WR, Poulos TL (1995) Crystal structure of recombinant pea cytosolic ascorbate peroxidase. Biochemistry 34:4331–4341
21. Kunishima N, Fukuyama K, Matsubara H, et al. (1994) Crystal structure of the fungal peroxidase from *Arythromyces ramosus* at 1.9Å resolution. J Mol Biol 235:331–344
22. Kunishima N, Amada F, Fukuyama K, et al. (1996) Pentacoordination of the heme iron of *Arthromyces ramosus* peroxidase shown by a 1.8Å resolution crystallographic study at pH 4.5. FEBS Lett 378:291–294

23. Perutz MF, Rossmann MG, Cullis AF, et al. (1960) Structure of haemoglobin: a three-dimensional Fourier synthesis at 5.5-Å resolution, obtained by X-ray analysis. Nature 185:416–422

24. Nagano S, Tanaka M, Watanabe Y, et al. (1995) Putative hydrogen bond network in the heme distal site of horseradish peroxidase. Biochem Biophys Res Commun 207:417–423

25. Nagano S, Tanaka M, Ishimori K, et al. (1996) Catalytic roles of the distal site asparagine-histidine couple in peroxidases. Biochemistry 35:14251–14258

26. Mukai M, Nagano S, Tanaka M, et al. (1997) Effects of concerted hydrogen bonding of distal histidine on active site structures of horseradish peroxidase: resonance Raman studies with Asn-70 mutants. J Am Chem Soc 119:1758–1766

27. Teraoka J, Kitagawa T (1981) Structural implication of the heme-linked ionization of horseradish peroxidase probed by the Fe-histidine stretching Raman line. J Biol Chem 256:3969–3977

28. Dunford HB (1982) Peroxidases. Adv Inorg Biochem 4:41–68

29. Ortiz de Montellano PR (1987) Control of the catalytic activity of prosthetic heme by the structure of hemoproteins. Account Chem Res 20:289–294

30. Job D, Dunfold HB (1976) Substituent effect on the oxidation of phenols and aromatic amines by horseradish peroxidase compound I. Eur J Biochem 66:607–614

31. Ozaki S, Matsui T, Watanabe Y (1996) Conversion of myoglobin into a highly stereospecific peroxygenase by the L29H/H64L mutation. J Am Chem Soc 118:9784–9785

32. Ozaki S, Ortiz de Montellano PR (1995) Molecular engineering of horseradish peroxidase: thioether sulsfoxidation and styrene epoxidation by Phe-41 leucine and threonine mutants. J Am Chem Soc 117:7056–7064

33. Miller VP, DePillis GD, Ferrer JC, et al. (1992) Monooxygenase activity of cytochrome c peroxidase. J Biol Chem 267:8936–8942

34. Newmyer SL, Ortiz de Montellano PR (1995) Horseradish peroxidase His-42->Ala, His-42→Val, and Phe-41→Ala mutants—histidine catalysis and control of substrate access to the heme iron. J Biol Chem 270:19430–19438

35. Fruetel J, Collins JR, Camper DL, et al. (1992) Caluculated and experimental absolute stereochemistry of the styrene and β-methylstyrene epoxides formed by cytochrome P-450cam. J Am Chem Soc 114:6987–6993

36. Ortiz de Montellano PR, Catalano CE (1985) Epoxidation of styrene by hemoglobin and myoglobin: transfer of oxidizing equivalent to the protein surface. J Biol Chem 260:9265–9271

37. Rao SI, Wilks A, Ortiz de Montellano PR (1993) The roles of His64, Tyr103, Tyr146, and Tyr151 in the epoxidation of styrene and β-methylstyrene by recombinant sperm whale myoglobin. J Biol Chem 268:803–809

38. Kelman DJ, DeGray JA, Mason RP (1994) Reaction of myoglobin with hydrogen peroxide forms a peroxyl radical which oxidizes substrate. J Biol Chem 269:7458–7463

39. Collonna S, Gagerro N, Casella L, et al. (1992) Chloroperozidase and hydrogen peroxide: an efficient system for enzymatic enantioselective sulfoxidations. Tetrahedron Asymmetry 3:95–106

40. Fu H, Kondo H, Ichikawa Y, et al. (1992) Chloroperoxidases-catalyzed asymmetric synthesis: enantioselective reactions of chiral hydroperoxides with sulfides and bromohydration of glycals. J Org Chem 57:7265–7270

41. Allain EJ, Hager LP, Deng L, et al. (1993) Highly enantioselective epoxidation of disubstituted alkenes with hydrogen peroxide catalyzed by chloroperoxidase. J Am Chem Soc 115:4415–4416

# Catalytic Roles of the Distal Site Hydrogen Bond Network of Peroxidases

Shingo Nagano[1,3], Motomasa Tanaka[1], Koichiro Ishimori[1], Isao Morishima[1], Yoshihito Watanabe[2], Masahiro Mukai[2], Takashi Ogura[2], and Teizo Kitagawa[2]

*Summary.* There are highly conserved hydrogen bonds between the distal histidine and adjacent asparagine in many peroxidases. To investigate the functional roles of the hydrogen bond between the distal histidine and Asn-70, Asn-70 in horseradish peroxidase (HRP) was replaced with Val or Asp. The disruption of the Asn-70–His-52 couple decreases the rate of compound I formation to less than 10% of that of native enzyme. Based on resonance Raman spectroscopy, the midpoint pH value of the Fe(II)-His stretching frequency in the acid–base transition was decreased by the mutation of Asn-70, suggesting that the distal histidine became less basic. With a less basic distal histidine, proton abstraction from hydrogen peroxide is harder for the mutants, resulting in remarkable deceleration of compound I formation.

*Key words.* HRP—Heme protein—Resonance Raman—Hydrogen bond—Acid–base catalysis

## Introduction

Peroxidases are heme proteins that catalyze oxidation of substrates utilizing hydrogen peroxide as an oxidant [1]. The peroxidase reaction cycle involves two catalytic intermediates, compound I and compound II. In the first step, resting-state (ferric) peroxidase reacts with peroxide to give the first catalytic intermediate, compound I, which is two-electron oxidized from the resting-state enzyme. Poulos and Kraut proposed a critical role of the highly conserved histidine (distal His) in the heme distal pocket for compound I formation (Fig. 1) [2]. Binding of the peroxide to the heme iron is facilitated by transfer of one of its hydrogens to the $N^\varepsilon$ atom of the distal His. The hydrogen atom is then delivered by the distal His to the terminal oxygen of the peroxide as the O–O bond is cleaved to give compound I. Transfer of one electron to

[1] Division of Molecular Engineering, Graduate School of Engineering, Kyoto University, Kyoto 606-01, Japan
[2] Institute for Molecular Science, Okazaki National Research Institutes, 38 Nishigounaka, Myodaiji, Okazaki, Aichi 444, Japan
[3] Present address: Department of Biochemistry, School of Medicine, Keio University, 35 Shinanomachi, Shinjuku-ku, Tokyo 160, Japan

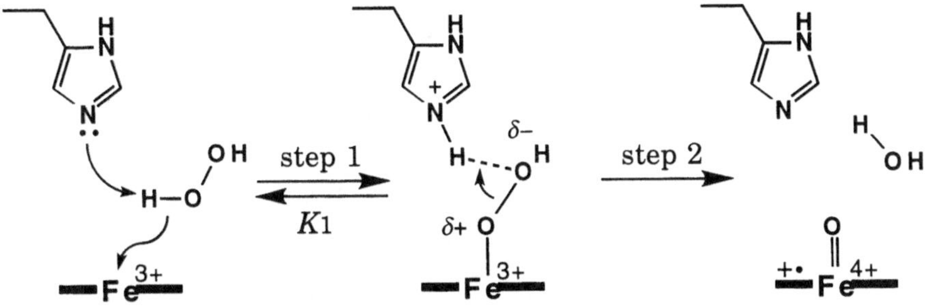

Fig. 1. Plausible mechanism of compound I formation

compound I gives compound II, in which the Fe(IV)=O species remains intact but the protein of the porphyrin radical has been reduced. Transfer of a second electron from a reducing substrate reduces compound II to the ferric enzyme and result in release of the ferryl oxygen as a molecule of water.

## Distal Histidine as a Base Catalyst

Erman et al. [3] and Newmyer et al. [4] have demonstrated a crucial role of the distal His for the rapid reaction with peroxide. Substitution of the distal His to Leu in cytochrome $c$ peroxidase (CcP) (H52L CcP) [3] and to Ala or Val in horseradish peroxidase (HRP) (H42A, H42V HRP) [4] greatly reduced compound I formation rates by five to six orders of magnitude. The value of the rate constant for the reaction between the distal His substituted peroxidases and hydrogen peroxide is similar to the values observed for the reaction of hydrogen peroxide and metalloporphyrins [5] and between metmyoglobin and hydrogen peroxide [6]. Replacement of the distal His with aliphatic residues has eliminated rate enhancement normally shown by this enzyme, suggesting that the distal His is one of the most important residues in the catalytic cycle of the peroxidase reaction.

## The Distal Histidine–Asparagine Couple

Inspection of the crystal structure of CcP [7] indicates the formation of a hydrogen bond between the distal His and Asn, an amino acid a further from the immediate vicinity of the heme (Fig. 2). Amino acid sequence alignments and X-ray crystal structures show that the hydrogen bond is highly conserved among many plant and fungal peroxidases. Contrary to peroxidases, the Asn–His couple could not be found in the distal site of myoglobin and hemoglobin, even though they also have the distal His in their heme pockets. When the Asn residue in the distal site of HRP was replaced with Val or Asp (N70V and N70D HRP), the rate constants of compound I formation were significantly depressed to less than 10% of native enzyme (Table 1) [8,9]. Because the distal His acts as a base catalyst in the reaction of the ferric resting HRP

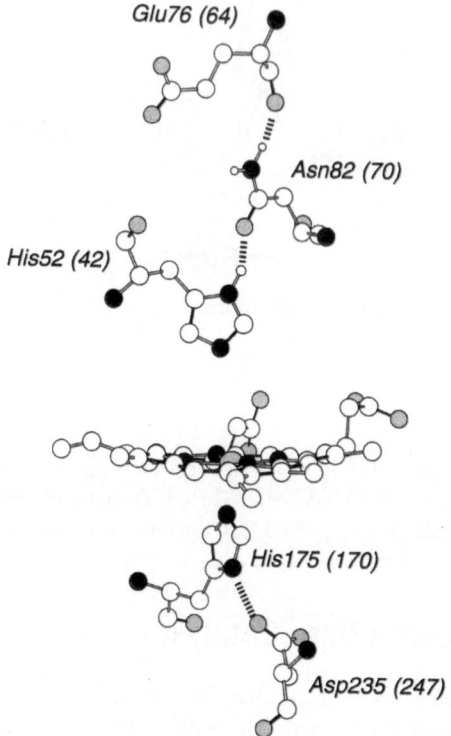

FIG. 2. X-ray crystallographic structures of the heme proximity of cytochrome *c* peroxidase (C*c*P). Hydrogen bond is expressed by a *hatched bond. Amino acid numbering* is for C*c*P but the *numbers in parentheses* denote the numbering for horseradish peroxidase (HRP)

TABLE 1. Rate constants of compound I Formation ($k_1$) and compound II reduction ($k_2$)

| Enzyme | $k_1 \times 10^{-7}/M^{-1} \cdot s^{-1}$ | $k_3 \times 10^{-5}/M^{-1} \cdot s^{-1 a}$ |
|---|---|---|
| Native | 1.5 | 5.2 |
| Wild type | 1.4 | 5.7 |
| N70V | 0.12, 0.03[b] | 0.20 |
| N70D | 0.15 | 0.58 |

[a] Calculated from the steady-state reaction rate.
[b] Ratio of fast and slow phases = 52:48.

with hydrogen peroxide (step 1 in Fig. 1) [2], it is reasonable to suppose that the reaction with hydrogen peroxide is decelerated by the decrease of basicity. Resonance Raman spectroscopy demonstrated that N70V and N70D have less basic distal His than that of native HRP (p$K_a$ of the distal His: native; 7.2; N70V; 5.9; N70D; 5.5) [10]. Based on the relationship between the basicity and rate constants, it is strongly suggested that the Asn in the distal site makes the distal His more basic, resulting in a larger equilibrium constant ($K_1$: step 1 of Fig. 1) and facile compound I formation, and when the Asn–His hydrogen bond is cleaved the distal His becomes less basic, leading to a smaller rate constant of compound I formation.

## Conformation of the Distal Histidine

In addition to the distal site Asn–His hydrogen bond, there is a notable difference between globins and peroxidases. The distance between the distal His and heme iron is 4.3 Å in sperm whale myoglobin, which may stabilize the oxygen molecule coordinated to the heme by making the hydrogen bond. On the other hand, the distal His of CcP lies 5.6 Å above the heme iron. Savenkova et al. have prepared the double-mutated HRP (distal His → Ala, Phe-41 → His), in which the distal His has moved from the original site [11]. The rate of compound I formation for the double mutant is $10^3$ times slower than that for native HRP [11]. This means that the conformation of the distal His also contributes to governing the rate of compound I formation.

## Proximal Ligand

Strong electron donation from the anionic proximal His to the heme iron has also been regarded as a predominant structural and functional factor that discriminates peroxidases from other hemoproteins (Push effect) [12]. Especially, the critical role of the Cys ligand of P-450s in the peroxide O–O bond cleavage step has been demonstrated [13]. However, replacing the proximal ligand of CcP, His-175, with Gln has only a minor effect on the rate of compound I formation (wild type. $3.9 \times 10^7$; H175Q, $1.2 \times 10^7$ $M^{-1}s^{-1}$), demonstrating that the anionic nature of proximal His is not critical in high rates of compound I formation although it has direct connection to the heme iron [14].

## Compound II

The reactivity of HRP compound II is linked to a group with a p$K_a$ of 8.6, probably the distal His, which is hydrogen bonded to the ferryl oxygen at neutral condition (His$\cdots$O=Fe(IV)). At basic conditions the distal His is not hydrogen bonded to the ferryl oxygen (His O=Fe(IV)) and the compound II is inactive [15]. From resonance Raman studies, Mukai et al. [10] have found that, when the His–Asn hydrogen bond is absent, the hydrogen bond between the distal His and ferryl oxygen (His$\cdots$O=Fe(IV)) is significantly weaker than that of native HRP at neutral condition. N70V and N70D HRPs in which the His–Asn hydrogen bond is disrupted have smaller rate constants of compound II reduction (Table 1) and have lower activities in the steady state [8,9]. As a result of these observations, it is postulated that the Asn residue in the distal site regulates the reactivity of compound II through the concerted hydrogen bond network (Asn$\cdots$His$\cdots$O=Fe(IV)).

# Conclusions

Rapid reactions of the resting peroxidases with peroxides to form compound I discriminate the peroxidases from other classes of hemoproteins. This characteristic reactivity is related to the distinct protein structural features, especially to the heme distal site. The Asn in the distal site is a key residue that regulates the basicity of the distal His to facilitate compound I formation and also affects the reactivity of compound II through the concerted hydrogen bond network.

# References

1. Everse J, Everse KE, Grisham MB (eds) (1991) Peroxidases in chemistry and biology. CRC, Boca Raton
2. Poulos TL, Kraut J (1980) The stereochemistry of peroxidase catalysis. J Biol Chem 255:8199–8205
3. Ermam JE, Vitello LB, Miller MA, et al. (1993) Histidine-52 is a critical residue for rapid formation of cytochrome-$c$ peroxidase compound-I. Biochemistry 32:9798–9806
4. Newmyer SL, Ortiz de Montellano PR (1995) Horseradish peroxidase His-42→Ala, His-42→Val, and Phe-41→Ala mutants—histidine catalysis and control of substrate access to the heme iron. J Biol Chem 270:19430–19438
5. Bruice TC (1991) Reactions of hydroperoxides with metallotetraphenylporphyrins in aqueous solutions. Acc Chem Res 24:243–249
6. Yonetani T, Schleyer H (1967) Studies on cytochrome $c$ peroxidase IX: the reaction of ferrimyoglobin with hydroperoxides and a comparison of peroxide-induced compounds of ferrimyoglobin and cytochrome $c$ peroxidase. J Biol Chem 242:1974–1979
7. Poulos TL, Finzel BC, Gunzalus IC, et al. (1985) The 2.6-angstrom crystal structure of *Pseudomonas putida* cytochrome P-450. J Biol Chem 260:16122–16130
8. Nagano S, Tanaka M, Watanabe Y, et al. (1995) Putative hydrogen bond network in the heme distal site of horseradish peroxidase. Biochem Biophys Res Commun 207:417–423
9. Nagano S, Tanaka M, Watanabe Y, et al. (1996) Catalytic roles of the distal site asparagine-histidine couple in peroxidases. Biochemistry 35: 14251–14258
10. Mukai M, Nagano S, Tanaka M, et al. (1997) Effects of concerted hydrogen bonding of distal histidine on active site structures of horseradish peroxidase. Resonance Raman Studies with Asn70 mutants. J Am Chem Soc 119:1758–1766
11. Savenkova MI, Newmyer SI, Ortiz de Montellano PR (1996) Rescue of His-42→Ala horseradish peroxidase by a Phe-41→His mutation. J Biol Chem 271:24598–24603
12. Yamaguchi K, Watanabe Y, Morishima I (1993) Direct observation of the push effect on the oxygen–oxygen bond cleavage of acylperoxoiron(III) porphyrin complexes. J Am Chem Soc 115:4058–4065
13. Adachi S, Nagano S, Ishimori K, et al. (1993) Roles of proximal ligand in heme proteins—replacement of proximal histidine of human myoglobin with cysteine and tyrosine by site-directed mutagenesis as models for P-450, chloroperoxidase, and catalase. Biochemistry 32:241–252
14. Sundaramoorthy M, Choudhury K, Edwards SL, et al. (1991) Crystal structure and preliminary function analysis of the cytochrome $c$ peroxidase His175Gln proximal ligand mutant. J Am Chem Soc 113:7755–7757
15. Makino R, Uno T, Nishimura Y, et al. (1986) Coordination structure and reactivities of compound II in iron and manganese horseradish peroxidase. J Biol Chem 261:8376–8382

# Catalytic Intermediates of Polyethylene-Glycolated Horseradish Peroxidase in Benzene

SHIN-ICHI OZAKI[1], YUJI INADA[2], and YOSHIHITO WATANABE[1]

*Summary.* Horseradish peroxidase was homogeneously solubilized in benzene by the covalent modification of lysine residues on the protein surface with polyethylene glycol. The addition of a stoichiometric quantity of hydrogen peroxide to polyethylene-glycolated horseradish peroxidase in benzene gave the compound I chromophore. In the presence of a large excess of guaiacol, compound I was reduced first to compound II and then to the ferric enzyme. Thus, catalytic intermediates in benzene were the ferryl species as established in aqueous buffer. The slower reduction of compound I with guaiacol in benzene than in buffer suggested that the accessibility of the active site for the substrates decreases in benzene.

*Key words.* Peroxidase—Polyethylene glycol conjugate—Compound I—Compound II—P-450

## Introduction

Horseradish peroxidase (HRP) catalyzes the oxidation of a variety of aromatic substrates utilizing hydrogen peroxide [1], and it is found to be catalytically active in not only aqueous but also nonaqueous media [2–5]. The enzyme can be solubilized homogeneously by the covalent modification of lysine residues on the protein surface with polyethylene glycol [2–4] or by the formation of noncovalent micellar complexes with bis(2-ethylhexyl)sodium sulfosuccinate [5]. Recent optical spectroscopic studies on the ferric resting state of polyethylene-glycolated HRP (PEG-HRP) have suggested that the heme active site remains structurally intact in benzene [4].

It has been established that the reactions in aqueous buffer normally proceed by the following mechanism [6,7]:

$$HRP + H_2O_2 \rightarrow \text{compound I} \tag{1}$$

[1] Institute for Molecular Science, Okazaki National Research Institutes, 38 Nishigounaka, Myodaiji, Okazaki, Aichi 444, Japan
[2] Toin Human Science and Technology Center, Department of Materials Science and Technology, Toin University of Yokohama, 1614 Kurogane-cho, Aoba-ku, Yokohama 225, Japan

$$\text{Compound I} + AH_2 \rightarrow \text{compound II} + AH\cdot \qquad (2)$$

$$\text{Compound II} + AH_2 \rightarrow HRP + AH\cdot \qquad (3)$$

where compound I ($Fe^{IV\cdot+}=O$) and II ($Fe^{IV}=O$) represent the ferryl intermediates and $AH_2$ the peroxidase substrate. However, the reaction intermediates in organic solvents have not been clearly defined. Thus, we report here the UV-visible spectra of compound I and II for PEG-HRP in benzene as well as in sodium phosphate buffer by rapid scanning spectroscopy. Furthermore, the conversion of the intermediates into the ferric state by the addition of a typical peroxidase substrate provides clear evidence that compounds I and II of PEG-HRP are the catalytic species in both organic and aqueous media.

The absorption spectra of the ferric state for PEG-HRP in benzene and sodium phosphate buffer are essentially identical to the spectra previously reported [2,4]. The addition of a stoichiometric quantity of hydrogen peroxide to PEG-HRP in benzene causes a decrease in the absorbance of the Soret band to give a compound I-like spectrum ($\lambda_{max} = 405\,nm$). The spectrum is subsequently converted into compound II [4] under the xenon lamp of rapid scanning spectroscopy, as Stillman et al. previously reported for the reductive photolysis of compound I for native HRP in buffer [8]. However, the further photoreduction of compound II is not observed. The Soret bands of PEG-HRP for ferric, compound I, and compound II in sodium phosphate buffer are also observed at 403, 403, and 420 nm, respectively. The Soret band of compound II in benzene is 413 nm, which shifts to a shorter wavelength by 7 nm with respect to $\lambda_{max}$ of compound II in the buffer (Table 1).

The decrease in absorbance at 405 nm fits the single exponential curve $k_{obs1} = 7.2 \pm 0.4\ (s^{-1})$; however, the systematic deviation in the residuals suggests that a single-step first-order formation of a compound I-like intermediate might be oversimplified. Factor analysis of the rapid scan data set revealed that the formation of compound I of PEG-HRP in benzene could consist of two steps. The initial first step might be the formation of an iron-peroxy complex like compound 0 [9], and the subsequent slow step would be the O–O bond cleavage process (Fig. 1).

To confirm that the detected species are compounds I and II in benzene, guaiacol was added after the formation of the first intermediate. Adding a large excess of guaiacol to the intermediate solution gives a compound II absorption spectrum, followed by the ferric PEG-HRP chromophore. The observed spectral changes clearly indicate the first intermediate to be compound I, and two ferryl species, compounds I and II, are catalytic intermediates in organic media. On the other hand, only the twofold excess of guaiacol with respect to the PEG-HRP is required to transform compound I efficiently into the ferric enzyme in sodium phosphate buffer. It appears that the conversion of compound II to the ferric enzyme is the rate-determining step in the peroxidase cycle in both benzene and aqueous buffer because compound II is observed during turnover.

The reaction of PEG-HRP with hydrogen peroxide to form compound I is faster in aqueous buffer than in benzene, and the oxidation of guaiacol by compound I of PEG-HRP proceeds faster in buffer than in benzene. Thus, the reaction media partly controls the accessibility of the heme center for guaiacol. The amphiphilic polyethylene glycol chains surrounding the PEG-HRP surface might be less flexible in organic

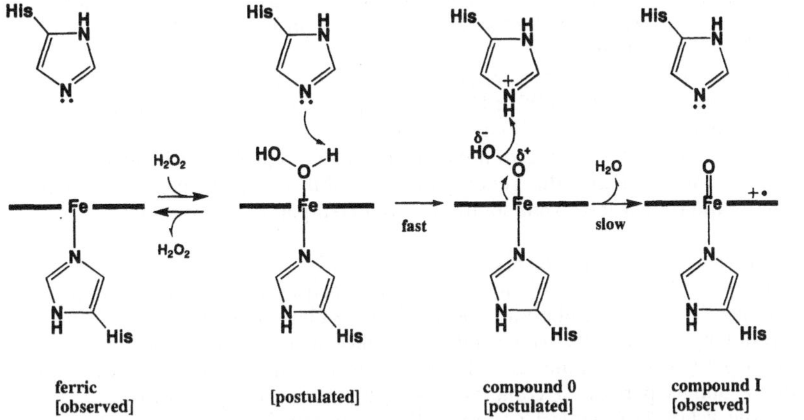

FIG. 1. Formation of compound I of polyethylene-glycolated horse radish peroxidase (PEG-HRP) may consist of two steps (*black bars* represent heme)

TABLE 1. Electronic absorption maxima for polyethylene-glycolated horse radish peroxidase in phosphate buffer and benzene

| Compound | Soret band in buffer (nm) | Soret band in benzene (nm) |
|---|---|---|
| Ferric | 403 | 405 |
| Compound I | 403 | 405 |
| Compound II | 420 | 413 |

solvent than in aqueous buffer and close the substrate channel to the heme, but this hypothesis remains to be proved.

The results presented here indicate that the catalytic process in aqueous media can be reproducible even in organic solvents without altering the mechanism. Thus, the chemical modification of enzymes to dissolve in appropriate organic solvents might be a potentially promising approach to observe transient enzymatic reaction intermediates such as the ferryl species of cytochrome $P450_{cam}$ at a temperature below the freezing point of the buffer [10].

*Acknowledgments.* We thank Mr. Ted King (Hi-Tech Scientific) for his helpful discussion. This work was supported by grant-in-aid for scientific research (no. 07458147) and for priority areas, molecular biometallics, for Y.W.

# References

1. Dunford HB (1991) In: Everse JE, Everse KE, Grisham MB (eds) Peroxidases in chemistry and biology, vol II. CRC, Boca Raton, pp 1–24

2. Takahashi K, Nishimura H, Yoshimoto T, et al. (1984) A chemical modification to make horseradish peroxidase soluble and active in benzene. Biochem Biophys Res Commun 121:261–265

3. Yang L, Murray RW (1994) Spectrophotomeric and electrochemical kinetic studies of poly(ethylene glycol)-modified horseradish peroxidase reactions in organic solvents and aqueous buffers. Anal Chem 66:2710–2718

4. Mabrouk PA (1995) The use of nonaqueous media to probe biochemically significant enzyme intermediates: the generation and stabilization of horseradish peroxidase compound II in neat benzene solution at room temperature. J Am Chem Soc 117:2141–2146

5. Kudryashova EV, Galperin IB, Gladilin AK, et al. (1996) Stabilization of peroxidase by non-covalent complex formation with polyelectrolysis in water cosolvent mixtures. In: Obinger C, Burner U, Ebermann R, et al. (eds) Plant peroxidases, biochemistry and physiology. University of Geneva Press, Geneva, Swizerland, pp 390–395

6. Chance B (1952) The kinetics and stoichiometry of the titration from the primary to the secondary peroxidase complexes. Arch Biochem Biophys 41:416–424

7. George P (1952) Chemical nature of the secondary hydrogen peroxide compound formed by cytochrome-c peroxidase and horseradish peroxidase. Nature 169:612–613

8. Stillman JS, Stillmann MJ, Dunford HB (1975) Horseradish peroxidase. XIX. A photo-chemical reaction of compound I at 5°K. Biochem Biophys Res Commun 63:32–35

9. Baek HK, Van Wart HE (1989) Elementary steps in the formation of horseradish peroxidase compound. I: Direct observation of compound 0, a new intermediate with a hyperporphyrin spectrum. Biochemistry 28:5714–5719

10. Douzou P, Sireix R, Travers F (1970) Temporal resolution of individual steps in an enzymic reaction at low temperature. Proc Natl Acad Sci USA 66:787–792

# Formation of a Hydroperoxy Complex of Heme During the Reaction of Ferric Myoglobin with $H_2O_2$

Tsuyoshi Egawa, Hideo Shimada, and Yuzuru Ishimura

*Summary.* Reaction of ferric myoglobin with hydrogen peroxide ($H_2O_2$) was studied by following the changes in absorption spectra of myoglobin during its conversion to a ferryl form. Analyses of the data by global analysis revealed that an intermediate species, which has not been hitherto described, occurred prior to the formation of ferryl myoglobin. At a pHs below 7, the new species exhibits the Soret absorption maximum at 408 nm, which shifts to 414 nm on raising the pH to above 8. By an analogy with the pH-dependent transition between the acidic and alkaline forms of ferric myoglobin, we suggest that the intermediate species is a mixture of a high-spin $Fe^{3+}(H_2O_2)$ form and a $Fe^{3+}(HO_2)^-$ form, the latter of which is in a thermal equilibrium between the high- and low-spin forms.

*Key words.* Heme–peroxide complex—Myoglobin—Peroxidase—Global analysis—Rapid-scan spectrophotometry

## Introduction

Myoglobin is a heme protein which reversibly binds with molecular oxygen ($O_2$). Under physiological conditions, the heme iron of myoglobin is usually in a ferrous state ($Fe^{2+}$), but is easily oxidized to a ferric state [1] by a variety of oxidants. Ferric myoglobin can react with $H_2O_2$, and it cleaves the oxygen-oxygen bond of $H_2O_2$ to give $H_2O$ and an oxoferryl ($Fe^{4+}=O$) species of the hemoprotein, which is referred to as ferryl myoglobin [2,3]. The reaction has therefore been used as a model of the $H_2O_2$ decomposing reactions such as those catalyzed by catalase and peroxidases [2–4].

In the above reaction, it is expected that a complex of heme and hydrogen peroxide (hydroperoxy-heme), which may be similar in structure to the peroxy intermediate in peroxidase-catalyzed reactions [5], is formed as an obligatory intermediate. However, such a hydroperoxy-heme species of myoglobin has never been observed nor detected in the reaction of ferric myoglobin with $H_2O_2$. By employing a method called "global

Department of Biochemistry, School of Medicine, Keio University, 35 Shinanomachi, Shinjuku-ku, Tokyo 160, Japan

analysis [6]", we succeeded in this study in extracting an absorption spectrum of such an intermediate species formed prior to the ferryl myoglobin formation. Although the intermediate is too unstable to be accumulated in a large quantity during the reaction, the present method enabled us to obtain absorption spectrum of the intermediate, which is most probably the hydroperoxy-heme species of myoglobin.

## Materials and Methods

Horse heart myoglobin was purchased from Sigma. Hydrogen peroxide was obtained from Tokyo Kasei Kogyo (Tokyo, Japan) and used without further purification. Transient optical spectra were measured by an RSP-601 stopped flow-rapid scan system (UNISOKU, Osaka, Japan). Optical changes during the reaction were analyzed by global analysis [6] and, as a result, the absorption spectrum of the intermediate species that formed before ferryl myoglobin was obtained. Details of our methods will be described elsewhere (T. Egawa et al, in manuscript).

## Results and Discussion

Changes in the optical spectrum of ferric myoglobin during the reaction with $H_2O_2$ were followed at pH 5.0, 6.0, 6.5, 7.0, and 8.0, and were analyzed by the methods described under Materials and Methods. Figure 1 shows absorption spectra of the intermediate species, which was obtained at various pH prior to the formation of ferryl myoglobin. As seen, Soret absorption spectrum of the intermediate changed markedly as pH was changed; at least two isoforms which are in equilibrium appear to exist with Soret absorption maxima at 408 and 423 nm.

We have recently suggested that one isoform of the intermediate formed at acidic pH is a $Fe^{3+}(H_2O_2)$ derivative of myoglobin (T. Egawa et al. in manuscript). The basis for the proposal depends on 1) the empirical rule for the relationships of the spin- and the oxidation-states to the absorption spectrum of heme [1,7,8], 2) the results from quantum chemical calculations of the absorption spectrum of the $Fe^{3+}(H_2O_2)$ derivative of protoheme [9], and 3) the fact that the intermediate species of myoglobin is formed prior to ferryl myoglobin.

As seen in Fig. 1, the Soret peak at 408 nm decreases as pH is raised, and a new band at 423 nm appears as pH increases. The inset of Fig. 1 plots the relative values for the optical absorption at 423 and 408 nm against pH, indicating that there is a $pK_a$ for the transition between the two isoforms at about pH 6.5.

It is well known that ferric myoglobin (metmyoglobin) exhibits pH-dependent changes, very similar to those found with the intermediate species described above. At an acidic pH, ferric myoglobin is predominantly in the acidic form, which has an $H_2O$ molecule at the sixth ligation position. This acidic form is in the equilibrium with an alkaline form with a hydroxyl anion at sixth coordination position (Fig. 2a). The acidic form of ferric myoglobin is in a high-spin state exhibiting the Soret maximum at 408 nm, while the alkaline form is a thermal mixture of high- and low-spin states showing an apparent Soret maximum at 414 nm [1].

On analogy with the pH-dependent changes in the absorption spectrum of ferric myoglobin, we propose that ferric myoglobin forms a hydrogen peroxide complex,

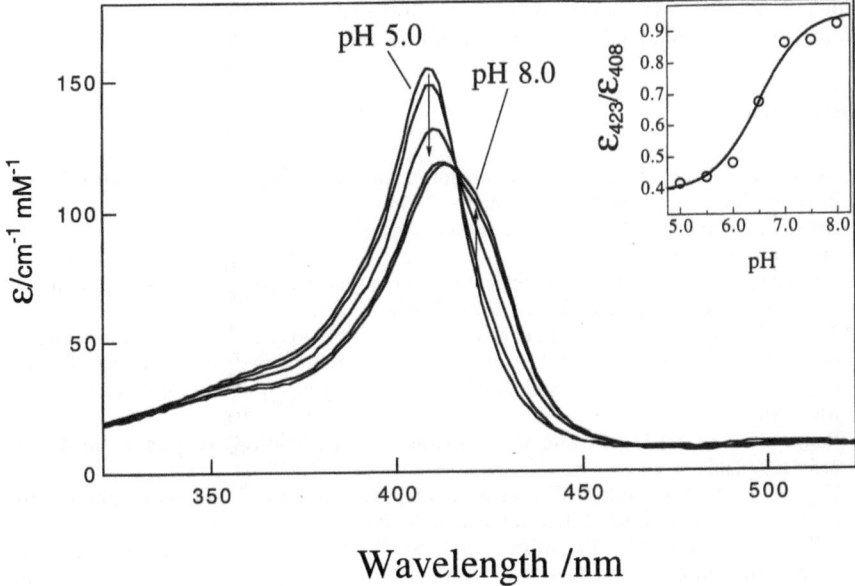

FIG. 1. Calculated absorption spectra of the intermediate species formed during the reaction of ferric myoglobin with $H_2O_2$ at various pH values (5.0, 6.0, 6.5, 7.0, 8.0). *Arrows* indicate directions of changes in absorption spectrum with increasing pH. The *inset* plots the relative absorption at 423 and 408 nm against pH. *Solid line* is a theoretical curve expected for dissociation of one proton with $pK_a$ 6.5

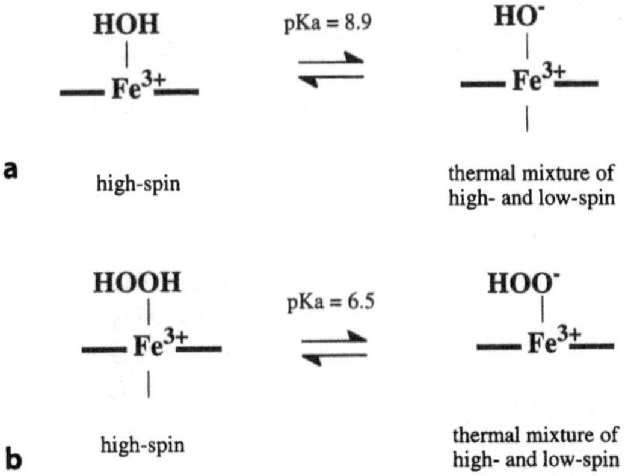

FIG. 2a,b. Schematic for the acid–base transition of ferric myoglobin (a) and for proposed acid–base transition of heme-peroxide complex of myoglobin (b)

which is a mixture of a high-spin $Fe^{3+}(H_2O_2)$ form and a $Fe^{3+}(HO_2)^-$ form, the latter of which is in a thermal equilibrium between the high- and low-spin forms (Fig. 2b).

# References

1. Antonini E, Brunori M (1971) Hemoglobin and myoglobin in their reactions with ligands. North-Holland, Amsterdam, pp 13–53
2. George P, Irvine DH (1952) The reaction between metmyoglobin and hydrogen peroxide. Biochem J 52:511–517
3. George P, Irvine DH (1956) A kinetic study of the reaction between ferrimyoglobin and hydrogen peroxide. J Colloid Sci 11:327–339
4. Adachi S, Nagano S, Ishimori K, et al. (1993) Roles of proximal ligand in heme proteins: replacement of proximal histidine of human myoglobin with cysteine and tyrosine by site-directed mutagenesis as model for P-450, chloroperoxidase, and catalase. Biochemistry 32:241–252
5 Poulos TL, Kraut J (1980) The stereochemistry of peroxidase catalysis. J Biol Chem 255:8199–8205
6. Hug SJ, Lewis JW, Einterz CM, et al. (1990) Nanosecond photolysis of phodopsin: evidence for a new, blue-shifted intermediate. Biochemistry 29:1475–1485
7. Selke M, Sisemore MF, Valentine JS (1996) The diverse reactivity of peroxy ferric porphyrin complexes of electron-rich and electron-poor porphyrins. J Am Chem Soc 118:2008–2012
8. Tajima K (1989) A possible model of a hemoprotein-hydrogen peroxide complex. Inorg Chim Acta 163:115–122
9. Harris DL, Loew GH (1996) Identification of putative peroxide intermediates of peroxidases by electronic structure and spectra calculations. J Am Chem Soc 118:10588–10594

# Photooxidation in Ternary System Human Serum Albumin–Chlorin e$_6$–Tryptophan

E.V. Petrotchenko[1], G.A. Kochubeev[2], S.A. Usanov[1], and P.A. Kiselev[1]

*Summary.* Light-induced oxidation reactions in ternary system tryptophan–human serum albumin–chlorin e$_6$ (Trp-HSA-Cle$_6$) were investigated. Solutions with different components content were illuminated under red light. Reagent conversion kinetics were estimated by reverse-phase HPLC. Under reaction, photoproducts are formed both free and covalently bound to protein Trp. Their distribution correlates with the initial binding ratio of Trp to albumin. Data obtained were considered in several aspects. (1) The involvement of Trp as a naturally occurring ligand of albumin in photooxidation reactions points out a possibility for targeted in situ generation of physiologically active products of albumin ligands. (2) Production of chemically modified HSA molecules also may have an impact on cell-mediated and vascular responses of photooxidation. (3) Ternary system Trp-HSA-Cle$_6$ may serve as a prototype of a light-driven oxidation enzyme with the HSA ligand (Trp) as the substrate, Cle$_6$ as the prosthetic group, and HSA as the apoenzyme.

*Key words.* Chlorin e$_6$—Photooxidation—Human serum albumin—Tryptophan

## Introduction

Nature uses light for the energy supply of living systems by photosynthesis. Energy equivalents are accumulated in macroenergetic bonds of substances and are released as needed for the accomplisment of various life processes. Although being universal and existing since time in the remote past, such routes are quite complicated and have a low overall quantum yield. Alternatively, light energy can be utilized directly in chemical reactions. Some porphyrins can transfer energy of absorbed light to oxygen, thus generating activated species of oxygen molecules, that are capable of oxidation reactions.

---

[1] Institute of Bioorganic Chemistry of the Academy of Sciences of Belarus, Zhodinskaya Str., 5, 220141 Minsk, Belarus
[2] Institute of Molecular and Atomic Physics of the Academy of Sciences of Belarus, Skoryny Av., 70, 220072 Minsk, Belarus

We have tried to construct such a system on protein matrices and to model the photochemical reaction assembly as a prototype of a photooxidation enzyme. We chose tryptophan (Trp), chlorin $e_6$ (Cle$_6$), and human serum albumin (HSA) as model components and investigated the photooxidation reactions in this ternary system.

## Materials and Methods

Two hundred-milliliter mixtures of human serum albumin (HSA) (Sigma, St. Louis, MO, USA), Cle$_6$ (Dialek, Minsk, Belarus), and Trp (Reanal, Budapest, Hungary) in 0.05 M phosphate buffer pH 7.4 with varying component ratios were illuminated by light above 560 nm under intensive mixing. Aliquots of 20 μl were taken at 15-min intervals and analyzed by reversed-phase HPLC. The column was an Ultrasphere ODS, 5 μm, 0.46 × 25 cm (Beckman, Palo-Alto, CA, USA) in 0.1% TFA acetonitrile. A gradient of 0–100% of acetonitrile during 10 min, 1 ml/min, was used. Absorbance at 254 and 280 nm was simultaneously monitored. Reagent conversion kinetics were estimated by measuring the heights of corresponding peaks on the chromatograms.

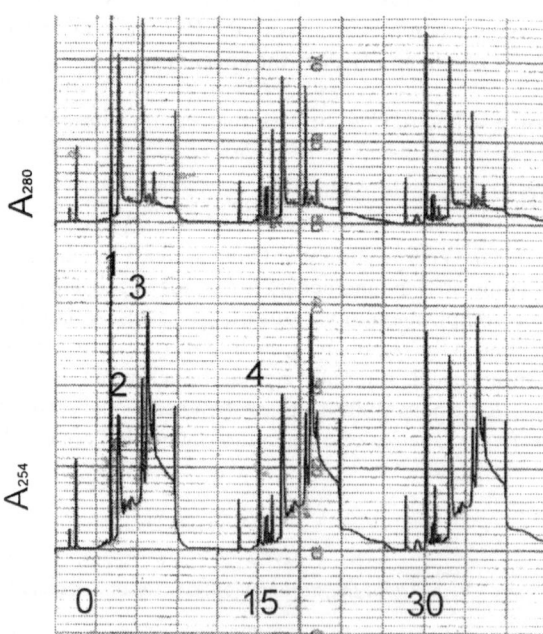

Fig. 1. Chromatograms of the mixture tryptophan–human serum albumin–chlorin $e_6$ (Trp-HSA-Cle$_6$) under 0, 15, and 30 min of illumination. 1, Trp; 2, HSA; 3, Cle$_6$; 4, Trp photoproduct

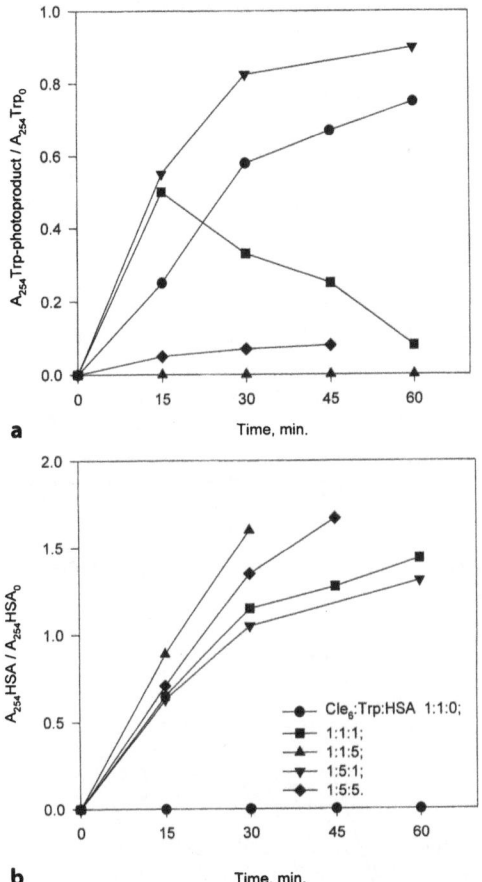

FIG. 2a,b. Accumulation of free (a) and conjugated with HSA (b) photoproducts under different reagent ratios

# Results

In the binary system Trp-$Cle_6$, predominant accumulation of one product occurs, as reflected in the emergence of a major peak on chromatograms of the reaction mixture. The peak is characterized by approximately equal absorbance at 254 and 280 nm and by minimal retention among other chromatographic zones appearing during the reaction. Addition to the system of sodium azide almost completely prevents loss of Trp during the reaction and principally does not influence on the loss of $Cle_6$.

In the ternary system, Trp-HSA-$Cle_6$, photoconversion of Trp differs to some extent. Under a fivefold excess of Trp over HSA, the distribution of reaction products follows those in a protein-free system (Fig. 1). With increasing protein content in the reaction mixture, redistribution of free tryptophan photoproducts occurs. The amplitude of the dominant peak is decreased, and at the ratio HSA/Trp of 5:1, that is, under

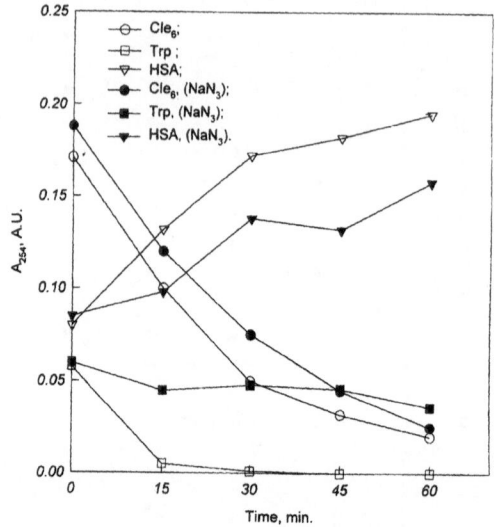

FIG. 3. Reagent conversion kinetics in ternary system Trp-HSA-Cle$_6$

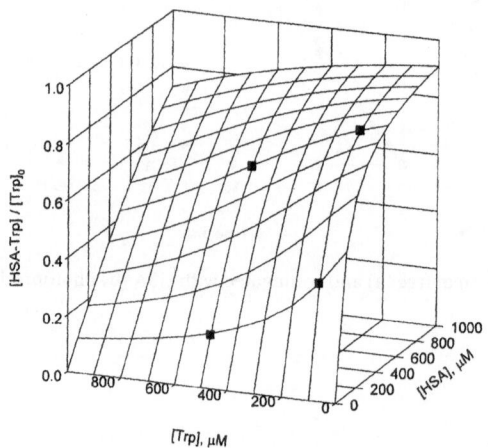

FIG. 4. Dependence of proportion of HSA-bound Trp on component ratio

conditions approaching those in vivo, free products of the photoconversion of Trp are completely absent on chromatograms (Fig. 2).

In all reactions in the presence of protein, an increase of the HSA peak of 2- to 3 fold at 254 nm and of 1.2- to 1.5 fold at 280 nm is observed. Monitoring of chromatograms at 400 nm reveals the presence of fractions coeluting with the HSA peak. However, carrying out the reaction under different HSA/Cle$_6$ ratios (1:1, 5:1) did not influence

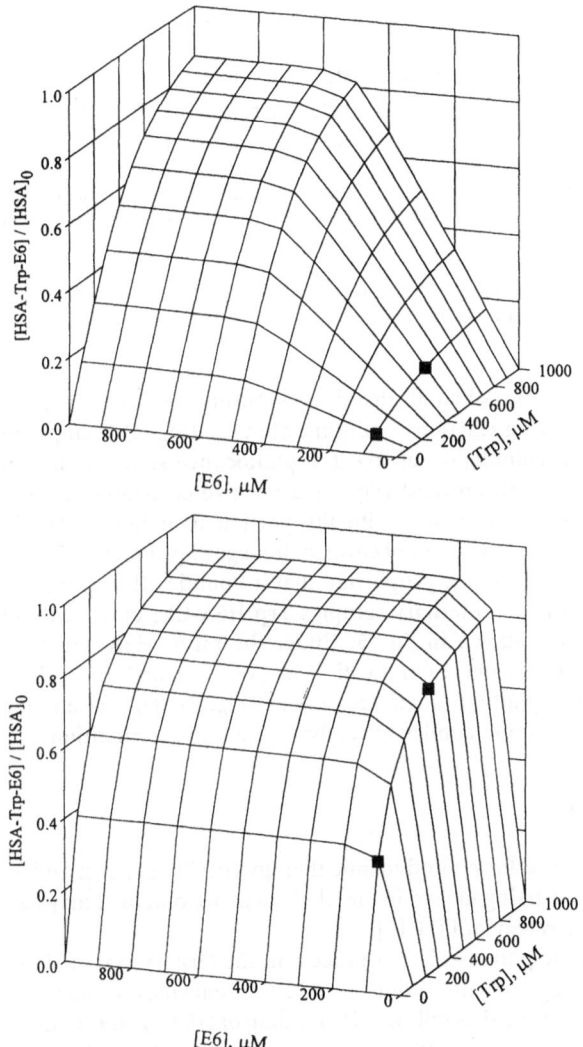

FIG. 5. Ternary complex Trp-HSA-Cle$_6$ content in relation to components ratio. *Top*, [HSA] = 500 μM; *bottom*, [HSA] = 100 μM

the order of HSA peak changes, but varying the Trp concentration resulted in a change in height of the HSA peak. Addition of sodium azide also inhibited HSA peak increase (Fig. 3). These data show that during the reaction, covalent conjugates of HSA with Cle$_6$ and Trp photooxidation products are formed.

Considering the ternary system Trp-HSA-Cle$_6$, it might be postulated that binding of Trp and Cle$_6$ with HSA occurs in two distinct sites. Trp interacts with the indole-binding site ($K_s \sim 10^{-4}$ M) and Cle$_6$ appears to be bound to the bilirubin-binding site

TABLE 1. Complexes of human serum albumin (HSA) with ligands in relation to component ratios

| P:T:E | $[P]_0$, µM | $[T]_0$, µM | $[E]_0$, µM | $[P*T]$, µM | $[P*E]$, µM | $[P*T*E]$, µM | $[P*T]/$ $[T]_0$ | $[P*T]/$ $[P]_0$ | $[P*E]/$ $[P]_0$ | $[P*T*E]/$ $[P]_0$ |
|---|---|---|---|---|---|---|---|---|---|---|
| 1:1:1 | 100 | 100 | 100 | 38.2 | 90.5 | 34.6 | 0.38 | 0.38 | 0.91 | 0.35 |
| 1:5:1 | 100 | 500 | 100 | 80.7 | 90.5 | 73.0 | 0.16 | 0.81 | 0.91 | 0.73 |
| 5:5:1 | 500 | 500 | 100 | 321. | 99.8 | 64.2 | 0.64 | 0.64 | 0.20 | 0.13 |
| 5:1:1 | 500 | 100 | 100 | 80.7 | 99.8 | 16.1 | 0.81 | 0.16 | 0.20 | 0.03 |
| 10:1:1 | 500 | 50 | 50 | 41.1 | 49.9 | 4.1 | 0.82 | 0.08 | 0.10 | 0.01 |

P, T, and E are HSA, Trp, and chlorin $e_6$, respectively; P*T and P*E, binary complexes; P*T*E, ternary complex; $[T]_0$, $[E]_0$, $[P]_0$, total concentration of Trp, $Cle_6$, and HSA, respectively.

$(K_s \sim 10^{-6} M)$ [1,2]. It is obvious, that different proportions of Trp are bound with the protein when varying reaction conditions are used (Fig. 4). Comparing these values with curves of accumulation of free Trp photoconversion products (i.e., those not covalently bound with protein) (Fig. 2), a positive correlation of the degree of the generation of these substances with the portion of unbound Trp in the reaction mixture can be seen. The extent of covalent linkage of products of Trp photooxidation to HSA also depends on the proportion that is bound with the protein ligand.

Calculated portions of ternary complex Trp-HSA-$Cle_6$ under different component concentrations are shown in Fig. 5. Under the ratios HSA:Trp:$Cle_6 = 1:1:1$ and $1:5:1$, almost all chlorin is bound with HSA (Table 1), and the initial rates of generation of free Trp photoproducts are similar (Fig. 2). This result suggests that the system acts in a manner similar to catalysis under these conditions.

## Discussion

As chlorin $e_6$ is used for photodynamic therapy (PDT) of cancer and Trp is a natural ligand of HSA occurring in vivo in blood, the data we obtained may have a relation to the theory and strategy of PDT [3].

Considering the photooxidation process in the ternary system Trp-HSA-$Cle_6$ as a functional analogy with the action of oxidation enzymes, elementary stages of the process could be imaged as follows: (1) binding of HSA ligands to the complex HSA–$Cle_6$; (2) absorption of a quantum of light by $Cle_6$ and its transition to the triplet state; (3) diffusion collision of an oxygen molecule with $Cle_6$, energy transfer, and transition of the oxygen into the singlet state; (4) diffusion of singlet oxygen and its interaction with target molecules; and (5) secondary reactions of oxidized substances. This chain of events resembles the catalytic cycle of oxidase with HSA ligands as substrate, $Cle_6$ as heme, HSA as apoenzyme, and substitution of the energy supply from reduction equivalents to light energy.

The critical moment for such a model to be valid is at the covalent linkage of HSA ligand photoproducts to protein. The question is: Is a covalent linkage an inalienable feature of such oxidation reactions or it can be avoided? To obtain the answer, the role of the chemical structure of the ligand, its oxidation chemistry, the mutual orientation and proximity of porphyrin and bound ligand to the protein, and the path of activated oxygen must be elucidated.

# References

1. McMenamy RH (1977) Albumin binding sites. In: Rosenoer VM, Oratz M, Rotschild MA (eds) Albumin structure, function and uses. Pergamon, Oxford, pp 143–158
2. Kochubeev GA, Frolov AA, Zenkevich EI, Gurinovich GP (1988) Features of complex formation between chlorin $e_6$ and human serum albumin. Biofizyka 32:175–178
3. Henderson BW, Dougherty TJ (1992) How does photodynamic therapy work? Photochem Photobiol 55:145–157

# Part 4
# Oxygen Sensing and Regulation of Blood Flow

# Tissue Oxygen Pressure and Oxygen Sensing by the Carotid Body

David F. Wilson[1], Sergei A. Vinogradov[1], Anil Mokashi[2],
Anna Pastuszko[1,3], Sukhamay Lahiri[2], and Mark W. Dewhirst[4]

*Summary.* A new optical technique for measuring oxygen, oxygen-dependent quenching of phosphorescence, has been developed for the study of tissue oxygenation in vivo and in vitro. Using phosphorescence imaging techniques it has been possible to obtain preliminary three-dimensional maps of oxygen distribution within growing tumors and the surrounding host tissue. These maps show that the oxygen levels in tumors have high inter- and intratumor heterogeneity but that tumors are generally more hypoxic than the host tissue. The lowest oxygen pressures are found in the growing edge where new tissue is being formed. Sensing of oxygen pressure in tissues is central to the homeostatic regulation of oxygen supply, and the carotid body is one of the premier sensory tissues of the body. The afferent neural activity increases manyfold as the arterial oxygen pressure falls from about 100 to near 0 torr. In the isolated cat carotid body perfused with a medium equilibrated with a gas containing $CO:O_2:CO_2$ at approximately 72%, 23%, and 5%, respectively, afferent neural activity increased approximately 20 fold above that for $N_2:O_2:CO_2$ of 72%, 23% and 5%, respectively, to near the maximal values obtained as $O_2$ approached zero. The effect of CO on afferent activity was reversed by bright light. The decrease in activity induced by nonsaturating intensities of monochromatic light were corrected to equal light intensity and plotted against the wavelength of the light. The resulting "action spectrum" is the absorption spectrum of the CO compound of mitochondrial cytochrome $a_3$. The oxygen-sensing activity attributable to cytochrome $a_3$ accounts for at least 80% of the total activity of the carotid body.

*Key words.* Carotid body—Oxygen—Tissue oxygenation—Phosphorescence—Tumors

[1] Departments of Biochemistry and Biophysics, University of Pennsylvania, 3260 Hamilton Walk, Philadelphia, PA 19104-6059, USA
[2] Physiology, University of Pennsylvania, 3260 Hamilton Walk, Philadelphia, PA 19104-6059, USA
[3] Pediatrics, University of Pennsylvania, 3260 Hamilton Walk, Philadelphia, PA 19104-6059, USA
[4] Department of Radiation Oncology, Duke University, Durham, NC 27710, USA

# Introduction

Tissue maintenance and function are critically dependent on delivery of oxygen in sufficient amounts and with an appropriate distribution (local oxygen pressures) to support the metabolic requirements of that tissue. A major barrier to effective study of the oxygen delivery system has been an inability to quantitate its performance through direct measurements of the oxygen pressure distribution within living tissue. This inability to directly measure the oxygen levels has caused investigators to substitute one or more indirect measurements, such as blood flow and oxygen extraction, as the measure of oxygen delivery. Although these measurements are very useful, the conclusions are often ambiguous because alterations in metabolic requirements or distribution of the delivered oxygen can occur that are not correctly identified. In this chapter, we use a new optical method for measuring oxygen, oxygen-dependent quenching of phosphorescence, to demonstrate that direct measurements of the oxygen distributions in living tissue are possible and that it is feasible obtain three-dimensional maps of these distributions.

Regulation of oxygen delivery to a tissue is dependent on biochemical mechanisms within the tissue for detecting and measuring oxygen pressure at local areas. These oxygen-measuring systems are generally referred to as oxygen sensors. The oxygen sensors must then communicate information about oxygen pressure to the appropriate vascular or neural elements, presumably through a messenger cascade(s). Currently available data indicate there may be several different functional sensors for oxygen, depending on the tissue, but in most cases the identity and roles of these sensors remain to be established. The carotid body, an organ located at the bifurcation of the carotid artery, detects the levels of oxygen and carbon dioxide in the blood of the carotid artery and translates this information into afferent electrical impulses to the brain [1,2]. The sensing of oxygen and carbon dioxide occurs through different detectors, but there are interactions between the sensory systems. Strong evidence is now available that the primary oxygen sensor of the carotid body is mitochondrial oxidative phosphorylation through mitochondrial cytochrome $c$ oxidase.

# Materials and Methods

Oxygen-dependent quenching of phosphorescence has been previously described in detail [3–7].

## Basic Principles

When light is absorbed by a chromophore, the electron in the lowest lying energy level is promoted into the excited singlet state. In the excited singlet state, the electron has the same electronic spin as it had in the ground state. The singlet state may return to the ground state with emission of a photon of light (fluorescence), a process that is very rapid (nanoseconds), or by releasing the energy to its environment as heat. In some chromophores, the singlet state can also undergo "intersystem crossing," in which the electron spin changes to the opposite of that in the ground state and converts the singlet state into a triplet state. The triplet state can, like the singlet state, return to the ground state with emission of a photon (phosphorescence) or by releasing the energy to its environment as heat. For the triplet state, however, return of the electron to the ground state requires that the two events, release of a photon and flip of the electron spin back to that of the ground state, occur simultaneously. Such a simultaneous double event is improbable and as a result the lifetime of the triplet state is much longer than that of the singlet state ($10^{-3}$ s vs. $10^{-8}$ s).

The excited-state molecules may also return to the ground state by transfer of the energy to some other molecule that has appropriate electronic energy levels to accept the energy "package." For efficient energy transfer, the molecules must be close together, such as occurs during the random collisions between molecules in solution. The degree of quenching is dependent of the number of triplet-state molecules that collide with quencher molecules and therefore is a direct measure of the concentration of quencher. This energy transfer results in a decrease in the phosphorescence lifetime and an attendant decrease in total light emitted (quenching). Triplet excited states, with their long lifetimes, have a high probability of colliding with potential quenching agents. Molecular oxygen is an efficient quencher of phosphorescence, and in biological fluids such as blood, it is the only quenching agent present in significant concentrations.

We have selected phosphors that (1) emit from 6% to 20% as many photons as they absorb, resulting in easily measured phosphorescence intensities; (2) have phosphorescence lifetimes of about $1 \times 10^{-3}$ s in the absence of quencher and nearly $20 \times 10^{-5}$ s in media saturated with air (20% oxygen), providing a wide dynamic range for oxygen determinations. The phosphorescence decay occurs at rates (lifetimes) that are easy to measure.

## Advantages of Phosphorescence Quenching for Oxygen Measurements

Phosphorescence quenching is an optical method and the only oxygen measuring system capable of noninvasive, quantitative determination of oxygen pressure in the vasculature of tissue in vivo. This is important because the performance of the pulmonary–cardiovascular system is directly related to its ability to sustain local tissue oxygen pressures at appropriate levels throughout the body. Other advantages include the following:

1. Quenching of phosphorescence by oxygen behaves according to a well-defined mathematical relationship, the Stern–Volmer equation:

$$T^{\circ}/T = 1 + k_Q T^{\circ} PO_2 \qquad (1)$$

where $T^{\circ}$ is the lifetime in the absence of oxygen, $T$ is the intensity and lifetime at an oxygen pressure PO$_2$, and $k_Q$ is a constant describing the frequency of quenching collisions between the probe molecules in the triplet state and molecular oxygen. The constant, $k_Q$, is a function of the diffusion constants for probe and oxygen, temperature, and probe environment.

2. So long as the phosphor remains in a well-defined physical environment, calibrations of $k_Q$ and $T^{\circ}$ are absolute and calibrations in vitro are fully valid for studies in vivo. Moreover, once the values of $k_Q$ and $T^{\circ}$ have been determined, they can be used indefinitely; that is, the calibration is only dependent on the chemical structures

involved. Values of $k_Q$ and $T°$ determined in one laboratory are equally valid in any other part of the world.

3. The method is highly specific for oxygen. We have not detected any agents in blood, other than oxygen, that affect the measured phosphorescence lifetime. Thus calibration constants measured in vitro are valid for measurements in vivo.

4. The response time is only a few milliseconds even at low oxygen pressures, and this means oxygen measurements are very rapid down to less than 0.1 torr.

5. This method is an optical technique and can be used to measure two- and three-dimensional distribution of oxygen within objects. Using an intensified video camera, phosphorescence quenching has been used to provide high-resolution digital maps of the distribution of oxygen pressure in the vasculature of a wide range of tissues in vitro and in vivo.

6. There is no evidence for toxicity of the Pd-porphyrins or any of the other phosphors in current use, including the new ones operating in the near-infrared region of the spectrum. Injection of 5 mg of Green 2W [6] per mouse into six mice produced no evidence of toxicity in the following 10 days, although less than 0.3 mg/ mouse is sufficient for imaging the oxygen pressure and light guide measurements require less than 0.15 mg/mouse.

7. Measurements of phosphorescence lifetime are independent of phosphor concentration so long as it is bound to albumin or otherwise remains in the same local environment.

8. The calibration parameters of some of the phosphors are independent of pH and ionic strength over wide ranges of these parameters.

9. Measurements of phosphorescence lifetimes are independent of any other chromophores, such as hemoglobin or myoglobin, that may be present in the system so long as these changes do not occur during the phosphorescence decay (<1 ms). Thus, the method is particularly effective for measuring oxygen in the vasculature of intact tissue.

## Tumor Growth in Rats in a Dorsal Skin Flap Window Chamber

Fischer rats were surgically fitted with dorsal flap window chambers, as previously described [8]. The skin on the back was anesthetized and abraded to produce an injury approximately 1.2 cm in diameter, and the injured skin area was placed in a metal frame that holds two glass windows, one on either side of the skin flap, with the space between the two being approximately 250 μm. R3230AC mammary carcinoma cells were transplanted into the chamber on one side of the host tissue and the tumors were allowed to grow to 2–3 mm in diameter. The growing tumor and host tissue can be observed through the windows from either the side of the growing tumor tissue or the side of the host tissue.

## Isolated Perfused Carotid Body from the Adult Cat

The carotid bifurcation was prepared from adult female cats for in vitro perfusion as described by Lahiri and co-workers [9,10]. Single-pass perfusion and superfusion were established using hydrostatic pressures of 80 torr (mm Hg). The perfusates and superfusates were modified Tyrode solution containing 154 mM $Na^+$, 123 mM $Cl^-$,

4.7 mM K$^+$, 2.2 mM Ca$^{2+}$, 1.1 mM Mg$^{2+}$, 22 mM glutamate, 5 mM glucose, and 5 mM hydroxyethylpiperazine ethanesulfonic acid (HEPES) at a final pH of 7.4. The perfusate was gassed with either 21% O$_2$:5% CO$_2$ or a mixture of 20% O$_2$, 75% CO with 5% CO$_2$. The temperature of the perfusion chamber was maintained at 36.5° ± 0.5°C. Paraffin oil was layered over the superfusate to a depth of about 4 mm.

## Measurements of the Photochemical Action Spectrum

The photochemical action spectrum was measured as the light-induced response in afferent neural activity in the isolated carotid body (carotid bifurcation) perfused with a mixture of CO and O$_2$ [10]. The theory is essentially that presented by Warburg and Negelein [11], who used this technique to identify the primary oxidase of yeast and mammalian cells. It has since been the primary method for identifying most of the heme oxidases and showing the participation of other enzymes with reduced heme in metabolic pathways (such as cytochrome P-450 in biological hydroxylations) [12].

# Results

## The Oxygen Dependence of Phosphorescence Lifetimes

The values of the phosphorescence lifetime at zero oxygen and the quenching constant for oxygen-dependent quenching of phosphorescence for a phosphor allow the oxygen pressure to be calculated from each measured phosphorescence lifetime. Each phosphor needs to be calibrated for a given set of experimental conditions, but the calibration then can be used whenever the measurement conditions match those used for calibration. For best results, the phosphorescence lifetimes are measured at many different oxygen pressures for each experimental condition (pH, temperature, and solvent) and the values of $k_Q$ and $T^\circ$ calculated from a best fit to the Stern–Volmer equation (Eq. 1) [7]. The lifetimes of the phosphor most often used for oxygen measurement, Pd-*meso*-tetra-(4-carboxyphenyl) porphyrin in buffer containing 0.5% or more of bovine serum albumin, show an excellent fit to Eq. 1. The values of $k_Q$ and $T^\circ$ of this phosphor are not dependent on the pH of the medium in the physiological range of from 6.8 to 7.8. Both $k_Q$ and $T^\circ$ are temperature dependent, however, increasing about 3% and decreasing about 0.8% per degree increase in temperature, respectively, between 23° and 38°C.

## Observing the Oxygen Distribution in Tissue in Three Dimensions

Phosphorescence emission can be readily measured using an intensified video camera [5,13,14]. With appropriate gating of the intensifier, a series of images can be collected with phosphorescence collection beginning with different delay times after a flash of excitation light. The phosphorescence lifetime can then be calculated for each pixel of the images (see [5,13,14]). This generates a two-dimensional map of the distribution of phosphorescence lifetimes and therefore of oxygen distribution in the tissue. Phosphors based on Pd-*meso*-tetra-(4-carboxyphenyl) porphyrin have absorption bands at about 420 nm (near-ultraviolet) and 524 nm (green), and the phosphorescence can be equally well excited at either wavelength. On the other hand, the intrinsic chromophores of tissue limit the depth of penetration of 420 nm light to about 50 μm while the depth of penetration of 524 nm light, which is absorbed less, can be as much as about 500 μm. This means that the oxygen measurements are for a superficial layer of tissue (420 nm excitation) and for a deeper layer (524-nm excitation), allowing identification of tissue volumes with different oxygen pressures within this thickness of tissue. We have applied this approach to tumors grown in dorsal window chambers in rats as described by Dewhirst and co-workers [8,15].

Figure 1 shows a composite of the oxygen pressure maps of a growing tumor. The oxygen pressure maps were taken from the tumor side of the window with blue excitation (tumor side, superficial) and green excitation (tumor side, deep), and then from the host tissue side with blue excitation (host side, superficial) and green excitation (host side, deep). Because of the increase in tissue mass and decreased oxygen pressures relative to the host tissue, the tumor has a much greater phosphorescence intensity than the surrounding host tissue. As a result, the oxygen maps essentially depict the tumor tissue. Blue excitation (superficial tissue layer) shows lower oxygen pressures than green excitation (deeper tissue layer), indicating the oxygen pressures

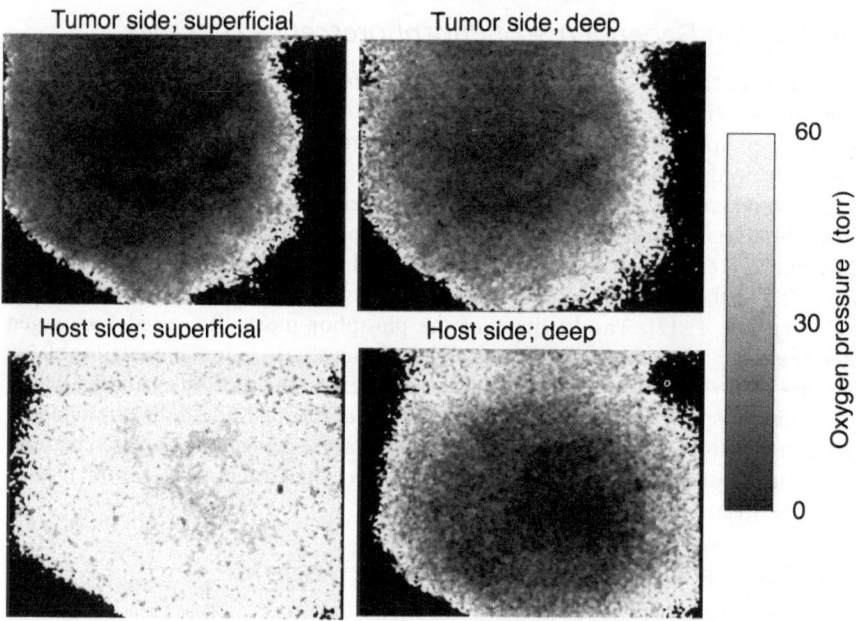

FIG. 1. Oxygen pressure distributions (maps) of a tumor growing in the window chamber as measured by phosphorescence imaging. Animals were anesthetized and 7 mg of Oxyphor-R2 (2.3 mg Pd-porphyrin) was injected into the tail vein. The phosphorescence was imaged with an intensified CCD camera gated to collect the phosphorescence beginning at 20, 40, 70, 120, 220, 400, 720, and 2500 μs after the flash of excitation light. The phosphorescence lifetime and oxygen pressure were then calculated for each pixel (480 × 512) of the image set. The measurements were made with blue excitation (420 nm; *superficial* penetration) and green excitation (524 nm; *deep* penetration) and from the side of the window with the growing tumor (*tumor side*) or from the side of the host tissue (*host side*)

Oxygen histogram

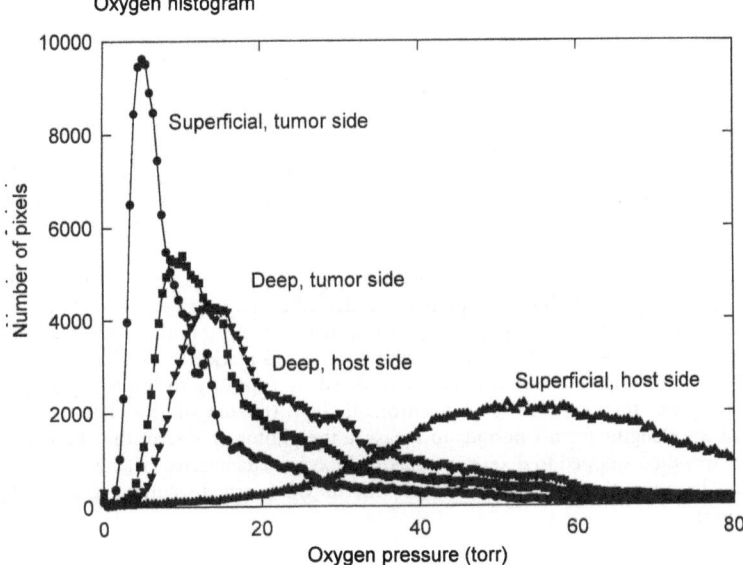

FIG. 2. Histograms of the oxygen distributions in the tumor shown in Fig. 1. In calculating the phosphorescence lifetimes and oxygen pressures, the data are first calculated in 16 bits and then reduced to the 8 most significant bits for presentation. The histograms were calculated for the 16-bit images before reduction to the 8-bit images shown in Fig. 1 by summing the number of pixels with each oxygen pressure and plotting that number against the oxygen pressure

in the growing edge of the tumor are more hypoxic than in the internal, older parts of the tumor. This is also clearly seen in Fig. 2, where the distributions of oxygen are presented as histograms in which the number of pixels with a given oxygen pressure is plotted against the oxygen pressure. In these histograms, the two-dimensional information is lost but quantitation of differences between oxygen distributions is easier. In this experiment, blue excitation from the host side produced oxygen pressures indistinguishable from those of host tissue without an implanted tumor and measurements of regions of host tissue in the window chamber outside the tumor area.

## Identification of the Oxygen Sensor of the Carotid Body

Measurements of the afferent neural activity of the isolated perfused-superfused carotid body show that the activity is low when perfused with normoxic media (air saturated) but that this increases as the oxygen pressure is decreased. Figure 3 shows the low activity when the perfusate contains 130 torr oxygen, with the remainder nitrogen except for 5% carbon dioxide. The afferent activity increased dramatically when 560 torr of carbon monoxide was added instead of the nitrogen, leaving the oxygen and carbon dioxide unchanged, indicating that carbon decreases the oxygen pressure "sensed" by the oxygen sensor. This is a competitive relationship, and the

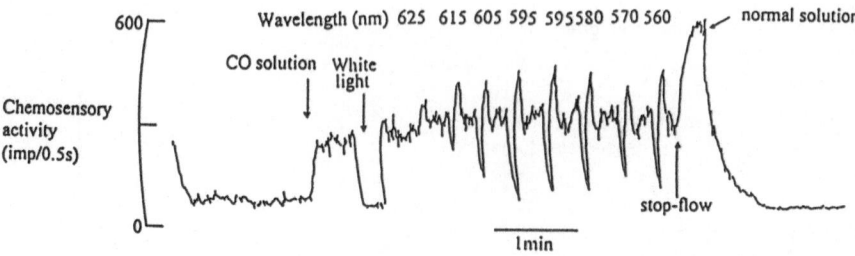

FIG. 3. Photosensitivity of the chemosensory discharge of carotid body perfused with a medium containing CO. The carotid body was prepared as described in Materials and Methods. After the initial period of perfusion with no CO, the perfusion was changed to that with $O_2$ and CO (130 and 560 torr, respectively). The afferent activity increased markedly with the CO-containing medium, but this increase was completely reversed by exposure to bright white light. The carotid body was then exposed to monochromatic light (6.7-nm width at half-height) of the indicated wavelengths for 6-s periods to measure the photoresponse of the afferent activity. The flow was then stopped to determine maximal oxygen chemosensory activity and the perfusion medium was then restored to 21% $O_2$ and no CO (normal solution). (From [10], with permission)

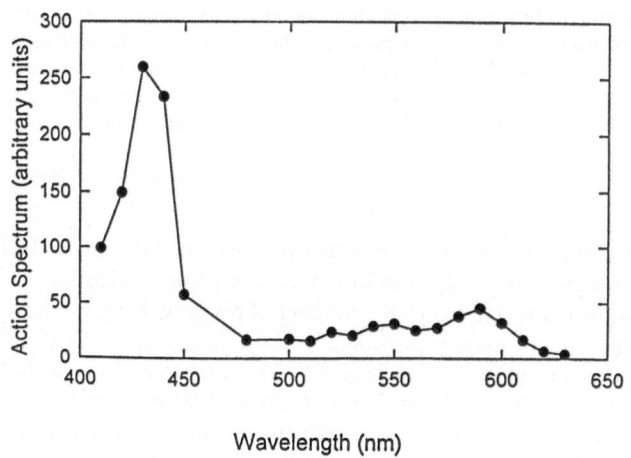

FIG. 4. The photochemical action spectrum for oxygen sensing by the cat carotid body. Isolated perfused carotid body was treated as shown in Fig. 3. When the perfusate was changed to that containing CO, the afferent activity increased approximately 20 fold. The carotid body was then subjected to 6-s periods of illumination, separated by brief recovery periods (Fig. 3), using light passed through a monochromator with a bandwidth at half-height of 10 nm. The extent of the decrease in afferent electrical activity of the carotid body induced by illumination was corrected to equal intensity of light quanta (light flux) at each measurement wavelength. The height of the response is proportional to the absorption of the CO complex of the oxygen sensor at that wavelength. (From [10], with permission)

effect depends on the ratio of O$_2$ pressure to CO pressure. Most importantly, the effect of CO is completely reversed by bright white light, showing that the inhibitory CO complex is reversibly dissociated by absorption of a photon. At less than saturating light intensities, this light-induced reversal is dependent on light intensity (number of photons per second) and on the wavelength of light. As the wavelength is stepped from 625 nm to 595 nm, the size of the photoinduced alteration in afferent activity increases severalfold and then decreases again at shorter wavelengths. In other experiments, two clear maxima, one near 430 nm and the other near 590 nm, were obtained. When the measurements were adjusted to the same light intensity at each wavelength (same number of photons per second), the spectrum shown in Fig. 4 was obtained. The two maxima were more accurately located using multiple local measurements as $432 \pm 2$ nm and $589 \pm 2$ nm with relative absorption values of about $6:1$.

## Discussion

Oxygen-dependent quenching of phosphorescence shows great promise for measuring oxygenation of tissue in vivo. As a noninvasive optical method, the measurements can be made in real time and without interfering with the tissue function or metabolism. In this chapter, we have focused on use of phosphorescence quenching for obtaining high-resolution, two- and three-dimensional maps of oxygen distribution in tissue. The oxygen pressure distributions obtained are very similar to those reported by Dewhirst and co-workers [8,15] using microelectrodes to measure the perivascular oxygen pressure. With microelectrodes, of course, the oxygen measurements had to be made one point at a time, whereas phosphorescence imaging can collect complete high-resolution sets of data for each tumor. The ability to measure oxygen distributions in three dimensions provides a unique, new opportunity to study oxygen delivery to developing tissue and other tissue accessible to optical measurements.

The carotid body is an organ that alters its afferent neural activity rapidly and precisely in response to alterations in the oxygen pressure of the blood in the carotid artery. Oxygen sensing by the carotid body occurs by means of a sensor for which CO is a competitive inhibitor and for which the effect of CO is reversed by bright light. This behavior has allowed the absorption spectrum of the CO complex of the sensor to be determined. Because only photons that are absorbed by the sensor–CO complex can result in reversal (by photodissociation) of the inhibitory effect of CO, the efficacy of photons of monochromatic light in inducing that reversal is directly proportional to the absorption of the CO compound at that wavelength. This "photochemical action spectrum" has been determined by measuring the release of the CO-induced increase in afferent electrical activity by monochromatic light. At each wavelength the response was measured at light intensities well below saturating intensity and the response was then corrected for differences in light intensity at the different wavelengths. The spectrum obtained shows the light-induced response obtained after correcting the intensity of the monochromatic light to the same number of photons per second at all wavelengths. The action spectrum is therefore a measure of the probability of absorption of light photon as a function of the wavelength of light, that is, the absorption spectrum of the CO–sensor complex.

The photochemical action spectrum obtained for the carotid body is the same as that originally reported by Warburg and Negelein [11] for yeast cells (see also [16,17]) and as the absorption spectrum of the cytochrome $a_3$-CO complex of the mitochondrial respiratory chain. This is unambiguous evidence that the mitochondrial respiratory chain is the primary oxygen sensor for the carotid body. The effects of CO on the cytochrome $c$ oxidase of the respiratory chain are presumably to alter the afferent electrical activity secondary to alterations in cellular energy metabolism, although the mechanism(s) of this communication remains to be established.

*Acknowledgments.* This work was supported by grants NS-31465 and HL-43413 from The U.S. National Institutes of Health.

# References

1. Lahiri S, Rumsey WL, Wilson DF, Iturriaga R (1993) Contribution of in vivo microvascular PO$_2$ in the cat carotid body chemotransduction. J Appl Physiol 75:1035–1043
2. Lahiri S, Wilson DF, Osanai S, et al. (1996) Photochemical action spectra, not absorption spectra, allow identification of the oxygen sensor of the carotid body. In: Zapata P, Eyzaguirre C, Torrance RW (eds) Frontiers in arterial chemoreception. Plenum, New York, pp 65–71
3. Vanderkooi JM, Maniara G, Green TJ, Wilson DF (1987) An optical method for measurement of dioxygen concentration based on quenching of phosphorescence. J Biol Chem 262:5476–5482
4. Wilson DF, Rumsey WL, Green TJ, Vanderkooi JM (1988) The oxygen dependence of mitochondrial oxidative phosphorylation measured by a new optical method for measuring oxygen concentration. J Biol Chem 263:2712–2718
5. Shonat RD, Wilson DF, Riva CE, Pawlowski M (1992) Oxygen distribution in the retinal and choroidal vessels of the cat as measured by a new phosphorescence imaging method. Appl Optics 33:3711–3718
6. Vinogradov SA, Wilson DF (1995) Metallotetrabenzoporphyrins. New phosphorescent probes for oxygen measurements. J Chem Soc Perkin Trans II 2:103–111
7. Lo L-W, Koch CJ, Wilson DF (1996) Calibration of oxygen dependent quenching of the phosphorescence of Pd-*meso*-tetra (4-carboxyphenyl) porphine: a phosphor with general application for measuring oxygen concentration in biological systems. Anal Biochem 236:153–160
8. Dewhirst MW, Ong ET, Klitzman B, et al. (1992) Perivascular oxygen tensions in a transplantable mammary tumor growing in a dorsal flap window chamber. Radiat Res 130:171–182
9. Iturriaga R, Rumsey WL, Mokashi A, et al. (1991) In vitro perfused-superfused cat carotid body for physiological and pharmacological studies. J Appl Physiol 70:1393–1400
10. Wilson DF, Mokashi A, Chugh D, et al. (1994) The primary oxygen sensor of the cat carotid body is cytochrome $a_3$ of the mitochondrial respiratory chain. FEBS Lett 351:370–374
11. Warburg O, Negelein E (1928) Uber die photochemische Dissoziation bei intermittierender Belichtung und das absolute Absorptionsspektrum des Atmungsferments. Biochem Z 202:202–228
12. Cooper DY, Scheyer S, Rosenthal O (1970) Some chemical properties of cytochrome P-450 and its carbon monoxide compound (P$_{450}$-CO). Ann NY Acad Sci 174:205–217
13. Rumsey WL, Pawlowski M, Lejavardi N, Wilson DF (1994) Oxygen pressure distribution in the heart in vivo and evaluation of the ischemic "border zone." Am J Physiol 266:H1676–H1680

14. Vinogradov SA, Lo L-W, Jenkins WT, et al. (1996) Non invasive imaging of the distribution of oxygen in tissue in vivo using near infra-red phosphors. Biophys J 70:1609–1617

15. Dewhirst MW, Secomb TW, Ong ET, et al. (1994) Determination of local oxygen consumption rates in tumors. Cancer Res 54:3333–3336

16. Melnick JL (1942) The photochemical spectrum of cytochrome oxidase. J Biol Chem 146:385–390

17. Caster LN, Chance B (1955) Photochemical action spectra of carbon monoxide-inhibited respiration. J Biol Chem 217:453–464

# Molecular Mechanisms of Hypoxia-Induced Angiogenesis

Eiji Ikeda[1], Annette Damert[2], and Werner Risau[2]

*Summary.* Tissues respond to hypoxia with several compensatory mechanisms to maintain a microenvironment in which the cells can survive. These compensatory mechanisms include the formation of new blood vessels from preexisting vessels, a process termed angiogenesis, which restores the reduced oxygen supply. Hypoxia-induced angiogenesis plays an important role in physiological and pathological situations such as embryogenesis, diabetic retinopathy, and solid tumor growth. Increasing evidence suggests that upregulation of the expression of vascular endothelial growth factor (VEGF) is a key step in hypoxia-induced angiogenesis. Investigation into this hypoxic induction of VEGF should therefore lead to elucidation of the molecular mechanisms involved in hypoxia-induced angiogenesis. Using C6 glioma cells, which generate a highly vascularized glioma in vivo, the nuclear run-on assay and determination of mRNA half-life demonstrate that hypoxic induction of VEGF is the result of both transcriptional activation and increased mRNA stability. Reporter gene studies revealed that the hypoxia-responsive transcription-activating element was present in the 5′-flanking region of the VEGF gene. This element contains a consensus binding site for hypoxia-inducible factor-1 (HIF-1), which was originally found to bind the hypoxia-responsive enhancer sequence in the erythropoietin gene.

*Key words.* Hypoxia—Angiogenesis—VEGF—HIF-1—Tumor angiogenesis

## Introduction

When mammals are exposed to hypoxia, several compensatory mechanisms respond to restore the reduced oxygen supply. One mechanism is hyperventilation, which increases the respiratory volume to obtain more oxygen from the atmosphere. Another mechanism is erythrocytosis, to transport more oxygen to the tissue; erythrocytosis

[1] Department of Pathology, School of Medicine, Keio University, 35 Shinanomachi, Shinjuku-ku, Tokyo 160, Japan
[2] Max-Planck-Institut für Physiologische und Klinische Forschung, W.G. Kerckhoff-Institut, Abteilung Molekulare Zellbiologie, Parkstraße 1, 61231 Bad Nauheim, Germany

is regulated by erythropoietin (EPO) production [1]. Another mechanism, which is a focal response, is angiogenesis.

Angiogenesis is defined as new blood vessel formation from preexisting vessels [2], and angiogenesis could restore reduced oxygen by increasing the blood volume in the tissue. Angiogenesis plays important roles in several physiological and pathological situations, such as normal embryonic development, corpus luteum formation, wound healing, solid tumor growth, rheumatoid arthritis, and diabetic retinopathy [3]. There are several known triggers of angiogenesis, among which hypoxia is one of the most important. Angiogenesis involved in normal embryonic development, solid tumor growth, or diabetic retinopathy is believed to be hypoxia regulated [4–7]. To analyze the molecular mechanisms of hypoxia-induced angiogenesis, we have focused on angiogenesis as involved in solid tumor growth.

We have investigated tumor angiogenesis as it occurs during the process of astrocytoma progression. Through histopathological observation, astrocytomas are classified into four grades. Astrocytomas of grades 1 and 2 are regarded in general as histologically benign tumors, and astrocytomas grades 3 and 4 are malignant. One of the important criteria of astrocytomas grades 3 and 4 is proliferation of tumor vessels. Several polypeptide growth factors such as fibroblast growth factors (FGFs), vascular endothelial growth factor (VEGF), and transforming growth factor-$\beta$ were shown to be involved, directly or indirectly, in tumor angiogenesis [3,8]. Increasing evidence suggests that VEGF is a key regulator of tumor angiogenesis, especially angiogenesis that accurs during the progression of astrocytoma [4,5,9].

VEGF is an endothelial cell-specific mitogen secreted from various cell types including tumor cells and is angiogenic in vivo [10–12]. VEGF has four different isoforms, $VEGF_{121}$, $VEGF_{165}$, $VEGF_{189}$, and $VEGF_{206}$ [12]. Up until the present, two high-affinity receptors of VEGF, Flt-1 [13] and Flk-1 (KDR) [14], have been recognized and cloned. In situ hybridization studies have shown that expression of the genes encoding VEGF and its receptor correlates both spatially and temporally with tumor angiogenesis involved in astrocytoma progression [4]. Furthermore, tumor angiogenesis in vivo is shown to be inhibited by either injecting a neutralizing antibody against VEGF [15] or overexpressing a dominant-negative receptor mutant [9].

These results have strongly suggested the essential role of VEGF and its receptors in tumor angiogenesis in astrocytoma. Another important finding by in situ hybridization studies is that induction of VEGF gene expression is most prominent in the palisading cells surrounding necrotic areas [4,5], which suggests that upregulation of the VEGF gene expression in tumors is hypoxia induced. Based on these results, we have concentrated on investigation into the mechanisms underlying hypoxic induction of VEGF to elucidate the molecular mechanisms of hypoxia-induced angiogenesis.

## Materials and Methods

### Cell Culture and Hypoxic Incubation

Rat C6 glioma cells (American Type Culture Collection) were cultured in Dulbecco's modified Eagle's Medium (GibcoBRL, Life Technologies, Berlin, Germany) supplemented with 10% fetal bovine serum. The cells were incubated at either 21% $O_2$, 5%

$CO_2$ for normoxia or 1% $O_2$, 5% $CO_2$ balanced with $N_2$ in an oxygen-regulated incubator (Model 3015 Forma Scientific/Labotect, Göttingen, Germany) for hypoxia. For inhibition of protein synthesis, cycloheximide was added to the culture medium (10 µg/ml) 90 min before hypoxic treatment.

Several genes are known to be upregulated by hypoxia. Among them, hypoxia-induced expression of the EPO gene has been intensively studied. EPO gene expression is shown to be induced by cobalt chloride treatment as well as hypoxia. We examined whether VEGF gene expression could also be induced by cobalt chloride ($CoCl_2$) treatment. We cultured C6 cells in medium containing $CoCl_2$ at 25, 100, or 200 µM for 38 h under normoxic condition, and secretion of VEGF into the culture medium was detected by Western blot analysis.

## Western Blotting

C6 cells were cultured in a 60-mm-dia culture dish with 3 ml of culture medium under normal conditions. Five hours before incubation either under normoxia with or without $CoCl_2$ or under hypoxia, the serum concentration was decreased to 5%. The culture medium was changed to fresh medium containing 1% serum, and then cell incubation was begun under a specific culture condition. After incubation, the conditioned medium was collected and concentrated ten times with Centricon-10 (Amicon, MA, USA) for Western blot analysis. At the same time, the number of cells in the culture dish was counted. The conditioned medium was analyzed by electrophoresis in 12.5% polyacrylamide-sodium dodecyl sulfate (SDS) gel under denaturing conditions and transferred to a nitrocellulose membrane. The membrane was incubated with rabbit antiserum against murine VEGF [7], and sequentially with peroxidase-conjugated goat anti-rabbit secondary antibodies (Dianova, Hamburg, Germany). Immobilized antibodies were detected with ECL (Amersham, UK).

## Northern Blotting

Total RNA was prepared from the cells by acid guanidinium thiocyanate:phenol:chloroform extraction [16], and 20 µg of total RNA was electrophoresed in a lane of 1% agarose gel containing 2.2 M formaldehyde. After the gel was stained with ethidium bromide, RNA was transferred to a nylon membrane (Hybond N$^+$, Amersham) by the capillary method. The membrane was cross-linked with UV light and prehybridized in a solution containing 50% formamide, 5× SSC, 5× Denhardt's solution, 1% SDS, and 100 µg/ml yeast tRNA. Then the membrane was hybridized with cDNA for murine $VEGF_{164}$ labeled with [$\alpha$-$^{32}$P]dCTP([$\alpha$-$^{32}$P]deoxycytidine) at 42°C overnight in the same solution that was used for prehybridization. The membrane was washed at 42°C once in 2× SSC, once in 2× SSC with 0.5% SDS, and twice in 0.3× SSC with 0.5% SDS, and then exposed to X-ray film. The membrane was subsequently hybridized with [$\alpha$-$^{32}$P]dCTP-labeled cDNA for chick $\beta$-actin for standardization. Quantitative analysis was performed with a PhosphorImager SF (Molecular Dynamics, CA, USA).

## Nuclear Run-On Assay

Nuclei were isolated from the C6 cells under normoxia or hypoxia for 3 or 15 h, and in vitro transcription was performed as described previously [17,18]. The labeled RNA of equal counts per minute (cpm) was hybridized at 42°C for 48 h to the murine VEGF$_{164}$ cDNA and the chick β-actin cDNA, both of which were cloned in pBluescript KS (Stratagene, CA, USA) and immobilized on a nylon membrane (GeneScreen Plus, DuPont NEN, MA, USA). The membrane was washed at 42°C twice in 2× SSC with 0.5% SDS, twice in 0.3× SSC with 0.5% SDS, and incubated at 37°C for 30 min with RNaseA (10 μg/ml) in 2× SSC. The membrane was further washed at 37°C in 2× SSC and finally in 0.3× SSC, then exposed to X-ray film.

## Determination of mRNA Half-Life

The half-life of VEGF mRNA in C6 cells cultured under normoxia or hypoxia for 3 or 15 h was determined. Actinomycin D was added to the culture medium of the C6 cells after normoxia or hypoxia culturing to block transcription, and the cells were then returned to the same culture conditions. Total RNA was then prepared every hour for 7 h, and the amount of VEGF mRNA remaining in the cells was quantified by Northern blot analysis. The amount of VEGF mRNA was normalized to the amount of β-actin mRNA.

## Reporter Gene Constructs and Transfection

A *KpnI-NheI* fragment of the human VEGF gene containing 2.3 kb of 5′-flanking region and 0.9 kb of 5′-untranslated region [19], which was provided by Dr. Judith A. Abraham (Scios Nova, CA, USA), was cloned upstream of the luciferase reporter gene in the pGL2-Basic Vector (Promega, WI, USA) [18]. From this construct, several deletion derivatives were generated and tested for their hypoxic responsiveness by transient transfection into C6 cells. Transfection of these constructs into C6 cells, measurement of reporter gene products, and determination of the degree of induction of luciferase activity by hypoxia were performed as described previously [18].

# Results

## Upregulation of VEGF Production by Hypoxia and Cobalt Chloride

Expression of the VEGF gene in C6 cells under normoxia or hypoxia was determined at both RNA and protein levels. Secretion of the VEGF protein into the culture medium was shown to be amplified in cells grown under hypoxia for 18 h as compared to cells cultured under normoxia (Fig. 1). The major isoform of the secreted VEGF protein was VEGF$_{165}$. Northern blot analysis revealed that the VEGF mRNA level had already increased after 3 h of hypoxic incubation and further increased after 15 h of hypoxia (Fig. 2). Several repeated experiments have shown that the VEGF mRNA level reaches the steady-state level after about 15 h of hypoxic incubation [18].

When cycloheximide, a protein synthesis inhibitor, was added to the culture medium, induction of VEGF mRNA was blocked in the cells under hypoxia for both 3 and 15 h (Fig. 2). From these results, we concluded that hypoxic induction of VEGF synthesis is regulated mainly at the mRNA level and that new protein synthesis after stimulation by hypoxia seems to be required for the VEGF mRNA level to increase. As

# Hypoxia

**0    18   hours**

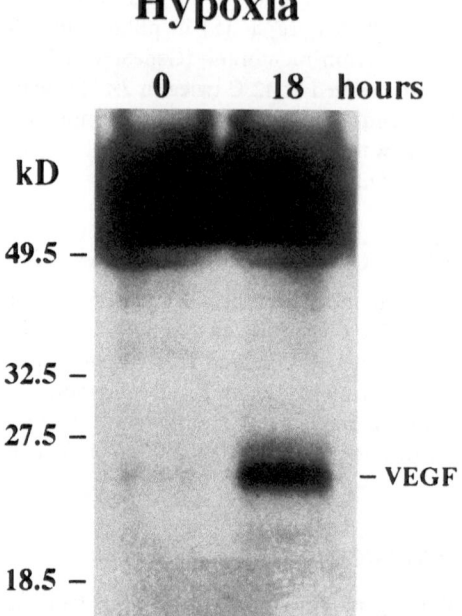

FIG. 1. Secretion of vascular endothelial growth factor (*VEGF*) from C6 cells exposed to hypoxia for 18h into the culture medium was detected by Western blot analysis

shown in Fig. 3, secretion of VEGF protein from the C6 cells into the culture medium was also amplified by $CoCl_2$ treatment.

## Transcriptional Activation of the VEGF Gene by Hypoxia

Transcriptional rates of the VEGF gene in C6 cells under normoxia or hypoxia were quantified by nuclear run-on assay, which revealed that transcription of the VEGF gene was already activated after 3h of hypoxia and further upregulated after 15h of hypoxia (Fig. 4).

## Stabilization of VEGF mRNA by Hypoxia

Half-lives of VEGF mRNA in C6 cells under normoxia or hypoxia were measured. C6 cells that had been cultured under normoxia or hypoxia for 3 or 15h were treated with actinomycin D to block new transcription, and the amounts of mRNA remaining in the cells were quantified every hour by Northern blot analysis and plotted (Fig. 5). These results revealed that the half-life of VEGF mRNA was lengthened in the cells after hypoxic treatment for 15h. As compared with long exposure to hypoxia, hypoxic incubation for 3h did not influence the half-life of VEGF mRNA. Together with the results of nuclear run-on assays, we concluded that the initial induction of VEGF mRNA is caused by transcriptional activation only, whereas the upregulation under longer lasting hypoxia is caused by both transcriptional activation and stabilization of mRNA.

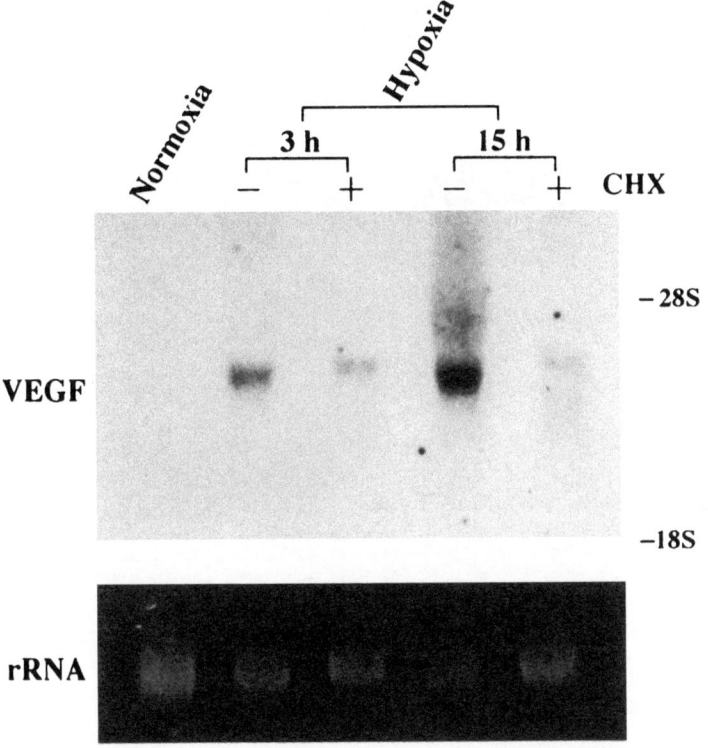

Fɪɢ. 2. Induction of VEGF mRNA by hypoxia in C6 cells cultured under normoxia or hypoxia for 3 or 15 h in the presence (+) or absence (−) of cycloheximide (*CHX*). Total RNA was analyzed by Northern blot analysis. *Top panel*, VEGF mRNA; *bottom panel*, 28S ribosomal RNA stained with ethidium bromide

## Sequences of the VEGF Gene Responsible for Hypoxic Induction

The results of nuclear run-on assays suggested that *cis*-acting elements in the VEGF gene may be responsible for hypoxic induction. The 5′-flanking (2.3-kb) and 5′-untranslated (1-kb) region of the human VEGF gene was cloned upstream of luciferase reporter gene in the promoterless luciferase reporter vector. From this construct, we generated several deletion derivatives, including the construct in which 1 kb of 5′-untranslated region was deleted (Fig. 6). Luciferase activity in C6 cells under normoxia or hypoxia was measured, and the degree of induction by hypoxia was calculated. The degree of induction by hypoxia was shown to be abruptly eliminated when the *SacI-BanI* (−2218 to −1925) fragment was deleted from the construct (Fig. 6). On the other hand, deletion of the 0.9 kb of the 5′-untranslated region (*NheI-NarI*) did not influence hypoxic responsiveness. These results suggested that the *SacI-BanI* fragment (293 bp) contains the hypoxia-responsive elements, and that the 5′-untranslated region, which is relatively long, does not contribute to hypoxia-induced transcriptional activation.

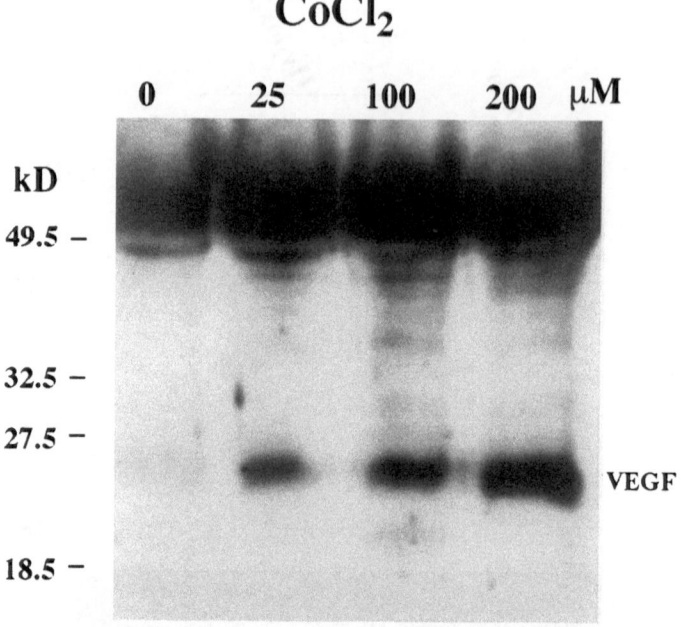

FIG. 3. Secretion of VEGF from C6 cells cultured under normoxia in culture medium containing 25, 100, or 200 μM CoCl₂ for 38 h. VEGF in the medium was detected by Western blot analysis

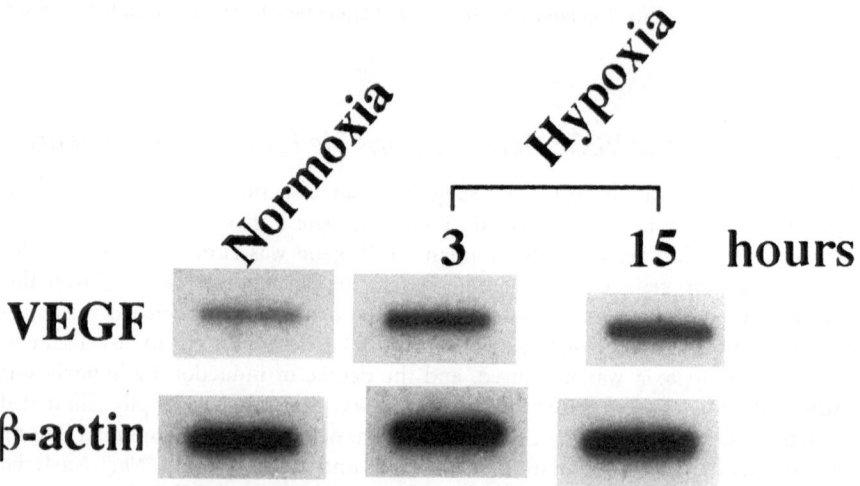

FIG. 4. Transcriptional rates of the genes for VEGF and β-actin in nuclei prepared from C6 cells cultured under normoxia or hypoxia for 3 or 15 h. The labeled RNA transcribed in vitro was hybridized to the VEGF and β-actin cDNA immobilized on a nylon membrane

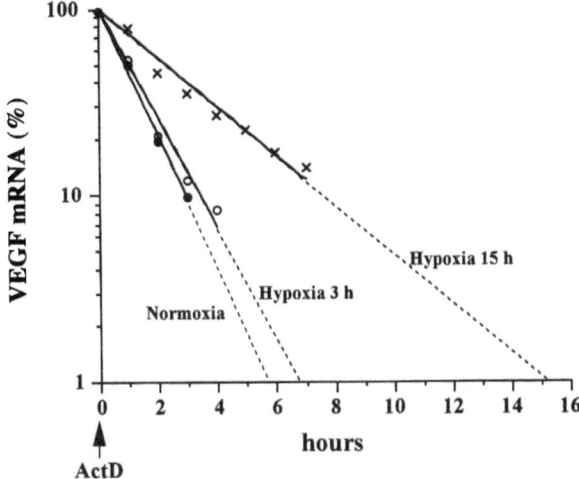

FIG. 5. Stability of VEGF mRNA in C6 cells cultured under normoxia or hypoxia for 3 or 15 h and treated with actinomycin D (*ActD*) to block transcription. The amount of VEGF or β-actin mRNA was determined every hour by Northern blot analysis; the amount of VEGF mRNA was normalized by that of β-actin and plotted. (From [18], with permission)

Sequence analysis of the *SacI-BanI* fragment revealed a consensus binding site for hypoxia-inducible factor 1 (HIF-1). HIF-1 was originally reported as a transcription factor that was indispensable to hypoxic induction of erythropoietin (EPO) by binding to the hypoxia-responsible enhancer sequence in the 3′-flanking region of the EPO gene [20,21]. To examine the role of HIF-1 in VEGF induction by hypoxia, we deleted the HIF-1 binding site from the hypoxia-responsive construct. Deletion of the HIF-1 binding site led to complete loss of hypoxia responsiveness (Damert et al, manuscript submitted), which means the HIF-1 binding site is crucial for transcriptional activation of VEGF gene by hypoxia.

Another interesting finding was that deletion of the 5′-end (*SacI-PvuII*) of the *SacI-BanI* fragment caused hypoxic responsiveness to decrease to some degree (Damert et al, manuscript submitted). This result suggested that a potentiating sequence is necessary for full induction, and binding of AP-1 to this potentiating sequence was demonstrated in both normoxic and hypoxic nuclear extracts by a electrophoretic mobility shift assay that included supershift analysis with specific monoclonal antibodies (Damert et al, manuscript submitted).

## Discussion

Angiogenesis is one of the most important compensatory mechanisms for restoration of the oxygen supply to hypoxic tissue. This hypoxia-induced angiogenesis is observed in several physiological and pathological situations such as normal embryonic development, solid tumor growth, and diabetic retinopathy [3]. Increasing evidence shows that VEGF is a key regulator of the angiogenesis involved in these situations

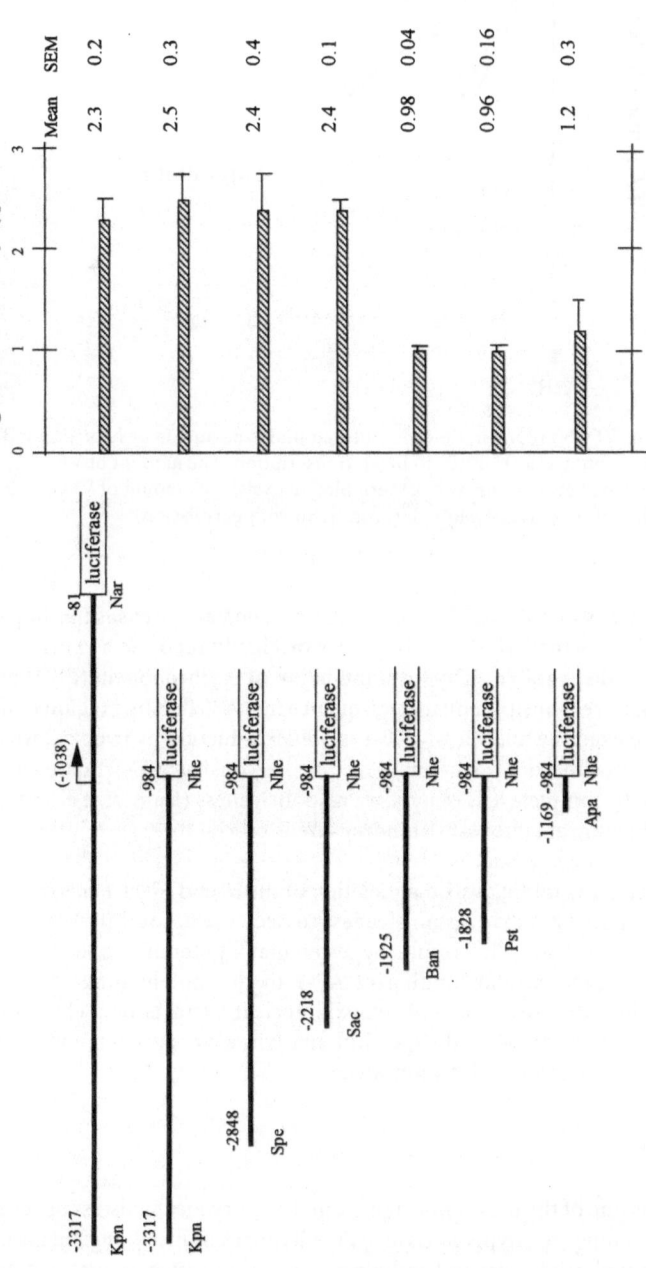

FIG. 6. Functional analysis of human VEGF 5′-flanking sequences for responsiveness to hypoxia. VEGF-luciferase deletion constructs were transfected into C6 cells, and the degree of induction by hypoxia was determined from three independent experiments. Nucleotides are numbered from the translation start site, and transcription start site is indicated with an *arrow*. *Kpn, KpnI; Spe, SpeI; Sac, SacI; Ban, BanI; Pst, PstI; Apa, ApaI; Nhe, NheI; Nar, NarI*. (From [18], with permission)

and that hypoxia-induced upregulation of the VEGF gene expression seems to be crucial [4,5]. Therefore, we focused on hypoxic induction of VEGF gene expression to analyze the molecular mechanism of hypoxia-induced angiogenesis. We used C6 glioma cells as an in vitro model of human glioblastoma, which is known to be a highly vascularized tumor [22]. When C6 cells were cultured under hypoxic condition, their production of VEGF was amplified, and the induction of VEGF was shown to be regulated mainly at the mRNA level. VEGF mRNA level started to increase after 3 h of hypoxic incubation and reached the steady-state level after about 15 h of hypoxic incubation.

Nuclear run-on assay and determination of half-lives of VEGF mRNAs revealed that transcriptional activation of the VEGF gene and stabilization of the VEGF mRNA were differentially regulated during the course of hypoxic incubation. Transcription of the VEGF gene was already activated after 3 h of hypoxia and continued to be activated during further incubation under hypoxia. On the other hand, stabilization of the VEGF mRNA was only detectable after longer incubation under hypoxia. From these results, we concluded that the increase in VEGF mRNA level after 3 h of hypoxia is caused by transcriptional activation only, whereas the increase in VEGF mRNA level at the steady-state level is mediated by both transcriptional activation and stabilization of the mRNA. This result suggests the involvement of several distinct molecular mechanisms in the hypoxic induction of VEGF.

Nuclear run-on assays suggested that *cis*-acting elements in the VEGF gene regulate the transcriptional rate of the VEGF gene in response to hypoxia. Several reporter constructs containing 5'-flanking and 5'-untranslated sequences of the human VEGF gene were tested for their hypoxic responsiveness by transient transfection into C6 cells. Hypoxic responsiveness of the reporter construct was abruptly eliminated when a 0.3-kb *SacI-BanI* fragment was deleted, in contrast to the reported result that hypoxic responsiveness was gradually lost in Hep 3B cells by serial deletion of the 5'-end of the human VEGF 5'-flanking region [23]. This discrepancy might suggest a cell-type-specific response to hypoxia. Our results revealed that a 0.3-kb *SacI-BanI* fragment contains the hypoxia-responsive elements that confer hypoxic responsiveness to C6 glioma cells. Sequence analysis of this *SacI-BanI* fragment revealed a consensus binding site for hypoxia-inducible factor 1 (HIF-1), which was originally found to bind a hypoxia-responsive enhancer in the 3'-flanking region of the EPO gene [20,21].

We examined the role of HIF-1 in hypoxic induction of VEGF expression by deleting a consensus binding site for HIF-1 from a hypoxia-responsive reporter construct, and showed that the HIF-1 binding site is crucial for hypoxia-induced transcriptional activation of the VEGF gene in C6 glioma cells (Damert et al, manuscript submitted), as reported for PC12 pheochromocytoma [24] and endothelial cells [25]. Through more detailed examination of the *SacI-BanI* fragment, we found a potentiating sequence that is necessary for full hypoxic induction of transcriptional activation but is not hypoxia inducible by itself. Binding of AP1 to this potentiating sequence was demonstrated in both normoxic and hypoxic nuclear extract from C6 cells (Damert et al, manuscript submitted).

The fact that HIF-1 is a key factor to regulate hypoxic induction of both EPO and VEGF provides strong evidence that hypoxic induction of these two molecules shares the same molecular mechanisms including the same oxygen-sensing system.

Heme proteins have been suggested to be involved as an oxygen-sensor in the hypoxic induction of EPO. One piece of circumstantial evidence to support this idea is the induction of EPO by $CoCl_2$ treatment, which could interact with a heme [26]. We examined whether VEGF production is also enhanced in cells treated with $CoCl_2$, and our data demonstrated that expression of the VEGF gene was also upregulated by $Cocl_2$ treatment. Together with the reported result that hypoxia-induced expression of the VEGF gene was inhibited by exposing the cells to carbon monoxide (CO) [27], heme proteins are likely to be implicated in the signaling that leads to the VEGF gene expression in response to hypoxia. Further studies are needed to clarify in more detail the molecular mechanisms of hypoxia-induced angiogenesis, including the cellular oxygen-sensing system and the factors involved in stabilization of VEGF mRNA.

*Acknowledgment.* We are grateful to Dr. Judith A. Abraham (Scios Nova) for the generous gift of a genomic clone for human VEGF.

# References

1. Krantz SB (1991) Erythropoietin. Blood 77:419–434
2. Folkman J, Shing Y (1992) Angiogenesis. J Biol Chem 267:10931–10934
3. Folkman J, Klagsbrun M (1987) Angiogenic factors. Science 235:442–447
4. Plate KH, Breier G, Weich HA, et al. (1992) Vascular endothelial growth factor is a potential tumour angiogenesis factor in human gliomas in vivo. Nature 359:845–848
5. Shweiki D, Itin A, Soffer D, et al. (1992) Vascular endothelial growth factor induced by hypoxia may mediate hypoxia-initiated angiogenesis. Nature 359:843–845
6. Millauer B, Wizigmann-Voos S, Schnürch H, et al. (1993) High affinity VEGF binding and developmental expression suggest Flk-1 as a major regulator of vasculogenesis and angiogenesis. Cell 72:835–846
7. Breier G, Albrecht U, Sterrer S, et al. (1992) Expression of vascular endothelial growth factor during embryonic angiogenesis and endothelial cell differentiation. Development (Camb) 114:521–532
8. Risau W (1990) Angiogenic growth factors. Prog Growth Factor Res 2:71–79
9. Millauer B, Shawver LK, Plate KH, et al. (1994) Glioblastoma growth inhibited in vivo by a dominant-negative Flk-1 mutant. Nature 367:576–579
10. Keck PJ, Hauser SD, Krivi G, et al. (1989) Vascular permeability factor, an endothelial cell mitogen related to PDGF. Science 246:1309–1312
11. Leung DW, Cachianes G, Kuang W-J, et al. (1989) Vascular endothelial growth factor is a secreted angiogenic mitogen. Science 246:1306–1309
12. Ferrara N (1993) Vascular endothelial growth factor. Trends Cardiovasc Med 3:244–250
13. Shibuya M, Yamaguchi S, Yamane A, et al. (1990) Nucleotide sequence and expression of a novel human receptor-type tyrosine kinase gene (flt) closely related the fms family. Oncogene 5:519–524
14. Terman BI, Carrion ME, Kovacs E, et al. (1991) Identification of a new endothelial cell growth factor receptor tyrosine kinase. Oncogene 6:1677–1683
15. Kim KJ, Li B, Winer J, et al. (1993) Inhibition of vascular endothelial growth factor-induced angiogenesis suppresses tumour growth in vivo. Nature 362:841–844
16. Chomczynski P, Sacchi N (1987) Single-step method of RNA isolation by acid guanidinium thiocyanate-phenol-chloroform extraction. Anal Biochem 162:156–159

17. Dignam JD, Lebovitz RM, Roeder RG (1983) Accurate transcription initiation by RNA polymerase II in a soluble extract from isolated mammalian nuclei. Nucleic Acids Res 11:1475–1489
18. Ikeda E, Achen MG, Breier G, et al. (1995) Hypoxia-induced transcriptional activation and increased mRNA stability of vascular endothelial growth factor in C6 glioma cells. J Biol Chem 270:19761–19766
19. Tischer E, Mitchell R, Hartman T, et al. (1991) The human gene for vascular endothelial growth factor. Multiple protein forms are encoded through alternative exon splicing. J Biol Chem 266:11947–11954
20. Wang GL, Jiang BH, Rue EA, et al. (1995) Hypoxia-inducible factor 1 is a basic-helix-loop-helix-PAS heterodimer regulated by cellular $O_2$ tension. Proc Natl Acad Sci USA 92:5510–5514
21. Semenza GL, Wang GL (1992) A nuclear factor induced by hypoxia via de novo protein synthesis binds to the human erythropoietin gene enhancer at a site required for transcriptional activation. Mol Cell Biol 12:5447–5454
22. Plate KH, Breier G, Millauer B, et al. (1993) Up-regulation of vascular endothelial growth factor and its cognate receptors in a rat glioma model of tumor angiogenesis. Cancer Res 53:5822–5827
23. Forsythe JA, Jiang BH, Iyer NV, et al. (1996) Activation of vascular endothelial growth factor gene transcription by hypoxia-inducible factor 1. Mol Cell Biol 16:4604–4613
24. Levy AP, Levy NS, Wegner S, et al. (1995) Transcriptional regulation of the rat vascular endothelial growth factor gene by hypoxia. J Biol Chem 270:13333–13340
25. Liu Y, Cox SR, Morita T, et al. (1995) Hypoxia regulates vascular endothelial growth factor gene expression in endothelial cells. Circ Res 77:638–643
26. Goldberg MA, Dunning SP, Bunn HF (1988) Regulation of the erythropoietin gene: evidence that the oxygen sensor is a heme protein. Science 242:1412–1415
27. Goldberg MA, Schneider TJ (1994) Similarities between the oxygen-sensing mechanisms regulating the expression of vascular endothelial growth factor and erythropoietin. J Biol Chem 269:4355–4359

# Mechanisms of Pulmonary Vasodilatation and Ductus Arteriosus Constriction by Normoxia

E. Kenneth Weir[1], Helen L. Reeve[1], Simona Tolarova[1],
David N. Cornfield[2], Daniel P. Nelson[1], and Stephen L. Archer[1]

*Summary.* In the fetus, the lungs are not ventilated and arterial oxygen tension is approximately 22 mmHg. Under these conditions, the ductus arteriosus is open and the pulmonary vasculature is constricted. At birth, the ductus constricts and the pulmonary vessels relax. The resting membrane potential in vascular smooth muscle cells is largely determined by potassium current. It appears that oxygen increases the potassium current in the smooth muscle cells of small pulmonary arteries, thus causing membrane hyperpolarization and inhibition of the voltage-gated calcium channel, which in turn leads to a reduction in cytosolic calcium and to relaxation. In the adult, the oxygen-sensitive potassium channels are delayed rectifiers ($K_v$). In the ductus arteriosus the oxygen-sensitive potassium channel is also in the $K_v$ group but oxygen inhibits, rather than activates, the potassium current and thus causes depolarization, calcium entry, and contraction. The mechanism by which the gating of potassium channels changes in response to oxygen is not clear. Transgenic mice that lack the 91-kDa subunit of NADPH oxidase still have normal hypoxic pulmonary vasoconstriction. Thus, while a role for NADPH oxidase is unlikely, it seems plausible that the oxygen sensor may be part of the channel or be coexpressed with it.

*Key words.* Oxygen—Potassium channel—Ductus arteriosus—Smooth muscle—NADPH oxidase

## Introduction

In the adult pulmonary vasculature, hypoxia causes localized vasoconstriction. Teleologically, this is advantageous as it prevents desaturated pulmonary arterial blood from flowing past poorly ventilated alveolae, which would otherwise result in systemic hypoxemia. The phenomenon of hypoxic pulmonary vasoconstriction (HPV) was described more than 100 years ago, but the first detailed report of the hemodynamic changes was given by von Euler and Liljestrand in 1946 [1]. It appears that HPV is,

[1] Veterans Affairs, Medical Center, One Veterans Drive, Minneapolis, MN 55417, USA
[2] Department of Pediatrics, University of Minnesota, Variety Heart and Research Center, 401 East River Road, Minneapolis, MN 55455, USA

from a phylogenetic perspective, an ancient mechanism in that it can be demonstrated in fish. When the gills of the dogfish are superfused with water that has a low oxygen content, the arteries supplying the gills constrict [2].

The mechanism of HPV at a cellular level is being elucidated. The greatest constriction in response to hypoxia occurs in pulmonary arteries that have a luminal diameter of about 200–300 μm, although pulmonary veins and larger pulmonary arteries also show some hypoxia-initiated constriction [3]. Acute HPV appears to be a property of individual pulmonary arterial smooth muscle cells [4], although this constriction is modulated by endothelium-derived substances and may indeed be mediated by such substances after the first few minutes.

It has been known for more than 10 years that hypoxia causes depolarization of pulmonary arterial smooth muscle and an increase in the frequency of action potentials related to calcium influx [5]. The increase in cytosolic calcium occurs, at least in part, as a result of calcium influx through the L-type voltage-gated calcium channels. The calcium influx can be inhibited by the L-type calcium channel blocker verapamil [6] and enhanced by the calcium channel agonist BAY K8644 [7,8]. HPV can be diminished by removing calcium from the perfusate of isolated, perfused lungs [9]. In the case of small pulmonary artery rings, hypoxic contraction is reduced by the removal of bath calcium [5]. While there is evidence that HPV may also involve the release of calcium from the sarcoplasmic reticulum [10,11], the results just summarized indicate that membrane depolarization and calcium influx are important components.

## The Role of Potassium Currents

If HPV involves smooth muscle membrane depolarization, how is this initiated? The resting membrane potential of vascular smooth muscle is determined, in large part, by potassium channel activity. A number of potassium channels have been identified in pulmonary arterial smooth muscle cells, including calcium-sensitive ($K_{Ca}$), ATP-sensitive ($K_{ATP}$), and voltage-sensitive ($K_v$) channels [12,13]. Inhibition of $K_{ATP}$ channels by glibenclamide has little effect on pulmonary vascular tone, and the response to tetraethylammonium (TEA; a blocker of $K_{Ca}$ channels) is modest [14,15]. However, 4-aminopyridine (4-AP; a $K_v$ blocker), causes significant pulmonary vasoconstriction in the isolated perfused rat lung, indicating that, in this species, $K_v$ channels are probably important in the control of smooth muscle membrane potential. This conclusion is supported by the finding that TEA and charybdotoxin (CTX) ($K_{Ca}$ blockers) do not change membrane potential, while 4-AP causes depolarization (Fig. 1).

Hypoxia reduces whole-cell potassium current ($I_K$) and causes membrane depolarization in freshly dispersed pulmonary arterial smooth muscle cells of the dog but has no effect on $I_K$ of renal artery smooth muscle cells [15]. Similarly, hypoxia/dithionite (a reducing agent) causes an inhibition of $I_K$ in rat pulmonary artery smooth muscle cells in primary culture but not in mesenteric artery smooth muscle [13]. These observations suggest that hypoxic inhibition of potassium channels may be a characteristic of pulmonary but not systemic arterial smooth muscle cells. This conclusion is strengthened by the fact that oxygen is sensed in a similar manner in the carotid body type 1 cell [16] and in the neuroepithelial body (NEB) [17].

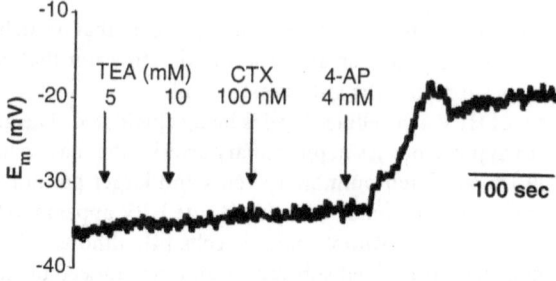

FIG. 1. Membrane potential ($E_m$) of a pulmonary artery smooth muscle cell dispersed from small pulmonary arteries of an adult rat. $E_m$ is not altered by tetraethylammonium *(TEA)* or charybdotoxin *(CTX)*, both $K_{Ca}$ channel blockers, but is depolarized by 4-aminopyridine *(4-AP)* ($K_v$ channel blocker), suggesting that $K_v$ channels control membrane potential. (From [19], with permission)

In the carotid body type 1 cell, hypoxia blocks an oxygen-sensitive potassium channel leading to membrane depolarization, calcium entry, dopamine release, enhanced sinus nerve activity, and an increase in respiratory rate [16]. In these cells, the rise in intracellular calcium caused by hypoxia is almost completely dependent on the presence of external calcium [16,18]. Suppression of the calcium response to hypoxia can be demonstrated within a minute of the removal of external calcium, suggesting that the suppression is not caused by depletion of calcium from the sarcoplasmic reticulum [18]. Calcium channel blockers inhibit the hypoxic increase in cytosolic calcium in the type 1 cell [18] and inhibit HPV [6], while the calcium channel agonist BAY K 8644 enhances both.

Subsequent work has suggested that, in the rat pulmonary vasculature, smooth muscle cells in which oxygen-sensitive $K_v$ channels predominate are more common in the small resistance pulmonary arteries. Conversely, cells in which $I_K$ flows mainly through $K_{Ca}$ channels are more commonly observed in conduit arteries [19]. It appears that the distribution of cells with different electrophysiological characteristics may determine the functional response of vessels. This is probably true of systemic as well as pulmonary vessels and may apply to the localized expression of receptors and enzymes, as well as ion channels. In the case of adult rat pulmonary arteries, the potassium channel most commonly blocked during hypoxia is a $K_v$ channel with a conductance of 37 pS [19].

## Pulmonary Vascular Responses to Oxygen in the Fetus

In the fetus, the lungs are not ventilated and pulmonary vascular resistance is greater than systemic levels. The high pulmonary vascular resistance may in part be caused by hypoxia. In the fetal lamb, marked pulmonary vasodilatation can be induced by ventilating the maternal sheep with 100% oxygen [20], which increases fetal aortic oxygen tension from 18 to 25 mmHg. Most of this vasodilatation could be prevented by pretreatment of the fetus with either TEA or iberiotoxin ($K_{Ca}$ channel blockers). Pretreatment with glibenclamide did not alter the vasodilatation caused by an in-

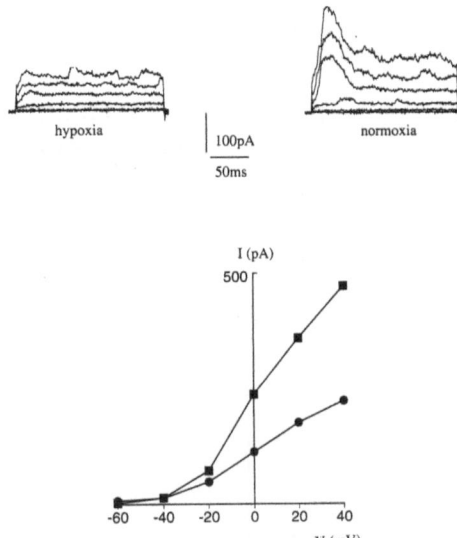

FIG. 2. The transition from hypoxia to normoxia in a fetal lamb pulmonary artery smooth muscle cell increases $I_K$, even at relatively negative membrane potentials (amphotericin-perforated patch-clamp technique). Spontaneous transient outward currents (STOCs) can be seen in the normoxic currents. Current, $I(pA)$; membrane potential, $V(mV)$. *Circles*, hypoxia; *squares*, normoxia

crease in oxygen tension. These findings suggested that $K_{Ca}$ channels might be involved in the mechanism of oxygen sensing in the fetal pulmonary vasculature. The conclusion was supported by whole-cell patch-clamp data showing that switching freshly dispersed fetal pulmonary artery smooth muscle cells from hypoxia to normoxia dramatically increases $I_K$ and hyperpolarizes the cell membrane. CTX reverses that increase in $I_K$ and depolarizes the cell membrane [20]. An example of the increase in $I_K$ caused by normoxia in pulmonary artery smooth muscle cells of the fetal lamb at term is shown in Fig. 2. Consequently, it seems likely that in the fetal lamb pulmonary vasculature, the $K_{Ca}$ channel is oxygen sensitive, and in the adult rat, one or more $K_v$ channels are oxygen sensitive. In the case of the full-term rabbit ductus arteriosus, again a $K_v$ channel is oxygen sensitive but in this instance normoxia closes rather than opens the channel.

## Ductus Arteriosus Response to Oxygen

The ductus arteriosus in the fetus permits oxygenated blood returning from the placenta to bypass the unventilated lungs and to reach the systemic circulation. At birth, exposure to increased oxygen tension leads to constriction of the ductus. This normoxic constriction, together with the normoxic vasodilatation of the pulmonary vasculature, results in the transition from the fetal to the adult circulation pattern. The mechanism of the normoxic ductal constriction has proved elusive.

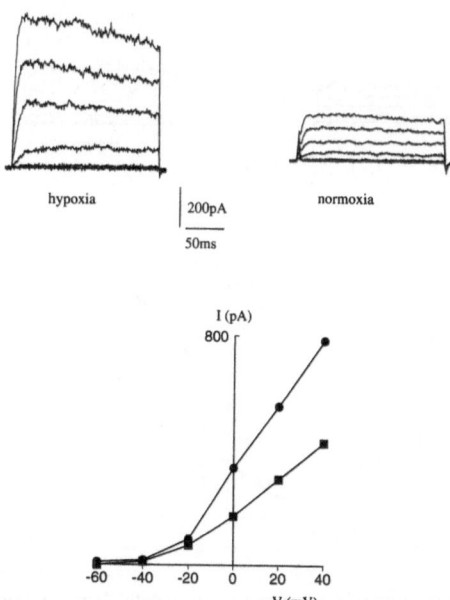

FIG. 3. Normoxia inhibits whole-cell $I_K$ in a freshly dispersed smooth muscle cell from the ductus arteriosus of the rabbit at term (amphotericin-perforated patch-clamp technique). *Circles*, hypoxia; *squares*, normoxia

As in the pulmonary vasculature, the potassium current is largely responsible for maintaining membrane potential in the ductus. While glibenclamide ($K_{ATP}$ blocker) and TEA ($K_{Ca}$ blocker) have relatively little effect on the tension of ductal rings and membrane potential in ductal smooth muscle cells, 4-AP ($K_v$ blocker) causes ring contraction and membrane depolarization [21]. These observations suggest that the gating of the $K_v$ channel might determine ductal tone. In fact, both normoxia and 4-AP inhibit whole-cell potassium current at negative membrane potentials. Figure 3 illustrates the reduction in $I_K$ caused in ductal smooth muscle cells by the transition from hypoxia to normoxia. In addition, normoxia has been shown to inhibit a 59-pS, 4-AP-sensitive channel in cell-attached patches [21]. The normoxia-induced contraction of ductal rings can be reversed by nisoldipine (an L-type calcium channel blocker). It seems likely that the initial, normoxic contraction of the ductus at birth is the result of $K_v$ channel inhibition, membrane depolarization, and calcium entry through the voltage-gated calcium channel. Calcium-induced calcium release from the sarcoplasmic reticulum may enhance this contraction.

## Gating of Oxygen-sensitive Channels

It has been known for 15 years that oxygen radicals and sulfhydryl oxidants, such as diamide will inhibit HPV [22,23]. The pulmonary vasodilator response to diamide (Fig. 4) is very reproducible and is rapidly reversible. Because of these observations, and by analogy with the role of sulfhydryl groups in the activation of pancreatic β-cells, it was suggested that changes in oxygen tension might be sensed by changes in sulfhydryl redox status, which might alter potassium channel gating and membrane potential [24]. This concept was made more plausible by the later observation that, in some mammalian potassium channels, the redox status of a cysteine residue in the N-terminal region can be controlled by the redox status of glutathione and may determine the gating of the channel [25]. It has been demonstrated that the critical cysteine is present in a β-subunit that can attach to the channel and determine its gating [26]. Oxidation, for instance as the result of an increase in oxygen radicals, would eliminate β-subunit activity and thereby enhance potassium efflux. It is possible that a redox-sensing subunit could be a part of different channels in different species or at different stages of ontological development. The effects of oxidized and reduced glutathione on $I_K$ in rat pulmonary artery smooth muscle cells are shown in Fig. 5 [27]. Diamide, which oxidizes glutathione, increases $I_K$ in pulmonary artery smooth muscle cells, while antioxidants N-acetylcysteine, dithionite and duroquinone reduce IK [13,28,29].

Although it is clear that redox changes can alter potassium channel gating, it is not certain that this is the mechanism by which variation in oxygen tension is sensed. Recent observations on the activity of hypoxia-inducible factor (HIF) make it more plausible [30]. Hypoxic activation of HIF to promote production of factors such as erythropoietin, tyrosine hydroxylase, and vascular endothelial growth factor involves redox-dependent stabilization of HIF-1 α protein. If the cells producing HIF in response to hypoxia are exposed to hydrogen peroxide or diamide, the level of HIF activity is markedly diminished. These findings recall the reports of a reduction of HPV by peroxides and diamide, as discussed earlier.

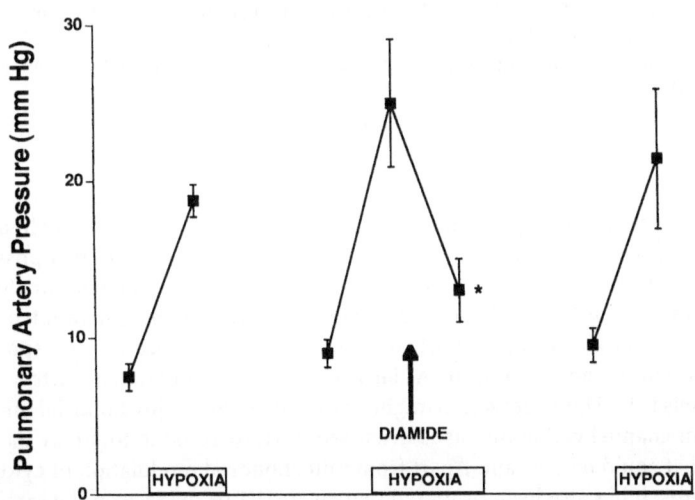

FIG. 4. The bolus injection of *diamide* (1 mg) markedly reduces the hypoxic pressor response in the isolated perfused rat lung ($n = 5$). *, $P < .05$ compared to value pre diamide. *Error bars* show SEM

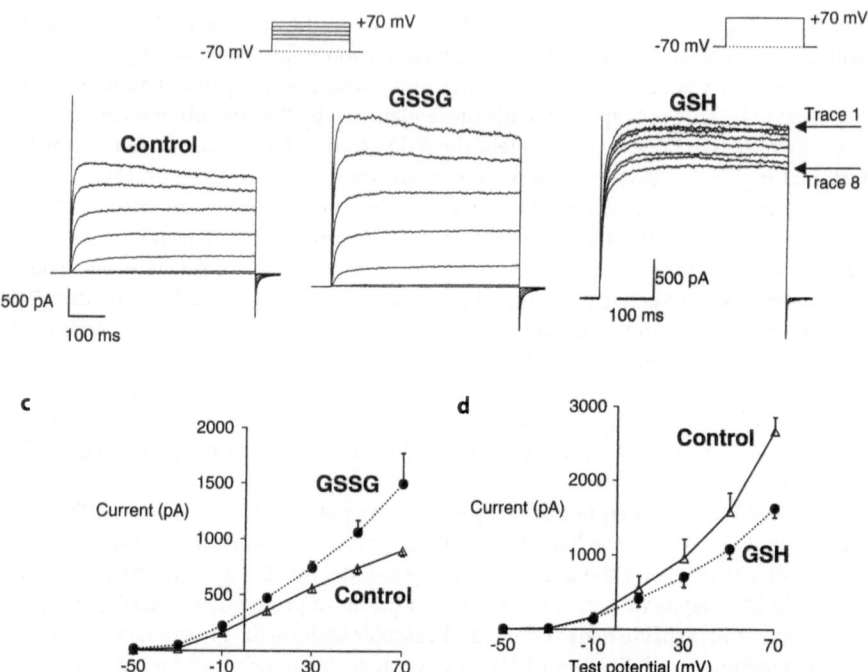

FIG. 5a–d. Effect of intracellular oxidized and reduced glutathione on whole-cell potassium currents in rat pulmonary artery smooth muscle cells. **a** Family of potassium currents elicited by voltage steps before (*control*) and after 5-min dialysis with oxidized glutathione (*GSSG*, 2 mM) into the cell, showing an increase in outward current. **b** Series of $K^+$ currents elicited by a repetitive voltage step from −70 to +70 mV, every 20 s after start of dialysis with reduced glutathione (GSH, 2 mM) into the cell, showing a decrease in outward current. **c** Current–voltage relationship for mean (+SEM) potassium current in the presence and absence of GSSG (2 mM). **d** Current–voltage relationship in presence and absence of GSH (2 mM). (From [27], with permission)

If the oxygen-sensitive potassium channel is redox controlled, how might a change in oxygen tension alter the redox status of the channel? There are two possibilities related to mitochondrial function. Hypoxia can increase the accumulation of reduced compounds such as NADH. In addition, hypoxia can reduce the generation of oxygen radicals from the electron transport chain. Inhibitors of the electron transport chain, such as rotenone and antimycin A, also decrease $I_K$ in pulmonary artery smooth muscle cells [31]. This observation might be thought to link mitochondrial function to potassium channel gating but, as is discussed next, these inhibitors may have other actions that could be relevant. Apart from mitochondrial modulation of cytoplasmic redox status, it has also been proposed that an NAD(P)H oxidase on the sarcolemma of the smooth muscle cell might "sense" the oxygen tension. The production of superoxide anion, and consequently of hydrogen peroxide, would be in proportion to the level of oxygen [32,33].

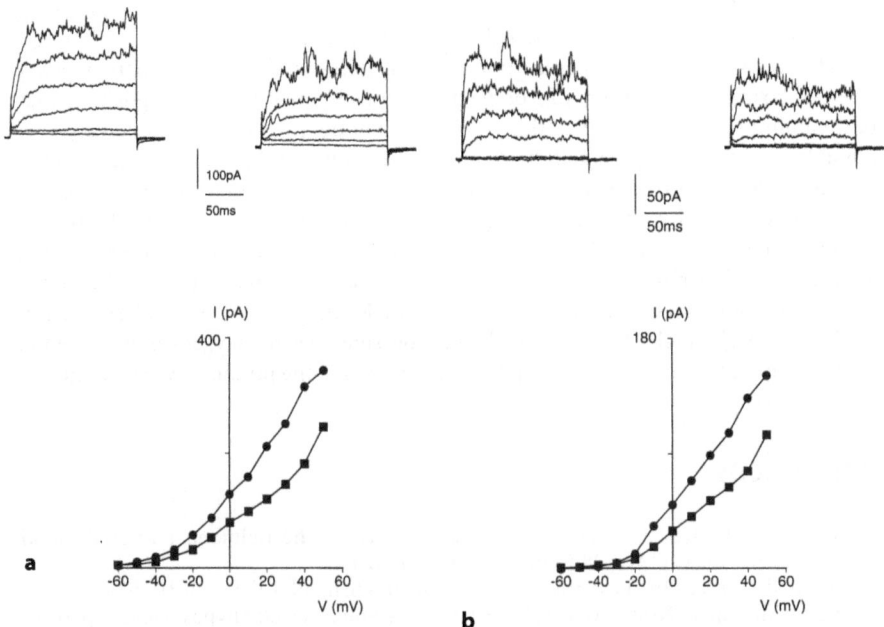

FIG. 6. **a** Hypoxia causes inhibition of whole cell $I_K$ in a freshly dispersed smooth muscle cell from the pulmonary artery of a wild-type mouse (amphotericin-perforated patch-clamp technique). Current, $I$ ($pA$); membrane potential, $V$ ($mV$). **b** Hypoxia causes inhibitin similar to that shown in **A** of $I_K$ in a smooth muscle cell from a transgenic mouse lacking the 91-kDa subunit of NADPH oxidase. *Circles*, normoxia; *squares*, hypoxia

It seems unlikely that the sensing of oxygen requires the action of NADPH oxidase. The inhibitor of NADPH oxidase, diphenyleneiodonium (DPI), does reduce $I_K$ in pulmonary vascular smooth muscle, which would be compatible with a role for the enzyme, but it also blocks calcium channels [34]. The blocking of calcium channels by DPI makes it more difficult to discern the physiological function of the enzyme because it causes smooth muscle relaxation. Transgenic mice that lack the 91-kDa subunit of NADPH oxidase still have a normal acute pulmonary pressor response to hypoxia [35], indicating that the entire enzyme is not necessary for oxygen sensing. In addition, hypoxia inhibits $I_K$ to the same extent in both transgenic and wild-type mice (Fig. 6). It is possible that a truncated portion of the enzyme is still capable of oxygen sensing; however, the transgenic mice show virtually no chemiluminescence from the surface of their perfused lungs [35]. The chemiluminescence is a measure of oxygen radical production [36], and the absence of chemiluminescence in the transgenic mice implies that NADPH oxidase is not generating oxygen radicals. The fact that rotenone and antimycin A markedly diminish chemiluminescence [31] indicates that they inhibit NADPH oxidase in addition to their well-known inhibition of the electron transport chain in the mitochondria. The persistence of HPV in the NADPH oxidase-deficient mouse does not exclude a role for NADH oxidase [37].

Regardless of the source of radicals, if redox regulation of potassium channels initiates hypoxic constriction of the pulmonary arteries, in the ductus, where hypoxia causes relaxation, the redox gating of the potassium channels must be opposite to that in the pulmonary arteries, or perhaps another channel might also be involved. It is possible that the redox gating of the calcium channel is an additional influence in the ductus, such that normoxia increases calcium entry through the voltage-gated calcium channel. This hypothesis is compatible with the observation that DPI blocks calcium channels [34], so long as DPI is assumed to inhibit the oxygen-sensing enzyme as well as NADPH oxidase. In the resistance pulmonary arteries, the dominant redox-modulated channel appears to be the $K_v$ channel. Differential expression of these two channels, both of which may be sensitive to oxygen-induced redox changes, could help to explain the opposite responses of the pulmonary artery and the ductus to oxygen.

# References

1. von Euler U, Liljestrand G (1946) Observations on the pulmonary arterial blood pressure in the cat. Acta Physiol Scand 12:301–320
2. Satchell G (1962) Intrinsic vasomotion in the dogfish gill. J Exp Biol 39:503–512
3. Shirai M, Sada K, Ninomiya I (1986) Effects of regional alveolar hypoxia and hypercapnia on small pulmonary vessels in cats. J Appl Physiol 61:440–448
4. Madden J, Vadula M, Kurup V (1992) Effects of hypoxia and other vasoactive agents on pulmonary and cerebral artery smooth muscle cells. Am J Physiol 263:L384–L393
5. Harder D, Madden J, Dawson C (1985) Hypoxic induction of $Ca^{2+}$-dependent action potentials in small pulmonary arteries of the cat. J Appl Physiol 59:1389–1393
6. McMurtry I, Davidson B, Reeves J, Grover R (1976) Inhibition of hypoxic pulmonary vasoconstriction by calcium antagonists in isolated rat lungs. Circ Res 38:99–104
7. McMurtry I (1985) Bay K8644 potentiates and A23187 inhibits hypoxic vasoconstriction in rat lungs. Am J Physiol 249:H741–H746
8. Tollins M, Weir E, Chesler E, et al. (1986) Pulmonary vascular tone is increased by a voltage-dependent calcium channel potentiator. J Appl Physiol 60:942–948
9. Farrukh I, Michael J (1992) Cellular mechanisms that control pulmonary vascular tone during hypoxia and normoxia. Am Rev Respir Dis 145:1389–1397
10. Salvaterra C, Goldman W (1993) Acute hypoxia increases cytosolic calcium in cultured pulmoncary arterial myocytes. Am J Physiol 264:L323–L328
11. Vadula M, Kleinman J, Madden J (1993) Effect of hypoxia and norepinephrine on cytoplasmic freed $Ca^{2+}$ in pulmonary and cerebral arterial myocytes. Am J Physiol 265:L591–L597
12. Clapp L, Gurney A (1991) Outward currents in rabbit pulmonary artery cells dissociated with a new technique. Exp Physiol 76:677–693
13. Yuan X-J, Goldman W, Tod M, et al. (1993) Hypoxia reduces potassium currents in cultured rat pulmonary but not mesenteric arterial myocytes. Am J Physiol 264:L116–L123
14. Hasunuma K, Rodman D, McMurtry I (1991) Effects of $K^+$-channel blockers on vascular tone in the perfused rat lung. Am Rev Respir Dis 144:884–887
15. Post J, Hume J, Archer S, Weir E (1992) Direct role for potassium channel inhibition in hypoxic pulmonary vasoconstriction. Am J Physiol 262:C882–C890
16. Lopez-Barneo J, Benot A, Urena J (1993) Oxygen sensing and the electrophysiology of arterial chemoreceptor cells. News Physiol Sci 8:191–195

17. Youngson C, Nurse C, Yeger H, Cutz E (1993) Oxygen sensing in airway chemorecep-
    tors. Nature 365:153-155
18. Buckler K, Vaughan-Johnes R (1994) Effects of hypoxia on membrane potential and
    intracellular calciumn in rat neonatal carotid body type 1 cells. J Physiol 476:423-428
19. Archer S, Huang J, Reeve H, et al. (1996) Differential distribution of electrophysiolog-
    ically distinct myocytes in conduit and resistance arteries determines their response to
    nitric oxide and hypoxia. Circ Res 78:431-442
20. Cornfield D, Reeve H, Tolarova S, et al. (1996) Oxygen causes fetal pulmonary vasodi-
    lation through activation of a calcium-dependent potassium channel. Proc Natl Acad
    Sci USA 93:8089-8094
21. Tristani-Firouzi M, Reeve H, Tolarova S, et al. (1996) Oxygen-induced constriction of
    rabbit ductus arteriosus occurs via inhibition of a 4-aminopyridine-, voltage-sensitive
    potassium channel. J Clin Invest 98:1959-1965
22. Weir E, Will J (1982) Oxidants: a new group of pulmonary vasodilators. Bull Eur
    Physiopathol 18:81-85
23. Weir E, Will J, Lundquist L, et al. (1983) Diamide inhibits pulmonary vasoconstriction
    induced by hypoxia or prostaglandin F2. Proc Soc Exp Biol Med 173:96-103
24. Archer S, Will J, Weir E (1986) Redox status in the control of pulmonary vascular tone.
    Herz 11:127-141
25. Ruppersberg J, Stocker M, Pongs O, et al. (1991) Regulation of fast inactivation of
    cloned mammalian $I_K(A)$ channels by cysteine oxidation. Nature 352:711-714
26. Rettig J, Helnemann S, Wunder F, et al. (1994) Inactivation properties of voltage-gated
    $K^+$ channels altered by presence of $\beta$-subunit. Nature 369:289-294
27. Weir E, Archer S (1995) The mechanism of acute hypoxic pulmonary vasoconstriction:
    the tale of two channels. FASEB J 9:183-189
28. Post J, Weir E, Archer S, Hume J (1993) Redox regulation of $K^+$ channels and hypoxic
    pulmonary vasoconstriction. In: Weir E, Hume J, Reeves J (eds) Ion flux in pulmonary
    vascular control. Plenum, New York, pp 189-204
29. Reeve H, Weir E, Nelson D, et al. (1995) Opposing effects of oxidants and antioxidants
    on $K^+$ channel activity and tone in rat vascular tissue. Exp Physiol 80:825-834
30. Huang L, Arany Z, Livingston D, Bunn H (1996) Activation of hypoxia-inducible
    transcription factor depends primarily upon redox-sensitive stabilization of its $\delta$ sub-
    unit. J Biol Chem 271:32253-32259
31. Archer S, Huang J, Henry T, et al. (1993) A redox-based O2 sensor in rat pulmonary
    vasculature. Circ Res 73:1100-1112
32. Cross A, Henderson L, Jones O, et al. (1990) Involvement of an NAD(P)H oxidase as a
    $pO_2$ sensor protein in the rat carotid body. Biochem J 272:743-747
33. Gorlach A, Jelkmann W, Hancock J, et al. (1993) Photometric characteristics of heme
    proteins in erythropoietin producing hepatoma cells (HepG2). Biochem J 290:771-
    776W
34. Weir E, Wyatt C, Reeve H, et al. (1994) Diphenyleneiodonium inhibits both potassium
    and calcium currents in isolated pulmonary artery smooth muscle cells. J Appl Physiol
    76:2611-2615
35. Weir E, Dinauer M, Nelson D, Archer S (1996) Hypoxic pulmonary vasoconstriction is
    unchanged in NADPH-oxidase "knock-out" mice. FASEB J 10:A99
36. Archer S, Nelson D, Weir E (1989) Simultaneous measurement of oxygen radicals and
    pulmonary vascular reactivity in the isolated rat lung. J Appl Physiol 67:1903-1911
37. Mohazzab-H K, Wolin M (1994) Properties of a superoxide anion-generating microso-
    mal NADH oxidoreductase, a potential pulmonary artery Po2 sensor. Am J Physiol
    267:L823-L831

# Biological Impediment to Oxygen Sensing in Injured Pulmonary Microcirculation Exposed to a High-Oxygen Environment

Kazuhiro Yamaguchi, Koichi Suzuki, Kazumi Nishio,
Takuya Aoki, Yukio Suzuki, Atsushi Miyata, Nagato Sato,
Katsuhiko Naoki, and Hiroyasu Kudo

*Summary.* To resolve the issue whether the $O_2$-sensing mechanism of pulmonary circulation is impaired in lungs exposed to hyperoxia, we prepared isolated perfused lungs from rats exposed to either a normoxic or hyperoxic environment and estimated hypoxia-elicited pressor changes in the pulmonary arteries as well as diameter changes in microvessels in the acini. To assess the important role of vasodilator prostaglandins as well as nitric oxide (NO) in blunting hypoxic pulmonary vasoconstriction (HPV) in hyperoxia-exposed lungs, we examined the effects of inhibition of constitutive and inducible forms of cyclooxygenase (COX-1 and COX-2) and of constitutive and inducible forms of NO synthase (eNOS and iNOS). Indomethacin was used as the inhibitor of both COX-1 and COX-2, while NS-398 was applied as a selective inhibitor of COX-2. Simultaneous restraint of eNOS and iNOS was attained by $N^{\omega}$-nitro-L-arginine methyl ester (L-NAME); restraint of iNOS was achieved by aminoguanidine. By addition of fluorescein isothiocyanate-dextran to the perfusion circuit, the architecture and diameter of the intraacinar microvessel were precisely determined in terms of a real-time confocal laser luminescence microscope. The results derived therefrom are (1) exposure to hyperoxia causes overall HPV and contractility of precapillary arterioles responding to hypoxia to be attenuated; (2) the attenuated overall HPV was significantly restored by adding L-NAME but not by adding aminoguanidine, indomethacin, or NS-398; and (3) hypoxia-induced constriction of precapillary arterioles was ameliorated by either L-NAME, aminoguanidine, or indomethacin but not by NS-398. The extent of recovery of arteriolar constriction in the presence of L-NAME was greater than that in the presence of aminoguanidine. We concluded that overall HPV would be blunted by the enhanced eNOS expression along the arteries rather than those in the acini, while contractility of intraacinar precapillary arterioles would be impaired by augmentation of eNOS and iNOS as well as COX-1 expression.

*Key words.* Hyperoxia—Hypoxic pulmonary vasoconstriction—COX-1—COX-2—eNOS—iNOS

Department of Medicine, School of Medicine, Keio University, 35 Shinanomachi, Shinjuku-ku, Tokyo 160, Japan

# Introduction

In the normal lung, pulmonary vascular $O_2$-sensing mechanisms, which induce hypoxic pulmonary vasoconstriction (HPV), have been considered to be exceedingly important in regulating the distribution of pulmonary blood flow, allowing the lung to maintain a reasonable match between ventilation and blood flow [1]. Active HPV shifts the pulmonary blood flow in a hypoxic region to a better oxygenated region of the lung, thus avoiding the occurrence of serious hypoxemia. If HPV is reduced, gas exchange efficiency in the lung is impaired and causes significant hypoxemia because of augmentation of the blood flow entering hypoxic regions. Impaired HPV has been reported to occur in various lung diseases, including acute respiratory distress syndrome induced by endotoxin [2], instillation of a toxic agent such as oleic acid [3,4] or ethchlorvynol [5], severe pneumonia caused by *Pneumococcus* [6] or *Pseudomonas aeruginosa* [7], air embolism [8], diffuse granulomatous lung disease [9], and oxidant lung injury induced by exposure to a high-$O_2$ environment [10]. Among these, hyperoxia-exposed lung injury should be cautiously handled from a clinical point of view, because hyperoxic gas breathing is generally provided, through an artificial ventilator, for treating patients with severe hypoxemia. This implies that hyperoxic gas breathing, that is, the procedure introduced for improving the hypoxemia, may negatively affect the patient situation.

Excess of exogenous reactive oxygen species (ROS), generally encountered under a condition in which the lung is exposed to a hyperoxic environment, has been suggested to be a pathological factor that diminishes the responsiveness of the pulmonary vasculature to various stimuli, especially alveolar hypoxia [11,12]. We have recently shown that antioxidant enzymes such as superoxide dismutase, catalase, and glutathione peroxidase in the lung tissue as well as those in the erythrocyte play a significant role in preserving HPV under conditions with excessive oxidant stress generated by xanthine and xanthine oxidase [12]. The complex effects of ROS on vascular reactivity have been extensively discussed by Gurtner and Burke-Wolin [11]. Summarizing the experimental findings reported to date, Gurtner and Burke-Wolin [11] suggested that ROS would act as a double-edged sword for modifying vascular reactivity in the pulmonary circulation. ROS can stimulate production of vasoconstrictive arachidonate mediators including thromboxane $A_2$ and peptide leukotrienes in the presence of endothelial cells. On the other hand, ROS may cause vasodilation through either direct activation of guanylate cyclase or interference with signal transduction within the smooth muscle cells.

However, several groups of investigators including ourselves [2–5] confirmed that increased production of vasodilating prostaglandin such as prostacyclin ($PGI_2$) would also be one of the decisive causes for blunted HPV in diseased lungs injured by endotoxin or oleic acid administration, both of which are suggested to involve ROS, as well. These findings indicate that, in addition to the mechanisms proposed by Gurtner and Burke-Wolin [11], augmented generation of vasodilator substances should also be taken into account as one of the crucial factors impairing HPV in oxidant-induced lung injury. Although nitric oxide (NO) has a vasodilating capability comparable to that of $PGI_2$, conspicuous attention has not been paid, in the previous studies [3–5], to the importance of NO for blunted HPV in oxidant-injured lungs.

In view of these facts as well as the significance of applying hyperoxic gas to ill patients, the current study was undertaken to solve the following issues. (1) Does ROS as yielded by prolonged exposure to hyperoxia actually diminish HPV in the pulmonary circulation? (2) If so, what kinds of pulmonary microvessels lose reactivity to hypoxic stimulation? (3) Does the hyperoxia-exposed lung significantly enhance the vasodilator substances produced by cyclooxygenase (COX) or NO synthase (NOS)? (4) If so, what types of COX or NOS are responsible for the increased genesis of vasodilators? We attempted to shed light on potential effects of not just the constitutive forms of COX (COX-1) and NOS (eNOS) but also the inducible forms of COX (COX-2) and NOS (iNOS) on impaired HPV in hyperoxia-induced lung injury.

Because pulmonary microvascular networks are exceedingly intricate and arterioles, venules, and capillaries are densely convoluted in the acini [13], classical epiluminescence microscopy, having ordinarily been used for observing blood cell kinetics in the systemic microcirculation with its much simpler architecture, may not be suitable for examining the detailed architecture of the pulmonary microcirculation with its interwoven structure. To overcome this obstacle, we have developed a real-time confocal laser scanning luminescence optical microscope allowing precise discrimination of individual microvessels from neighboring vessels [14]. Applying this novel method, we attempted to systematically investigate the effect of hypoxic stimulation on alterations in the vascular size of pulmonary microvessels injured by long-term exposure to hyperoxia under conditions in which the respective forms of COX or NOS were thoroughly inhibited.

## Materials and Methods

### Experimental Protocols

To assess whether $O_2$-sensing mechanisms would be significantly impaired in the pulmonary microvasculature injured by exposure to a hyperoxic environment, we estimated the effects of hypoxic stimulation on changes in mean pulmonary arterial pressures ($P_{PA}$) and intraacinar microvascular diameter in hyperoxia-exposed lungs. Male pathogen-free Sprague–Dawley rats (SD rats, 8 weeks old) weighing 250–300 g ($n = 130$) were exposed to either normoxia (21% $O_2$) or hyperoxia (90% $O_2$) for 48 h. Animals exposed to normoxia and hyperoxia were defined as the N-group and H-group, respectively.

Preparing isolated perfused lungs from these animals, we estimated the effects of inhibiting COX or NOS on HPV elicited by 2% $O_2$. Each animal group was subdivided into five groups on the basis of the agents applied.

1. In the no-medications group, measurements were made without administration of any agents.
2. In the indomethacin group, indomethacin was used for restraint of both constitutive and inducible forms of cyclooxygenase (COX-1 and COX-2). Perfusate concentration of indomethacin was adjusted to 20 μM.
3. In the NS-398 group, N-(2-cyclohexyloxy-4-nitrophenyl) methane sulphonamide (NS-398) was applied as a specific inhibitor for inducible COX-2. The high specifi-

city of NS-398 for suppressing COX-2 functions was confirmed by Futaki and colleagues [15]. Perfusate concentrations of NS-398 were adjusted to $1\mu M$.

4. In the L-NAME group, the constitutive form of NO synthase (eNOS) and inducible form of NO synthase (iNOS) were concomitantly inhibited with $N^{\omega}$-nitro-L-arginine methyl ester (L-NAME) at a perfusate concentration of $100\mu M$.

5. In the aminoguanidine group, aminoguanidine at a perfusate concentration of 4mM was administered as a specific inhibitor for iNOS. Unlike the majority of NO synthase inhibitors currently available, recent evidence has suggested that aminoguanidine is a selective inhibitor for iNOS [16].

## Preparation of Isolated Perfused Rat Lungs

We used isolated perfused rat lungs in which hematocrit (Ht) and overall perfusion rate could easily be adjusted to meet our requirements. A detailed description of the isolated perfused lung preparation has been provided elsewhere [12,14,17]. The isolated lung was fixed on a microscopic stage in the supine position and perfused at a constant flow rate of 10ml/min in a recirculating manner using a roller pump with Krebs–Henseleit solution with 3% bovine serum albumin. The perfusate Ht was adjusted to 5% by adding fresh blood obtained from donor rats. Gas exchange was adequately maintained with an extracorporeal membrane oxygenator (ECMO, Merasilox-S, Senko, Tokyo, Japan) inserted between the isolated lung and the roller pump. A gas mixture containing 21% $O_2$ and 5% $CO_2$ was used as the gas flowing into the ECMO. A warmed and humidified gas mixture containing the same composition of gases as those used for the ECMO was supplied continuously to the lung surface so as to maintain a temperature of $37° \pm 0.5°C$ and to avoid desiccation of the lung surface.

After stable $P_{PA}$ had been attained, the gas flowing into the ECMO and that blown onto the lung surface were switched from the mixture containing 21% $O_2$ to that containing 2% $O_2$, allowing the HPV to be elicited. Increment of $P_{PA}$ from the baseline value during hypoxic gas exposure was employed as the measure of overall HPV, that is, vasoconstriction occurring in all the pulmonary vessels. $P_{PA}$ was continuously monitored by force displacement of pressure transducers.

## Measurements of Pulmonary Microvessel Diameters

To obtain images adequate for precise estimation of the events taking place in pulmonary microvessels, we used a confocal luminescence microscope, equivalent to the instrument known as a tandem scanning reflected light microscope [14]. Our confocal unit yields a resolution velocity approximately 1000 fold greater than does a conventional confocal scanning optical microscope with a unitary beam such as the galvanometer type and is designed to generate a continuous image at a rate of 1ms/frame. The reflected light or fluorescent emission from the specimen was imaged onto a high-sensitivity charge-coupled-device (CCD) camera with an image intensifier (EktaPro Intensified Imager VSG, Kodak, San Diego, CA, USA) that can detect even very low fluorescence signals. By incorporating an excitation wavelength of 488nm emitted from a low-power air-cooled argon (Ar) ion laser (532-BSA04; output power, 10mW/ V; Omnichrome, Chin, CA, USA) with appropriate fluoresceins, the present confocal system allowed us to obtain apparently instantaneous images at 1000 frames/s, with an optical sectioning depth of 0.75 μm and a two-point spatial resolution of 0.2 μm.

The final magnifying power of our system reached ×968 (with the 40× objective) on the video screen. The size of the resulting field of view was 180 × 210 μm, corresponding roughly to a diameter of the single pulmonary microvessel running beside the terminal bronchiole [18]. We registered confocal images at a rate of 250 frames/s by means of a high-speed video analysis system (EktaPro 1000 Processor, Kodak) connected with the image-intensified CCD camera (EktaPro Intensified Imager VSG, Kodak). All images obtained were monitored on color video television (PVM-1444Q, Sony, Tokyo, Japan) and stored in a videocassette recorder (SVQ-260, Sony). To precisely determine the diameter and architecture of microvessels, we had added 200 μl of 5% fluorescein isothiocyanate-dextran (FITC-dextran) with a molecular weight of 145 000 (Sigma, St. Louis, MO, USA) to the reservoir. Vessel diameters were estimated by processing a confocal video image with the computer-assisted digital image analyzing system (Quadra 840AV/Image 1.58, Apple, Cupertino, CA, USA).

## Statistical Evaluation

Statistical differences in the results obtained for varied experimental conditions were generally judged in terms of one-way analysis of variance (ANOVA) followed by multiple comparison analysis of the Scheffe examination. Values are presented as means ± SD; $P < .05$ was taken to be statistically significant.

# Results

## HPV in Normoxia- and Hyperoxia-Exposed Lungs

Increments of $P_{PA}$ by hypoxic stimulation in the absence of any medications averaged 4.1 Torr for the N-group but 1.1 Torr for the H-group, with a distinct difference between the two values (Fig. 1). Baseline diameters of intraacinar precapillary arterioles in the N-group were 25 μm but were reduced by 2.6 μm, corresponding to 11% of the baseline value, after hypoxic stimulation. Although arteriolar diameters in the H-group before stimulation with hypoxic gas did not differ from those of the N-group, they were not constricted, even dilated, by hypoxic stimulus. HPV did not alter venular and capillary diameters in either group.

## Effects of COX Inhibition on HPV in Normoxia- and Hyperoxia-Exposed Lungs

In the N-group, neither indomethacin nor NS-398 exerted any significant influence on the extent of hypoxia-induced vasoconstriction, including increments of $P_{PA}$ and microvascular diameter changes.

Although the extent of overall HPV in the H-group was not improved by treatment with indomethacin, hypoxia-elicited constriction of intraacinar precapillary arterioles having a diameter of about 20–30 μm was obviously restored; that is, the decrease in arteriolar diameter by hypoxic stimulation was 2.9 μm, consistent with the value observed in the N-group treated with indomethacin. Venular as well as capillary

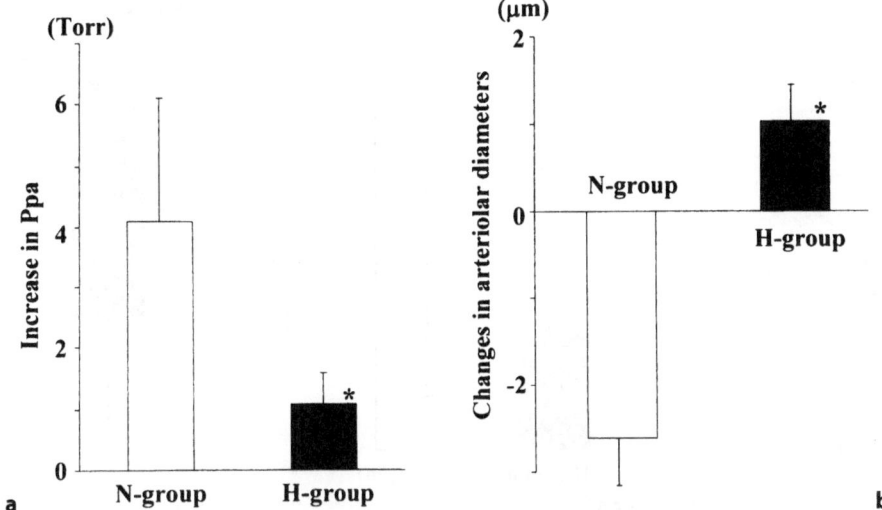

FIG. 1a,b. Hypoxic pulmonary vasoconstriction (HPV) in normoxia-exposed lungs (*N-group, open bars*) and in hyperoxia-exposed lungs (*H-group, solid bars*). Values are given as means ± SD. **a** Overall HPV represented by increment of pressor changes in the main pulmonary artery. **b** HPV represented by diameter changes in precapillary arterioles in the acini. *, Significantly different from the value obtained for the N-group

diameter changes induced by hypoxic gas exposure did not differ between the H- and N-group receiving indomethacin. NS-398, a specific inhibitor for COX-2, ameliorated neither overall HPV nor microvascular contractility of hyperoxia-injured lungs during hypoxic stimulation.

## Effects of NOS Inhibition on HPV in Normoxia- and Hyperoxia-Exposed Lungs

In the N-group, inhibition of both eNOS and iNOS by L-NAME significantly augmented hypoxia-induced pressor changes of pulmonary circulation, but exerted little influence on diameter changes in any microvessels including precapillary arterioles, postcapillary venules, and capillaries (Fig. 2). Aminoguanidine, an inhibitor of iNOS, had no significant effect on either overall pressor changes or microvascular diameter changes in intact lungs during hypoxic stimulation (Fig. 3).

In the H-group, L-NAME obviously increased hypoxia-induced $P_{PA}$ changes, the maximum $P_{PA}$ during hypoxic stimulation reaching a value comparable to that observed in the N-group (Fig. 2). In parallel with this, L-NAME significantly ameliorated the contractility of intraacinar precapillary arterioles responding to hypoxic gas exposure (Fig. 2). The extent of vasoconstriction of the precapillary arterioles in the H-group treated with L-NAME was statistically the same as that obtained in the N-group with L-NAME. However, L-NAME did not have any significant influence on hypoxia-elicited diameter changes in postcapillary venules and capillaries of the H-group. Aminoguanidine did not alter the overall extent of HPV in the H-group, but signifi-

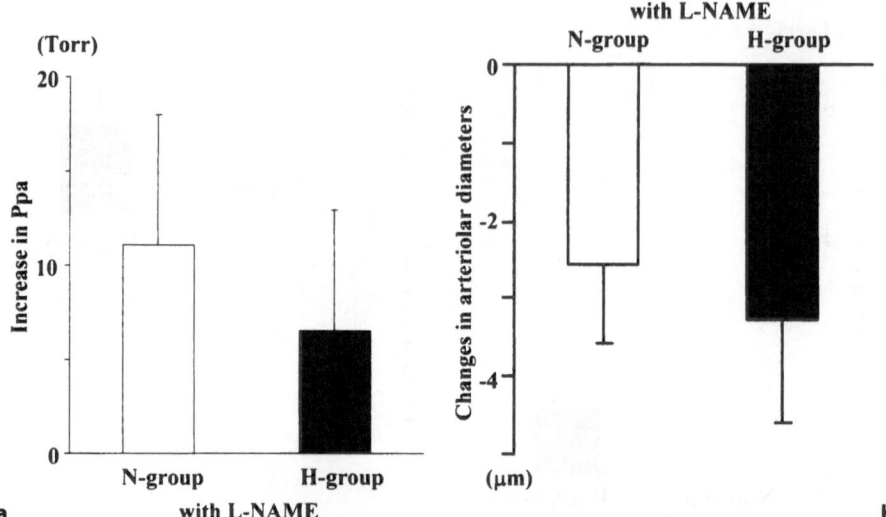

FIG. 2a,b. Effects of $N^{\omega}$-nitro-L-arginine methylester (*L-NAME*) on overall HPV and hypoxia-induced constriction of intraacinar arterioles in N- and H-groups. Values are presented as means ± SD. **a** Overall HPV. **b** Intraacinar arterioles

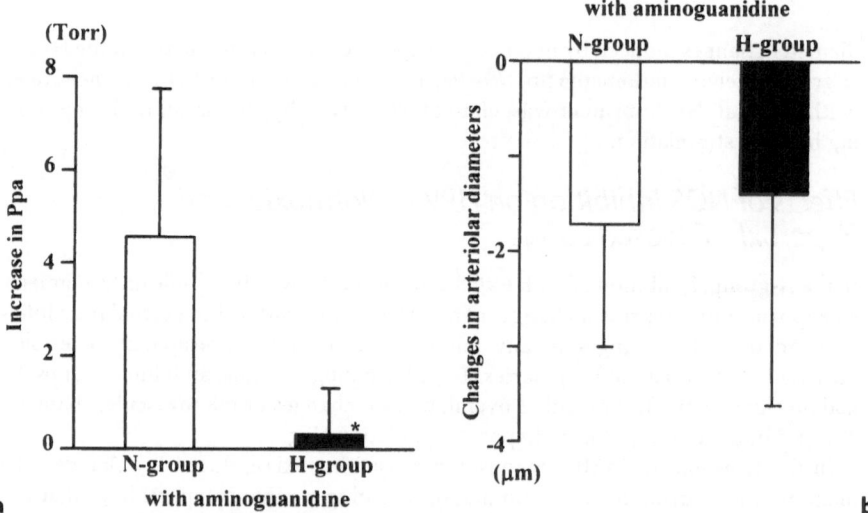

FIG. 3a,b. Effects of aminoguanidine on overall HPV and hypoxia-elicited constriction of intraacinar arterioles in N- and H-groups. Values are means ± SD. **a** Overall HPV. **b** Intraacinar arterioles. Recovery of intraacinar arteriolar contractility of the H-group by aminoguanidine is significantly less than that by L-NAME as shown in Fig. 2. *, Significantly different from the value observed in the N-group

cantly improved arteriolar contractility of hyperoxia-exposed lungs during hypoxic stimulation (Fig. 3). The extent of recovery of arteriolar contractility responding to hypoxia in the H-group with aminoguanidine was appreciably less than that observed for the H-group with L-NAME (Figs. 2 and 3). Aminoguanidine exerted no remarkable influence on the contractility of postcapillary venules and capillaries in the H-group.

## Discussion

### Hypoxia-Induced Constriction of Intraacinar Arterioles in Normoxia- and Hyperoxia-Exposed Lungs

Hypoxic stimulation distinctly increased $P_{PA}$ associated with significant constriction of intraacinar precapillary arterioles in normoxia-exposed intact lungs (see Fig. 1). However, neither postcapillary venules nor capillaries in intact lungs were constricted by hypoxic stimulus, with the result that hypoxic gas could only evoke vasoconstriction of intraacinar arterioles. Although previous histological and radiological studies [18,19] indicated that, in intact lungs, hypoxia causes significant vasoconstriction in pulmonary arteries with diameters ranging from 100 to 1000 µm, the question whether much smaller intraacinar arterioles are actually constricted when exposed to hypoxia has not been conclusively settled. To the best of our knowledge, this may be the first report demonstrating direct evidence for hypoxia-induced vasoconstriction of very small intraacinar arterioles connecting directly with capillaries under conditions resembling in vivo circumstances (Fig. 1).

The overall extent of hypoxia-elicited pressor changes in lungs injured by hyperoxia exposure was about a quarter of that observed in normal lungs, indicating the significant attenuation of HPV in hyperoxia-injured lungs (Fig. 1). Concomitantly, constriction of precapillary arterioles in the acini conspicuously faded out, suggesting that intraacinar arterioles would be one of the responsible portions for impaired HPV in hyperoxia-induced lung injury (Fig. 1).

### Effects of COX and NOS on HPV in Normoxia-Exposed Lungs

The overall extent of hypoxia-induced pressor responses in normoxia-treated lungs was significantly enhanced by simultaneous inhibition of eNOS and iNOS, but was not altered by restraining either iNOS alone, COX-2 alone, or both COX-1 and COX-2, suggesting that NO generated through eNOS would function as the important modulator coping with an excessive vasoconstriction of intact pulmonary vessels during hypoxic stimulation (see Figs. 2 and 3). Our experimental results may additionally suggest that vasodilator prostaglandins have little effect in opposition to HPV occurring in intact lungs. Although inhibition of eNOS by L-NAME enhanced pressor changes of pulmonary vessels responding to hypoxia, it did not alter the extent of hypoxia-elicited constriction of precapillary arterioles in the intact acini, leading to the conclusion that eNOS may be predominantly expressed along the relatively larger arteries rather than intraacinar arterioles with a diameter of 20–30 µm (Fig. 2).

## Effects of COX and NOS on Blunted HPV in Hyperoxia-Exposed Lungs

Although simultaneous inhibition of COX-1 and COX-2 by indomethacin did not amend the extent of overall HPV in hyperoxia-injured lungs, it led intraacinar precapillary arterioles to regain the contractility lost during the hypoxic period. On the other hand, inhibition of COX-2 alone by NS-398 improved neither overall HPV nor responsiveness of precapillary arterioles to hypoxic stimulation. Combining the results obtained for normoxic-treated lungs, these findings may indicate that constitutive COX-1 expression is conspicuously augmented at least along the intraacinar arteriolar walls of hyperoxia-injured lungs, although enhanced COX-1 expression may not be a key factor explaining impairment of their overall HPV. Reactive oxygen species such as superoxide as well as hydrogen peroxide were reported to be potent inducers of COX-2 gene expression in a variety of cells or tissues [20], indicating the possibility that prolonged hyperoxia exposure could significantly induce COX-2 mediated by reactive oxygen species. Therefore, we attempted to elucidate the effect of inhibiting COX-2 on impaired HPV in hyperoxia-injured lungs. Our experimental results seem to support the conception that COX-2 is not sufficiently induced or vasoactive effects of the prostaglandins generated through the pathway relating to COX-2 are canceled out by one another and do not cause paralytic HPV in hyperoxia-injured lungs.

In hyperoxia-injured lungs, restraint of both eNOS and iNOS by L-NAME evidently restored the responsiveness of intraacinar arterioles to hypoxia as well as the overall HPV (Fig. 2). This finding qualitatively differed from that obtained for normoxia-exposed intact lungs in which the responsiveness of intraacinar arterioles to hypoxic stimulation was not augmented by administration of L-NAME, suggesting that exposure to hyperoxia would enhance the expression of eNOS or iNOS in intraacinar arterioles (Fig. 2). Restraint of iNOS alone by aminoguanidine did not improve the overall HPV but ameliorated the contractile response of intraacinar arterioles to hypoxic stimulation, although the extent was much less than that obtained under a condition in which eNOS and iNOS were concomitantly inhibited (Figs. 2 and 3). Taken together, these findings may suggest that expressions of both eNOS and iNOS are significantly augmented in intraacinar precapillary arterioles exposed to hyperoxic environments. However, enhanced iNOS induction in intraacinar arterioles may not be a crucial factor responsible for depression of overall HPV observed in lungs injured by hyperoxia exposure. Instead, increased eNOS expression may function as one of the important factors blunting overall HPV in hyperoxia-injured lungs.

## Conclusions

1. Hypoxic stimulation was able to constrict precapillary arterioles with diameters of about 20–30 µm but neither postcapillary venules nor capillaries in intact acini.

2. NO generated through eNOS was one of the important substances modulating HPV occurring in normal lungs.

3. Prolonged exposure to hyperoxia substantially attenuated pressor changes of pulmonary arteries and contractility of precapillary arterioles responding to hypoxic stimulation.

4. The attenuated overall pressor responses to hypoxia in hyperoxia-injured lungs were significantly restored by treating the lung with L-NAME but not with indomethacin, NS-398, or aminoguanidine. These findings may indicate that enhanced expression of eNOS (preferentially along arterioles with relatively lager diameters) significantly impairs hypoxic-elicited overall pressor responses of pulmonary circulation in hyperoxia-injured lungs.

5. On the other hand, contractility of precapillary arterioles was appreciably amended by either L-NAME, aminoguanidine, or indomethacin but not by NS-398.

6. The extent of recovery of arteriolar constriction responding to hypoxic stimulation in hyperoxia-injured lungs treated with L-NAME was much greater than that in lungs treated with aminoguanidine, suggesting that eNOS expression and iNOS induction are distinctly augmented by exposure to hyperoxic environments. However, COX-2 may not be induced abundantly in hyperoxia-exposed lungs, although COX-1 expression is expected to increase.

# References

1. Fishman AP (1976) Hypoxia and pulmonary circulation. Circ Res 38:221–231
2. Frank DU, Lowson SM, Roos CM, Rich GF (1996) Endotoxin alters hypoxic pulmonary vasoconstriction in isolated rat lungs. J Appl Physiol 81:1316–1322
3. Leeman M, Delcroix M, Vachiery J, et al. (1992) Blunted hypoxic vasoconstriction in oleic acid lung injury: effect of cyclooxygenase inhibitors. J Appl Physiol 72:251–258
4. Yamaguchi K, Mori M, Kawai A, et al. (1994) Regulation of blood flow in pulmonary microcirculation by vasoactive arachidonic acid metabolites—analysis in acute lung injury. Adv Exp Med Biol 345:113–120
5. Yagi K, Baudenditel LW, Dahms TE (1992) Ibuprofen reduces ethchlorvynol lung injury: possible role of blood flow distribution. J Appl Physiol 72:1156–1165
6. Light RB (1986) Indomethacin and acetylsalicylic acid reduce intrapulmonary shunt in experimental pneumococcal pneumonia. Am Rev Respir Dis 134:520–525
7. Graham LM, Vasil A, Vasil ML, et al. (1990) Decreased pulmonary vasoreactivity in an animal model of chronic pseudomonas pneumonia. Am Rev Respir Dis 142:221–229
8. Cheney FW, Eisenstein BL, Overand PT, Bishop MJ (1989) Regional alveolar hypoxia does not affect air embolism-induced pulmonary edema. J Appl Physiol 66:2369–2373
9. Schulman LL, Wood JA, Enson Y (1989) Control of shunt pathway perfusion in diffuse granulomatous lung disease. J Appl Physiol 67:1717–1726
10. Truog WE, Redding GJ, Standert TA (1987) Effects of hyperoxia on vasoconstriction and $\dot{V}_A/\dot{Q}$ matching in the neonatal lung. J Appl Physiol 63:2536–2541
11. Gurtner GH, Burke-Wolin T (1991) Interactions of oxidant stress and vascular reactivity. Am J Physiol 260:L207–L211
12. Yamaguchi K, Asano K, Takasugi T, et al. (1996) Modulation of hypoxic pulmonary vasoconstriction by antioxidant enzymes in red blood cells. Am J Respir Crit Care Med 153:211–217
13. Weibel ER (1991) Design and morphometry of the pulmonary gas exchange. In: Crystal RG, West JB, Barnes PJ, et al. (eds) The lung: scientific foundations. Raven, New York, pp 795–805
14. Yamaguchi K, Nishio K, Sato N, et al. (1997) Leukocyte kinetics in the pulmonary microcirculation: observations using real-time confocal luminescence microscopy coupled with high-speed video analysis. Lab Invest 76:809–822

15. Futaki N, Arai I, Hamasaka Y, et al. (1993) Selective inhibition of NS-398 on prostanoid production in inflamed tissue in rat carrageenan-air-pouch inflammation. J Pharm Pharmacol 45:753–755
16. Griffiths MJD, Messent M, Curzen NP, Evans TW (1995) Aminoguanidine selectively decreases cyclic GMP levels produced by inducible nitric oxide synthase. Am J Respir Crit Care Med 152:1599–1604
17. Yamaguchi K, Takasugi T, Fujita H, et al. (1996) Endothelial modulation of pH-dependent pressor response in isolated perfused rabbit lungs. Am J Physiol (Heart Circ Physiol 39):H252–H258
18. Kato M, Staub N (1966) Response of small pulmonary arteries to unilobar hypoxia and hypercapnia. Circ Res 19:426–440
19. Shirai M, Sada K, Ninomiya I (1986) Effects of regional alveolar hypoxia and hypercapnia on small pulmonary vessels in cats. J Appl Physiol 61:440–448
20. Feng L, Xia Y, Garcia GE, et al. (1995) Involvement of reactive oxygen intermediates in cyclooxygenase-2 expression induced by interleukin-1, tumor necrosis factor-$\alpha$, and lipopolysaccharide. J Clin Invest 95:1669–1675

# Transcriptional Responses Mediated by Hypoxia-Inducible Factor 1

GREGG L. SEMENZA, FATON AGANI, NARAYAN IYER,
BING-HUA JIANG, ERIK LAUGHNER, SANDRA LEUNG, RICK ROE,
CHARLES WIENER, and AIMEE YU

*Summary.* Hypoxia-inducible factor 1 (HIF-1) is a basic helix–loop–helix protein that activates transcription of hypoxia-inducible genes, including those encoding erythropoietin, vascular endothelial growth factor, heme oxygenase-1, inducible nitric oxide synthase, and the glycolytic enzymes aldolase A, enolase 1, lactate dehydrogenase A, phosphofructokinase L, and phosphoglycerate kinase 1. Hypoxia response elements from these genes consist of a HIF-1-binding site as well as additional DNA sequences that are required for function, which in some elements include a second HIF-1-binding site. HIF-1 is a heterodimer: the HIF-1α subunit is unique to HIF-1, while HIF-1β (ARNT) can dimerize with other proteins. Cotransfection of HIF-1α and HIF-1β (ARNT) expression vectors and a reporter gene containing a wild-type hypoxia response element resulted in increased transcription in nonhypoxic cells and a superinduction of transcription in hypoxic cells, whereas HIF-1 expression vectors had no effect on transcription of reporter genes containing a HIF-1 binding-site mutation. In HeLa cells, HIF-1α and HIF-1β protein levels and HIF-1 DNA-binding activity increased exponentially as oxygen tension decreased, with maximum values at 0.5% oxygen and half-maximal values at 1.5%–2% oxygen. HIF-1α and HIF-1β (ARNT) mRNAs were detected in all human and rodent organs assayed. HIF-1α protein levels were induced in vivo when animals were subjected to anemia or hypoxia.

*Key words.* Erythropoietin—Glycolysis—Hypoxia—Transcription—Vascular endothelial growth factor

## Introduction

Hypoxia is a significant pathophysiological factor in a variety of cardiovascular, endocrine, hematological, oncologic, and pulmonary disorders (reviewed in [1]). This chapter summarizes recent investigations in our laboratory that have demonstrated a key role for the transcription factor hypoxia-inducible factor 1 (HIF-1) in mediating

Center for Medical Genetics, The Johns Hopkins University School of Medicine, 600 North Wolfe Street Baltimore, MD 21287-3914, USA homeostatic responses to hypoxia in mammals. HIF-1 was first identified [2] as a DNA-binding activity that was induced when cultured cells were exposed to hypoxic conditions (1% $O_2$) (reviewed in [3–5]). We have continued to utilize cultured cells to study the role of HIF-1 and have now also begun the more demanding process of analyzing HIF-1 expression in vivo.

## Identification of Target Genes for Activation by HIF-1 in Hypoxic Cells

HIF-1 is a basic helix—loop—helix transcription factor that appears to be involved in many important homeostatic responses to hypoxia. Genes whose transcription is activated by HIF-1 include the following: (a) EPO, encoding erythropoietin, the primary regulator of erythropoiesis and thus a major determinant of blood $O_2$-carrying capacity [4,6]; (b) VEGF, encoding vascular endothelial growth factor, the primary regulator of angiogenesis and thus a major determinant of tissue perfusion [7–10]; (c) ALDA, ENO1, LDHA, PFKl, and PGK1, encoding the glycolytic enzymes aldolase A, enolase 1, lactate dehydrogenase A, phosphofructokinase L, and phosphoglycerate kinase 1, respectively, which provide a metabolic pathway for ATP generation in the absence of $O_2$ [11–15]; and (d) HO1 and iNOS, encoding heme oxygenase 1 and inducible nitric oxide synthase, which are responsible for the synthesis of the vasoactive molecules carbon monoxide and nitric oxide, respectively [16,17].

## Analysis of Hypoxia Response Elements

In the case of the EPO and VEGF genes, cis-acting DNA sequences of 33 base pairs (bp) and 35 bp, respectively, have been identified that are sufficient for hypoxia-induced transcription of reporter genes and thus constitute hypoxia response elements (HREs) (reviewed in [1,4]). These HREs have in common the presence of a HIF-1 site and flanking sequences that are essential for function. Mutation of the EPO sequence 5'-CACAG-3' located downstream of the HIF-1 site resulted in a complete loss of HRE function [2]. The sequence 5'-(C/A)ACAG-3' is also present both immediately upstream and downstream of the HIF-1 site in the VEGF gene. The factor that recognizes this sequence has not been identified.

For other genes such as ENO1, HO1, LDHA, and PGK1, the minimal functional HRE sequence has not been precisely determined. However, these HREs are all characterized by the presence of two HIF-1 sites within less than 30 bp. In the case of ENO1 and LDHA, mutation of one site completely eliminated HRE function, whereas mutation of the other site severely diminished but did not eliminate the transcriptional response to hypoxia, suggesting that the two sites did not function equivalently [12,14]. In the ALDA and ENO1 genes, sequences outside the functional HRE were identified that bound HIF-1 in vitro but did not contribute to the hypoxic response, based on mutagenesis analysis in transfection assays, indicating that a HIF-1 site is necessary but not sufficient for HRE function [14]. Comparison of the ten HIF-1 binding sites identified within the HREs just described revealed an invariant core sequence 5'-CGTG-3'.

# Structural Analysis of HIF-1

Protein purification [18] and isolation of cDNA sequences [19] revealed that HIF-1 is composed of two subunits. The HIF-1α subunit is an 826-amino-acid protein whose sequence was not previously reported, and the HIF-1β subunit is the protein product of the (aryl hydrocarbon receptor nuclear translocator protein) (*ARNT*) gene, which was previously shown to encode 774- and 789-amino-acid isoforms of the aryl hydro-carbon receptor (AHR) nuclear translocator protein [20]. ARNT can therefore dimer-ize with HIF-1α in cells subjected to hypoxia to form HIF-1 and can also dimerize with AHR in cells exposed to aryl hydrocarbons such as dioxin to form the AHR complex. Additional transcription factors that may utilize ARNT as a dimerization partner include three *Drosophila* bHLH-PAS proteins: single-minded (SIM) [21], similar (SIMA) [22], and trachealess (TRH) [23,24].

The HIF-1α and HIF-1β (ARNT) proteins share in common the following structural motifs [6,25]. (a) The bHLH, or basic helix–loop–helix domain, is the hallmark of an extensive superfamily of transcription factors. The HLH domains mediate protein dimerization, which is necessary for DNA binding mediated by the basic domains. (b) Although the HLH domain is sufficient for dimerization of most bHLH proteins, HIF-1α and HIF-1β (ARNT) contain a second required dimerization domain, PAS, which was originally identified by the presence of related sequences in the *PER* (which does not contain a bHLH domain), *ARNT*, and *SIM* proteins [20]. All PAS domains contain two internal homology units of approximately 50 amino acids, the A and B repeats, each of which contains an invariant HXXD motif (H, histidine; X, any amino acid; D, aspartate) [19]. For HIF-1α and HIF-1β (ARNT), the HLH and PAS domains together create a functional interface for subunit protein–protein dimeriza-tion [6,25–27]. (c) The carboxyl half of the HIF-1α and HIF-1β (ARNT) proteins contains one or more potent transactivation domains, which are presumed to interact directly or indirectly with components of the transcription initiation complex and thus affect the rate of transcription of genes to which they have bound [6,15,25,28–30].

# Functional Analysis of HIF-1

To demonstrate directly that HIF-1 functions as a transcriptional activator, cultured cells were cotransfected with (a) reporter plasmids containing an HRE from the *ENO1*, *EPO*, or *VEGF* gene and (b) effector plasmids that allow constitutive expression of HIF-1α and HIF-1β (ARNT) from a cytomegalovirus promoter [6,9,14]. Forced expression of HIF-1 activated reporter gene transcription in nonhypoxic cells and caused superactivation in hypoxic cells. Reporter genes that contained an HRE with a mutation which disrupted HIF-1 binding did not respond to hypoxia or expression of recombinant HIF-1, demonstrating that transcriptional activation required sequence-specific binding of HIF-1. Although reporter gene transcription was activated by cotransfected HIF-1α expression vector in a dose-dependent manner, there was no effect of added HIF-1β (ARNT) expression vector, suggesting that HIF-1β (ARNT) is present in excess relative to HIF-1α in both nonhypoxic and hypoxic cells [14].

To identify HIF-1α sequences required for transactivation of reporter genes in hypoxic cells, we cotransfected a reporter plasmid, an effector plasmid encoding full-length HIF-1β (ARNT) and either full-length or deletion mutants of HIF-1α. Expression of full-length HIF-1α (aa 1–826) resulted in 7- and 29-fold higher levels of reporter gene transcription at 20% and 1% $O_2$, respectively, than in the absence of expression vectors [6]. Expression of the mutant HIF-1α (aa 1–390) resulted in only 4- and 6-fold increases over control levels at 20% and 1% $O_2$, respectively. These results indicate that the carboxyl-terminal half of HIF-1α is required for transactivation in hypoxic cells, whereas transactivation in nonhypoxic cells may be partially mediated by other sequences such as the HIF-1β (ARNT) transactivation domain. Analysis of nuclear extracts from transfected cells by gel shift and immunoblot assays revealed that the mutant HIF-1α (aa 1–390) protein was expressed at higher levels than full-length HIF-1α, indicating that the deletion specifically affected transactivation [6].

We also constructed an effector plasmid that constitutively expressed a mutant form of HIF-1α which lacked both the basic domain and transactivation domain, such that it could dimerize with HIF-1β (ARNT) but the resulting heterodimer could not bind to DNA or activate transcription. Cotransfection of this dominant-negative mutant form of HIF-1α resulted in a dose-dependent repression of reporter gene transcription in hypoxic cells, thus demonstrating that the endogenous activation of reporter genes in hypoxic cells was also mediated by HIF-1 [6,9,14].

## HIF-1 Expression as a Function of Cellular Oxygen Concentration

We performed a series of experiments in collaboration with H. Marti and C. Bauer (Institute of Physiology, Zurich) to determine the relationship between HIF-1 expression and cellular oxygen concentration. HeLa cells were incubated over a range of physiological and pathophysiological $O_2$ concentrations, in either the absence or presence of 1 mM KCN to block oxidative phosphorylation and eliminate any intracellular or extracellular $O_2$ gradients resulting from $O_2$ consumption, and nuclear extracts were prepared for gel shift and immunoblot assays. HIF-1α protein, HIF-1β (ARNT) protein, and HIF-1 DNA-binding activity all increased exponentially as cellular $O_2$ concentrations decreased from 20% to 0.5% $O_2$ both in the presence and in the absence of KCN [31]. For all three parameters, the curves showed a point of inflection at 4%–5% $O_2$; the response was half-maximal at 1.5%–2% $O_2$ and maximal at 0.5% $O_2$. The magnitude of the HIF-1α induction was much greater than that of HIF-1β, as might be expected given that HIF-1β is present in excess and that HIF-1α is the subunit specific to HIF-1. In vivo measurements have determined that $O_2$ concentrations in most tissues under normal physiological conditions are in the range of 2%–5%, indicating that any decrease in tissue oxygenation would occur along the steep portion of the HIF-1 response curve [31]. If cells in vivo respond in a manner similar to HeLa cells in culture, then the induction of HIF-1 expression could provide a means to activate homeostatic transcriptional responses that are proportional to the degree of the inciting hypoxic stimulus.

# Expression of HIF-1 In Vivo

HIF-1$\alpha$ and HIF-1$\beta$ mRNAs were detected by blot hybridization in all human tissues assayed, including brain, heart, kidney, lung, liver, pancreas, placenta, and skeletal muscle [32]. In addition, a BLAST search of the expressed sequence tag database (dbEST) identified HIF-1$\alpha$ expression in bone, fetal and adult brain, pancreatic islets, placenta, retina, uterus, and white blood cells. The ubiquitous expression of HIF-1$\alpha$ and HIF-1$\beta$ mRNA is consistent with the proposed role of HIF-1 in coordinating homeostatic responses to hypoxia throughout the body. To determine whether HIF-1 mRNA expression was induced by hypoxia in vivo, rats were exposed to 21% or 7% $O_2$ for 1 h before RNA isolation. HIF-1$\alpha$ and HIF-1$\beta$ mRNAs were detected in brain, heart, kidney, liver, lung, and spleen [32]. Modest increases in HIF-1 mRNA levels were demonstrated in brain, kidney, and lung tissue from hypoxic rats. In C57BL/6J mice, the basal levels of HIF-1$\alpha$ mRNA were lower than in rats, allowing a clearer demonstration of the induction by hypoxia in brain, kidney, and lung [32]. However, recent studies utilizing an isolated perfused and ventilated ferret lung preparation suggest that in vivo the level of HIF-1 activity is determined by precise regulation of the steady-state levels of HIF-1$\alpha$ protein as a function of oxygen tension, in agreement with the results obtained in tissue culture cells [14,31]. Taken together, there is now a considerable body of experimental evidence supporting the hypothesis that HIF-1 coordinates transcriptional responses to hypoxia that underlie essential aspects of cellular and systemic oxygen homeostasis.

# References

1. Semenza GL (1996) Transcriptional regulation by hypoxia-inducible factor 1: molecular mechanisms of oxygen homeostasis. Trends Cardiovasc Med 6:151–157
2. Semenza GL, Wang GL (1992) A nuclear factor induced by hypoxia via de novo protein synthesis binds to the human erythropoietin gene enhancer at a site required for transcriptional activation. Mol Cell Biol 12:5447–5454
3. Semenza GL (1994) Regulation of erythropoietin production: new insights into molecular mechanisms of oxygen homeostasis. Hematol Oncol Clinics N Am 8:863–884
4. Wang GL, Semenza GL (1996) Molecular basis of hypoxia-induced erythropoietin expression. Curr Opin Hematol 3:156–162
5. Wang GL, Semenza GL (1996) Oxygen sensing and response to hypoxia by mammalian cells. Redox Rep 2:89–96
6. Jiang B-H, Rue E, Wang GL, et al. (1996) Dimerization, DNA binding, and transactivation properties of hypoxia-inducible factor 1. J Biol Chem 271:17771–17778
7. Levy AP, Levy NS, Wegner S, et al. (1995) Transcriptional regulation of the rat vascular endothelial growth factor gene by hypoxia. J Biol Chem 270:13333–13340
8. Liu Y, Cox SR, Morita T, et al. (1995) Hypoxia regulates vascular endothelial growth factor gene expression in endothelial cells. Circ Res 77:638–643
9. Forsythe JA, Jiang B-H, Iyer NV, et al. (1996) Activation of vascular endothelial growth factor gene transcription by hypoxia-inducible factor 1. Mol Cell Biol 16:4604–4613
10. Ikeda E, Achen MG, Breier G, et al. (1995) Hypoxia-induced transcriptional activation and increased mRNA stability of vascular endothelial growth factor in C6 glioma cells. J Biol Chem 270:19761–19766

11. Firth JD, Ebert BL, Pugh CW, et al. (1994) Oxygen-regulated control elements in the phosphoglycerate kinase 1 and lactate dehydrogenase A genes: similarities with the erythropoietin 3' enhancer. Proc Natl Acad Sci USA 91:6496–6500

12. Firth JD, Ebert BL, Ratcliffe PJ (1995) Hypoxic regulation of lactate dehydrogenase A: interaction between hypoxia-inducible factor 1 and cAMP response elements. J Biol Chem 270:21021–21027

13. Semenza GL, Roth PH, Fang H-M, et al. (1994) Transcriptional regulation of genes encoding glycolytic enzymes by hypoxia-inducible factor 1. J Biol Chem 269:23757–23763

14. Semenza GL, Jiang B-H, Leung SW, et al. (1996) Hypoxia response elements in the aldolase A, enolase 1, and lactate dehydrogenase A gene promoters contain essential binding sites for hypoxia-inducible factor 1. J Biol Chem 271:32529–32537

15. Li H, Ko HP, Whitlock JP Jr (1996) Induction of phosphoglycerate kinase gene expression by hypoxia: roles of ARNT and HIF1α. J Biol Chem 271:21262–21267

16. Lee PJ, Jiang B-H, Chin BY, et al. (1997) Hypoxia-inducible factor 1 mediates transcriptional activation of the heme oxygenase-1 gene in response to hypoxia. J Biol Chem 272:5375–5381

17. Melillo G, Musso T, Sica A, et al. (1995) A hypoxia-responsive element mediates a novel pathway of activation of the inducible nitric oxide synthase promoter. J Exp Med 182:1683–1693

18. Wang GL, Semenza GL (1995) Purification and characterization of hypoxia-inducible factor 1. J Biol Chem 270:1230–1237

19. Wang GL, Jiang B-H, Rue EA, et al. (1995) Hypoxia-inducible factor 1 is a basic-helix-loop-helix-PAS heterodimer regulated by cellular $O_2$ tension. Proc Natl Acad Sci USA 90:5510–5514

20. Hoffman EC, Reyes H, Chu F-F, et al. (1991) Cloning of a factor required for activity of the Ah (dioxin) receptor. Science 252:954–958

21. Nambu JR, Lewis JO, Wharton KA, et al. (1991) The *Drosophila* single-minded gene encodes a helix-loop-helix protein that acts as a master regulator of CNS midline development. Cell 67:1157–1167

22. Nambu JR, Chen W, Hu S, et al. (1996) The *Drosophila melanogaster*-similar bHLH-PAS gene encodes a protein related to human hypoxia-inducible factor 1α and *Drosophila* single-minded. Gene 172:249–254

23. Isaac DD, Andrew DJ (1996) Tubulogenesis in *Drosophila*: a requirement for the trachealess gene product. Genes Dev 10:103–117

24. Wilk R, Weizman I, Shilo B-Z (1996) *trachealess* encodes a bHLH-PAS protein that is an inducer of tracheal cell fates in *Drosophila*. Genes Dev 10:93–102

25. Reisz-Porszasz S, Probst MR, Fukunaga BN, et al. (1994) Identification of functional domains of the aryl hydrocarbon receptor nuclear translocator protein (ARNT). Mol Cell Biol 14:6075–6086

26. Lindebro MC, Poellinger L, Whitelaw ML (1995) Protein-protein interaction via PAS domains: role of the PAS domain in positive and negative regulation of the bHLH/PAS dioxin receptor-Arnt transcription factor complex. EMBO J 14:3528–3539

27. Wood SM, Gleadle JM, Pugh CW, et al. (1996) The role of the aryl hydrocarbon receptor nuclear translocator (ARNT) in hypoxic induction of gene expression: studies in ARNT-deficient cells. J Biol Chem 271:15117–15123

28. Whitelaw ML, Gustafsson J-A, Poellinger L (1994) Identification of transactivation and repression functions of the dioxin receptor and its basic helix-loop-helix/PAS partner factor Arnt: inducible versus constitutive modes of regulation. Mol Cell Biol 14:8343–8355

29. Li H, Dong L, Whitlock JP Jr (1994) Transcriptional activation function of the mouse Ah receptor nuclear translocator. J Biol Chem 269:28098–28105

30. Rowlands JC, McEwan IJ, Gustafsson J-A (1996) *trans*-activation by the human aryl hydrocarbon receptor and aryl hydrocarbon receptor nuclear translocator proteins: direct interactions with basal transcription factors. Mol Pharmacol 50:538–548
31. Jiang B-H, Semenza GL, Bauer C, et al. (1996) Hypoxia-inducible factor 1 levels vary exponentially over a physiologically relevant range of $O_2$ tension. Am J Physiol 271:C1172–C1180
32. Wiener CM, Booth G, Semenza GL (1996) In vivo expression of mRNAs encoding hypoxia-inducible factor 1. Biochem Biophys Res Commun 225:485–488

# Differences in Particle Size and Oxygen-Binding Affinity Between Cross-Linked Hemoglobin and Hemoglobin Vesicle

SHINJI TAKEOKA, YUICHI MANO, and EISHUN TSUCHIDA

*Summary.* Modified hemoglobins (Hb) such as intramolecular cross-linked Hb are currently undergoing clinical tests. On the other hand, a Hb vesicle that encapsulates a purified and concentrated Hb solution with the bilayer membrane of phospholipids is expected to overcome the problems of the modified Hb. The differences in size and oxygen-binding affinity of the Hb vesicles and intramolecular cross-linked Hb were compared to discuss the difference in the cellular and acellular types of Hb-based oxygen carriers, which have oxygen-binding and dissociation equilibrium curves similar to those of red blood cells (RBCs). The particle sizes of the cross-linked Hb, the Hb vesicle, and RBC are 5, 250, and 8000 nm, respectively. The Hb solution shows the lowest viscosity, whereas the viscosity of the Hb vesicle is relatively high and similar to the RBC. The permeability of the membrane filters depends on the particle size: cross-linked Hb > Hb vesicles > RBC. The oxygen-binding and -releasing rates of the Hb vesicles are between those of the acellular Hb and RBC, which could be explained in terms of the total surface area of Hb and the oxygen diffusion in the concentrated Hb solution (36 g/dl) inside the Hb vesicle.

*Key words.* Hemoglobin vesicle—Cross-linked hemoglobin—Particle size—Oxygen-binding affinity—Oxygen carriers

## Introduction

Various efforts have been made to develop red cell substitutes, especially those utilizing hemoglobin (Hb), to overcome the problems associated with blood transfusions such as the necessity for blood typing and cross-matching, blood-borne infections, and difficulties of storage [1]. Acellular Hb solutions such as chemically modified Hb (intermolecular cross-linked Hb, intramolecular cross-linked Hb, polymer-conjugated Hb) are currently undergoing clinical tests. However, some acellular Hb issues include vasoconstriction, toxicity enhancement complexed with endotoxin, and autoregulation based on the differences in oxygen transport and solution proper-

Department of Polymer Chemistry, Advanced Research Institute of Science and Engineering, Waseda University, 3-4-1 Ohkubo, Shinjuku-ku, Tokyo, 169 0072, Japan

ties [2]. On the other hand, liposome-encapsulated hemoglobin or hemoglobin vesicles (Hb vesicles), which have a cellular structure composed of a phospholipid bilayer and encapsulate a concentrated Hb solution (~38 g/dl) is considered to be one candidate for a red blood cell substitute [3,4]. An Hb vesicle with excellent physiochemical properties was prepared [5,6] and a 90% blood exchange transfusion with the Hb vesicles was carried out to evaluate their efficacy [7]. Hb vesicles are expected to overcome such issues related to cross-linked Hb. In this chapter, the oxygen transport and physical properties of Hb vesicles, cross-linked Hb, and RBCs are compared to discuss their physiological influences.

## Materials and Methods

### Purification of Hemoglobin and Synthesis of Intramolecular Cross-Linked Hb

Hemoglobin was easily isolated from outdated red blood cells (Hokkaido Red Cross Blood Center, Sapporo, Japan) using our original processes [8]. The profile of the obtained HbCO solution is as follows: [Hb], 40 g/dl; [metHb], <1%; protein purity, 99.95%; residual phospholipids, <0.2%; residual organic solvent, <0.1 ppm.

The reaction of bis(3,5-dibromosalicyl)fumarate with Lys-$\alpha_1$99 and Lys-$\alpha_2$99 of deoxyhemoglobin was carried out in 100 mM HEPES buffer (pH 7.3) for 2 h at 37°C [9]. The reaction mixture was heated for 1.5 h at 75°C to denature the non-cross-linked Hb. The precipitate was removed by centrifugation (1900 × $g$, 20 min), and the supernatant was purified by filtration, dialysis, and then ultrafiltration.

### Preparation of Hb Vesicles

The Hb vesicle was prepared as previously reported [10]. The lipid bilayer was composed of Presome PPG-I [a mixture of 1,2-dipalmitoyl-$sn$-glycero-3-phosphatidylcholine (DPPC), cholesterol, 1,2-dipalmitoyl-$sn$-glycero-3-phospatidylglycerol (DPPG) (Nippon Fine Chemicals, Osaka, Japan), and $\alpha$-tocopherol (Merck, Darmstadt, Germany) at a molar composition of 5:5:1:0.1)]. To the outer surface of the Hb vesicle was introduced 0.3 mol% of PEG-DPPE, of which the PEG (polyethylene glycol) weight was about 5000. The Hb vesicle encapsulated ~38 g/dl of Hb containing pyridoxal 5′-phosphate (18 mM) as an allosteric effector and homocystine (5 mM) as a reductant. The average diameter of the Hb vesicles was 251 ± 76 nm, and the Hb concentration was 10 g/dl.

### Stopped-Flow Measurements

A 150-W xenon lamp was used as a monitor light source equipped with a cut-off filter (HOYA L-39, HOYA, Tokyo, Japan) to eliminate light in the ultraviolet region. To measure the oxygen-binding rate constant [$k_{on}(O_2)$] [11], equal volumes ($1 \times 10^{-4}$ l) of deoxy-form oxygen carriers ([heme]: $2 \times 10^{-5}$ mol l$^{-1}$) and buffered solution equilibrated with various $P(O_2)$s were rapidly mixed in a mixing cell (Unisoku TSP-601, Unisoku, Osaka, Japan). For measurement of the oxygen dissociation rate constant [$k_{off}(O_2)$], equal volumes ($1 \times 10^{-4}$ l) of oxy-form oxygen carriers and a buffered solution containing sodium dithionite in an nitrogen atmosphere ($8.0 \times 10^{-2}$ mol$\,$l$^{-1}$) were rapidly mixed in a mixing cell at $37 \pm 0.2$°C. The reactions were monitored at 415 nm (oxy Hb) and 430 nm (deoxy Hb).

## Results and Discussion

### Characterization of Intramolecular Cross-Linked Hb

Intramolecular cross-linking was confirmed by electrophoresis (sodium dodecyl phosphate-polyacrylamide gel electrophoresis, SDS-PAGE). Tetrameric Hb without cross-linking dissociated into monomer units after treatment with sodium dodecyl sulfate. The band was confirmed around 16 100. On the other hand, the $\alpha,\alpha$-cross-linked Hb showed two bands attributed to the $\alpha$–$\alpha$ dimer and $\beta$ monomer, both of which had the same band intensity. The UV-visible spectrum of the cross-linked Hb was exactly the same as that of the stroma-free Hb, and there was no sign of metHb formation during the modification of Hb. From differential scanning calorimetry, the endothermic peak of the stroma-free Hb was observed at 62.2°C, corresponding to the denaturation of Hb by heating. On the other hand, intramolecularly cross-linked Hb was observed at 76.1°C.

Figure 1 shows the oxygen-binding and dissociation equilibrium curves of these oxygen carriers. The degree of oxygen affinity of the stroma-free Hb was 9 Torr but became 32 Torr after being cross-linked. The oxygen affinity of the Hb vesicles was controlled at 32 Torr by coencapsulating pyridoxal 5′-phosphate with threefold Hb, which was lower than that (27 Torr) of RBCs for more effective oxygen transport.

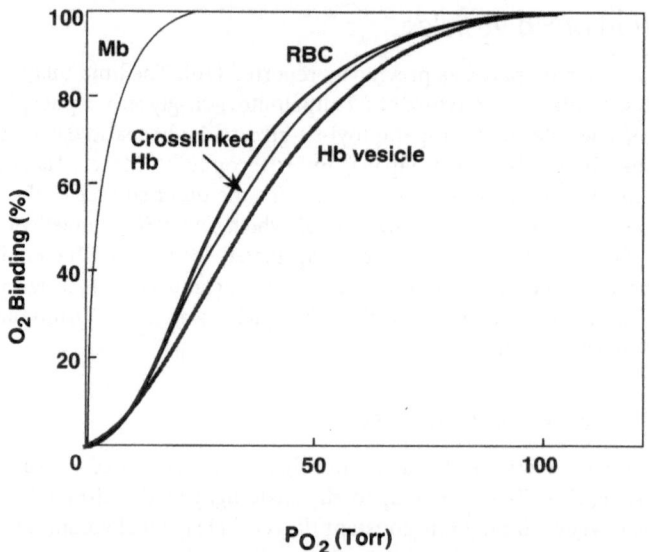

FIG. 1. Oxygen-binding equilibrium curves of oxygen carriers at 37°C. Hemoglobin (*Hb*) vesicle; cross-linked Hb; red blood cell (*RBC*); myoglobin (*Mb*)

However, the appropriate oxygen affinity of the oxygen carriers in vivo has not yet been determined, especially during critical care use.

## Solution Properties of Hb-Based Oxygen Carriers

Intramolecular cross-linked Hb is the same size as Hb (~5 nm). The size of the Hb vesicles was determined from the final pore size of the membrane filter during the extrusion procedure. In this experiment, the size of the vesicles is controlled to 250 nm in diameter (the diameter of the RBCs is known to be 8 μm). As shown in Fig. 2, such a difference was clearly reflected on the filter permeability of the isopore membrane filters. The RBC can penetrate the membrane filter up to a pore size of 5 μm. Hb vesicles can penetrate pores of 0.2 μm, whereas cross-linked Hb can penetrate pores even as small as at 0.05 μm in diameter.

The Hb vesicles tend to aggregate in the albumin solution, resulting in high viscosity (8.9 cP). This is prevented by modification with polyoxyethylene chains. The solution viscosity of the modified Hb vesicle (3.6 cP) is as low as that of RBC (4.0 cP), while that of the observed Hb solution is low (1.6 cP) because of the small size; however, it increases with increasing Hb concentration, especially greater than 30 g/dl (>10 cP).

## Kinetics of Oxygen Binding

To elucidate the oxygen-binding properties of the oxygen carriers, the dynamics of binding were studied using stopped-flow measurements. The oxygen-binding rate constants of the Hb vesicle and RBC were $2.8 \times 10^5$ and $7.5 \times 10^4 \, M^{-1} s^{-1}$, respectively.

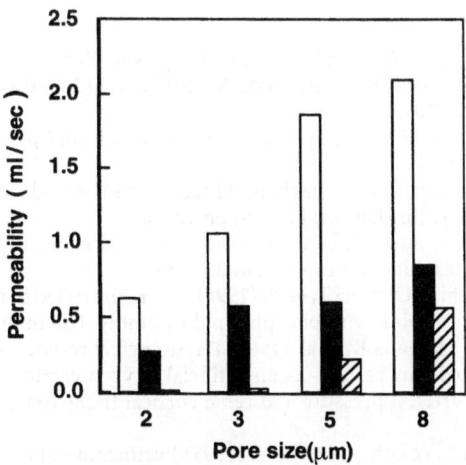

FIG. 2. Permeability through isopore membrane filters at the applied pressure of 0.3 kg/cm² at 37°C. Cross-linked hemoglobin (*Hb*), *white areas*; Hb vesicle, *black areas*; red blood cell (*RBC*), *striped areas*; Pore size, 2–8 μm; membrane diameter, 25 mm

These were significantly less than that of cross-linked Hb ($4.5 \times 10^6 M^{-1} s^{-1}$). The Hb concentration in the inner aqueous phases of the encapsulated Hb is 36 g/dl, which is $10^3$ times higher when compared with that of the modified Hb solution used in the experiment. Coin and Olson [11] showed that the cellular reaction was almost entirely limited by the velocity of the oxygen diffusion. Furthermore, because oxygen transfer of the Hb vesicle and RBC occurs through the membrane shell, the total surface area is the other important parameter used to determine the oxygen-binding property. It is reasonable that the Hb vesicles have an oxygen-binding rate 10 fold faster than RBC because of a lower total surface area (23 fold).

## Conclusion

The properties of the Hb vesicle and cross-linked Hb were compared with Hb and RBC. All of these have similar oxygen-transporting affinities, comparable to RBC; however, the solution properties such as viscosity and filter permeability and the kinetic parameters of oxygen binding are quite different because the particle diameters of the cross-linked Hb, Hb vesicle, and RBC are 5, 200, and 8000 nm, respectively. The high binding rate of oxygen and the high permeability of cross-linked Hb compared with the Hb vesicle and RBC would be reasons for the autoregulation [12] or vasoconstriction [13] caused by cross-linked Hb.

*Acknowledgment.* This work was supported in part by grants-in-aid from the Ministry of Education, Science, Sports and Culture of Japan (no. 07508005 and 08680940) and the Kawakami Memorial Foundation.

## References

1. Tsuchida E (ed) (1995) Artificial red cells. Wiley, New York
2. Winslow RM, Vandegriff KD, Intaglietta M (eds) (1995) Blood substitutes: new challenges. Birkhauser, Boston
3. Djordjevich L, Miller IF (1980) Synthetic erythrocytes from lipid encapsulated hemoglobin. Exp Hematol 8:584–592
4. Winslow RM, Vandegriff KD, Intaglietta M (eds) (1995) Blood substitutes: physiological basis of efficacy. Birkhauser, Boston, pp 90–104
5. Sakai H, Takeoka S, Park SI, et al. (1997) Rheological behavior of surface-modified hemoglobin vesicles. Bioconjugate Chem 8:23–30
6. Takeoka S, Ohgushi T, Ohmori T, et al. (1996) Layer controlled hemoglobin vesicles by interaction of hemoglobin with phospholipid assembly. Langmuir 12:1775–1779
7. Izumi Y, Sakai H, Hamada K, et al. (1996) Physiological responses to exchange transfusion with hemoglobin vesicles as an artificial oxygen carrier in anesthetized rats: changes in mean arterial pressure and renal cortical tissue oxygen tension. Crit Care Med 24:1869–1873
8. Sakai H, Takeoka S, Yokohama H, et al. (1993) Purification of concentrated hemoglobin using organic solvent and heat treatment. Protein Expr Purif 4:563–569
9. Winslow RM, Chapman KW (1994) Methods in enzymology, vol 231, Hemoglobin, part B. Academic Press, New York, pp 3–16
10. Sakai H, Hamada K, Takeoka S, et al. (1996) Physical properties of hemoglobin vesicles as red cell substitutes. Biotechnol Prog 11:119–125

11. Coin JT, Olson JC (1979) The rate of oxygen uptake by human red blood cells. J Biol Chem 254:1178–1190
12. Winslow RM, Vandegriff KD, Intaglietta M (eds) (1995) Blood substitutes: physiological basis of efficacy. Birkhauser, Boston, pp 143–154
13. Farmer MC, Przybelski RJ, Mckenzie JE, Burhop KE (1995) Preclinical data and clinical trials with diaspirin cross-linked hemoglobin. In: Tsuchida E (ed) Artificial red cells. Wiley, Chichester, pp 177–185

# Activation of Gene Expression of Collagenase and ICAM-1 by UVA Radiation and by Exposure to Singlet Oxygen

Karlis Briviba[1], Meinhard Wlaschek[2], Karin Scharffetter-Kochanek[2], Susanne Grether-Beck[3], Jean Krutmann[3], and Helmut Sies[1]

*Summary.* UVA irradiation alone or in conjunction with photosensitizers substantially modulates gene expression. The biological effect of UVA irradiation can be mediated by electron transfer or by a pathway involving singlet oxygen ($^1O_2$). UVA irradiation or $^1O_2$, generated in a dark reaction by thermal decomposition of the endoperoxide of 3,3'-(1,4-naphthylidene) dipropionate, induces gene expression of collagenase (MMP-1) in human skin fibroblasts and of the intercellular adhesion molecule-1 (ICAM-1) in human keratinocytes, respectively. Further supporting evidence for the role of singlet oxygen in the UVA induction of collagenase and ICAM-1 was obtained by examining the modulation of the effects by employing deuterium oxide as a singlet oxygen enhancer or sodium azide as a quencher of singlet oxygen. Using ICAM-1 based luciferase reporter gene constructs, it was shown that ICAM-1 promoter activation induced by UVA radiation as well as by singlet oxygen generated from the endoperoxide of 1,4-naphthylidene dipropionate required the activation of the transcription factor AP-2.

*Key words.* UVA irradiation—Singlet oxygen—Collagenase—ICAM-1—AP-2

## Introduction

Ultraviolet A (UVA) (320–400 nm) is the major component of the UV solar spectrum that reaches the earth. UVA irradiation of skin causes severe damage of connective tissue in several photodermatological disorders including induced phototoxicity, porphyria, and photoaging. Exposure of human cells to UVA radiation modulates gene expression. However, the mechanism by which UVA radiation induces transcriptional activation is not yet known. In contrast to UVB (280–320 nm), which can be ab-

[1] Institut für Physiologische Chemie I, Heinrich-Heine-Universität Düsseldorf, Postfach 10 10 07, D-40001 Düsseldorf, Germany
[2] Department of Dermatology, University of Cologne, Joseph-Stelzmann-Str. 9, D-50931 Cologne, Germany
[3] Clinical and Experimental Photodermatology, Department of Dermatology, Heinrich-Heine-Universität Düsseldorf, Postfach 10 10 07, D-40001 Düsseldorf, Germany

sorbed by DNA or proteins and resembles UVC as a DNA-damaging agent, UVA radiation is weakly absorbed by most biomolecules but can initiate the generation of reactive oxygen species via a variety of chromophores such as porphyrins or flavins [1].

Absorption of UVA light can lead to the formation of triplet-excited states of endogenous or exogenous photosensitizers, which can transfer an electron to a substrate leading to the formation of a substrate radical (type I reaction) or, alternatively, the energy is transferred to oxygen yielding singlet molecular oxygen, $^1O_2$ (type II reaction) (Fig. 1). Evidence for the involvement of singlet oxygen in UVA-mediated cytotoxicity [2] and expression of heme oxygenase [3] was obtained using the enhancing effect of deuterium oxide and the quenching effect of sodium azide.

We present here our own recent work on the role of singlet oxygen in UVA radiation-induced expression of collagenase, MMP-1, in human fibroblasts and of the intercellular adhesion molecule-1, ICAM-1, in human keratinocytes.

## Generation of Singlet Oxygen In Vitro

Singlet oxygen can be produced by photoexcitation, for example, by irradiation of photosensitizers like rose bengal or methylene blue with visible light in the presence of molecular oxygen (type II reaction). However, the formation of type I reactants such as hydroxyl or superoxide anion radicals can also occur in such systems. The generation of $^1O_2$ by chemiexcitation, that is, formation of singlet oxygen without light, for example by thermodecomposition of the endoperoxide of (1,4-naphthylidene) dipropionate (NDPO$_2$) as shown in the following reaction [4], is more selective because type I reactions do not occur.

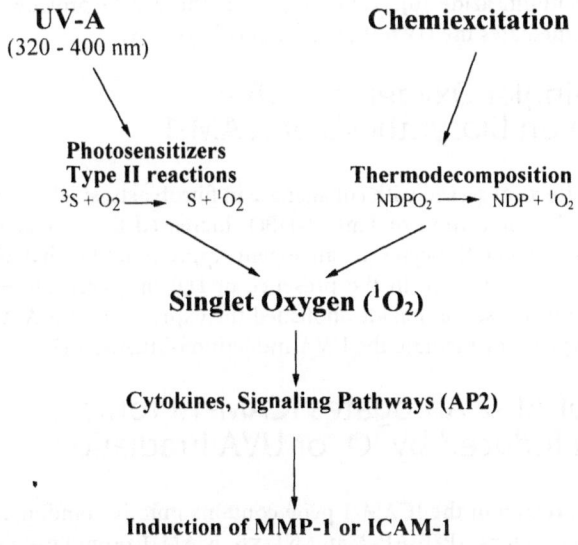

FIG. 1. Scheme of UVA radiation-induced gene expression

$$2NDPO_2 \xrightarrow{37°C} 2NDP + {}^1O_2 + O_2$$

NDPO$_2$ as well as NDP, the stable endproduct formed during this reaction, are non-toxic as tested up to 40 mM in various cell culture systems. For induction of collagenase or ICAM-1, up to 3 mM NDPO$_2$ was used, which generates about 9 μM ${}^1O_2$/min, and the calculated steady-state concentration of singlet oxygen in H$_2$O-based phosphate-buffered saline (PBS) is about $10^{-12}$ M.

## Effect of Singlet Oxygen and UVA Irradiation on Collagenase Expression

Exposure of human skin fibroblasts to ${}^1O_2$ generated by thermodecomposition of NDPO$_2$ for 30 min induced collagenase mRNA levels in a dose-dependent manner (0.1 mM up to 3 mM NDPO$_2$) [5]. There was no induction of collagenase by NDP. The effect on collagenase expression with 3 mM NDPO$_2$ was equivalent to that observed with 200–300 kJ/m$^2$ UVA, a dose that is acquired easily during 3 h of sun exposure in June at latitude 40°N (Rome). In contrast to collagenase MMP-1, the mRNA level of the tissue inhibitor of matrix-metalloproteinases, TIMP-1, remained unchanged [5,6].

To assess the role of singlet oxygen in the induction of genes upon exposure to UVA, the effects of deuterium oxide, an enhancer of the lifetime of singlet oxygen, and of sodium azide, an effective quencher of singlet oxygen, are widely used. When fibroblasts were exposed to UVA in the presence of 95% D$_2$O, there was an increase in steady-state levels of collagenase mRNA compared to that in H$_2$O, while sodium azide abolished the UVA-induced increase in the collagenase mRNA level [6].

An induction of biosynthesis and secretion of collagenase protein of as much as twofold was found after exposure of cells to NDPO$_2$ or to UVA irradiation compared to the nontreated controls. The induction was more pronounced in D$_2$O and was decreased by sodium azide [6]. Taken together, these data provide evidence that singlet oxygen mediates the UVA induction of collagenase.

## Effect of Singlet Oxygen and UVA Irradiation on Biosynthesis of ICAM-1

Similar to the effects observed with collagenase in fibroblasts, singlet oxygen generated by thermal decomposition of 1 mM NDPO$_2$ increased the surface expression of ICAM-1 in human keratinocytes to an extent approximating that obtained with 300 kJ/m$^2$ UVA [7]. Exposure in the presence of D$_2$0 increased the effect of UVA radiation. In contrast, sodium azide abolished the expression of ICAM-1, indicating that singlet oxygen may mediate the UVA induction of ICAM-1 [7].

## Deletion of AP-2 Abrogates ICAM-1 Promoter Activation Induced by ${}^1O_2$ or UVA Irradiation

The 5′-flanking region of the ICAM-1 gene contains putative binding sites for transcription factors such as AP-1, AP-2, and NF-κB. ICAM-1 promoter-based luciferase reporter gene constructs with or without consensus binding sequences for these transcriptions factors were transfected into 293 cells. The deletion of the putative AP-2 binding site in the pIC277 construct resulted in the loss of the UVA- or $^1O_2$-induced, but not the UVB-induced, ICAM-1 promoter activation. In contrast, deletion of the NF-κB or putative AP-1 sites affected neither UVA nor UVB radiation-induced ICAM-1 promoter activation [7]. Thus, AP-2 is identified as the singlet oxygen-responsive element of the human ICAM-1 gene.

## Conclusions

1. Singlet oxygen induces expression of collagenase in human skin fibroblasts and ICAM-1 in human skin keratinocytes.
2. Singlet oxygen may mediate the UVA radiation-induced expression of collagenase and ICAM-1.
3. ICAM-1 promoter activation induced by UVA radiation as well as by singlet oxygen generated from NDPO$_2$ required the activation of the transcription factor AP-2.

*Acknowledgments.* The work reported here was supported by the Deutsche Forschungsgemeinschaft, projects Scha 411/2-2 and Kr 558/8-1, and by the Sonderforschungsbereich 503, projects B1 and B2, as well as by the National Foundation for Cancer Research, Bethesda, MD, U.S.A.

## References

1. Tyrrell RM (1991) UVA (320–380 nm) radiation as an oxidative stress. In: Sies H (ed) Oxidative stress: oxidants and antioxidants. Academic Press, London, pp 57–83
2. Tyrrell RM, Pidoux M (1989) Singlet oxygen involvement in the inactivation of cultured human fibroblasts by UVA (334 nm, 365 nm) and near visible (405 nm) radiations. Photochem Photobiol 49:407–412
3. Basu-Modak S, Tyrrell RM (1993) Singlet oxygen: primary effector in the ultraviolet A/near visible light induction of the human heme oxygenase. Cancer Res 53:4505–4510
4. Di Mascio P, Sies H (1989) Quantification of singlet oxygen generated by thermolysis of 3,3′-(1,4-naphthylidene) dipropionate. Monomol and dimol photoemission and the effects of 1,4-diazabicyclo [2.2.2] octane. J Am Chem Soc 111:2909–2914
5. Scharffetter-Kochanek K, Wlaschek M, Briviba K, Sies H (1993) Singlet oxygen induces collagenase expression in human skin fibroblasts. FEBS Lett 331:304–306
6. Wlaschek M, Briviba K, Stricklin GP, et al. (1995) Singlet oxygen may mediate the ultraviolet A-induced synthesis of interstitial collagenase. J Invest Dermatol 104:194–198
7. Grether-Beck S, Olaizola-Horn S, Schmitt H, et al. (1996) Activation of transcription factor AP2 mediates ultraviolet-A radiation- and singlet oxygen-induced expression of the human ICAM-1 gene. Proc Natl Acad Sci USA 93:14586–14591

# Redox Regulation of the Nuclear Factor Kappa B (NF-κB) Signaling Pathway and Disease Control

Takashi Okamoto, Shinsaku Sakurada, Yang Jian-Ping, and Naoko Takahashi

*Summary.* Nuclear factor kappa B (NF-κB) is an inducible cellular transcription factor that activates various cellular and viral genes. NF-κB usually exists as a molecular complex with an inhibitory molecule, IκB, in the cytosol. On stimulation of the cells, such as by proinflammatory cytokines IL-1 and tumor necrosis factor (TNF), IκB is dissociated and NF-κB is translocated to the nucleus and activates the expression of the target genes. We found that a redox control mechanism is involved in the DNA-binding activity of NF-κB and that a cellular-reducing catalyst thioredoxin (Trx), together with kinases, is primarily involved as an effector molecule in this signaling pathway. Trx was recently demonstrated to associate with the redox-sensitive cysteine within the DNA-binding loop of NF-κB. Effects of antioxidants in blocking NF-κB activation can be explained by the involvement of radical oxygen intermediates (ROI) in this pathway. These findings support the idea that redox regulation involving ROI and Trx plays a crucial role in the signal transduction pathway leading to NF-κB activation, thus contributing substantially to understanding of the pathogenetic processes of various diseases including AIDS, hematogenic cancer cell metastasis, and rheumatoid arthritis (RA).

*Key words.* Thioredoxin—Radical oxygen intermediates—AIDS—Cancer—Rheumatoid arthritis—Transcription—Signal transduction

## Introduction

Oxygen is an essential molecule for all aerobic life forms including humans. Oxygen appears to play contradictory roles in the biological sphere of this planet: while oxygen is indispensable for the cell to obtain essential chemical energy as a form of ATP, it is often transformed into highly reactive forms, radical oxygen intermediates (ROI), which are often toxic to the cell. During the long period of evolution cells have acquired a multiplicated endogenous antioxidant system that has been a prerequisite for the maintenance of a stable form of life. These defense mechanisms include reducing enzyme systems such as glutaredoxin and thioredoxin (Trx) [1,2]. However,

Department of Molecular Genetics, Nagoya City University Medical School, 1 Kawasumi, Mizuho-cho, Mizuho-ku, Nagoya 467, Japan

438

previous studies have revealed that these reducing enzymes together with ROI are involved in cell signaling [1–4]. The term redox regulation has thus been proposed to indicate the active role of oxidoreductive modifications of proteins in regulating their activities. Oxidation and reduction of biomolecules are now considered to be "signals" in certain instances and are utilized for the maintenance of cellular homeostasis. This chapter describes the nature of redox regulation of one of the cellular transcription factors, nuclear factor kappa B (NF-κB) [5–9]. The possible pathophysiological as well as therapeutic implications are discussed.

## Transcription Factor NF-κB and Its Involvement in Various Pathological Conditions

NF-κB regulates expression of a wide variety of cellular and viral genes (see [5–8], for reviews). These genes include cytokines such as interleukins IL-2, IL-6, and IL-8, granulocyte-macrophage-colony-stimulating factor (GM-CSF) and tumor necrosis factor (TNF), cell adhesion molecules such as ICAM-1 and E-selectin, inducible nitric oxidase synthase (iNOS), and viruses such as human immunodeficiency virus (HIV) and cytomegalovirus [10–20]. Through the causal relationship with these genes, NF-κB is considered to be causally involved in currently intractable diseases such as acquired immunodeficiency syndrome (AIDS), hematogenic cancer cell metastasis, and rheumatoid arthritis (RA) (Fig. 1). Although the genes induced by NF-κB are variable according to the context of cell lineage and are also under the control of other

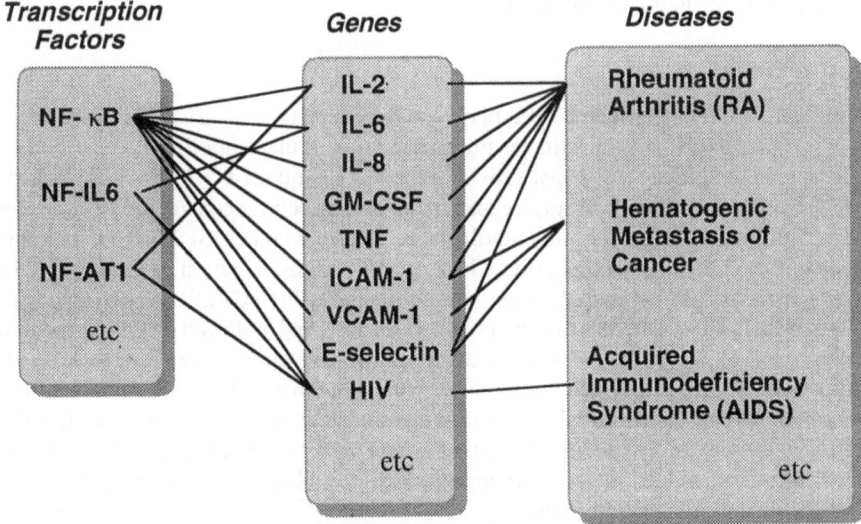

FIG. 1. Relationships between transcription factors and diseases: a pathogenetic paradigm regarding nuclear factor kappa B (NF-κB). Transcription factors are causally associated with various genes that are known to be involved in pathogenetic processes of the relevant diseases. This diagram shows cross-correlations between transcription factors and diseases with attention to the transcription factor NF-κB; indicating that some disease processes may be controlled by modifying the actions of transcription factors

transcription factors, NF-κB plays a major role in regulation of expression of these genes and thus contributes markedly to the pathogenesis. Therefore, biochemical intervention of NF-κB conceivably interferes with the pathogenic process and would be effective for treatment.

## AIDS

The pivotal role of NF-κB in the HIV life cycle, especially in virus reactivation within latently infected cells, has been widely accepted. After activation through intracellular signaling pathways such as those elicited by T-cell receptor–antigen complex or by receptors for IL-1 or TNF, NF-κB initiates HIV gene expression by binding to the target DNA element within the promoter region of HIV LTR [11,20–22]. Then, the virus-encoded *trans*-activator Tat is produced and triggers explosive viral replication [23–25]. Because activation of HIV gene expression by cellular transcription factor NF-κB conceptually precedes the production of Tat, NF-κB can be ascribed as being a critical determinant between the maintenance and breakdown of viral latency. Various attempts have been carried out to control the signaling pathways to NF-κB activation in view of therapeutic intervention of AIDS pathology. For example, the antioxidant compounds known to block NF-κB activation cascade have been suggested to be effective in preventing the clinical development of AIDS by blocking HIV replication [26–28]. As viral transcription is the only step in the amplification of viral genetic information, a clinically relevant anti-NF-κB compound should be effective in blocking viral replication from the latent status and in substantially decreasing the viral load in virus-infected individuals.

## Cancer Metastasis

Another situation in which NF-κB plays a role is hematogenic cancer cell metastasis [29]. NF-κB induces E-selectin on the surface of vascular endothelial cells [30,31]. Because some cancer cells constitutively express a ligand for E-selectin, called sialyl-Lewis$^x$ antigen, on their cell surface, induction of E-selectin is considered to be a rate-determining step of cancer cell–endothelial cell interaction [32,33]. We examined this phenomenon in detail with regard to the role of NF-κB in induction of E-selectin [29]. When primary human umbilical venous endothelial cells (HUVEC) were treated with IL-1 or TNF, nuclear translocation of NF-κB was observed, followed by the augmented expression of E-selectin. We examined the cell-to-cell interaction between HUVEC and QG90 cell, a tumor cell line derived from human small cell carcinoma of the lung expressing sialyl-Lewis$^x$ antigen, and found that IL-1 was able to induce the attachment of cancer cells to HUVEC. However, pretreatment of HUVEC with *N*-acetylcysteine, aspirin, or pentoxyphillin efficiently blocked the cell-to-cell attachment in a dose-dependent manner.

## Rheumatoid Arthritis

Similarly, recent evidence has indicated the involvement of NF-κB in the pathogenesis of RA [34–36]. In fact, Handel et al. [37] have demonstrated the presence of the NF-κB subunit proteins, p65 and p50, in the nuclei of the lining cells of fresh synovial tissue obtained from patients with RA, indicating activation of NF-κB in situ. Because of its regulatory role in gene expression of IL-1, TNF, IL-2, IL-6, IL-8, GM-CSF, and chemokines such as RANTES and MIP-1 α, ICAM-1, E-selectin and iNOS, which are known to be overexpressed in the rheumatoid synovium, NF-κB is considered to be a major transcriptional regulator in the expansion and maintenance of chronic inflammatory response in the affected joints. For example, sustained NF-κB activation would induce production of cytokines and thus activate maturation of B lymphocytes to produce antibodies, while GM-CSF and chemokine production together with over-expression of cell adhesion molecules would support recruitment of leukocytes from the blood, thus augmenting the local inflammatory response. Additionally, some of the effective antirheumatic drugs, including corticosteroids, aspirin, and gold compounds, are now known to block the NF-κB cascade. Because these features of inflammatory responses are not specific for RA and are found in other chronic inflammatory processes irrespective of the affected organs or tissues, the pathological roles of NF-κB are probably universal. Therefore, clinical applicability of anti-rheumatic drugs should be evaluated in other diseases in which NF-κB plays a role.

## Signal Transduction Pathway to Activate NF-κB

NF-κB consists of two subunit molecules, p65 and p50, and usually exists as a molecular complex with an inhibitory molecule, IκB, in the cytosol [19,20,38–42]. On stimulation of the cells, such as that by proinflammatory cytokines, IL-1, and TNF, IκB is

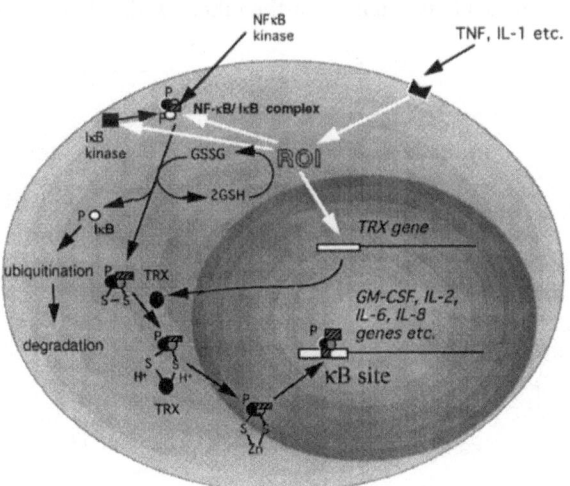

FIG. 2. Signal transduction pathways for NF-κB activation. The first step involves kinase pathways such as by NF-κB and IκB kinases. The second step involves "redox regulation" by thioredoxin (*TRX*). After stimulation of the cells by tumor necrosis factor (*TNF*) or IL-1, for example, radical oxygen intermediates (*ROI*) are produced. ROI induce not only activation of kinase cascade but also production of thioredoxin (Trx). TNF receptor-associated factor (TRAF) is known to be associated with the TNF receptor and is known to stimulate NF-κB activation. Similarly, IRAK (IL-1 receptor-associated kinase) is considered to participate in the NF-κB activation through its ability to phosphorylate IκB. Phosphorylation of NF-κB or IκB will release NF-κB. However, NF-κB must go through Trx-mediated reduction of the "redox-sensitive" cysteine to recognize the target DNA sequence (*κB site*). The phosphorylated IκB will be ubiquitinated and then degraded by proteasome or other proteases

dissociated and NF-κB is translocated to the nucleus, activating expression of the target genes (Fig. 2). Thus, activity of NF-κB itself is regulated by the upstream regulatory mechanism. Not much is known about what happens immediately downstream of the cell-surface receptor or about what triggers these NF-κB signaling cascades. The TNF receptor-associated factors (TRAFs) have been identified and shown to be involved in the TNF-mediated NF-κB activation [43,44].

Moreover, involvement of ROI is suggested at least in one of the upstream steps of the NF-κB activation pathway because the signaling was efficiently blocked by pretreatment of the cells with antioxidants such as $N$-acetyl-L-cysteine (NAC) or α-lipoic acid [26,27,28,45–49]. Therefore, antioxidants are now considered to be effective NF-κB inhibitors. Moreover, we found that NAC could also block the induction of Trx [50]. Therefore, anti-NF-κB actions of antioxidants are considered to be twofold: (1) blocking the signaling immediately downstream of the signal elicitation, and (2) suppressing the induction of the redox effector Trx.

## Origin of Redox Signaling: Involvement of Radical Oxygen Intermediates

Figure 3 illustrates the intracellular redox cascade involving successive reductions of oxygen by addition of four electrons and the redox regulatory cascade in the cell. Among these ROI, hydrogen peroxide has the longest half-life and is considered to be a major mediator of oxidative signal. On the other hand, cellular reducing systems

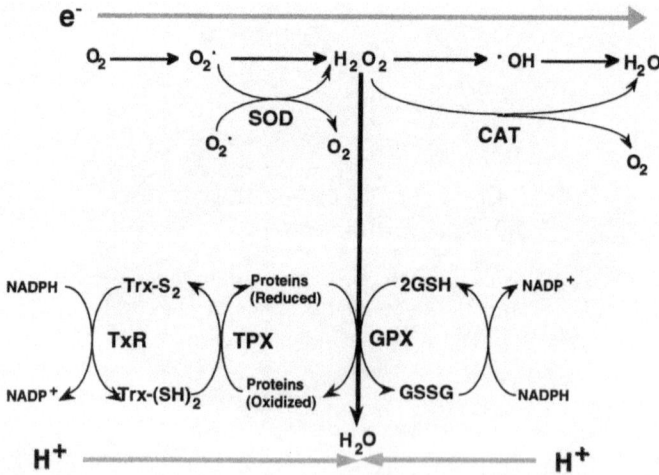

FIG. 3. Cellular redox system. Successive reduction of oxygen by addition of four electrons generates reactive oxygen species that are called ROI. Among these ROI, hydrogen peroxide has the longest half-life and is considered to be a mediator of oxidative signal. To maintain redox homeostasis, there are multiplicated antioxidant defense mechanisms within the cells, including superoxide dismutase (*SOD*), catalase (*CAT*), glutathione (*GSH*), glutathione peroxidase (*GPX*), thioredoxin (*Trx*), and Trx peroxidase (*TPX*). Unlike other antioxidant enzymes, the Trx system may be more specifically involved in the redox repairment of the oxidized protein molecules. The oxidized or reduced status of the protein confers biological information be regulating the activity of the relevant protein molecule

such as Trx and glutathione (GSH) counteract the action of hydrogen peroxide. While GSH is directly involved in scavenging ROI, Trx appears to participate in this cascade by repairing the oxidized proteins through its reducing activity. This reversible oxidation and reduction of a functional protein determine its activity. Therefore, GSH itself may not be directly involved in the redox signaling but the intensity of the oxidative signal may be determined by the internal GSH level. Thus, total GSH/GSSG (glutathione disulfide) content could be a useful indicator for the responsiveness of cellular redox signaling. This redox regulatory pathway is clearly illustrated in the NF-κB activation pathway.

## Activation of NF-κB by Kinase and Trx-Mediated Redox Regulation

There are at least two independent steps in the NF-κB activation cascade: kinase pathways and the redox-signaling pathway. These two distinct pathways are involved in the NF-κB activation cascade in a coordinate fashion, which may contribute to a fine-tuned, as well as fail-safe, regulation of NF-κB activity. At least two distinct types of kinase pathways are known to be involved in NF-κB activation: NF-κB kinase and IκB kinase (see Fig. 2). We found that a 43 kDa serine kinase, NF-κB kinase, is associated with NF-κB [51]. This kinase phosphorylates the subunits, particularly p65, of NF-κB and dissociates it from IκB. Molecular cloning of the gene for this kinase is now under way. There is another kinase (or kinases) that is known to phosphorylate IκB [52–59]. As a candidate of IκB kinases, IRAK (IL-1 receptor-associated kinase) was recently cloned and found to share structural similarity to Pelle, a protein kinase known to be involved in the activation of a NF-κB homologue in *Drosophila* [58]. Consistent with these findings, NF-κB was shown to be phosphorylated in some cell lines and IκB was phosphorylated in others in response to stimulation with TNF or IL-1 [58–61]. In most cases, NF-κB dissociation by the kinase cascade is a primary step of NF-κB activation.

After dissociation from IκB, however, NF-κB must go through redox regulation by the cellular reducing catalyst, thioredoxin (Trx) [62,63], to recognize the target DNA sequence and induce transcription. Trx is a cellular reducing catalyst and is known to participate in redox reactions through reversible oxidation of its active center dithiol to a disulfide (see Figs. 2 and 3). Interestingly, human Trx has been initially identified as a factor responsible for induction of the α-subunit of interleukin-2 receptor, which has been revealed to be under the control of NF-κB [64]. We and others have demonstrated in vitro that NF-κB cannot bind to the κB DNA sequence of the target genes until it is reduced [62,63,65,66]. Based on the estimation of high local pI value near one of the conserved cysteine residues, we have assigned the cysteine residue at the 62nd amino acid position of p50 subunit as a target of redox regulation [63], which was confirmed by a site-directed mutagenesis study by others [67] in which the cysteine-62 substitution abolished the DNA-binding activity.

## Structural Basis for the Trx-Mediated Redox Regulation of NF-κB

Structural biological approaches have proven quite powerful in explaining the molecular mechanism of the redox regulation of NF-κB by Trx. First, in early 1995 two independent groups demonstrated the three-dimensional structure of the NF-κB sub-unit p50 homodimer cocrystallized with the target DNA [68,69]. NF-κB appears to have a novel DNA-binding structure called a beta-barrel, a group of β-sheets stretch-ing toward the target DNA. There is a loop in the tip of the beta-barrel structure that intercalates with the nucleotide bases and is considered to make a direct contact with the DNA. This DNA-binding loop contains the cysteine-62 that we predicted to be the target of redox regulation as a proton acceptor from Trx. Although in both studies this cysteine was replaced with alanine, presumably for technical reasons of crystalliza-tion, yet these observations confirmed our earlier speculations [62,63]. Additionally, an National Institutes of Health (NIH) group recently solved the three-dimensional structure of the Trx molecule that is associated with the DNA-binding loop of p50 by using nuclear magnetic resonance (NMR) [70]. A boot-shaped hollow on the surface of Trx containing the redox-active cysteines could stably recognize the DNA-binding loop of p50 [68,69] and is likely to reduce the oxidized cysteine by donating protons in a structure-dependent fashion. The reduction of NF-κB by Trx therefore is consid-ered to be specific and dependent on the structural compatibility between the target protein and Trx. However, the disulfide bridge formation between Trx and NF-κB might be transient, as the binding of Trx to the NF-κB DNA-binding loop would block the recognition of DNA because of the apparent competition of the same cysteine residue (Fig. 4). In favor of this model, we have demonstrated that NF-κB and Trx concomitantly migrated to the nucleus in the rheumatoid synovial cells during the first phase of the NF-κB activation process and that Trx was relocated in the cyto-plasm after 30 min of stimulation while the NF-κB nuclear predominance was still observed for several hours (Fig. 5).

   In addition to these findings, our in vitro binding study has demonstrated that zinc is required for the DNA-binding activity of NF-κB as well as the reduction of the redox-sensitive cysteine [63,70]. We found that even the fully activated NF-κB could still be blocked by gold ion by a redox mechanism [71]. We found that the zinc ion is a necessary component of the active NF-κB and that addition of monovalent gold ion could efficiently block its activity by oxidizing the redox-active cysteines on NF-κB. Because gold did not appear to replace zinc [71], it is likely that gold ion oxidizes these thiolate anions on NF-κB into disulfides and thus abrogates the DNA-binding activity

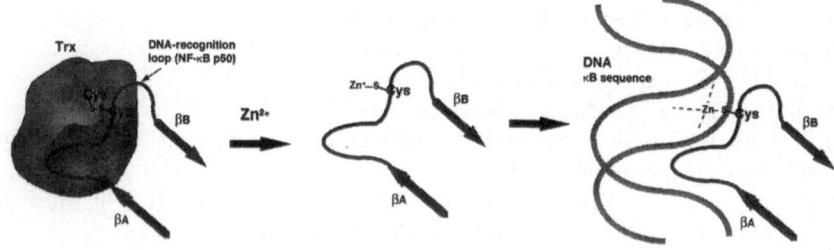

Fig. 4. Reduction of the "redox-sensitive" cysteine on NF-κB by Trx. A boot-shaped hollow on the surface of Trx, also containing the redox-active cysteines, could stably recognize the DNA-binding loop of p50 and reduce the oxidized cysteine by donating protons in a structure-dependent way (based on the three-dimensional structure of the Trx molecule with the DNA-binding loop peptide of p50 [70]). Zinc is required to make NF-κB competent for the DNA binding [71]. We thus assume that the active intermediate of NF-κB is associated with zinc.

# NF-κB (p65)          Trx

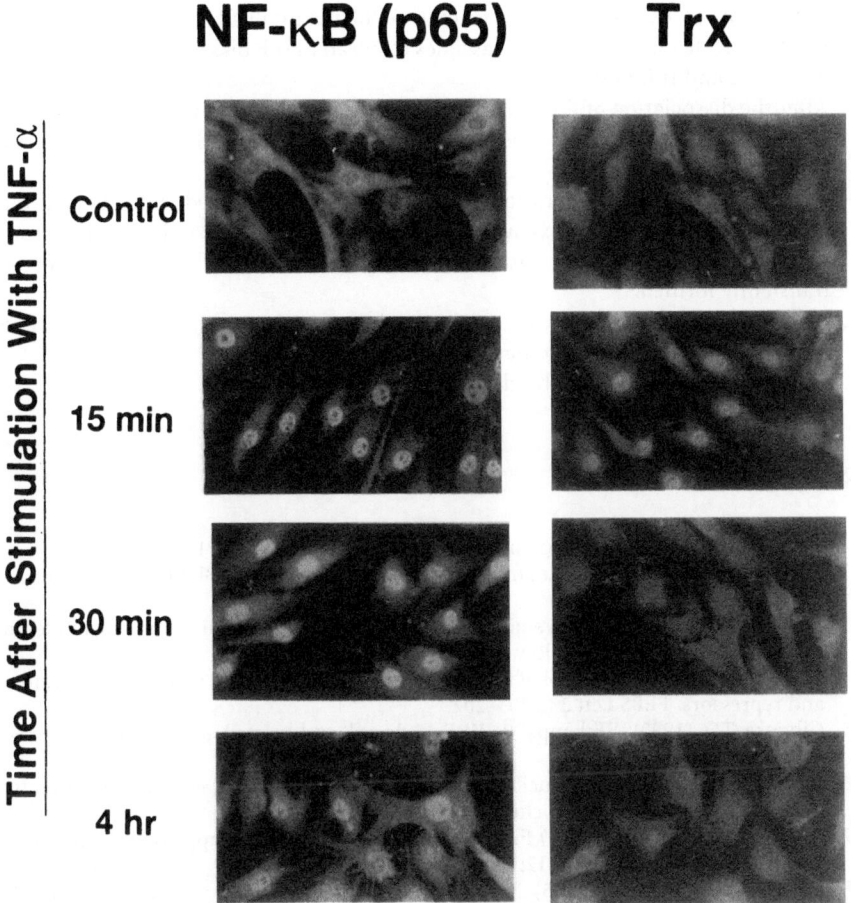

FIG. 5. Nuclear translocation of NF-κB and Trx in rheumatoid synovial fibroblasts (RSF). RSFs were treated with 10ng/ml of IL-1β for various time periods (0, 15min, 30min, and 4h). The cells were then reacted with rabbit antibodies against p65 of the NF-κB subunit or Trx and subsequently stained with fluorescein isothiocyanate (FITC) conjugated goat antirabbit IgG. Note that the concomitant nuclear translocation of p65 and Trx occurs only at the stage after 15 min of TNF stimulation. This colocalization of NF-κB (p65) and Trx suggests the interaction between these two proteins in the cultured cells during the NF-κB activation cascade

because of its higher oxidation potential over zinc ion. It is notable that gold compounds have been successfully used for the treatment of RA [72,73]. Our finding could explain why gold is effective in RA and suggests that NF-κB might have a crucial role in the disease process [37,74]. It may be that a gold compound is potentially effective in other diseases in which NF-κB plays a pathological role.

## Conclusion: Redox Regulation of NF-κB

Taken together, we would like to propose the following model with regard to the redox regulation of NF-κB:

1. Generation of ROI in response to extracellular stimuli (TNF, irradiation, etc.) would activate one of the kinase cascades that lead to IκB dissociation (such as NF-κB kinase and IκB kinases).
2. After the dissociation of IkB from NF-κB, NF-κB will move to the nucleus because of the exposure of its nuclear localization signal (NLS), which has been covered by IκB.
3. During the nuclear translocation process, NF-κB is associated with Trx and the redox-active cysteine (for example, the cysteine-62 of p50) is reduced in a structure-dependent manner. A disulfide bridge between NF-κB and Trx might be transiently formed.
4. Association of zinc ion with the redox-sensitive cysteine of NF-κB will dissociate Trx, and Trx will relocate to the cytoplasm.
5. Zinc-associated NF-κB will bind to the target DNA.

# References

1. Holmgren A (1985) Thioredoxin. Annu Rev Biochem 54:237–271
2. Holmgren A (1989) Thioredoxin and glutaredoxin systems. J Biol Chem 264:13963–13966
3. Ziegler DM (1985) Role of reversible oxidation-reduction of enzyme thios-disulfides in metabolic regulation. Annu Rev Biochem 54:305–329
4. Allen JF (1993) Redox control of transcription: sensors, response regulators, activators and repressors. FEBS Lett 332:203–207
5. Gilmore TD (1990) NF-kappa B, KBF-1, dorsal and related matters. Cell 62:841–843
6. Baeuerle PA. (1991) The inducible transcription activator NF-kappa B: regulation by distinct protein subunits. Biochim Biophys Acta 1072:63–80
7. Baeuerle PA, Henkel T (1994) Function and activation of NF-kappa B in the immune system. Annu Rev Immunol 12:141–179
8. Thanos D, Maniatis T (1995) NF-kappa B: a lesson in family values. Cell 80:529–532
9. Maekawa T, Itoh F, Okamoto T, et al. (1989) Identification and purification of the enhancer-binding factor of human immunodeficiency virus-1. Multiple proteins and binding to other enhancers. J Biol Chem 264:2826–2831
10. Schindler U, Baichwal VR (1994) Three NF-kappa B binding sites in the human E-selectin gene required for maximal tumor necrosis factor alpha-induced expression. Mol Cell Biol 14:5820–5831
11. Okamoto T, Matsuyama T, Mori S, et al. (1989) Augmentation of human immunodeficiency virus type 1 gene expression by tumor necrosis factor alpha. AIDS Res Hum Retroviruses 5:131–138
12. Stade BG, Messer G, Riethmuller G, Johnson JP (1990) Structural characteristics of the 5' region of the human ICAM-1 gene. Immunobiology 182:79–87
13. Mukaida N, Mahe Y, Matsushima K (1990) Cooperative interaction of nuclear factor-kappa B- and cis-regulatory enhancer binding protein-like factor binding elements in activating the interleukin-8 gene by pro-inflammatory cytokines. J Biol Chem 265:21128–21133
14. Roebuck KA, Rahman A, Lakshminarayanan V, et al. (1995) $H_2O_2$ and tumor necrosis factor-alpha activate intercellular adhesion molecule 1 (ICAM-1) gene transcription through distinct cis-regulatory elements within the ICAM-1 promoter. J Biol Chem 270:18966–18974

15. Donnelly RP, Crofford LJ, Freeman SL, et al. (1993) Tissue-specific regulation of IL-6 production by IL-4. Differential effects of IL-4 on nuclear factor-kappa B activity in monocytes and fibroblasts. J Immunol 151:5603–5612
16. Schreck R, Baeuerle PA (1990) NF-kappa B as inducible transcriptional activator of the granulocyte-macrophage colony-stimulating factor gene. Mol Cell Biol 10:1281–1286
17. Staynov DZ, Cousins DJ, Lee TH (1995) A regulatory element in the promoter of the human granulocyte-macrophage colony-stimulating factor gene that has related sequences in other T-cell-expressed cytokine genes. Proc Natl Acad Sci USA 92:3606–3610
18. Xie Q-W, Kashiwabara Y, Nathan C (1994) Role of transcription factor NF-kappa B/Rel in induction of nitric oxide synthase. J Biol Chem 269:4705–4708
19. Sen R, Baltimore D (1986) Inducibility of kappa immunoglobulin enhancer-binding protein NF-kappa B by a posttranslational mechanism. Cell 46:705–716
20. Nabel G, Baltimore D (1987) An inducible transcription factor activates expression of human immunodeficiency virus in T cells. Nature 326:711–713
21. Bohnlein E, Lowenthal JW, Siekevitz M, et al. (1988) The same inducible nuclear protein regulates mitogen activation of both the interleukin-2 receptor-alpha gene and type 1 HIV. Cell 53:827–836
22. Okamoto T, Benter T, Josephs SF, et al. (1990) Transcriptional activation from the long-terminal repeat of human immunodeficiency virus in vitro. Virology 177:606–614
23. Arya SK, Guo C, Josephs SF, Wong-Staal F (1985) trans-Activator gene of human T-lymphotropic virus type III (HTLV-III). Science 229:69-73
24. Sodroski J, Patarca R, Rosen C (1985) Location of the trans-activating region on the genome of human T-cell lymphotropic virus type III. Science 229:74-77
25. Okamoto T, Wong-Staal F (1986) Demonstration of virus-specific transcriptional activator(s) in cells infected with HTLV-III by an in vitro cell-free system. Cell 47:29–35
26. Roederer M, Staal FJT, Raju PA, et al. (1990) Cytokine-stimulated human immunodeficiency virus replication is inhibited by N-acetyl-L-cysteine. Proc Natl Acad Sci USA 87:4884–4888
27. Suzuki YJ, Aggarwal BB, Packer L (1992) Alpha-lipoic acid is a potent inhibitor of NF-kappa B activation in human T cells. Biochem Biophys Res Commun 189:1709–1715
28. Merin JP, Matsuyama M, Kira T, et al. (1996) α-Lipoic acid blocks HIV-1 LTR-dependent expression of hygromycin resistance in THP-1 stable transformants. FEBS Lett 394:9–13
29. Tozawa K, Sakurada S, Kohri K, Okamoto T (1995) Effects of anti-nuclear factor kappa B reagents in blocking adhesion of human cancer cells to vascular endothelial cells. Cancer Res 55:4162–4167
30. Montgomery KF, Osborn L, Hession C, et al. (1991) Activation of endothelial-leukocyte adhesion molecule 1 (ELAM-1) gene transcription. Proc Natl Acad Sci USA 88:6523–6527
31. Whelan J, Ghersa P, Huijsduijnen RH, et al. (1991) An NF kappa B-like factor is essential but not sufficient for cytokine induction of endothelial leukocyte adhesion molecule 1 (ELAM-1) gene transcription. Nucleic Acids Res 19:2645–2653
32. Dejana E, Bertocci F, Bortolami MC (1988) Interleukin 1 promotes tumor cell adhesion to cultured human endothelial cells. J Clin Invest 82:1466-1470.
33. Takada A, Ohmori K, Yoneda T, et al. (1993) Contribution of carbohydrate antigens sialyl Lewis A and sialyl Lewis X to adhesion of human cancer cells to vascular endothelium. Cancer Res 53:354–361
34. Alvaro-Gracia JM, Zvaifler NJ, Brown CB, et al. (1991) Cytokines in chronic arthritis. VI. Analysis of the synovial cells involved in granulocyte macrophage colony-stimulating factor production and gene expression in rheumatoid arthritis and its regulation by IL-1 and TNF-α. J Immunol 146:3365–3372

35. Ulfgren AK, Lindblad S, Klareskog L, et al. (1995) Detection of cytokine producing cells in the synovial membrane from patients with rheumatoid arthritis. Ann Rheum Dis 54:654–659
36. Arend WP, Dayer JM (1995) Inhibition of the production and effects of interleukin-1 and tumor necrosis factor a in rheumatoid arthritis. Arthritis Rheum 38:151–157
37. Handel ML, McMorrow LB, Gravallese EM (1996) Nuclear factor-κB in rheumatoid synovium. Localization of p50 and p60. Arthritis Rheum 38:1762–1770
38. Baeuerle PA, Baltimore D (1988) Activation of DNA-binding activity in an apparently cytoplasmic precursor of the NF-kappa B transcription factor. Cell 53:211–217
39. Baeuerle PA, Baltimore D (1988) I-kappa B: a specific inhibitor of the NF-kappa B transcription factor. Science 242:540–546
40. Ghosh S, Gifford AM, Riviere LR, et al. (1990) Cloning of the p50 DNA binding subunit of NF-kappa B: homology to Rel and dorsal. Cell 62:1019–1029
41. Ghosh S, Baltimore D (1990) Activation in vitro of NF-kappa B by phosphorylation of its inhibitor I-kappa B. Nature 344:678–682
42. Read MA, Whitley MZ, Williams AJ, Collins T (1994) The proteasome pathway is required for cytokine-induced endothelial-leukocyte adhesion molecule expression. J Exp Med 179:503–512
43. Rothe M, Wong SC, Henzel WJ, Goeddel DV (1994) A novel family of putative signal transducers associated with the cytoplasmic domain of the 75 kDa tumor necrosis factor receptor. Cell 78:681–692
44. Hsu H, Shu HB, Pan MG, Goeddel DV (1996) TRADD-TRAF2 and TRADD-FADD interactions define two distinct TNF receptor 1 signal transduction pathways. Cell 84:299–308
45. Schreck R, Rieber P, Baeuerle PA (1991) Reactive oxygen intermediates as apparently widely used messengers in the activation of the NF-kappa B transcription factor and HIV-1. EMBO J 10:2247–2258
46. Meyer M, Schreck R, Baeuerle PA (1993) $H_2O_2$ and antioxidants have opposite effects on activation of NF-kappa B and AP-1 in intact cells: AP-1 as secondary antioxidant-responsive factor. EMBO J 12:2005–2015
47. Biswas DK, Dezube BJ, Ahlers CM, Pardee AB (1993) Pentoxifylline inhibits HIV-1 LTR-driven gene expression by blocking NF-kappa B action. J AIDS 6:778–786
48. Suzuki YJ, Packer L (1994) Signal transduction for nuclear factor-kappa B activation. Proposed location of antioxidant-inhibitable step. J Immunol 153:5008–5015.
49. Packer L, Witt EH, Tritschler HJ (1995) α-Lipoic acid as a biological antioxidant. Free Radical Biol Med 19:227–250
50. Sachi Y, Hirota K, Masutani H, et al. (1995) Three NF-kappa B binding sites in the human E-selectin gene required for maximal tumor necrosis factor alpha-induced expression. Immunol Lett 44:189–193
51. Hayashi T, Sekine T, Okamoto T (1993) Identification of a new serine kinase that activates NF-kappa B by direct phosphorylation. J Biol Chem 826:26790–26795
52. Shirakawa F, Mizel SB (1989) In vitro activation and nuclear translocation of NF-kappa B catalyzed by cyclic AMP-dependent protein kinase and protein kinase C. Mol Cell Biol 9:2424–2430
53. Meichle A, Schutze S, Hensel G, et al. (1990) Protein kinase C-independent activation of nuclear factor κB by tumor necrosis factor. J Biol Chem 265:8339–8343
54. Feuillard J, Gouy H, Bismuth G, et al. (1991) NF-Kappa B activation by tumor necrosis factor alpha in the Jurkat T cell line is independent of protein kinase A, protein kinase C, and Ca(2+)-regulated kinase. Cytokine 3:257–265
55. Ostrowski J, Sims JE, Sibley CH, et al. (1991) A serine/threonine kinase activity is clsely associated with a 65-kDa phosphoprotein specifically recognized by the kappa B enhancer element. J Biol Chem 266:12722–12733

56. Schutze S, Potthoff K, Machleidt T, et al. (1992) TNF activates NF-kappa B by phosphatidylcholine-specific phospholipase C-induced "acidic" sphingomyelin breakdown. Cell 71:765–776

57. Brown K, Gerstberger S, Carlson L, et al. (1995) Control of I-kappa B-alpha proteolysis by site-specific, signal-induced phosphorylation. Science 267:1485–1488

58. Cao Z, Henzel WJ, Gao X (1996) IRAK: a kinase associated with the interleukin-1 receptor. Science 271:1128–1131

59. Chen ZJ, Parent L, Maniatis T (1996) Site-specific phosphorylation of IκBα by a novel ubiquitination-dependent protein kinase activity. Cell 84:853–862

60. Naumann M, Scheidereit C (1994) Activation of NF-kappa B in vivo is regulated by mutiple phosphorylations. EMBO J 13:4597–4607

61. Li C-C H, Dai R-M, Chen E, Longo DL (1994) Phosphorylation of NF-KB1-p50 is involved in NF-kappa B activation and stable DNA binding. J Biol Chem 269:30089–30092

62. Okamoto T, Ogiwara H, Hayashi T, et al. (1992) Human thioredoxin/adult T cell leukemia-derived factor activates the enhancer binding protein of human immunodeficiency virus type 1bt thiol redox control mechanism. Int Immunol 4:811–819

63. Hayashi T, Ueno Y, Okamoto T (1993) Oxidoreductive regulation of nuclear factor kappa B. Involvement of a cellular reducing catalyst thioredoxin. J Biol Chem 268:11380–11388

64. Tagaya Y, Maeda Y, Mitsui A, et al. (1989) ATL-derived factor (ADF), an IL-2 receptor/ Tac inducer homologous to thioredoxin; possible involvement of dithiol-reduction in the IL-2 receptor induction. EMBO J 8:757–764

65. Molitor JA, Ballard DW, Greene WC (1991) Kappa-B-specific DNA binding proteins are differentially inhibited by enhancer mutations and biological oxidation. New Biol 3:987–996

66. Toledano MB, Leonard WJ (1991) Modulation of transcription factor NF-kappa B binding activity by oxidation-reduction in vitro. Proc Natl Acad Sci USA 88:4328–4332

67. Matthews JR, Wakasugi N, Virelizier JL, et al. (1992) Thioredoxin regulates the DNA binding activity of NF-kappa B by reduction of a disulfide bond involving cysteine 62. Nucleic Acids Res 20:3821–3830

68. Ghosh G, Van Duyne G, Ghosh S, Sigler PB (1995) Structure of NF-kappa B p50 homodimer bound to a kappa B site. Nature 373:303–310

69. Müller CW, Rey FA, Sodeoka M, et al. (1995) Structure of the NF-kappa B p50 homodimer bound to DNA. Nature 373:311–317

70. Qin J, Clore GM, Kennedy WMP, et al. (1995) Solution structure of human thioredoxin in a mixed disulfide intermediate complex with its target peptide from the transcription factor NF-kappa B. Structure (Lond) 3:289–297

71. Yang JP, Merin JP, Nakano T, et al. (1995) Inhibition of the DNA-binding activity of NF-kappa B by gold compounds in vitro. FEBS Lett 361:89–96

72. Skosey JL (1993) Treatment of rheumatoid arthritis. In: McCarty DJ, Koopman WJ (eds) Arthritis and allied conditions. Lea & Febiger, Philadelphia, pp 603–614

73. Insel PA (1996) Analgesic-antipyretic and anti-inflammatory agents and drugs employed in the treatment of gout. In: Hardman JG, et al. (eds) The pharmacological basis of therapeutics. Macmillan, New York, pp 670–681

74. Sakurada S, Kato T, Okamoto T (1996) Induction of cytokines and ICAM-1 by proinflammatory cytokines in primary rheumatoid synovial fibroblasts and inhibition by N-acetyl-L-cysteine and aspirin. Int Immunol 8:1483–1493

# Thioredoxin and Its Involvement in the Redox Regulation of Transcription Factors, NF-κB and AP-1

Tetsuya Ohno[1], Kiichi Hirota[2], Hajime Nakamura[3],
Hiroshi Masutani[1], Tetsuro Sasada[1], and Junji Yodoi[1]

*Summary.* Thioredoxin (TRX) is a cellular factor that has a disulfide-reducing activity and plays important roles in regulation of cellular processes. Activity of a number of transcription factors is posttranslationally altered by redox modification(s). One such factor is NF-κB, whose activity is altered by the intracellular redox state. The DNA-binding activity of AP-1 is modified by a nuclear protein, reducing factor-1 (Ref-1). Ref-1 activity is in turn modified by various redox-active compounds, including TRX. This short review summarizes the role of TRX as an important intracellular reducing factor and its involvement in the redox regulation of transcription factors NF-κB and AP-1.

*Key words.* Thioredoxin (TRX)/adult T-cell leukemia- (ATL-) derived factor (ADF)—Redox regulation—Transcription factors—NF-κB—AP-1

## The Role of TRX as a Reducing Factor

Thioredoxin (TRX), also known as adult T-cell leukemia- (ATL-) derived factor (ADF) [1], has a redox-active disulfide (Cys-Gly-Pro-Cys) within its active site and operates in concert with NADPH and thioredoxin reductase as a general protein disulfide-reducing system [2]. Regulation of the intracellular oxidoreductive milieu is critical for living cells, and TRX plays an important role in maintaining their redox environment [3].

TRX has both intracellular and extracellular activities (Fig. 1). TRX gene expression is enhanced through a novel cis-acting regulatory element responsive for the oxidative stress (oxidative-responsive element, ORE) [4]. Thus, TRX is an inducible antioxidative stress protein. Oxidative stresses including UV irradiation promptly induce

---

[1]Institute for Virus Research, Kyoto University, 53 Shogoin-Kawaharacho, Sakyo-ku, Kyoto 606-01, Japan
[2]Department of Anesthesia, Kyoto University Hospital, 54 Shogoin-Kawaharacho, Sakyo-ku, Kyoto 606-01, Japan
[3]Department of Gastroenterological Surgery, Kyoto University Graduate School of Medicine, 54 Shogoin-Kawaharacho, Sakyo-ku, Kyoto 606-01, Japan

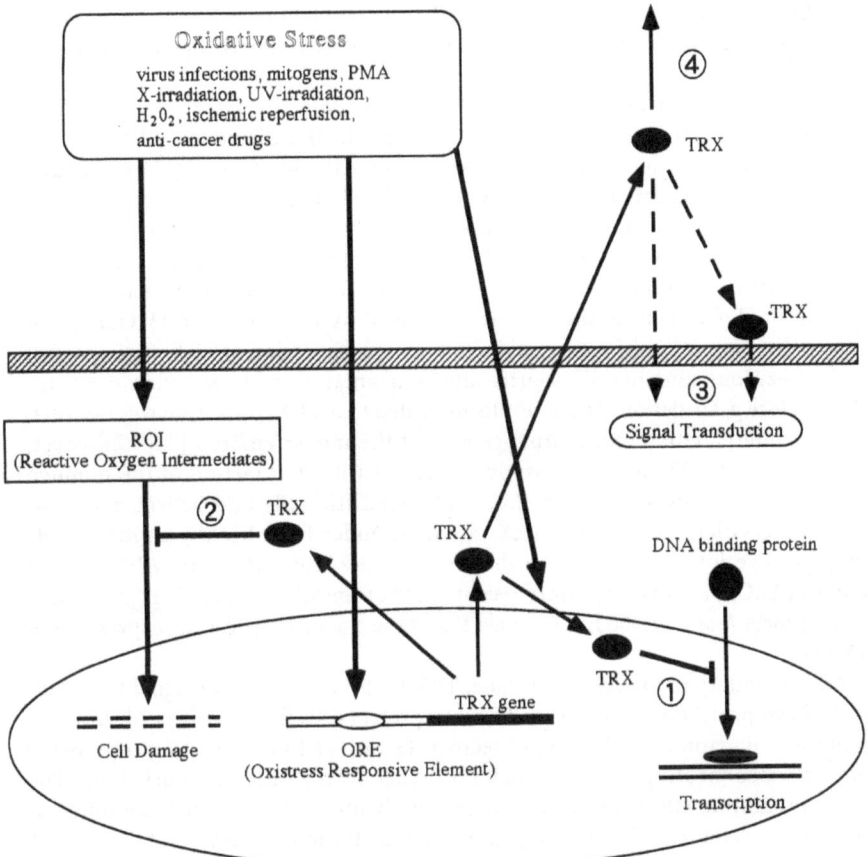

FIG. 1. The role of thioredoxin (TRX) as a reducing factor. 1, regulation of protein-nucleotide interaction; 2, cytoprotection against ROI; 3, signal transduction; 4, cytokine-like effect

the translocation of TRX into the nucleus (Hirota et al, in preparation]. The DNA-binding activity of several transcription factors, such as NF-κB [5,6], AP-1 (Jun and Fos complex) [7,8], Sp-1 [9], Ets-1 [10], Myb [11], and p53 [12,13], requires their reduced form. TRX creates the reduced condition in the nucleus and contributes to the augmentation of the DNA-binding activity of the transcription factors. This regulation of protein–nucleotide interaction is one of the important intracellular roles of TRX. In the case of NF-κB, direct molecular interactions between NF-κB and TRX have been demonstrated by nuclear magnetic resonance (NMR) structure analysis [14]. The binding of the Jun and Fos complex to the AP-1 site is enhanced by a nuclear protein, reducing factor 1 (Ref-1) [8], and TRX has been proved to interact directly with Ref-1 [15]. Our recent study showed that the DNA-binding domain of PEBP2/AML1 was also activated by Ref-1 or TRX (Akamatsu et al, in preparation). Interestingly, the intracellular expression of TRX is also regulated by the cell cycle

[16]. This evidence and the fact that a cell-cycle-regulating protein, p53, is under redox regulation suggest the possible involvement of TRX in redox regulation of the cell cycle.

The second function of intracellular TRX is a cytoprotective activity against oxidative stress. Although glutathione is a major intracellular antioxidant, TRX is an endogenously inducible and protective protein againt oxidative damage [17,18]. We and others previously reported that TRX-transfected cells were more resistant to treatment with anticancer agents such as cisplatin (CDDP) than the control cells [19,20]. In the tissues of the pregnant uterus, TRX is overexpressed in decidua and trophoblast cells, suggesting that TRX may be beneficial in protecting the fertilized egg and placental trophoblasts from the cytotoxic effects of oxygen radicals [21]. Our recent study on the phenotype of mice carrying a targeted disruption of TRX gene revealed that homozygous mutants die shortly after implantation and the concepti are resorbed before gastrulation. These results indicated that TRX expression was essential for early differentiation and morphogenesis of the mouse embryo [22]. Moreover, dysregulation of TRX may be possibly associated with dysfunction of the immune system. Earlier study has shown that while dendritic cells and activated macrophages express large amount of TRX in lymph nodes from healthy controls, cells highly positive for TRX are lost in the lymph nodes from AIDS and AIDS-related complex (ARC) patients [23]. More recent study of immunosuppressive agents such as cyclosporin A and FK506 has revealed that these agents suppress the expression of TRX [24].

At the boundary to the cell membrane, TRX is also a regulator of signal transduction and cytoprotection. Exogenous TRX can protect cells from anti-Fas Ab-induced apoptosis and cytotoxicity by tumor necrosis factor (TNF) [25], hydrogen peroxide from activated neutrophils [16], and postischemic reperfusion injury [26]. The growth-promoting effects of other cytokines on lymphocytes as well as on nonlymphocytes are reinforced by the exogenous administration of TRX [27,28]. TRX recently was demonstrated to be a potent costimulator of cytokine expression [29]. Human TRX is secreted by living cells, for example, lymphocytes, hepatocytes, and fibroblasts, through a unique leaderless pathway [30], and is identified in human plasma or serum [31]. Plasma levels of human TRX are indicative of the inflammatory response against oxidative stress in the course of progression of several degenerative diseases. One of the important examples is that HIV-infected individuals show significantly elevated plasma TRX levels and those with higher levels of plasma TRX tend to have lower CD4 cell numbers [32]. This suggests that plasma TRX is one of the important parameters in the terminal stage of HIV infection.

# Redox Regulation of Transcription Factors

Until now, the activity of a growing number of transcription factors has been demonstrated to be controlled under physiological oxidant-antioxidant homeostasis. Although in the early 1950s the possibility that protein activity might be reversibly regulated by redox changes in response to environmental changes was already suggested, iron-responsive element-binding protein (IRE-BP) was showed for the first time in 1989 to require free sulfhydryl groups for its specific interaction with the IRE

[33]. Thereafter there have been accumulating examples of redox-responsive tran-scription factors such as NF-κB [5,6], AP-1 [7,8], Myb [11], Ets-1 [10], p53 [12,13], glucocorticoid receptor [34], Sp-1 [9], Egr-1 [35], USF [36], PEBP2/AML1 (Akamatsu et al, in preparation), and HoxB5 [37]. Except for HoxB5, which is activated in its oxidative state, in most of these transcription factors the reduced state of conserved cysteine residues in the DNA-binding domain is critical for their DNA binding. Based on a series of these reports, the concept of redox (reduction/oxidation) regulation in eukaryotic cells became a novel subdiscipline.

In mammalian cells, there are two main thiol-antioxidant systems, the glutathione (GSH) -glutathione disulfide (GSSG) system and TRX system. These two physiologi-cally relevant systems appear to have synergistic but distinct roles to each other in the regulation of the intracellular redox condition and the activity of transcription fac-tors. Living cells may sense environmental cues and transduce stress signals into a change in gene expression with cooperation between these two systems. Figure. 2 shows the regulation of two well-known transcription factors, NF-κB and AP-1.

In the case of NF-κB, oxidative stress induces its activation via changes of GSSG levels in the cytosol [38]. TRX may act adversely in this stage [39,40], but as men-tioned earlier, TRX itself is probably translocated from the cytosol to the nucleus at the cue of various kinds of stress (Hirota et al., in preparation). In the nucleus, oxidized NF-κB cannot bind to the κB-site effectively, and this lost DNA-binding activity is restored after reduction by TRX [5,6]. For the specific binding between NF-κB and the κB site, one of the conserved cysteine residues in the DNA-binding domain of NF-κB, Cys-62, is essential and its oxidation or mutation deteriorates the DNA-binding activity of NF-κB [41]. This critical cysteine residue lies in the vicinity of the basic amino acids, arginine and lysine, which make a cationic environment and render the thiol group quite susceptible to oxidation. During the activation of NF-κB in the cytosol, oxidation of this Cys-62 is unavoidable and thus any reducing agents,

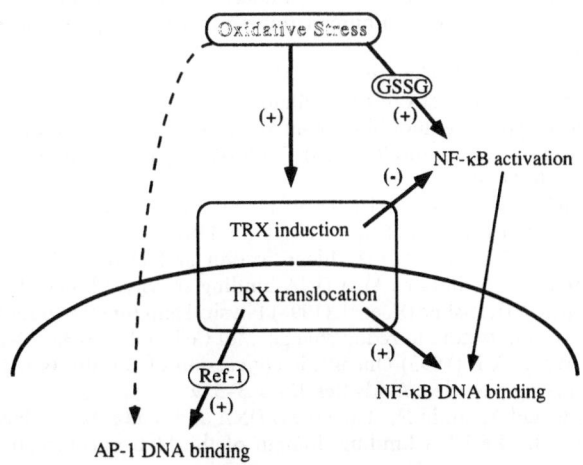

FIG. 2. Involvement of thioredoxin in the redox regulation of transcription factors NF-κB and AP-1. *GSSG*, glutathione–glutathione disulfide

in this case TRX, are indispensable to the antioxidant-induced signal transduction of the cells. As NF-κB activation and its DNA binding are sequential events, cells can potentially acquire the most effective NF-κB activity by changing intracellular redox states.

In contrast to NF-κB, AP-1 activation is strongly enhanced by transient expression or exogenous application of TRX and hydrogen peroxide is only a weak inducer of AP-1 [39]. As AP-1 is almost exclusively localized in the nucleus, AP-1 is less sensitive to changes of cytosolic GSSG levels induced by oxidative stress [38]. The in vitro DNA-binding activity of AP-1 is stimulated by the reducing agents including Ref-1 and TRX [8]. A single conserved cysteine residue closely flanked by basic amino acids in the DNA-binding domains of Fos and Jun is responsible for this regulation. In this case, this cysteine residue is not essential as Cys-62 in NF-κB, because its mutation to serine enhances the DNA-binding activity and decreases the redox sensitivity of this factor [42]. It is possibly suggested that the conversion of the cysteine to reversible oxidation products such as sulfenic (RSOH) or sulfinic ($RSO_2H$) acids could contribute to the regulation of the DNA binding of this transcription factor [7]. In the light of these data, physiologically relevant intracellular antioxidants will exhibit their own distinct effects on transcription factors in the course of their activation.

# References

1. Yodoi J, Uchiyama T (1992) Human T-cell leukemia virus type I (HTLV-I) associated diseases; virus, IL-2 receptor dysregulation and redox regulation. immunol Today 13:405–411
2. Holmgren A, Björnstedt M (1995) Thioredoxin and thioredoxin reductase. Methods Enzymol 252:199–208
3. Nakamura H, Nakamura K, Yodoi J (1997) Redox regulation of cellular activation. Annu. Rev. Immunol 15:351–369
4. Taniguchi Y, Taniguchi-Ueda Y, Mori K, et al. (1996) A novel promoter sequence is involved in the oxidative stress-induced expression of the adult T-cell leukemia-derived factor (ADF)/human thioredoxin (Trx) gene. Nucleic Acids Res 24:2746–2752
5. Okamoto T, Ogiwara H, Hayashi T, et al. (1992) Human thioredoxin/adult T cell leukemia-derived factor activates the enhancer binding protein of human immuno-deficiency virus type 1 by thiol redox control mechanism. Int Immunol 4:881–819
6. Hayashi T, Ueno Y, Okamoto T (1993) Oxidoreductive regulation of NF-κB. J Biol Chem 268:11380–11388
7. Abate C, Patel L, Rauscher FJ III, et al. (1990) Redox regulation of Fos and Jun DNA-binding activity in vitro. Science 249:1157–1161
8. Xanthoudakis S, Curran T (1992) Identification and characterization of Ref-1, a nuclear protein that facillitates AP-1 DNA-binding activity. EMBO J 11:653–665
9. Wu X, Bishopric NH, Disher DJ, et al. (1996) Physical and functional sensitivity of zinc finger trasncription factors to redox change. Mol Cell Biol 16:1035–1046
10. Wasylyk C, Wasylyk B (1993) Oncogenic conversion of Ets affects redox regulation in-vivo and in-vitro. Nucleic Acids Res 221:523–529
11. Myrset AH, Bostad A, Jamin N, et al. (1993) DNA and redox state induced conformational changes in the DNA-binding domain of the Myb oncoprotein. EMBO J 12: 4625–4633
12. Hainaut P, Milner J (1993) Redox modulation of p53 conformation and sequence-specific DNA binding in vitro. Cancer Res 53:4469–4473

13. Rainwater R, Parks D, Anderson ME, et al. (1995) Role of cysteine residues in regulation of p53 function. Mol Cell Biol 15:3862–3903

14. Qin J, Clore GM, Kennedy WM, et al. (1995) Solution structure of human thioredoxin in a mixed disulfide intermediate complex with its target peptide from the transcription factor NF-κB. Structure 3:289–297

15. Hirota K, Matsui M, Iwata S, et al. (1997) AP-1 transcriptional activity is regulated by a direct association between thioredoxin and Ref-1. Proc Natl Acad Sci VSA 94: 3633–3638

16. U-Taniguchi Y, Furuke K, Masutani H, et al. (1995) Cell cycle inhibition of HTLV-1 transfomed T cell lines by retinoic acid: the possible therapeutic use of thioredoxin reductase inhibitors. Oncol Res 7:183–189

17. Nakamura H, Matsuda M, Furuke K, et al. (1994) Adult T cell leukemia-derived factor/ human thioredoxin protects endothelial F-2 cell injury caused by activated neutrophils or hydrogen peroxide. Immunol Lett 42:75–80

18. Tomimoto H, Akiguchi I, Wakita H, et al. (1993) Astriglial expression of ATL-derived factor, a human thioredoxin homologue, in the gerbil brain after transient global ischemia. Brain Res 625:1–8

19. Yokomizo A, Ono M, Nanri H, et al. (1995) Cellular levels of thioredoxin associated with drug sensitivity to cisplatin, mitomycin C, doxorubicin and etoposide. Cancer Res 55:4293–4296

20. Sasada T, Iwata S, Sato N, et al. (1996) Redox control of resistance to cis-diamminedichloroplatinum (II) (CDDP) protective effect of human thioredoxin against CDDP-induced cytotoxicity. J Clin Invest 97:2268–2276

21. Kobayashi F, Sagawa N, Nanbu Y, et al. (1995) Biological and topological analysis of adult T-cell leukemia-derived factor, homologous to thioredoxin, in the pregnant human uterus. Hum Reprod 10:1603–1608

22. Matsui M, Oshima M, Oshima H, et al. (1996) Early embryonic lethality caused by targeted disruption of the mouse thioredoxin gene. Dev Biol 178:179–185

23. Masutani H, Naito M, Takahashi K, et al. (1992) Dysregulation of adult T-cell leukemia-derived factor (ADF)/thioredoxin in HIV infection: loss of ADF high-producer cells in lymphoid tissues of AIDS patients. AIDS Res Hum Retroviruses 8:1707–1715

24. Furuke K, Nakamura H, Hori T, et al. (1995) Suppression of adult T cell leukemia-derived factor/human thioredoxin induction by FK506 and cyclosporin A: a new mechanism of immune modulation via redox control. Int Immunol 7:985–993

25. Matsuda M, Masutani H, Nakamura H, et al. (1991) Protective activity of adult T cell leukemia-derived factor (ADF) against tumor necrosis factor-dependent cytotoxicity on U937 cells. J Immunol 147:3837–3841

26. Yokomise H, Fukuse T, Hirata T, et al. (1994) Effect of recombinant human adult T cell leukemia-derived factor on rat lung reperfusion injury. Respiration 61:99–104

27. Wakasugi N, Tagaya Y, Wakasugi H, (1990) Adult T cell leukemia-derived facto/ thioredoxin, produced by both human T-lymphotropic virus type-I and Epstein-Barr virus-transformed lymphosytes, acts as an autocrine growth factor and synergizes with interleukin 1 and inerleukin 2. Proc Natl Acad Sci USA 87:8282–8286

28. Namamura H, Masutani H, Tagaya Y, et al. (1992) Expression and growth-promoting effect of adult T-cell leukemia-derived factor, a human thioredoxin homologue in hepatocellular carcinoma. Cancer (Phila) 69:2091–2097

29. Schenk H, Vogt M, Droege W, et al. (1996) Thioredoxin as a potent costimulus of cytokine expression. J Immunol 156:765–771

30. Rubartelli A, Bajetto A, Allavena G, et al. (1992) Secretion of thioredoxin by normal and neoplastic cells through a leaderless secretory pathway. J Biol Chem 267:24161–24164

31. Kitaoka Y, Sorachi KI, Nakamura H, et al. (1994) Detection of adult T-cell leukemia-derived factor, human thioredoxin in human serum. Immunol Lett 41:155–161

32. Nakamura H, DeRosa S, Roederer M, et al. (1996) Elevation of plasma thioredoxin levels in HIV infected individuals. Int Immunol 8:603–611
33. Hentze MW, Rouault TA, Harford JB, et al. (1989) Oxidation-reduction and the molecular mechanism of a regulatory RNA-protein interaction. Science 244:357–359
34. Makino Y, Okamoto K, Yoshikawa N, et al. (1996) Thioredoxin: a redox-regulationg cellular cofactor for glucocorticoid hormone action. Cross talk between endocrine control of stress response and cellular antioxidant defense system. J Clin Invest 98:2469–2477
35. Huang RP, Adamson ED (1993) Characterization of the DNA-binding properties of the early growth response-1 (Egr-1) transcription factor: evidence for modulation by a redox mechanism. DNA Cell Biol 12:265–273
36. Pognonec P, Kato H, Roeder RG (1992) The helix-loop-helix/leucine repeat transcription factor USF can be functionally regulated in a redox-dependent manner. J Biol Chem 267:24563–24567
37. Galang CK, Hauser CA (1993) Cooperative DNA binding of the human HoxB5 (Hox-2.1) protein is under redox regulation in vitro. Mol Cell Biol 13:4609–4617
38. Galter D, Mihm S, Dröge W (1994) Distinct effects of glutathione disulphide on the nuclear transcription factor kappa B and the activator protein-1. Eur J Biochem 221:639–648
39. Meyer M, Schrek R, Baeuerle PA (1993) $H_2O_2$ and antioxidants have opposite effects on activation of NF-κB and AP-1 in intact cells: AP-1 as secondary antioxidant-responsive factor. EMBO J 12:2005–2015
40. Schenk H, Klein M, Erdbrügger W, et al. (1994) Distinct effects of thioredoxin and antioxidants on the activation of transcription factors NF-κB and AP-1. Proc Natl Acad Sci USA 91:1672–1676
41. Matthews JR, Wakasugi N, Virelizier J-L, et al. (1992) Thioredoxin regulates the DNA binding activity of NF-κB by reduction of a disulphide bond involving cysteine 62. Nucleic Acids Res 20:3821–3830
42. Okuno H, Akahori A, Sato H, et al. (1993) Escape from redox regulation enhances the transforming activity of Fos. Oncogene 8:695–701

# Inhibition of Cytokines and ICAM-1 Induction in Rheumatoid Fibroblasts by anti-NF-κB Reagents

Shinsaku Sakurada, Tetsuji Kato, Kohichi Mashiba, Yang Jian-Ping, and Takashi Okamoto

*Summary.* The role of the transcription factor, nuclear factor NF-κB, in the induction of cytokines and intercellular adhesion molecule-1 (ICAM-1) on stimulation with interleukin-1 (IL-1) and tumor necrosis factor-α (TNF-α) was studied in primary rheumatoid synovial fibroblasts (RSF) obtained from patients with rheumatoid arthritis (RA). The production of GM-CSF, IL-6, and IL-8 and the expression of ICAM-1 were augmented after nuclear translocation of NF-κB induced by treatment with IL-1 or TNF-α. We examined the effects of *N*-acetyl-L-cysteine (NAC) and acetylsalicylic acid (aspirin) on the induction of proinflammatory cytokines and ICAM-1. Pretreatment of RSF with NAC inhibited nuclear translocation of NF-κB completely, and the induction of these cytokines and ICAM-1 was markedly suppressed. On the other hand, the effect of aspirin was only partial. These observations indicate the pivotal role of NF-κB in RA pathogenesis, thus highlighting the possibility of a novel therapeutic strategy.

*Key words.* Rheumatoid arthritis—Synovial fibroblasts—NF-κB—*N*-Acetyl-L-cysteine—Aspirin

## Introduction

Rheumatoid arthritis (RA) is characterized as chronic and progressive inflammatory processes in synovium with systemic immunological abnormalities. The active involvement of proinflammatory cytokines and cell adhesion molecules (CAMs) has been implicated in these processes [1]. Tumor necrosis factor-α (TNF-α) and interleukin-1 (IL-1) have been known to be important in inducing the expression of other cytokines and CAMs. Furthermore, it is well established that TNF-α and IL-1 stimulate gene expression of proinflammatory cytokines and CAMs through a signal transduction pathway leading to nuclear factor κB (NF-κB) activation [2]. Although NF-κB is by no means a sole determinant for inducible expression of these genes, these observations led us to investigate the role of NF-κB in primary rheumatoid

Department of Molecular Genetics, Nagoya City University Medical School, 1 Kawasumi Mizuho-cho Mizuho-ku, Nagoya 467, Japan

synovial fibroblasts (RSF). We attempted to block the NF-κB activation pathway using NF-κB inhibitiors and then examined the suppression of induction of the RA-associated cytokines and CAMs.

# Materials and Methods

## Cells

RSF were isolated from fresh synovial tissue biopsy samples from three RA patients. The cells obtained were subcultured in F-12 (HAM) medium supplemented with 10% fetal calf serum (FCS) and maintained in the same medium.

## Immunofluorescence

RSF were cultured in eight-well-chamber slides and then stimulated with 10 ng/ml of IL-1β or TNF-α for different time periods. Other sets of cells were pretreated for 1 h with 10 mM N-acetyl-L-cysteine (NAC) or 2.5 mM aspirin before treatment with these cytokines. Rabbit polyclonal anti-NF-κB (p65) (Santa-Cruz Biotechnology, Santa-Cruz, CA, USA) and mouse monoclonal anti-human intercellular adhesion molecule-1 (ICAM-1) (Becton & Dickinson, San Jose, CA, USA) antibodies were used for indirect immunofluorescence and flow cytometrical analysis after fixation.

## Preparation of Nuclear Extracts and Electrophoretic Mobility Shift Assay

Nuclei from $4 \times 10^6$ cells were isolated by treatment with hypotonic lysis buffer. Released nuclei were collected by microcentrifugation and suspended in a different buffer. Nuclear proteins were extracted by sonication and microcentrifugation. DNA-binding activity was examined by electrophoretic mobility shift assay (EMSA) using the specific κB DNA probe (5′-TTTCTAGGGACTTTCCGCCTGGGGACTTTCCAG-3′). For competion experiments, a 50-fold molar excess of unlabeled wild-type κB or mutated κB was preincubated with the proteins before the radioactive probe was added.

## Detection of Cytokines in RSF Culture Medium by ELISA

The concentration of cytokines in RSF culture supernatants was determined using cytokine-specific ELISA kits for IL-1α, IL-1β, and TNF-α (Otsuka Pharmaceutical,

---

Fig. 1. a Nuclear translocation of the nuclear factor NF-κB in rheumatoid synovial fibroblasts (RSF) treated with 10 ng/ml of interleukin 1 (IL-1β) at different time periods (0.5, 1, 3, 6, 9, 12, or 24 h). The cells were reacted with rabbit antibody against p65 of NF-κB subunit and stained with fluorescein isothiocyanate- (FITC-) conjugated goat antirabbit immunoglobulin (IgG) after fixation and permealization. b Electrophoretic mobility shift assay (EMSA) using the κB DNA probe demonstrated the activation of NF-κB induced by IL-1β in RSF. Nuclear extracts were prepared from either untreated (lanes 1-3 and 7-9) or IL-1β-treated (10 ng/ml) RSF (lanes 4-6 and 10-12) and tested for κB DNA-binding activity by EMSA. Specifity of the binding was assessed by an excess amount of cold competitor oligonucleotides (ratios to radiolabeled κB DNA probe are indicated). Positions of the DNA–protein complex "Bound" (closed arrowhead), "Unbound" (open arrowhead), and "Non-specific" (asterisk) are shown

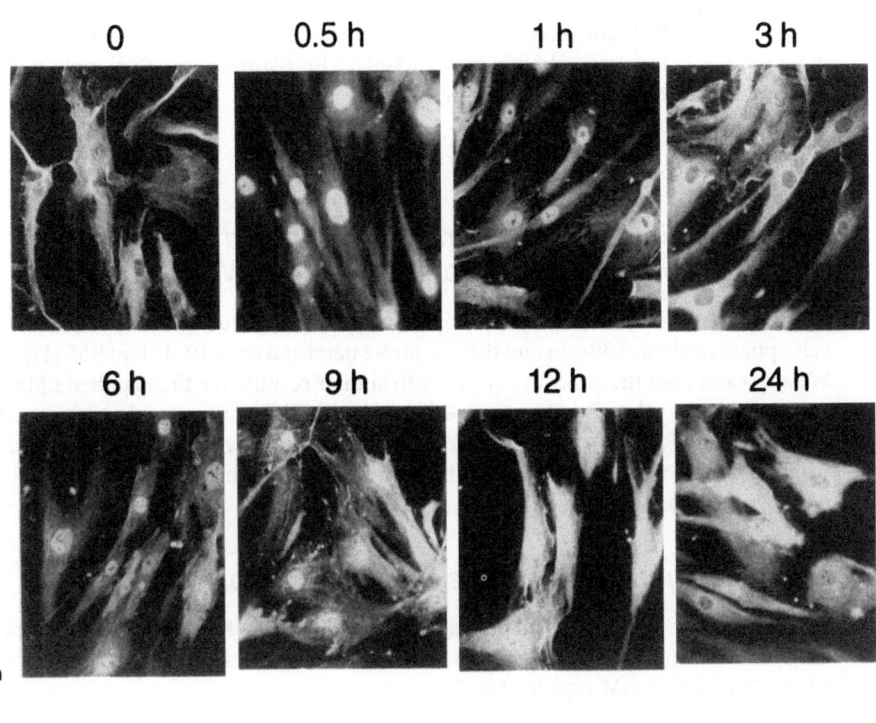

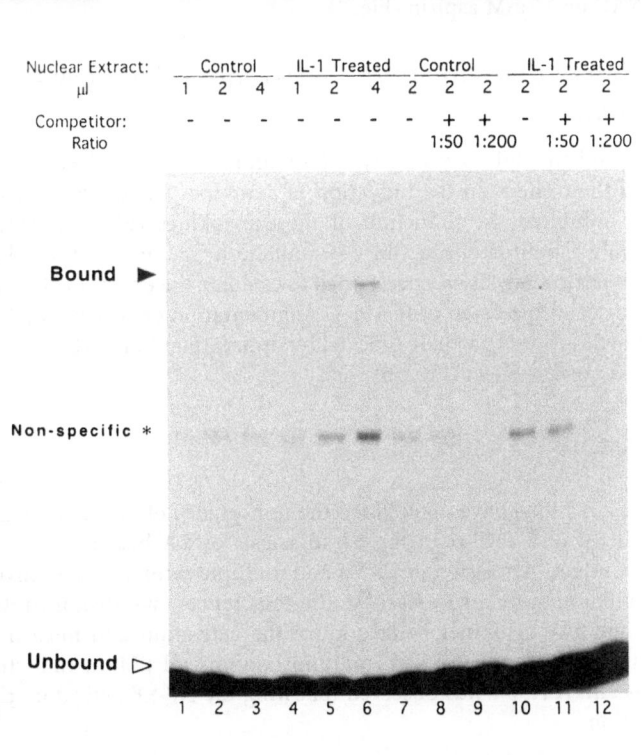

Tokyo, Japan), IL-6 and IL-8 (Toray Fuji Bionics, Tokyo, Japan), and granulocyte-macrophage colony-stimulating factor (GM-CSF) (Amersham, Buckinghamshire, UK) as recommended by the manufacturers.

## Results

### Demonstration of Nuclear Translocation and Activation of NF-κB in RSF by Immunofluorescence and EMSA

The treatment with IL-1β induced a biphasic nuclear translocation of p65 in RSF; the early phase peaked at 30 min and the late phase persisted until 10–12 h in RSF (Fig. 1a). We also examined the effect of TNF-α, with similar results. We then applied EMSA to examine whether the κB sequence-specific DNA-binding activity could be induced in the nucleus of RSF after IL-1β treatment. The nuclear extract from RSF stimulated with IL-1β demonstrated κB DNA-binding activity while that of the control RSF did not have such activity (Fig. 1b).

### NAC and Aspirin Blocked the NF-κB Activation of RSF

We examined the effect of reagents known to modulate the NF-κB activation cascade. The nuclear translocation of the NF-κB was effectively blocked by pretreatment with 10 mM NAC or 2.5 mM aspirin (Fig. 2).

### Inhibition of Cytokine and ICAM-1 Induction by NAC and Aspirin

Having found an inhibitory activity of NAC and aspirin on NF-κB activation, we then examined their effect on the induction of cytokines elicited by IL-1β in RSF. NAC strikingly inhibited the induction of these cytokines (Fig. 3a). However, aspirin showed only a limited effect; GM-CSF induction was not at all blocked by aspirin. Flow cytometric analysis was performed to evaluate the effects of NAC and aspirin on the cell-surface expression of ICAM-1. Augmentation of surface ICAM-1 expression induced by IL-1β was partially blocked by pretreatment of NAC. In contrast, aspirin showed no notable effect (Fig. 3b).

## Discussion

Recent observations have highlighted the importance of proinflammatory cytokines such as IL-1β and TNF-α in the pathogenesis of RA because these cytokines are known to induce expression of CAMs and multiple cytokine genes involved in rheumatoid inflammatory processes [3]. In this report, we demonstrated that these proinflammatory cytokines could induce the activation and nuclear translocation of NF-κB in RSF as revealed by immunostaining and EMSA. Induction of various cytokines and ICAM-1 was observed following the NF-κB activation and its nuclear translocation.

IL-1 (-)

IL-1 (+)          **No Pretreatment**

IL-1 (+)          **NAC 10 mM**

IL-1 (+)          **Aspirin 2.5mM**

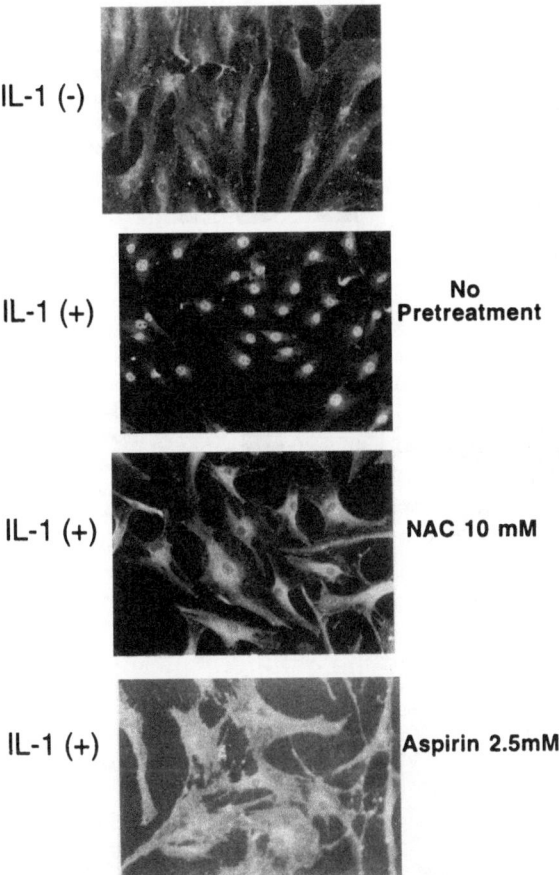

FIG. 2. Effects of N-acetyl-L-cysteine (NAC) and aspirin on the nuclear translocation of p65 NF-κB subunit in RSF. Cells were treated with 10 ng/ml of IL-1β in the presence or absence of 10 mM NAC or 2.5 mM aspirin. After 30 min of treatment the cells were stained by the immunofluorescence technique as described in Fig. 1a

We then examined the effects of reagents reported to inhibit the NF-κB activation cascade, such as NAC and aspirin in RSF. In particular, antioxidant NAC was shown to be strikingly effective in blocking the NF-κB translocation and subsequent induction of target genes. We also observed the apparent differences in the effects of NAC and aspirin in blocking of the NF-κB activation cascade [4]. Experimental observations demonstrated here support the idea that the apparent complex nature of rheumatoid inflammatory processes involving multiple cytokines and CAMs could be explained, at least in part, by the role of a single transcription factor, NF-κB, in RSF obtained from patients with RA.

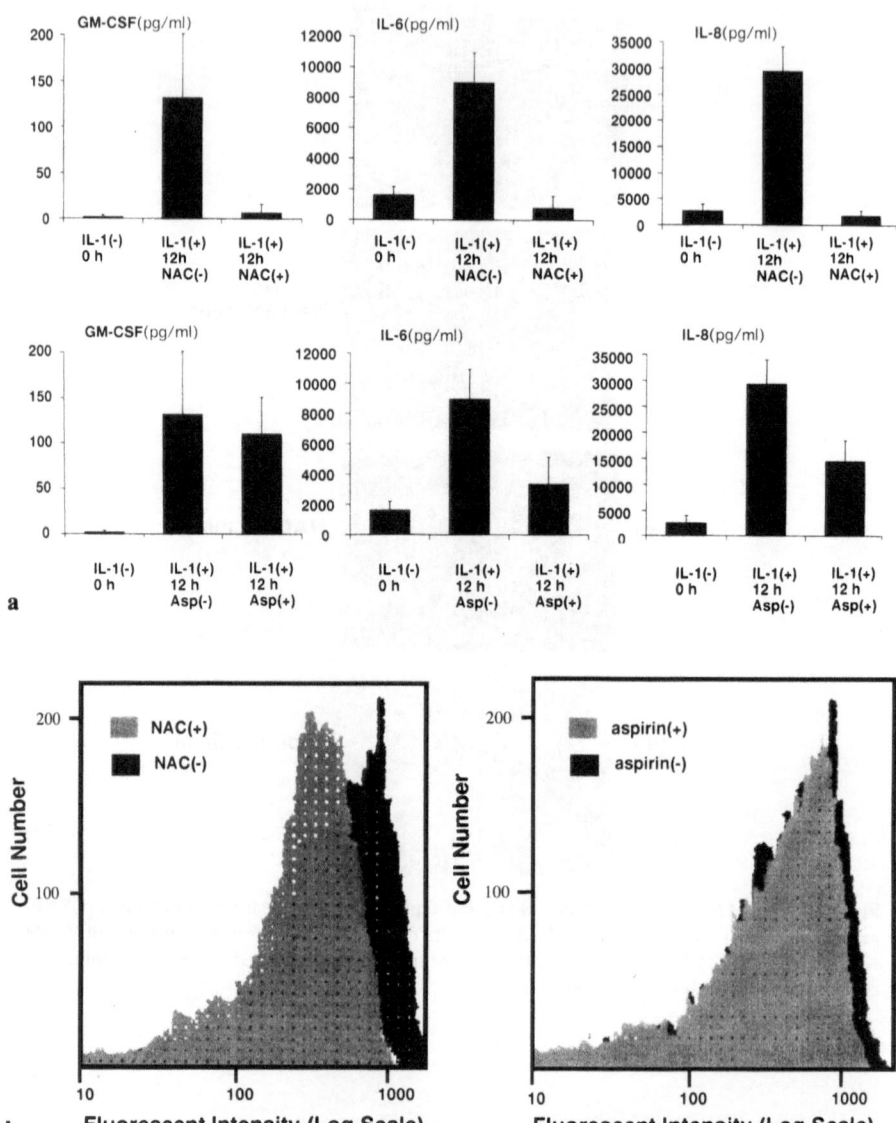

FIG. 3. **a** Effects of NAC and aspirin on the induction of various cytokines by IL-1β. Cells were treated with 10 ng/ml of IL-1β in the presence or absence of 10 mM NAC or 2.5 mM aspirin. The concentration of cytokines was determined using ELISA after 12 h of treatment. Each value indicates the mean ($n = 3$); *error bar* shows SD. **b** Flow cytometric analysis demonstrates the inhibitory effects of NAC and aspirin on the induction of cell-surface intercellular adhesion molecule-1 (ICAM-1) induced by IL-1β. Cells were pretreated with 10 mM NAC or 2.5 mM aspirin for 2 h before the addition of 10 ng/ml IL-1β and incubated with mouse monoclonal antibody to human ICAM-1 and FITC-conjugated rabbit antimouse IgG. Nonspecific fluorescence was assessed by substitution of anti-ICAM-1 antibody by mouse IgG1

# References

1. Alvaro-Gracia JM, Zvaifler NJ, Brown CB, et al. (1991) Cytokines in chronic arthritis. VI. Analysis of the synovial cells involved in granulocyte macrophage colony-stimulating factor production and gene expression in rheumatoid arthritis and its regulation by IL-1 and TNF-α. J Immunol 146:3365–3371
2. Baeuerle PA (1991) The inducible transcription activator NF-κB: regulation by distinct protein subunits. Biochim Biophys Acta 1072:63–80
3. Wickes IP, Leizer T, Wawryk S, et al. (1992) The effect of cytokines on the expression of MHC antigens and ICAM-1 by normal and transformed synoviocytes. Autoimmunity 12:13–19
4. Weber C, Erl W, Pietsh A, et al. (1995) Aspirin inhibits nuclear factor-κB mobilization and monocyte adhesion in stimulated human endothelial cells. Circulation 91:1914

# Detection of a Nuclear 60-kDa Protein Coimmunoprecipitated with Human Thioredoxin

Akira Nishiyama[1], Keizo Furuke[2], Kiichi Hirota[3],
Hiroshi Masutani[1], and Junji Yodoi[1]

*Summary.* Thioredoxin (TRX) has potent dithiol-reducing activity produced by two cysteine residues in its active site, and its reducing activity carries thiol-redox control functions. In unicellular life, TRX and its family proteins play an important role in the cell cycle, DNA replication, and protein secretion. In the mammalian system, however, the involvement of TRX in these cellular functions remains to be elucidated. To investigate the cellular functions of TRX in the mammalian system, we analyzed its binding protein. By an immunoprecipitation study using anti-TRX monoclonal antibody, a 60-kDa (p60) protein was coprecipitated with TRX. The p60 protein was also shown to bind to the glutathione S-transferase–TRX fusion protein. The p60 protein was detected in various human cell lines as well as peripheral blood lymphocytes. A subcellular fractionation study indicated that p60 binding to TRX is distributed in the nuclear fraction, suggesting its possible involvement in TRX-dependent redox regulation in the nuclear compartment.

*Key words.* Human thioredoxin—Protein–protein interaction—Nuclear protein—Immunoprecipitation—Redox regulation

## Introduction

Thioredoxin (TRX) is a ubiquitous protein that has a reducing activity produced by two cysteine residues in its active site [1]. This consensus active site sequence is conserved from bacteria to higher eukaryotes. TRX has important thiol-redox regulative functions by its reducing activity.

In unicellular organisms, TRX and its family proteins are involved in various cellular functions. TRX was originally discovered as a hydrogen donor for ribonucleotide reductase in *Escherichia coli* [1]. In the cell cycle, TRX gene deficiency affects the

[1] Department of Biological Responses, Institute for Virus Research, Kyoto University, 53 Shogoin Kawaharacho, Sakyo-ku, Kyoto 606-01, Japan
[2] Division of Cell and Gene Therapies, Center for Biologics Evaluation and Research, Food and Drug Administration, Rockville Pike, Bethesda, MD 20892, USA
[3] Department of Anesthesia, Kyoto University Hospital, 54 Shogoin Kawaharacho, Sakyo-ku, Kyoto 606-01, Japan

timing of the S and $G_1$ phases in yeast [2]; the S phase is threefold longer and $G_1$ is nearly absent. One of the TRX family proteins, Dsb proteins, is distributed in bacterial periplasm and catalyzes the formation of the disulfide bond in exported proteins [3]. Moreover, another TRX-like membrane protein is essential for biogenesis of a cytochrome oxidase [4]. In *Drosophila* and mouse, TRX gene deletion causes embryonic lethality [5,6]. These data suggest that TRX plays an important role in cellular responses by reducing thiol residues in mammalian cells. For example, TRX is known to have an enhancing effect on binding of transcriptional factors to DNA [7–9]. In our study to investigate unknown cellular functions of mammalian TRX, we have analyzed TRX-binding proteins in human cells. We report here a nuclear 60-kDa protein as a candidate of TRX-binding proteins.

# Methods

## Cell Culture and Reagents

Jurkat cells were maintained in RPMI1640 medium (Gibco BRL, Gaithersburg, MD, USA) containing 10% fetal calf serum (Gibco) and antibiotics (100 U/ml penicillin and 100 µg/ml streptomycin) in 5% $CO_2$ at 37°C. Anti-human thioredoxin monoclonal antibodies, 11 Ab and 21 Ab, were established and provided by Fujirebio (Tokyo, Japan). Mouse monoclonal immunoglobulin G (MOPC 21 antibody) was purchased from Sigma (St. Louis, MO, USA). Protein G sepharose was purchased from Zymed (San Francisco, CA, USA).

## Immunoprecipitation and Subcellular Fractionation

The cells were collected and washed twice with ice-cold phosphate-buffered saline (PBS). The cell lysates were obtained with NP-40 buffer containing 150 mM NaCl, 1.0% Nonidet P-40, and 50 mM Tris-HCl pH 7.5. After centrifugation for 15 min at $12\,000 \times g$, the supernatants containing 1 mg of protein were subjected to immunoprecipitation. After preclearing by protein G sepharose, samples were rotated with antibodies for 2 h at 4°C and incubated with protein G sepharose for an additional 20 min. Then, samples were centrifuged and washed five times with NP-40 buffer. The precipitated proteins were separated on sodium dodecyl sulfate-polyacrylamide gel electrophoresis (SDS-PAGE) and visualized by silver staining (silver stain kit, Wako, Osaka, Japan). Subcellular fractionation was performed by the procedure described by Hockenbery et al. [10]. In each collected fraction, immunoprecipitation with anti-TRX antibody 11Ab was conducted.

## Bacterial Expression of the Fusion Protein and Its Binding Assay

The glutathione S-transferase (GST) gene fusion system (Pharmacia Biotech, Uppsala, Sweden) was used for the expression of GST fusion protein. TRX cDNA was inserted into the pGEX-4T vector and transformed *E. coli* strain BL21. GST fusion proteins were prepared according to the manufacturer's instructions. Binding assays using GST fusion proteins were conducted under the same conditions as immunoprecipitation. Briefly, Jurkat cells were lysed by NP-40 buffer and centrifuged at $12\,000 \times g$. The supernatants were precleared by using glutathione-sepharose (Pharmacia) and incubated with each GST fusion protein. Then, samples were added to glutathione- . sepharose and rotated additionally. The samples were centrifuged and washed with NP-40 buffer. Collected samples were analyzed by SDS-PAGE.

## Results

To analyze TRX-binding protein in mammalian cells, we performed immunoprecipitation experiments using the anti-TRX monoclonal antibodies 11Ab and 21Ab. A 60-kDa protein (p60) was specifically coimmunoprecipitated using anti-TRX antibodies (Fig. 1A), while other proteins were also detected by immunoprecipitation with control MOPC 21 antibody. These two anti-TRX antibodies recognize different sites of TRX. Although p60 was coimmunoprecipitated by using either of the anti-TRX antibodies, both antibodies failed to recognize p60 in Western blotting analysis (data not shown). We used bacterial expressed GST-TRX fusion protein to analyze whether p60 directly binds to TRX. As shown in Fig. 1B, p60 bound to GST-TRX but not to GST, suggesting that p60 can directly interact with TRX. We also detected p60 in other cell

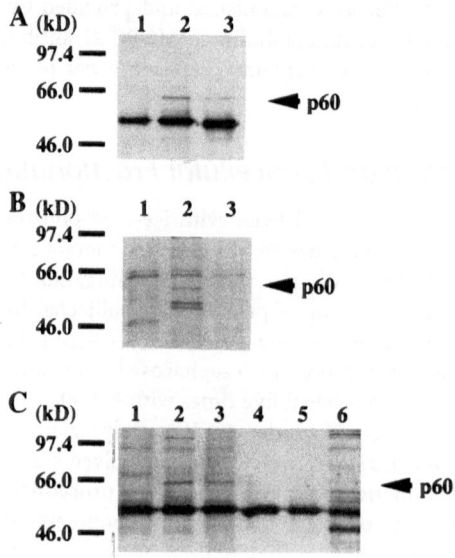

Fig. 1A–C. Detection and analysis of coimmunoprecipitated proteins with thioredoxin (TRX). A Immunoprecipitation using anti-TRX antibodies. Antibodies used in this experiment are MOPC-21 antibody (*lane 1*) as negative control, and anti-TRX antibodies 11Ab (*lane 2*) and 21Ab (*lane 3*). B Binding assay with glutathione S-transferase- (GST-) TRX fusion protein was conducted using 5 μg GST-TRX (*lane 2*). As negative controls, 5 μg GST (*lane 1*) and glutathione-sepharose (*lane 3*) were used. C Localization of the TRX–p60 interaction. In total lysate (*lane 2*) and each subcellular fraction, nuclear (*lane 3*), organelle (*lane 4*), membrane (*lane 5*), and cytoplasm (*lane 6*), immunoprecipitations were conducted by anti-TRX antibody 11Ab. As a negative control, MOPC-21 antibody was used (*lane 1*)

lines, including U937, HeLa, and human peripheral blood T lymphocytes (data not shown).

Finally, to determine the localization of this interaction, we performed subcellular fractionation of Jurkat cells. In each collected fractions, the interaction was detected only in the nuclear fraction (Fig. 1C).

## Discussion

In this study, we identified a 60-kDa protein that binds directly and stably to TRX. Interactions between TRX and other proteins have been reported in yeast [11]. In this report, the cooperative function of TRX and a 10-kDa protein is involved in lysosome division of the yeast cell cycle. This 10-kDa protein and TRX are copurified as protein with lysosome vesicle fusion activity. It is revealed that TRX is required in organelle division by TRX gene disruption. This is one example of a complexed mechanism that requires TRX.

The result of the subcellular fractionation study suggested that the p60–TRX complex is mainly localized in the nuclear compartment. There is accumulating evidence indicating the roles of TRX in the nuclear compartment. TRX was reported to be a hydrogen donor of ribonucleotide reductase in *E. coli* [1]. TRX is known to enhance the DNA-binding activity of several transcriptional factors by its reducing activity in mammalian cells [7–9]. On UV or phorbol myristate acetate stimulation, TRX is translocated from cytoplasm to nucleus (Hirota et al, manuscript in preparation). As TRX lacks a nuclear localization signal and C-terminal-truncated TRX failed to translocate, TRX-binding protein may be involved in this phenomenon.

From these findings, it is possible that, in the nucleus or nuclear compartment of mammalian cells, TRX acts synergistically with p60 and regulates nuclear processes in which p60 is involved. Through analysis of p60 function and its regulation by TRX, novel functions of TRX in the nuclear compartment would be made clear.

## References

1. Holmgren A (1985) Thioredoxin. Annu Rev Biochem 54:237–271
2. Muller EGD (1991) Thioredoxin deficiency in yeast prolongs S phase and shortens the $G_1$ interval of the cell cycle. J Biol Chem 266:9194–9202
3. Bardwell JC, McGovern K, Beckwith J (1991) Identification of a protein required for disulfide bond formation in vivo. Cell 67:581–589
4. Loferer H, Bott M, Hennecke H (1993) *Bradyrhizobium japonicum* TlpA, a novel membrane-anchored thioredoxin-like protein involved in the biogenesis of cytochrome $aa_3$ and development of symbiosis. EMBO J 9:3373–3383
5. Salz HK, Flickinger TW, Mittendorf E, et al. (1994) The *Drosophila* maternal effect locus deadhead encodes a thioredoxin homolog required for female meiosis and early embryonic development. Genetics 136:1075–1086
6. Matsui M, Oshima M, Oshima H, et al. (1996) Early embryonic lethality caused by targeted disruption of the mouse thioredoxin gene. Dev Biol 178:179–185
7. Grippo J, Holmgren A, Pratt WB (1985) Proof that the endogenous, heat-stable glucocorticoid receptor-activating factor is thioredoxin. J Biol Chem 260:93–97
8. Xanthoudakis S, Miao G, Wang F, et al. (1992) Redox activation of Fos-Jun DNA binding activity is mediated by a DNA repair enzyme. EMBO J 11:3323–3335

9. Matthews JC, Wakasugi N, Virelizier J, et al. (1992) Thioredoxin regulates the DNA binding activity of NF-kappa B by reduction of a disulphide bond involving cysteine 62. Nucleic Acids Res 20:3821–3830
10. Hockenbery D, Nuñez G, Milliman C, et al. (1990) Bcl-2 is an inner mitochondrial membrane protein that blocks programmed cell death. Nature 348:334–336
11. Xu Z, Wickner W (1996) Thioredoxin is required for vacuole inheritance in *Saccharomyces cerevisiae*. J Cell Biol 132:787–794

# Ito Cells: A Putative Cellular Component Responsible for Carbon Monoxide-Mediated Microvascular Relaxation in the Rat Liver

Yoshiyuki Wakabayashi, Satoshi Kashiwagi, Nobuhito Goda, Yuzuru Ishimura and Makoto Suematsu

*Summary.* Ito cells are microvascular pericytes occurring specifically in the liver. They are characterized by abundant fat droplets and constitute a major storage pool of vitamin A in the liver. These cells encircle the outer surface of microvascular walls and constitute a well-organized meshwork of intercellular connection by using their unique neuron-like dendritic structure. Ito cells have thus been considered to be a putative machinery controlling sinusoidal blood flow. Carbon monoxide (CO) generated by the heme oxygenase reaction serves as an endogenous relaxing factor that actively relaxes hepatic sinusoids. Although such a CO-dependent vasorelaxing mechanism seems to involve cyclic guanosine monophosphate- (cGMP-) dependent relaxation of Ito cells, it is still unknown whether Ito cells can exhibit cell relaxation in response to CO through unidentified cGMP-independent mechanisms. This chapter provides an overview of mechanisms for CO-dependent Ito cell relaxation in vivo and in vitro.

*Key words.* Heme oxygenase—Liver-specific pericyte—Ion channel—Guanylate cyclase

## Introduction

Ito cells, the liver-specific microvascular pericytes [1], have attracted great interest as a regulatory mechanism of sinusoidal blood flow. These cells line the outer surface of sinusoidal walls and thus analogous to vascular smooth muscle cells (VSMCs) in larger vessels (Fig. 1). Like VSMCs, Ito cells can exhibit contractile phenotypes in response to a variety of vasoactive substances; vasoconstrictive agonists such as angiotensin II and endothelin-1 (ET-1) evoke an increase in intracellular calcium concentration and cellular contraction in Ito cells [2]. Ito cells can relax in response to a variety of vasorelaxants such as prostaglandin $E_2$ and nitric oxide (NO). In addition, considerable experimental evidence has been provided that suggests the possibility that carbon monoxide (CO) serves as an endogenous modulator of sinusoidal tone

Department of Biochemistry, School of Medicine, Keio University, 35 Shinanomachi, Shinjuku-ku, Tokyo 160, Japan

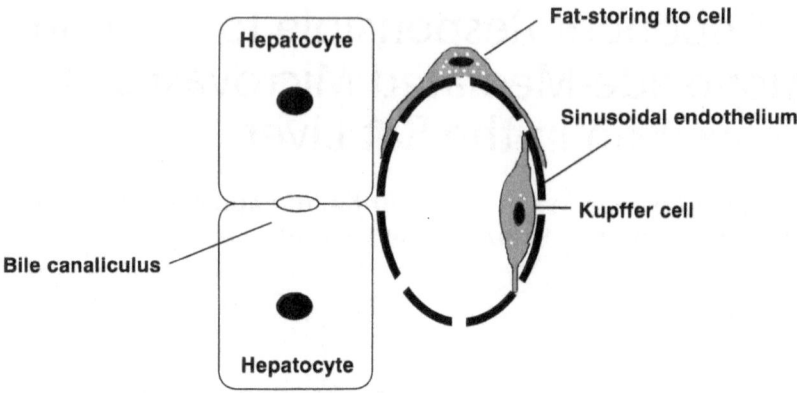

FIG. 1.   Anatomical orientation of liver cells

through its vasorelaxing action on Ito cells [3,4]. This chapter summarizes recent aspects of the Ito cell-mediated regulatory mechanism of sinusoidal perfusion through this novel gaseous mediator.

## Hepatic Microcirculation and Fat-Storing Ito Cells

Ito cells are characterized by abundant fat droplets that serve as a major vitamin A storage pool in the liver [5]. Because of their anatomical orientation in the sinusoidal compartment, Ito cells have attracted great interest by investigators as a machinery controlling sinusoidal blood flow. Kawada et al. [2] showed that cultured Ito cells can cause wrinkling of silicone rubber membranes when exposed to vasoconstrictors such as thromboxane $A_2$ analogues, prostaglandin (PG) $F_{2\alpha}$, and ET-1, while vasorelaxants such as $PGI_2$ analogue, $PGE_2$, and NO relax the cells. Intravital evidence suggesting the ability of Ito cells to constrict was also provided by Zhang et al. [6], who applied in vivo microscopy to the rat liver and showed that transportal administration of ET-1 can induce a discrete pattern of constriction in local sinusoidal segments colocalized with Ito cells.

## CO: A Novel Microvascular Regulator in the Liver

Carbon monoxide (CO) is a by-product of the heme oxygenase reaction that oxidatively degrades protoheme IX into a bile pigment biliverdin. This reaction serves as a rate-limiting step for protoheme IX metabolism. CO has recently been recognized as a major gaseous substance that is endogenously produced in the liver (see the chapter by M. Suematsu et al, this volume). Using the isolated perfused liver preparation, we demonstrated that suppression of endogenous CO production by administration of zinc protoporphyrin IX (ZnPP), a heme oxygenase inhibitor, elicited discontinuous and discrete patterns of narrowing of sinusoidal vessels in the perfused rat liver [4]. Morphometric analysis based on digital microfluorography revealed a significant correlation between the location of Ito cells and local sinusoidal segments, representing a narrowing response.

## Guanylate Cyclase: A CO-Dependent Relaxing Mechanism in Ito Cells?

Like NO, CO has the ability to activate soluble guanylate cyclase [7]. However, the potential of CO to upregulate cGMP is known to be as small as only a few percent of that of NO [8]. Using the cultured system, we showed that exogenously applied CO at micromolar levels increased intracellular cGMP levels. In addition, the same doses of CO downregulated actin polymerization in the same cells. These lines of evidence suggest that CO serves as a relaxing mediator for Ito cells. On the other hand, experimental data collected from the isolated perfused liver preparation raised the possibility that CO functions as a relaxant through an unidentified cGMP-independent mechanism. First, exogenous cGMP analogue did not completely reverse the ZnPP-elicited increase in the vascular resistance and sinusoidal narrowing. Second, the levels of CO concentrations detected in the venous perfusate were not so high as those that induced a relaxing response in the cultured cells.

Although such a cGMP-independent cell-relaxing mechanism is quite unknown in Ito cells, our recent electrophysiological evaluation using a whole-cell configuration of patch-clamp techniques has revealed the presence of outward potassium rectifier channels in Ito cells [9]. Because this type of $K^+$ channel present in rabbit corneal epithelial cells or human jejunal circular smooth muscle is known to increase its opening probability in response to CO and to be hyperpolarized [10,11], it is not reasonable to speculate that endogenously produced CO activates this type of channel and hyperpolarizes Ito cells to help relax the sinusoids.

## Conclusion

Although the reactivity of Ito cells in response to gaseous monoxides such as NO and CO has been studied in cultured systems in vitro as well as in vivo, the cellular components responsible for endogenous CO production in the liver are not yet well understood. Furthermore, the signal-transducing molecules (presumably proteins or enzymes possessing a heme moiety) that serve as receptors for CO are quite unknown in Ito cells. Further efforts to specify such heme proteins are necessary to establish the role of Ito cells in CO-mediated control of sinusoidal blood perfusion.

*Acknowledgments.* This work was supported by a grant-in-aid for scientific research from the Ministry of Education, Science, Sports and Culture of Japan, and by grants from School of Medicine, Keio University.

## References

1. Ito T (1951) Cytological studies on stellate cells of Kupffer and fat-storing cells in the capillary wall of human liver. Acta Anat Jpn 26:42

2. Kawada N, Tran-Thi TA, Klein H, et al. (1993) The contraction of hepatic stellate (Ito) cells stimulated with vasoactive substances: possible involvement of endothelin 1 and nitric oxide in the regulation of the sinusoidal tonus. Eur J Biochem 213:815–823
3. Suematsu M, Kashiwagi S, Sano T, et al. (1994) Carbon monoxide as an endogenous modulator of hepatic vascular perfusion. Biochem Biophys Res Commun 205:1333–1337
4. Suematsu M, Goda N, Sano T, et al. (1995) Carbon monoxide: an endogenous modulator of sinusoidal tone in the perfused rat liver. J Clin Invest 96:2431–2437
5. Suematsu M, Oda M, Suzuki H, et al. (1993) Intravital and electron microscopic observation of Ito cells in rat hepatic microcirculation. Microvasc Res 46:28–42
6. Zhang JX, Pegoli W Jr, Clemens MG (1994) Endothelin-1 induces direct constriction of hepatic sinusoids. Am J Physiol 266:G624–G632
7. Schmidt HW (1992) NO, CO and OH: endogenous soluble guanylyl cyclase activating factors. FEBS Lett 307:102–107
8. Kharitonov V, Sharma VS, Pilz RB, et al. (1995) Basis of guanylate cyclase activation by carbon monoxide. Proc Natl Acad Sci USA 92:2568–2571
9. Kashiwagi S, Suematsu M, Wakabayashi Y, et al. (1997) Electrophysiological characterization of cultured hepatic stellate cells in rats. Am J Physiol 272:G742–G750
10. Farrugia G, Irons WA, Rae JL, et al. (1993) Activation of whole cell currents in isolated human jejunal circular smooth muscle cells by carbon monoxide. Am J Physiol 264:G1184–G1189
11. Adam R, Farrugia G, Rae JL (1994) Carbon monoxide stimulates a potassium-selective current in rabbit corneal epithelial cells. Am J Physiol 267:C435–C442

# K$^+$ Channels and the Normoxic Constriction of the Rabbit Ductus Arteriosus

HELEN L. REEVE[1], MARTIN TRISTANI-FIROUZI[2], SIMONA TOLAROVA[1], STEPHEN L. ARCHER[1], and E. KENNETH WEIR[1]

*Summary.* The ductus arteriosus (DA) is a vital fetal structure that acts as a right-to-left shunt to divert blood flow away from the constricted pulmonary circulation in the developing fetus. At birth, the DA constricts as a direct result of the increase in oxygen ($O_2$) tension that occurs. The mechanism for this $O_2$-mediated constriction remains controversial. We have shown that the smooth muscle of the DA contains at least two types of potassium (K$^+$) channel, a 4-aminopyridine-sensitive, delayed rectifier (Kv) channel and a tetraethlyammonium-sensitive, calcium- (Ca$^{2+}$-) dependent K$^+$ channel. Increased levels of $O_2$ appear to selectively inhibit the activity of the Kv channel. Because this channel controls the resting membrane potential of DA smooth muscle cells, this inhibition results in the depolarization of the cell membrane, opening of voltage-gated L-type Ca$^{2+}$ channels, influx of Ca$^{2+}$, and hence constriction. We suggest that, under normal conditions, this mechanism may initiate the normoxic constriction of the DA.

*Key words.* Ductus arteriosus—Smooth muscle—Oxygen:potassium channel—Calcium channel

The ductus arteriosus (DA) is a shunt pathway that diverts right ventricular output directly into the descending aorta in the developing fetus. Because the lung vasculature is constricted during development, the DA acts to prevent unnecessary blood flow to the pulmonary circulation. At birth, the increase in blood oxygen ($O_2$) that results from ventilation of the lungs causes the DA to constrict and thus removes this right-to-left shunt pathway [1].

The mechanism by which constriction of the DA occurs at birth remains controversial. There is evidence that the DA is actively maintained patent (dilated) during development by the presence of dilator prostaglandins, in particular prostaglandin $E_2$ [2,3], as well as a lesser contribution from nitric oxide [4]. Indeed, the most common and most successful treatment for patent DA in premature infants is administration of

[1] Veterans Affairs, Medical Center, One Veterans Drive, Minneapolis, MN 55417, U.S.A.
[2] Pediatric Cardiology, Primary Children's Medical Center, 100 N. Medical Drive, Salt Lake City, UT 84113, USA

the prostaglandin synthetase inhibitor indomethacin [5,6]. However, normoxic constriction of the DA can still be demonstrated in DA rings treated with indomethacin, suggesting that normal DA constriction at birth is not caused simply by the decrease in production and potency of dilator prostaglandins [7].

The mechanism responsible for normoxic constriction of the DA is known to be intrinsic to the smooth muscle, as it can be demonstrated in strips of DA from which the adventital and intimal layers have been removed [8]. It is also known to be directly caused by the change in $O_2$ that occurs at birth, because lung ventilation with nitrogen is not a sufficient stimulus to induce DA constriction [9,10].

There have been many studies of the mechanism by which $O_2$ mediates DA constriction. Coceani et al. found that carbon monoxide could relax the normoxia-constricted DA in the fetal lamb and that it did so by interfering with a cytochrome P-450 enzyme system [11–14]. This suggested that DA constriction might be mediated by the activation of a cytochrome P-450 mechanism, resulting in the production of a constrictor. It was also shown that endothelin-1 (ET-1) was a potent constrictor of the DA, through activation of $ET_A$ receptors [15]. Because $O_2$ increased the synthesis of ET-1, this could be a plausible mechanism for DA constriction, with the high levels of $O_2$ associated with birth activating ET-1 synthesis through a cytochrome P-450 system. On closer examination, however, this mechanism seems unlikely to be the only pathway by which normoxic constriction of the DA occurs. The time course of $O_2$-induced ET-1 synthesis is much slower than that of normoxic constriction, perhaps suggesting a role for ET-1 in the maintenance of DA constriction, rather than the initiation. It is also evident that normoxic constriction cannot be completely blocked by the presence of endothelin antagonists, with up to 50% constriction to 95% $O_2$ still occurring in the presence of the endothelin antagonist BQ123 [16].

A second mechanism for normoxic DA constriction was recently proposed by our laboratory [17]. Hypoxic isolated DA rings constricted in response to 4-aminopyridine (4AP; 10mM), a blocker of delayed rectifier (Kv) potassium ($K^+$) channels. In the presence of 4AP, there was no additional constriction of the rings to normoxia. In contrast, there was no consistent constriction of the rings to either tetraethylammonium (TEA), a blocker of calcium- ($Ca^{2+}$-) dependent $K^+$ channels (KCas) or glibenclamide (a blocker of ATP-dependent $K^+$ channels), even at high concentration. Because there is substantial evidence in other $O_2$-sensitive tissues such as the pulmonary artery [18–20] and carotid body [21,22] that $K^+$ channels are involved in the $O_2$ response associated with these vessels, the role of $K^+$ channels in the DA was further investigated. Amphotericin-perforated patch-clamp recordings [23] of whole-cell $K^+$ channel currents from single, isolated, DA smooth muscle cells indicated that there were at least two types of $K^+$ channel present, a 4AP-sensitive Kv channel and a TEA-sensitive KCa channel. Single-channel studies showed these channels to have conductances of 58 and 150pS, respectively, in the cell-attached configuration (Fig. 1).

Recordings of membrane potential under hypoxic conditions indicated that the cell membrane could only be depolarized by 4AP (1mM) and not by TEA (5mM) or glibenclamide (10µM), suggesting that the $K^+$ channel setting the resting membrane potential in DA smooth muscle (and therefore controlling tone) was a Kv channel. While this provided an explanation for the constriction observed in the DA rings to 4AP, it did not provide evidence for an effect of normoxia on $K^+$ channel activity. Further studies, however, showed that whole-cell $K^+$ channel currents recorded from

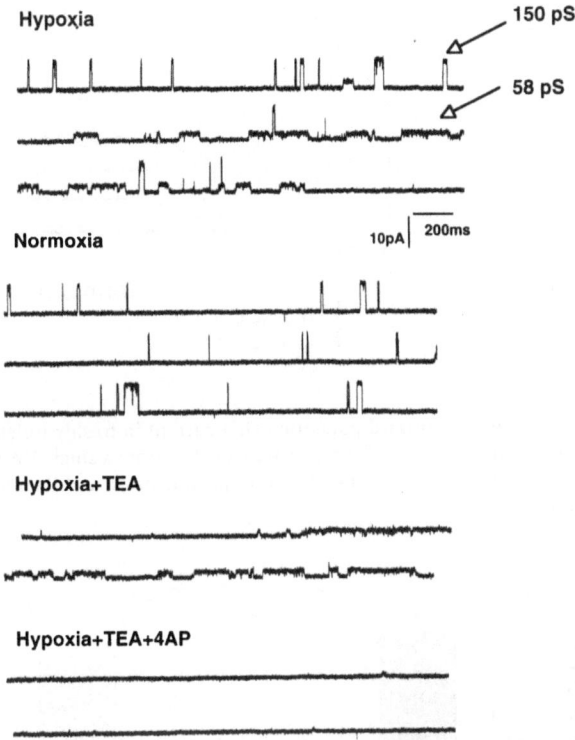

**Hypoxia**

150 pS

58 pS

**Normoxia**

10pA | 200ms

**Hypoxia+TEA**

**Hypoxia+TEA+4AP**

FIG. 1. Normoxia inhibits a 58-pS, 4-aminopyridine (4AP-) sensitive K$^+$ channel, shown by recordings of single-channel activity in cell-attached patches at +40 mV. Traces recorded from a single ductus arteriosus (DA) smooth muscle cell during hypoxia and normoxia show the presence of two types of K$^+$ channel with conductances of 58 and 150 pS (*arrows*). Normoxia decreases the activity of the 58-pS channel. Pharmacological characterization of the single channels using tetraethylammonium (*TEA*; 5 mM) and 4AP (1 mM), shown in *bottom traces*, indicate the 58-pS channel to be inhibited by 4AP and the 150-pS channel to be inhibited by TEA. (Redrawn from [17], with permission)

the DA smooth muscle could be partially inhibited by exposure to normoxia (Fig. 2). Single-channel studies indicated that this normoxic inhibition was specific to the 58-pS Kv channel (see Fig. 1). In addition, cellular membrane potential was depolarized by normoxic exposure, suggesting a physiological role for this inhibition. These data indicated that normoxia may cause DA constriction by inhibition of the activity of a 4AP-sensitive Kv channel in the smooth muscle membrane, resulting in membrane depolarization and Ca$^{2+}$-influx through voltage-dependent CA$^{2+}$ channels. This influx in Ca$^{2+}$ would act as the stimulus for DA constriction. Indeed, a normoxia-induced rise in Ca$^{2+}$ was shown to occur in isolated DA rings loaded with the Ca$^{2+}$-sensitive fluorescent dye fura-2, and this rise could be mimicked by 4AP. Further studies in isolated DA rings appeared to indicate that this rise in Ca$^{2+}$ was the result of influx through voltage-dependent Ca$^{2+}$ channels rather than release from intracellular

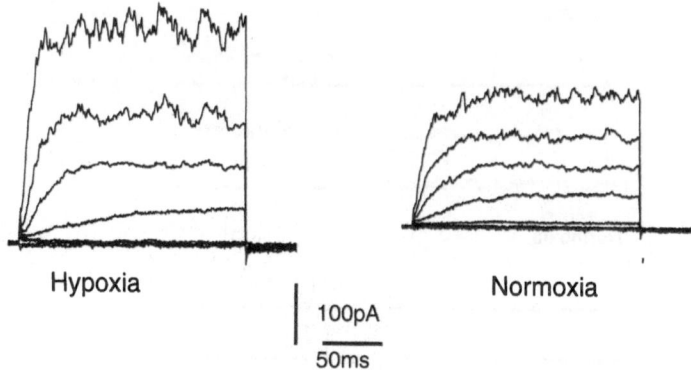

FIG. 2. Normoxia inhibits the outward potassium ($K^+$) current in freshly isolated DA smooth muscle cells. Representative $K^+$ channel currents were evoked from a single DA smooth muscle cell from a holding potential of −70 mV in +20 mV steps in hypoxia and then following 2-min exposure to normoxia

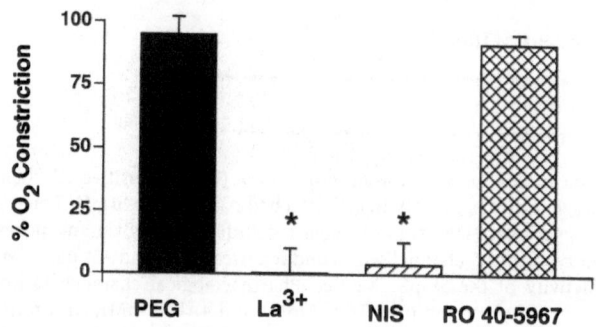

FIG. 3. Normoxic constriction is completely reversed by L-type calcium channel blockers. $La^{3+}$ (5 mM) and nisoldipine (*NIS*; 0.5 μM) completely relaxed the normoxic-constricted DA rings while RO 40-5967 and polyethylene glycol (*PEG*; vehicle for nisoldipine) had no effect. *, P < .001 compared with PEG. (From [17], with permission)

stores. Normoxia-induced constriction could be rapidly and completely reversed by the nonspecific $Ca^{2+}$-channel-blocking anion $La^{3+}$ (5 mM) and low concentrations of the L-type $Ca^{2+}$ channel blocker nisoldipine (0.5 μM). The T-type $Ca^{2+}$ channel blocker RO 40-5967 had no effect on normoxic DA constriction (Fig. 3).

While at present there are no data characterizing the source(s) of the $Ca^{2+}$ rise initiating ET-1 constriction in the DA, in other vascular tissue the activation of $ET_A$ receptors initiates an $IP_3$-mediated release of intracellular $Ca^{2+}$ and influx through nonselective cation channels. Membrane depolarization may then open L-type $Ca^{2+}$ channels and thus maintain constriction [24]. Because the normoxic response in the isolated DA rings can be completely inhibited by low concentrations of L-type $Ca^{2+}$

channel blockers, this may provide further evidence that an ET-1-mediated mechanism is not the sole mediator of normoxic DA constriction.

We have provided a complete characterization of a pathway through which normoxia may cause constriction of the DA. Inhibition of a 58-pS Kv channel by increased $O_2$ results in membrane depolarization and influx of $Ca^{2+}$ through L-type $Ca^{2+}$ channels. This theory is not necessarily in conflict with the extensive data provided by the work of Coceani and colleagues [11–14]. Recent evidence shows that ET-1 is an inhibitor of Kv channels in vascular smooth muscle [25]. It is plausible that the *initiating* event for the functional closure of the DA occurs through the effect of $O_2$ on K⁺ channel activity. Because both increases in $O_2$ [2] and a decrease in prostaglandin production [26] are known to initiate synthesis of ET-1, the conditions associated with birth lead to an increase in production of the constrictor. ET-1 may then maintain DA constriction before its permanent anatomical closure (necrosis and fibrosis of the intimal and medial layers) through its effect on K⁺ channel activity and/or release of intracellular $Ca^{2+}$ through $ET_A$ receptor activation.

*Acknowledgments.* Dr. Reeve is the 1997 recipient of the Giles F. Filley Memorial Award for Excellence in Respiratory Physiology and Medicine. She is supported, in part, by grant funding from the American Heart Association (Minnesota Affiliate) and by VA Merit Review Grant funding through Drs. E.K. Weir and S.L. Archer.

# References

1. Heymann MA, Rudolph AM (1975) Control of the ductus arteriosus. Physiol Rev 55:62
2. Coceani F, Olley PM. (1973) The response of the ductus arteriosus to prostaglandins. Can J Physiol 51:220–223
3. Coceani F, Olley PM (1988) The control of cardiovascular shunts in the fetal and perinatal period. Can J Physiol Pharmacol 66:1129–1134
4. Coceani F, Kelsey L, Seidlitz E (1994) Occurrence of endothelium-derived relaxing factor/nitric oxide in the lamb ductus arteriosus. Can J Physiol Pharmacol 72:82–88
5. Friedman WF, Hirschklau MJ, Printz MP, et al. (1976) Pharmacologic closure of patent ductus arteriosus in the premature infant. N Engl J Med 295:526–529
6. Heymann MA, Rudolph AM, Silverman NH (1976) Closure of the ductus arteriosus in premature infants by inhibition of prostglandin synthesis. N Engl J Med 295:530–533
7. Clyman RI, Mauray F, Roman C, et al. (1983) Effect of gestational age on ductus arteriosus response to circulating prostaglandin E₂. J Pediatr 102:907–911
8. Fay FS (1971) Guinea pig ductus arteriosus. I. Cellular and metabolic basis for oxygen sensitivity. Am J Physiol 221:470–479
9. Kennedy JA, Clark SL (1941) Observations on the ductus arteriosus of the guinea pig in relation to its method of closure. Anat Rec 79:349–371
10. Kennedy JA, Clark SL (1942) Observations on the physiological reactions of the ductus arteriosus. Am J Physiol 136:140
11. Coceani F, Hamilton NC, Labuc J, Olley PM (1984) Cytochrome P-450-linked monooxygenase: involvement in the lamb ductus arteriosus. Am J Physiol 246:H640–H643
12. Coceani F, Breen CA, Lees JG, (1988) Further evidence implicating a cytochrome P-450-mediated reaction in the contractile tension of the lamb ductus arteriosus. Circ Res 62:471–477
13. Coceani F, Wright J, Breen C (1989) Ductus arteriosus: involvement of a sarcolemmal cytochrome P-450 in $O_2$ constriction? Can J Physiol Pharmacol 67:1448–1450

14. Coceani F, Kelsey L, Ackerley C, et al. (1994) Cytochrome P-450 during ontogenic development: occurrence in the ductus arteriosus and other tissues. Can J Physiol Pharmacol 72:217-226
15. Coceani F, Armstrong C, Kelsey L (1989) Endothelin is a potent constrictor of the lamb ductus arteriosus. Can J Physiol Pharmacol 67:902-904
16. Coceani F, Kelsey L, Seidlitz E (1992) Evidence of an effector role of endothelin in closure of the ductus arteriosus at birth. Can J Physiol Pharmacol 70:1061-1064
17. Tristani-Firouzi M, Reeve HL, Tolarova S, et al. (1996) Oxygen-induced constriction of rabbit ductus arteriosus occurs via inhibition of a 4-aminopyridine-, voltage-sensitive potassium channel. J Clin Invest 98:1959-1965
18. Post JM, Hume JR, Archer SL, Weir EK (1992) Direct role for potassium channel inhibition in hypoxic pulmonary vasoconstriction. Am J Physiol 262:C882-C890
19. Yuan X-Y, Goldman WF, Tod ML, et al. (1993) Hypoxia reduces potassium currents in cultured rat pulmonary but not mesenteric arterial myocytes. Am J Physiol 262:C882-C890
20. Archer SL, Huang JMC, Reeve HL, et al. (1996) The differential distribution of electro-physiologically distinct myocytes in conduit and resistance arteries determines their response to nitric oxide and hypoxia. Circ Res 78:431-442
21. Lopez-Barneo J, Lopez-Lopez JR, Urena J, Gonzalez C (1988) Chemotransduction in the carotid body: $K^+$ current modulated by $PO_2$ in type I chemoreceptor cells. Science 241:580-582
22. Peers C (1990) Hypoxic suppression of $K^+$ currents in type I carotid body cells: selective effect on the $Ca^{2+}$-activated $K^+$ current. Neurosci Lett 119:253-256
23. Rae J, Cooper K, Gates G, Watsky M (1991) Low access resistance perforated patch recordings using amphotericin B. J Neurosci Methods 37:15-26
24. Wong AY, Klassen GA (1996) Endothelin-induced electrical activity and calcium dynamics in vascular smooth muscle cells: a model study. Ann Biomed Eng 24:547-560
25. Shimoda LA, Booth GM, Undem BJ, et al. (1997) Endothelin-1 (ET-1) causes inhibition of delayed rectifier $K^+$ (KDR) current and constriction in human pulmonary artery (PA). FASEB J 11:1994
26. Prins B, Hu R-M, Nazario B, et al. (1994) Prostaglandin $E_2$ and prostacyclin inhibit the production and secretion of endothelin from cultured endothelial cells. J Biol Chem 16:11938-11944

# Modulation of Adhesion Molecule Expression in Pulmonary Vascular Endothelium by Oxygen

Yukio Suzuki[1], Takuya Aoki[3], Kazumi Nishio[3], Osamu Takeuchi[2], Kyoko Toda[2], Koichi Suzuki[3], Atsushi Miyata[3], Nagato Sato[3], Katsuhiko Naoki[3], Hiroyasu Kudo[3], and Kazuhiro Yamaguchi[2]

*Summary.* Pulmonary oxygen toxicity (POT) is an important clinical problem that occurs in patients on long-term mechanical ventilation requiring a high inspired-oxygen concentration. Although the histological evidence of POT is pulmonary edema with neutrophil infiltration into the lung parenchyma, the pathogenesis of POT is not fully understood. To elucidate the mechanism of the development of POT, we investigated the effect of hyperoxia (90% $O_2$, 5% $CO_2$) on adhesion molecule expression in cultured human endothelial cells. The level of intercellular adhesion molecule-1 (ICAM-1) expression had increased in hyperoxia-exposed endothelial cells at 48h and at 72h as compared with normoxic control (21% $O_2$, 5% $CO_2$). In contrast, the levels of P-selectin and E-selectin expression were unchanged during hyperoxic exposure. These hyperoxia-induced ICAM-1 expressions were dose dependently attenuated by a protein kinase C inhibitor (H-7). The levels of ICAM-1 mRNA and the numbers of adherent neutrophils were increased at 48h and at 72h of hyperoxia-exposed endothelial cells. These results suggest that increased ICAM-1 expression in endothelial cells plays an important role in neutrophil accumulation during POT.

*Key words.* Intercellular adhesion molecule-1—Pulmonary oxygen toxicity—Endothelial cell—Neutrophil–Protein kinase C

## Introduction

Neutrophil adhesion to the vascular endothelium is one of the first and essential inflammatory events at the site of injury. Adhesion of neutrophils is mediated by the increased expression of leukocyte integrins or endothelial cell adhesion molecules including P-selectin, E-selectin, and intercellular adhesion molecule (ICAM) -1.

---

[1] Department of Internal Medicine, [2] Biomedical Laboratory, Kitasato Institute Hospital, 5-9-1 Shirokane, Minato-ku, Tokyo 108, Japan
[3] Department of Internal Medicine, School of Medicine, Keio University, 35 Shinanomachi, Shinjuku-ku, Tokyo 160, Japan

Pulmonary oxygen toxicity (POT) is an important clinical problem that occurs in patients on long-term mechanical ventilation requiring a high inspired-oxygen concentration. Although the histological evidence of POT is pulmonary edema with neutrophil infiltration into the lung parenchyma [1], the mechanisms responsible for neutrophil accumulation in hyperoxia-exposed lungs are not fully understood. To clarify the mechanism of the development of POT, we investigated the effect of hyperoxia (90% $O_2$, 5% $CO_2$) on adhesion molecule expressions in cultured human endothelial cells.

## Materials and Methods

Human pulmonary artery endothelial cells (Kurabo, Osaka, Japan) were cultured to confluence and cell monolayers were then exposed to normoxic (21% $O_2$, 5% $CO_2$) or hyperoxic (90% $O_2$, 5% $CO_2$) conditions for various periods at 37°C in a humidified multigas incubator (APM-36, ASTEC, Fukuoka, Japan). To determine the levels of adhesion molecule expression, cells were incubated with either antihuman P-selectin (WGA-1, Takara Biomedicals, Kyoto, Japan), E-selectin (BB1G-E5, British Biotechnology, Oxon, UK), or ICAM-1 (LB-2, Becton Dickinson, San Jose, CA, USA) monoclonal antibody followed by fluorescein isothiocyanate-(FITC-) labeled antimouse IgG monoclonal antibody (Becton Dickinson). The intensity of fluorescence was determined with a flow cytometry (FACScan, Becton Dickinson), and results were expressed as percent intensity of fluorescence against the control. In some experiment, cell monolayers were treated with various concentrations of the protein kinase C inhibitor 1-(5-isoquinolinesulphonyl)-2-methylpiperazine dihydrochloride (H-7, Seikagaku Kogyo, Tokyo, Japan) and then exposed to hyperoxia for 48 h.

Neutrophil adhesion assays were carried out by applying isolated human neutrophils ($5 \times 10^5$ cells/ml) to cell monolayers that had previously been exposed to either normoxia or hyperoxia for various periods in a six-well culture plate. Plates were then incubated for 60 min under 5% $CO_2$, and nonadherent neutrophils were removed by gently washing the plates. Ten randomly selected fields were read at ×200 magnification. Neutrophil adhesion was evaluated by counting the number of neutrophils adhering to endothelial cell monolayers per high-power field (HPF).

ICAM-1 mRNA expression was analyzed by the reverse transcriptase-polymerase chain reaction (RT-PCR) method. Total RNA was extracted from endothelial cells that had been exposed to either normoxia or hyperoxia for various periods by the method of Chomczynski and Sacchi [2]. First-strand cDNA was synthesized by SuperScript RT (GIBCO-BRL, Gaithersburg, MD, USA). We amplified synthesized first-strand cDNA by PCR (Perkin Elmer Cetus, Norwalk, CT, USA) with the 5'- and 3'-primers with Taq polymerase (Takara Biomedicals). PCR cycles were allowed to run for 30 s at 94°C, followed by 30 s at 55°C and 1 min at 72°C. The 5'- and 3'-primers were as follows: 5'-TGACCATCTACAGCTTTCCGCC-3', 5'-GTCTGAGGTTACACGGTCCGA-3' [3]. Human glyceraldehyde-3-phosphate dehydrogenase (GAPDH) primers (Clontech Laboratories, Palo Alto, CA, USA) were used as an internal control. The 5'- and 3'-primers were as follows: 5'-TGAAGGTCGGAGTCAACGGATTTGGT-3', 5'-CATGTGGGCCATGAGGTCCACCAC-3'. A 10 µl of aliquot of the amplified DNA reaction mixture was fractionated by 2.0% agarose gel electrophoresis, and the amplified product was visualized by ultraviolet fluorescence after staining with ethidium bromide.

## Results

The ICAM-1 expression levels in hyperoxia-exposed human pulmonary artery endothelial cells (HPAEC) at 24 h were unchanged as compared with normoxic conditions. Exposure of endothelial cells to hyperoxia induced increases in ICAM-1 expression at 48 h and at 72 h (Fig. 1). The levels of ICAM-1 expression at 96 hours subsequently returned to baseline values and were unchanged as compared with normoxic conditions. On the other hand, exposure of HPAEC to hyperoxia induced no significant changes in P-selectin and E-selectin expression during the experiment.

To address the mechanism by which hyperoxia leads to ICAM-1 expression, we examined the effect of protein kinase C inhibitor (H-7) on hyperoxia-induced ICAM-1 expression in HPAEC. Although 1 or $5 \mu M$ of H-7 did not affect the levels of hyperoxia-induced ICAM-1 expression, 10 or $20 \mu M$ of H-7 attenuated the levels of hyperoxia-induced ICAM-1 expression by 26% or 34%, respectively.

To assess whether upregulated ICAM-1 expression is followed by increased neutrophil adhesion to endothelial cells, neutrophil adhesion assays were performed. The numbers of adherent neutrophils at 24 h were not significantly changed. Adherent neutrophils were increased at 48 and 72 h of hyperoxic exposure ($176 \pm 21$ and $144 \pm 32$ cells/HPF, respectively, $P < 0.01$) compared with normoxic conditions ($57 \pm 5$ cells/HPF).

By using RT-PCR method, we examined whether increased ICAM-1 expression is regulated at the transcriptional level. The level of ICAM-1 mRNA expression was unchanged under normoxic conditions. The level of ICAM-1 mRNA

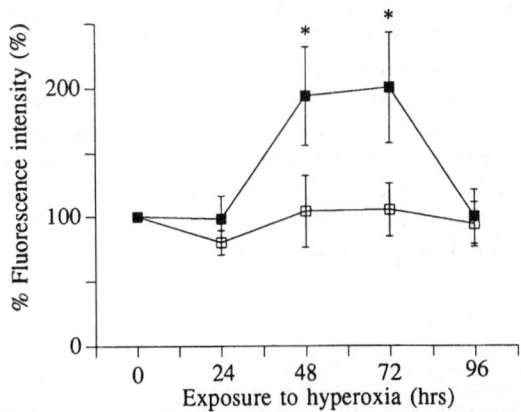

FIG. 1. Changes in intracellular adhesion molecule (ICAM)-1 expression in human pulmonary artery endothelial cells. Data are expressed as percent intensity of fluorescence compared with time 0 (mean $\pm$ SD). *Open squares*, normoxia; *closed squares*, hyperoxia; *, $P < 0.001$ compared with normoxic conditions

expression was unchanged at 24 h of hyperoxic exposure but was upregulated at 48 h and 72 h.

## Discussion

The CD18/ICAM-1 adhesion pathway has been shown to be involved in several models of lung inflammation including sepsis, reperfusion injury, bronchial asthma, and immune complex-mediated lung injury [4]. However, POT seems to be different from these inflammatory diseases because the injury is relatively subacute and the toxic route of entry is inhalational. Therefore, we investigated the effect of hyperoxia (90% $O_2$) on adhesion molecule expressions in cultured human endothelial cells to clarify the mechanism of the development of POT.

We demonstrated that exposure of endothelial cells to hyperoxia for 48–72 h increased the expression of ICAM-1, but not P-selectin or E-selectin. This suggests that hyperoxia may selectively induce ICAM-1 expression in endothelial cells. Bradley and colleagues reported that hydrogen peroxide increased endothelial ICAM-1 expression but not E-selectin or vascular cell adhesion molecule-1 expression in human umbilical vein endothelial cells [5].

We have demonstrated that hyperoxia-induced ICAM-1 expression is attenuated by H-7. This suggests that hyperoxia-induced ICAM-1 expression in human endothelial cells is at least partly mediated by protein kinase C activation. We have also shown increased neutrophil adhesion to hyperoxia-exposed HPAEC. The increases in ICAM-1, ICAM-1 mRNA, and neutrophil adhesion suggest that ICAM-1 plays an important role in the pathogenesis of POT. In other words, reducing the level of either ICAM-1 mRNA or ICAM-1 may attenuate POT.

## Conclusions

We demonstrated that exposure of endothelial cells to hyperoxia for 48–72 h induced an increased level of ICAM-1, but not P-selectin or E-selectin, expression as compared with normoxic control. Treatment with a protein kinase C inhibitor downregulated hyperoxia-induced ICAM-1 expression of endothelial cells, suggesting that hyperoxia induces ICAM-1 expression through the protein kinase C pathway. Neutrophil adhesion to endothelial cells was increased at 48 and 72 h of hyperoxic exposure. In addition, ICAM-1 mRNA expression in endothelial cells was upregulated at 48 and 72 h of hyperoxic exposure. The current study suggested that ICAM-1 plays an important role in neutrophil-dependent endothelial cell injury during the development of POT.

## References

1. Royston BD, Webster NR, Nunn JF (1990) Time course of changes in lung permeability and edema in the rat exposed to 100% oxygen. J Appl Physiol 69:1532–1537
2. Chomczynski P, Sacchi N (1987) Single-step method of RNA isolation by acid guanidium thiocyanate-phenol-chloroform extraction. Anal Biochem 162:156–159

3. Saito I, Terauchi K, Shimura M, et al. (1993) Expression of cell adhesion molecules in the salivary and lacrimal glands of Sjögren's syndrome. J Clin Lab Anal 7:180–187
4. Hamacher J, Schaberg T (1994) Adhesion molecules in lung diseases. Lung 172:189–213
5. Bradley JR, Johnson DR, Pober JS (1993) Endothelial activation by hydrogen peroxide: selective increases of intercellular adhesion molecule-1 and major histocompatibility complex class I. Am J Pathol 142:1598–1609

# Cross Talk Between Nitric Oxide and Cyclooxygenase Pathways in Glomerular Mesangial Cells

Toshifumi Tetsuka[1] and Aubrey R. Morrison[2]

*Summary.* Both nitric oxide (NO) and prostaglandins (PGs) are produced in glomerular inflammatory processes. Thus, the interactions between these two pathways were studied in cultured rat mesangial cells. Two NO donors, sodium nitroprusside (SNP) and $S$-nitroso-$N$-acetylpenicillamine (SNAP), amplified interleukin-1β (IL-1β) -induced mRNA and protein expression of an inducible isoform of cyclooxygenase (COX-2), followed by an increase in $PGE_2$ production. The effect of NO is likely to be mediated by cyclic guanosine monophosphate (cGMP), because (i) 8-Br-cGMP mimicked the effect of NO on COX-2 expression and $PGE_2$ formation; and (ii) methylene blue, an inhibitor of guanylate cyclase, reversed the effect of NO donors on COX-2 mRNA expression. On the other hand, a cyclooxygenase inhibitor, indomethacin, potentiated IL-1β-stimulated inducible NO synthase (iNOS) mRNA expression and NO production. The stimulatory effect of indomethacin on iNOS expression and NO production was reversed by the addition of exogenous $PGE_2$. These data indicate that (1) NO can amplify IL-1β-induced COX-2 gene expression, possibly via the activation of guanylate cyclase; and (2) endogenous $PGE_2$ negatively modulates IL-1β-induced iNOS expression. The cross talk between these pathways may play a role in modulating glomerular inflammatory processes.

*Key words.* Nitric oxide—Prostaglandins—Cyclooxygenase—Glomerular—Mesangium—Interleukin-1

## Introduction

At the site of glomerular inflammation, infiltrating macrophages and activated mesangial cells release cytokines, growth factors, prostaglandins (PGs), nitric oxide (NO), and reactive oxygen species [1–3]. These stimuli may act in concert to modify the glomerular inflammatory processes and the function of glomerular cells. Thus, in

[1] Department of Molecular Genetics, Nagoya City University Medical School, Mizuho-cho, Mizuho-ku, Nagoya 467, Japan
[2] Department of Molecular Biology and Pharmacology, Washington University Medical School, Box 8103, 660 S.Euclid Ave., St. Louis, MO 63110, USA

this study, the interactions between NO and PGs pathways were examined in cultured rat mesangial cells.

# Methods

Primary mesangial cell cultures were prepared from male Sprague–Dawley rats [4]. $PGE_2$ in the medium was determined by stable isotope gas chromatography–mass spectrometry (GC–MS). Concentration of nitrite, that is, the stable metabolite of NO, in the medium was measured by the Griess reaction.

For Northern blot analysis, total RNA was isolated using the acid guanidium thiocyanate–phenol–chloroform method; 20 μg of total RNA was fractionated by 1% agarose-formaldehyde gel electrophoresis and transferred onto nylon membrane. The membrane was hybridized with radiolabeled cDNA for murine cyclooxygenase-1 (COX-1), cyclooxygenase-2 (COX-2), inducible nitric oxide synthase (iNOS), and rat glyceraldehyde-3-phosphate dehydrogenase (GAPDH) probes.

For Western blot analysis, cells were solubilized in hypotonic lysis buffer containing 1% NP-40. 30–60 μg of protein was fractionated by sodium dodecyl sulfate-polyacrylamide gel electrophoresis (SDS-PAGE) and transferred to polyvinylidene difluoride (PVDF) membrane. The membranes were blotted with polyclonal rabbit IgG antibody against murine COX-2 or iNOS and visualized by an enhanced chemiluminescence method.

# Results

To determine the effect of NO on PG production, the effect of two NO donors, S-nitroso-N-acetylpenicillamine (SNAP) and sodium nitroprusside (SNP), were studied; 0.1–100 μM of SNAP potentiated interleukin-1β (IL-1β) -induced $PGE_2$ formation in a dose-dependent manner (Fig. 1a), although SNAP by itself did not increase basal $PGE_2$ formation. Similarly, SNP potentiated IL-1β-induced $PGE_2$ production but not basal $PGE_2$ formation. The stimulatory effect of SNP on IL-1β-induced $PGE_2$ production was maximal at 1–10 μM and was attenuated at 100 μM (data not shown).

Because the inducible isoform of cyclooxygenase (COX-2) is one of the major determinants of PGs production [4,5], we evaluated whether NO modulates COX-2 expression. SNAP amplified IL-1β-induced COX-2 mRNA (Fig. 1b) and COX-2 protein expression (Fig. 1c), although SNAP by itself did not induce COX-2 mRNA or COX-2 protein. Similarly SNP, another NO donor, also potentiated COX-2 mRNA expression (data not shown). Neither IL-1β nor NO donors affected the constitutive isoform of cyclooxygenase (COX-1) mRNA expression.

The endothelium-derived NO and micromolar range of NO donors SNAP and SNP are known to increase cellular cyclic guanosine monophosphate (cGMP) in mesangial cells [6–8]. To test the possibility that the NO effect was mediated through the stimulation of the guanylate cyclase, we examined the effect of an inhibitor of soluble guanylate cyclase, methylene blue [9], on COX-2 expression. Methylene blue reversed the stimulatory effect of SNAP on IL-1β-induced COX-2 mRNA expression (Fig. 2a). To further confirm the effect of the cGMP pathway, we determined the effect of the

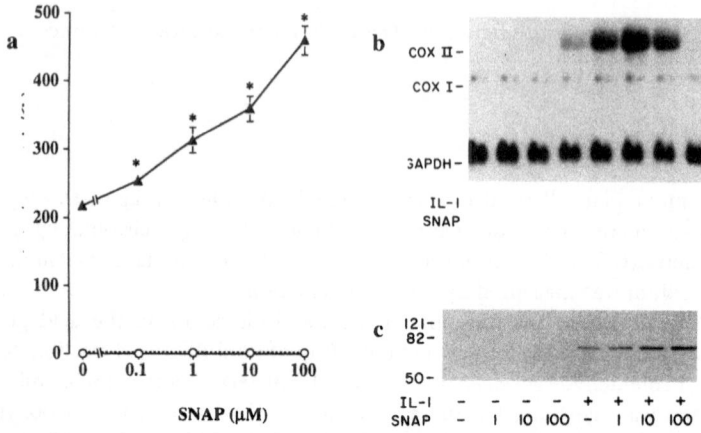

FIG. 1a–c. Effect of S-nitroso-N-acetylpenicillamine (*SNAP*) on prostaglandin-E₂ (PGE₂) production (a), cyclooxygenase-2 (*COX-2*)mRNA (b), and COX-2 protein (c). a Cells were stimulated with interleukin-1β (*IL-1β*) (50 U/ml) and/or SNAP for 6 h. PGE₂ in the medium was determined by gas chromatography–mass spectrometry (GC-MS). *, $P < .05$ (vs. SNAP, $0 \mu M$). *Open circles*, basal; *filled triangles*, IL-1β. b Cells were stimulated with IL-1β (50 U/ml) and/or SNAP for 3 h and harvested; 20 μg of total RNA was used for Northern blot analysis. c Cells were stimulated with IL-1β (50 U/ml) and/or SNAP for 6 h and harvested. The protein samples were subjected to Western blot analysis. *GAPDH*, glyceraldehyde-3-phosphate dehydrogenase

membrane-permeable cGMP analogue 8-Br-cGMP on IL-1β-induced COX-2 mRNA expression and PGE₂ production. Similarly to NO, 8-Br-cGMP potentiated IL-1β-induced COX-2 mRNA expression and PGE₂ production (Fig. 2b,c). These data support the notion that cGMP mediates the action of NO and indicates that cGMP can influence COX-2 gene expression.

Next, to determine the effect of endogenous PGs on the NO pathway, the effect of a nonselective cyclooxygenase inhibitor, indomethacin, on NO production was examined. Indomethacin enhanced the effect of IL-1β on nitrite production (Fig. 3a), and indomethacin inhibited IL-1β-induced PGE₂ release by less than 99% (375.3 ± 14.9 to 1.1 ± 0.2 ng; $P < .001$). The replacement of endogenous PGE₂ by exogenous PGE₂ reversed the effect of indomethacin in a dose-dependent manner (Fig. 3a, inset).

To determine whether the alterations in NO production by PGs occur because of changes in iNOS mRNA levels, the steady-state level of iNOS mRNA was determined by Northern blot analysis. As the peak steady-state level of iNOS mRNA occurred 12 h after IL-1β stimulation, the effect of indomethacin and PGE₂ on IL-1β-induced iNOS mRNA was determined at 12 h. As with nitrite production, indomethacin enhanced the IL-1β-induced iNOS mRNA level (Fig. 3b). This stimulatory effect of indomethacin was reversed by addition of exogenous PGE₂.

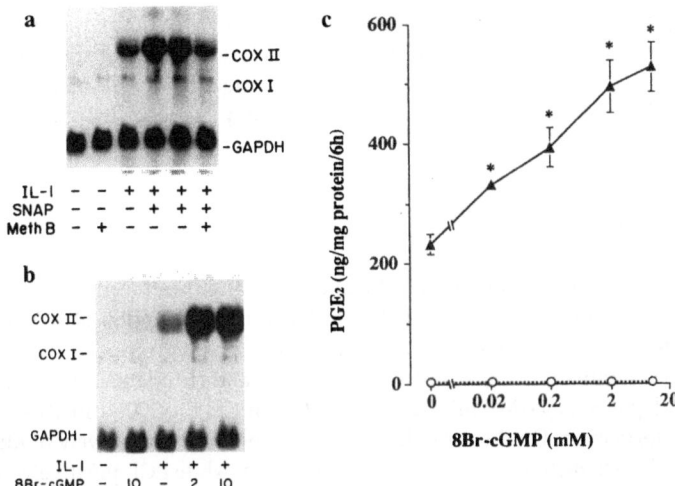

FIG. 2a,b. Effects of methylene blue (*Meth B*) and 8-Br-cyclic guanosine monophosphate (*8-Br-cGMP*) on *COX-2* mRNA and PGE$_2$ production. **a** Cells were preincubated with Meth B (10 μM) for 40 min. Cells were then stimulated with *IL-1β* (50 U/ml) and/or *SNAP* (10 μM) for 3 h and harvested; 20 μg of total RNA was used for Northern blot analysis. **b** Cells were stimulated with *IL-1β* (50 U/ml) and/or 8-Br-cGMP for 3 h and harvested; 20 μg of total RNA was used for Northern blot analysis. **c** Cells were stimulated with IL-1β (50 U/ml) and/or 8-Br-cGMP for 6 h, and PGE$_2$ in the medium was determined by GC-MS. $P < .02$ (vs. 8-Br-cGMP, 0 μM). *Open circles*, basal; *filled triangles*, IL-1β

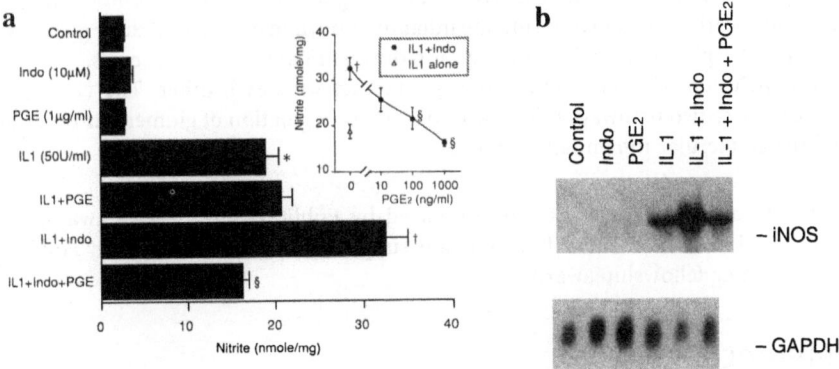

FIG. 3a,b. Effect of indomethacin (*Indo*) and exogenous PGE$_2$ on IL-1β-induced nitrite production (**a**) and inducible nitric oxide synthase (*iNOS*) mRNA expression (**b**). **a** Cells were stimulated with IL-1β (50 U/ml), Indo (10 μM) and/or PGE$_2$ for 24 h. Results are mean ± SEM ($n$, 9–12). *, $P < .001$ vs. control; †, $P < .001$ vs. IL-1β alone; §, $P < .01$ vs. IL-1 + Indo (without PGE$_2$). **b** Cells were stimulated with IL-1β (50 U/ml), Indo (10 μM), and/or PGE$_2$ (1 μg/ml) for 12 h and harvested. Total RNA was subjected to Northern blot analysis

In contrast to PGE₂, a stable analogue of PGI₂, carbaprostacyclin, increased IL-1β-induced nitrite production and iNOS mRNA expression (data not shown). Forskolin, an activator of adenylate cyclase, mimicked the effect of carbaprostacyclin but not PGE₂. Thus, the activation of adenylate cyclase may mediate the effect of PGI₂ but not PGE₂.

## Discussion

This study has demonstrated that NO can amplify IL-1β-induced COX-2 expression and PGE₂ formation. It was likely that cGMP mediated the action of NO, because (a) an inhibitor of soluble guanylate cyclase, methylene blue, reversed the stimulatory effect of NO donors on COX-2 mRNA expression; and (b) the membrane-permeable cGMP analogue 8-Br-cGMP mimicked the effect of NO on COX-2 mRNA expression and PGE₂ formation. On the other hand, endogenous PGE₂ negatively modulated IL-1β-induced iNOS mRNA expression and NO release, while PGI₂ positively modulated IL-1β-induced iNOS mRNA expression and NO release.

NO may synergize not only with IL-1β but also with a variety of stimuli, because NO donors also potentiated COX-2 mRNA expression induced by lipopolysaccharide and by a Ca²⁺ ionophore (data not shown). Peunova and Enikolopov [10] have demonstrated that NO amplifies c-fos gene expression stimulated by a Ca²⁺ signal in PC12 cells. Mühl et al. [11] have shown that NO amplifies IL-1β-induced iNOS expression in rat mesangial cells. Thus, NO may synergize with cytokine-like stimuli and amplify gene expression including iNOS, COX-2, and c-fos. The effector molecules of these gene products, such as PGs, NO, and AP-1 transcription factor, may further modulate subsequent long-term changes in cellular function and gene expression. In fact, the current study has demonstrated that the products of the COX pathway, PGE₂ and PGI₂, have a negative and positive effect, respectively, on IL-1β-induced iNOS expression. Thus the loop of positive and negative feedback regulation for these proinflammatory genes may work to sustain or terminate inflammation, and products of these genes can be targets for pharmacological intervention during inflammatory processes.

In conclusion, NO and COX pathways interact with each other. The cross talk between these two pathways might be important in regulation of glomerular function and the glomerular inflammatory process.

*Acknowledgments.* This work was supported by Public Health Service awards HL 20787 and DK PO-38111 to A.R.M. T.T. was supported by a National Kidney Foundation matching fellowship award.

## References

1. Sedor JR (1992) Cytokines and growth factors in renal injury. Semin Nephrol 12: 428–440
2. Baird NR, Morrison AR (1993) Amplification of the arachidonic acid cascade: implications for pharmacologic intervention. Am J Kidney Dis 21:557–564
3. Pfeilschifter J, Kuntz D, Mühl H (1993) Nitric oxide: an inflammatory mediator of glomerular mesangial cells. Nephron 64:518–525

4. Rzymkiewicz D, Leingang K, Baird N, et al. (1994) Regulation of prostaglandin endoperoxide synthase gene expression in rat mesangial cells by interleukin-1 beta. Am J Physiol 266:F39–F45
5. Reddy ST, Hershman HR (1994) Ligand-induced prostaglandin synthesis requires expression of the TIS10/PGS-2 prostaglandin synthase gene in murine fibroblasts and macrophages. J Biol Chem 269:15473–15480
6. Garg UC, Hassid A (1989) Inhibition of rat mesangial cell mitogenesis by nitric oxide-generating vasodilators. Am J Physiol 257:F60–F66
7. Marsden PA, Brock TA, Ballermann BJ (1990) Glomerular endothelial cells respond to calcium-mobilizing agonists with release of EDRF. Am J Physiol 258:F1295–F1303
8. Shultz PJ, Schorer AE, Raij L (1990) Effects of endothelium-derived relaxing factor and nitric oxide on rat mesangial cells. Am J Physiol 258:F162–F167
9. Gruetter CA, Barry BK, McNamara DB, et al. (1979) Relaxation of bovine coronary artery and activation of coronary arterial guanylate cyclase by nitric oxide, nitroprusside and a carcinogenic nitrosamine. J Cyclic Nucleotide Res 5:211–224
10. Peunova N, Enikolopov G (1993) Amplification of calcium-induced gene transcription by nitric oxide in neuronal cells. Nature 364:450–453
11. Mühl H, Pfeilschifter J (1995) Amplification of nitric oxide synthase expression by nitric oxide in interleukin 1β-stimulated rat mesangial cells. J Clin Invest 95:1941–1946

# Part 5
## Pathophysiology and Physiology of Gaseous Monoxides

# Pathophysiological Reactivities of Nitric Oxide

Andrew J. Gow[1], Raymond Foust III[1], Molly McClelland[1], Stuart Malcolm[3], and Harry Ischiropoulos[1,2]

*Summary.* Nitric oxide (·NO) is a free radical signal transducing agent that mediates a variety of physiological processes. The cellular reactivity of ·NO may be regulated by reaction with heme proteins, metal complexes, thiols, $O_2$ and superoxide ($O_2^-$). To evaluate the pathophysiological role of ·NO in biological systems, its reactivity with thiols, which regulates physiological responses, and with $O_2^-$, which mediates pathological actions, was examined. Experimental evidence supporting a mechanism for the formation of *S*-nitrosothiols under physiological conditions indicates that ·NO reacts directly with reduced thiol to form an intermediate that is converted to *S*-nitrosothiol by the reduction of an electron acceptor. This novel mechanism for the formation of *S*-nitrosothiols can explain the presence of *S*-nitrosylated proteins in vivo. The nearly diffusion limited reaction of ·NO with $O_2^-$ forms peroxynitrite ($ONOO^-$), a relatively long-lived, highly reactive species. Peroxynitrite exhibits selective reactivity with proteins to form bioactive nitrotyrosine residues that have been detected in atherosclerosis, sepsis, inflammation, and neurodegenerative diseases. Tyrosine nitration results in inactivation of protein function and interferes with tyrosine phosphorylation, a key event in cellular signal transduction. These data indicate that the reactivity of ·NO with target molecules is critical in regulating its biological function.

*Key words.* Peroxynitrite—Superoxide—*S*-Nitrosothiols—Nitration

## Nitric Oxide as a Signal Transduction Molecule

Nitric oxide (˙NO) has been shown to play a critical role in a diverse range of physiological functions including vasodilation, inhibition of platelet aggregation, neutrophil adherence, and neurotransmission [1–4]. Nitric oxide is synthesized by nitric oxide

[1] Institute for Environmental Medicine, University of Pennsylvania, 3620 Hamilton Walk, Philadelphia, PA 19104-6068, USA
[2] Department of Biochemistry and Biophysics, University of Pennsylvania, School of Medicine, 3620 Hamilton Walk Philadelphia, PA 19104-6068, USA
[3] Biokinetics Laboratory, Pearson Hall, Broak and Montgomery Str., Temple University, PA 19104, USA

synthases. These enzymes catalyze the five-electron oxidation of one of the two equivalent guanidino groups of L-arginine to citrulline and nitric oxide. The oxidation is carried out by using electrons from NADPH through the flavins, flavin mononucleotide and flavin adenine dinucleotide (FMN and FAD) and by incorporating two oxygen molecules. Nitric oxide is a free radical (contains an unpaired electron in its outer electron sphere) that can freely diffuse isotropically among cells where it principally reacts with soluble guanylate cyclase. The relatively high affinity of ·NO for binding to reduced iron in iron–heme proteins facilities the activation of soluble guanylate cyclase to produce cyclic guanosine monophosphate cGMP [2]. However, ·NO also reacts rapidly with a number of molecules such as superoxide anion and heme and nonheme iron-containing proteins. These reactants are readily available within the cytoplasm and thus may contribute to the short half-life of NO in vivo. Therefore, it has been suggested that ·NO reacts with thiols to form S-nitrosothiols, which are relatively longer lived species capable of donating nitric oxide [5]. S-nitrosothiols are vasodilators, inhibit platelet aggregation, and may also play other yet unrecognized roles [5–8]. A number of S-nitrosothiol-containing proteins have been identified in vivo [5–8]; however, the mechanism of the synthesis of S-nitrosothiol in vivo remains unknown.

## Potential Mechanisms for the Formation of S-Nitrosothiols In Vivo

The direct reaction of NO with free thiol would be an unbalanced one:

$$\cdot NO + R\text{-}SH \rightarrow RS\text{-}NO + e^- + H^+ \qquad (1)$$

It has been suggested that S-nitrosothiol is formed by the reaction of free sulfhydryl groups with the higher oxides of nitrogen ($NO_x$) formed by the reaction between ·NO and molecular oxygen [9–12]. However, the reaction of ·NO with molecular oxygen follows second-order kinetics with respect to nitric oxide. As a result, the rate of production of $NO_x$ is slow, approximately 3–300 pmole $s^{-1}$, at physiological concentrations of ·NO (0.1–1.0 μM). Therefore, the formation of nitrosothiol via this pathway is unlikely in tissue under normal physiological conditions, although it will become significant under higher ·NO concentrations. Experimental evidence has revealed that ·NO–iron dinitrosyl complexes react with thiols to form S-nitrosothiol [13]; however, this pathway necessitates that the thiols be in close proximity to the relatively bulky ·NO–iron dinitrosyl complexes. Moreover, the concentration of free iron in cells and tissues under physiological conditions is kept relatively low by iron-binding proteins and metal chelation.

Therefore, we proposed a novel mechanism for the formation of S-nitrosothiols by the reaction of reduced thiol with physiological concentrations of nitric oxide [14]. The major thesis of the proposed mechanism is that the reaction proceeds only in the presence of an electron acceptor. The reaction of a reduced thiol with ·NO will form a radical intermediate that will then reduce oxygen under aerobic conditions to form S-nitrosothiol plus superoxide (reactions 2 and 3). A second ·NO molecule will react with superoxide to form peroxynitrite (reaction 4) (see following):

$$\cdot NO + R\text{-}SH \rightarrow RS\text{-}\cdot \dot{N}OH \tag{2}$$

$$RS\text{-}\cdot \dot{N}OH + O_2 \rightarrow RS\text{-}NO + O_2^- + H^+ \tag{3}$$

$$\cdot NO + O_2 \rightarrow ONOO^- \tag{4}$$

$$\overline{2\cdot NO + R\text{-}SH + O_2 \rightarrow RS\text{-}NO + ONOO^- + H^+} \tag{5}$$

From this mechanism, certain predictions can be made, and experimental evidence for each prediction can verify the proposed mechanism:

1. $S$-nitrosothiols will be formed aerobically by the addition of excess (more than 200 fold) reduced thiol to physiological concentrations of $\cdot NO$ ($0.1$–$5\,\mu M$), or anaerobically in the presence of an electron acceptor such as $NAD^+$. The product formed either aerobically or anaerobically will release $\cdot NO$ on addition of reduced equivalents. It is important that the concentration of the reduced thiol is higher than the concentration of $\cdot NO$, similar to in vivo physiological conditions. Concentrations of $\cdot NO$ greater than the concentration of reduced thiol will oxidize the thiol to sulfenic acid and will not form $S$-nitrosothiol [15].

2. Addition of reduced thiol will increase the rate of $\cdot NO$ decomposition.

3. The reaction will consume oxygen, and in the presence of sufficient superoxide dismutase the reaction will produce $H_2O_2$.

Experimental evidence in support of this mechanism can be found in a recent publication [14]. Briefly, we found that a $\cdot NO$ donor with the characteristic absorbance of $S$-nitrosothiol was formed by the addition of reduced thiol to a nitric oxide-containing buffer system either under aerobic or anaerobic conditions. Under an anaerobic environment, the data clearly showed that the formation of $S$-nitrosothiol is dependent on the presence of an electron acceptor, $NAD^+$. Moreover, addition of $0.75\,mM$ cysteine accelerated by 1.7 fold the rate of decomposition of $6\,\mu M\cdot NO$ under aerobic conditions. In addition, cysteine increased the consumption of $O_2$ by the decomposition of $\cdot NO$ from 0.15 to $0.59\,\mu M\ O_2/\mu M\cdot NO$. This implies that the reaction is unlikely to proceed via a $NO_x$ intermediate and that oxygen is consumed by the reaction of nitric oxide with cysteine. The production of superoxide by the reaction of $\cdot NO$ and cysteine was demonstrated by measuring $H_2O_2$ production in the presence of Cu/Zn superoxide dismutase. The reaction of $6\,\mu M\cdot NO$ with $0.25\,mM$ cysteine yields $5\,\mu M$ $H_2O_2$ in the presence of Cu/Zn superoxide dismutase. Overall, these results demonstrate that under physiological conditions $\cdot NO$ reacts directly with cysteine to form $S$-nitrosocysteine in the presence of an electron acceptor and provides a biochemically reasonable pathway for the formation of $S$-nitrosothiols in vivo.

Although $\cdot NO$ mediates important physiological functions such as long-term potentiation, vasorelaxation, and immune responses, numerous studies have provided strong evidence that implicates the formation of $\cdot NO$-derived oxidants as part of the pathogenic mechanism for tissue toxicity.

## Nitric Oxide-Derived Oxidants in Tissue Injury

Recent publications have provided evidence that production of $\cdot NO$-derived oxidants are part of the pathogenic mechanism associated with tissue injury in major organs (Table 1). The development of pulmonary injury after lungs were challenged with

TABLE 1. Published data implicating ·NO-derived oxidants with tissue injury: the double-edged role of nitric oxide

| Organ | Physiological function | Pathological mechanism |
|---|---|---|
| Lung | Vasodilation<br>Bronchodilation | Injury mediated by activated inflammatory cells<br>Ischemia-reperfusion injury<br>Sepsis, ARDS |
| Heart Vasculature | Vasodilation<br>Prevention of platelet and neutrophil adherence to endothelium | Ischemia-reperfusion injury<br>Sepsis, hemorrhagic shock |
| Brain | Vasodilation<br>Long-term potentiation | Glutamate toxicity<br>Ischemia-reperfusion MPTP, 3-NP toxicity |

ARDS, adult respiratory distress syndrome; MPTP, 1-methyl -4-phenyl-1,2,3,6-tetrahydropyridine; 3-NP, 3-nitroproprionic acid.

immunecomplexes, or smoke, conditions that activate inflammatory cells (macrophages, neutrophils), was inhibited by a competitive inhibitor of nitric oxide synthesis [16,17]. Nitric oxide-derived oxidants are implicated in the pathogenic mechanisms of lung injury derived from sepsis, respiratory distress syndrome, and ischemia-reperfusion [18–21]. Inhibition of nitric oxide synthesis was shown to be protective in the global myocardial ischemia-reperfusion piglet model [22] and reduced infarct size in the in situ rabbit heart [23]. Moreover, ·NO-derived oxidants were detected in rat heart after ischemia-reperfusion injury [24]. A clear indication for the involvement of ·NO in ischemic injury was provided by Huang et al, who showed that mice lacking neuronal nitric oxide synthase (NOS) were protected from ischemic injury [25], and by data showing that antisense oligonucleotides targeting the inducible form of NOS protected rat kidney against ischemia [26]. Neuronal injury from hypoxia-reoxygenation and N-methyl-D-aspartate (NMDA) receptor activation was blocked by NOS inhibition [27–30]. Neurotoxicity derived from injections of 1-methyl-4-phenyl-1,2,3,6-tetrahydropyridine (MPTP), a model of Parkinson's disease, 3-nitropropionic acid (3-NP), and malonate are associated with ·NO-derived oxidants [31–33]. It was recently shown that apoptotic death of PC12 cells following downregulation of Cu/Zn superoxide dismutase proceeds via the formation of ·NO-derived oxidants [34], and ·NO-derived oxidants inhibit DOPA synthesis in the same cell system [35]. In addition, upregulation of nitric oxide synthesis by endotoxin has been associated with tissue injury in liver ischemia reperfusion [36].

Nitric oxide is a weak one-electron oxidant and, similarly to superoxide, it must be converted to another reactive species to be toxic. The reaction of ·NO with $O_2^-$ to form $ONOO^-$ can explain the ·NO-mediated toxicity. Because both ·NO and $O_2^-$ are free radicals, they react rapidly to form peroxynitrite:

$$\cdot NO + O_2^- \rightarrow ONOO^- \qquad (6)$$

The second-order rate constant of the reaction between nitric oxide and superoxide is $6.7 \times 10^9 \, M^{-1} \, s^{-1}$ [37]. This is an extremely fast rate that is approximately 30 times faster than the reaction of ·NO with oxyhemoglobin or guanylate cyclase and 3 times faster than the reaction of superoxide with superoxide dismutase ($2.9 \times 10^9$ $M^{-1}$ $s^{-1}$). This implies that the formation of peroxynitrite can outcompete the major scavenging pathways for ·NO and $O_2^-$. Moreover, the rate of $ONOO^-$ formation increases by 100 times with only a 10-fold increase in the concentration of ·NO and $O_2^-$. Activated rat alveolar macrophages, human neutrophils, and endothelial and smooth muscle cells generate $ONOO^-$ [38–42], thus providing evidence for the formation of $ONOO^-$ in vivo. Furthermore, the above published studies have provided evidence that peroxynitrite may be the critical mediator of nitric oxide-derived toxicity (see Table 1) because, unlike ·NO and $O_2^-$, $ONOO^-$ is a strong oxidant capable of reacting with many cellular targets. To understand the molecular mechanisms of peroxynitrite-mediated toxicity we have examined the biochemistry of peroxynitrite. These studies have revealed the unique reactivities of peroxynitrite and have led to the development of methodologies for its detection in biological systems. The major findings regarding the biochemistry of peroxynitrite are outlined next.

## Oxidative Biochemistry of Peroxynitrite

The $pK_a$ of $ONOO^-$ is 6.8 and at pH 7.4, approximately 25% will be protonated to form peroxynitrous acid (ONOOH):

$$ONOO^- + H^+ \rightarrow ONOOH \rightarrow NO_3^- \tag{7}$$

Peroxynitrous acid isomerizes to nitrate at a rate of $0.6 s^{-1}$ in phosphate buffer and 37°C [43]. Although ONOOH does not appear to physically separate to form hydroxyl radical and nitrogen dioxide (·OH . . . ·$NO_2$), it will oxidize biological molecules to give the same oxidized products as the hydroxyl radical [43]. Hydroxyl radical-like reactivity derived from peroxynitrite may not be important in vivo because the rate of $ONOO^-$ dissociation to hydroxyl radical-like reactivity is slower than other direct reactions of peroxynitrite in biological systems (Fig. 1).

The most rapid and direct reactions of $ONOO^-$ described to date are the oxidation of zinc- and iron-thiolate centers and nitration of proteins that can mediate the

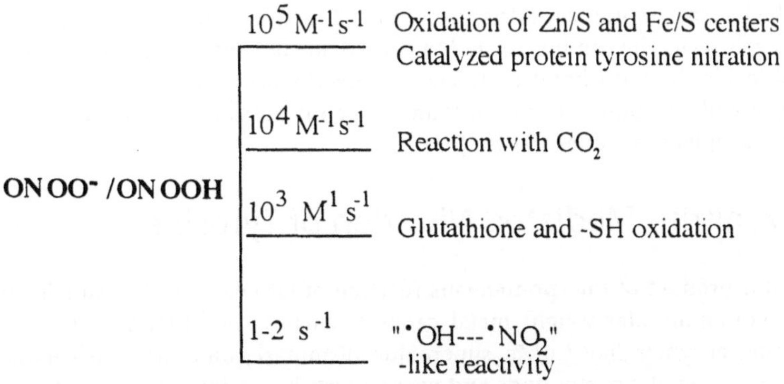

**ON OO⁻ /ON OOH**

$10^5 M^{-1}s^{-1}$    Oxidation of Zn/S and Fe/S centers
             Catalyzed protein tyrosine nitration

$10^4 M^{-1}s^{-1}$    Reaction with $CO_2$

$10^3 M^{-1}s^{-1}$    Glutathione and -SH oxidation

$1\text{-}2\ s^{-1}$    "·OH---·$NO_2$"
         -like reactivity

FIG. 1. Potential reactive pathways of peroxynitrite based on rate constants

inactivation of key enzymes in vivo [44–46]. Peroxynitrite will oxidize cellular thiols and lipids [47,48]. Peroxynitrite-mediated oxidation of glutathione and protein cysteine residues, which occurs at rates 1000 times faster than with equimolar concentrations of $H_2O_2$ would deplete cells of antioxidant defenses [47]. Recent data are starting to elucidate other potential intracellular targets of peroxynitrite, including proteins, antioxidant defense pathways, transcription factors, and DNA [49–56]. Overall, the net result of peroxynitrite reactivity is the induction of oxidative stress and ultimately cell death. The nature of peroxynitrite-induced cell death is dependent on the concentration of peroxynitrite; high levels of peroxynitrite result in necrotic death whereas relatively low levels result in the induction of apoptosis [49–52]. Exposure of type II pneumocytes to $ONOO^-$ significantly decreased cyanide-sensitive $O_2$ consumption [57]. Exposure to peroxynitrite resulted in the inactivation of succinate-cytochrome $c$ reductase and cytochrome $c$ oxidase of the mitochondrial electron transport chain in neuronal cells but not in astrocytes [58]. These data indicate that peroxynitrite can establish a diffusion gradient that will allow its diffusion inside cells where it can react with cellular targets such as mitochondria.

## Mitochondria as a Potential Cellular Target of Peroxynitrite

In addition to studies utilizing intact cells [57,58], published data from studies using isolated mitochondria preparations have revealed important mechanistic information regarding the interaction of peroxynitrite with mitochondria. Cassina and Radi have reported that while ·NO induces reversible injury to mitochondrial electron transport components, $ONOO^-$ induces irreversible injury that is consistent with the changes observed in intact cells [59]. These observations led the authors to speculate that peroxynitrite is the ultimate species responsible for mitochondrial dysfunction under pathological conditions. Exposure of isolated liver mitochondria to peroxynitrite results in an efflux of calcium and depolarization [60,61]. Another event that is associated with the reaction of peroxynitrite with isolated mitochondria is the uncoupling of the electron transport chain. Uncoupling of the mitochondrial electron transport chain and an increase in hydrogen peroxide production have been shown to occur after exposure of isolated rat heart mitochondria to peroxynitrite [62]. Moreover, the increase in mitochondria-derived reactive species may be further amplified because peroxynitrite can inactivate Mn superoxide dismutase, which is strategically located inside the mitochondria, to account for the production of superoxide [45]. Mn superoxide dismutase has been found to be nitrated and inactivated in rejected human transplanted kidney tissues [63].

## Peroxynitrite-Mediated Nitration of Tyrosine

The major product of the spontaneous reaction of $ONOO^-$ with tyrosine is nitrotyrosine. Low molecular weight metal catalysts such as $Fe^{+3}$-EDTA or $Cu^{+2}$, metal-containing enzymes like Cu/Zn superoxide dismutase, and $CO_2$ can catalyze the nitration of phenolic compounds and protein tyrosine residues to give nitrophenols and nitrotyrosine [45,46,64,65]. The reaction of peroxynitrite with $CO_2$ may provide the necessary nitrating agent that can explain the formation of nitrotyrosine in vivo.

Recent data from our laboratory as well as from others independently provided experimental evidence that $CO_2$ reacts in a catalytic manner with peroxynitrite to form a potent nitrating agent. Not only does the yield of protein nitration increase more than twofold by the reaction of peroxynitrite with $CO_2$ but all other reactivities of peroxynitrite such as oxidation of cysteine and tryptophan are partially inhibited [64,65]. Endogenous tyrosine nitration is almost certainly derived via enzymatically produced ·NO, although ·NO itself is not a nitrating agent. Potential in vivo nitrating agents include $ONOO^-$, metal nitrates, alkyl and acyl nitrates, nitryl halides, nitrogen oxides, and acid-catalyzed reactions of nitrite.

Using human plasma as a model, data revealed that nitrotyrosine could not be detected even at concentrations of ·NO that resulted in as much as 1 mM nitrogen dioxide or after exposure to ·NO in the presence of other biological oxidants such as $H_2O_2$ and metal catalysts [64]. Therefore, under plausible pathophysiological concentrations, the intermediate formed by the reaction of $ONOO^-$ with $CO_2$ is the most reasonable nitrating agent. However, there is no single mechanism for the nitration of tyrosine but rather a continuum of reactions that are a function of the reactive nitrating species and the nitration conditions. As such, the mechanism of nitration by $ONOO^-$ and other nitrating agents is complicated by the presence of more than one oxidizing species in the reaction mixture. Overall, in the presence of $CO_2$ the reaction of peroxynitrite with proteins will preferentially yield nitrotyrosine. Consistent with these data is the detection of extensive tyrosine nitration in pathological conditions known to increase the levels of both bicarbonate/$CO_2$ and $ONOO^-$, such as inflammation and ischemia-reperfusion injury (Table 2).

The direct evidence for $ONOO^-$ formation in human atherosclerotic plaques is consistent with the observations that $ONOO^-$, but not ·NO, induced lipid oxidation of β-very low density lipoprotein (β-VLDL) and LDL [67,75] and that hypercholesterolemia induced the release of both nitrogen oxides and $O_2^-$ from rabbit aorta [76,77]. The evidence for $ONOO^-$ formation in human lungs is consistent with the detection of

TABLE 2. Detection of nitrotyrosine in human and animal diseases

Human:
    Atherosclerotic plaques of coronary vessels [66]
    Low density lipoprotein isolated from atherosclerotic lesions [67]
    Lungs of infants with sepsis or respiratory disease [18,19]
    Synovial fluid of patients with arthritis [68]
    Multiple sclerosis plaques [69]
    Chronic renal failure in septic patients [70]
    Rejected renal allografts [63]

Animal:
    Rabbit lungs following exposure to hyperoxia [18]
    Lungs of endotoxin-treated rats [20]
    Ischemia-reperfusion-injured rat lungs [21]
    Ischemia-reperfusion-injured rat heart [24]
    Aorta of septic rats [71]
    Inflammatory bowel disease [72]
    CO-poisoned rats [73]
    Rat skeletal muscle SERC2a isoform [74]
    Brain lesions in MPTP, 3-nitropropionic acid, and malonate neurotoxicity [31–33]

nitrotyrosine in animal models of these diseases, and the detection of nitrotyrosine in inflammatory disease is consistent with the ability of macrophages and neutrophils to generate peroxynitrite [38,39]. Moreover, in all animal models of injury increase in the levels of nitrotyrosine was prevented by inhibition of ·NO synthesis, indicating that nitrotyrosine is in part derived from ·NO.

## Consequences of Protein Nitration

The body of published data suggests that nitrotyrosine represents a protein modification specific for ONOO⁻ formation in vivo. Nitration of tyrosine residues has been utilized as a protein modification to study the role of specific tyrosine residues in the function of proteins. Chemically induced tyrosine nitration using tetranitromethane (TNM) as the nitrating agent has been shown to inactivate nearly 140 mammalian proteins whose activity is dependent on tyrosine residues. Thus far peroxynitrite-mediated nitration of tyrosine residues has been shown to inactivate mitochondrial Mn superoxide dismutase [45,63], the lipid aggregatory activity of surfactant protein A [78], and the activity of glutamine synthase [79]. We have also provided experimental evidence to indicate that tyrosine nitration may have a significant impact in cellular function, in particular on the degradation of proteins and on signal transduction [80]. Tyrosine phosphorylation is an important regulator of signal transduction in cells and has been implicated in cellular responses to growth factors, cytokines, and calcium ionophores. We have shown that nitrated peptides are not readily phosphorylated in vitro by purified tyrosine kinases, a result that was confirmed by the work of Kong et al, who showed that ONOO⁻-mediated nitration of a single tyrosine residue in the cell cycle kinase CDC2 prevents tyrosine phosphorylation [80,81]. Therefore, nitrotyrosine formation may also interfere with normal signal transduction pathways.

Overall, the inherent ability of tissues and cells to withstand oxidative stress is dependent on their antioxidant capacity, the ability to derive energy from alternate pathways and to repair oxidatively modified biomolecules, and the availability of trophic support to maintain energy requirements, ionic homeostasis, and structural integrity. Nitric oxide can be part of the physiological cellular response as well as the pathogenic mechanism of the disease. The availability of target reactive molecules such as thiols and superoxide may provide biochemically reasonable pathways for understanding the pathophysiological functions of nitric oxide.

## References

1. Furchgott RF (1996) The discovery of endothelium-derived relaxing factor and its importance in the identification of nitric oxide. JAMA 276:1186–1188
2. Murad F (1996) Signal transduction using nitric oxide and cyclic guanosine monophosphate. JAMA 276:1189–1192
3. Ignarro LJ (1990) Biosynthesis and metabolism of endothelium-derived nitric oxide. Annu Rev Pharmacol Toxicol 30:535–560
4. Moncada S, Palmer RMJ, Higgs E (1991) Nitric oxide: physiology, pathophysiology, and pharmacology. Pharmacol Rev 43:109–142

5. Myers PR, Minor RL Jr, Guerra R Jr, et al. (1990) Vesorelaxant properties of the endothelium-derived relaxing factor more closely resemble S-nitrosocysteine than nitric oxide. Nature 345:161–163
6. Stamler JS, Simon DI, Osborne JA, et al. (1992) S-Nitrosylation of proteins with nitric oxide: synthesis and characterization of biologically active compounds. Proc Natl Acad Sci USA 89:444–448
7. Jia L, Bonaventura C, Bonaventura J, Stamler JS (1996) S-Nitrosohaemoglobin: a dynamic activity of blood involved in vascular control. Nature 380:221–226
8. Lander HM, Milbank AJ, Tauras JM, et al. (1996) Redox regulation of cell signalling. Nature 381:380–381
9. Pryor WA, Church DF, Govindian CK, Crank G (1982) Oxidation of thiols by nitric oxide and nitrogen dioxide: synthetic utility and toxicological implications. J Org Chem 47:157–161
10. Wink DA, Nims RW, Darbyshire JF, et al. (1994) Reaction kinetics for nitrosation of cysteine and glutathione in aerobic nitric oxide solutions at neutral pH. Insights into the fate and physiological effects of intermediates generated in the NO/O$_2$ reaction. Chem Res Toxicol 7:519–525
11. Kharitonov VG, Sundquist AR, Sharma VS (1995) Kinetics of nitrosation of thiols by nitric oxide in the presence of oxygen. J Biol Chem 270:28158–28164
12. Hogg N, Singh RJ, Kalyanaraman B (1996) The role of glutathione in the transport and catabolism of nitric oxide. FEBS Lett 382:223–228
13. Boese M, Mordvintcev PI, Vanin AF, (1995) S-Nitrosation of serum albumin by dinitrosyl-iron complex. J Biol Chem 270:29244–29249
14. Gow A, Buerk DG, Ischiropoulos H (1997) A novel reaction mechanism for the formation of S-nitrosothiol in vivo. J Biol Chem 272:2841–2844.
15. DeMaster EG, Quast BJ, Redfern B, Nagasawa HT (1995) Reaction of nitric oxide with the free sulfhydryl group of human serum albumin yields a sulfenic acid and nitrous oxide. Biochem J 34:11494–11499
16. Mulligan MS, Hevel JM, Marletta MA, Ward PA (1991) Tissue injury caused by deposition of immune complexes is L-arginine dependent. Proc Natl Acad Sci USA 88:6338–6342
17. Ischiropoulos H, Mendiguren I, Fisher D, et al. (1994) Role of neutrophils and nitric oxide in lung alveolar injury from smoke inhalation. Am J Respir Crit Care Med 150:337–341
18. Haddad IY, Pataki G, Hu P, et al. (1994) Quantitation of nitrotyrosine levels in l ung sections of patients and animals with acute lung injury. J Clin Invest 94:2407–2413
19. Kooy NW, Royall JA, Ye Y-Z, et al. (1995) Evidence for in vivo peroxynitrite production in human acute lung injury. Am J Respir Crit Care Med 151:1250–1254
20. Wizemann TM, Gardner CR, Laskin JD, et al. (1994) Production of nitric oxide and peroxynitrite in the lung during acute endotoxemia. J Leukocyte Biol 56:759–768
21. Ischiropoulos H, Al-Mehdi AB, Fisher AB (1995) Reactive species in rat lung injury: contribution of peroxynitrite. Am J Physiol 269:L185–L164
22. Matheis G, Sherman MP, Buckberg GD, et al. (1992) Role of L-arginine-nitric oxide pathway in myocardial reoxygenation injury. Am J Physiol 262:H616–620
23. Patel VC, Yellon DM, Singh KJ, et al. (1993) Inhibition of nitric oxide limits infarct size in the in situ rabbit heart. Biochem. Biophys Res Commun 194:234–238
24. Wang P, Zweir JL (1996) Measurement of nitric oxide and peroxynitrite generation in the postischemic heart. J Biol Chem 271:29223–29230
25. Huang Z, Huang PL, Panahian N, et al. (1994) Effects of cerebral ischemia in mice deficient in neuronal nitric oxide synthase. Science 265:1883–1885
26. Noiri E, Peresleni T, Miller F, Goligorsky MS (1996) In vivo targeting of inducible NO synthase with oligodeoxynucleotides protects rat kidney against ischemia. J Clin Invest 97:2377–2383

27. Dawson V, Dawson T, London E, et al. (1991) Nitric oxide mediates glutamate neuro-toxicity in primary cortical cultures. Proc Natl Acad Sci USA 88:6368–6371
28. Cazevieille C, Muller A, Meynier F, Bonne C (1993) Superoxide and nitric oxide cooperation in hypoxia/reoxygenation-induced neuronal injury. Free Radical Biol Med 14:389–395
29. Beckman JS (1991) The double-edged role of nitric oxide in brain function and superoxide-mediated injury. J Dev Physiol (Eynsham) 15:53–59
30. Lipton SA, Choi Y-B, Pan Z-H, et al. (1993) A redox-based mechanism for the neuro-protective and neurodestructive effects of nitric oxide and related nitroso-compounds. Nature 364:626–632
31. Schulz JB, Matthews RT, Muqit MK, et al. (1995) Inhibition of neuronal nitric oxide synthase by 7-nitroindazole protects against MPTP-induced neurotoxicity in mice. J Neurochem 64:936–939
32. Schulz JB, Matthews RT, Jenkins BG, et al. (1995) Blockade of neuronal nitric oxide synthase protects against excitotoxicity in vivo. J Neurosci 15:8419–8429
33. Schulz JB, Huang PL, Matthews RT, et al. (1996) Striatal malonate lesions are attenu-ated in nitric oxide oxide synthase knockout mice. J Neurochem 67:430–433
34. Troy CM, Derossi D, Prochiantz A, et al. (1996) Down-regulation of copper/zinc superoxide dismutase leads to cell death via the nitric oxide-peroxynitrite pathway. J Neurosci 16:253–261
35. Ischiropoulos H, Duran D, Horwitz J (1995) Peroxynitrite-mediated inhibition of DOPA synthesis in PC12 cells. J Neurochem 65:2366–2372
36. Ma TT, Ischiropoulos H, Brass CA (1995) Endotoxin stimulated nitric oxide produc-tion increases injury and reduces rat liver chemiluminescence during reperfusion. Gastroenterology 108:463–469
37. Huie RE, Padjama S (1993) The reaction of NO with superoxide. Free Radical Res Commun 18:195–199
38. Ischiropoulos H, Zhu L, Beckman JS (1992) Peroxynitrite formation from macroph-age-derived nitric oxide. Arch Biochem Biophys 298:446–445
39. Carreras MC, Pargament GA, Catz SD, et al. (1994) Kinetics of nitric oxide and hydro-gen peroxide production and formation of peroxynitrite during the respiratory burst of human neutrophils. FEBS Lett 341:65–68
40. Kooy NW, Royall JA (1994) Agonist-induced peroxynitrite production from endothe-lial cells. Arch Biochem Biophys 310:352–359
41. Phelps DT, Ferro TJ, Higgins PJ, et al. (1995) Tumor necrosis factor induces the peroxynitrite-mediated depletion of lung endothelial glutathione via protein kinase C activation. Am J Physiol 269:L551–L559
42. Boota A, Zar H, Kim Y-M, et al. (1996) IL-1$\beta$ stimulates superoxide and delayed peroxynitrite production by pulmonary vascular smooth muscle cells. Am J Physiol 271:L932–L938
43. Beckman JS, Beckman TW, Chen J, et al. (1990) Apparent hydroxyl radical production by peroxynitrite: implications for endothelial injury from nitric oxide and superoxide. Proc Natl Acad Sci USA 87:1620–1624
44. Crow JP, Beckman JS, McCord JM (1995) Sensitivity of the essential zinc-thiolate moiety of yeast alcohol dehydrogenase to hypoclorite and peroxynitrite. Biochemistry 34:3544–3522
45. Ischiropoulos H, Zhu L, Chen J, et al. (1992) Peroxynitrite-mediated tyrosine nitration catalyzed by superoxide dismutase. Arch Biochem Biophys 298:431–437
46. Beckman JS, Ischiropoulos H, Zhu L, et al. (1992) Kinetics of superoxide dismutase and iron-catalyzed nitration of phenolics by peroxynitrite. Arch Biochem Biophys 298:438–445
47. Radi R, Beckman JS, Bush KM, Freeman BA (1991) Sulfhydryl oxidation by peroxyni-trite: the cytotoxic potential of superoxide and nitric oxide. J Biol Chem 266:4244–4250

48. Radi R, Beckman JS, Bush KM, Freeman BA (1991) Peroxynitrite-induced membrane lipid peroxidation: the cytotoxic potential of superoxide and nitric oxide. Arch Biochem Biophys 288:481–487
49. Ischiropoulos H (1995) Exposure of endothelial cells to peroxynitrite inhibits tyrosine phosphorylation and induces apoptosis. Endothelium 3S:47–48
50. Bonfoco E, Krainc D, Ankarrona M, et al. (1995) Apoptosis and necrosis: two distinct events induced, respectively, by mild and intense insults with $N$-methyl-D-aspartate or nitric oxide/superoxide in cortical cell cultures. Proc Natl Acad Sci USA 92:1044–1048
51. Estévez AG, Radi R, Barbeito L, et al. (1995) Peroxynitrite-induced cytotoxicity in PC12 cells: evidence for an apoptotic mechanism differentially modulated by neurotrophic factors. J Neurochem 65:1543–1550
52. Lin K-T, Xue J-Y, Nomen M, et al. (1995) Peroxynitrite-induced apoptosis in HL-60 cells. J Biol Chem 70:16487–16490
53. Szabo C, Zignarelli B, O'Connor M, Salzman AL (1996) DNA strand breakage, activation of poly(ADP-ribose) synthetase, and cellular energy depletion are involved in the cytotoxicity in macrophages and smooth muscle cells exposed to peroxynitrite. Proc Natl Acad Sci USA 93:1753–1758
54. Salgo MG, Stone K, Squadrito GL, et al. (1995) Peroxynitrite causes DNA nicks in plasmid pBR322. Biochem Biophys Res Commun 210:1025–1030
55. Yermilov V, Rubio J, Ohshima H (1995) Formation of 8-nitroguanine in DNA treated with peroxynitrite in vitro and its rapid removal from DNA by depurination. FEBS Lett 376:207–210
56. Roussyn I, Briviba K, Masumoto H, Sies H (1996) Selenium-containing compounds protect DNA from single-strand breaks caused by peroxynitrite. Arch Biochem Biophys 330:216–218
57. Hu P, Ischiropoulos H, Beckman JS, Matalon S (1994) Peroxynitrite inhibition of oxygen consumption and sodium transport in alveolar type II cells. Am J Physiol 266:L628–L634
58. Bolãnos JP, Heals SJR, Land JM, Clark JB (1995) Effect of peroxynitrite on the mitochondrial respiratory chain: differential susceptibility of neurones and astrocytes in primary culture. J Neurochem 64:1965–1972
59. Cassina A, Radi R (1996) Differential inhibitory action of nitric oxide and peroxynitrite on mitochondrial electron transport. Arch Biochem Biophys 328:309–316
60. Packer MA, Murphy MP (1995) Peroxynitrite formed by simultaneous nitric oxide and superoxide generation causes a cyclosporin A-sensitive mitochondrial calcium efflux and depolarisation. Eur J Biochem 234:231–239
61. Schweizer M, Richter C (1996) Peroxynitrite stimulates the pyridine nucleotide-linked calcium release from intact rat liver mitochondria. Biochemistry 35:4524–4528
62. Radi R, Rodriguez M, Castro L, Telleri R (1994) Inhibition of mitochondrial transport chain by peroxynitrite. Arch Biochem Biophys 308:89–95
63. MacMillan-Crow LA, Crow JP, Kerby JD, et al. (1996) Nitration and inactivation of manganese superoxide dismutase in chronic rejection of human renal allografts. Proc Natl Acad Sci USA 93:11853–11858
64. Gow A, Duran D, Thom SR, Ischiropoulos H (1996) Carbon dioxide catalyzed protein tyrosine nitration by peroxynitrite. Arch Biochem Biophys. 333:42–48
65. Denicola A, Trujillo M, Freeman BA, Radi R (1996) Peroxynitrite reaction with carbon dioxide/bicarbonate: kinetics and influence on peroxynitrite-mediated oxidation reactions. Arch Biochem Biophys 333:49–58
66. Beckman JS, Ye Y-Z, Anderson PG, et al. (1994) Extensive nitration of protein tyrosine in human atherosclerosis detected by immunohistochemistry. Biol Chem Hoppe-Seyler 375:81–88

67. Leeuwenburgh C, Hardy MM, Hazen SL, et al. (1997) Reactive nitrogen intermediates promote low density lipoprotein oxidation in human atherosclerotic intima. J Biol Chem 272:1433–1436
68. Kaur H, Halliwell B (1994) Evidence for nitric oxide-mediated oxidative damage in chronic inflammation. Nitrotyrosine in serum and syniovial fluid from rheumatoid patients. FEBS Lett 350:9–12
69. Basarga O, Michaels FH, Zheng YM, et al. (1995) Activation of the inducible form of nitric oxide synthase in the brains of patients with multiple sclerosis. Proc Natl Acad Sci USA 92:12041–12045
70. Fukuyama N, Takebayashi Y, Hida M, et al. (1997) Clinical evidence of peroxynitrite formation in chronic renal failure patients with septic shock. Free Radical Biol Med 22:771–774
71. Szabo C, Salzman AL, Ischiropoulos H (1995) Endotoxin triggers the expression of an inducible isoform of nitric oxide synthase and the formation of peroxynitrite in the rat aorta in vivo. FEBS Lett 363:235–238
72. Miller MJS, Thompson JH, Zhang X-J, et al. (1995) Role of inducible nitric oxide synthase expression and peroxynitrite formation in the guinea pig ileitis. Gastroenterology 109:1475–1483
73. Ischiropoulos H, Beers MF, Ohnishi ST, et al. (1996) Nitric oxide production and perivascular tyrosine nitration in brain following carbon monoxide poisoning in the rat. J Clin Invest 97:2260–2267
74. Viner RI, Ferrington DA, Huhmer AFR, et al. (1996) Accumulation of nitrotyrosine on the SERCA2a isoform of SR Ca-ATPase of rat skeletal muscle during aging: a peroxynitrite-mediated process? FEBS Lett 379:286–290
75. White CR, Brock TA, Chang, L-Y, et al. (1994) Superoxide and peroxynitrite in atherosclerosis. Proc Natl Acad Sci USA 91:1044–1048
76. Minor RL, Myers PR, Guerra R, et al. (1990) Dietinduced atherosclerosis increases the release of nitrogen oxides from rabbit aorta. J Clin Invest 86:2109–2116
77. Ohara Y, Peterson TE, Harrison DG (1993) Hypercholesterolemia increases endothelial superoxide anion production. J Clin Invest 91:2546–2551
78. Haddad IY, Ischiropoulos H, Holm BA, et al. (1993) Mechanisms of peroxynitrite induced injury to pulmonary surfactants. Am J Physiol 265:L555–L564
79. Barlett BS, Friguet B, Yim MB, et al. (1996) Peroxynitrite-mediated nitration of tyrosine residues in *Escherichia coli* glutamine synthetase mimics adenylation: relevance to signal transduction. Proc Natl Acad Sci USA 93:1776–1780
80. Gow A, Duran D, Malcom S, Ischiropoulos H (1996) Effects of peroxynitrite induced modifications to signal transduction and protein degradation. FEBS Lett 385:63–66
81. Kong S-K, Yim MB, Stadtman ER, Chock PB (1996) Peroxynitrite disables the tyrosine phosphorylation regulatory mechanism: lymphocyte-specific tyrosine kinase fails to phosphorylate nitrated cdc2(6-20)NH2 peptide. Proc Natl Acad Sci USA 93:3377–3382

# Defenses Against Peroxynitrite

HELMUT SIES, HIROSHI MASUMOTO, VICTOR SHAROV, and KARLIS BRIVIBA

*Summary.* Stimulated inflammatory cells produce superoxide and nitric oxide radicals, which react, in a fast reaction, to generate peroxynitrite. Biological systems require protection against this reactive species, which leads to DNA damage by oxidizing guanine and by causing single-strand breaks and to tyrosine nitration, potentially interfering with phosphorylation—dephosphorylation signaling pathways. So far, there is no known enzymatic defense against peroxynitrite. Our recent work showed that peroxynitrite appears to react preferentially with selenium compounds. Ebselen, an antiinflammatory selenoorganic compound, reacts with peroxynitrite, forming the corresponding selenoxide at a second-order rate constant of $2 \times 10^6$ $M^{-1}s^{-1}$, which is about 100 fold higher than the rate constant observed with low molecular mass compounds such as ascorbate, cysteine, and methionine. The selenoxide can be reduced back to the parent compound at the expense of glutathione (GSH) or other thiols, so that a steady-state line of defense can be maintained. Selenium-containing compounds protect plasmid DNA from single-strand breaks better than the corresponding sulfur-containing compounds; e.g., selenomethionine protected DNA from peroxynitrite-induced damage more efficiently than methionine. Likewise, nitration reactions were suppressed efficiently by seleno-compounds. We postulate, based on preliminary evidence, that defense against peroxynitrite is a novel function of selenoproteins such as GSH peroxidase.

*Key words.* Peroxynitrite—Peroxynitrite reductase—Nitrotyrosine—Ebselen—Selenomethionine

## Introduction

There are several enzymatic sites in biological systems where the superoxide anion radical ($O_2^{-}$) and nitric oxide, a biological messenger, are produced. These two relatively stable radicals react with each other to form a stronger biological oxidant,

Institut für Physiolgische Chemie I, Heinrich-Heine-Universität Düsseldorf, Postfach 10 10 07, D-40001 Düsseldorf, Germany.

peroxynitrite [1]. The reaction rate constant, $6.7 \times 10^9$ $M^{-1}s^{-1}$ [2], is larger than that for the dismutation of $O_2^{-}$ catalyzed by superoxide dismutase (SOD). Indeed, the formation of peroxynitrite in vitro by endothelial cells, Kupffer cells, macrophages, and neutrophils has been demonstrated [3–6].

## Peroxynitrite-Mediated Damage

Peroxynitrite damages biological molecules by oxidation and nitration reactions and can initiate lipid peroxidation in biological membranes or low-density lipoproteins. A variety of enzymes such as myeloperoxidase, SOD, or alcohol dehydrogenase can be inactivated by peroxynitrite. Peroxynitrite hydroxylates phenolic compounds and oxidizes thiols like other reactive species, e.g., the hydroxyl radical or singlet oxygen [7,8]. Nitration of tyrosine is more characteristic for peroxynitrite and is used to detect the formation of peroxynitrite. The nitration of tyrosine residues has been even shown in vivo in the human in atherosclerotic lesions [9] and in plasma in chronic renal patients with septic shock [10].

Peroxynitrite is genotoxic, can cause strand breaks, and can oxidize guanine residues to 8-hydroxyguanine in isolated bacteriophage PM2 DNA, as characterized by means of several repair enzymes with defined substrate specificities [11]. Peroxynitrite is also capable of generating single-strand breaks in viable DNA in rat thymocytes [12]. Peroxynitrite reacts preferentially with guanine in DNA [13]. The majority of mutations caused by peroxynitrite occurred at G:C base pairs with GC→TA transversion when a plasmid pretreated with peroxynitrite was replicated in bacteria or mammalian cells [14].

Thus, biological systems require a defense system against peroxynitrite. Low molecular mass compounds such as ascorbate, cysteine, glutathione, $CO_2$, and methionine have been shown to react with peroxynitrite. $CO_2$ rapidly reacts with peroxynitrite; however, another reactive species, nitrosoperoxycarbonate, capable of nitrating even more effectively than peroxynitrite is formed [15]. There is some doubt whether glutathione, ascorbate, or α-tocopherol is capable of protecting sufficiently against damage generated by peroxynitrite. So far, there is no known enzymatic defense against peroxynitrite, such as SOD against $O_2^{-}$ or catalase and glutathione peroxidase against $H_2O_2$ and organic hydroperoxides.

## Selenium-Containing Compounds Protect Against Peroxynitrite-Mediated Reactions

Recently, we showed that low molecular mass selenium-containing compounds such as ebselen, selenomethionine, or selenocystine protect against peroxynitrite-mediated DNA damage [16], as well as oxidizing and nitration reactions [17], about 100 fold more effectively than their corresponding sulfur analogs or low molecular antioxidants like glutathione or ascorbate (Table 1). Ebselen, a glutathione peroxidase mimic, reacts with peroxynitrite yielding ebselen selenium-oxide at stoichiometry 1:1 [18]. The second-order rate constant was estimated to be $2 \times 10^6$ $M^{-1}s^{-1}$, i.e., two or three orders of magnitude higher than the rate constants for ascorbate or glutathione [19,20]. Ebselen selenium oxide is reduced readily back to ebselen at the expense of

TABLE 1. Half-maximal inhibitory concentrations of the peroxynitrite-mediated oxidation of dihydrorhodamine-123 (DHR-123) and of the peroxynitrite-mediated nitration of 4-hydroxyphenylacetate (4-HPA) for some selenoorganic compounds and their sulfur analogs[a]

| Compound | Half-maximal inhibitory concentration (μM) | |
|---|---|---|
| | DHR-123 oxidation | 4-HPA nitration |
| Glutathione peroxidase (GPx)[b] | 0.15 | n.d. |
| Ebselen | 0.2 | 60 |
| Selenium-methionine | 0.3 | 50 |
| Selenium-cystine | 2.5 | 30 |
| Ebsulfur | 15 | 300 |
| Methionine | 20 | 750 |
| Cystine | $>10^3$ | $>10^3$ |
| Sodium selenite | $>10^4$ | $>10^4$ |

n.d., not determined
[a] From [17] and [b] unpublished work

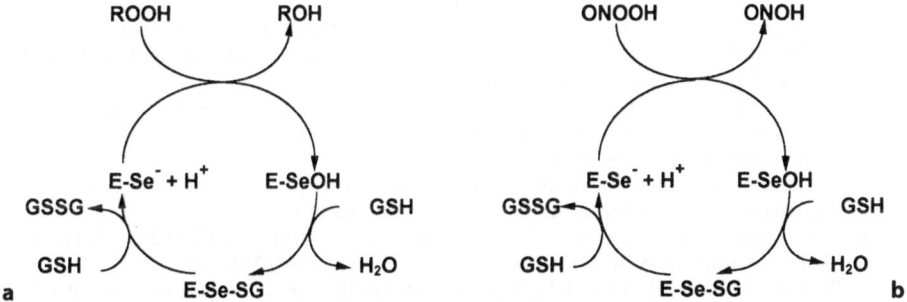

FIG. 1a,b. Proposed catalytic mechanism of selenoperoxidases in their function of reduction of (a) hydroperoxide to the corresponding alcohol (b) peroxynitrous acid to ONOH (or peroxynitrite to nitrite). E, enzyme; GSSG, glutathione disulfide; GSH, glutathione

reducing equivalents such as glutathione, N-acetylcysteine, or dihydrolipoate, so that ebselen is available for further use in peroxynitrite reduction. This catalytic cycle is similar to that described for glutathione peroxidase. It is interesting that a selenium-containing enzyme, glutathione peroxidase (GPx), in the reduced but not in the oxidized form protected against the peroxynitrite-mediated oxidation of dihydro-rhodamine 123 more effectively than ebselen (Sies et al., unpublished work).

We propose that GPx reduces peroxynitrite, yielding nitrite and oxidized GPx that can be reduced by GSH. Furthermore, reduced GPx as well as selenomethionine protected against 3-nitrotyrosine formation in human fibroblast lysate from human fibroblasts, showing a high efficiency in competition with endogenous antioxidants or proteins capable of reacting with peroxynitrite. Moreover, selenomethionine oxide formed by peroxynitrite can be recovered to selenomethionine by GSH (Briviba et al., unpublished results). A number of different selenopeptides and selenoproteins, many of them with functions so far unknown, have been described in vivo [21]. The presence of one of them, selenoprotein P, has been associated recently with protection against liver damage in two oxidant injury models [22]. Thus, selenoproteins containing selenocysteine or selenomethionine may fulfill a function in defense against peroxynitrite via a GPx-like catalytic cycle (Fig. 1).

*Acknowledgments*. This study was supported by the Deutsche Forschungsgemeinschaft, SFB 503, Project B1. V.S. Sharov was a Research Fellow of the Alexander von Humboldt Foundation, Bonn, Germany.

# References

1. Beckman JS, Beckman TW, Chen J, et al. (1990) Apparent hydroxyl radical production by peroxynitrite: implications for endothelial injury from nitric oxide and superoxide. Proc Natl Acad Sci USA 87:1620–1624
2. Huie RE, Padmaja S (1993) The reaction of NO with superoxide. Free Radical Res Commun 18:195–199
3. Kooy NW, Royall JA (1994) Agonist-induced peroxynitrite production from endothelial cells. Arch Biochem Biophys 310:352–359
4. Wang JF, Komarov P, Sies H, de Groot H (1991) Contribution of nitric oxide synthase to luminol-dependent chemiluminescence generated by phorbol-ester-activated Kupffer cells. Biochem J 279:311–314
5. Carreras MC, Pargament GA, Catz SD, et al. (1994) Kinetics of nitric oxide and hydrogen peroxide production and formation of peroxynitrite during the respiratory burst of human neutrophils. FEBS Lett 341:65–68
6. Ischiropoulos H, Zhu L, Beckman JS (1992) Peroxynitrite formation from macrophage-derived nitric oxide. Arch Biochem Biophys 298:446–451
7. Ramezanian MS, Padamaja S, Koppenol W (1996) Nitration and hydroxylation of phenolic compounds by peroxynitrite. Chem Res Toxicol 9:232–240
8. Radi R, Beckman JS, Bush KM, Freeman BA (1991) Peroxynitrite oxidation of sulfhydryls. The cytotoxic potential of superoxide and nitric oxide. J Biol Chem 266: 4244–4250
9. Beckman JS, Ye YZ, Anderson PG, et al. (1994) Extensive nitration of protein tyrosines in human atherosclerosis detected by immunohistochemistry. Biol Chem Hoppe-Seyler 375:81–88
10. Fukuyama N, Takebayashi Y, Hida M, et al. (1997) Clinical evidence of peroxynitrite formation in chronic renal failure patients with septic shock. Free Radical Biol Med 22:771–774
11. Epe B, Ballmaier D, Roussyn I, et al. (1996) DNA damage by peroxynitrite characterized with DNA repair enzymes. Nucleic Acids Res 24:4105–4110
12. Salgo MG, Bermudez E, Squadrito GL, Pryor WA (1995) DNA damage and oxidation of thiols peroxynitrite causes in rat thymocytes. Arch Biochem Biophys 322:500–505
13. Douki T, Cadet J, Ames BN (1996) An adduct between peroxynitrite and 2'-deoxyguanosine: 4,5-dihydro-5-hydroxy-4-(nitrosooxy)-2'-deoxyguanosine. Chem Res Toxicol 9:3–7
14. Juedes MJ, Wogan GN (1996) Peroxynitrite-induced mutation spectra of pSP189 following replication in bacteria and in human cells. Mutat Res 349:51–61
15. Radi R, Cosgrove TP, Beckman JS, Freeman BA (1993) Peroxynitrite-induced luminol chemiluminescence. Biochem J 290:51–57
16. Roussyn I, Briviba K, Masumoto H, Sies H (1996) Selenium-containing compounds protect DNA from single-strand breaks caused by peroxynitrite. Arch Biochem Biophys 330:216–218

17. Briviba K, Roussyn I, Sharov VS, Sies H (1996) Attenuation of oxidation and nitration reactions of peroxynitrite by selenomethionine, selenocystine and ebselen. Biochem J 319:13–15
18. Masumoto H, Sies H (1996) The reaction of ebselen with peroxynitrite. Chem Res Toxicol 9:262–267
19. Masumoto H, Kissner R, Koppenol WH, Sies H (1996) Kinetic study of the reaction of ebselen with peroxynitrite. FEBS Lett 398:179–182
20. Sies H, Masumoto H (1997) Ebselen as a glutathione peroxidase mimic and as a reactant with peroxynitrite. Adv Pharmacol 38:229–246
21. Behne D, Hilmert H, Scheid S, et al. (1988) Evidence for specific selenium target tissues and new biologically important selenoproteins. Biochim Biophys Acta 966:12–21
22. Burk RF, Hill KE, Awad JA, et al. (1995) Pathogenesis of diquat-induced liver necrosis in selenium-deficient rats: assessment of the roles of lipid peroxidation and selenoprotein P. Hepatology 21:561–569

# Bioactive 6-Nitronorepinephrine Formation Requiring Nitric Oxide Synthase in Mammalian Brain

Toshio Nakaki[1], Futoshi Shintani[2], Shigenobu Kanba[3], Eiji Suzuki[2], and Masahiro Asai[2]

*Summary.* The possibility of an interaction between norepinephrine and NO˙ or NO˙-related molecules in vivo was explored. Bubbling of NO˙ gas in norepinephrine aqueous solution yielded 6-nitronorepinephrine (6-nitroNE). We also identified 6-nitroNE in the mammalian brain. Amounts of 6-nitroNE in the rat brain were attenuated by administration of an inhibitor of nitric oxide synthase. $N^G$-nitro-L-arginine methyl ester. This was reversed by coadministration of L-arginine, suggesting that nitric oxide synthase participated in the formation of 6-nitroNE. Moreover, we found that 6-nitroNE inhibits the activity of catechol-*O*-methyl transferase, as well as norepinephrine transport into rat synaptosomes. At values to $100\,\mu M$, 6-nitroNE did not displace the binding of $\alpha 1$-, $\alpha 2$-, or $\beta$-adrenergic receptor ligand. A rat brain microdialysis experiment showed that perfusion of 6-nitroNE into the rat hypothalamic paraventricular nucleus significantly elevated norepinephrine while decreasing 3-methoxy-4-hydroxyphenylglycol, a metabolite of norepinephrine. We propose that 6-nitroNE is a potential signal molecule linking the actions of NE and NO˙.

*Key words.* Nitric oxide—Norepinephrine—6-Nitronorepinephrine—Brain—COMT inhibitor

## Introduction

Norepinephrine (NE) functions as a neurotransmitter in the nervous system [1]. Nitric oxide (NO˙), a gaseous radical, is also thought to function as a neurotransmitter [2,3]. We previously found that NE levels detected by microdialysis of the rat hypothalamic paraventricular nucleus [4] were diminished by perfusion of a solution containing NO˙. We hypothesized that NE and NO˙ or NO˙-related molecules may interact.

[1] Department of Pharmacology, Teikyo University School of Medicine, 2-11-1 Kaga, Itabashi-ku, Tokyo 173, Japan
[2] Department of Neuro-Psychiatry, School of Medicine, Keio University, 35 Shinanomachi, Shinjuku-ku, Tokyo 160, Japan
[3] Department of Neuro-Psychiatry, Yamanashi Medical University, Yamanashi 409-38, Japan

Therefore, we attempted to identify a putative product of their reaction in mammalian brain.

## Materials and Methods

### Identification of 6-NitroNE

One hundred porcine brains (150–200 g/brain) were purchased from Tokyo Shibaura Zoki (Tokyo, Japan). Catechol derivatives were extracted from the brain tissues by a previously described method [5]. The samples were analyzed by means of high performance liquid chromatography (HPLC) with electrochemical detection (ECD) employing a previously described method [6]. A peak with a retention time identical to that of 6-nitroNE synthesized in vitro was obtained. The fractions were pooled and subjected to ultraviolet (UV) spectrometry, mass spectrometry (MS), and nuclear magnetic resonance (NMR) spectroscopy.

Absorbance was measured at a wavelength between 190 and 500 nm using a UV photometer (Waters 990, Waters, Tokyo, Japan). The mobile phase was a water solution that consisted of 5.5 mM formate and 6.7 mM ammonium formate (pH 3.7), and a strong cation-exchange column (Whatman Japan, Tokyo, Japan) was used for sample separation at 25°C. Further purification was performed by using a $C_{18}$ reversephase column (Whatman Japan) and a mobile phase that consisted of 0.09% trifluoroacetic acid (vol/vol water) and 0.07% acetonitrile (vol/vol water).

Mass spectral analysis was carried out on a JMS-SX/SX102A (BEBE configuration) tandem mass spectrometer (JEOL, Tokyo, Japan) by Dr. Takeshi Kinoshita (Analytical and Metabolic Research Laboratories, Sankyo, Tokyo, Japan). Positive- and negative-ion antigen-binding fragment mass spectroscopy (Fab MS), employing 3-nitrobenzyl alcohol as the matrix, produced $[M + H]^+$ ions at m/z 215.0669 (calculated for $C_8H_{11}N_2O_5$, 215.0668) and $[M - H]^-$ ions at m/z 213. Collisionally activated dissociation (CAD) was performed using argon as the collision gas at a pressure sufficient to reduce the precursor ion signal by 80%. The accelerating voltage was 10 kV and the collision cell potential was set at 5 kV. The $^1$H-NMR spectrum was recorded on a JEOL GX-400 NMR spectrometer (JEOL) using tetramethylsilane as an internal reference.

Groups of four male Sprague–Dawley rats (Sankyo Laboservice, Tokyo, Japan) were given both 200 mg/kg of $N^G$-nitro-L-arginine methyl ester or $N^G$-nitro-D-arginine methyl ester and 800 mg/kg of L- or D-arginine hydrochloride intraperitoneally 60 min before decapitation under ether anesthesia. The drugs were dissolved in saline, and the solution was adjusted to pH 7.4 before use. The brain contents of 6-nitroNE were then extracted by the method described previously. The peak areas of samples were compared with those of standard 6-nitroNE to allow quantiation. All experiments conformed to the standards put forth in the Handbook for the Use of Animals in Neuroscience Research.

### NE Uptake

Uptake of $[^3H]$L-NE into synaptosomes was measured by the method of Richelson and Pfenning [7].

## Microdialysis

Surgery and the microdialysis experiments were performed employing previously described methods [8]. To examine the effects of 6-nitroNE, Hanks' solution containing 100 µM 6-nitroNE was perfused into the microdialysis probe without stopping the aforementioned sampling; perfusion was continued for as long as 60 min. All drugs were freshly prepared and adjusted to pH 7.4 on the day of use. The rats were given free access to food and water throughout the microdialysis procedure.

## HPLC Analysis

The HPLC-ECD system was set up based on a previously described method [6] with minor modification. In brief, the chromatographic system consisted of a dual-plunger pump (EP-10, Eicom, Kyoto, Japan), an electrochemical detector with a carbon graphite electrode (Coulochem II, ESA, Chelmsford, MA, USA), an auto-sampling injector (231–401, Gilson Medical Electronics, Villiers le Bel, France) with a 100-ml loop, and an ODS column. We used a CA-5ODS column (150 mm × 4.6 mn i.d., Eicom), and the mobile phase consisted of 95% 0.1 M phosphate buffer solution (consisting of 0.0862 M $NaH_2PO_4 \cdot 2H_2O$ and 0.0138 M $Na_2HPO_4 \cdot 12H_2O$; pH 6.0), 2% methanol, 200 mg/l sodium 1-octanesulfonate, and 50 mg/l ethylenediamine tetraacetic acid · 2Na. The buffer was sonicated to degas it before use. Separation was achieved at 25°C using a flow rate of 1 ml/min. An electrochemical detector with a carbon graphite working electrode was set at +400 mV for the guard cell, at +50 mV for the oxidation electrode, and at −200 to −300 mV for the reduction electrode. Data were recorded on an integrator (chromatocorder 12, JASCO, Tokyo, Japan), and peak heights/area of dialysis samples were compared with those of standards determined each day for quantitation. The detection limit was 0.5 pg/100 ml.

## Binding Assays and Catechol-O-Methyl Transferase

Several concentrations of 6-nitroNE were incubated in a receptor-binding preparation containing [³H]prazosin (1.10 Tbq/mmol, New England Nuclear, Boston, MA, USA) [9], [³H]yohimbine (2.59 Tbq/mmol, New England Nuclear) [9], or [³H]CGP12177 (1.11 Tbq/mmol, New England Nuclear) [10], which are $\alpha_1$-, $\alpha_2$-, and nonselective β-adrenergic receptor ligands, respectively. Catechol-O-methyl transferase (COMT) activity was determined by measuring the amount of normetanephrine formed from NE [11].

## 6-NitroNE

The 6-nitroNE used in these experiments was synthesized in vitro from L-NE and NO˙. After 10 mg of L-NE hydrochloride had been dissolved in 100 ml of distilled water, 200–300 ml of NO˙ gas was bubbled in a gastight glass tube at 37°C at atmospheric pressure for 2 min. The reaction products were separated and purified with the HPLC system described previously. The structure was confirmed using both MS and NMR spectroscopy (data not shown). The purity of 6-nitroNE exceeded 95%.

## Results and Discussion

Catechol-containing molecules were extracted from porcine brain tissue and separated by HPLC with ECD. The presence of 6-nitroNE was then detected in porcine brain using a UV detection method, MS (Fig. 1a), and NMR spectroscopy (Fig. 1b). The 6-nitroNE content of the whole brain constitutes about 10% of total NE.

Next, we tested whether NOS is involved in the formation of 6-nitroNE in the brain. The 6-nitroNE levels, as measured by HPLC-ECD in rat brains minus the olfactory bulbs, fell following intraperitoneal injection of *NG*-nitro-L-arginine methyl ester (200 mg/kg), but not *NG*-nitro-D-arginine methyl ester, which is not an inhibitor of nitric acid synthase (NOS). This decrease was reversed by coadministration of

a

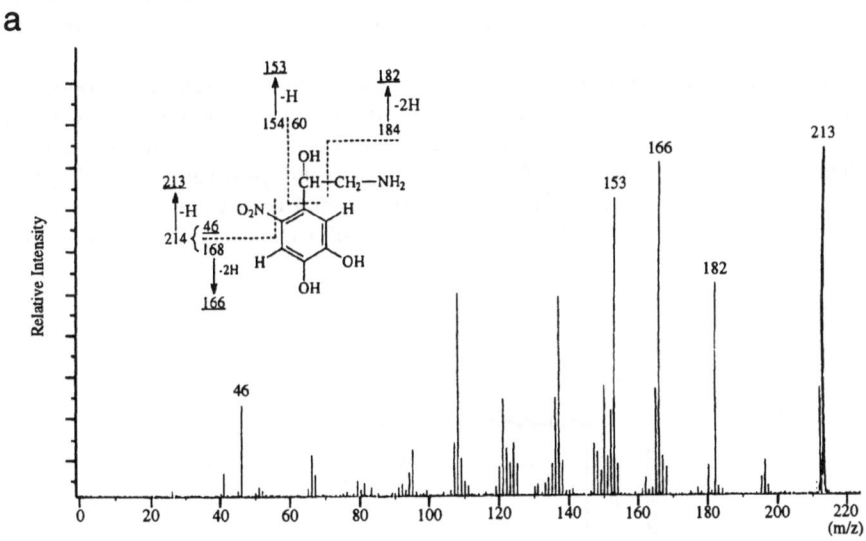

b

FIG. 1a,b. Identification of 6-nitronorepinephrine (6-nitroNE) in porcine brain. **a** Negative-ion antigen-binding fragment mass spectroscopy (delayed after depolarization [Fab MS/MS (DAD)] spectrum of 6-nitroNE and fragmentation patterns. **b** Structure of 6-nitroNE and the ¹H-NMR (nuclear magnetic resonance) data (400 MHz in D₂O, s [ppm]). (Based on [14], with permission)

800 mg/kg of L-arginine hydrochloride but not by the D-isomer (data not shown). These results suggest that NOS is involved in the formation of 6-nitroNE. In fact, NOS-containing and NE-containing neurons coexist in both the locus coeruleus [12] and the hypothalamic paraventricular nuclei [3], where 6-nitroNE may be formed. Ischiropoulos et al. showed that peroxynitrite and $Cu^{2+}$ mediate nitration of the electrophilic center of the phenyl moiety of tyrosine [13]. Although the nitration of NE may be mediated through peroxynitrite, the mechanisms governing nitration of 6-nitroNE in vivo remain to be determined.

At levels to 100 μM, 6-nitroNE did not displace the binding of [³H]prazosin, [³H]yohimbine, or [³H]CGP12177, which are $\alpha_1$-, $\alpha_2$-, and nonselective β-adrenergic receptor ligands, respectively (data not shown). Likewise, the activity of monoamine oxidase (MAO) was unaffected (data not shown). However, transport of 8 nM [³H]L-NE into rat brain synaptosomes prepared from frontal and occipital cortices was inhibited by 6-nitroNE with an $IC_{50}$ of 31 μM. 6-NitroNE inhibited the activity of COMT with an $IC_{50}$ of 7.5 μM (Fig. 2). It is still an open question whether endogenous 6-nitroNE inhibits catecholamine reuptake and COMT activities. We are currently investigating this issue.

To examine the effects of 6-nitroNE on central noradrenergic neurotransmission in vivo, we conducted a microdialysis experiment to determine the content of NE and 3-methoxy-4-hydroxyphenyl glycol in the rat paraventricular nucleus. Significant elevations of NE and decreases in 3-methoxy-4-hydroxyphenyl glycol were observed during the perfusion of 6-nitroNE, and both were dose dependent (Fig. 3a,b). The inhibition by 6-nitroNE both of COMT activities and of the

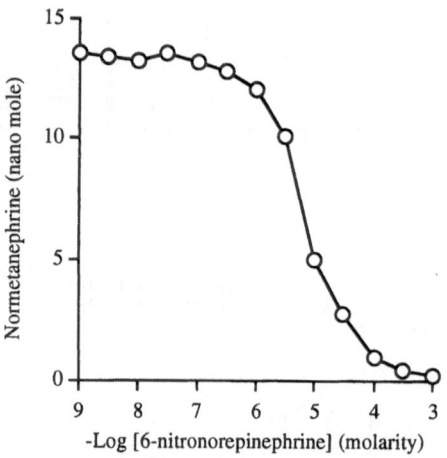

FIG. 2. Effects of 6-nitroNE on catechol-O-methyl transferase activity. Inhibitory effects of 6-nitroNE on catechol-O-methyl transferase activity catalyzing the reaction from L-NE to normetanephrine. The ordinate represents amounts of normetanephrine generated from norepinephrine by catechol-O-methyl transferase; the abscissa represents the molarity of 6-nitroNE. (Based on [14], with permission)

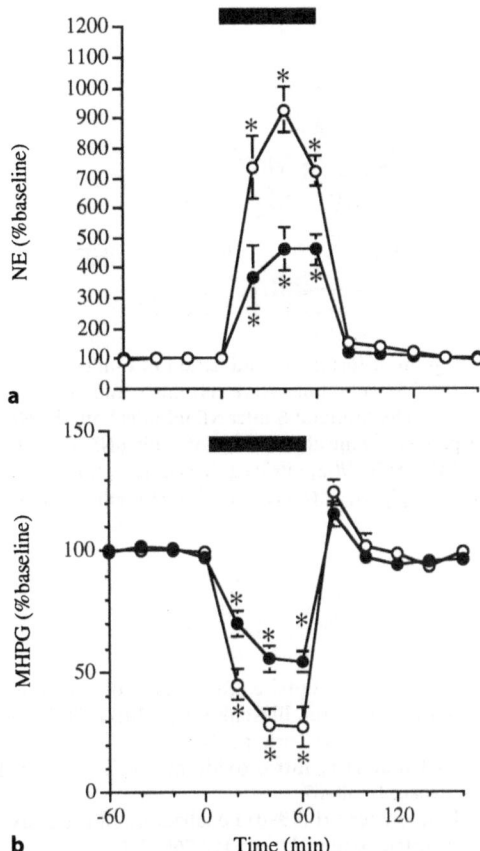

FIG. 3a,b. Effects of 6-nitroNE perfused directly into the rat paraventricular nucleus on NE and 3-methoxy-4-hydroxyphenyl glycol (MHPG). The amine was administered through a microdialysis tube at the concentrations of NE (a) or MHPG (b) recovered in dialysates from the same region. *Circles* and *vertical bars* represent means and SE, respectively; *horizontal bars* represent the infusion period for each concentration of 6-nitroNE (*open circles*, 100 µM; *closed* circles, 10 µM). The 100% values of NE and MHPG are $1.22 \pm 0.11$ and $6.75 \pm 0.29$ (pg/20 min, defined as means $\pm$ SE of four samples), respectively. *, significantly different ($P < .05$) from baseline value. (From [14], with permission).

reuptake of NE into presynaptic neurons may account for the in vivo results (Fig. 4).

## Conclusion

Our results suggest that NO' or NO'-derived molecules react with NE in vivo, and that the resultant 6-nitroNE has distinct biological activities which may modulate synaptic transmissions.

516    T. Nakaki et al.

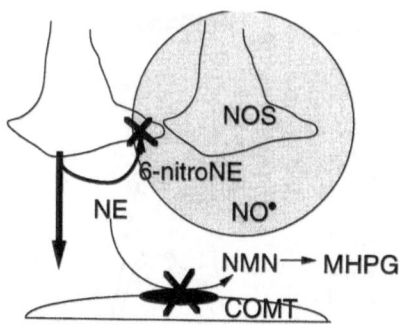

FIG. 4. Hypothetical diagram of formation and action of 6-nitroNE at synapses. Nitric oxide (NO·) diffuses from the NOS-containing nerve terminals and reacts with NE released from adjacent adrenergic nerves. The resultant 6-nitroNE inhibits both the NE reuptake process and COMT activities. The precise chemical reactions of 6-nitroNE formation are unknown. *NE*, norepinephrine; *NO.*, nitric oxide; *NOS*, nitric oxide synthase; *NMN*, normetanephrine; *MHPG*, 3-methoxy-4-hydroxyphenylglycol; *COMT*, catechol-*O*-methyl transferase; 6-nitroNE, 6-nitronorepinephrine

# References

1. von Euler US (1972) Synthesis, uptake and storage of catecholamines in adrenergic nerves, the effects of drugs. In: Blaschko H, Muscholl E (eds) Catecholamines. Springer, Berlin Heidelberg New York, pp 186–230
2. Garthwaite J (1991) Glutamate, nitric oxide and cell-cell signaling in the nervous system. Trends Neurosci 14:60–67
3. Bredt DS, Hwang PM, Snyder SH (1990) Localization of nitric oxide synthase indicating a neural role for nitric oxide. Nature 347:768–770
4. Shintani F, Kato R, Kinoshita N, et al. (1995) Nitric oxide and superoxide anion involved in hypothalamic stress response. In: Proceedmgs, Satellite symposium, 4th International Brain Research Organization world congress in neuro science, July 8, 1995, Otsu
5. Nakadate T, Kubota K, Muraki T, et al. (1980) Mechanism of chlorpromazine action on plasma glucose. Eur J Pharmacol 64:107–113
6. Shintani F, Kanba S, Nakaki T, et al. (1993) Interleukin-1β augments release of norepinephrine, dopamine and serotonin in the rat anterior hypothalamus. J Neurosci 13:3574–3581
7. Richelson E, Pfenning M (1984) Blockade by antidepressants and related compounds of biogenic amine uptake into rat brain synaptosomes: most antidepressants selectively block norepinephrine uptake. Eur J Pharmacol 104:277–286
8. Shintani F, Nakaki T, Kanba S, et al. (1995) Involvement of interleukin-1 in immobilization stress-induced increase in plasma adrenocorticotropic hormone and in release of hypothalamic monoamines in the rat. J Neurosci 15:1961–1970
9. Richelson E, Nelson A (1984) Antagonism by neuroleptics of neurotransmitter receptors of normal human brain in vitro. Eur J Pharmacol 103:197–204
10. Porzig H, Becker C, Reuter H (1982) Competitive and non-competitive interaction between specific ligands and beta-adrenoceptors in living cardiac cells. Naunyn-Schmiedeberg's Arch Pharmacol 321:89–99
11. Axelrod J, Tomchick R (1958) Enzymatic O-methylation of epinephrine and other catechols. J Biol Chem 233:702–705

12. Xu ZQ, Pieribone VA, Zhang X, et al. (1994) A functional role for nitric oxide in locus coeruleus: immunohistochemical and electrophysiological studies. Exp Brain Res 98:75–83

13. Ischiropoulos H, Zhu L, Chen J, et al. (1992) Peroxynitrite-mediated tyrosine nitration catalyzed by superoxide dismutase. Arch Biochem Biophys 298:431–437

14. Shintani F, Kinoshita T, Kanba S, et al. (1996) Bioactive 6-nitronorepinephrine identified in mammalian brain. J Biol Chem 271:13561–13565

# Role of Nitric Oxide in the Regulation of Cerebral Blood Flow in Conscious Rats

Shin-ichi Takahashi and Louis Sokoloff

*Summary.* Effects of inhibition of brain nitric oxide (NO) synthase on the cerebral vasodilation evoked by (1) functional activation, (2) autoregulatory response to hypotension, and (3) 5% $CO_2$ inhalation were studied in conscious rats with the [$^{14}$C]iodoantipyrine method for measurement of local cerebral blood flow (CBF). Enzymatic assays confirmed that three regimens of $N^G$-nitro-L-arginine methyl ester (L-NAME) administration, that is, (1) single intravenous injection (30 mg/kg), (2) intracisternal infusion (approximately 10 fold the total estimated content of L-arginine in rat brain), and (3) chronic intraperitoneal administration (50 mg/kg twice daily for 4 days) reduced NO synthase activity in whole brain to 47%, 12%, and 16% of control levels, respectively. Percent increases in local CBF in the stations of the whisker-to-cortical barrel sensory pathway elicited by unilateral vibrissal stimulation was unchanged by any of the three regimens of NO synthase inhibition. After almost complete inhibition of NO synthase activity by chronic intraperitoneal injection of L-NAME, neither global nor local CBF was affected when arterial blood presure was lowered by controlled withdrawal of blood, indicating that the normal autoregulatory vasodilator response was preserved. Finally, the percent enhancement of average blood flow in the brain as a whole by inhalation of 5% $CO_2$ in the inspired air was also unaltered following NO synthase inhibition by chronic intraperitoneal injection of L-NAME. In most of these experimental paradigms the NO synthase inhibition resulted in reduced baseline CBF despite increased systemic blood pressure. These results indicate that NO plays a role in the tonic regulation of cerebral vascular tone, but in conscious rats it does not mediate the increases in CBF produced by functional activation or increased blood $CO_2$ tension or the autoregulatory cerebral vasodilation in response to hypotension.

*Key words.* Autoradiographic [$^{14}$C]iodoantipyrine method—$N^G$-nitro-L-arginine methyl ester—Whisker barrel—Autoregulation—Carbon dioxide

Laboratory of Cerebral Metabolism, National Institute of Mental Health, Building 36, Room 1A-05, Bethesda, MD 20892-4030, USA

# Introduction

The gaseous molecule nitric oxide (NO) has been identified [1,2] as the endothelium-derived relaxing factor (EDRF) that is produced in vascular endothelium and mediates the vasodilator effects of acetylcholine and other vasodilator drugs [3,4]. NO is formed from the oxidation of L-arginine catalyzed by the mixed-function oxidase, nitric oxide synthase, an enzyme present in vascular endothelium. The biological half-life of NO is very short, usually several seconds, but it must be normally produced continuously because it exerts a tonic vasodilator influence on blood vessels, including those of the brain; inhibition of its production results in cerebral vasoconstriction and reduced cerebral blood flow (CBF) despite an associated elevation of systemic blood pressure [5,6].

Roy and Sherrington [7] postulated more than 100 years ago that CBF was regulated during altered functional activity by mechanisms that adjusted it to the associated changes in brain tissue metabolic demands. It was presumed that the rate of energy metabolism is raised during increased functional activity and that products or consequences of enhanced energy metabolism, for example, increased $Pco_2$ and $H^+$ concentration and decreased $Po_2$ in the tissue, dilate the cerebral blood vessels and increase the rate of blood flow. That local CBF is adjusted up and down in association with corresponding levels of functional activity has been observed and proven during the subsequent 100 years, but the fundamental mechanisms underlying these adjustments remain obscure.

Increased $Pco_2$ and $H^+$ concentration and decreased $Po_2$ do, indeed, raise CBF and are tonically active because when these are changed in the opposite direction, CBF decreases [8]. Experimental findings, however, have ruled out each of these chemical factors as the sole or even an essential factor in the enhancement of CBF during functional activation. Autoregulation of CBF is another characteristic of the regulation of the cerebral circulation; it is manifested as the maintenance of relatively constant CBF over a wide range of arterial blood pressures [9,10]. Maintenance of constant CBF in the face of changed perfusion pressures requires adjustment of cerebrovascular resistance and cerebral vascular tone to compensate for the altered pressure gradient. The mechanisms underlying the phenomenon of autoregulation are still unknown [11], but recent findings suggest that the cerebral vascular endothelium may play a role [12].

Nitric oxide synthase has been demonstrated immunohistochemically to reside also in neurons of the central nervous system [13] and in astrocytes in vivo as well as in primary cultures [14]. Its presence in these cell types has led to the hypothesis that nitric oxide plays a role in intercellular signaling in the nervous system [15]. Such signaling need not be confined to neurons and glia but could include signaling from neurons, glia, and endothelial cells to the smooth muscle of the cerebral blood vessels. Many of the physiological properties of NO are those to be expected of the putative mediator of the regulation of CBF.

In the current study we have employed three modes of administration of $N^G$-nitro-L-arginine methyl ester (L-NAME), a potent inhibitor of NO synthase, that is, (1) single intravenous injection, (2) intracisternal infusion, and (3) chronic intraperitoneal administration, to examine in conscious rats the effects of inhibition of NO synthesis on the increases in local CBF in the neural components of the whisker-to-cortical barrel sensory pathway elicited by vibrissal stimulation [16,17], on the autoregulatory cerebral vasodilator response to systemic hypotension, and on the enhancement of CBF by $CO_2$ inhalation.

## Materials and Methods

### Chemicals

4-Iodo-[N-methyl-$^{14}$C]antipyrine ([$^{14}$C]iodoantipyrine) (Spec. Act., 54 mCi/mmol) was purchased from Du Pont NEN (Boston, MA, U.S.A.). $N^G$-Nitro-L-arginine methyl ester hydrochloride (L-NAME) was obtained from Sigma Chemical (St. Louis, MO, U.S.A.).

### Animals

All procedures performed on animals were in strict accordance with the National Institutes of Health "Guide for the Care and Use of Laboratory Animals" and were approved by the local Animal Care and Use Committee. Normal adult male Sprague–Dawley rats were obtained from Taconic Farms (Germantown, NY, USA) and maintained in animal quarters with a standard 12-h light/dark cycle and normal humidity and temperature. Water and food were allowed ad libitum. Those rats that received chronic treatment with L-NAME were prepared for several days before the experimental procedure; they were injected intraperitoneally twice daily with 50 mg/kg of L-NAME dissolved in normal saline (16.5 mg/ml, pH 7) for 4 consecutive days just before the determination of CBF. Control rats were injected similarly with equivalent volumes of normal saline.

On the day of the experiment the rats were prepared by the insertion of polyethylene catheters (PE-50, Clay-Adams, Parsippany, NJ, USA) into both femoral arteries and one femoral vein under halothane (5% for induction and 0.8%–1.5% for maintenance)/70% $N_2O$/30% $O_2$ anesthesia. In those rats that were to receive infusions of L-NAME intracisternally, a midline incision was made in the upper cervical region, and soft tissues were dissected to expose the atlantooccipital membrane. A PE-10 polyethylene catheter (Clay-Adams, Parsippany, NJ, USA) was then inserted into the cisterna magna through a small incision to a depth of about 2 mm and fixed in place with acrylic cement (Super Binder, Loctite Corporation, Newington, CT, USA). In all animals surgical wounds were treated with 5% lidocaine ointment and sutured. In rats in which the effects of functional activation of the whisker-to-cortical barrel pathway were studied, the vibrissae on the right side of the face were cut close to the skin to minimize spurious stimulation of the unstimulated control side. Following the surgery a loose-fitting plaster cast was applied to the lower abdomen and hips and taped to a lead brick to prevent locomotion. The animals were then allowed 3–4 h to recover from the surgery and anesthesia. Body temperature was maintained at 37°C throughout the operative and postoperative periods by a thermostatically controlled infrared lamp coupled to a rectal temperature probe.

## Physiological Variables

Several physiological variables were measured repeatedly after the 3-h period of recovery from the surgery and anesthesia to assess the physiological status of the animal during the control state and during the experimentally induced conditions. Mean arterial blood pressure (MABP) was monitored through one of the femoral arterial catheters with a pressure transducer (Model 4-327-I, SensorMedics, Anaheim, CA, USA) and recorded on a Beckman Model R-611 polygraph (Beckman Instruments, Fullerton, CA, USA). Arterial blood $Po_2$, $Pco_2$, and pH were measured with a Corning pH/Blood Gas Analyzer (model 170, Corning Medical, Medfield, MA, USA). Hematocrit was determined in samples of arterial blood centrifuged in a Microfuge B (Beckman).

## Experimental Conditions

### Effects of NO Synthase Inhibition on Stimulation of CBF by Functional Activation of the Whisker-to-Sensory Cortex Pathway

Unilateral stroking of whiskers in rats produces selective increases in local CBF in the whisker-to-cortical barrel sensory pathway [16,17]. To identify the specific brain structures in which CBF was altered by whisker stimulation and to establish a baseline for comparison with the effects in inhibitor-treated animals, local CBF was measured in a group of control rats that had received only normal saline intravenously (Fig. 1). The effects of unilateral whisker stimulation on local CBF were examined in rats treated with three different regimens of L-NAME administration to inhibit NO synthase activity.

#### Acute Intravenous L-NAME Administration

A dose of 30 mg/kg of L-NAME dissolved in 1.2–1.5 ml of normal saline was infused intravenously during a 1- to 1.5-min period in four rats; five control rats received similar infusions of normal saline alone. Ten minutes after completion of the infusion, when MABP had reached a stable although elevated level, the procedure for measurement of local CBF was initiated. The vibrissae on the left side of the face were manually stroked forward and backward with a small paint brush at a rate of two to four strokes per second throughout the period of measurement of CBF.

#### Intracisternal Infusion of L-NAME

L-NAME, 15 mg/ml in artificial cerebrospinal fluid (CSF) adjusted to pH 7.25, was infused intracisternally at a rate of 2 µl/min for 20 min in seven rats. The total dose of L-NAME was about 600 µg (i.e., 2.2 µmol or about ten times the total estimated content of L-arginine in rat brain). Six control rats were similarly infused with equivalent volumes of artificial CSF alone. Unilateral vibrissal stimulation as described and measurement of CBF were initiated immediately after completion of the infusion.

#### Chronic Intraperitoneal L-NAME Administration

Seven rats were injected intraperitoneally twice daily with 50 mg/kg of L-NAME in 1.2–1.5 ml of normal saline for 4 days; four control rats were injected similarly

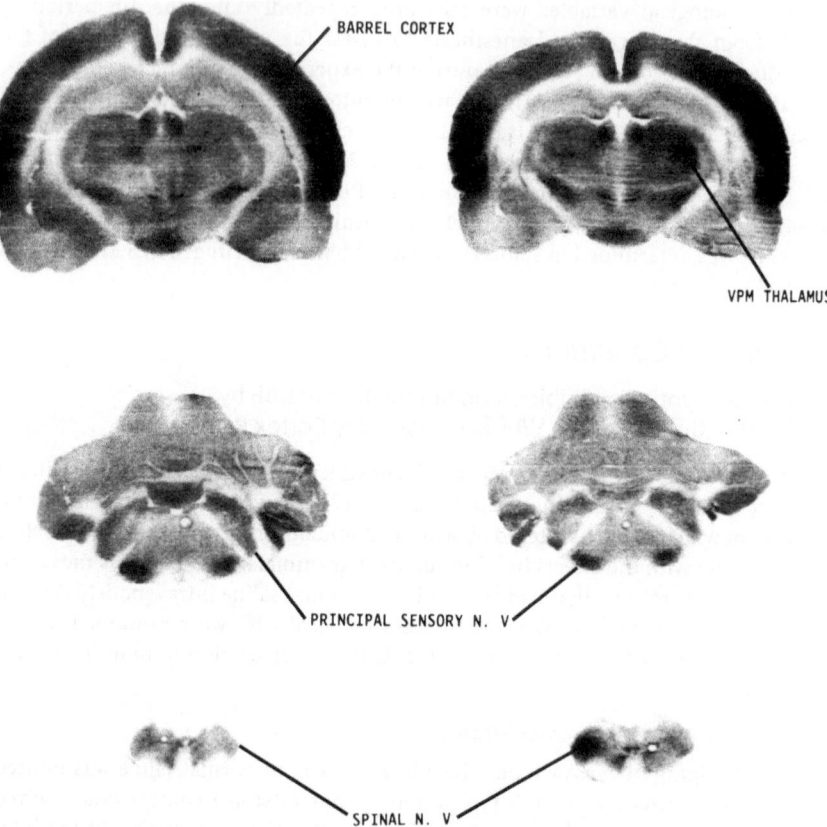

FIG. 1. [$^{14}$C]Iodoantipyrine autoradiographs of coronal sections of rat brain illustrating effects of vibrissal stimulation on rates of blood flow in stations of the whisker-to-sensory cortical barrel pathway; the darker the region, the higher the rate of blood flow. Sections are at levels of barrel cortex and ventral posteromedial nucleus (*VPM*) of thalamus (*top*), trigeminal principal sensory nucleus (*middle*), and trigeminal spinal subnucleus caudalis (*bottom*). *Left*: autoradiographs representative of those obtained from control rats in which whiskers were not stimulated during [$^{14}$C]iodoantipyrine infusion. *Right*: autoradiographs representative of those from rats with left unilateral whisker-stroking

with normal saline alone. CBF was measured approximately 14h after the last intraperitoneal injection. Unilateral vibrissal stimulation as described previously was carried out throughout the 1-min period of measurement of local CBF.

## Effects of NO Synthase Inhibition on Autoregulatory Vasodilator Response to Systemic Hypotension

In these experiments we employed the regimen of chronic intraperitoneal L-NAME administration, which was shown to achieve almost complete inhibition of NO synthase activity in brain [18,19]. CBF was measured in some of these L-NAME-treated rats (group a) and the saline-treated controls with their resting blood pressures left unaltered. The other L-NAME-treated rats were made hypotensive by withdrawal of 1.5–4.5 ml of arterial blood during a 3- to 5-min until the desired level of MABP was obtained; two levels of lowered blood pressure were obtained, moderate hypotension (group b) and severe hypotension (group c). Approximately 10 min after establishment of the MABP level at a stable desired level and immediately before initiation of the procedure for determination of CBF, arterial blood gas tensions, pH, and hematocrit were redetermined. MABP was monitored continuously throughout the measurement of CBF.

### Effects of NO Synthase Inhibition on enhancement of CBF by 5% $CO_2$ Inhalation

Fifteen rats that had been chronically pretreated with L-NAME and 14 control rats that had been pretreated with normal saline alone were administered either compressed room air (8 L-NAME-treated and 7 saline-treated rat) or 5% $CO_2$ in room air (7 L-NAME-treated and 7 saline-treated rat) via a plastic bag enveloping the animals. Five minutes after the onset of administration of the gases, the infusion of [$^{14}$C]iodoantipyrine was initiated and continued for the 1-min period of measurement of CBF.

## Assay of Brain NO Synthase Activity

NO synthase activity was measured in brain homogenates by assay of the conversion of [$^{14}$C]arginine to [$^{14}$C]citrulline according to a modification [20] of the procedure of Bredt and Snyder [21]. The enzyme assays were performed on the brains of two rats representative of each of the four conditions in which local CBF was measured: (1) normal controls receiving vehicle, (2) acute intravenous injections of L-NAME, (3) intracisternal infusion of L-NAME, and (4) intraperitoneal injections of L-NAME twice daily for 4 days.

## Measurement of Local CBF

Local CBF was determined by the [$^{14}$C]iodoantipyrine method of Sakurada et al. [22], slightly modified by the use of a programmed intravenous infusion of [$^{14}$C]iodoantipyrine designed to achieve a constantly rising tracer concentration in the arterial blood (i.e., ramp input function) [23]. The [$^{14}$C]iodoantipyrine (40 μCi in 0.8 ml of normal saline) was infused continuously through the femoral venous catheter by a programmed computer-driven infusion pump (Model 2400-003, Harvard Apparatus, South Natick, MA, USA) for about 1 min during which precisely timed arterial blood samples were collected at approximately 3-s intervals onto previously weighed filter paper disks in plastic beakers. The beakers were immediately capped after collection of blood to avoid evaporation.

At a precisely recorded time, approximately 1 min after onset of the infusion, the rat was decapitated, and the brain was rapidly removed and frozen in isopentane chilled to −40° to −50°C with dry ice. The filter paper disks and beakers were then reweighed. The weights of the blood samples were determined by difference, and their volumes were calculated from their weights on the basis of an assumed blood density of 1.05 g/ml. The filter paper disks were then transferred to scintillation counting vials; 10 ml of scintillation phosphor solution (Aquasol, Du Pont NEN, Boston, MA, USA) were added, which precipitated the blood proteins on the filter paper disks, and then 1 ml of water was added. After a 24-h period for complete elution of the labeled tracer into the phosphor solution, the samples were assayed for [$^{14}$C]iodoantipyrine concentration by liquid scintillation counting (Packard Instrument, Downers Grove, IL, USA) with external standardization.

The frozen brains were cut into 20-μm-thick coronal sections in a cryostat (Bright Instrument, Huntingdon, UK). Sets of four contiguous sections were taken every 200 μm throughout the entire rostrocaudal span of the brain and autoradiographed on Kodak EMC-1 X-ray film (Kodak, Rochester, NY, USA) together with a set of calibrated [$^{14}$C]methylmethacrylate standards. Local tissue concentrations of $^{14}$C were determined by quantitative autoradiography and image processing of the autoradiographs with a McIntosh-based image-processing system (Image 1.08, W. Rasband, NIMH, Bethesda, MD, USA). Local CBF was computed by means of the operational equation of the [$^{14}$C]iodoantipyrine method [22] with appropriate corrections for the lag and washout of the dead space in the arterial sampling catheter [24]; catheter flow rates had been adjusted to be at least 40 times the dead space volume (40 μl) to minimize the magnitude of the corrections.

The local CBF for each structure was calculated from the mean of the tracer concentrations found in six brain sections in which it appeared in the autoradiographs. Average CBF in the brain as a whole, weighted for the relative sizes of its component structures, was also determined from the individual rates of blood flow, pixel by pixel, with the computer program developed by G. Mies (Max Planck Institut für Neurologische Forschung, Köln, FRG) for use with the McIntosh personal computer and Image program.

## Data Analyses

### Effects of NO Synthase Inhibition on Stimulation of CBF by Functional Activation of the Whisker-to-Sensory Cortex Pathway

Four structures in the vibrissal sensory pathway were selected for determination of local CBF. These were the trigeminal spinal subnucleus caudalis, trigeminal principal sensory nucleus, ventral posteromedial nucleus (VPM) of the thalamus, and barrel field of the sensory cortex [16,25]. Vibrissal stroking was previously shown to stimulate glucose utilization in these stations of the pathway [17,25,26] and also to increase blood flow in the barrel field of the somatosensory cortex [17,26]. All these regions were clearly visible in the images reconstructed from the digitized autoradiographs and readily identified by comparison of the brain sections, which were stained after autoradiography, with an atlas of the rat brain [27] (see Fig. 1). Local blood flow was determined in each of these structures on both the stimulated and unstimulated sides of the brain, and side-to-side differences were compared statistically by Student's t-test for paired comparison. Percent increases in blood flow resulting from the vibrissal stimulation were calculated as the percent difference between the rates in the stimulated and unstimulated sides. The nonparametric Wilcoxon–Mann–Whitney test was used to evaluate the statistical significance of the differences in stimulus-induced percent increases in blood flow in controls and L-NAME-treated rats.

### Effects of NO Synthase Inhibition on Autoregulatory Vasodilator Response to Systemic Hypotension

In rats that were subjected to controlled blood withdrawal, paired $t$-tests were applied to compare the physiological parameters before and after blood withdrawal. A two-group $t$-test was used to compare the values of the physiological variables in the L-NAME-treated rats before blood withdrawal with those in the saline control rats.

Local CBF was determined in 32 representative brain structures. Two-group $t$-tests with Bonferroni corrections for four comparisons (i.e., saline control vs. group a, group a vs. group b, group a vs. group c, group b vs. group c) were used to compare the values for average and local CBF in the saline-treated controls and the three L-NAME-treated groups with the different MABP levels; in this analysis group variances were assumed to be homogeneous and a pooled variance obtained from all four groups was used. Similar statistical analysis was used to compare the physiological variables in the three L-NAME-treated groups.

### Effects of NO Synthase Inhibition on CBF Enhancement by 5% $CO_2$ Inhalation

In rats subjected to $CO_2$ inhalation, paired $t$-tests were applied to compare the physiological parameters before and after $CO_2$ inhalation. A two-group $t$-test was used to compare the physiological variables during $CO_2$ inhalation in the L-NAME-treated rats with those in the saline control rats.

Inasmuch as $CO_2$ inhalation is known to cause global increases in CBF, average blood flow in the brain as a whole was determined. A two-group $t$-test was applied to compare CBF in the groups breathing $CO_2$ and room air in both the saline-treated and L-NAME-treated animals. Percent increases in blood flow resulting from $CO_2$ inhalation were calculated as the percent difference between CBF in rats breathing $CO_2$ and in those breathing air.

## Results

### *Effects of L-NAME Administration on NO Synthase Activity in Brain*

Acute intravenous administration of L-NAME was found to reduce NO synthase activity in brain to 47% of the control level at 10 min after its injection when local CBF was measured (Table 1). When L-NAME was infused intracisternally or administered intraperitoneally in repeated doses for 4 days, brain NO synthase activity was reduced to 12% and 16% of control levels, respectively (Table 1).

### *Systemic Effects of L-NAME*

Acute intravenous administration of L-NAME (30 mg/kg) raised MABP in every animal from a mean of $130 \pm 2$ to $151 \pm 2$ ($\pm$SE) mmHg ($n = 4$; $P < .001$, paired $t$-test).

TABLE 1. Effects of various treatments with $N^G$-nitro-L-arginine methylester hydrochloride (L-NAME) on nitric oxide (NO) synthase activity in rat brain

| Treatment | Relative activity |
|---|---|
| Controls | 100 |
| L-NAME | |
| Acute intravenous injection (30 mg/kg) | 47 |
| Intracisternal infusion over 20 min (600 µg) | 12 |
| Intraperitoneal injection 2 times/day × 4 days (50 mg/kg) | 16 |

Value are means of values obtained when NO synthase activity was assayed in parallel in individual whole-brain homogenates of two rats for each listed condition. Enzyme activity was determined as nanomoles of citrulline produced/mg protein/per minute.

MABP at the time of CBF measurement in the L-NAME-treated rats was about 20% greater than that of the saline controls. Arterial pH and $P_{CO_2}$ were unaffected by intravenous L-NAME injections. Intracisternal infusions of L-NAME had no effect on MABP but resulted in a fall in arterial $P_{CO_2}$ from $37 \pm 0.8$ to $33 \pm 0.3$ ($\pm$SE) ($n = 7$; $P < .05$) and a rise in pH from $7.44 \pm 0.01$ to $7.48 \pm 0.01$ ($\pm$SE) ($n = 7$; $P < .05$, paired $t$-test).

The twice-daily intraperitoneal injections for 4 days caused no obvious behavioral changes in the rats given either L-NAME or saline [28,29]. The chronic intraperitoneal administration of L-NAME was without effect on arterial pH and $P_{CO_2}$ but produced a sustained significant increase in MABP to levels comparable to those after a single intravenous dose. MABP in these animals was $154 \pm 3$ ($n = 6$) compared to $125 \pm 5$ ($\pm$SE) mmHg ($n = 4$) in the saline-treated controls ($P < .0001$, $t$-test for group comparison).

The 24 L-NAME-treated rats were further divided into three groups of 8 rats each. In group a, CBF was determined with blood pressure left at its resting level (MABP range; 135–170 mmHg); in group b the rats were made moderately hypotensive (MABP range; 105–133 mmHg) by controlled withdrawal of blood before measurement of CBF; and the rats in group c were made more severely hypotensive by greater blood withdrawal (MABP range; 78–98 mmHg) (Table 2). There were no statistically significant differences in MABP, arterial $P_{CO_2}$, $P_{O_2}$, pH, plasma glucose, and hematocrit before blood withdrawal among these three groups. After the blood withdrawal, slight decreases were observed in hematocrit in groups b and c (Table 2).

Inhalation of 5% $CO_2$ raised $P_{CO_2}$ from $39 \pm 1$ to $49 \pm 1$ ($\pm$SE) mmHg ($n = 7$; $P < .01$, paired $t$-test) in L-NAME-treated animals and from $38 \pm 1$ to $49 \pm 1$ ($\pm$SE) mmHg ($n = 7$; $P < .01$, paired $t$-test) in saline-treated animals. The arterial $P_{CO_2}$ levels were statistically significantly higher in rats breathing 5% $CO_2$ than the value of $39 \pm 1$ ($\pm$SE) mmHg in air-breathing rats ($n = 7$; $P < .01$, grouped $t$-test) at the time of CBF measurement.

## Effects of L-NAME on Stimulus-Induced Increases in Local CBF

Stroking of vibrissae on the left side of the face resulted in marked increases in local CBF in the left trigeminal subnucleus caudalis and principal sensory nucleus and the right VPM of the thalamus and barrel cortex above the levels in the homologous structures on the opposite side of the brain (Fig. 2).

TABLE 2. Effects of chronic treatment with L-NAME on physiological variables

| MABP range at time of CBF Determination | Before blood withdrawal | | | | | Just before CBF measurement | | | | |
|---|---|---|---|---|---|---|---|---|---|---|
| | Mean arterial blood pressure (mmHg) | Hematocrit (%) | $Po_2$ (mmHg) | $Pco_2$ (mmHg) | pH | Mean arterial blood pressure (mmHg) | Hematocrit (%) | $Po_2$ (mmHg) | $Pco_2$ (mmHg) | pH |
| *Saline-treated:* 120–138 mmHg ($n = 8$) | — | — | — | — | — | 128 ± 6 | 49 ± 2 | 89 ± 6 | 35 ± 2 | 7.47 ± 0.02 |
| *L-NAME-treated:* | | | | | | | | | | |
| Group a, 135–170 mmHg ($n = 8$) | — | — | — | — | — | 151 ± 11‡ | 53 ± 3‡ | 92 ± 4 | 36 ± 4 | 7.47 ± 0.01 |
| Group b, 105–133 mmHg ($n = 8$) | 149 ± 7‡ | 52 ± 2‡ | 90 ± 5 | 38 ± 2 | 7.47 ± 0.02 | 118 ± 9†,* | 49 ± 3†,*** | 93 ± 6 | 35 ± 2 | 7.47 ± 0.01 |
| Group c, 78–98 mmHg ($n = 8$) | 153 ± 9‡ | 53 ± 2‡ | 90 ± 3 | 36 ± 1 | 7.49 ± 0.02 | 88 ± 8†,*** | 50 ± 2† | 91 ± 6 | 35 ± 3 | 7.48 ± 0.02 |

MABP, mean arterial blood pressure; CBF, cerebral blood flow.

Values are means ± SD in number of rats in parentheses.

‡ $P < .001$, when pooled values for all L-NAME-treated rats before blood withdrawal were compared with saline-treated group.

† $P < .01$, compared to the values before induction of hypotension in the same animal (paired $t$-test).

* $P < .005$, compared to L-NAME-treated group a.

** $P < .005$, compared to L-NAME-treated group b.

*** $P < .05$, compared to L-NAME-treated group a.

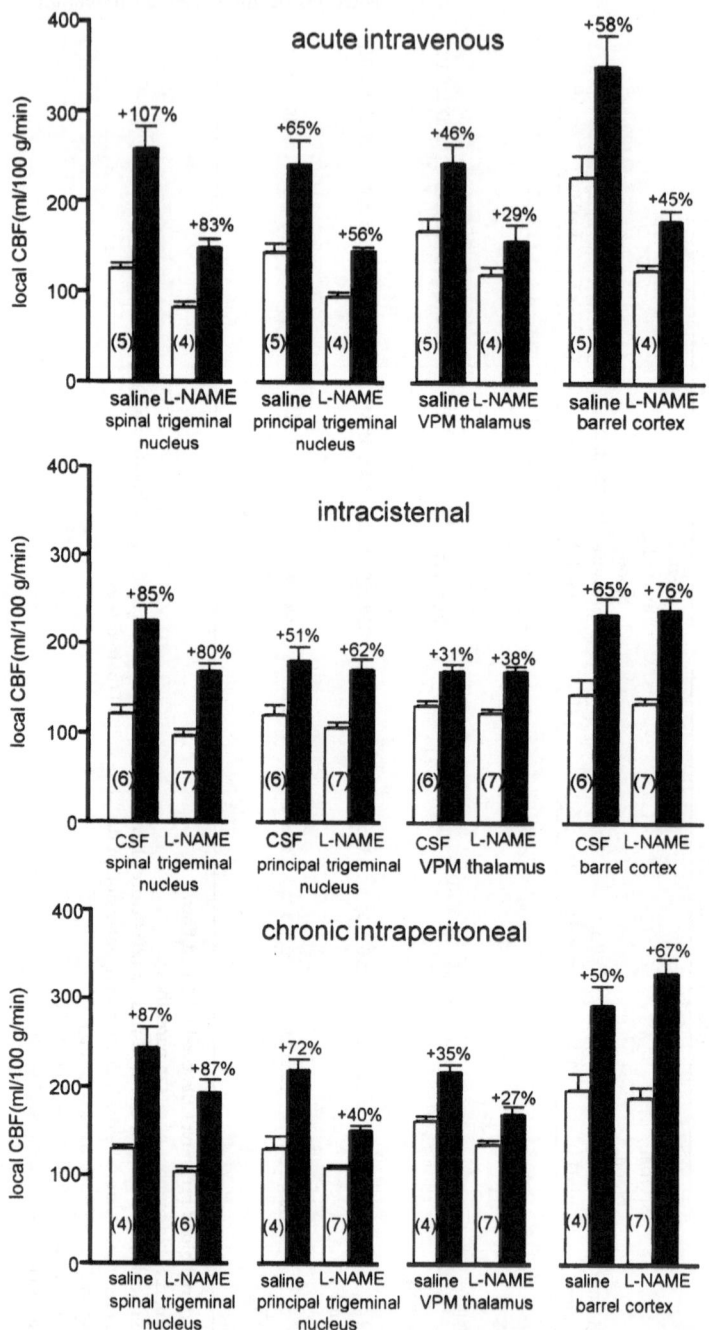

FIG. 2. Effects of single intravenous injection (*upper*), of intracisternal infusion (*middle*), and twice-daily injections for 4 days (*lower*) of $N^{G}$-nitro-L-arginine methyl ester (*L-NAME*) on baseline rates of local cerebral blood flow and on the increases in blood flow induced by unilateral vibrissal stimulation in four stations of the whisker-to-cortical barrel pathway. Data are mean rates of local blood flow ± SE obtained in the numbers of rats indicated in parentheses. *Solid bars*, stimulated side; *open bars*, control side

Although intravenous administration of L-NAME raised MABP, it significantly lowered local CBF in all brain structures below the values in saline-treated rats, including the stations of the whisker-to-cortical barrel pathway. Despite the approximately 53% inhibition of brain NO synthase activity (see Table 1), unilateral whisker stroking increased CBF unilaterally in all the stations of the whisker-to-cortical barrel pathway. The absolute increments in CBF caused by stimulation were less than those in the saline-treated control rats, but the baseline values in the contralateral unstimulated side were lower, and the percent increases, therefore, were not statistically significantly less than those in the saline-treated controls. (Fig. 2, upper panel).

When L-NAME was infused intracisternally to bypass the blood–brain barrier, there was also a tendency toward lower rates of blood flow throughout the brain, but the percent increases in CBF in the stations of the whisker-to-cortical barrel pathway elicited by functional activation were essentially unchanged from those in control rats infused with artificial CSF (Fig. 2, middle panel), despite the 88% inhibition of NO synthase activity in the brain.

Inhibition of brain NO synthase activity by L-NAME has been reported to be a relatively slow process but to be almost complete after twice-daily intraperitoneal injection of 50 mg/kg of L-NAME for 4 days [18]. In the present study this treatment was found to inhibit total brain NO synthase activity, including parenchymal and endothelial enzyme activities, by 84% (see Table 1). Nevertheless, despite NO synthase inhibition, whisker-stroking resulted in percent increases in blood flow in all stations of the whisker-to-cortical barrel pathway that were as great as those in the saline-treated control rats (Fig. 2, lower panel).

## Effects of L-NAME on Autoregulatory Vasodilator Response to Systemic Hypotension

Global CBF in the rats in group a (i.e., those treated chronically with L-NAME but studied with their elevated MABP left unaltered) was significantly lower than the levels in the saline-treated control rats ($P < .005$, two-group $t$-test) (Table 3). There were, however, no statistically significant differences in global CBF among the three L-NAME-treated groups (i.e., groups a, b, and c), despite their different levels of MABP. Changes in local CBF were similar to those in global CBF; it was statistically significantly lower in 29 of 32 representative brain structures examined in the L-NAME-treated group (a) than in the saline-treated group (two-group $t$-test) (Table 3). None of the 32 structures, however, showed statistically significant decreases in blood flow in the L-NAME-treated animals made hypotensive compared to those with the elevated MABP, except for the auditory cortex in which CBF was significantly reduced in the severely hypotensive group c ($P < .005$).

A plot of values of global CBF versus MABP obtained in all the individual rats clearly demonstrates that CBF in not significantly correlated with MABP ($r_{xy} = .18$; $P > .4$) and is maintained within a narrow range despite wide changes in arterial blood pressure in the rats treated with L-NAME (Fig. 3a). On the other hand, a similar plot of the values for whole-brain cerebrovascular resistance (CVR) (i.e., MABP in mmHg/global CBF in ml/100 g per min) versus MABP demonstrates a highly significant positive correlation between them ($r_{xy} = .85$; $P < .001$), indicating progressively

TABLE 3. Effects of chronic treatment with L-NAME on autoregulation of local CBF (ml/100 g per min)

| Structure | Saline: controls (n = 8) | L-NAME: group a (n = 8) | L-NAME: group b (n = 8) | L-NAME: group c (n = 8) |
|---|---|---|---|---|
| Whole–brain CBF | 130 ± 17 | 95 ± 15*** | 97 ± 12 | 89 ± 10 |
| Medulla | | | | |
| Nucleus tractus solitarius | 134 ± 33 | 101 ± 18* | 99 ± 19 | 101 ± 17 |
| Pons | | | | |
| Pontine reticular nucleus | 124 ± 24 | 92 ± 15*** | 91 ± 13 | 88 ± 14 |
| Cerebellum | | | | |
| Cerebellar cortex | 81 ± 9 | 62 ± 15*** | 63 ± 7 | 61 ± 8 |
| Cerebellar white matter | 54 ± 6 | 41 ± 7*** | 43 ± 7 | 42 ± 5 |
| Fastigial nucleus | 171 ± 21 | 127 ± 26*** | 132 ± 17 | 130 ± 20 |
| Interpositus nucleus | 186 ± 20 | 129 ± 31*** | 134 ± 16 | 140 ± 19 |
| Dentate nucleus | 206 ± 21 | 146 ± 33*** | 156 ± 29 | 156 ± 21 |
| Mesencephalon | | | | |
| Substantia nigra (reticulata) | 109 ± 19 | 86 ± 14** | 85 ± 7 | 85 ± 12 |
| Red nucleus | 158 ± 32 | 117 ± 19*** | 119 ± 14 | 110 ± 15 |
| Inferior colliculus | 224 ± 28 | 171 ± 35** | 182 ± 34 | 168 ± 26 |
| Superior colliculus (superficial layer) | 137 ± 22 | 109 ± 18 | 108 ± 15 | 102 ± 15 |
| Diencephalon | | | | |
| Medial geniculate body | 204 ± 34 | 163 ± 29* | 159 ± 18 | 139 ± 16 |
| Dorsal lateral geniculate body | 151 ± 22 | 116 ± 22*** | 119 ± 14 | 114 ± 15 |
| Lateral habenular nucleus | 192 ± 30 | 160 ± 34 | 156 ± 23 | 156 ± 21 |
| Medial habenular nucleus | 146 ± 24 | 115 ± 20** | 110 ± 16 | 108 ± 11 |
| Lateral posterior mediorostral thalamus | 150 ± 18 | 114 ± 23*** | 119 ± 17 | 113 ± 13 |
| Ventral posteromedial thalamus | 144 ± 24 | 111 ± 24* | 119 ± 22 | 116 ± 19 |
| Subthalamic nucleus | 179 ± 27 | 141 ± 28* | 144 ± 25 | 150 ± 28 |
| Ventromedial hypothalamus | 93 ± 16 | 72 ± 11** | 67 ± 4 | 67 ± 12 |
| Telencephalon | | | | |
| Lateral amygdala | 118 ± 19 | 88 ± 17*** | 80 ± 6 | 77 ± 11 |
| Medial amygdala | 96 ± 16 | 72 ± 13*** | 68 ± 7 | 64 ± 11 |
| Caudate-putamen | 158 ± 24 | 112 ± 15*** | 117 ± 13 | 105 ± 16 |
| Globus pallidus | 89 ± 14 | 65 ± 8*** | 66 ± 8 | 65 ± 9 |
| Nucleus Accumbens | 205 ± 36 | 119 ± 25*** | 119 ± 24 | 104 ± 14 |
| Motor cortex | 162 ± 22 | 108 ± 24*** | 114 ± 25 | 99 ± 11 |
| Sensory-motor cortex | 183 ± 31 | 135 ± 23*** | 137 ± 24 | 124 ± 20 |
| Sensory cortex | 196 ± 50 | 142 ± 46* | 147 ± 31 | 138 ± 26 |
| Auditory cortex | 268 ± 33 | 233 ± 43 | 206 ± 27 | 172 ± 20**** |
| Visual cortex | 168 ± 29 | 123 ± 17*** | 121 ± 15 | 117 ± 16 |
| Cingulate cortex | 211 ± 40 | 135 ± 23*** | 139 ± 26 | 119 ± 16 |
| Myelinated fiber tracts | | | | |
| Internal capsule | 54 ± 7 | 46 ± 9 | 43 ± 7 | 45 ± 5 |
| Genu of corpus callosum | 40 ± 7 | 31 ± 7* | 33 ± 4 | 29 ± 5 |

The values are means ± SD of the numbers of animals indicated in the parentheses.
$*P < .05$, $**P < .01$, $***P < .005$, compared to saline-treated group (two-group $t$-test); $****$, $P < .005$, compared to L-NAME-treated group a (two-group $t$-test).

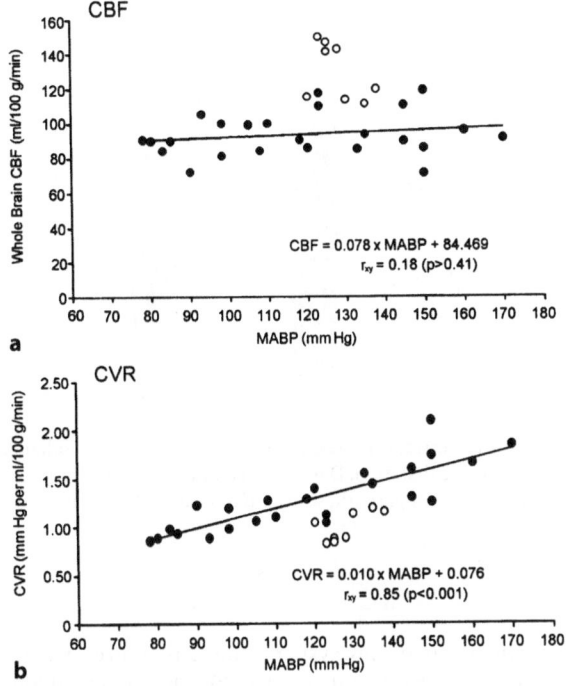

FIG. 3A,B. Relationships between global brain blood flow (*CBF*) and mean arterial blood pressure (*MABP*) (a) and between total brain cerebrovascular resistance (*CVR*) and mean arterial blood pressure (b). Each point represents results obtained in one rat. *Open circles*, saline-treated controls; *closed circles*, $N^G$-nitro-L-arginine methyl ester (L-NAME)-treated animals. The lines in Figs. 1a and 1b are the least-squares best fits to the values obtained in the L-NAME-treated animals. The slope of the line in Fig. 1a is not statistically significantly different from zero, indicating that autoregulation of CBF was preserved following L-NAME treatment

decreasing cerebral vascular tone with decreasing blood pressure (Fig. 3b). The maintenance of constant CBF in the face of decreasing arterial pressure because of a compensating reduction in CVR is strong evidence that autoregulation of the cerebral circulation is preserved after almost complete blockade of NO synthesis.

## Effects of L-NAME on $CO_2$ Enhancement of CBF

The inhalation of 5% $CO_2$ statistically significantly increased global CBF in both saline-treated rats (+34%; $P < .05$, grouped $t$-test) and L-NAME-treated rats (+29%; $P < .01$, grouped $t$-test) (Fig. 4).

## Discussion

The present results agree with those of previous studies, which demonstrated that intravenous administration of the NO synthase inhibitor L-NAME constricts cerebral

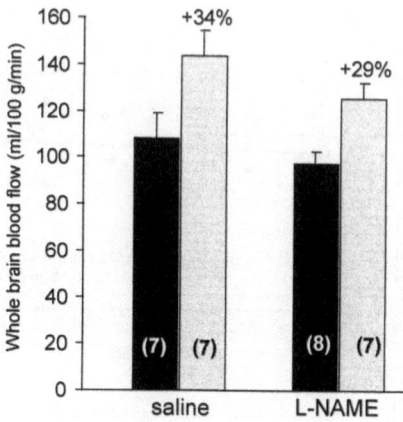

FIG. 4. The effects of chronic intraperitoneal injection of L-NAME on enhancement of global blood flow induced by 5% $CO_2$ inhalation. Data are mean rates of local blood flow ± SE obtained from the numbers of rats shown in parentheses. *Solid bars*, air inhalation control; *hatched bars*, $CO_2$ inhalation

vessels and reduces CBF while raising systemic blood pressure, evidence of increased cerebral vascular resistance [5,6,19,30] (see Fig. 2). The increased blood pressure is presumably the result of widespread vasoconstriction, and the increased cerebral vascular resistance indicates that the cerebral vessels are also affected. These changes indicate that NO synthase activity is effectively inhibited in, at least, the vascular endothelium and that NO normally plays a role in the regulation of systemic as well as cerebrovascular tone. Enzymatic assay of NO synthase activity in the homogenized brain after an acute intravenous injection of L-NAME, however, showed only about 53% inhibition of total brain activity (see Table 1).

It was uncertain how this inhibition was distributed between the endothelial and parenchymal or neuronal enzyme species. It was possible that the blood–brain barrier limited access of L-NAME to the neuronal or astroglial enzymes, which could conceivably have been the species most involved in the coupling of blood flow to functional activity. Therefore, to bypass the blood–brain barrier, L-NAME was infused directly into the cisterna magna and was, indeed, found to inhibit brain NO synthase activity by 88% (Table 1). Despite the greater degree of NO synthase inhibition with intracisternal administration of L-NAME, the effects on resting CBF and vascular resistance, although qualitatively similar, were less than those following intravenous administration, and there were no significant reductions in the percent increases in local blood flow as a result of stimulation in the stations of the whisker-to-cortical barrel pathway (see Fig. 2). The comparative effects of these two modes of administration indicate that NO does have a role in the normal regulation of cerebral vascular resistance and that it is the vascular endothelial NO synthase activity which contributes more to this function than the neuronal enzyme, but neither appears to be required for the enhancement of blood flow by neural functional activation.

Dwyer et al. [18] reported that the inhibition of NO synthase activity by L-NAME is not immediate but occurs only after a delay. Therefore, the possibility was considered that inhibition of NO synthase activity in the brain in vivo after the intravenous or intracisternal administration of L-NAME might not yet have been fully established during the 10- to 20-min period before the measurement of CBF; it was conceivable that the inhibition observed in the in vitro assay of brain NO activity had occurred after homogenization of the brain or during the assay itself. For this reason, we applied the procedure used by Dwyer et al. [18] to achieve almost complete inhibition of NO synthase activity in the rat brain in vivo, that is, twice-daily intraperitoneal injections of 50 mg/kg of L-NAME for 4 days, and found the brain NO activity to be inhibited by 84%. Presumably, the prolonged systemic administration of L-NAME inhibited both the endothelial and neuronal enzymes in the brain. In fact, baseline blood flow was reduced throughout the brain by approximately the same degree as that after an acute intravenous injection (Fig. 2). Nevertheless, there were no significant reductions in the percent enhancement of blood flow by stimulation in any of the stations of whisker-to-cortical barrel pathway (Fig. 2), further evidence that NO does not mediate the coupling of local CBF to functional activity.

We have also examined the effects of this regimen of NO inhibition on local cerebral energy metabolism in conscious rats by means of the [$^{14}$C]deoxyglucose method and found that blockade of NO synthesis does not alter local cerebral glucose utilization in any of a large number of structures representing almost all functional systems of the rat brain [28].

Our study also provides cogent evidence that in unanesthetized animals NO does not mediate the autoregulatory cerebral vasodilatation that occurs when arterial blood pressure is reduced. Our results are in accord with those of several recent studies conducted in anesthetized animals [31–33]. In these studies, however, NO synthase inhibitors were administered by a single intravenous dose, which we found by direct enzymatic analysis to inhibit NO synthase activity in whole brain by only 50%, and it was uncertain how much of the inhibition was of the endothelial enzyme and how much of the neuronal enzyme. Systemic blood pressure is moderately elevated and cerebral blood flow reduced following acute intravenous injections of L-NAME, probably because of inhibition of NO production in the vascular endothelium. Gotoh et al. [12] have emphasized the role of the endothelium in the mechanisms of autoregulation, and it seems conceivable that the NO produced in the vascular endothelium is more important for autoregulation than the NO produced by the intracerebral NO synthase. Nevertheless, the failure of acute single intravenous doses of NO synthase inhibitors to block autoregulation in the reports cited might have been attributed to incomplete inhibition of endothelial or parenchymal NO synthase activity in the brain. In our study, however, the animals received 50 mg/kg of L-NAME twice a day for 4 days, a regimen that inhibits brain NO synthase activity at least 84%, providing much stronger evidence against any important role for NO in the mechanisms of the autoregulatory cerebral vasodilatation that accompanies lowering of arterial blood pressure.

In our study both average and local CBF remained unchanged when MABP was lowered to about 80 mm Hg. It has been reported that the blood pressure range over which autoregulation is effective in the normal anesthetized rat is between 80 and 160 mmHg [9], and this range appears to apply also to conscious animals [34]. Because in the current studies it proved difficult to carry out the blood sampling required for measurement of CBF without also inducing in the hypotensive animals additional and uncontrollable falls in blood pressure, we could not determine the full range of autoregulation. Wang et al. [33], however, have reported that the range of autoregulation following NO synthesis inhibition does not differ from that in control animals.

The current study also provides evidence that almost complete inhibition of NO synthesis by chronic systemic L-NAME administration does not alter the percent enhancement of CBF by $CO_2$. These results are consistent with the results of our previous study [30] in which acute intravenous injections of NO synthase inhibitors were employed. Irikura et al. [35] recently reported that the increase in CBF elicited by $CO_2$ inhalation is preserved in mutant mice lacking neuronal NO synthase gene expression. They also found that the $CO_2$ effect on CBF flow was not fully blocked even after application of L-NAME to the mutant mice in which endothelial NO synthase activity was preserved. These results clearly indicate that NO produced in either vascular endothelium or brain tissue is not essential for the enhancement of CBF by $CO_2$.

The results of our study support a role for NO in the tonic regulation of the cerebral circulation, but they provide no evidence that NO is the mediator of the dynamic changes in CBF associated with alterations in neural functional activity, blood $CO_2$ tension, or arterial blood pressure in conscious rats.

# References

1. Ignarro LJ, Buga GM, Wood KS, et al. (1987) Endothelium-derived relaxing factor produced and released from artery and vein is nitric oxide. Proc Natl Acad Sci USA 84:9265–9269
2. Palmer RMJ, Ferrige AG, Moncada S (1987) Nitric oxide release accounts for the biological activity of endothelium-derived relaxing factor. Nature 327:524–526
3. Furchgott RF, Zawadzki JV (1980) The obligatory role of endothelial cells in the relaxation of arterial smooth muscle by acetylcholine. Nature 228:373–376
4. Moncada S, Palmer RM, Higgs EA (1991) Nitric oxide: physiology, pathophysiology, and pharmacology. Pharmacol Rev 43:109–142
5. Iadecola C, Pelligrino DA, Moskowitz MA, et al. (1994) Nitric oxide synthase inhibition and cerebrovascular regulation. J Cereb Blood Flow Metab 14:175–192
6. Tanaka K, Gotoh F, Gomi S, et al. (1991) Inhibition of nitric oxide synthesis induces a significant reduction in local cerebral blood flow in the rat. Neurosci Lett 127: 129–132
7. Roy CW, Sherrington CS (1890) On the regulation of the blood-supply of the brain. J Physiol (Lond) 11:85–108
8. Kety SS, Schmidt CF (1948) The effects of altered arterial tensions of carbon dioxide and oxygen on cerebral blood flow and oxygen consumption of normal young men. J Clin Invest 27:484–492
9. Hernàndez MJ, Brennan RW, Bowman GS (1978) Cerebral blood flow autoregulation in the rat. Stroke 9:150–154
10. Lassen NA (1959) Cerebral blood flow and oxygen metabolism in man. Physiol Rev 39:183–238
11. Edvinsson L, MacKenzie ET, McCulloch J (1993) Autoregulation: arterial and intra-cranial pressure. In: Edvinsson L, McKenzie ET, McCulloch J (eds) Cerebral blood flow and metabolism. Raven, New York, pp 553–580
12. Gotoh F, Fukuuchi Y, Amano T, et al. (1987) Role of endothelium in responses of pial vessels to change in blood pressure and to carbon dioxide in cats. J Cereb Blood Flow Metab 7(supp 1):S275

13. Bredt DS, Hwang PM, Snyder SH (1990) Localization of nitric oxide synthase indicating a neural role for nitric oxide. Nature 347:768–770
14. Murphy S, Simmons ML, Agullo L, et al. (1993) Synthesis of nitric oxide in CNS glial cells. Trends Neurosci 16:323–328
15. Garthwaite J (1991) Glutamate, nitric oxide and cell-cell signalling in the nervous system. Trends Neurosci 14:60–67
16. Woolsey TA, Van Der Loos H (1970) The structure organization of layer IV in the somatosensory region (SI) of mouse cerebral cortex. Brain Res 17:205–242
17. Greenberg J, Hand P, Sylvestro A, et al. (1979) Localized metabolic-flow couple during functional activity. Acta Neurol Scand 60(suppl 72):12
18. Dwyer MA, Bredt DS, Snyder SH (1991) Nitric oxide synthase: irreversible inhibition by L-$N^G$-nitroarginine in brain in vitro and in vivo. Biochem Biophys Res Commun 176:1136–1141
19. Adachi K, Takahashi S, Melzer P, et al. (1994) Increases in local cerebral blood flow associated with somatosensory activation are not mediated by nitric oxide. Am J Physiol 267:H2155–H2162
20. Giovanelli J, Campos KL, Kaufman S (1991) Tetrahydrobiopterin, a cofactor for rat cerebellar nitric oxide synthase, does not function as a reactant in the oxygenation of arginine. Proc Natl Acad Sci USA 88:7091–7095
21. Bredt DS, Snyder SH (1990) Isolation of nitric oxide synthase, a calmodulin-requiring enzyme. Proc Natl Acad Sci USA 87:682–685
22. Sakurada O, Kennedy C, Jehle J, et al. (1978) Measurement of local cerebral blood flow with ido[$^{14}$C]antipyrine. Am J Physiol 234:H59–H66
23. Patlak CS, Pettigrew KD (1976) A method to obtain infusion schedules for prescribed blood concentration time courses. J Appl Physiol 40:458–463
24. Freygang WH Jr, Sokoloff L (1958) Quantitative measurement of regional circulation in the central nervous system by the use of radioactive inert gas. Adv Biol Med Phys 6:263–279
25. Sharp FR, Gonzalez MF, Morgan CW, et al. (1988) Common fur and mystacial vibrissae parallel sensory pathways: $^{14}$C 2-deoxyglucose and WGA-HRP studies in the rat. J Comp Neurol 270:446–469
26. Ginsberg MD, Dietrich WD, Busto R (1987) Coupled forebrain increases of local cerebral glucose utilization and blood flow during physiologic stimulation of a somatosensory pathway in the rat: demonstration by double labeled autoradiography. Neurology 37:11–19
27. Paxinos G, Watson (1986) The rat brain in stereotaxic coordinates, 2nd edn. Academic Press, New York
28. Takahashi S, Cook M, Jehle J, et al. (1995) Lack of inhibition of nitric oxide synthesis on local glucose utilization in the rat brain. Brain Res 65:414–419
29. Takahashi S, Cook M, Jehle J, et al. (1995) Preservation of autoregulatory vasodilator response to hypotension after inhibition of nitric oxide synthesis. Brain Res 678: 21–28
30. Sokoloff L, Kennedy C, Adachi K, et al. (1992) Effects of inhibition of nitric oxide synthase on resting local cerebral blood flow and on changes induced by hypercapnia or local functional activity. In: Kriegelstein J, Oberpichler-Schwenk H (eds) Pharmacology of cerebral ischemia 1992. Wissenschaftliche Verlagsgesellschaft, Stuttgart, pp 371–381
31. Buchanan JE, Phillis JW (1993) The role of nitric oxide in the regulation of cerebral blood flow. Brain Res 610:248–255
32. DeWitt DS, Prough DS, Colonna DM, et al. (1992) Effects of nitric oxide synthase inhibitors on cerebral blood flow and autoregulation in rats. Anesthesiology 77:A689
33. Wang Q, Paulson OB, Lassen NA (1992) Is autoregulation of cerebral blood flow in rats influenced by nitro-L-arginine, a blocker of the synthesis of nitric oxide? Acta Physiol Scand 145:297–298

S. Takahashi and L. Sokoloff

34. Niwa K, Shinohara Y, Takizawa S, et al. (1993) Regional differences in autoregulation of cerebral blood flow, a study using hypotensive rats by exsanguination in the awake state. Jpn J Stroke 15:303-309
35. Irikura K, Huang PL, Ma J, et al. (1995) Cerebrovascular alterations in mice lacking neuronal nitrix oxide synthase gene expression. Proc Natl Acad Sci USA 92:6823-6827

# Physiological Implication of Induction of Heme Oxygenase-1 Gene Expression

Shigeki Shibahara

*Summary.* Heme oxygenase-1 (HO-1) is an essential enzyme in heme catabolism that cleaves heme to form biliverdin, iron, and carbon monoxide. HO-1 is induced by various environmental factors, including its substrate heme and heavy metals, and is considered as a member of the defense system against oxidative stress. Recently, we showed that HO-1 expression was remarkably increased in various human cell lines by treatment with nitric oxide (NO) donors. Furthermore, in rat brain, HO-1 mRNA was detectable only in the scattered neuron-like cells within the dentate gyrus hilus, while transient forebrain ischemia caused a remarkable increase in HO-1 mRNA levels in both neuronal and glia-like cells present in the neocortex, hippocampus, and a part of the thalamus. This induction of HO-1 mRNA was accompanied by an increase in HO-1 protein levels, except for the CA1 subfield of the hippocampus, in which pyramidal neurons are particularly vulnerable to ischemic insult. It is therefore conceivable that induction of HO-1 protein may represent a neuronal defense under certain conditions. In vivo and in vitro studies on the regulation of HO-1 gene expression are discussed.

*Key words.* Heme—Bilirubin—Carbon monoxide—Transcription—Stress protein

## Introduction

Heme oxygenase (E.C.1.14.99.3) catalyzes an essential step in heme catabolism to form carbon monoxide, iron, and biliverdin [1,2]. Biliverdin is subsequently converted by biliverdin reductase to bilirubin in mammals [3]. Heme oxygenase activity is higher in liver, spleen, and bone marrow, where large amounts of heme, derived from hemoglobin of senescent erythrocytes, are degraded by resident macrophages [4–6]. However, because of the essential role of hemoproteins, all cells and tissues possess the systems for both heme biosynthesis and heme catabolism. There are two isozymes of heme oxygenase, heme oxygenase-1 (HO-1) and heme oxygenase-2 (HO-2) [7],

Department of Applied Physiology and Molecular Biology, Tohoku University School of Medicine, 2-1 Seiryo-machi, Aoba-ku, Sendai, Miyagi 980–77, Japan

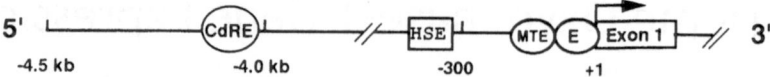

Cd-responsive element (CdRE): TTTTGCTAGATTT (- 4.1 kb)

Heat shock element (HSE): CTGGAACCTTCTGG (- 381 to -368)

Macrophage-specific TPA-responsive element (MTE):
GT<u>CATAT</u>GAC (-156 to -147)

USF-binding site (E)<u>CACGTG</u> (-44 to -39)

Fig. 1. The *cis*-regulatory elements of the human heme oxygenase-1 gene. The 5′-flanking region and the exon 1 of the human HO-1 gene are shown; an *arrow* indicates the direction of transcription. The nucleotide sequence of each element is shown at the *bottom*, and the numbers shown within parentheses are the nucleotide residues from the transcriptional initiation site. The function of heat-shock element (*HSE*) may be repressed in vivo [44]. The E-box motifs, CANNTG, are underlined. *TPA*, 12-O-tetradecanoylphorbol-13-acetate; *USF*, upstream stimulatory factor

each of which is encoded by a separate gene [8,9]. HO-1 shares about 43% similarity in its amino acid sequence with HO-2.

The HO-1 gene is induced by various environmental factors, such as its substrate heme [4,6,10,11], cadmium [12–16], heat shock [17], 12-O-tetradecanoylphorbol-13-acetate (TPA) [18,19], and nitric oxide (NO) donors [20–22]. In contrast, HO-2 is not inducible under the conditions in which HO-1 expression is noticeably increased [7,23–25]. We have identified several *cis*-regulatory elements in the human HO-1 gene promoter (Fig. 1), such as the cadmium-responsive element (CdRE) [15,16], the macrophage-specific TPA-responsive element (MTE) [18,19], and a upstream stimulatory factor- (USF-) binding element [26,27]. MTE consists of a 10-bp sequence, GTCATATGAC (nucleotide positions −156 to −147), and is required for the TPA-mediated activation in a myelomonocytic cell-specific manner [18]. The general properties of HO-1 and HO-2 were discussed in a recent review [28].

## Possible Functions of Heme Breakdown Products

Heme breakdown yields unique products: biliverdin/bilirubin, iron, and carbon monoxide (CO). Bilirubin has been considered as a toxic waste product, but it was shown that biliverdin or bilirubin produced locally may work as a physiological antioxidant [29,30]. It is thus proposed that HO-1 constitutes a member of the defense system against oxidative stress. Iron is an essential metal for living organisms, and iron released from heme is effectively reused for heme biosynthesis. However, in the presence of iron, superoxide anion radical and hydrogen peroxide can be involved in the formation of hydroxyl radical, a highly reactive species of radical. Thus, free iron concentration must be strictly regulated within cells. On the other hand, CO has been suggested to possess NO-like activities [31]. Indeed, CO was proposed to act as retrograde messages for long-term potentiation (LTP) in the hippocampus [32,33]. However, targeted disruption of the HO-2 gene in mice had no noticeable effects on the gross neuronal structure and the basal hippocampal synaptic transmission [34], suggesting that CO produced by HO-2 may not be a regulator for LTP in the hippocampus. It remains to be investigated whether CO produced by HO-1 functions as a signaling gas under certain conditions. CO has recently been shown to function as a modulator of sinusoidal tone in the liver [35,36].

## Expression of HO-1 mRNA in Human Brain and Its Overexpression in Brain Tumors

To assess the functional significance of HO-1 in the brain, we determined the expression levels of HO-1 mRNA in human brain by Northern blot analysis. Both HO-1 and HO-2 mRNAs were expressed in every region of the human brain examined, with the highest levels found in the frontal cortex, temporal cortex, occipital cortex, and hypothalamus [21]. Subsequently, we have shown that both HO isozyme mRNAs are expressed in the primary brain tumors examined, including two glioblastomas multiforme, five astrocytomas, and one choroid plexus carcinoma [22]. The expression levels of HO-1 mRNA are higher in these brain tumors compared to normal brain tissues, with the highest levels of HO-1 mRNA detected in one case of glioblastoma multiforme. It is therefore conceivable that expression of HO-1 mRNA is increased in the human brain in vivo under certain conditions.

## Induction of HO-1 in Rat Brain Following Transient Forebrain Ischemia

We have established that rat HO-1 is a heat-shock protein (HSP) by showing that the rat HO-1 gene contains a heat-shock element (HSE) [9] and is transcriptionally activated by heat shock [17,37]. Consistent with this, hyperthermia was shown to lead to remarkable induction of HO-1 mRNA and protein in the rat brain [23], suggesting the significance of this induction as a neuronal defense against heat-shock stress. On the other hand, the heat-mediated induction of HO-1 in rat brain may be harmful to brain tissues because of the potential toxicity of heme degradation products.

To study the regulation of HO-1 expression in the brain, we analyzed, by Northern blot analysis, the expression levels of HO-1 and HSP-70 mRNAs in rat brain following 20-min forebrain ischemia [38]. Expression of neither mRNA was detectable in sham-operated rats or in rats after 0.5h of recirculation. The levels of both mRNAs increased to the detectable level at 1.5h of recirculation, reached the maximum at 12h, and then decreased. In situ hybridization analysis showed that HO-1 mRNA was expressed in certain neuron-like cells in the dentate hilus of the control rat brain, suggesting that HO-1 may play a hitherto unknown role in these neuronal cells [38].

The expression level of HO-1 mRNA was increased in both neuronal and glia-like cells following transient ischemic insult. The increased expression of HO-1 mRNA in glia-like cells is of particular significance, because HSP-70 mRNA is induced mainly in neuronal cells such as hippocampal pyramidal cells and dentate granule cells after transient ischemia in the same rat model. These results suggest that each gene is regulated in a different manner between neuronal and glia-like cells. The separate regulation of HO-1 and HSP-70 mRNAs was also reported in the rat heart subjected to hemodynamic stress [39]. Furthermore, regional differences in the induction of HO-1 protein were noted in the rat brain following transient ischemic insult [40]. In the cortical mantle, both neurons and astrocytes expressed increased HO-1 protein, while in hippocampal CA2 and CA3 subfields prominent induction was mainly observed in astrocytes. In contrast, HO-1 protein was undetectable in both neurons and astrocytes in the CA1 subfield [40], although HO-1 mRNA was evident in these cell types [38]. These observations are of interest in view of the pathogenesis of the delayed CA1 neuronal death.

## Regulation of the Human HO-1 Gene Under Thermal Stress

HO-1 is not necessarily induced by heat shock in cultured cells derived from human, monkey, pig, and mouse [41], while rat HO-1 has been generally accepted as an HSP. Because of such species-specific regulation of HO-1 expression by heat shock, it is sometimes difficult to compare data obtained from different species. The human HO-1 gene promoter contains a single copy of HSE identical to the functional HSE [42] (see Fig. 1), although transient expression assays suggested that the human HO-1 HSE is unable to confer the heat-mediated transcriptional activation of a reporter gene in the transfected cells. On the other hand, the HSE of the human HO-1 gene was shown to be bound in vitro by heat-shock factor [43,44], suggesting that this HSE is potentially functional.

Consistent with this, the synthetic HSE of the human HO-1 gene is sufficient to confer the heat-mediated inducibility of the fusion gene carrying this HSE [44]. We are assuming that the sequence flanking the HO-1 HSE may act as a silencer that prevents the heat-mediated activation of the HO-1 gene. Such a silencing effect is of physiological significance in the brain, because heme degradation products possess potential toxic effects that may damage the brain tissues. Thus, if the human HO-1 gene lacks such a silencer element, HO-1 is easily induced in the human brain under various conditions, as seen in rat brain [23]. It is therefore conceivable that the human HO-1 gene may gain the silencer element during evolution to protect its harmful induction caused by heat shock. In addition, transcriptional activation of the human HO-1 gene by cadmium was repressed under thermal stress (42°C), whereas expression of the HSP70 gene was always increased at 42°C [44]. These results indicate that the regulation of HO-1 gene expression is different from that of HSP70 gene expression in humans. It is however noteworthy that the property of HO-1 as a stress protein is conserved in humans.

## Conclusion

As summarized in Fig. 1, there are multiple *cis*-regulatory elements in the 5′-flanking region of the HO-1 gene. For example, CdRE and MTE are responsible for transcriptional activation caused by cadmium and TPA, respectively, although neither reagent is a physiological substance. Possibly, cadmium or TPA may mimic or modify a naturally occurring mediator that is involved in a signal transduction system, leading to activation of the HO-1 gene transcription. Therefore, the HO-1 gene provides a good system to study a regulatory network of transcription factors involved in gene activation caused by various environmental factors.

# References

1. Tenhunen RH, Marver HS, Schmid R (1968) The enzymatic conversion of heme to bilirubin by microsomal heme oxygenase. Proc Natl Acad Sci USA 61:748–755
2. Tenhunen RH, Marver HS, Schmid R (1969) Microsomal heme oxygenase. Characterization of the enzyme. J Biol Chem 244:6388–6394
3. Tenhunen RH, Ross ME, Marver HS, et al. (1970) Reduced nicotinamide-adenine dinucleotide phosphate dependent biliverdin reductase: partial purification and characterization. Biochemistry 9:298–303
4. Tenhunen RH, Marver HS, Schmid R (1970) The enzymatic catabolism of hemoglobin: stimulation of microsomal heme oxygenase by hemin. J Lab Clin Med 75:410–421
5. Pimstone NR, Tenhunen RH, Seitz PT, et al. (1971) The enzymatic degradation of hemoglobin to bile pigments by macrophages. J Exp Med 133:1264–1281
6. Shibahara S, Yoshida T, Kikuchi G (1978) Induction of heme oxygenase by hemin in cultured pig alveolar macrophages. Arch Biochem Biophys 188:243–250
7. Maines MD, Trakshel GM, Kutty RK (1986) Characterization of two constitutive forms of rat liver microsomal heme oxygenase. Only one molecular species of the enzyme is inducible. J Biol Chem 261:411–419
8. Cruse I, Maines MD (1988) Evidence suggesting that the two forms of heme oxygenase are products of different genes. J Biol Chem 263:3348–3353
9. Muller RM, Taguchi H, Shibahara S (1987) Nucleotide sequence and organization of the rat heme oxygenase gene. J Biol Chem 262:6795–6802
10. Pimstone NR, Engel P, Tenhunen RH, et al. (1971) Inducible heme oxygenase in the kidney: a model for the homeostatic control of hemoglobin catabolism. J Clin Invest 50:2042–2050
11. Yoshida T, Biro P, Cohen T, et al. (1988) Human heme oxygenase cDNA and induction of its mRNA by hemin. Eur J Biochem 171:457–461
12. Ishizawa S, Yoshida T, Kikuchi G (1983) Induction of heme oxygenase in rat liver. J Biol Chem 258:4220–4225
13. Shibahara S, Muller R, Taguchi H, et al. (1985) Cloning and expression of cDNA for rat heme oxygenase. Proc Natl Acad Sci USA 82:7865–7869
14. Taketani S, Kohno H, Yoshinaga T, et al. (1989) The human 32-kDa stress protein induced by exposure to arsenite and cadmium ions is heme oxygenase. FEBS Lett 245:173–176
15. Takeda K, Ishizawa S, Sato M, et al. (1994) Identification of a cis-acting element that is responsible for cadmium-mediated induction of the human heme oxygenase gene. J Biol Chem 269:22858–22867
16. Takeda K, Fujita H, Shibahara S (1995) Differential control of the metal-mediated activation of the human heme oxygenase-1 and metallothionein IIA genes. Biochem Biophys Res Commun 207:160–167
17. Shibahara S, Muller RM, Taguchi H (1987) Transcriptional control of rat heme oxygenase by heat shock. J Biol Chem 262:12889–12892
18. Muraosa Y, Shibahara S (1993) Identification of a cis-regulatory element and putative trans-acting factors responsible for 12-O-tetradecanoylphorbol-13-acetate (TPA) - mediated induction of heme oxygenase expression in myelomonocytic cell lines. Mol Cell Biol 13:7881–7891
19. Muraosa Y, Takahashi K, Yoshizawa M, et al. (1996) cDNA cloning of a novel protein with the two zinc-finger domains that may function as a transcription factor for the human heme oxygenase-1 gene. Eur J Biochem 235:471–479

20. Kim Y-M, Bergonia HA, Muller C, et al. (1995) Loss and degradation of enzyme-bound heme induced by cellular nitric oxide synthesis. J Biol Chem 270:5710–5713
21. Takahashi K, Hara E, Suzuki H, et al. (1996) Expression of heme oxygenase isozyme mRNAs in the human brain and induction of heme oxygenase-1 by nitric oxide donors. J Neurochem 67:482–489
22. Hara E, Takahashi K, Tominaga T, et al. (1996) Expression of heme oxygenase and inducible nitric oxide synthase mRNA in human brain tumors. Biochem Biophys Res Commun 225:153–158
23. Ewing JF, Maines MD (1991) Rapid induction of heme oxygenase 1 mRNA and protein by hyperthermia in rat brain: heme oxygenase 2 is not a heat shock protein. Proc Natl Acad Sci USA 88:5364–5368
24. Shibahara S, Yoshizawa M, Suzuki H, et al. (1993) Functional analysis of cDNAs for two types of human heme oxygenase and evidence for their separate regulation. J Biochem (Tokyo) 113:214–218
25. Trakshel GM, Kutty RK, Maines MD (1986) Purification and characterization of the major constitutive form of testicular heme oxygenase. The noninducible isoform. J Biol Chem 261:11131–11137
26. Sato M, Fukushi Y, Ishizawa S, et al. (1989) Transcriptional control of the rat heme oxygenase gene by a nuclear protein that interacts with adenovirus 2 major late promoter. J Biol Chem 264:10251–10260
27. Sato M, Ishizawa S, Yoshida T, et al. (1990) Interaction of upstream stimulatory factor with the human heme oxygenase gene promoter. Eur J Biochem 188:231–237
28. Shibahara S (1994) Heme oxygenase—regulation of and physiological implication in heme catabolism. In: Fujita H (ed) Regulation of heme protein synthesis. AlphaMed Press, Dayton, OH, pp 103–116
29. Stocker R, Glazer AN, Ames BN (1987) Antioxidant activity of albumin-bound bilirubin. Proc Natl Acad Sci USA 84:5918–5922
30. Stocker R, Yamamoto Y, McDonagh AF, et al. (1987) Bilirubin is an antioxidant of possible physiological importance. Science 235:1043–1046
31. Marks GS, Brien JF, Nakatsu K, et al. (1991) Does carbon monoxide have a physiological function? Trends Pharmacol Sci 12:185–188
32. Stevens CF, Wang Y (1993) Reversal of long-term potentiation by inhibitors of haem oxygenase. Nature 364:147–149
33. Zhuo M, Small SA, Kandel ER, et al. (1993) Nitric oxide and carbon monoxide produce activity-dependent long-term synaptic enhancement in hippocampus. Science 260:1946–1950
34. Poss KD, Thomas MJ, Ebralidze AK, et al. (1995) Hippocampal long-term potentiation is normal in heme oxygenase-2 mutant mice. Neuron 15:867–873
35. Suematsu M, Kashiwagi S, Sano T, et al. (1994) Carbon monoxide as an endogenous modulator of hepatic vascular perfusion. Biochem Biophys Res Commun 205:1333–1337
36. Suematsu M, Goda N, Sano T, et al. (1995) Carbon monoxide: an endogenous modulator of sinusoidal tone in the perfused rat liver. J Clin Invest 96:2431–2437
37. Okinaga S, Shibahara S (1993) Identification of a nuclear protein that constitutively recognizes the sequence containing a heat-shock element. Its binding properties and possible function modulating heat-shock induction of the rat heme oxygenase gene. Eur J Biochem 212:167–175
38. Takeda A, Onodera H, Sugimoto A, et al. (1994) Increased expression of heme oxygenase mRNA in rat brain following transient forebrain ischemia. Brain Res 666:120–124
39. Katayose D, Isoyama S, Fujita H, et al. (1993) Separate regulation of heme oxygenase and heat shock protein 70 mRNA expression in the rat heart by hemodynamic stress. Biochem Biophys Res Commun 191:587–594

40. Takeda A, Kimpara T, Onodera H, et al. (1996) Regional difference in induction of heme oxygenase-1 protein following rat transient forebrain ischemia. Neurosci Lett 205:169–172
41. Shibahara S (1988) Regulation of heme oxygenase gene expression. Semin Hematol 25:370–376
42. Shibahara S, Sato M, Muller RM, et al. (1989) Structual organization of the human heme oxygenase gene and the function of its promoter. Eur J Biochem 179:557–563
43. Mitani K, Fujita H, Sassa S, et al. (1991) A heat-inducible nuclear factor that binds to the heat-shock element of the human haem oxygenase gene. Biochem J 277:895–897
44. Okinaga S, Takahashi K, Takeda K, et al. (1996) Regulation of human heme oxygenase-1 gene expression under thermal stress. Blood 87:5074–5084

# Carbon Monoxide: Toxic Waste or Endogenous Modulator of Hepatobiliary Function?

Makoto Suematsu[1], Tsuyoshi Sano[1], Shinji Takeoka[2], Yoshiyuki Wakabayashi[1], Takashi Yonetani[3], Eishun Tsuchida[2], and Yuzuru Ishimura[1]

*Summary.* Carbon monoxide (CO) is widely known to be a toxic gas that interferes with oxygen transport by the red blood cells. However, this "toxin" is continuously generated in the body by a heme-degrading enzyme called heme oxygenase. Our findings revealed that CO plays a crucial role in keeping liver blood vessels actively in a relaxed state and acts as a physiological mechanism for maintaining ample blood flow in the liver. CO produced in the liver tissue can reach its capillary vessels, called sinusoids, and relax the Ito cells, the microvascular pericytes covering the sinusoidal wall. The potential importance of this double-edged substance in the field of life sciences has emerged much like nitric oxide, another gaseous molecule that was established as a neurovascular mediator including vascular cell relaxation. Our discovery may, therefore, attract the wider interests of not only vascular biologists but also researchers in neuroscience, immunology, and pharmacology among others. With careful and persistent investigation, the heme oxygenase–CO pathway could serve as a potential therapeutic target for control of various disease conditions.

*Key words.* Carbon monoxide—Heme oxygenase—Nitric oxide—Hemoglobin—Sinusoids

## Introduction

Carbon monoxide, a product of heme oxygenase, [1] is thought to upregulate cyclic guanosine monophosphate (cGMP) via activation of guanylyl cyclase, and thus shares several biological actions with nitric oxide (NO) such as smooth muscle relaxation or inhibition of platelet aggregation [2]. There are, however, few experimental data suggesting that CO generated endogenously by heme oxygenase activity participates in modulation of cell function under physiological conditions, except for recent

[1] Department of Biochemistry, School of Medicine, Keio University, 35 Shinanomachi, Shinjuku-ku, Tokyo, 160 Japan
[2] Department of Polymer Chemistry, Waseda University, Toyama, Shinjuku-ku, Tokyo 169, Japan
[3] Department of Biophysics and Biochemistry, University of Pennsylvania, Philadelphia, PA 19104-6089, USA

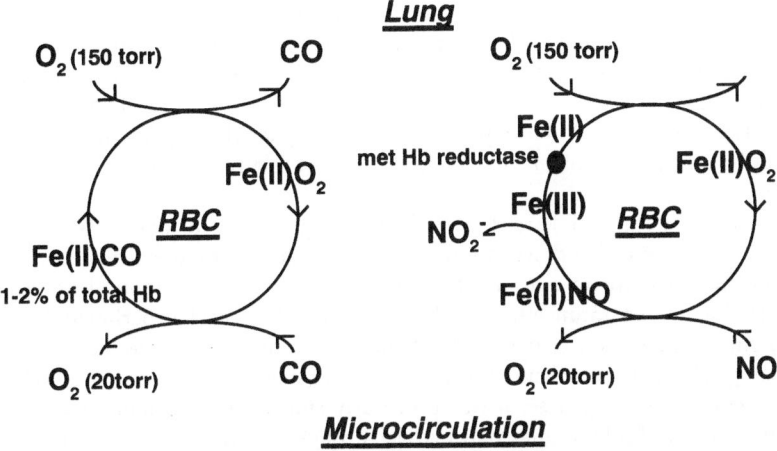

FIG. 1. Fate of CO and NO in vivo

observations that provide evidence for the contribution of CO as a neural messenger in brain [3]. As to its vascular actions, no definite evidence has been provided to establish CO as an endogenous modulator of vascular tone. The major source of CO production in the body is from physiological degradation of heme, and most CO generated in vivo is known to be transferred by hemoglobin in erythrocytes and then evacuated from the lung to the airway (Fig. 1).

Heme is metabolized to biliverdin and CO by heme oxygenase [1]. In humans, nearly 80% of the bilirubin excreted in bile is derived from hemoglobin heme. Cytochrome P-450 is a major contributor to the bilirubin derived from nonhemoglobin sources in the liver [1]. Two isoforms have been characterized: heme oxygenase-1, which is inducible by a variety of stressors such as hyperthermia, cytokines, or hypoxia-reoxygenation, and heme oxygenase-2, which is thought to constitute the major enzyme activity in the liver parenchyma [4]. The liver is, like the brain and spleen, one of the most abundant sources of heme oxygenase among organs [4]. We have recently demonstrated that suppression of endogenous CO generation elicits a marked increase in vascular resistance, which is concurrent with an elevation of bile formation in the liver, suggesting the biological action of CO as a modulator of hepatobiliary function. This chapter overviews recent research advance supporting the roles of CO in regulation of organ function in the liver.

## CO and NO: Differences in Biological Activities

Considerable evidence has been accumulated to indicate that exogenously applied CO markedly increases the intracellular cGMP content in a variety of cells such as platelets [2], corneal epithelial cells [5], neural cells [3], and Ito cells, liver-specific microvascular pericytes [6]. On the other hand, recent studies indicated that CO is not so potent as NO in activating soluble guanylate cyclase in its isolated purified enzyme preparation [7]: to be activated, this enzyme requires critical conformational changes involving dislocation of the five-coordinate interaction of the heme molecule with the distal histidine, and NO, but not CO, readily evokes such structural changes. In other words, CO cannot function as a potent activator of guanylate cyclase under circumstances in which the local concentration of NO is at relatively higher levels than that of CO.

It is therefore impossible to demonstrate the significance of CO in cGMP-dependent regulation of cell and organ function without discussing the actual amounts of the gaseous mediator in situ and the relationship of microanatomical orientation between the CO-generating effector cells and target cells in each experimental system. Furthermore, there is a criticism about the methodology for collecting data from physiological experimental systems, i.e., the use of metalloporphyrins as a "specific" inhibitor of heme oxygenase: it was previously shown that metalloporphyrins can alter the activity of soluble guanylate cyclase through their direct interactions with the heme moiety [8]. For example, Sn-protoporphyrin IX, one of widely used inhibitors of heme oxygenase, serves as an activator of soluble guanylate cyclase. Zn protoporphyrin IX (ZnPP), another heme oxygenase inhibitor with a greater inhibitory coefficient, inhibits the guanylate cyclase activity. It is therefore necessary, in physiological experimental models, to confirm the CO hypothesis by using pharmacological tools other than metalloporphyrins.

Another important difference in biological action between NO and CO is the reactivity with radical species: different from NO, CO is a nonradical monoxide, and its reactivity with biological reagents that can interact with NO such as SH compounds and nonheme iron is far less than that of NO [9]. For example, the rate of NO binding with oxyhemoglobin is about 3000 fold greater than that of CO. In addition, NO, but not CO, can interact with oxygen radical species such as triplet oxygen and superoxide anion to convert into nitrate and nitrite. These facts indicate that regional concentrations of hemoglobin, oxygen, and glutathione may highly influence the local concentration of NO as compared with that of CO [10].

## Hepatic Microcirculation: Its Spontaneous Relaxation by CO

The hepatic microcirculation is a unique vascular system that is composed of parenchymal hepatocytes and a variety of nonparenchymal cells such as sinusoidal endothelial cells, Kupffer cells, and Ito cells [11]. Liver sinusoids, the capillaries of the liver, share functions in common with capillaries elsewhere in the body, that is, exchange of oxygen and nutrients with the parenchyma and removal of metabolites and waste products. However, unlike other capillaries, they lack a basement membrane and have a discontinuous endothelium, forming small pores named fenestrae. It is well recognized that, despite the absence of vascular smooth muscle cells, the liver sinusoids are able to control blood flow. However, among the nonparenchymal cells, Ito cells (fat-storing cells, stellate cells) have attracted great interest as a contractile machinery analogous to the microvascular pericytes that control sinusoidal blood perfusion.

These cells, lining the outer surface of sinusoidal walls, are characterized by their abundant fat droplets containing vitamin A [12] and constitute a well-organized mesh of intercellular connection using their unique neuron-like dendritic structure [13]. Ito cells can exhibit contractile phenotypes like vascular smooth muscle cells in response to a variety of vasoactive substances; vasoconstrictive agonists such as angiotensin II and endothelin-1 evoke an increase in the intracellular calcium concentration and cellular contraction in Ito cells [14]. Ito cells can also exhibit relaxation in response to a variety of vasorelaxants such as prostaglandin $E_2$ and nitric oxide (NO) [15]. Detailed information about CO-mediated responses in this cell is available in chapter 4 in this volume (by Y. Wakabayashi et al).

We have shown that suppression of endogenous CO generation by ZnPP elicits a marked increase in the vascular resistance in the isolated perfused rat liver preparation [16]. Table 1 illustrates the effects of ZnPP, a heme oxygenase inhibitor, on vascular resistance. Administration of $1\,\mu M$ ZnPP eliminated the baseline CO generation, and the vascular resistance increased by 30%. The increase in vascular resistance and sinusoidal constriction were attenuated significantly by adding CO ($1\,\mu M$) or a cGMP analogue, 8-bromo-cGMP ($1\,\mu M$), in the perfusate. Of interest was that NO synthase inhibitors such as nitro-$N^{\omega}$-L-arginine methyl ester (L-NAME) and aminoguanidine neither induced a significant change in the vascular resistance nor altered the CO level in the venous effluents. The ZnPP-induced increase in the hepatic vascular resistance can be reproduced by administration of oxyhemoglobin at $1.5\,g/dl$ (Table 1). On the other hand, the same dose of HbCO did not alter the baseline vascular resistance (data not shown). Taking into account the fact that CO hemoglobin can scavenge NO [17], these results suggest that CO rather than NO plays the greater role in modulation of the vascular tone in a steady-state condition.

Because the potency of NO to activate soluble guanylate cyclase is about 50 fold

TABLE 1. Increase in the vascular resistance in isolated perfused rat liver preparation by ZnPP, a heme oxygenase inhibitor, and the effects of pretreatment with CO and 8-bromo-cGMP (8Br-cGMP)

| Groups | Vascular resistance (cm $H_2O$ min ml$^{-1}$) | CO concentrations (nM) |
| --- | --- | --- |
| Control | $0.21 \pm 0.02$ | $250 \pm 80$ |
| ZnPP | $0.32 \pm 0.05$* | <20 (undetectable) |
| ZnPP + CO | $0.32 \pm 0.03$** | $560 \pm 20$ |
| ZnPP + 8Br-cGMP | $0.24 \pm 0.04$** | <20 (undetectable) |
| HbO$_2$ | $0.27 \pm 0.03$* | Not determined |

Data depict the values measured at 10 min after the start of administration of each reagent (mean ± SD of 6–8 experiments). CO ($1\,\mu M$) or 8Br-cGMP ($1\,\mu M$) was infused simultaneously with ZnPP. Note that oxyhemoglobin (HbO$_2$, $1.5\,g/dl$ at a final concentration) reproduces a similar vascular response in the perfused liver, suggesting the involvement of endogenously produced CO. In these experiments, CO concentrations in the perfusate could not be measured because of the interference with the light absorption by Hb.
*, $P < .05$ as compared with the control values; **, $P < .05$ as compared with the values in the ZnPP-treated group.

greater than that of CO in vitro, as described previously [7], CO-mediated vasorelaxation through the cGMP-dependent mechanism seems hardly to occur when NO is abundantly generated in the same vascular compartment. In the liver, however, there are several circumstances in which CO serves as a modulator of sinusoidal tone. First, the actual flux in the liver is different between the two mediators: the level of CO flux from rat liver is about 600–700 pmol min$^{-1}$g$^{-1}$ liver [6], while that of NO is at most about 30 pmol min$^{-1}$g$^{-1}$ liver [18]. Second, because of its chemical properties as a radical species, NO can interact with superoxide anion canceling each other out [19] or bind to glutathione excreted from the liver parenchyma into the circulation. Finally, intraorgan compartments responsible for NO or CO generation should be taken into account: the main sources of NO in the unstimulated liver are endothelial cells (and Kupffer cells, if the liver is stimulated), which confront the luminal side of sinusoids. By contrast, hepatocytes that are located in the extravascular space account for a major source of endogenous CO generation because a major fraction of heme oxygenase activity in the control liver occurs in the liver parenchyma [4]. Under these conditions, the actual concentration of NO in and around the sinusoidal compartment in vivo is likely to be much smaller than that of CO. These circumstances are helpful in explaining why CO can preferentially modulate the resistance of hepatic sinusoids in the steady state.

Although it is evident that, at least in the perfused liver, CO plays a modulatory role in regulation of sinusoidal tone, whether cGMP-mediated mechanisms explain the whole picture of CO-dependent vasorelaxion in sinusoids is not known. Several findings in the current study support the possible involvement of unidentified cGMP-independent mechanisms: First, the minimum CO concentration that was required to elicit a significant increase in cGMP in cultured Ito cells was at least 100 fold higher than that detected in the effluent of the control perfused liver. Second, the effects of 8Br-cGMP on the ZnPP-induced increase in the vascular resistance was limited only to about 60% inhibition, suggesting the presence of alternative mechanisms for CO-mediated vasorelaxation. Such an unknown action of CO, which involves a cGMP-independent signal transduction mechanism, has recently been postulated by Kourembanas et al. [20], who reported that exogenously applied CO can modulate the expression of endothelin and growth factors independently of cGMP. A recent observation by Rich et al. [5] has provided evidence that CO elicits an increase in potassium current and thereby hyperpolarizes the resting membrane potential in rabbit corneal epithelial cells. We have recently revealed that the membrane potential of Ito cells is determined specifically by the two distinct K$^+$ channels, that is, outward and inward rectifier K$^+$ channels [13]. Because the hyperpolarizing effect of CO has also been reported in other types of cells such as corneal epithelial cells [5], electrophysiological properties of cultured Ito cells and the effect of CO on transmembrane currents should further be evaluated to reveal such a cGMP-independent novel vasorelaxing mechanism.

*Acknowledgments.* This work was supported by the International Research Program of Grants-in-Aid for Scientific Research from the Ministry of Education, Science, Sports and Culture of Japan, and by grants from School of Medicine, Keio University, from Kanae Research Foundation, and from Takeda Research Foundation.

# References

1. Maines MD (1988) Heme oxygenase: function, multiplicity, regulatory mechanisms, and clinical applications. FASEB J 2:2557–2568
2. Brüne B, Ullrich V (1987) Inhibition of platelet aggregation by carbon monoxide mediated by activation of guanylate cyclase. Mol Pharmacol 32:497–504.
3. Verma A, Hirsch DJ, Glatt CE, et al. (1993) Carbon monoxide: a putative neural messenger. Science 259:381–384
4. Maines MD, Trakshel GM, Kutty RK (1986) Characterization of two constitutive forms of rat liver microsomal heme oxygenase: only one molecular species of the enzyme is inducible. J Biol Chem 261:11131–11137
5. Rich A, Farrugia G, Rae JL (1994) Carbon monoxide stimulates a potassium-selective current in rabbit corneal epithelial cells. Am J Physiol 267:C435–C442
6. Suematsu M, Goda N, Sano T, et al. (1995) Carbon monoxide: an endogenous modulator of sinusoidal tone in the perfused rat liver. J Clin Invest 96:2431–2437
7. Kharitonov V, Sharma VS, Pilz RB, et al. (1995) Basis of guanylate cyclase activation by carbon monoxide. Proc Natl Acad Sci USA 92:2568–2571
8. Ignarro LJ, Barrot B, Wood KS (1984) Regulation of soluble guanylate cyclase activity by porphyrins and metalloporphyrins. J Biol Chem 259:6201–6207
9. Moncada S, Pallmer RMJ, Higgs EA (1991) Nitric oxide: physiology, pathophysiology, and pharmacology. Pharmacol Rev 43:109–142
10. Suematsu M, Wakabayashi Y, Ishimura Y (1996) Gaseous monoxides: a new class of microvascular regulator in the liver microcirculation. Cardiovasc Res 32:679–686
11. McCuskey RS, Urbaschek R, McCuskey PA, Urbascheck B (1982) In vivo microscopic responses of the liver to endotoxins. Klin Wochenschr 60:749–751
12. Suematsu M, Oda M, Suzuki H, et al. (1993) Intravital and electron microscopic observation of Ito cells in rat hepatic microcirculation. Microvasc Res 46:28–42
13. Kashiwagi S, Suematsu M, Wakabayashi Y, et al. (1997) Electrophysiological characterization of cultured hepatic stellate cells in rats. Am J Physiol (Gastrointest Liver Physiol) 272:G742–G750
14. Pinzani M, Failli P, Ruocco C, et al. (1992) Fat-storing cells as liver-specific pericytes. Spatial dynamics of agonist-stimulated intracellular calcium transients. J Clin Invest 90:642–646
15. Kawada N, Tran-Thi T-A, Klein H, Decker K (1993) The contraction of hepatic stellate (Ito) cells stimulated with vasoactive substances: possible involvement of endothelin 1 and nitric oxide in the regulation of the sinusoidal tonus. Eur J Biochem 213:815–823
16. Suematsu M, Kashiwagi S, Sano T, et al. (1994) Carbon monoxide as an endogenous modulator of hepatic vascular perfusion. Biochem Biophys Res Commun 205: 1333–1337
17. Kosaka H, Watanabe M, Yoshihara H, et al. (1992) Detection of nitric oxide production in lipopolysaccharide-treated rats by ESR using carbon monoxide hemoglobin. Biochem Biophys Res Commun 184:1119–1124
18. Obolenskaya MY, Vanin AF, Mordvintcev PI, et al. (1994) EPR evidence of nitric oxide production by the regenerating rat liver. Biochem Biophys Res Commun 202:571–576
19. Beckman JS, Beckman TW, Chen J, et al. (1990) Apparent hydroxyl radical production by peroxynitrite: implications for endothelial injury from nitric oxide and superoxide. Proc Natl Acad Sci USA 87:1620–1624
20. Kourembanas S, McQuilan LP, Leung GK, Faller DV (1993) Nitric oxide regulates the expression of vasoconstrictors and growth factors by vascular endothelium under both normoxia and hypoxia. J Clin Invest 92:99–104

# Microspectrophotometry of Nitric Oxide-Dependent Changes in Hemoglobin in Single Red Blood Cells Incubated with Stimulated Macrophages

Takuo Shiraishi, Kazuhiro Tsujita, and Katsuko Kakinuma

*Summary.* A highly sensitive microspectrophotometer was developed to measure spectral changes of oxyhemoglobin (oxy Hb) in single red blood cells (RBCs) incubated with stimulated macrophages as a model of nitric oxide- (NO-) dependent cytotoxicity. Our microspectrophotometer, using a modified acousto-optic tunable filter (AOTF), allows fast spectrophotometric data acquisition. This spectrophotometer examined single human RBCs treated with various concentrations of NO that exhibited absorption spectral changes due to the conversion of oxy Hb to methemoglobin (met Hb). The change in absorption differences at $\alpha$- (557–590 nm) and $\beta$- (542–525 nm) bands showed a linear relationship with the concentration of NO within 100 $\mu$M. We next examined RBCs incubated with murine macrophages with and without lipopolysaccharides (LPS) in the presence of glucose for 24 and 40 h. The RBCs incubated with LPS-stimulated macrophages showed significant spectral changes because of NO-dependent conversion of oxy Hb to met Hb, which suggests that RBC trap NO with very high efficiency by direct cell–cell interaction with macrophages. This spectrophotometric system is available for the use of a few drops of samples to study NO-specific cytotoxicity as a model of RBC without use of any chemical reagent in parallel with microscopic observations on cell structures such as membrane damage leading to hemolysis, adherence, and phagocytosis.

*Key words.* AOTF—Macrophages—Microspectrophotometry—NO—Red blood cells

## Introduction

Nitric oxide (NO) is now known to play a role in a number of physiological and pathological processes [1,2]. Nitric oxide in the blood is well maintained at a steady-state level (2–3 $\mu$M) by the dynamic balance between the continuous supply of NO from endothelial NO syntheses and the rapid scavenging of NO by oxyhemoglobin (oxy Hb) in red blood cells (RBCs) [3,4]. Nevertheless, several studies have shown that

Biophotonics Information Laboratories, Yamagata Advanced Technology Research Center, Matsuei 2-2-1, Yamagata 990, Japan

macrophage-mediated cytotoxicity is linked to the release of large amounts of NO ($10$–$100\,\mu M$) in response to certain microbes or microbial products such as bacterial lipopolysaccharides (LPS) and inflammatory cytokines [5,6].

In this investigation, we developed a highly sensitive microspectrophotometer using an acousto-optic tunable filter (AOTF) [7,8] to measure the spectral changes in RBC caused by NO-dependent conversion of oxy Hb to methemoglobin (met Hb) by stimulated macrophages. The RBCs incubated with stimulated macrophages showed significant spectral changes as the result of NO-dependent conversion of oxy Hb to met Hb, which corresponds to the spectral changes of RBC treated with concentrations of NO several times higher than that in the culture medium as determined by Griess reaction method [9]. The results are discussed in terms of the cell–cell interaction between RBCs and macrophages.

# Materials and Methods

## Reagents

$NaNO_2$, sulfanilamide, and $N$-(1-naphtyl) ethylenediamine dihydrochloride were obtained from Wako Pure Chemical Industries (Tokyo, Japan); interferon-$\gamma$ (INF-$\gamma$) and LPS were obtained from Sigma Chemical (St. Louis, MO, USA); thioglycollate medium and culture medium (RPMI-1640) were obtained from Kanto Chemical (Tokyo, Japan) and Gibco BRL, Gaithersburg, MD, USA, respectively; and $N_2$ gas (99.999% pure), NO gas (99.7%), and CO gas (99.95%) were obtained from Sumitomo Seika Chemicals (Tokyo, Japan).

## AOTF-Based Microspectrophotometric System

This microspectrophotometric system consists of an AOTF (M-IM100, Brimrose, Baltimore, MD, USA) unit, mounted at the side port of a microscope (TMD 300, Nikon, Tokyo, Japan), a two-dimensional charge-coupled device (CCD) array (CS8310, Tokyo Electric Industry, Tokyo, Japan), an image processor (DVS3000, Hamamatsu Photonics, Shizuoka, Japan), a personal computer controller, and a RF drive unit (Fig. 1). This system was developed to obtain absorption spectra in the range of 400–650 nm (original range, 450–600 nm) with 166 increments at 1.5-nm intervals.

## Cell Preparation

Fresh RBCs were collected from 20 ml of human venous blood and diluted tenfold with Krebs ringer phosphate (KRP) buffer containing 5 mM glucose. Macrophages were prepared from peritoneal exudates of CD-1 mice 4 days after injecting thioglycollate into the peritoneal cavity [10].

## Microspectrophotometry of RBCs

Spectrophotometric measurements of RBCs were performed using a sampling area ($4 \times 4\,\mu m^2$) for a single RBC and an adjacent extracellular area as reference, both of which were selected directly from the CRT screening of the image. The data acquisi-

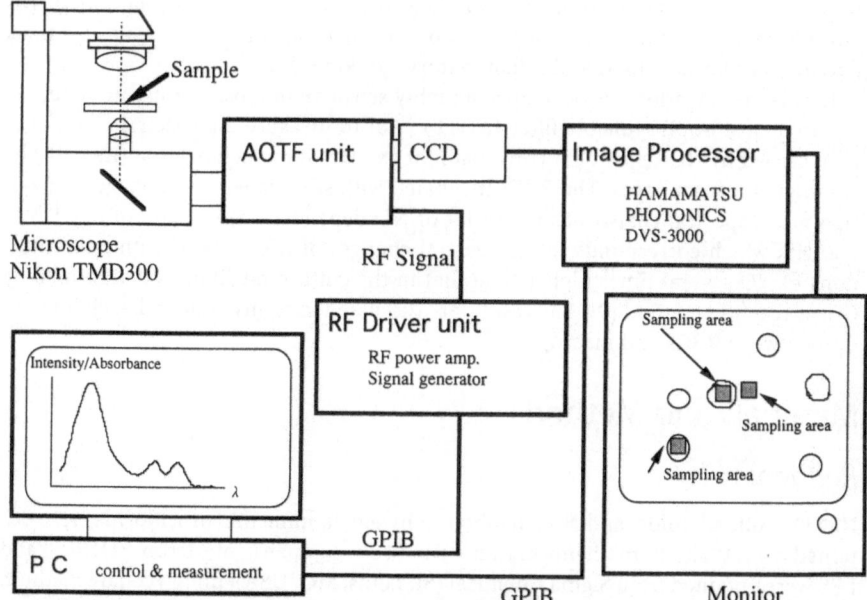

FIG. 1. The acousto-optic tunable filter- (*AOTF-*) based microspectrophotometric system consists of an AOTF unit mounted at the side port of a microscope, a two-dimensional charge-coupled device (*CCD*) array, an image processor, a personal computer controller, and a RF drive unit. *GPIB*, general purpose interface bus

tion time was approximately 90 s. In all measurements, the spectra of individual RBCs (2–5 cells) were averaged.

## NO-Trapping RBC

KRP buffer containing 5 mM glucose was deoxygenated in an airtight tube by bubbling with pure $N_2$ gas and then continuously bubbled with NO gas for 30 min to obtain a 3 mM NO solution [11]. Various concentrations of NO solution (0, 30, 70, and 300 μM) were prepared in deoxygenated airtight tubes under $N_2$ gas flow to which a 1:100 volume of RBCs was injected with a microsyringe.

## RBCs Incubated with Macrophages

Macrophages were suspended in a concentration of $8 \times 10^5$ cells per milliliter of the RPMI-1640 culture medium, an aliquot (0.2 ml) of which was placed in each of 96 wells of microtiter plates and preincubated for 1.5 h at 37°C in air containing 5% $CO_2$. The macrophage suspension was mixed with 10 μg/ml of LPS and 84 U/ml of INF-γ as stimulators. An equal volume of the culture medium was added instead of the stimulators to other macrophages for resting cells. To both the macrophage suspensions of stimulated and resting cells, an aliquot of RBCs was added to the final $8 \times 10^6$ cells per milliliter. All the macrophage suspensions containing 5 mM glucose were incubated at 37°C for 40 h without shaking. The microspectrophotometry on the RBCs in those macrophage suspensions was carried out at various time intervals.

## Results and Discussion

Because the AOTF unit has an optimal wavelength range from 500 to 600 nm, the optical absorption maxima at α- and β-bands were used for evaluation of oxy Hb and met Hb, and the peak at the γ-band was measured for characterization of the spectral shift throughout our investigation.

The effect of NO on the absorption spectra of RBC was examined under various concentrations. The difference spectra of RBC plus or minus NO (see inset in Fig. 2) reflected the NO-dependent conversion of oxy Hb to met Hb, in good agreement with the difference spectra previously reported [12]: the absorption bands at 542 and 577 nm and the isosbestic points at 525 and 590 nm. The spectral changes at the α- and β-bands, as expressed as the absorption differences between 577 and 590 nm and between 542 and 525 nm, respectively, were plotted against the concentration of NO (Fig. 2). The absorption differences at both α- and β-bands exhibited a linear relation to the concentration of NO within 100 μM. A great deviation, however, was found at 300 μM of NO: some RBCs showed Heinz bodies in the membranes (data not shown), and the number of RBCs markedly decreased. It should be noted that $NO_2^-$, which is derived from rapid conversion of NO in aqueous medium [13], had no effect on the spectrum of RBC in the presence of 5 mM glucose (data not shown).

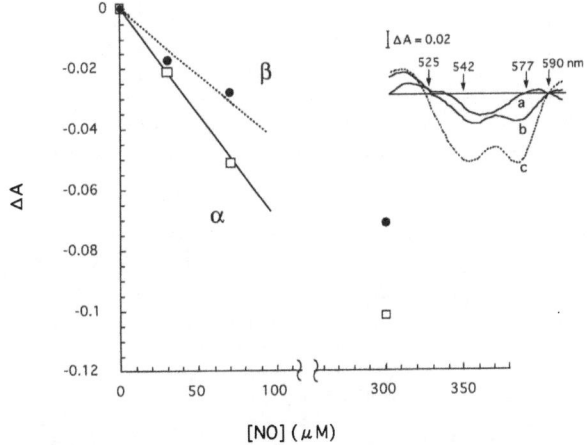

FIG. 2. Spectral changes of oxyhemoglobin in red blood cells (RBCs) treated with various concentrations of nitric oxide (*NO*) and subjected to microspectrophotometry. The spectral changes of RBC were recorded against the spectrum of NO-free RBC as reference (baseline) to obtain the difference spectra of 30 (*a*), 70 (*b*), and 300 (*c*) μM NO (*inset*). The absorption differences (*ΔA*) between 577 and 590 nm (α-band, *open squares and solid line*) and between 542 and 525 nm (β-band, *closed circles and dotted line*) were then plotted against the concentration of NO

We next examined microspectrophotometry of RBCs incubated with stimulated and resting macrophages for 0, 24, and 40 h (Fig. 3). The absorption spectrum at zero time showed that of typical oxy Hb, while RBC incubated with stimulated macrophages (Fig. 3a) for 24 h showed a marked change in the spectrum; the γ-band shifted 10 nm toward a shorter wavelength and the β-band peak markedly decreased. Further incubation for 40 h resulted in greater spectral changes: the disappearance of the α- and β-bands and the appearance of a new band around 500 nm, with a marked shift of the γ-band, characteristics of met Hb. In contrast, RBCs incubated with resting macrophages showed no change in the oxy Hb spectrum for 40 h (Fig. 3b). The absorption differences at α- ($\Delta A_{577-590\,nm}$) and β- ($\Delta A_{542-525\,nm}$) bands in Fig. 3a were plotted against the incubation time (Fig. 4). Incubation for 24 h caused the absorption changes to be 0.05 and 0.037 at the α- and β-bands, respectively, which corresponded to the spectral changes of RBCs treated with 70 μM NO (see Fig. 2), three times as much as that determined by the Griess reaction (the concentration of $NO_2^-$: 23 μM) [9,13]. In addition, the microscopic image of the incubated mixture showed time-dependent hemolysis of RBCs: about 50% decrease in the number of RBC at 40 h as compared to that incubated with resting macrophages.

These results showed that RBCs trapped highly concentrated NO produced in situ by stimulated macrophages, as the spectral changes of the RBCs corresponded to those of RBC treated with a concentration of NO about threefold greater than that in

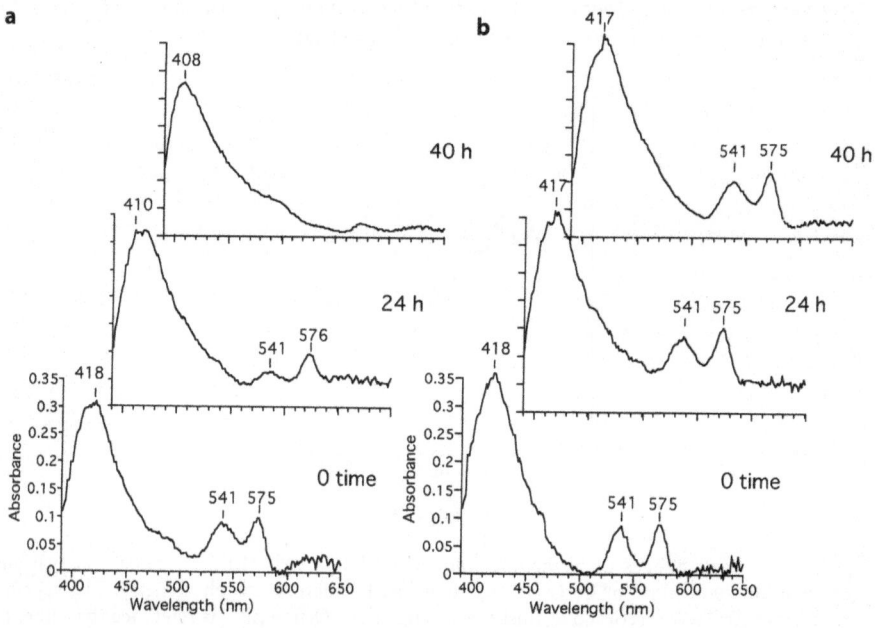

FIG. 3a,b. Absorption spectra of RBC incubated with stimulated (a) and resting (b) macrophages. Macrophages were mixed with lipopolysaccharide and interferon-γ for stimulation and with the same volume of the buffer for control (resting state) and incubated for 40 h. The absorption spectra of RBC were measured at 0, 24, and 40 h

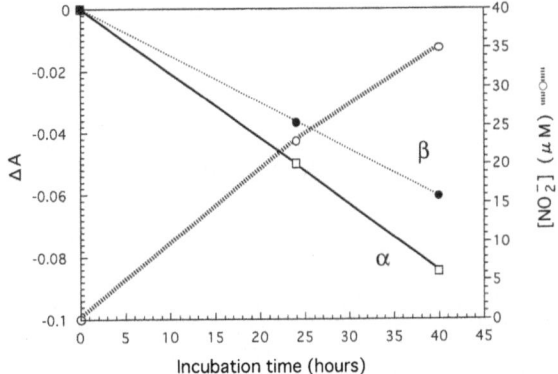

FIG. 4. Absorption changes of RBC incubated with stimulated macrophages and the concentration of NO in the culture medium. From the spectra of RBC incubated with stimulated macrophages (Fig. 3), the absorption differences at α ($\Delta A_{577-590\,nm}$) (*open squares* and *solid line*) and β ($\Delta A_{542-525\,nm}$) (*closed circles* and *dotted line*) bands were plotted, respectively, against the incubation time as reference at 0 time. The concentration of NO produced by stimulated macrophages was measured with the same culture medium as stimulated macrophages, except for the absence of RBC, by the Griess reaction method as $NO_2^-$ (*open circles* and *broken line*)

the culture medium. The reactivity of NO with RBCs is a complex function of its diffusibility, half-life in aqueous medium, membrane permeability, and culture condition [14]. The culture medium contained RBCs and macrophages in a ratio of 10:1 to obtain a high trapping efficiency by the direct cell–cell interaction. For consideration of NO-dependent cytotoxicity, a localized high concentration of NO produced in situ by a cell or a tissue seems more important than that diluted in a reaction medium. This microspectrophotometer with AOTF, which allows fast collection of spectrophotometric data, could be available in future for ex vivo monitoring of NO-dependent spectral changes in RBCs in a vein or in an inflammatory lesion.

*Acknowledgments.* We thank Prof. Takashi Yonetani, Department of Biochemistry, University of Pennsylvania, and Dr. Tetsuhiko Yoshimura, Institute for Life Support Technology, Yamagata Technopolis Foundation, for their valuable suggestions. We thank Dr. Teruo Kirikae and Prof. Masayasu Nakano, Department of Microbiology, Jichi Medical School, for their kind guidance for the preparation of macrophages and the assay of NO production.

# References

1. Furchgott RF, Zawadzki JV (1980) The obligatory role of endothelial cells in the relaxation of arterial smooth muscle by acetylcholine. Nature 288:373–376
2. Moncada S, Palmer RMJ, Higgs EA (1991) Nitric oxide: physiology, pathophysiology, and pharmacology. Pharmacol Rev 43:109–142
3. Ignarro LJ, Buga GM, Wood KS, et al. (1987) Endothelium-derived relaxing factor produced and released from artery and vein is nitric oxide. Proc Natl Acad Sci USA 84:9265–9269

4. Kosaka H, Uozumi M, Tyuma I (1989) The interaction between nitrogen oxides and hemoglobin and endothelium-derived relaxing factor. Free Radical Biol Med 7:653–658
5. Stuehr DJ, Marletta MA (1985) Mammalian nitrate biosynthesis: mouse macrophages produce nitrite and nitrate in response to *Escherichia coli* lipopolysaccharide. Proc Natl Acad Sci USA 82:7738–7742
6. Nathan C (1992) Nitric oxide as a secretory product of mammalian cells. FASEB J 6:3051–3064
7. Chang IC (1981) Acousto-optic tunable filters. Opt Eng 20:824–829
8. Tran CD, Simianu V (1992) Multiwavelength thermal lens spectrophotometer based on an acousto-optic tunable filter. Anal Chem 64:1419–1425
9. Schmidt HHHW, Kelm M (1996) Determination of nitrite and mitrate by the Griess reaction. In: Methods in nitric oxide research. Wiley, New York, pp 491–497
10. Kirikae F, Kirikae T, Qureshi N, et al. (1995) CD14 is not involved in *Rhodobacter sphaeroides* diphosphoryl lipid, a inhibition of tumor necrosis factor alpha and nitric oxide induction by taxol in murine macrophages. Infect Immunol 63:486–497
11. Tracey WR, Linden J, Peach MJ, et al. (1990) Comparison of spectrophotometric and biological assays for nitric oxide (NO) and endothelium-derived relaxing factor (EDRF): nonspecificity of the diazotization reaction for NO and failure to detect EDRF. J Pharmacol Exp Ther 252:922–928
12. Feelisch M, Kubitzek D, Werringloer J (1996) The oxyhemoglobin assay. In: Methods in nitric oxide research. Wiley, New York, pp 455–477
13. Ignarro LJ, Fukuto JM, Griscavage JM, et al. (1993) Oxidation of nitric oxide in aqueous solution to nitrite but not nitrate: comparison with enzymatically formed nitric oxide from L-arginine. Proc Natl Acad Sci USA 90:8103–8107
14. Vanderkooi JM, Wright WW, Erecinska M (1994) Nitric oxide diffusion coefficients in solutions, proteins and membranes determined by phosphorescence. Biochim Biophys Acta 1207:249–254

# An Antioxidant Role of Nitric Oxide in Modulation of Oxidative Stress in Human Placental Trophoblastic Cells

Nobuhito Goda[1], Michiya Natori[3], Makoto Suematsu[2], Kaoru Kiyokawa[1], Yuzuru Ishimura[2], Yasunori Yoshimura[1], and Shiro Nozawa[1]

*Summary.* Nitric oxide (NO) has recently attracted great interest as a mediator of cell and organ functions. During pregnancy, larger amounts of NO are known to be produced in systemic circulation on the maternal side than those in nonpregnant females. Placental trophoblastic cells (PTCs), which are major constituents of placenta and are in direct contact with maternal blood, can produce NO via the reaction of endothelial form of NO synthase during pregnancy. Endogenously produced NO in placenta can relax vascular tone, inhibit platelet aggregations, and attenuate adherent reactions of leukocytes and platelets. By contrast, the decrease of NO generation in the placenta is correlated with impairment of uteroplacental perfusion, which elicits severe perinatal complications such as preeclampsia (PE) and intrauterine growth retardation (IUGR). Although the mechanisms through which endogenous NO modulates a variety of cellular functions in the placenta have not been revealed yet, our results provide new insight into molecular mechanisms of endogenous NO in PTCs, suggesting that alterations in placental NO formation may not only play a role in the physiological changes of advancing gestation but may also contribute to the pathophysiology of PE and IUGR.

*Key words.* NO—Placenta—Mitochondria—Preeclampsia—IUGR

## Introduction

Regulation of placental circulation is pivotal in ensuring adequate delivery of oxygen and nutrients from the maternal to the fetal side. Impairment of such an integrated regulation is associated with severe perinatal complications, including intrauterine growth retardation (IUGR) and preeclampsia (PE), which are well known to be major causes of high mortality and morbidity during pregnancy. Despite these clinical interests, the mechanisms responsible for maintaining an ample blood supply to the

[1]Departments of Obstetrics and Gynecology and [2]Biochemistry, School of Medicine, Keio University, 35 Shinanomachi, Shinjuku-ku, Tokyo 160, Japan
[3]Department of Clinical Research, National Ohkura Hospital, 2-10-1 Ohkura, Setagaya-ku, Tokyo 157, Japan

uteroplacental circulation are poorly understood. Because of the absence of neural innervation, the placental circulation is likely to depend on locally produced vasoactive factors such as endothelin [1], prostacyclin [2], and nitric oxide (NO) [3].

NO has been identified as an endothelium-derived relaxing factor and is known to play an indispensable role in regulating a variety of cellular functions. The physiological roles of endogenously produced NO are discussed mostly in the regulation of microvascular functions such as arteriole tone, macromolecular permeability [4], and adhesion of platelets [5] and leukocytes to endothelial cells [6]. These biological actions of NO are explained mainly by two different biochemical properties of NO: activation of soluble guanylate cyclase and cancellation of superoxide anion ($O_2^-$) as an antioxidant. Recent data showing that, in normal human placenta, NO synthase is constitutively expressed in trophoblast cells that are major constituents of placenta indicate the involvement of endogenously produced NO in the maintenance of placental functions [7]. However, the mechanisms through which NO generated in placenta exerts its biological action are not yet completely studied. This chapter intends to summarize the roles of this gaseous substance in regulation of placental functions in view of pathophysiological relevance.

## Role of Endogenously Produced NO in the Placenta

Considerable evidence has indicated the biological significance of NO in the maintenance of pregnancy. Conrad and his co-workers showed the involvement of NO in pregnancy: urinary excretion of nitrates and cyclic guanosine monophosphate (cGMP) were more elevated in pregnant compared to nonpregnant rats [8]. In addition, nitrosylhemoglobin, a complex of NO with hemoglobin, increased in the blood of pregnant rats but not in nonpregnant ones. Recent immunohistochemical and biochemical investigations disclosed that a calcium-dependent NO synthase (endothelial form of NO synthase) is constitutively expressed in the trophoblastic layer of normal human placenta [7,9]. These lines of information suggest that endogenously generated NO is attributable to alterations in physiological vascular function so as to guarantee ample blood supply to uteroplacental circulation, resulting in fetal well-being throughout pregnancy.

On the other hand, attenuation of NO production by administration of a specific NO synthase inhibitor has been reported to cause PE and IUGR in pregnant rats [10–12]. Sooranna et al. [9] showed that NO synthase activity in placenta from these complications is significantly lower than those from normal pregnancy in humans. Although the etiology of such complications is not yet completely disclosed, the decrease in uteroplacental blood perfusion is thought to be associated with the development of these conditions. In addition, histochemical analyses have shown pathological thrombus formation with fibrin deposition in an intervillous space in placentas, suggesting the involvement of disruption of anticoagulant systems in PE and IUGR [13–15]. Because endogenously produced NO can exert a variety of microvascular functions and anti-coagulant properties, these experimental and clinical investigations indicate that NO released in uteroplacental circulation can function as an important mediator under pathophysiological conditions.

# NO as an Endogenous Antioxidant Molecule in the Placenta

The biological actions of NO have been discussed mainly on two different biochemical properties of this gaseous monoxide: first, NO can activate soluble guanylate cyclase and thus produce cGMP, resulting in relaxation of vascular smooth muscle cells and inhibition of platelet aggregation. Because findings such as vasospasm of uterine spiral arteries and pathological accumulation of thrombus and fibrin deposition in the intervillous space are often encountered in patients suffering from PE and IUGR [13–15], it is quite reasonable to consider that significant suppression of cGMP production evoked by reduction of NO formation in the placenta is a possible mechanism of these complications.

Second, NO can react with and cancel out $O_2^-$, resulting in formation of relatively stable substances such as nitrite and nitrate. $O_2^-$ is known to enhance oxidative stress via formation of lipid peroxides and other oxygen radical species such as hydroxyl radical and hydrogen peroxide. In this respect, NO can modulate oxidative stress on the basis of the respective quantitative relationship between NO and $O_2^-$. On the other hand, the interaction of NO with $O_2^-$ is known to generate peroxynitrite ($ONOO^-$), a highly reactive oxidant that may jeopardize physiological cell and tissue function [16]. However, we believe that, under physiological conditions, NO produced intracellularly is not necessarily present at levels adequate to react with $O_2^-$ for $ONOO^-$ generation: our recent results demonstrated that endogenously produced NO is likely to react

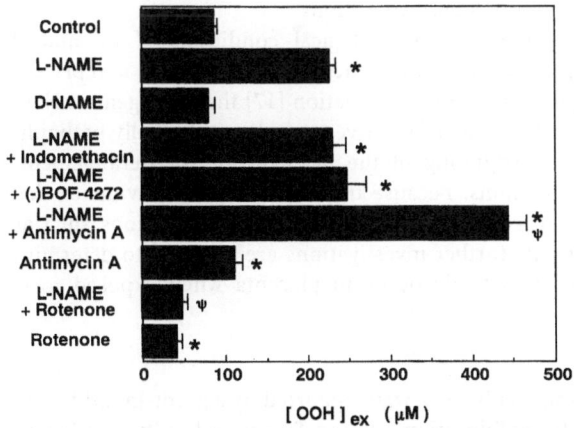

FIG. 1. Effects of inhibitors of mitochondrial respiratory chain on $N^G$-nitro-L-arginine methyl ester- (*L-NAME-*) induced oxidative stress in cultured human placental trophoblastic cells (HPTCs). HPTCs were incubated with 100 μM indomethacin, 10 nM (−) BOF-4272, 30 μM antimycin A, or 20 μM rotenone for 1 h before application of 1 mM L-NAME. Oxidative impacts were measured after 30 min of treatment with L-NAME or D-NAME (1 mM). Values are means ± SE of seven experiments. *, Statistical significance at $P < .05$ compared with control group; ψ, statistical significance at $P < .05$ compared with group treated with L-NAME. (Modified from [17], with permission of the American Physiological Society)

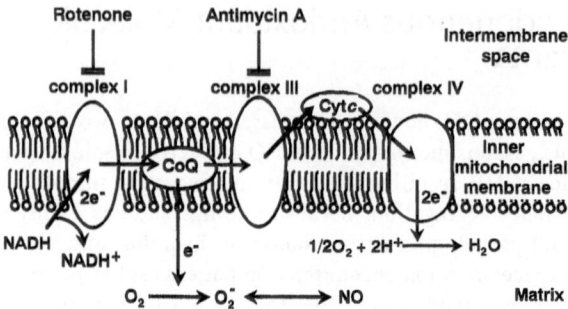

FIG. 2. Schema of possible mechanism in regulation of intracellular oxidative stress by nitric oxide (*NO*) in HPTCs. Endogenously produced NO is likely to react with $O_2^-$ released from the respiratory chain of mitochondria in HPTCs, leading to the modulation of intracellular oxidative stress. *CoQ*, coenzyme Q; *Cyt c*, cytochrome *c*

with $O_2^-$ released from the respiratory chain of mitochondria in human placental trophoblastic cells (Fig. 1) [17]. In other words, intracellular oxidative stress is enhanced by suppression of NO production in these cells (Fig. 2). In addition, NO has been reported to regulate microvascular endothelial function in vivo [18] and in vitro [19] through attenuation of intracellular peroxide formation. Collectively, endogenous NO is capable of modulating intracellular oxidative stress as an important antioxidant in the placenta and may play a key role in maintenance of placental functions under physiological conditions.

When considering pathophysiological conditions of PE and IUGR in which production of placental lipid peroxides is elevated [20,21] and placental NO synthase activity is suppressed [9], our observation [17] that endogenous NO is an indispensable modulator of placental oxidative stress has potentially pathophysiological relevance to the understanding of the mechanisms for derangement of intervillous microvascular functions. Because there is little direct evidence that endogenously produced NO can regulate a variety of placental functions such as antiadhesivity and anticoagularity, further investigations are required to determine whether these cellular alterations actually occur in placenta when exposed to endogenous NO suppression.

*Acknowledgments.* This work was supported by a grant-in-aid for scientific research from the Ministry of Education, Science, Sports and Culture of Japan, by grants from Keio University School of Medicine, and by a grant from the Japan Society for the Promotion of Science.

## References

1. Fried G, Liu YA (1994) Effects of endothelin, calcium channel blockade and EDRF inhibition on the contractility of human uteroplacental arteries. Acta Physiol Scand 151:477–484

2. Gerber JG, Payne NA, Murphy RC, et al. (1981) Prostacyclin produced by pregnant uterus in the dog may act as a circulating vasodepressor substance. J Clin Invest 67:632–636
3. Buttery LD, McCarthy A, Springall DR, et al. (1994) Endothelial nitric oxide synthase in the human placenta: regional distribution and proposed regulatory role at the feto-maternal interface. Placenta 15:257–265
4. Kubes P, Granger DN (1992) Nitric oxide modulates microvascular permeability. Am J Physiol 262 (Heart Circ Physiol 31):H611–H615
5. Sneddon JM, Vane JR (1988) Endothlium-derived relaxing factor reduces platelet adhesion to bovine endothelial cells. Proc Natl Acad Sci USA 85:2800–2804
6. Kubes P, Suzuki M, Granger DN (1991) Nitric oxide: an endogenous modulator of leukocyte adhesion. Proc Natl Acad Sci USA 88:4651–4655
7. Conrad KP, Vill M, McGuire PG, et al. (1993) Expression of nitric oxide synthase by syncytiotrophoblast in human placental villi. FASEB J 7:1269–1276
8. Conrad KP, Joffe GM, Kruszyna H, et al. (1993) Identification of increased nitric oxide biosynthesis during pregnancy in rats. FASEB J 7:566–571
9. Sooranna SR, Morris NH, Steer PJ (1995) Placental nitric oxide metabolism. Reprod Fertil Dev 7:1525–1531
10. Diket AL, Pierce MR, Munshi UK, et al. (1994) Nitric oxide inhibition causes intra-uterine growth retardation and hind-limb disruptions in rats. Am J Obstet Gynecol 171:1243–1250
11. Molnar M, Sutu T, Toth T, et al. (1994) Prolonged blockade of nitric oxide synthesis in gravid rats produces sustained hypertension, proteinuria, thrombocytopenia, and intrauterine growth retardation. Am J Obstet Gynecol 170:1458–1466
12. Seligman SP, Buyon JP, Clancy RM, et al. (1994) The role of nitric oxide in the pathogenesis of preeclampsia. Am J Obstet Gynecol 171:944–948
13. Khong TY (1987) Immunohistological study of the leukocytic infiltrate in maternal uterine tissues in normal and preeclamptic pregnancies at term. Am J Reprod Immunol Microbiol 15:1–8
14. Russell P (1980) Inflammatory lesions of the placenta. III. The histopathology of villitis of unknown aetiology. Placenta 1:227–244
15. Wallenburg HCS, Stolte LAM, Janssens J (1973) The pathogenesis of placental infarction. I. A morphologic study in the human placenta. Am J Obstet Gynecol 116:835–840
16. Beckman JS, Beckman TW, Chen J, et al. (1990) Apparent hydroxyl radical production from peroxynitrite: implications for endothelial injury by nitric oxide and superoxide. Proc Natl Acad Sci USA 87:1620–1624
17. Goda N, Suematsu M, Mukai M, et al. (1996) Modulation of mitochondrion-mediated oxidative stress by nitric oxide in human placental trophoblastic cells. Am J Physiol 271 (Heart Circ Physiol 40):H1893–H1899
18. Suematsu M, Tamatani T, Delano FA, et al. (1994) Microvascular oxidative stress preceding leukocyte activation elicted by in vivo nitric oxide suppression. Am J Physiol 266 (Heart Circ Physiol 35):H2410–H2415
19. Niu X, Smith CW, Kubes P (1994) Intracellular oxidative stress induced by nitric oxide synthesis inhibition increases endothelial cell adhesion to neutrophils. Circ Res 74:1133–1140
20. Cester N, Staffolani R, Rabini RA, et al. (1994) Pregnancy-induced hypertension: a role for peroxidation in microvillous plasma membranes. Mol Cell Biochem 131:151–155
21. Walsh SW, Wang Y (1993) Secretion of lipid peroxides by the human placenta. Am J Obstet Gynecol 169:1462–1466

# Liberation of Nitric Oxide from S-Nitrosothiols

Misato Kashiba-Iwatsuki, Keiko Kitoh, Manabu Nishikawa, Eisuke F. Sato, and Masayasu Inoue

*Summary.* Nitric oxide (NO) has a wide variety of functions. NO generates S-nitroso-thiols (RS-NO), which exhibit various activities attributable to NO. The metabolism of RS-NO, however, remains to be elucidated. RS-NO are considerably stable under physiological conditions. However, when incubated with rat liver homogenate, they rapidly disappeared from the mixture. Because such an effect was not observed with plasma, some compound(s) present in the liver might enhance the metabolism of RS-NO. Because levels of ascorbic acid and reduced glutathione (GSH) are significantly higher in cells than in plasma, effects of ascorbic acid, GSH, and other compounds were tested on the stability of RS-NO. S-Nitrosoglutathione (GS-NO) decomposed very slowly, which was accelerated by various compounds with reducing activity (ascorbic acid = cysteine > GSH). Because the cellular levels of ascorbic acid are fairly high, this compound might be an important modulator of the metabolism of RS-NO. Both NO and RS-NO reversibly inhibited the respiration of ascites tumor cells (NO > RS-NO). The inhibitory effect of RS-NO on the respiration was enhanced strongly by ascorbic acid, suggesting that ascorbic acid releases NO from RS-NO. These results suggested that ascorbic acid, cysteine, and related thiols might play important roles as modulators for RS-NO metabolism and NO action.

*Key words.* Nitric oxide—S-Nitrosothiols—Ascorbic acid—Cysteine—Thiols

## Introduction

Nitric oxide (NO) has received much attention not only as an endothelium-derived relaxing factor (EDRF) [1] but also as a key molecule for a wide variety of biological events. One important aspect of the chemistry of NO is its high reactivity with various molecules such as heme iron, nonheme iron, oxygen, and superoxide anion. The reduced form of glutathione (GSH) and related thiol compounds also react with the oxides of nitrogen and reversibly form S-nitrosothiols (RS-NO). Although the

Department of Biochemistry, Osaka City University Medical School, 1-4-54 Asahimachi, Abeno, Osaka 545, Japan lifetime of NO is short, RS-NO are fairly stable in vitro. For example, under physiological conditions, the in vitro half-lives of S-nitrosoglutathione (GS-NO) and S-nitrosoalbumin are longer than 10h. The in vitro half-life of S-nitrosocysteine (Cys-NO) is about 90 min. GS-NO, Cys-NO, and S-nitrosoalbumin have been reported to possess EDRF-like activity [2,3]. Stamler et al. [4] reported that plasma and bronchial lavage fluid from healthy humans contained about 7 and 0.3 μM of RS-No and GS-NO, respectively. However, the stability and fate of RS-NO in vivo remain to be elucidated.

Because the rate of NO release from GS-NO and S-nitroso-N-acetylpenicillamine (SNAP) is enhanced by platelet lysate and vascular smooth muscles [5], their constituent(s) have been postulated to affect the metabolism of RS-NO. $Cu^+$ has been shown to decompose RS-NO [6] by some mechanism that is enhanced by reducing compounds such as GSH [7]. However, the concentrations of free copper ion in plasma and tissues are extremely low. RS-NO is also decomposed by cysteine, GSH, and ascorbic acid [5,8–11]. Because NO reacts with molecular oxygen, the biological effects of NO and RS-NO would be affected by the local concentration of oxygen. In fact, recent studies revealed that NO and related metabolites inhibited respiration in mitochondria and in ascites tumor cells, particularly under low oxygen tension [12]. Although physiological concentrations of oxygen in and around cells are fairly low (~40μM), most in vitro experiments have been carried out under air atmospheric conditions in which oxygen tension is extremely high (~220μM). Thus, the in vivo fate of NO and RS-NO should be determined under physiological concentrations of reducing compounds, metals, and oxygen. However, only a limited information is available for the fate of RS-NO in the presence of these molecules.

This work describes the mechanism for the decomposition of RS-NO by reducing compounds and the effect of oxygen tension. The effect of these reducing agents on the RS-NO-suppressed respiration of ascites tumor cells is also reported.

# Experimental Procedures

## Chemicals

GSH, ascorbic acid, cysteine, N-acetylcysteine (NAC), and sodium nitrite were obtained from Wako (Osaka, Japan). Diethylenetriamine-N,N,N',N'',N''-pentaacetic acid (DETAPAC), bathocuproine disulfonic acid, and ethylenediaminetetraacetic acid (EDTA) were obtained from Dojin Chemical (Kumamoto, Japan). Other chemicals used were of the highest grade commercially available.

## Synthesis of RS-NO

RS-NO was synthesized as reported by Saville [13].

## Preparation of Rat Liver Homogenate

Under ether anesthesia, rats were exsanguinated and the liver was perfused with ice-cold saline. The liver was excised and homogenized in 5 vol of ice-cold 10 mM Tris-HCl buffer (pH 7.4) containing 0.1 M mannitol and 10 mM $MgCl_2$. After centrifugation at $10\,000 \times g$ for 30 min, the supernatant fraction was obtained. The pellet was washed twice with the mannitol solution by repeated centrifugation. To eliminate macromolecules with molecular weight larger than 10000, the supernatant fraction was subjected to ultrafiltration using a centricut filter (Kurabou, Osaka, Japan). Proteins were eliminated in the supernatant by treating with 5% trichloroacetic acid (TCA), and the acid-soluble fraction was neutralized by NaOH. The four fractions thus obtained were used for experiments after adding 20 mM EDTA.

## Preparation of Human Plasma Samples

Human blood was obtained from healthy donors. Blood samples were collected in succinate-containing tubes. After centrifugation of the blood, fresh plasma samples were obtained and used for experiments.

## Preparation of SDS Micelle

α-Tocopherol micelles were prepared by vigorously mixing appropriate amounts of α-tocopherol and 0.5 M sodium dodecyl sulfate (SDS) solution for 2 min.

## Preparation of Tumor Cells

Ehrlich ascites tumor cells (EATC) were supplied by the Japanese Cancer Research Resources Bank (Tokyo) and inoculated in the peritoneal cavity of male DDY mice. After 7–9 days of inoculation, EATC-containing ascites was obtained. After washing three times by centrifugation in calcium-free Krebs-Ringer phosphate buffer (KRP), EATC were resuspended in calcium-free KRP at $1 \times 10^8$ cells/ml.

## Analysis

Oxygen consumption was measured polarographically at 37°C using a Clark-type oxygen electrode fitted to a 2-ml water-jacketed closed chamber. This system was kept at 37°C and equipped with a magnetic stirrer. EATC were used at a final concentration of $1 \times 10^7$/ml in KRP containing 1 mM calcium. Concentration of GS-NO was determined spectrophotometrically ($\varepsilon_{545} = 13.0\,M^{-1}cm^{-1}$) and by the method of Saville [13]. NO was measured by a NO monitor (NO-501, Inter Medical).

# Results

## Effect of Liver Homogenate and Plasma on the Stability of GS-NO

Because GS-NO is fairly stable in vitro, it disappeared from the incubation mixture at a half-life longer than 10 h. As shown in Fig. 1, in the presence of either plasma or the particulate fraction of the liver, the rate of GS-NO disappearance was enhanced only slightly. However, the supernatant fraction markedly enhanced the disappearance of GS-NO in the mixture. The enhancing effect of the supernatant was fully replaced by the filtered fraction of the supernatant. This activity also remained unaffected after removing proteins by treating the supernatant with TCA. These results suggested that

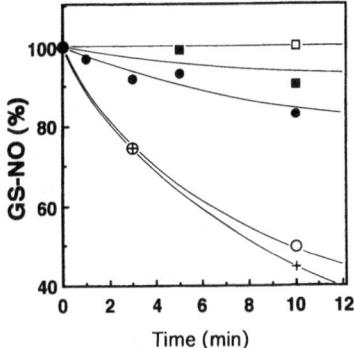

Fɪɢ. 1. Effect of plasma and liver fractions on the fate of *GS-NO*. Incubation mixtures contained, in a final volume of 1 ml of 50 μM GS-NO and 20 mM EDTA, either 900 μl of plasma (*closed squares*), the particulate fraction of rat liver homogenate (12 mg protein/ml) (*closed circles*), supernatant (10 mg protein/ml) (*open circle*), or an equal volume of the filtered fraction of the supernatant (+). During incubation at 37°C in air atmospheric conditions, changes in the concentration of GS-NO were determined. A control experiment was performed in the absence of plasma and liver fractions (*open square*)

low molecular weight compound(s) in the supernatant enhanced the disappearance of GS-NO. The contribution of proteins and other macromolecules in the decomposition of GS-NO might be small in the liver.

## Effect of Various Compounds on the Stability of GS-NO

The reactivity of GS-NO with various reducers was also studied. In the presence of either ascorbic acid or cysteine, the concentration of GS-NO in the mixture rapidly decreased. GSH also decreased the concentration of GS-NO although its activity was significantly lower than those of cysteine and ascorbic acid. α-Tocopherol-containing micelles did not affect the rate of GS-NO disappearance. The enhancing effect of ascorbic acid, cysteine, and GSH increased in a dose-dependent manner. It should be noted that plasma concentrations of ascorbic acid, GSH, and cysteine are approximately 80, 25, and 1 μM, respectively. In contrast, intracellular concentrations of ascorbic acid, GSH, and cysteine are 1–2 mM, 1–10 mM, and ~50 μM, respectively. Although these reducers also decompose GS-NO under anaerobic conditions, their enhancing effects are almost the same as those in air under atmospheric conditions. Hence, oxygen concentration might not affect the rate of GS-NO disappearance in the presence of either ascorbic acid, cysteine, or GSH.

The metabolite(s) formed by the reaction between GS-NO and ascorbic acid was also studied by using an NO analyzer. When ascorbic acid was mixed with GS-NO, the steady-state levels of NO rapidly increased. Within 15 min, the NO concentration reached a steady-state level, and this level remained for a fairly long time. This result indicated that NO was formed by the reaction between GS-NO and ascorbic acid.

## Effect of Metals on the Decomposition of GS-NO

The mechanism for the decomposition of GS-NO by these reducing compounds may involve contaminated metals. To test this possibility, the effect of chelating agents on the rate of GS-NO decomposition was tested. However, neither EDTA, diethylene triamino pentoacetic acid (DTPA), nor bathocuproine affected the rate of decomposition in the presence of three reducing agents. These observations suggested that the enhancing effect of the reducing compounds did not require the presence of metals. However, the presence of $10\,\mu M$ $CuCl_2$ increased the rate of decomposition of GS-NO either in the presence or absence of ascorbic acid and cysteine.

## Effect of Ascorbic Acid on RS-NO Decomposition

Concentrations of ascorbic acid in plasma ($\sim 80\,\mu M$) and cells ($\sim 5\,mM$) are significantly higher than those of cysteine. Furthermore, the reactivity of GS-NO with ascorbic acid was significantly higher than with GSH when the effect of ascorbic acid on the decomposition of RS-NO was studied with different nitrosothiols. Ascorbic acid also enhanced the decomposition of Cys-NO and NAC-NO.

## Effect of Reducing Agents on Cellular Respiration Inhibited by GS-NO

EATC rapidly consume oxygen without adding any respiratory substrate. Previous reports revealed that the respiration of tumor cells was transiently inhibited by NO in an oxygen-dependent manner. GS-NO also inhibited the respiration of EATC, particularly at low oxygen tension. The presence of either ascorbic acid or cysteine significantly enhanced the inhibitory effect of GS-NO. GSH also enhanced the inhibitory effect of GS-NO although its efficacy was lower than that of ascorbic acid and cysteine. Cys-NO also inhibited the respiration of EATC; the inhibitory effect of Cys-NO was stronger than that of GS-NO.

# Discussion

The present work demonstrates that small molecular weight compounds in the liver supernatant enhanced the decomposition of RS-NO. The liver homogenate used in the experiments contained about 200, 1, and $300\,\mu M$ of ascorbic acid, cysteine, and GSH, respectively. The present study also reports that ascorbic acid, cysteine, and GSH decomposed RS-NO (ascorbic acid > cysteine >> GSH). Considering the concentration of the three reducing compounds in the homogenates and their reactivities, ascorbic acid might be principally responsible for the enhanced decomposition of GS-NO. Ascorbate decomposed not only GS-NO but also Cys-NO and NAC-NO, suggesting that ascorbic acid generally releases NO from RS-NO. The experiments using a NO electrode revealed that NO was released as a reaction product on interaction of GS-NO and ascorbic acid.

Because the reducing agents enhanced the decomposition of RS-NO even in the presence of chelating agents, these compounds might directly react with RS-NO. The presence of trace amounts of copper ion enhanced the effects of ascorbic acid and cysteine, which suggested the synergistic effect of reducers and copper.

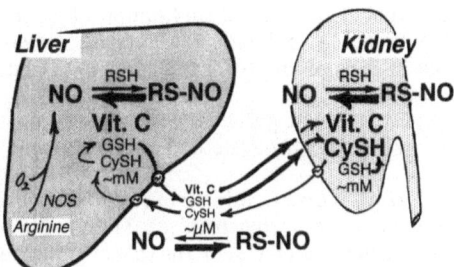

FIG. 2. Hypothetical scheme showing the metabolism of S-nitrosothiol in vivo. *NOS*, nitric oxide synthase; *RS-NO*, S-nitrosothiols; *GSH*, reduced glutathione; *CySH*, cysteine

NO inhibits the respiration of mitochondria and tumor cells [12]. RS-NO, such as GS-NO and Cys-NO, also inhibited the respiration of tumor cells by a mechanism that was enhanced by ascorbic acid and cysteine. This observation suggested that the reducing agents enhanced the inhibitory effect of RS-NO by releasing NO. It should be noted that the reactivity of ascorbic acid with GS-NO is higher than that of cysteine, while the enhancing effect of cysteine to inhibit that of tumor cell respiration was higher than that of ascorbic acid.

The metabolism of GSH occurs via intraorgan cycles including secretion of the tripeptide to the circulation and bile, degradation in the kidney and bile by γ-glutamyltransferase, and reabsorption of the constituent amino acid. Thus, local concentration of cysteine in the kidney and bile would be fairly high. Preliminary experiments in this laboratory revealed that GS-NO was rapidly decomposed in the kidney. This might reflect that a high concentration of renal cysteine enhanced the degradation of GS-NO. Because bile is the major route for the excretion of copper ion, efficient decomposition of RS-NO would occur in the bile. Stabilities of RS-NO in these organs should be studied further.

Thus, local concentrations of reducing agents, such as ascorbic acid, cysteine, and metals, might determine the fate of RS-NO and its activity in vivo (Fig. 2).

# References

1. Furchgott RF, Zawadzki JN (1980) The obligatory role of endothelial cells in the relaxation of arterial smooth muscle by acetylcholine. Nature 288:373–573
2. Keaney JF, Simon DI, Stamler JS, et al. (1993) NO forms an adduct with serum albumin that has endothelium-derived relaxing factor-like properties. J Clin Invest 91:1581–1589
3. Myers PR, Minor RL, Guerra R, et al. (1990) Vasorelaxant properties of the endothelium-derived relaxing factor more closely resemble S-nitrosocysteine than nitric oxide. Nature 345:161–163
4. Stamler JS, Simon DI, Osborne JA, et al. (1992) S-Nitrosylation of proteins with nitric oxide: synthesis and characterization of biologically active compounds. Proc Natl Acad Sci USA 89:444–448
5. Radomski MW, Rees DD, Dutra A, Moncada S (1992) S-Nitroso-glutathione inhibits platelet activation in vitro and in vivo. Br J Pharmacol 107:745–749

6. Macaninly J, Williams DLH, Askew SC, et al. (1993) Metal ion catalysis in nitrosothiol (RSNO) decomposition. J Chem Soc Chem Commun 1758–1759
7. Gorren ACF, Schrammel A, Schmidt K, Mayer B (1996) Decomposition of S-nitrosoglutathione in the presence of copper ions and glutathione. Arch Biochem Biophys 330:219–228
8. Ignarro L, Gruetter CA (1980) Requirement of thiols for activation of coronary arterial guanylate cyclase by glyceryl trinitrate and sodium nitrite. Biochim Biophys Acta 631:221–231
9. Pietraforte D, Mallozzi C, Scorza G, Minetti M (1995) Role of thiols in the targeting of S-nitrosothiols to red blood cells. Biochemistry 34:7177–7185
10. Hogg N, Singh RJ, Kalynaraman B (1996) The role of glutathione in the transport and catabolism of nitric oxide. FEBS Lett 382:223–228
11. Kashiba-Iwatsuki M, Yamaguchi M, Inoue M (1995) Role of ascorbic acid in the metabolism of S-nitroso-glutathione. FEBS Lett 389:149–152
12. Takehara Y, Kanno T, Yoshioka T, et al. (1995) Oxygen-dependent regulation of mitochondrial energy metabolism by nitric oxide. Arch Biochem Biophys 323:27–32
13. Saville B (1958) A scheme for the colorimetric determination of microgram amounts of thiols. Analyst 83:670–672

# Change of Nitric Oxide Concentration in Exhaled Gas After Lung Resection

HIROHISA HORINOUCHI, MITSUTOMO KOHNO, MASATOSHI GIKA,
ATSUSHI TAJIMA, KATSUYUKI KUWABARA, AKIRA YOSHIZU,
MASAO NARUKE, YOTARO IZUMI, MASAFUMI KAWAMURA,
KOJI KIKUCHI, and KOICHI KOBAYASHI

*Summary.* Nitric oxide (NO) has been shown to have a diversity of biological functions. In patients who undergo lung resection such as lobectomy and pneumonectomy, the vascular bed and its ventilation area are reduced. Few patients, however, develop pulmonary hypertension. Endogenous NO may play a role in stabilizing pulmonary circulation after major lung resection. To evaluate endogenous NO production, NO in exhaled gas was analyzed. From August to December 1996, eight patients with a mean age of 62 years who underwent lung resection (seven lobectomies and one pneumonectomy) were examined. NO concentration in their exhaled gas and minute ventilation volume was measured once in the preoperative period, on postoperative days 1 and 3, and once 1 week after surgery. Patients breathed pure oxygen, which contained no NO, through a mouthpiece; exhaled gas was introduced directly to the NO analyzer. NO concentration was analyzed continuously by the chemiluminescence method and recorded. Mean concentration of exhaled NO and total NO in exhaled gas in 1 min were calculated. For minute ventilation volume, there was a significant increase between preoperative examination and postoperative day 1. A significant increase in NO concentration in exhaled gas was observed at 1 day after lung resection. To standardize the change of NO concentration and total exhalation volume, the ratio compared to preoperative data was evaluated. The ratio of NO concentration compared to preoperative data changed 191% ± 109%, 296% ± 152%, and 93.7% ± 65.0% at postoperative days 1 and 3 and at 1 week postoperatively, respectively. The ratio of NO production changed 230% ± 119%, 363% ± 205%, and 119% ± 86.2% at postoperative days 1 and 3 and 1 week, respectively. There was a significant difference in NO concentration between preoperative data and postoperative day 3, and significant differences in total NO volume exhaled in 1 min between preoperative data and postoperative days 1 and 3. In conclusion, we showed that after lung resection, NO in the exhaled gas increases in concentration and total volume in unit time, which might be an important factor that prevents alteration of pulmonary circulation. Further careful study is necessary to clarify the mechanism and role of increased NO in exhaled gas after lung resection.

Department of Surgery, School of Medicine, Keio University, 35 Shinanomachi, Shinjuku-ku, Tokyo 160, Japan

*Key words.* Nitric oxide—Lung resection—Exhaled gas—Spontaneous breathing

## Introduction

Nitric oxide (NO) has been shown to have a diversity of biological functions [1–4]. Especially in the lung, this molecule is associated with not only vasodilatation but also with inflammation [1,5]. NO has been revealed to be a strong vasodilator when used as an inhalation agent [3]. In asthmatic patients, however, NO is considered to be an indicator of inflammatory change of the lower respiratory tract [6]. In patients who undergo lung resection such as lobectomy and pneumonectomy, the vascular bed and ventilation area are reduced. However, after lung resection, few patients develop pulmonary hypertension [7].

We hypothesize that endogenous NO may play a role in stabilizing pulmonary circulation after major lung resection. To evaluate endogenous NO production, NO in the exhaled gas was analyzed.

## Patients

From August to December 1996, eight patients who underwent lung resection (seven lobectomies and one pneumonectomy) were enrolled in this study. Informed consent to take part in this study was obtained from all patients. The mean age of these patients was 62 years (47–76 years). All patients showed no inflammatory process before surgery, with the exception of one patient who had right middle lobe atelectasis from lung carcinoma. No patient had a history of cardiac disease, asthma, or other disease except their current lung cancer (Table 1).

## Methods

NO concentration in exhaled gas was measured using an NO analyzer (NOA280, Sievers, Boulder, CO, USA) by the chemiluminescence method. Minute ventilation volume was measured with a respirometer (RM121, Ohmeda, Yamanashi, Japan). Patients were placed in a sitting position and breathed pure oxygen that contained no NO through a mouthpiece from a reservoir bag. To avoid nasal inhalation and to

TABLE 1. Characteristics of patients enrolled in study

| No. | Age (years) | Procedure |
| --- | --- | --- |
| 1 | 61 | Left lower lobectomy |
| 2 | 67 | Left upper lobectomy |
| 3 | 76 | Right middle lobectomy |
| 4 | 73 | Left lower lobectomy |
| 5 | 66 | Left pneumonectomy |
| 6 | 55 | Left lower lobectomy |
| 7 | 47 | Right upper sleeve lobectomy |
| 8 | 61 | Left lower lobectomy |

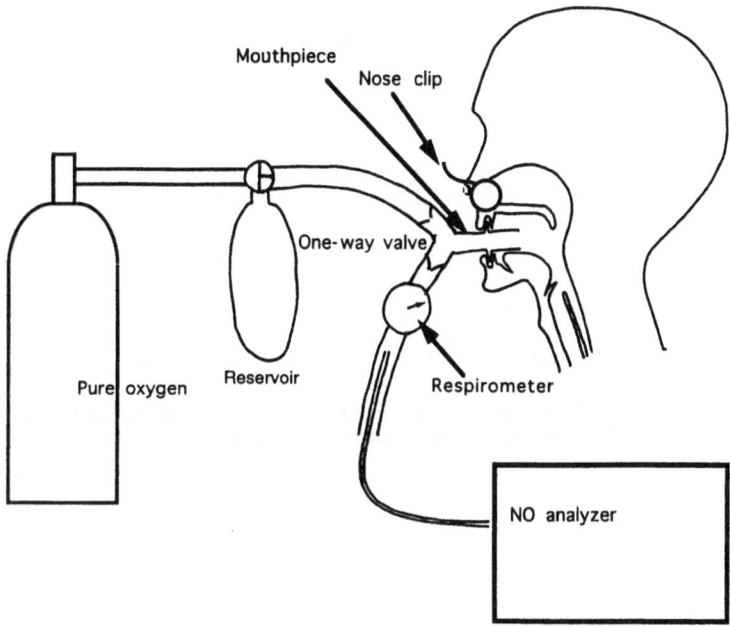

FIG. 1. Schematic depiction of measurement system. *NO*, nitric oxide

eliminate the influence of NO production in the nasal cavity, a nose clip was applied. Exhaled gas was introduced directly to the NO analyzer. NO concentration was analyzed continuously and recorded (Fig. 1).

Patients breathed oxygen freely for 1 min to minimize the influence of ambient air and after several deep breaths, patients continued breathing normally. NO concentration was recorded while the patient breathed pure oxygen. Mean concentration of exhaled NO and total NO in exhaled gas in 1 min were calculated.

Measurements were carried out preoperatively, on postoperative days 1 and 3, and once 1 week after the operation (examination was done on either the sixth, seventh, or eighth postoperative day). Because of the patient's condition, in three cases data collection was incomplete. Differences at each time point were tested by Student's paired *t*-test.

## Results

There was no operative mortality. No patients had major postoperative complications.

Because of difference of body mass and lung function, minute ventilation volume varied from 4.0 to 14.2 l/min (7.64 ± 6.49 l/min) at preoperative examination, from 4.9 to 14.2 l/min (7.49 ± 3.36 l/min) at postoperative day 1, from 4.5 to 10.0 l/min (6.40 ± 2.15 l/min) at postoperative day 3, and from 4.80 to 14.0 l/min (8.33 ± 3.75 l/min) at 1

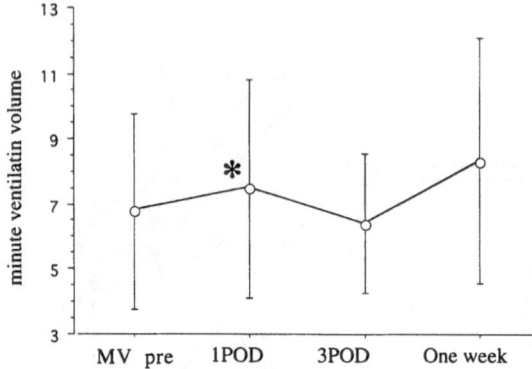

FIG. 2. Change of minute ventilation (*MV*) volume. *, *P* < .05 compared to preoperative value (*pre*). *1POD*, 1 day after operation; *3POD*, 3 days after operation

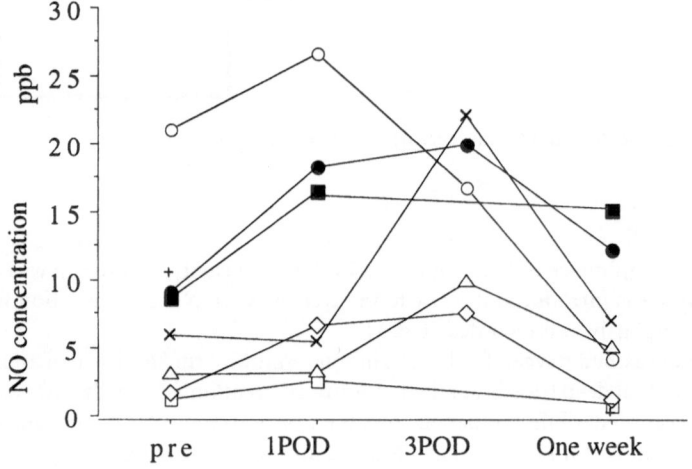

FIG. 3. Change of NO concentration in exhaled gas. Cases 1–8 are shown by *diamonds, triangles, open circles, solid squares, solid circles, plus signs, open squares,* and *multiplication* signs, respectively

week postoperative. There was a significant difference between the data at preoperative examination and postoperative day 1 (*P* = .048). (Fig. 2).

NO concentration varied from 1.6 to 21.1 ppb (7.639 ± 6.492 ppb) at preoperative examination, from 2.52 to 26.6 ppb (11.32 ± 9.23 ppb) at postoperative day 1, from 7.6 to 22.1 ppb (15.28 ± 6.31 ppb) at postoperative day 3, and from 0.10 to 15.1 ppb (8.33 ± 3.8 ppb) at 1 week postoperative. There was also a significant difference between the data at preoperative examination and on postoperative day 1 (*P* = .03) (Fig. 3).

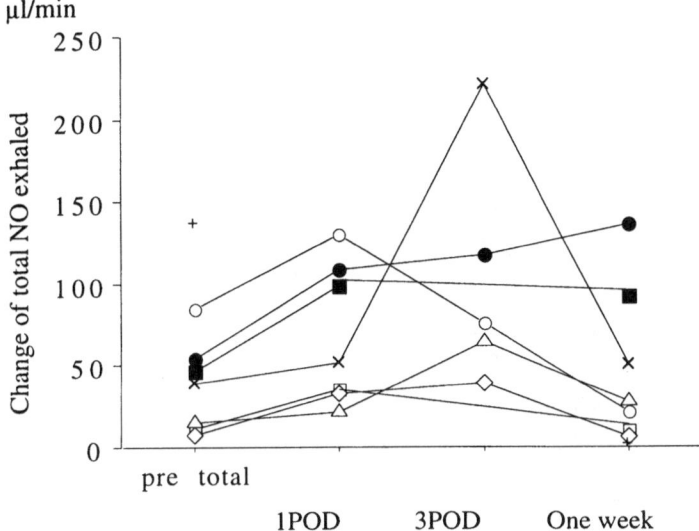

FIG. 4. Change in total NO exhaled in 1 min

Total NO in exhaled gas varied from 7.20 to 136.5 μl/min (49.52 ± 43.60 μl/min) at preoperative examination, from 21.8 to 130.3 μl/min (68.6 ± 43.1 μl/min) at postoperative day 1, from 38.8 to 221.0 μl/min (103.5 ± 71.6 μl/min) at postoperative day 3, and from 1.29 to 136.53 μl/min (43.2 ± 48.0 μl/min) at 1 week postoperative. There was a significant difference between the data at preoperative examination and at postoperative day 1 ($P = .049$) (Fig. 4).

To standardize the change of NO concentration and total exhalation volume, the ratio compared to preoperative data was evaluated. The ratio of NO concentration compared to preoperative data changed 191% ± 109%, 296% ± 152%, and 93.7% ± 65.0% at postoperative days 1 and 3 and at 1 week, respectively. The ratio of NO production changed 230% ± 119%, 363% ± 205%, and 119% ± 86.2% at postoperative days 1 and 3 and at 1 week, respectively. There were significant differences in NO concentration between preoperative data and postoperative day 3 ($P = .0447$) and between postoperative day 3 and 1 week after the operation ($P = .0363$). There were significant differences in total NO volume exhaled in 1 min between preoperative data and postoperative day 1 ($P = .0276$) and day 3 ($P = .0454$) (Fig. 5a,b).

## Discussion

In this study we showed the increase of NO concentration in exhaled gas after lung resection, as well as the increase of total NO volume exhaled in 1 min. This fact indicates that after lobectomy endogenous NO is induced and is produced at a higher level than in the preoperative state.

Kharitonov and colleagues [6] investigated NO exhaled from normal volunteers and asthma patients. They showed that the main component of NO in the exhaled gas

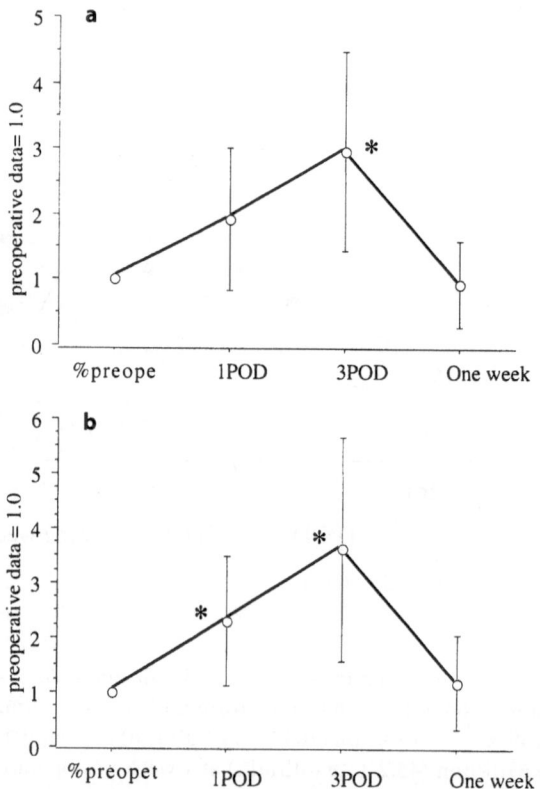

FIG. 5. **a** Change of ratio of NO concentration in exhaled gas *, $P < .05$ compared to preoperative value. **b** Change of ratio of total NO exhaled in 1 min. *, $P < .05$ compared to preoperative value

is derived from the lower respiratory tract. This results suggested that exhaled NO could influence vascular tone in pulmonary microvasculature.

Meanwhile Okamoto et al. [7] showed that in lobectomized patients, pulmonary arterial pressure (PAP) and pulmonary arterial resistance (PAR) are not significantly altered at resting state on the second postoperative day, although PAP and PAR rise considerably when exercise is added in the third postoperative week. They concluded that right ventricular dysfunction at rest compensates for the increase in right ventricular volume. They also speculated that the right ventricle may play an important role serving as a "reservoir" for increased afterload. Although we did not measure PAP and PVR in this series, the sequelae of those indices might be the same in our series. Therefore from our observation, increased NO could be one major factor that regulates the pulmonary vascular tone after lung resection.

On the other hand, the surgical procedure was performed under one-lung ventilation. Williams and his colleagues [8] described that one-lung anesthesia is thought to cause acute lung injury in the same manner as ischemia-reperfusion injury. In five cases of our series, NO concentration returned to preoperative level 1 week after operation. All these patients underwent lobectomy under one-lung ventilation. From the standpoint of lung injury, increase of NO production in the lung might be the result of the reperfusion effect. However, in one patient who underwent left pneumonectomy, total NO exhalation in 1 min rose 199% at postoperative day 1 and stayed high throughout this study. In this patient, right lung perfusion was maintained during surgery. Therefore the increase of total NO exhalation that is attributable to the right lung in this patient seemed an important factor to normalize pulmonary perfusion.

In conclusion, we showed that after lung resection, NO in exhaled gas increases in concentration and total volume in unit time, which might be an important factor to prevent alteration of pulmonary circulation. However, there are many humoral factors that are thought to regulate pulmonary artery pressure, such as angiotensin II, endogenous cathecholamins, prostaglandins, and altered fluid distribution. Further careful study is necessary to clarify the mechanism and role of increased NO in exhaled gas after lung resection.

# References

1. Ignarro LJ, Burga GM, Woods KS, et al. (1987) Endothelium-derived relaxing factor produced and released from artery and vein is nitric oxide. Proc Natl Acad Sci USA 84:9265–9269
2. Massaro AF, Gaston B, Kita D, et al. (1955) Expired nitric oxide levels during treatment of acute asthma. Am J Respir Crit Care Med 152:800–803
3. Roberts JD Jr, Polaner DM, Lang P, Zapol WM (1992) Inhaled nitric oxide in persistent pulmonary hypertention of the newborn. Lancet 340:818–819
4. Malinski T, Bailey F, Zang ZG, Chopp M (1993) Nitric oxide measures by porphyrinic microsensor in rat brain after transient middle cerebral artery occlusion. J Cereb Blood Flow Metab 13:355–358
5. Barnes PJ, Belvisi MG (1993) Nitric oxide and lung disease. Thorax 48:1034–1043
6. Kharitonov SA, Chung KF, Evans D, et al. (1996) Increased exhaled nitric oxide in asthma is mainly derived from the lower respiratory tract. Am J Respir Crit Care Med 153:1773–1780
7. Okada M, Ota T, Okada N (1994) Right ventricular dysfunction after major pulmonary resection. J Thorac Cardiovasc Surg 108: 503–511
8. Williams EA, Evans TW, Goldstraw P (1996) Acute lung injury following lung resection: is one-lung anaesthesia to blame? Thorax 51:114–116

# Induction of NADPH Cytochrome P-450 Reductase in Kupffer Cells After Chronic Ethanol Consumption Associated with Increase of Superoxide Anion Formation

Hirokazu Yokoyama, Yasutada Akiba, Masahiko Fukuda, Yukishige Okamura, Takeshi Mizukami, Michinaga Matsumoto, Hidekazu Suzuki, and Hiromasa Ishii

*Summary.* Using native polyacrylamide gel electrophoresis (PAGE) and diaphorase staining, NADPH cytochrome $c$ P-450 reductase was examined in the Kupffer cells of rats fed ethanol chronically as well as in controls. Formation of superoxide anion released from Kupffer cells into the hepatic sinusoid was estimated using the cytochrome $c$ method, which was applied to a liver perfusion model in both groups after an additional acute ethanol challenge. Kupffer cells were found to carry NADPH cytochrome $c$ P-450 reductase, and chronic ethanol consumption resulted in its induction being doubled. This change was associated with increase of superoxide anion release from Kupffer cells into the hepatic sinusoid after an acute ethanol challenge ($0.020 \pm 0.03$ O.D./g liver versus $0.012 \pm 0.002$ O.D./g liver; $P < .05$). In conclusion, release of superoxide anion from Kupffer cells into the hepatic sinusoid increases in rats chronically fed ethanol. Induction of NADPH reductase in Kupffer cells caused by chronic ethanol consumption may, at least in part, be involved in this mechanism.

*Key words.* NADPH cytochrome P-450 reductase—Superoxide anion—Kupffer cells—Chronic ethanol consumption

## Introduction

The microsomal ethanol-oxidizing system (MEOS) is a complex of several enzymes that participate in ethanol metabolism. In MEOS, cytochrome P-450 2E1 has been known to be one of the major enzymes for ethanol oxidation [1]. The current consensus is that various free radicals, including superoxide anion, are produced during ethanol oxidation via P-450 2E1 [2].

---

Department of Internal Medicine, School of Medicine, Keio University, 35 Shinanomachi Shinjuku-ku, Tokyo 160, Japan

MEOS requires a supply of electrons from cytochrome NADPH reductase when it functions [3]. Also, the superoxide anion has been shown to be derived from NADPH reductase when it transfers electrons from NADPH to various enzymes of the cytochrome system [4]. Thus, NADPH reductase is supposed to be an actual major source of superoxide anions during ethanol oxidation. Moreover, it was shown that chronic ethanol consumption resulted in induction of NADPH reductase in liver [5].

Recent studies have demonstrated that Kupffer cells carry ethanol-inducible cytochrome P-450 2E1, which participates in ethanol oxidation [6]. Therefore, it is conceivable that Kupffer cells carry NADPH reductase that is induced by chronic ethanol consumption. Furthermore, it is supposed that formation of superoxide anion during ethanol oxidation increases when NADPH reductase is induced.

In this chapter, our recent knowledge of NADPH reductase of Kupffer cells and the effect of its induction on formation of superoxide anion from Kupffer cells is summarized.

# NADPH Reductase in Kupffer Cells and Its Induction After Chronic Ethanol Consumption

Wistar male rats weighing about 150 g were chronically fed ethanol or were pair-fed without ethanol treatment for 4 weeks according to Lieber's method [7]. After this feeding, the livers were removed from the animals and the Kupffer cells were isolated with the metrizamide (Sigma, St. Louis, MO, USA) centrifugation equilibrium method. After sonication, protein concentration of each fraction was estimated by Lowry's method [8] using a kit purchase from Sigma.

Details of the method to detect NADPH reductase have appeared elsewhere [9]. Briefly, 50 μg of each sonicated sample was applied to native non-sodium dodecyl sulfate (non-SDS) polyacrylamide gel electrophoresis (PAGE) (% T = 7.5). After electrophoresis, the gel was dipped in diaphorase staining buffer, 0.1 mM Tris pH = 8.0 containing 1 mM β-NADPH and 0.5 mM nitroblue tetrazolium (Sigma).

The staining intensity of the band representing NADPH reductase was measured using a image scanner JX-250 (Sharp, Tokyo, Japan) and a Power Macintosh 8500/132 (Apple Computer, Cupertino, CA, USA) with software Photoshop (Adobe System, Mountain View, CA, USA). The band intensity was expressed as counts/pixel.

The band representing NADPH reductase was found in Kupffer cells, indicating that Kupffer cells carry NADPH reductase (data not shown). Because it has been already demonstrated that Kupffer cells can oxidize ethanol via their P-450 2E1 [6], it is not surprising that Kupffer cells carry NADPH reductase, which is essential for the functioning of P-450 2E1.

The intensity of the band as estimated by an image scanning system that represents the amount of NADPH reductase in Kupffer cells was significantly increased (twofold) in rats chronically fed ethanol when compared to controls (11.2 ± 0.1 versus 5.2 ± 0.2 counts/pixel; $P < .005$). This finding is consistent with the previous finding that NADPH reductase is induced in the liver of rats chronically fed ethanol [4].

## Superoxide Anion Release from Kupffer Cells in Rats Chronically Fed Ethanol

After the chronic ethanol feeding, rats were given continuous ethanol infusion; that is, ethanol was injected intravenously as a bolus at a dose of 1.75 g/kg body weight followed by continuous infusion of $250 \, mg \, kg^{-1} h^{-1}$ for 90 min. Then, the portal vein was cannulated with a 18-gauge Teflon catheter and the liver was perfused with Hanks' balanced salt solution (HBSS) (GIBCO BRL, Grand Island, NY, USA) at a rate of 2–3 ml/g wet liver. After blood was removed from the liver, HBSS containing $50 \, \mu M$ of oxidized ferricytochrome $c$ (Wakojunyaku, Tokyo, Japan) was perfused from the portal vein, and perfusate was collected from the inferior vena cava every 2 min. The change in absorbance ($\Delta ABS$) was measured at 550 nm, reflecting the formation of reduced ferricytochrome $c$ that is converted from the oxidized ferricytochrome $c$ in the presence of superoxide anion.

The formation of superoxide anion derived from Kupffer cells after an acute ethanol challenge significantly increased in rats chronically fed ethanol when compared to controls ($0.020 \pm 0.03$ O.D./g liver versus $0.012 \pm 0.002$ O.D./g liver; $\Delta ABS$ at $P < .05$). In that NADPH reductase is one of the acutal sources of superoxide anion during ethanol oxidation [3], the current finding that induction of NADPH reductase was associated with an increase of superoxide anion in Kupffer cells is also reasonable.

## Conclusion

Superoxide anion released from Kupffer cells increases in rats chronically fed ethanol. Induction of NADPH reductase in Kupffer cells caused by chronic ethanol consumption may be involved in this mechanism.

Because superoxide anions derived from Kupffer cells have been implicated in the pathogenesis of alcoholic liver disease, the induction of NADPH reductase by chronic ethanol consumption and the increase of superoxide anion formation may, at least in part, account for the mechanisms of progression of alcoholic liver disease.

## References

1. Lieber CS (1992) Medical and nutritional complication of alcoholism: mechanisms and management, 2nd edn. Plenum, New York
2. Wu D, Cederbaum AI (1996) Ethanol cytotoxicity to a transfected HepG2 cell line expressing human cytochrome P-450 2E1. J Biol Chem 271:23914–23919
3. Halliwell B, Gutteridge JMC (1989) Free radicals in biology and medicine, 2nd edn. Oxford University Press, Oxford
4. Lagrange P, Livertoux MH, Grassiot MC, Minn A (1994) Superoxide anion production during monoelectronic reduction of xenobiotics by preparations of rat brain cortex, microvessels, and choroid plexus. Free Radical Biol Med 17:355–359
5. Joly JG, Ishii H, Teshke R, et al. (1973) Effects of chronic ethanol feeding on the activities and submicrosomal distribution of nicotinamide adenine dinucleotide phosphate (NADPH)—cytochrome P-450 reductase and the demethylases for aminopyrine and ethylmorphine. Biochem Pharmacol 22:1532–1535

6. Koivisto T, Minshin VM, Kak KM, et al. (1996) Induction of cytochrome p-450 2E1 by ethanol in rat Kupffer cells. Alcohol Clin Exp Res 20:207–212
7. Lieber CS, Jones DP, DeCarli LM (1965) Effects of prolonged ethanol intake: production of fatty liver despite adequate diet. J Clin Invest 44:1009–1021
8. Lowry OH, Rosebrough NJ, Farr AL, Randall RJ (1951) Protein measurement with the folin phenol reagent. J Biol Chem 193:265–275
9. Yokoyama H, Akiba Y, Suzuki H, et al. (1996) Establishment of a new method to detect NADPH cytochrome P-450 reductase in rat liver by NADPH diaphorase staining after native polyacrylamide gel electrophoresis (PAGE): its induction by chronic ethanol consumption. Hepatology 24:440A (abstr)

# In Vivo Measurement of Superoxide in the Cerebral Cortex Utilizing *Cypridina luciferin* analog (MCLA) Chemiluminescence

Daisuke Uematsu[1], Yasuo Fukuuchi[2], Nobuo Araki[2], Shigeru Watanabe[2], Yoshiaki Itoh[2], and Keiji Yamaguchi[2]

*Summary.* Temproal profile of superoxide ($O_2^-$) generation following cerebral hypoxia and ischemia has been obscure, although it has been implicated in the progression of reperfusion injury. We have examined the time course of $O_2^-$ generation in the cat cortex following reversible hypoxia and forebrain incomplete ischomia. We used 20 cats anesthetized with halothane inhalation. A closed cranial window was placed over the exposed temporoparietal cortex, and *Cypridina luciferin* analog (MCLA), a chemiluminescence probe for the measurement of $O_2^-$, was dissolved in the artificial cerebrospinal fluid (ACSF) and superfused continuously throughout the experiments. We simultaneously monitored a reflectance (398 nm) from the cortex by means of an in vivo fluoromicroscope with two photomultiplier tubes. We subtracted a hemodynamic artifact from the MCLA chemiluminescence. Hypoxia was induced by pure $N_2$ inhalation for 1 min, and forebrain incomplete ischemia was induced by 30-min ligation of both common carotid arteries combined with hypotension (= 50 mmHg). As a result, the MCLA chemiluminescence increased following reoxygenation and reperfusion, indicating an enhanced $O_2^-$ generation. It was significantly reduced during hypoxia and severe ischemia, but fluctuated during mild ischemia. We speculate a breakdown of arachidonic acid and a biochemical interaction between endothelial cells and polymorphonuclear leukocytes as a source of $O_2^-$ generation.

*Key words.* Superoxide ($O_2^-$)—MCLA chemiluminescence—Brain—Hypoxia/ischemia—Reoxygenation/reperfusion

## Introduction

Excessive generation of oxygen-derived free radicals has been implicated in the pathogenesis of neuronal injury following cerebral ischemia and reperfusion [1,2], hypoxia and reoxygenation [3], and other pathological conditions such as neurode-

[1] Department of Neurology, Urawa Municipal Hospital, 2460 Mimuro, Urawa, Saitama 336, Japan
[2] Department of Neurology, School of Medicine, Keio University, 35 Shinanomachi, Shinjuku-ku, Tokyo 160, Japan

580

generative and Alzheimer diseases [4]. However, the exact temporal profile of oxidative stress following these insults has not been fully assessed in vivo. This study examined the time course of superoxide ($O_2^-$) generation in vivo in the cat cortex following hypoxic and ischemic insults by means of a newly developed *Cypridina luciferin* analog (MCLA) chemiluminescence technique [5].

## Materials and Methods

Studies were performed on 20 adult male cats anesthetized by 6% halothane inhalation. The respiration was controlled by a Harvard respirator. The skull was fixed in a stereotaxic head holder and a burr hole was made. The dura was incised carefully, and the pial surface was exposed. A quartz cranial window equipped with inlet/outlet tubes was placed in the burr hole and sealed with dental cement. Optical measurement was made from the cortex utilizing a photon-counting system with high-sensitivity bialkali photomultiplier tubes in the dark room. We utilized MCLA chemiluminescence to assess the level of oxidative stress represented by $O_2^-$. Artificial cerebrospinal fluid containing 100μM MCLA was superfused onto the pial surface continuously at a rate of 5 ml/h. Reversible hypoxic insult was induced by 100% $N_2$ inhalation for 60 s, which was followed by reoxygenation with air inhalation. In another series of experiments, forebrain incomplete ischemia was induced for 15 min by ligation of both carotid arteries in combination with hypotension (= 50 mmHg), which was followed by 1-h reperfusion. In these studies, we simultaneously monitored changes in cortical blood volume, which appreciably influences the MCLA photon counting, and corrected this hemodynamic artifact. In some animals, the cortical surface was serially superfused with solutions containing 100 mM MCLA plus (1) 500 u/ml superoxide dismutase (SOD), (2) 500 u/ml SOD + 5 mM β-carotene, and (3) 500 u/ml SOD + 5 mM β-carotene + 500 mM ($N^G$-monomethyl L-arginine (L-NAME)), each for 15 min.

## Results

Following MCLA superfusion, the chemiluminescence emitted from the cortical extracellular space increased gradually, reaching a steady level in 30 min (Fig. 1). Nitrogen inhalation caused a drastic decrease in the chemiluminescence, indicating a reduction of $O_2^-$ generation. Following reoxygenation, it rapidly recovered, exceeding the resting level (Fig. 2). The maximal reduction and overshoot of MCLA chemiluminescence during hypoxia and reoxygenation were −48% ± 27% and +19% ± 14%, respectively; both were statistically significant ($P < .05$).

Figure 3 shows a representative recording of the ischemia/reperfusion study. In this animal with severe ischemia, the MCLA chemiluminescence started to decrease immediately following ischemia and remained at a lower level. Following reperfusion, it recovered and then overshot, exceeding the resting level. In three animals with milder ischemia, it fluctuated and tended to increase even during the ischemic period. Superfusion with SOD suppressed the chemiluminescence by 32% (Fig. 4), whereas neither β-carotene nor L-NAME had any effect on chemiluminescence. Thus, we assume that the chemiluminescence measured in this model reflects mainly changes in $O_2^-$.

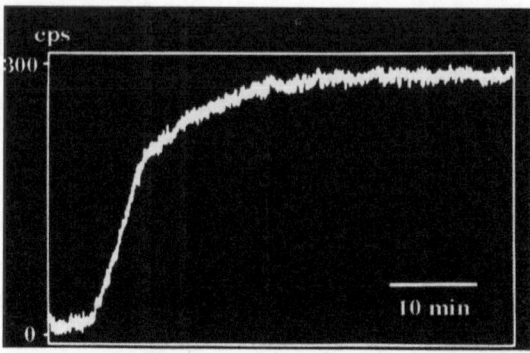

FIG. 1. Time course of cortical chemiluminescence following *Cypridina luciferin* analog (MCLA) superfusion. *cps*, counts per second

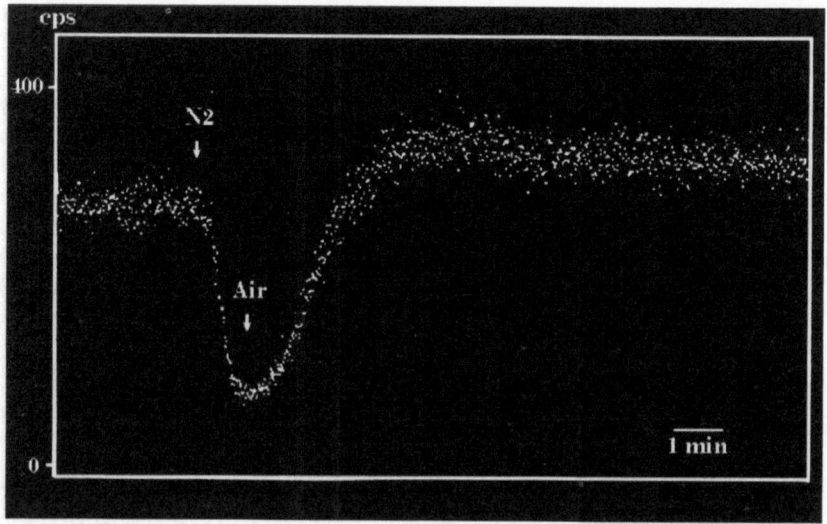

FIG. 2. Changes in MCLA chemiluminescence during hypoxia and reoxygenation

## Discussion

We have succeeded in monitoring the level of superoxide generation *in vivo* from the cerebral cortex. We demonstrated that the cortical superoxide level is suppressed during severe hypoxia and ischemia and enhanced following reoxygenation and the reperfusion phase. However, in milder ischemia the cortical $O_2^-$ tended to fluctuate and even increase transiently during the ischemic period. We presume that focal $O_2^-$ level should be a function of (1) $O_2$ supply, (2) severity of tissue injury, and (3) activity

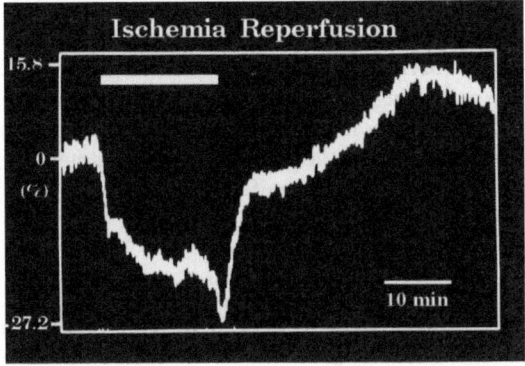

FIG. 3. Changes in MCLA chemiluminescence during severe ischemia and reperfusion

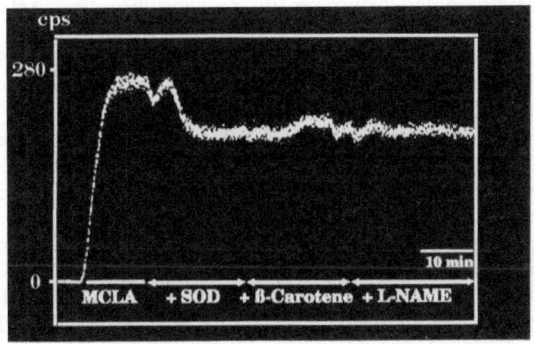

FIG. 4. Changes in MCLA chemiluminescence during serial superfusion with superoxide dismutase (*SOD*), β-*carotene*, and $N^G$-nitro-L-arginine methyl ester (*L-NAME*)

of free radical scavengers. A blood flow threshold, in other words, an $O_2$ supply threshold, may exist below which the $O_2^-$ generation was reduced.

We assessed basically the extracellular $O_2^-$ level in this study. As regards the source of $O_2^-$, we propose a breakdown of arachidonic acid in the cellular membrane and an interaction between endothelial cells and leukocytes in the capillary microcirculation. Although all types of cortical cells, including parenchymal cells (neurons and glias), vascular smooth muscle cells, and endothelial cells are possible sites of radical production, it is obscure at present which cells are mainly involved in our experiments. It has been postulated that neurons are more vulnerable to oxidative stress than glial cells because of their lower antioxidant capacity [6].

On the other hand, Kontos et al. [7] demonstrated, using a cytochemical technique, that following ischemia/reperfusion, superoxide was located primarily in the extracellular space, occasionally in the endothelial and vascular smooth muscle cells, and that no superoxide was detected in the brain parenchyma. A biochemical interaction

between endothelial cells and blood components such as polymorphonuclear leukocytes and platelets has been considered to play an important role in postischemic
reperfusion injury, generating several bioactive mediators including prostanoides/
leukotrienes, platelet-activating factors, and oxygen-derived free radicals [8]. We are
planning further studies utilizing a photon-counting camera system and radical
scavengers to ascertain the exact source of $O_2^-$ generation.

# References

1. Armsted WM, Mirro R, Busija DW, Leffer CW (1988) Postischemic generation of
   superoxide anion by newborn pig brain. Am J Physiol 255:H401–H403
2. Cao W, Carney JM, Duchon A, et al. (1988) Oxygen free radical involvement in ischemia
   and reperfusion to brain. Neurosci Lett 88:233–238
3. Uematsu D, K-Yatsugi S, Takahashi M, et al. (1996) Superoxide generation following
   hypoxia and reoxygenation in the cat cortex in vivo. J Cereb Blood Flow Metab 15:S280
4. Demopoulos HB, Flamm ES, Pietronigro DD (1980) The free radical pathology and
   the microcirculation in the major central nervous disorders. Acta Neurol Scand
   (Suppl) 492:91–119
5. Takahashi A, Nakano M, Mashiko S, Inaba H (1990) The first observation of $O_2^-$ genra-
   tion in situ lungs of rats treated with drugs to induce experimental acute respiratory
   distress syndrome. FEBS Lett 261:369–372
6. Makar TK, Nedergaard M, Preuss A, et al. (1994) Vitamin E, ascorbate, glutathione,
   glutathione disulfide, and enzymes of glutathone metabolism in cultures of chick astrocytes and neurons: evidence that astrocytes play an important role in antioxidative
   processes in the brain. J Neurochem 62:45–53
7. Kontos CH, Wei EP, Williams JI, et al. (1992) Cytochemical detection of superoxide in
   cerebral inflammation and ischemia in vivo. Am J Physiol 263:H1234–H1242
8. Mirro R, Armsted WM, Mirro J, et al. (1989) Blood-induced superoxide anion generation on the cerebral cortex of newborn pigs. Am J Physiol 257:H1560–H1564

# Effects of Zinc Protoporphyrin IX, a Heme Oxygenase Inhibitor, on Mitochondrial Membrane Potential in Rat Cultured Hepatocytes

Yuichi Shinoda, Makoto Suematsu, Yoshiyuki Wakabayashi, and Yuzuru Ishimura

*Summary.* Carbon monoxide (CO), a product of heme oxygenase, is released spontaneously from the liver and serves as an endogenous modulator of hepatobiliary function, although the whole mechanisms remain unknown. This study aimed to examine whether CO generated spontaneously might modulate hepatocellular function through its active interaction with the mitochondrial respiratory chain. Livers from male Wistar rats were used to provide isolated hepatocytes in culture using a collagenase perfusion technique. Endogenous CO production was abolished by application of 1 μM zinc protoporphyrin IX (ZnPP), a heme oxygenase inhibitor. When the hepatocytes contained in the suspension were observed by laser confocal digital microfluorography assisted by rhodamine 123, a fluorochrome sensitive to mitochondrial inner membrane potential ($\Delta\Psi$), the ZnPP application evoked a significant increase in the fluorescence, suggesting the increasing $\Delta\Psi$ values. These results suggest that CO endogenously generated by heme oxygenase modulates hepatocyte function, at least in part, through changing mitochondrial membrane potential.

*Key words.* Heme oxygenase—CO—ATP—Mitochondria

## Introduction

Carbon monoxide (CO) is a biological modulator that is produced endogenously by heme oxygenase (HO) [1]. CO is believed to upregulate cyclic guanosine monophosphate (cGMP) via activation of guanylate cyclase and to share several biological actions with nitric oxide (NO), such as smooth muscle relaxation or inhibition of platelet aggregation [2]. Our recent observation has shown that the basal flux of CO from the control perfused rat liver ex vivo is about $0.6 \, \text{nmol} \, \text{min}^{-1} \text{g}^{-1}$ liver [3]. Zinc protoporphyrin IX (ZnPP), a heme oxygenase inhibitor, induced marked alterations in sinusoidal tone [3] and bile acid-dependent bile transport [4]. These results suggest

Department of Biochemistry, School of Medicine, Keio University, 35 Shinanomachi, Shinjuku-ku, Tokyo 160, Japan

the physiological role of CO in regulation of hepatobiliary functions. On the other hand, because the ability of CO to activate guanylate cyclase is shown to be as small as a few percent of that of NO [5], another possibility suggests unknown cGMP-independent pathways as a putative mechanism for biological actions of CO. Actually, our recent report showed that exogenously applied CO at micromolar levels did not elicit a significant increase in cGMP in cultured rat hepatocytes in vitro [6]. Among several heme enzymes in hepatocytes, cytochrome oxidase in the mitochondrial respiratory chain or cytochrome P450 could serve as a candidate molecule that transduces intracellular signals. In this study, we have attempted to examine the effects of endogenous CO suppression on mitochondrial function and ATP synthesis in cultured rat hepatocytes.

## Materials and Methods

### Isolation of Couplet Hepatocytes

Hepatocytes were prepared from male Wistar rats (240–280 g) by type IV collagenase (Wako, Osaka, Japan) perfusion technique after anesthesia, as previously described [7].

### Rhodamine-123-Assisted Digital Confocal Microfluorography in Isolated Hepatocytes

Intracellular mitochondrial function in parenchymal hepatocytes was visually assessed in each experimental group by microfluorography using rhodamine-123 (Rh), a functional fluoroprobe that can bind mitochondria in proportion to the inner membrane potential [8]. The hepatocytes were loaded with Rh at 0.8 μM according to the previous experimental protocol [9]. The cells were then treated with medium containing desired concentrations of 1 μM ZnPP, with or without 2 μM CO, for 10 min at 37°C. The inverted intravital microscope (TMD-300, Nikon, Tokyo, Japan) equipped with a digital laser scanning confocal imaging system (Insight, Meridian, MI, USA) under epi-illumination at 488 nm using the argon laser power supply.

The confocal Rh microfluorographs were visualized by a line-scanning laser confocal system assisted by a cooled charge-coupled device (CCD) camera (Optronics TEC 470, Hamamatsu Photonics, Tokyo, Japan) every 1 min after the start of experiments. The fluorescence images were digitally processed by image processor (Image 1.58/Power Macintosh 8100, Apple Japan, Tokyo, Japan). To perform the data calibration, the relationship between the gray levels and the Rh concentrations was established by examining the gray levels of the solution containing known Rh concentrations. Based upon the calibration lines, the gray-level data measured from video images were interpreted into the apparent Rh concentrations that yielded gray levels equivalent to those measured in the cells [8]. Measurements of Rh concentrations were carried out using a variable window for the gray-level measurements according to our previous methods [8]. We analyzed more than 25 cells in each experimental group.

## Determination of ATP Contents in Isolated Hepatocytes

The rat hepatocyte suspension also served as samples for determination of ATP contents, as described elsewhere [10]. All the data in the current study were expressed as mean ± SE of measurements; $P$ values less than .05 were considered statistically statistically significant.

# Results and Discussion

## Endogenous CO Alters Mitochondria Function Assessed by Rh Microfluorograph

Figure 1 illustrates a representative laser confocal image showing Rh-loaded rat cultured hepatocytes inoculated on the culture dish. Most of the fluorescence activities occurred in cytoplasm but not in nuclei and displayed multiple pinpoint patterns, suggesting the presence of fluorescence activities in mitochondria. As seen, when the three-dimensional reconstruction was established, bile canalicular space between the adjacent hepatocytes was visualized. Based on fluorimetrical measurements using two-dimensional Rh images, the effects of ZnPP on the inner membrane potentials were examined (Fig. 2). To calibrate the fluorescence measurement of Rh in cultured hepatocytes, we attempted to establish a calibration line showing the relationship between gray levels and Rh dissolved in the buffer with known concentrations, indicating a linear relationship between the two parameters. According to this relationship, we estimated alterations in the inner membrane potential in mitochondria of cultured hepatocytes. In the cells exposed to ZnPP at a final concentration of 1 μM, the basal fluorescence intensity increased time dependently and reached a steady-state level, displaying approximately 25% elevation as compared with the control values. Because mitochondrial inner membrane potential is proportional to logarithmic values of the Rh intensity in hepatocytes, the net depletion of the membrane potential

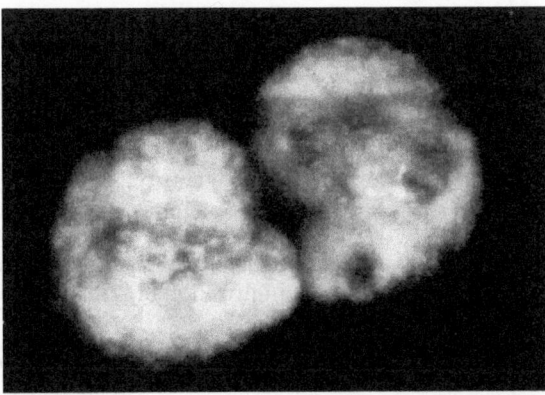

FIG. 1. Representative microfluorograph of three-dimensional features of couplet hepatocytes stained by Rh-123. The picture was reconstructed from serial two-dimensional confocal images of Rh-123. Note a single bile canalicular space between the adjacent hepatocytes

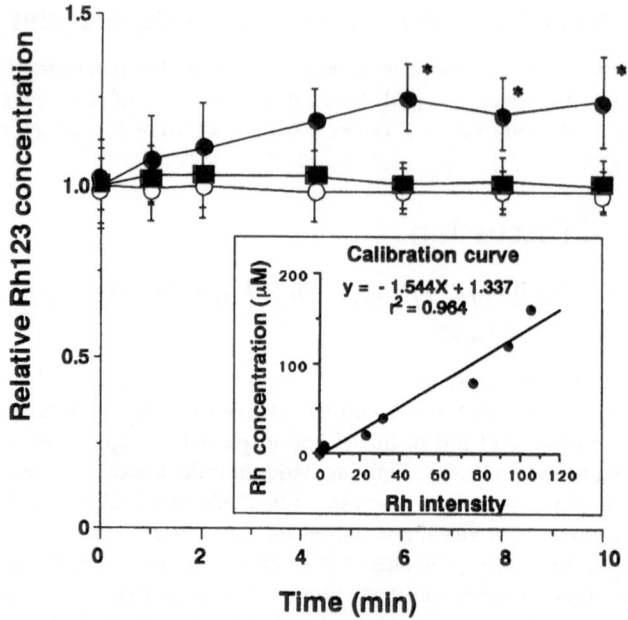

FIG. 2. Quantitative analysis of time-dependent alterations in the Rh-123 fluorescence intensity in the cells. *Inserts* show the relationship between known concentrations of Rh-123 solutions (0–126 μM) and their gray levels measured using the current microscopic system, indicating a linear correlation ($r^2 = .964$) between the two parameters. Using this calibration line, all gray-level measurements in the hepatocytes were interpreted to "apparent" Rh concentrations. These concentration values were then normalized by the initial Rh concentration at time 0 (relative Rh-123 concentration). *Open and closed circles with bars* denote mean ± SD of measurements from 12–17 different hepatocytes in control and zinc protoporphyrin IX-(ZnPP)-treated groups, respectively. *Closed squares and bars* represent data in cells treated with ZnPP + 2 μM CO ($n = 12$). *Open circles*, control. *, $P < .05$ as compared with the control

(d$\Delta\Psi$) could be estimated according to the following equation, as described elsewhere [11]:

$$d\Delta\Psi = 59\log\left[\text{relative Rh intensity}\right](mV)$$

On the basis of this formula, the ZnPP-induced increase in the potential was estimated to be approximately 5–6 mV. On the other hand, ZnPP-induced elevation of the Rh concentration values was represed by administration of 2 μM CO.

## Determination of ATP Content in Isolated Hepatocytes

As the action of ZnPP on mitochondrial function might result from interference with the interaction between the mitochondrial respiratory chain and endogenously generated CO, we inquired whether the intracellular ATP levels were altered by ZnPP (Table 1). In response to application of 1 μM ZnPP, hepatocellular ATP levels were

TABLE 1. ATP contents in isolated hepatocytes treated with $1\,\mu M$ ZnPP or with $1\,\mu M$ ZnPP + $2\,\mu M$ CO

| Group | ATP contents (n mol/$10^6$ cell) |
|---|---|
| Control | $2.39 \pm 0.08$ |
| ZnPP | $3.31 \pm 0.17^*$ |
| ZnPP + CO | $3.01 \pm 0.20$ |

Data represent mean $\pm$ SE of measurements using hepatocytes suspension isolated from six different livers.
*, $P < .05$ as compared with control CO concentrations.
ZnPP, zmc protoporphyrin IX.

significantly increased. The ZnPP-elicited elevation of the ATP contents was significantly attenuated by supplementation with $2\,\mu M$ CO. The same concentration of CO, however, did not alter the baseline ATP levels. These results suggest that the effect of ZnPP on mitochondrial respiration is attributable to its inhibitory action on heme oxygenase as a CO-generating system in hepatocytes [6].

*Acknowledgments.* The authors thank Chikara Oshio, M.D., for technical support to prepare the hepatocyte couplets. This work was supported by a grant-in-aid for scientific research from the Ministry of Education, Science, Sports and Culture of Japan, and by grants from Keio University School of Medicine, by the Sasakawa Scientific Research grant from The Japan Science Society, and grants from Takeda Research Foundation and Kanae Research Foundation.

# References

1. Maines MD (1988) Heme oxygenase: function, multiplicity, regulatory mechanisms, and clinical applications. FASEB J 2:2557–2568
2. Brüne B, Ullrich V (1987) Inhibition of platelet aggregation by carbon monoxide is mediated by activation of guanylate cyclase. Mol Pharmacol 32:497–504
3. Suematsu M, Goda N, Sano T, et al. (1995) Carbon monoxide: an endogenous modulator of sinusoidal tone in the perfused rat liver. J Clin Invest 69:2431–2437
4. Sano T, Shiomi M, Wakabayashi Y, et al. (1997) Carbon monoxide generated by heme oxygenase modulates bile acid-dependent biliary transport in perfused rat liver. Am J Physiol 272:G1268–G1275
5. Kharitonov VG, Sharma VS, Pilz RB, et al. (1995) Basis of guanylate cyclase activation by carbon monoxide. Proc Natl Acad Sci USA 92:2568–2571
6. Shinoda Y, Suematsu M, Wakabayashi Y, et al. (1996) Heme oxygenase-carbon monoxide pathway as a modulatory machinery for bile canalicular function. Hepatology 24:350A (abstract)
7. Oshio C, Phillips MJ (1981) Contractility of bile canaliculi: implications for liver function. Science 212:1041–1042
8. Suematsu M, Suzuki H, Ishii H, et al. (1992) Topographic dissociation between mitochondrial dysfunction and cell death during low-flow hypoxia in perfused rat liver. Lab Invest 67:434–442

9. Kurose I, Kato S, Ishii H, et al. (1993) Nitric oxide mediates lipopolysaccharide-induced alteration of mitochondrial function in cultured hepatocytes and isolated perfused liver. Hepatology 18:380–388

10. Yamaguchi T, Wakabayashi Y, Sano T, et al. (1996) Taurocholate induces directional transport of bilirubin into bile in the perfused rat liver. Am J Physiol 33:1028–1032

11. Suzuki H, Suematsu M, Ishii H, et al. (1994) Prostaglandin E1 abrogates early reductive stress and zone-specific paradoxical oxidative injury in hypoperfused rat liver. J Clin Invest 93:155–164

# Structure and Function of NO Reductase with Oxygen-Reducing Activity

Taketomo Fujiwara, Takanori Akiyama,
and Yoshihiro Fukumori

*Summary.* We succeeded in large-scale purification of NO reductase from *Paracoccus denitrificans* ATCC 35512 (formerly named *Thiosphaera pantotropha*) by using hydroxylapatite column chromatography. The spectral and enzymatic properties were the same as those of the enzyme purified by the previous method reported by Fujiwara and Fukumori. The enzyme was composed of two kinds of subunits with molecular masses of 34 and 15 kDa, respectively, and contained two hemes *b* and one heme *c* per molecule. We analyzed the metal content of NO reductase: the results suggested that NO reductase has no copper but does have 1–2 g·atoms of nonheme iron per mole. We also determined the structural gene of the enzyme. The *norC* and *norB* genes encoding the cytochrome *c* and cytochrome *b* subunits, respectively, showed considerable homology with those of *Pseudomonas stuzeri* NO reductase. Furthermore, the six invariant histidines in subunit I of the heme-copper oxidase superfamily were also conserved in the cytochrome *b* subunit of NO reductase.

*Key words.* NO reductase—Cytochrome *c* oxidase—Oxygen—Evolution—*Paracoccus denitrificans*

## Introduction

Nitric oxide reductase (NO reductase) purification was previously reported by Fujiwara and Fukumori [1]. Recent progress in determining the amino acid sequence of *Pseudomonas stuzeri* NO reductase [2] shows significant homology between NO reductase and *bc*-type cytochrome *c* oxidases that were found in microaerobic bacteria such as *Bradyrhizobium japoniocum* [3] and *Magnetospirillum magnetotacticum* [4], suggesting that the oxygen-reducing respiratory system developed from the anaerobic denitrifying respiratory system [5]. However, *P. stuzeri* NO reductase shows no oxygen-reducing activity [6]. We report here the large-scale purification and gene structure of a highly active NO reductase with oxygen-reducing activity from *Paracoccus denitrificans* ATCC 35512.

---

Department of Life Science, Faculty of Bioscience and Biotechnology, Tokyo Institute of Technology, 4259 Nagatsuta, Midori-ku, Yokohama 226, Japan

# Materials and Methods

## Organisms and Culture

*Paracoccus denitrificans* ATCC 35512 was anaerobically cultured as previously described by Robertson and Kuenen [7] with some modifications. The cells were harvested in the stationary phase of growth by centrifugation at $10\,000 \times g$ for 15 min.

## Physical Measurements

Absorption spectra were recorded with a Shimadzu MPS-2000 spectrophotometer (Kyoto, Japan) using cuvettes of 1-cm light path. The contents of hemes *b* and *c* were calculated on the basis of the pyridine ferrohemochrome spectra using millimolar extinction coefficients of $34.4\,\text{mM}^{-1}\text{cm}^{-1}$ at 557 nm for heme *b* and $29.1\,\text{mM}^{-1}\text{cm}^{-1}$ at 550 nm for heme *c* [8]. Protein concentrations were determined with a BCA protein assay reagent (Pierce Chemical, Rockford, IL, USA) using bovine serum albumin as a standard. The contents of copper and iron atoms in the purified enzyme preparation were measured by inductively coupled plasma atomic emission spectrometry using an SPS 1500 VR Plasma Spectrometer (Seiko Instruments, Tokyo, Japan).

## Enzyme Assay

NO consumption catalyzed by NO reductase was routinely measured by the method described in [1]. Cytochrome *c*-NO reductase activity of the enzyme was also measured by the method described in [1].

## Reagents

Sucrose monocaprate (SM-1080) was purchased from Mitsubishi-Kasei Foods (Tokyo, Japan). All other reagents used in this study were of the highest grade commercially available.

## Cloning and Sequencing of the Gene Encoding NO Reductase

The genomic DNA was extracted from *Paracoccus denitrificans* ATCC 35512. The 1.7-kbp fragment was amplified by a conventional polymerase chain reaction (PCR) method using for primer 1 the N-terminal sequence of the cytochrome *c* subunit (MSDIMTKNMA) and for primer 2 the partial amino acid sequence of the cytochrome *b* subunit (ALFYWMRF) and was then used as the probe for the Southern hybridization experiment. The nucleotide sequences of 2.5- and 4.0-kbp *Eco*RI fragments obtained by the southern hybridization experiment were determined from both directions according to the Sanger dideoxy method.

# Results and Discussion

## Large-Scale Purification of NO Reductase

All purifications were conducted at 4°C. The enzyme was purified by the method as described in [1] with slight modifications. The enzyme was solubilized with SM-1080

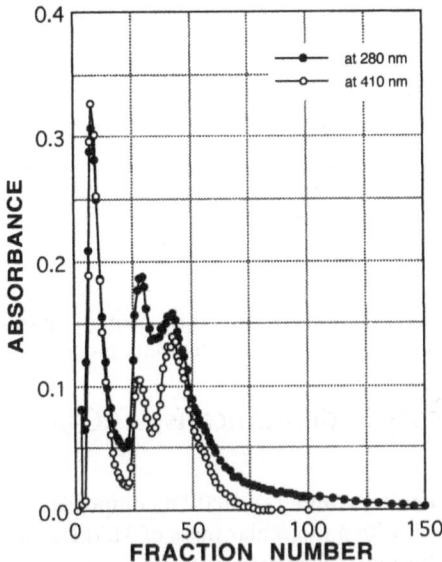

FIG. 1. Elution profiles of hydroxyapatite column chromatography for the purification of NO reductase from *Paracoccus denitrificans* ATCC35512. The column was equilibrated with 10 mM Tris-HCl buffer (pH8.0) containing 0.2% Tween 20, and NO reductase was eluted with the linear gradient of phosphate buffer. NO reductase was fractionated in fraction no. 4–14

and partially purified by two ion-exchange column chromatographies with DEAE-Toyopearl and DEAE-Biogel columns. In the previous method [1], the enzyme preparation obtained after column chromatography was finally purified by nondenaturing polyacrylamide gel electrophoresis (PAGE) in the presence of 0.2% (w/v) Tween-20. In the current study, we have found that the native PAGE could be replaced with hydroxyapatite column chromatography; the elution profile is shown in Fig. 1. NO reductase was eluted at a low concentration of sodium phosphate buffer. The contaminating proteins were separately eluted at different concentration of sodium phosphate.

We compared the spectral and enzymatic properties of NO reductase purified in the current study with those of the enzymes prepared by the previous method [1]. We could not find any differences in spectral and enzymatic properties. Furthermore, the subunit structure and heme contents were also the same as those of the previous enzyme (data not shown).

## Metal Content of Paracoccus denitrificans ATCC35512 NO Reductase

The contents of nonheme iron and copper in the NO reductase were analyzed by inductively coupled plasma atomic emission spectrometry. As summarized in Table 1, the enzyme has about 1–2 atoms of nonheme iron in the molecule, while copper was

TABLE 1. The copper and nonheme iron content of *Paracoccus denitrificans* ATCC31632 nitric acid (NO) reductase

| Sample no. | Heme (μM) | Fe (μM) | Cu (μM) | Nonheme iron (μM) | Cu (mol per mol of enzyme) | Nonheme iron (mol per mol of enzyme) |
|---|---|---|---|---|---|---|
| 1 | 4.08 | 5.89 | 0.370 | 1.802 | 0.08 | 1.32 |
| 2 | 1.08 | 1.84 | 0.091 | 0.756 | 0.10 | 2.09 |

scarcely detected in the preparation. These results strongly suggest that copper is not essential for the reduction of NO and $O_2$ by NO reductase.

## Structure of Paracoccus denitrificans ATCC35512 NO Reductase

The NO reductase of *P. denitrificans* ATCC35512 is composed of two kinds of subunit: a heme *b*-binding subunit with a molecular mass of 34.5 kDa (subunit I) and a heme *c*-binding subunit with the molecular mass of 15 kDa (subunit II) [1]. In this study, we determined the gene structure of *P. denitrificans* ATCC35512 NO reductase by the method described in Materials and Methods. Figure 2 shows the complete amino acid sequence of the enzyme deduced from nucleotide sequences of the genes *norC* and *norB*. The molecular masses of subunits I and II were determined to be 52 464 and 16 972, respectively. The molecular mass of the subunit I is much higher than that obtained by sodium dodecylsulfate- (SDS-) PAGE, which may be because of the hydrophobic property of the subunit I.

The amino acid sequences of the enzyme were compared with the protein sequence bank using the BLAST program. The subunit II has the heme *c*-binding motif, -Cys-X-Y-Cys-His-, and one transmembrane region in the molecule. These results indicate that subunit II is a membrane-anchored cytochrome *c*. On the other hand, the sequence analysis shows 54% identity and 72% homology with the subunit I of *Pseudomonas stuzeri* NO reductase; 39% homology with the subunit I of *cb*-type cytochrome *c* oxidase (FixN); and 21% homology with the subunit I of *aa₃*-type cytochrome *c* oxidase (*Rhodobacter sphaeroides*) [10]. Furthermore, the subunit I of NO reductase has 12 transmembrane segments and, interestingly, the six conserved metal-binding histidines that are the ligands for heme and copper in the heme-copper oxidase superfamily (Fig. 3). NO reductase has a low-spin heme *b* and a CO-reactive high-spin heme *b*, and the two histidines in helix VII and one histidine in helix VI that are copper ligands of the active site of the heme-copper oxidase superfamily were also conserved in the enzyme. Therefore, it seems likely that NO reductase has a bimetallic center with high-spin heme *b* and nonheme iron similar to that of the heme-copper oxidase superfamily.

The crystal structure of mitochondrial and bacterial cytochrome *c* oxidases were recently published by Yoshikawa et al [10] and Iwata et al [11]. The structural data provide helpful information to understand the mechanism of electron transport and proton pump in the molecule. In the current study, we succeeded in large-scale purification of NO reductase and determined the complete amino acid sequences of

*Amino acid sequence of cytochrome c subunit of*
*P.denitrificans ATCC35512 NO reductase*

MSDIMTKNMARNVFYGGSIFFILIFGALTVHSHIYARIKAVDESQLTPSVVEGKHVWERNAC**IDCHTL**
LGEGAYFAPELGNVMKRWGVQDDPETAVETLKGWMESMPTGIEGRRQMPRFELTDEEFRALSDFLLWT
GTINTQNWPPNDAG

*Amino acid sequence of cytochrome b subunit of*
*P.denitrificans ATCC35512 NO reductase*

MRYHSQRIAYAYFLVAMVLFAVQVTLGLIMGWIYVSPNFLSELLPFNIARMLHTNSLVVWLLLGFFGA
TYYILPEEAEREIHSPLLAWIQLGIFVLGTAGVVVTYLFDLFHGHWLLGKEGREFLEQPKWVKLGIAV
AAVIFMYNVSMTALKGRRTAVTNVLLMGLWGLVLLLWLFAFYNPANLVLDKQYWWWVIHLWVEGVWELI
MAAILAFLMLKLTGVDREVVEKWLYVIVATALFSGILGTGHHYYWIGLPAYWQWIGSIFSSFEIVPFF
AMMSFAFVMVWKGRRDHPNKAALVWSLGCTVLAFFGAGVWGFLHTLHGVNYYTHGTQITAAHGHLAFY
GAYVCLVLALVTYCMPLMKNRDPYNQVLNMASFWLMSSGMVFMTVTLTFAGTVQTHLQRVEGGFFMDV
QDGLALFYWMRFGSGVAVVLGALLFIYAVLFPRREVVTAGPVQAHKDGHLEAAE

FIG. 2.  The complete amino acid sequence of the cytochrome c (*top*) and cytochrome b (*bottom*) subunits of *P. denitrificans* ATCC35512 NO reductase

#### Helix II
PFNIARML**H**TNSLVVWLLLGFFGATYYI
PFNVARMV**H**TNLLIVWLLFGFMGAAYYL
SFGRLRPL**H**TSAVIFAFGGNVLIATSFY
TDLAIFAVH**L**SGASSILGAINMITTFLN
HYDQIFTA**H**GVIMIFFVAMPFVIGLMNL
TYNVFATN**H**GLIMIFFMVMPAMIGGFGN

#### Helix VI
WWWVI**H**LWV**E**GVWELIMAAILAFLM
WWFVVH**H**LWV**E**GVWELIMGAMLAFVL
QWWYG**H**NAV**G**FFLTAGFLAIMYYFI
LWFFG**H**PEV**Y**IIVLPAFGIVSHVIA
IWAWG**H**PEV**Y**ILILPVFGVFSEIAA
FWFFG**H**PEV**Y**ILILPGFGMISHVIS

#### Helix VII
WLYVIVATALFSGILGTG**H**HYYWIG
WLYVIIAMALITGIIGTG**H**HFFWIG
LSIIHFWALIFLYIWAGP**H**HLHYTA
PMVYAMVAIGVLGFVVWA**H**HM-YTA
SLVWATVCITVLSFIVWL**H**HFFTMG
GMAYAMVAIGGIGFVVWA**H**HMYTVG

#### Helix VIII
IGSIF**S**SFEIVPFFAMMSFAFVMV
VGSIF**S**ALEPLPFFAMVLFALNMV
LGMTF**S**IMLWMPSWGGMINGLMTL
YFMMATMVIAVPTGIKIFSWIATM
FFGITTMIIAIPTGVKIFNWLFTM
YFVAATMVIAVPTGVKVFSWIATM

#### Helix X
TAA**H**G**H**LAFYGAYVCLVLALVTY
TAA**H**G**H**LAFYGAYAMIVMTMISY
TIG**H**V**H**SGALGWVGFVSFGALYC
VVA**H**F**H**YVMSLGAVFGIFAGSTS
LIA**H**F**H**NVIIGGVVFGCFAGMTY
VIA**H**F**H**YVMGIAAVFAIFSGWYY

FIG. 3.  Sequence fragments of functionally important transmembrane segments of subunit I. *P. denitrificans nor B* (current study), *Paracoccus denitrificans*; *Ps. stutzerinor B* [2], *Pseudomonas stutzeri*; *B. japonicum fixN* [3], *Brady rhizobium japonicum*; *R. sphaeroides coxI* [9], *Rhodobacter sphaeroides*; *E. coli cyoB* [12], *Escherichia coli*; *N. winogradskyi coxB* [13], *Nitrobacter winogradskyi*

the enzyme. Further research should yield insight into the structure and function of NO reductase.

# References

1. Fujiwara T, Fukumori Y (1996) Cytochrome *cb*-type nitric oxide reductase with cytochrome *c* oxidase activity from *Paracoccus denitrificans* ATCC 35512. J Bacteriol 178:1866–1871
2. Zumft WG, Braun C, Cuypers H (1994) NO reductase from *Pseudomonas stutzeri*. Primary structure and gene organization of a novel bacterial cytochrome *bc* complex. Eur J Biochem 219:481–490
3. Preisig O, Anthamatten D, Hennecke H (1993) Genes for a microaerobically induced oxidase complex in *Bradyrhizobium japonicum* are essential for a nitrogen-fixing endosymbiosis. Proc Natl Acad Sci USA 90:3309–3313
4. Tamegai T, Fukumori Y (1994) Purification, and some molecular and enzymatic features of a novel *ccb*-type cytochrome *c* oxidase from a microaerobic denitrifier, *Magnetospirillum magnetotacticum*. FEBS Lett 347:22–26
5. Van der Oost J, de Boer APN, de Gier JWN, et al. (1994) The heme-copper oxidase family consists of three distinct types of terminal oxidases and its related to NO reductase. FEMS Microbiol Lett 121:1–10
6. Heiss B, Frunzke K, Zumft WG (1989) Formation of the N-N bond from nitric oxide by a membrane-bound cytochrome *bc* complex of nitrate-respiring (denitrifying) *Pseudomonas stutzeri*. J Bacteriol 171:3288–3297
7. Robertson LA, Kuenen JG (1983) *Thiosphaera pantotropha* gen. nov. sp. nov., a facultatively anaerobic, facultatively autotrophic sulfur bacterium. J Gen Microbiol 129:2847–2855
8. Yamanaka T (1988) Hemes. In: Otsuka S, Yamanaka T (eds) Metalloproteins; chemical properties and biological effects. Kodansha, Tokyo, pp 95–99
9. Shapleigh JP, Gennis RB (1992) Cloning, sequencing and deletion from the chromosome of the gene encoding subunit I of the $aa_3$-type cytochrome *c* oxidase of *Rhodobacter sphaeroides*. Mol Microbiol 6:635–642
10. Tsukihara T, Aoyama H, Yamashita E, et al. (1996) The whole structure of the 13-subunit oxidized cytochrome *c* oxidase at 2.8 Å. Science 272:1136–1144
11. Iwata S, Ostermeier C, Ludwig B, et al. (1995) Structure at 2.8 Å resolution of cytochrome *c* oxidase from *Paracoccus denitrificans*. Nature 376:660–669
12. Chepuri V, Lemieux L, Au DC, et al. (1990) The sequence of the cyo operon indicates substantial structural similarities between the cytochrome *o* ubiquinol oxidase of *Escherichia coli* and the $aa_3$-type family of cytochrome *c* oxidases. J Biol Chem 265:11185–11192
13. Berben GPR (1996) *N. winogradskyi* DNA for coxA, coxB and coxC genes. ACCESSION X89566

# Oxidative Modification of Apolipoprotein E3 and Its Biological Significance

SHIN-ICHI HARA[1], TATEHIKO TANAKA[2], MICHIO YAMADA[1], YUJI NAGASAKA[3], and KAZUYUKI NAKAMURA[3]*

*Summary.* Apolipoprotein E (apoE), in very low density lipoprotein (VLDL), formed aggregates and lost its heparin-binding activity with lipid peroxidation by an oxidation system consisting of $10\,\mu M$ ferrous sulfate in saline under aerobic conditions. Sodium dodecyl sulfate-polyacrylamide gel electrophoresis (SDS-PAGE) and amino acid analysis of the aggregated apoE indicated the oxidation of basic amino acid residues and the intermolecular cross-linking of apoE may be caused by the formation of 4-hydroxy-2-nonenal (HNE), which reacts with ε-amino groups of lysyl residues on apoE molecules. The presence of 1% heparin inhibited the oxidative modification of apoE to restore its heparin-binding activity. These findings suggest that the oxidative modification of apoE in VLDL causes the accumulation of lipid peroxides by the decrease in the rate of VLDL uptake via binding to heparin on the surface of cells. This may be a possible mechanism of the deposit of oxidized lipids in the vascular system and of oxidized apoE in senile plaques in the brain of Alzheimer's disease.

*Key words.* Apolipoprotein E—lipid peroxidation—Heparin—4-Hydroxy-2-nonenal (HNE)—Alzheimer's disease

## Introduction

Apolipoprotein E (apoE) is a component of lipoproteins and plays an important role in lipid transport in blood plasma [1]. Furthermore, apoE participates in restoration and regeneration of nervous tissue [2]. It has been reported that apoE4, an isoform of apoE, is a risk factor of Alzheimer's disease (AD) [3]. Immunoreactivity of apoE [4] or proteoglycan [5] has been shown in senile plaques in AD brains. ApoE has two heparin-binding sites rich in lysyl and arginyl residues [6,7]. These amino acid

[1] Department of Neuropsychiatry, [2] Central Laboratory for Research and Education, [3] First Department of Biochemistry, Yamaguchi University School of Medicine, Kogushi 1144, Ube, Yamaguchi 755, Japan
* To whom the correspondence should be addressed.

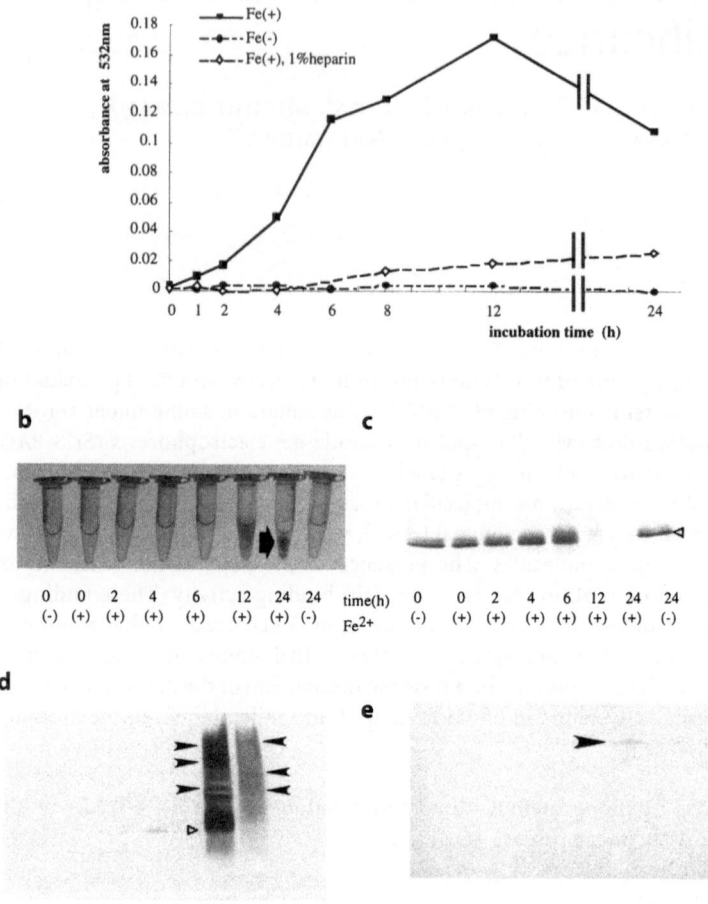

FIG. 1a–e. Time course of oxidation of very low density lipoprotein (VLDL) by ferrous ion-catalyzed oxidation system. **a** Thiobarbituric acid (TBA) reactivity of oxidized VLDL. **b** External appearance of oxidized apolipoprotein E (apoE). White precipitates were formed after 24 h incubation with 10 μMFeSO$_4$ (*arrow*). **c** ApoE immunoreactivity of oxidized VLDL (supernatant). *Triangular arrow*, monomeric ApoE. **d** ApoE immuno-reactivity of oxidized VLDL (precipitates). Aggregated apoE was found in the precipitates (*arrowheads*). **e** 4-Hydroxy-2-nonenal (HNE) immunoreactivity of oxidized VLDL (precipitates). *Arrow head* indicates the HNE-positive band of 50 kDa

residues are susceptible to metal ion-catalyzed oxidation [8]. In this chapter, we investigate the mechanism of oxidative modification of apoE in VLDL by a ferrous ion-catalyzed oxidation system and discuss its biological consequences related to Alzheimer's disease.

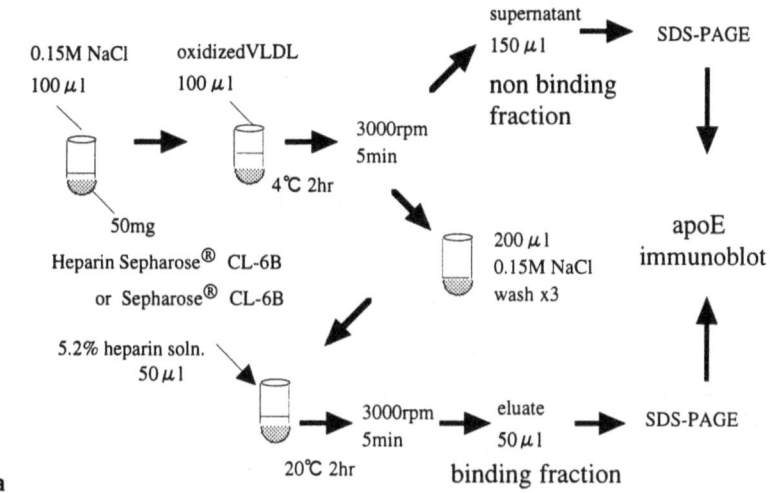

a

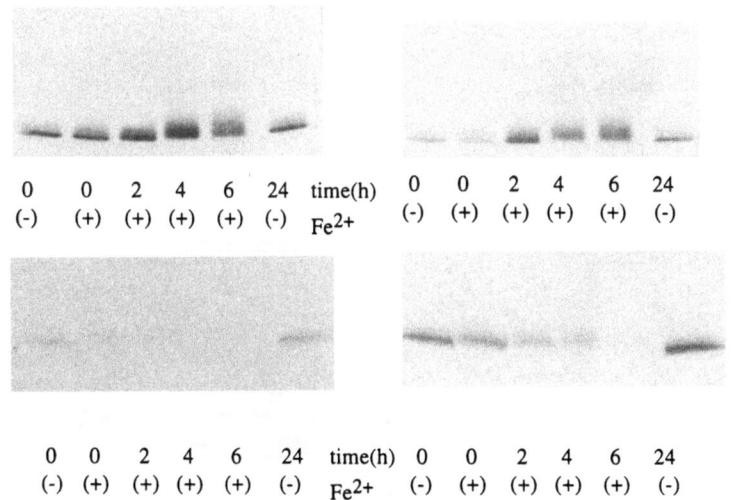

| 0 | 0 | 2 | 4 | 6 | 24 | time(h) | 0 | 0 | 2 | 4 | 6 | 24 |
| (-) | (+) | (+) | (+) | (+) | (-) | $Fe^{2+}$ | (-) | (+) | (+) | (+) | (+) | (-) |

b

FIG. 2a,b. Heparin-binding activity of oxidized apolipoprotein E (apoE). a Schematic diagram of the heparin-binding assay. b Immunoreactivity of apoE in nonbinding fractions (*upper panels*) and binding fractions (*lower panels*), obtained with sepharose CL-6B (*left panels*) and heparin sepharose CL-6B (*right panels*)

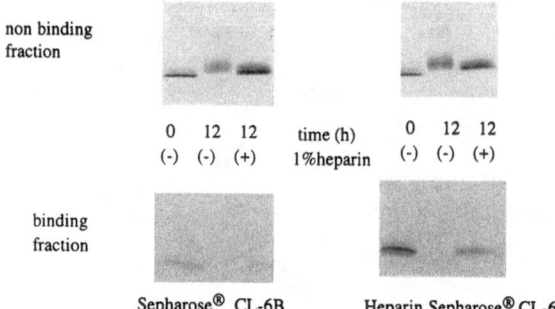

FIG. 3. Effect of 1% heparin on apoE heparin-binding activity. *Arrowhead* indicates the band of apoE

a

b

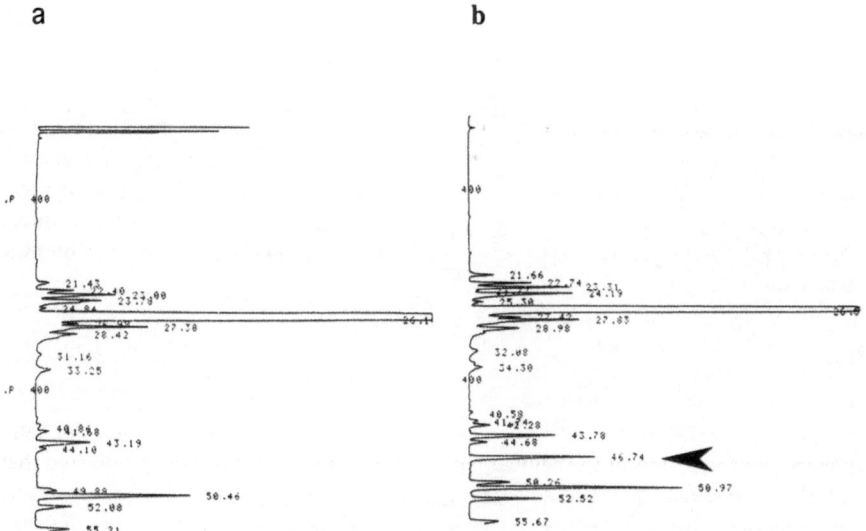

FIG. 4a,b. Amino acid analysis of oxidized apolipoprotein E (apoE). a Nontreated apoE. b Oxidized apoE (12h). An abnormal peak with high fluoresence was detected in the oxidized apoE (*arrowhead*)

## Materials and Methods

Very low density lipoprotein (VLDL) was isolated from healthy human blood plasma by ultracentrifugation in a fraction of $d < 1.006$ [9]. The phenotype of apoE was determined by isoelectric focusing followed by immunoblotting with antihuman apoE antisera. The VLDL was dialyzed against 0.15 M NaCl and incubated with 10 μM FeSO$_4$ at 37°C in aerobic conditions. The oxidation of VLDL was monitored by observing the appearance of the reaction mixture by lipid peroxide assay with thiobarbituric acid (TBA) [10] and by heparin-binding assay with heparin sepharose CL-6B [11]. ApoE in VLDL was separated by sodium dodecyl sulfate-polyacrylamide gel electrophoresis (SDS-PAGE) [12] using gels of 12.5% T and 2.7% C. The bands of apoE were visualized by immunoblotting with antihuman apoE antisera and anti-4-hydroxyl-2-nonenal (anti-HNE) antisera (a gift from Dr. K. Uchida, Nagoya University). For amino acid analysis the bands of apoE on polyvinylidene difluoride (PVDF) membrane stained with 1% amido black were cut out and hydrolyzed in the gas phase of 6 N HCl and 7% thioglycolic acid at 110°C for 24 h. The hydrolysate then was extracted with 30% methanol in 0.1 N HCl to be labeled with fluorescent reagent (AccQ, Waters,) in borate buffer, and the labeled amino acids were separated by HPLC with an AccQ·Tag (Millipore Milford, MA, USA) column.

## Results

The phenotype of apoE in VLDL was E3/3, the wild type of apoE. As the oxidation of VLDL in the reaction mixture with $10 \mu M$ $FeSO_4$ proceeded, TBA reactivity increased (Fig. 1a), and white precipitates were formed by incubation for 12 h. The precipitates became less soluble in SDS with incubation for 24 h (Fig. 1b,c ). In the precipitates, apoE with a molecular weight greater than 34 kDa and an HNE-positive band of 50 kDa were found (Fig. 1d,e). ApoE in oxidized VLDL lost its heparin-binding activity (Fig. 2), which was restored by the presence of 1% heparin in which TBA reactivity was not increased (Fig. 3). Amino acid analysis of the apoE showed an abnormal product with high fluorescence that was eluted from the HPLC column between methionine and lysine (Fig. 4).

## Discussion

ApoE is thought to be one of the pathological molecular chaperones that induce beta-pleated conformation in the senile plaques of AD [13], and it has been reported that a carboxyl-terminal fragment of apoE could form amyloyd-like fibrils [14]. Statistical analyses have shown that apoE4, an isoform of apoE, is a risk factor of AD. ApoE4 may play some role in the pathogenesis of AD, but no direct evidence for the cause of AD has yet been shown.

In our experiments, it was shown that apoE3 in VLDL that was oxidized by a ferrous ion-catalyzed oxidation system lost its heparin-binding activity and formed aggregates which were less soluble with SDS. Heparin is a potent antioxidant. By the loss of its heparin-binding activity, apoE can be more easily oxidized and forms aggregates, possibly because of the formation of intermolecular cross-linking between lysyl residues whose ε-amino groups react with 4-hydroxy-2-nonenal (HNE) in oxidized lipids [15]. The precipitation of VLDL caused by the aggregation of apoE leads to the accumulation of lipid peroxides in tissues and a decrease in the rate of VLDL uptake via binding to heparin on the surface of cells. This may be a possible mechanism of the accumulation of oxidized lipids in the vascular system and of the deposit of apoE in senile plaques in the brain of Alzheimer's disease patients.

# References

1. Mahley RD (1988) Apolipoprotein E: cholesterol transport protein with expanding role in cell biology. Science 240:622–630
2. Ignatius MJ, Gebicke-Harter PJ, Skene JH, et al. (1986) Expression of apolipoprotein E during nerve degeneration and regeneration. Proc Natl Acad Sci USA 83:1125–1129
3. Corder EH, Saunders AM, Strittmatter WJ, et al. (1993) Gene dose of apolipoprotein E type 4 allele and the risk of Alzheimer's disease in late onset families. Science 261:921–923
4. Namba Y, Tomonaga M, Kawasaki H, et al. (1991) Apolipoprotein E immunoreactivity in cerebral amyloid deposits and neurofibrillary tangles in Alzheimer's disease and kuru plaque amyloid in Creutzfeldt–Jakob disease. Brain Res 541:163–166
5. Snow AD, Lara S, Nochlin D, et al. (1989) Cationic dyes reveal proteoglycans structurally integrated within the characteristic lesions of Alzheimer's disease. Acta Neuropathol 78:113–123
6. Cardin AD, Demeter DA, Weintraab HJ, et al. (1991) Molecular design and modeling of protein-heparin interactions. Methods Enzymol 203:556–583
7. Weisgraber KH (1994) Apolipoprotein E: structure–function relationships. Adv Protein Chem 45:249–302
8. Stadtman ER (1990) Metal ion-catalyzed oxidation of proteins: biochemical mechanism and biological consequences. Free Radical Biol Med 9:315–325
9. Hatch FT, Lees RS (1968) Practical methods for plasma lipoprotein analysis. Adv Lipid Res 6:1–68
10. Kikugawa K, Kojima T, Yamaki S, et al. (1992) Interpretation of the thiobarbituric acid reactivity of rat liver and brain homogenates in the presence of ferric ion and ethylenediaminetetraacetic acid. Anal Biochem 202:249–255
11. Hara S (1997) Oxidative modification of apolipoprotein E3 and its modulation by heparin-binding (in Japanese). Yamaguchi Med J 46:53–63
12. Laemmli UK (1970) Cleavage of structural proteins during the assembly of bacteriophage T4. Nature 227:680–685
13. Wisniewski T, Frangione B (1992) Apolipoprotein E: a pathological chaperone protein in patients with cerebral and systemic amyloid. Neurosci Lett 135:235–238
14. Wisniewski T, Lalowski M, Golabek A, et al. (1995) Is Alzheimer's disease an apolipoprotein E amyloidosis? Lancet 345:956–958
15. Cohn JA, Tsai L, Friguet B, et al. (1996) Chemical characterization of a protein-4-hydroxy-2-nonenal cross-link: immunochemical detection in mitochondria exposed to oxidative stress. Arch Biochem Biophys 328(1):158–164

# Sequential Multistep Mechanisms for Leukocyte Adhesion: Applicable to Lung Microcirculation?

Takuya Aoki[1], Yukio Suzuki[2], Koichi Suzuki[1], Atsusi Miyata[1], Kazumi Nishio[1], Nagato Sato[1], Katsuhiko Naoki[1], Hiroyasu Kudo[1], Harukuni Tsumura[3], and Kazuhiro Yamaguchi[1]

*Summary.* This study was designed to examine how leukocyte traffic occurs in the pulmonary microcirculation under physiological shear rates. The leukocyte-endothelium interaction was visualized in perfused rat lungs using an intravital high-speed confocal laser video microscope by injecting fluorescence-tagged isolated leukocytes. In the control lung, transient cessation of leukocyte movement was observed in pulmonary capillaries. The percentage of leukocytes displaying such behavior increased in response to prestimulation by chemoattractants. As a consequence, leukocyte velocity decreased and the density of adherent leukocytes in pulmonary microvessels was markedly elevated. In contrast to observations in the mesenteric microcirculation, a major population of the adherent cells was observed in alveolar capillaries rather than in postcapillary venules. These results suggest that the pulmonary microvascular system is characterized by specific adhesive mechanisms for circulating leukocytes distinct from those previously reported in the mesenteric microcirculation.

*Key words.* Capillary entrapment—Wall shear rates—Adhesion—Rolling—Alveolar capillary

## Introduction

In inflammatory diseases, leukocyte accumulation is considered to occur via sequential multistep mechanisms, including rolling, adhesion, and transendothelial migration [1–3]. Different classes of adhesion molecules have specific functions in each process. Neutrophil adhesion and transendothelial migration have been shown to be mediated through CD11/CD18 and its endothelial ligand intercellular adhesion molecule-1 (ICAM-1) in vivo [4,5] and in vitro [6–8].

[1] Cardiopulmonary Division, Department of Internal Medicine, School of Medicine, Keio University, 35 Shinanomachi, Shinjuku-ku, Tokyo 160, Japan
[2] Department of Internal Medicine, Kitasato Institute Hospital, 5-9-1 Shirokane, Minato-ku, Tokyo 108, Japan
[3] Biomedical Department, Sankei Corporation, 2-77-7 Yushima, Bunkyo-ku, Tokyo 113, Japan

In the pulmonary microcirculation, however, the role of adhesion molecules in the mechanism underlying leukocyte accumulation is controversial, and whether sequential multistep leukocyte–endothelium interactions are applicable is not known. The important factors accounting for this controversy are possible differences in circulating blood leukocyte numbers and wall shear rates among the experimental models and unique features of the interaction between activated leukocytes and endothelium in the pulmonary microcirculation. We theorized that sequential multistep leukocyte–endothelium interactions are not applicable to the microvessels in the lung. The goal of the present study was to examine the behavior of leukocytes in rat pulmonary microcirculation perfused under physiological shear rates.

# Methods

## Animal Preparation

Specific-pathogen-free male Sprague–Dawley rats (Sankyo, Tokyo, Japan), 8 weeks of age and weighing 250–300 g, were used. All the following experimental protocols were approved by the Animal Committee of Keio University School of Medicine, Tokyo, Japan. After anesthesia by pentobarbital sodium (50 mg/kg i.p.), the trachea was cannulated and connected to a ventilator, then ventilated at a tidal volume of 10 ml/kg and a respiratory rate of 60/min. Lungs were exposed by median sternotomy and blood was withdrawn from the heart. The trachea was ligated at the level of one tidal volume above functional residual capacity and fixed on a microscope stage in the supine position. The main pulmonary artery and left atrium were catheterized. Pulmonary arterial pressure was measured with a pressure transducer (SEN-6102M, Nihon Koden, Tokyo, Japan) connected to the pulmonary artery cannula and monitored continuously during the experiment (AcqKnowledge III, BIOPAC Systems, Goleta, CA, USA). Krebs-Henseleit solution containing 3% albumin and autoerythrocytes was used as the perfusate (hematocrit = 6.5% ± 0.5%). An extracorporeal membrane oxygenator (Merasilox-S, Senkou-ikakougyou, Tokyo, Japan) was connected to the perfused lung circuit and equilibrated with 21% $O_2$–5% $CO_2$. The lungs were perfused at a rate of 10 ml/min with a peristaltic roller pump, and the perfusate from the left atrium was allowed to collect in the reservoir.

The in vivo microscopic system had three different sources of illumination lights. The first is a normal lamp light (Techno Light KTS-150, Kenko, Tokyo, Japan), the second a xenon lamp light (Nikon, Tokyo, Japan) for fluorescence imaging, and the third a laser power supply (Omnichrome, Chino, CA, USA) for confocal imaging. The image was displayed with a high-sensitivity charge-coupled device (CCD) camera (TEC-470, Optronics, San Diego, CA, USA) or a high-sensitivity intensified imager (II) camera (EktaPro Intensified Imager, Kodak, San Diego, CA, USA) and color video monitor (PVM-1444Q, Sony, Tokyo, Japan). The image was recorded on videotape with a tape recorder (SVQ-260, Sony). In some experiments, views of high-speed movements of the cells were displayed by using the confocal laser scanning microscope with the II camera and stored in a high-speed video recorder system (EktaPro TR6,000 System, Kodak, San Diego, CA, USA). Velocities of leukocytes and erythrocytes were recorded at a rate of 250 frames/s using the high-speed video system.

Leukocytes that remained in the same internal portions of capillaries were excluded from the speed analysis. Centerline blood flow velocity ($V_c$) was determined in arterioles and venules. The mean blood cell velocity ($V_{mean}$) was calculated from $V_{mean} = V_c/1.6$. The vessel diameters ($D_v$) of arterioles and venules were measured by processing the confocal video images with a computer-assisted digital image analyzing system (Apple Quadra 840-AV/Image 1.58, Apple, CA, USA). The vessel wall shear rates ($\gamma$) were calculated based on the definition for a Newtonian fluid: $\gamma = 8 \ (V_{mean}/D_v) \ (S^{-1})$, as described elsewhere [9]. We defined rolling as a caterpillar-like motion with deformation along vessel walls, which has been observed in mesenteric venules [10].

## Visualization of Vessel Networks, Erythrocytes, and Leukocytes

To visualize vessel networks, we administered fluorescein isothiocyanate- (FITC-) dextran (MW 145000), at a final concentration of 0.015%, into the perfusion circuit. Erythrocytes were labeled with FITC. Rat blood was diluted with phosphate buffered saline and centrifuged. The erythrocyte pellet was then diluted with phosphate buffered saline. FITC was added at a final concentration of 0.1 mg/ml. After a 30-min incubation at 37°C, the solution was centrifuged and diluted with 5 ml of phosphate buffered saline. Thereafter, 1 ml of the dilute solution was administered into the perfusion circuit when necessary. Leukocytes were labeled with carboxyfluorescein diacetate succinimidyl ester (CFSE) (Molecular Probes, Eugene, OR, USA) according to our previously described method [9].

## Experimental Groups

CFSE-labeled leukocytes were activated with 10 nM of cytokine-induced neutrophil chemoattractant (rat IL-8) at 37°C for 10 min just before administration into the perfusion circuit (IL-8 group; $n = 5$). Perfused rat lungs that were not pretreated with CFSE-labeled leukocytes served as a control group ($n = 5$). In separate sets of experiments, the behavior of CFSE-labeled leukocytes in the mesenteric microcirculation was visually examined by injecting the suspension of CFSE-labeled leukocytes into recipient rats. First, native leukocyte behavior was examined when mesenteries were superfused with 100 nM formyl-methionyl-leucyl-phenylalanine fMLP. Second, activated leukocyte behavior was examined when mesenteries were superfused with 100 nM fMLP. Observation of the mesenteric microcirculation under fluorescence in vivo microscopy with epi-illumination at 480 nm was carried as described elsewhere [11].

## Statistical Analysis

The results are presented as mean ± SEM. All $P$ values were determined using one-way analysis of variance (ANOVA) followed by the Scheffe-type multiple comparison test to detect differences among groups. A $P$ value less than .05 was considered statistically significant.

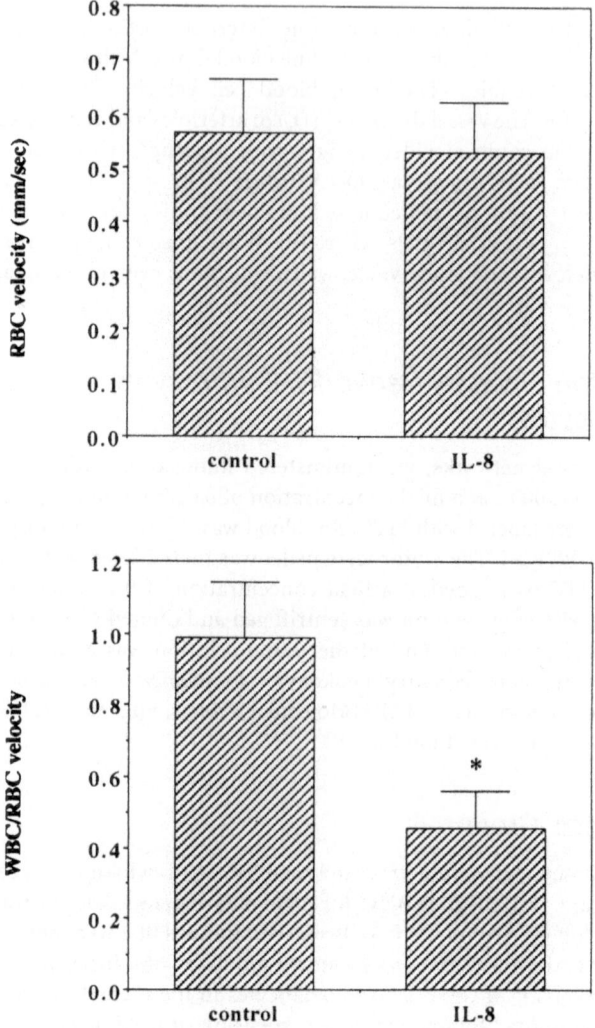

Fig. 1. The mean red blood cell (RBC) velocities (*upper graph*) and the relative leukocyte velocities (*lower graph*) in pulmonary capillaries of the control and the IL-8 groups. There were no differences in capillary RBC velocities, while relative leukocyte velocities in the capillaries of the IL-8 group were decreased as compared with that in the control group. Values are means ± SEM (each group; $n = 5$). *; $P < .01$ as compared with the control group

## Results

Mean pulmonary arterial pressure was unaffected by the injection of IL-8-activated CFSE-labeled leukocytes. Mean pulmonary arterial pressure was constant throughout the experiments. There were no significant differences in pulmonary arteriolar and venular wall shear rates between the control and the IL-8 groups.

No leukocyte rolling was observed in either pulmonary arterioles or venules, but rolling leukocytes were observed in mesenteric venules. Stationary adherent leukocytes were rarely observed in pulmonary arterioles and venules, and leukocytes remained within pulmonary capillaries in both the control and the IL-8 groups.

There were two distinct pulmonary leukocyte behavior patterns in the control capillaries: the majority of leukocytes were in "continuous" motion, varying relatively little in velocity. A minority, however, ceased motion transiently (0.04–0.2 s), then resumed moving within the capillary. The latter "discontinuous" motion pattern, with leukocytes stopping at least once for more than 0.04 s during observation of a $200 \times 200\,\mu m$ peripheral area of the lung, was also seen in the control group. When leukocytes were activated with rat IL-8, approximately half the total leukocytes showed discontinuous motion within the capillaries of the aforementioned area.

When prestimulated with rat IL-8, leukocyte behavior in pulmonary and mesenteric microcirculations is quite different. In the pulmonary microcirculation, IL-8 reduced relative leukocyte velocity in capillaries (Fig. 1), while the capillary leukocyte density was increased. Adherent leukocytes in pulmonary arterioles and venules were not observed. In the mesenteric microcirculation, however, superfusion of fMLP induced leukocyte adherence in venules but not in arterioles or capillaries. Activation with rat IL-8 diminished rolling leukocytes in mesenteric venules, and decreased leukocyte adherence, despite fMLP superfusion.

## Discussion

We visualized the behavior of unstimulated and stimulated leukocytes in the pulmonary microcirculation perfused ex vivo under physiologically controlled shear rates. Several lines of evidence support the concept that the sequential multistep interactions known to be involved in the leukocyte adhesion observed in mesenteric venules [1–3] are not applicable to adhesion mechanisms operating in the pulmonary microcirculation. First, no rolling leukocytes were seen in either capillaries or venules in the lungs. Second, when activated, leukocytes were observed to be trapped in pulmonary microvessels. The same leukocytes could not, however, adhere to mesenteric venules. Furthermore, most leukocyte sequestration took place in alveolar capillaries rather than postcapillary venules. These observations raise the possibility that leukocyte accumulation in pulmonary microvessels involves mechanisms distinct from those in the mesenteric microvessels.

Leukocyte activation induces several changes, including expression of adhesion molecules and cellular deformability. The mechanism of pulmonary capillary entrapment, especially the contributions of CD18–ICAM-1 interaction and mechanical properties, awaits clarification.

*Acknowledgments.* The authors thank Dr. Makoto Suematsu and Dr. Yuzuru Ishimura, Department of Biochemistry, School of Medicine, Keio University, Tokyo, Japan, for technical support and advice.

608    T. Aoki et al.

# References

1. Springer TA (1994) Traffic signals for lymphocyte recirculation and leukocyte emigration: the multistep paradigm. Cell 76:301–314
2. Butcher EC (1991) Leukocyte-endothelial cell recognition: three (or more) steps to specificity and diversity. Cell 67:1033–1036
3. Springer TA (1990) Adhesion receptors of the immune system. Nature 346:425–434
4. Von Andrian UH, Chambers JD, McEvoy LM, et al. (1991) Two-step model of leukocyte-endothelial cell interaction in inflammation: distinct roles for LECAM-1 and the leukocyte β2 integrins in vivo. Proc Natl Acad Sci USA 88:7538–7542
5. Hernandez LA, Grisham MB, Twohig B, et al. (1987) Role of neutrophils in ischemia-reperfusion-induced microvascular injury. Am J Physiol 22:H699–H703
6. Smith CW, Rothlein R, Hughes BJ, et al. (1988) Recognition of an endothelial determinant for CD18-dependent human neutrophil adherence and transendothelial migration. J Clin Invest 82:1746–1756
7. Smith CW, Marlin SD, Rothlein R, et al. (1989) Cooperative interactions of LFA-1 with intercellular adhesion molecule-1 in faciliating adherence and transendothelial migration of human neutrophils in vitro. J Clin Invest 83:2008–2017
8. Luscinskas FW, Cybulsky MI, Kiely JM, et al. (1991) Cytokine-activated human endothelial monolayers support enhanced neutrophil transmigration via a mechanism involving both endothelial-leukocyte adhesion molecule-1 and intercellular adhesion molecule-1. J Immunol 146:1617–1625
9. Suematsu M, Schmid-Schönbein GW, Chavez-Chavez RH, et al. (1993) In vivo visualization of oxidative changes in microvessels during neutrophil activation. Am J Physiol 264:H881–H891
10. Mayadas TN, Johnson RC, Rayburn H, et al. (1993) Leukocyte rolling and extravasation are severely compromised in P selectin-deficient mice. Cell 74:541–554
11. Suematsu M, Schmid-Schönbein GW, Chavez-Chavez RH, et al. (1993) In vivo visualization of oxidative changes in microvessels during neutrophil activation. Am J Physiol 264:H881–H891

# Role of Nitric Oxide in Autoregulation of Cerebral Blood Flow in the Rat

Kortaro Tanaka, Yasuo Fukuuchi, Toshitaka Shirai, Shigeru Nogawa, Hiroyuki Nozaki, Eiichiro Nagata, Taro Kondo, Satoshi Koyama, and Tomohisa Dembo

*Summary.* We examined the role of nitric oxide (NO) in autoregulation of cerebral blood flow (CBF) in the rat. Autoregulation of the CBF is defined as the physiological tendency of the brain to maintain a constant CBF despite changes in arterial blood pressure (ABP). Male Sprague–Dawley rats were divided into six groups: a saline group ($n = 9$), a saline + hypotension group ($n = 8$), an $N^G$-monomethyl-L-arginine (L-NMMA) group ($n = 7$), an L-NMMA + hypotension group ($n = 9$), a denervation group ($n = 6$), and a denervation + hypotension group ($n = 15$). We employed the [$^{14}$C]iodoantipyrine method to measure CBF. In each hypotension group, the ABP was lowered by withdrawing blood. In each L-NMMA group, 30 mg/kg of L-NMMA, a potent NO synthase (NOS) inhibitor, was injected intravenously before making the CBF measurement. In each denervation group, unilateral chronic transection of the perivascular NOS-containing nerve fibers was performed at 2 weeks before the CBF measurement. We found a significant impairment of autoregulation in the L-NMMA and L-NMMA + hypotension groups, whereas the saline and saline + hypotension groups as well as the denervation and denervation + hypotension groups did not show any definite disturbance of autoregulation. We infer that NO, probably derived from the vascular endothelium, may play an important role in autoregulation of CBF.

*Key words.* Nitric oxide—Cerebral blood flow—Autoregulation—Autoradiography—$N^G$-Monomethyl-L-arginine

## Introduction

Autoregulation of cerebral blood flow (CBF) has been defined as the physiological tendency of the brain to maintain a constant CBF despite changes in arterial blood pressure. The mechanisms postulated to be involved in autoregulation have long raised controversial issues [1]. We reported previously that inhibition of nitric oxide synthase (NOS) by $N^G$-monomethyl-L-arginine (L-NMMA) induces a significant

Department of Neurology, School of Medicine, Keio University, 35 Shinanomachi, Shinjuku-ku, Tokyo 160, Japan.

reduction in basal CBF [2], suggesting that NO may play an important role in maintenance of the basal CBF. On the other hand, the role of NO in autoregulation of the CBF has not yet been adequately investigated [3]. The present study was therefore undertaken to examine the role of NO in autoregulation of the CBF during hypotension.

In the brain, NOS is widely located in various cellular components such as neurons, perivascular nerve fibers, endothelial cells, and so on [4]. As the first part of this study, we assessed the effects of intravenous administration of L-NMMA on autoregulation. We then investigated the role of NOS-containing perivascular nerve fibers in the brain, of which major and minor sources have been found to be the sphenopalatine ganglion (SPG) and the trigeminal ganglion (TG), respectively [5]. Chronic transection of these NOS-containing nerve fibers was undertaken before examination of autoregulation.

## Materials and Methods

The experimental protocol has been approved as meeting the Animal Experimentation Guidelines of Keio University School of Medicine. Male Sprague–Dawley rats weighing 250–300 g were divided into six groups: a saline group ($n = 9$), a saline + hypotension group ($n = 8$), an L-NMMA group ($n = 7$), an L-NMMA + hypotension group ($n = 9$), a denervation group ($n = 6$), and a denervation + hypotension group ($n = 15$). In each group, on the day of CBF measurement, polyethylene catheters were inserted into the femoral artery and vein under anesthesia with pentobarbital sodium (25 mg/kg). The arterial blood pressure was monitored continuously, and arterial blood gases were determined immediately before the CBF measurement. After the animals had recovered from the anesthesia, CBF was measured by the [$^{14}$C]iodoantipyrine method [6]. The body temperature of each animal was kept at $37° ± 0.5°C$ during the entire experiment.

In both the saline and saline + hypotension groups, saline (0.5 ml) was injected intravenously at 6 min before making the CBF measurement. In the saline + hypotension group, the arterial blood pressure was gradually lowered by 30–40 mmHg by withdrawing arterial blood at 3 min before the CBF measurement. On the other hand, the arterial blood pressure of the saline group underwent no artificial manipulation.

In the L-NMMA and L-NMMA + hypotension groups, 30 mg/kg of L-NMMA was injected intravenously at 6 min before CBF measurement. At 3 min after L-NMMA administration, the arterial blood pressure was gradually lowered by 30–40 mmHg in the L-NMMA + hypotension group. On the other hand, the L-NMMA group underwent no artificial manipulation of blood pressure except that the intravenous administration of L-NMMA moderately increased arterial blood pressure, as observed previously [2].

In each denervation group, the left postganglionic parasympathetic nerve fibers from the SPG and the left nasociliary nerve from the TG were transected at the ethmoidal foramen 2 weeks before making the CBF measurement, as described previously [7]. On the day of CBF measurement, the CBF was measured in a similar manner to that outlined earlier in both the denervation and denervation + hypotension groups. In the denervation + hypotension group, the arterial blood pressure was lowered by 30–40 mmHg. In each denervation group, after completion of the CBF measurement, the circle of Willis and its major branches attached to the pial membrane were quickly removed and processed for NADPH-diaphorase staining according to the procedures used previously in our laboratory [5].

Total disappearance of NADPH-diaphorase-positive perivascular nerve fibers was confirmed on the left middle cerebral artery and its branches, while the right middle cerebral artery and the anterior and posterior cerebral arteries on both sides demonstrated abundant networks of NADPH-diaphorase-positive nerve fibers on their adventitia, indicating that our denervation procedure had efficiently removed the innervation of NOS-containing perivascular nerve fibers from the area of the left middle cerebral artery.

# Results

Table 1 summarizes the physiological parameters, which were determined immediately before the CBF measurement. The mean arterial blood pressure (MABP) was significantly decreased in each hypotension group as compared to that of each corresponding group that was not subjected to withdrawal of blood. The arterial blood gases did not reveal any significant differences among the groups, and were within the normal ranges.

Figure 1 illustrates the relationship between the values of MABP and CBF in the left somatosensory area of the frontoparietal cortex of the animals in both the saline and saline + hypotension groups. MABP expresses the value obtained during the CBF measurement. The somatosensory area represents one of the major areas within the territory of the middle cerebral artery. CBF did not show any definite correlation with MABP between 70 and 140 mmHg. Similar findings were obtained for the right somatosensory cortex as well as for other cerebral and cerebellar cortices, indicating that CBF autoregulation was maintained in these animals, which had received intravenous injection of saline.

Figure 2 illustrates the relationship between the values of MABP and CBF in the left somatosensory area of the frontoparietal cortex of the animals in both the L-NMMA

TABLE 1. Physiological parameters at cerebral blood flow (CBF) measurement (means ± SD)

| Group | MABP | $PaO_2$ | $PaCO_2$ | pH |
|---|---|---|---|---|
| Saline | 118 ± 12 | 91 ± 10 | 36 ± 3 | 7.41 ± 0.01 |
| Saline + hypotension | 78 ± 6* | 105 ± 9 | 34 ± 3 | 7.41 ± 0.04 |
| L-NMMA | 144 ± 15*·*** | 93 ± 10 | 38 ± 1 | 7.35 ± 0.05 |
| L-NMMA + hypotension | 102 ± 13** | 92 ± 7 | 41 ± 5 | 7.37 ± 0.06 |
| Denervation | 111 ± 4 | 105 ± 30 | 39 ± 12 | 7.41 ± 0.07 |
| Denervation + hypotension | 67 ± 11*** | 90 ± 14 | 37 ± 11 | 7.37 ± 0.07 |

MABP, mean arterial blood pressure; $PaO_2$, arterial oxygen tension; $PaCO_2$, arterial carbon dioxide tension.
*, $P < .01$ as compared to the saline group; **, $P < .01$ as compared to the $N^G$-monomethyl-L-arginine (L-NMMA) group; ***, $P < .01$ as compared to the denervation group.

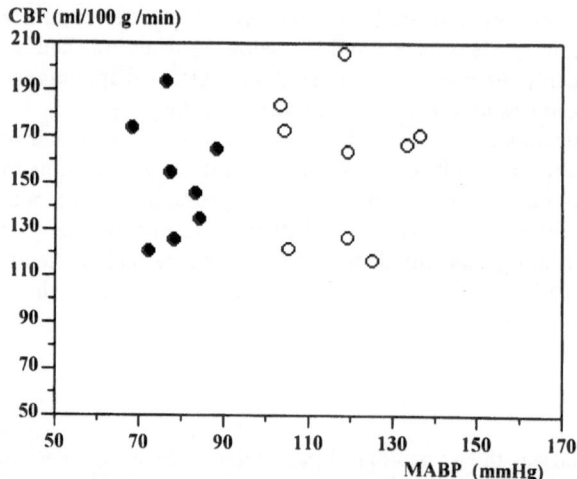

FIG. 1. Scatterplots of the relations between mean arterial blood pressure (*MABP*) and cerebral blood flow (*CBF*) in the left somatosensory area of the frontoparietal cortex of animals in both the saline (*open symbols*) and saline + hypotension (*closed symbols*) groups. Each data point on the figure represents an individual pair of measurements of CBF and MABP, where MABP expresses the value obtained during the CBF measurement

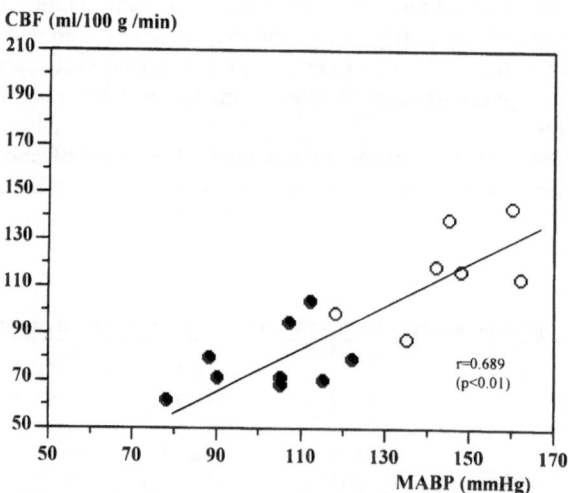

FIG. 2. Scatterplots of the relations between MABP and CBF in the left somatosensory area of the frontoparietal cortex of animals in both the $N^G$-monomethyl-L-arginine (*L-NMMA*) (*open symbols*) and L-NMMA + hypotension (*closed symbols*) groups. Each data point on the figure represents an individual pair of measurements of CBF and MABP, where MABP expresses the value obtained during the CBF meausrement. Note the significant linear correlation between the two parameters, indicating that autoregulation of CBF was abolished

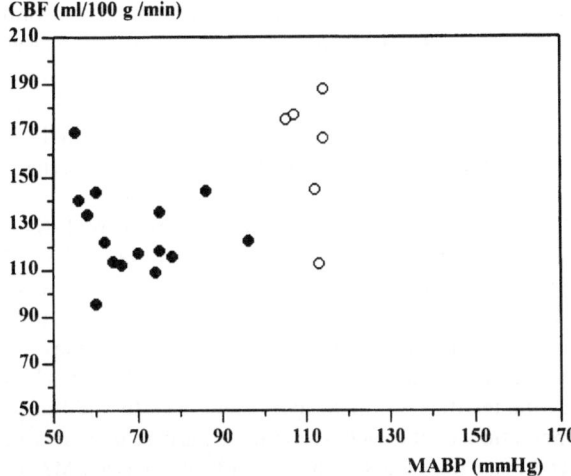

Fɪɢ. 3. Scatterplots of the relations between MABP and CBF in the left somatosensory area of the frontoparietal cortex of animals in both the denervation (*open symbols*) and denervation + hypotension (*closed symbols*) groups. Each data point on the figure represents an individual pair of measurements of CBF and MABP, where MABP expresses the value obtained during the CBF measurement

and L-NMMA + hypotension groups. In contrast to Fig. 1, a statistically significant linear correlation was noted between MABP and CBF, indicating that CBF autoregulation was clearly impaired after intravenous injection of L-NMMA. Similar findings were obtained for the right somatosensory area as well as for other cerebral and cerebellar cortices.

Figure 3 illustrates the relationship between the values of MABP and CBF in the left somatosensory area of the frontoparietal cortex of the animals in both the denervation and denervation + hypotension groups. As shown in Fig. 1, CBF did not reveal any definite correlation with MABP between 50 and 120 mmHg. Similar findings were obtained for the right somatosensory cortex as well for other cerebral and cerebellar cortices on both sides, indicating that the chronic denervation of the NOS-containing perivascular nerve fibers apparently did not affect CBF autoregulation.

## Discussion

The major findings of the current experiments can be summarized as follows: (1) intravenous administration of L-NMMA, a potent inhibitor of each isoform of NOS including neuronal and endothelial NOS, clearly impaired autoregulation of the CBF in each region of the cerebral cortices, and (2) complete denervation of NOS-containing perivascular nerve fibers on the middle cerebral artery did not affect CBF autoregulation in any region.

NOS can be classified into constitutive NOS and inducible NOS [8]. Under physiological conditions, only the constitutive NOS is present in the brain, whereas the expression of inducible NOS is identified only under pathological conditions such as ischemia and inflammation [8]. Constitutive NOS is now known to be widely localized not only in discrete neuronal populations of the brain such as neurons and glial cells, but also in the vascular endothelium as well as in the perivascular autonomic nerves of the cerebral arteries [9]. Indeed, NOS has been demonstrated in the endothelium of the cerebral vasculature of all categories including the major arteries, arterioles, and capillaries [10].

As we reported previously, intravenous administration of L-NMMA induces a significant and homogeneous reduction of CBF in the rat brain [2]. On the other hand, the neuronal NOS in the brain clearly demonstrates a heterogeneous distribution, such as with the highest content in the cerebellar cortex. Similarly, the distribution of perivascular NOS-containing nerve fibers is not uniform within the brain. These nerve fibers are known to be abundant in the anterior portion of the circle of Willis, while they are scarce in the territory of the basilar artery [11]. In addition, we have found that chronic denervation of the NOS-containing perivascular nerve fibers affected neither the basal CBF nor the reduction of CBF induced by L-NMMA in the rat [7]. On the basis of these findings, we speculated that endothelial NOS, which is located ubiquitously throughout the brain, may play a pivotal role in the control of the basal CBF.

According to our previous research [12], intravenous administration of L-NMMA (30 mg/kg) inhibits NOS catalytic activity in the brain by 70%–80% by 3 min after the injection. The current findings suggest therefore that NO produced in the brain may indeed play an important role in autoregulation of CBF. Impairment of autoregulation was uniformly found in each region of the brain, including the cerebellar cortex. Such homogeneous impairment of autoregulation appears to be compatible not with the heterogeneous distribution of neuronal NOS but with the endothelial NOS.

Chronic denervation of NOS-containing perivascular nerve fibers on the middle cerebral artery and its branches did not apparently affect autoregulation, suggesting that these nerve fibers may not be involved in the mechanisms underlying autoregulation. However, the data for the chronic experiment should be interpreted cautiously because 2 weeks of denervation could have elicited compensatory mechanisms such as upregulation of endothelial NOS for maintaining autoregulation. Accordingly, an acute denervation model is currently under investigation at our laboratory.

In conclusion, we infer that NO produced within the brain plays an important role in CBF autoregulation, and its major origin as involved in such autoregulation may be the vascular endothelium. In line with our speculation, a recent paper has reported that endothelial NOS knockout mice display an apparent impairment of autoregulation during hypotension [13].

# References

1. Gotoh F, Tanaka K (1988) Regulation of cerebral blood flow. In: Vinken PJ, Bruyn GW, Klawans HL (eds) Handbook of clinical neurology, vol 53. Elsevier, Amsterdam, pp 47–77
2. Tanaka K, Gotoh F, Gomi S, et al. (1991) Inhibition of nitric oxide synthesis induces a significant reduction in local cerebral blood flow in the rat. Neurosci Lett 127:129–132

3. Tanaka K (1996) Is nitric oxide really important for regulation of the cerebral circulation? Yes or no? Keio J Med 45:14–27
4. Iadecola C (1993) Regulation of the cerebral microcirculation during neural activity: is nitric oxide the missing link? Trends Neurosci 16:206–214
5. Suzuki N, Fukuuchi Y, Koto A, et al. (1994) Distribution and origins of cerebrovascular NADPH-diaphorase-containing nerve fibers in the rat. J Auton Nerv Syst 49(suppl):S51–S54
6. Sakurada O, Kennedy C, Jehle J, et al. (1978) Measurement of local cerebral blood flow with iodo[$^{14}$C]antipyrine. Am J Physiol 234:H59–H66
7. Tanaka K, Fukuuchi Y, Shirai T, et al. (1995) Chronic transection of post-ganglionic parasympathetic and nasociliary nerves does not affect local cerebral blood flow in the rat. J Auton Nerv Syst 53:95–102
8. Forstermann U, Kleinert H (1995) Nitric oxide synthase: expression and expressional control of the three isoforms. Naunyn-Schmiedeberg's Arch Pharmacol 352:351–364
9. Iadecola C, Pelligrino DA, Moskowitz MA, Lassen NA (1994) Nitric oxide synthase inhibition and cerebrovascular regulation. J Cereb Blood Flow Metab 14:175–192
10. Gabbott PLA, Bacon SJ (1993) Histochemical localization of NADPH-dependent diaphorase (nitric oxide synthase) activity in vascular endothelial cells in the rat brain. Neuroscience 57:79–95
11. Suzuki N, Fukuuchi Y, Koto A, et al. (1993) Cerebrovascular NADPH-diaphorase-containing nerve fibers in the rat. Neurosci Lett 151:1–3
12. Tanaka K, Fukuuchi Y, Shirai T, et al. (1996) Nitric oxide synthase-containing post-ganglionic parasympathetic and nasociliary nerves do not contribute to basal cerebral blood flow in the rat. In: Proceedings of symposium on biochemistry and molecular biology of nitric oxide, University of California, July 13–17, 1996
13. Huang Z, Huang PL, Ma J, et al. (1996) Enlarged infarcts in endothelial nitric oxide synthase knockout mice are attenutated by nitro-L-arginine. J Cereb Blood Flow Metab 16:981–987

# Key Word Index

*N*-Acetyl-L-cysteine   457
Acid–base catalysis   354
Acousto-optic tunable filter (AOTF)   550
ACTH   237
Adhesion   603
Adrenal cortex   231, 237, 244
Adult T-cell leukemia- (ATL-) derived factor
    (ADF)   450
AIDS   438
Alveolar capillary   603
Alzheimer's disease   597
Angeli's salt   199
Angiogenesis   388
Angiotensin II   237
AP-1   450
AP-2   434
Apolipoprotein E   597
L-Arginine   298
Artificial amino acid   161
Ascorbic acid   562
Aspirin   457
ATP   120, 585
Autoradiographic [$^{14}$C]iodoantipyrine
    method   518
Autoradiography   609
Autoregulation   518, 609
Axial ligand   181

*Bacillus megaterium*   166
Bilirubin   248, 537
Biliverdin   304, 322
Bioenergetics   102
Brain   510, 580

Calcium channel   473
Calcium signal   252

Calmodulin   289
Cancer   438
Capillary entrapment   603
Carbon dioxide   518
Carbon monoxide (CO)   304, 537, 544,
    585
Carotid body   377
Catalytic mechanism   161
Cavity mutant   172
Cerebral blood flow   609
Cerebral oxygenation   84
Chlorin e$_6$   367
Chronic ethanol consumption   576
Circular dichroism spectroscopy   204
Collagenase   434
Compound I   359
Compound II   359
COMT inhibitor   510
Control   289
Coupling   106
COX-1   410
COX-2   410
Cross-linked hemoglobin   428
Crystal structure   147, 276, 282
Crystallization   98
Crystallography   263
*c*-type cytochrome   112
Cybrid   72
Cyclooxygenase   484
Cysteine   562
Cytochrome *a$_3$*   120
Cytochrome *aa$_3$*   120
Cytochrome *bc*   112
Cytochrome *bo*   24
Cytochrome *c* oxidase   3, 13, 40, 57, 72, 84,
    98, 102, 106, 112, 120, 591
Cytochrome oxidase; *see* Cytochrome *c*
    oxidase

Cytochrome P-450   127, 139, 147, 161, 166, 172, 181, 199, 231, 252, 359
Cytochrome P-450$_{17\alpha,lyase}$   214
Cytochrome P-450$_{11\beta}$ (CYP11B1)   214, 221, 231, 237, 244
Cytochrome P-450$_{aldo}$ (CYP11B2)   221, 237
Cytochrome P-450$_{nor}$   156
Cytochrome P-450$_{scc}$ (CYP11A1)   204
Cytoplast   72

Denitrification   156
Denitrifying fungus   147
Dioxygen reduction   102
Dioxygenase   263, 282
Distal site   199
Ductus arteriosus   400, 473

Ebselen   505
Electron transfer   221
Enantioselective oxidation   340
Endothelial cell   479
eNOS   410
EPR   139, 298
Erythropoietin   421
Escherichia coli   33
Evolution   591
Exhaled gas   569
Extradiol-type dioxygenase   276

Fatty acid   166
Fluorescence imaging   252
Fusarium oxysporum   156

Gene complex   248
Glioblastoma   328
Global analysis   363
Glomerular   484
Glucocorticoid   244
Glycolysis   421
Guanylate cyclase   469
Gunn rats   248

Helicobacter pylori   112
Heme   33, 304, 315, 340, 537
Heme-copper oxidase   112
Heme oxygenase (HO)   304, 315, 322, 328, 469, 544, 585
Heme-peroxide complex   363
Heme protein   354

Heme-thiolate complex   172
Heme-thiolate protein   189
Hemoglobin   544
Hemoglobin vesicle   428
Hemoprotein   199, 304
Heparin   597
Heteroplasmy   72
Horseradish peroxidase (HRP)   354
Human   328
Human mitochondria   72
Human serum albumin   367
Human thioredoxin   464
Hydrogen bond   161, 354
$\alpha$-Hydroxyheme   304, 315
4-Hydroxy-2-nonenal (HNE)   597
$N^w$-hydroxy-L-arginine   298
Hyperoxia   410
Hypoxia   84, 388, 421
Hypoxia-inducible factor 1 (HIF-1)   388
Hypoxia/ischemia   580
Hypoxic pulmonary vasoconstriction   410

Immunoprecipitation   464
Inducible nitric oxide synthase (iNOS)   328, 410
Intercellular adhesion molecule-1 (ICAM-1)   434, 479
Interleukin-1   484
Intramolecular electron transfer   24
Intrauterine growth retardation (IUGR)   557
Ion channel   469
Iron-sulfur protein   139

Kupffer cells   576

Langmuir-Blodgett films   204
LDL   252
Ligand exchange   282
Ligand photolysis   47
Ligand shuttle   47
Lipid peroxidation   597
Liver-specific pericyte   469
Lung resection   569

Macrophages   550
Magnetic circular dichroism (MCD) spectroscopy   172
MCLA chemiluminescence   580
Membrane protein   13, 98
Membrane protein crystallization   3

Mesangium   484
Microscopic imaging   252
Microspectrophotometry   550
Mitochondria   557, 585
Mitochondrial energy state   84
$N^G$-Monomethyl-L-arginine   609
Monooxygenase   166
Monooxygenation   127
Mutagenesis   127, 161
Myoglobin   172, 340, 363

NAD(P)H   156
NADPH cytochrome P-450 reductase   576
NADPH oxidase   400
Near-infrared photometry   84
NF-κB   450, 457
Nitration   493
Nitric oxide (NO)   189, 199, 289, 298,
        484, 510, 544, 550, 557, 562, 569,
        609
Nitric oxide (NO) reductase   147, 591
Nitric oxide synthase (NOS)   289, 298
6-Nitronorepinephrine   510
S-Nitrosothiols   493, 562
$N^G$-nitro-L-arginine methyl ester   518
Nitrotyrosine   505
Nonheme iron   282
Nonionic detergent   98
Norepinephrine   510
Nuclear protein   464

O–O bond   181
$O_2$ reduction   13
Oxygen   139, 263, 377, 400, 591
   potassium channel   473
Oxygen activation   47, 57, 166, 340
Oxygen carriers   428
Oxygen radical   189, 333
Oxygenase   304
Oxygen-binding affinity   428

Paracoccus denitrificans   3, 591
PCB degradation   276
Peroxidase   40, 359, 363
Peroxy intermediate   57, 102
Peroxy state   33
Peroxygenase   340
Peroxynitrite   493, 505
Peroxynitrite reductase   505
Phosphorescence   377
Photooxidation   367

Placenta   557
Polyethylene glycol conjugate   359
Porphyrin   181
Potassium channel   400
Preeclampsia   557
Prostaglandins   484
Protein kinase C   479
Protein–protein interaction   127, 464
Proton pump   13, 47, 57
Proton translocation   40
Protonation   40
Protonation sites   106
Pulmonary oxygen toxicity   479
Putidaredoxin   139

Quinol oxidase   33
Quinol oxidation site   24
Quinone analogues   24

Radical oxygen intermediates   438
Rapid quenching method   214
Rapid-scan spectrophotometry   363
Red blood cells   550
Redox change   199
Redox potential   322
Redox regulation   450, 464
Reductase   127, 289
Reoxygenation/reperfusion   580
Reperfusion injury   333
Resonance Raman spectrophotometry   57,
        102, 354
Respiration   47
Rheumatoid arthritis   438, 457
Rolling   603

Selenomethionine   505
Semiquinone   33
Semiquinone radical   24
Signal transduction   438
Singlet oxygen   434
Sinusoids   544
Site-directed mutagenesis   156
Smooth muscle   400, 473
Sodium nitroprusside   328
Spin-state equilibrium   204
Spontaneous breathing   569
Stable expression   231
Steroid hormone   214
Steroid hydroxylases   231
Steroidogenesis   244

Stopped-flow spectrophotometry   322
Stress protein   537
Structure determination   3
Substrate specificity   221
Successive reaction   214
Superoxide anion ($O_2^-$)   333, 493, 576, 580
Synovial fibroblasts   457

*Thiobacillus novellus*   120
Thiolate   181
Thiols   562
Thioredoxin   438
Thioredoxin (TRX)   450
Tissue oxygenation   377
Transcription   421, 438, 537
Transcription factors   450
Transcriptional regulation   244
Tryptophan   367
Tumor angiogenesis   388
Tumors   377
Tyrosine nitration   189

UDP-glucuronosyltransferase   248
Unconjugated hyperbilirubinemia   248
UVA irradiation   434

V79 fibroblasts   231
Vascular endothelial growth factor (VEGF)
        388, 421
Vasoconstriction   189
Verdoheme   304, 322
Verdoheme $IX_\alpha$   315

Wall shear rates   603
Whisker barrel   518

X-ray crystal structure   13
X-ray crystallography   3
Xanthine dehydrogenase   333
Xanthine oxidase   333